Mechanical Engineer's Handbook

Academic Press Series in Engineering

Series Editor
J. David Irwin
Auburn University

This a series that will include handbooks, textbooks, and professional reference books on cutting-edge areas of engineering. Also included in this series will be single-authored professional books on state-of-the-art techniques and methods in engineering. Its objective is to meet the needs of academic, industrial, and governmental engineers, as well as provide instructional material for teaching at both the undergraduate and graduate level.

The series editor, J. David Irwin, is one of the best-known engineering educators in the world. Irwin has been chairman of the electrical engineering department at Auburn University for 27 years.

Published books in this series:

Control of Induction Motors
2001, A. M. Trzynadlowski

Embedded Microcontroller Interfacing for McoR Systems
2000, G. J. Lipovski

Soft Computing & Intelligent Systems
2000, N. K. Sinha, M. M. Gupta

Introduction to Microcontrollers
1999, G. J. Lipovski

Industrial Controls and Manufacturing
1999, E. Kamen

DSP Integrated Circuits
1999, L. Wanhammar

Time Domain Electromagnetics
1999, S. M. Rao

Single- and Multi-Chip Microcontroller Interfacing
1999, G. J. Lipovski

Control in Robotics and Automation
1999, B. K. Ghosh, N. Xi, and T. J. Tarn

Mechanical Engineer's Handbook

Edited by
Dan B. Marghitu

Department of Mechanical Engineering, Auburn University, Auburn, Alabama

ACADEMIC PRESS

A Harcourt Science and Technology Company

San Diego • San Francisco • New York • Boston • London • Sydney • Tokyo

Academic Press
A Harcourt Science and Technology Company
525 B Street, Suite 1900, San Diego, California 92101-4495, USA
http://www.academicpress.com

Academic Press
Harcourt Place, 32 Jamestown Road, London NW1 7BY, UK
http://www.academicpress.com

Library of Congress Catalog Card Number: 2001088196

International Standard Book Number: 0-12-471370-X

PRINTED IN THE UNITED STATES OF AMERICA
01 02 03 04 05 06 CO 9 8 7 6 5 4 3 2 1

Table of Contents

CHAPTER 2   *Dynamics*

Dan B. Marghitu, Bogdan O. Ciocirlan, and Cristian I. Diaconescu

---

CHAPTER 3   *Mechanics of Materials*

Dan B. Marghitu, Cristian I. Diaconescu, and Bogdan O. Ciocirlan

CHAPTER 6 *Theory of Vibration*

Dan B. Marghitu, P. K. Raju, and Dumitru Mazilu

CHAPTER 7 *Principles of Heat Transfer*

Alexandru Morega

CHAPTER 8   Fluid Dynamics

Nicolae Craciunoiu and Bogdan O. Ciocirlan

APPENDIX   *Differential Equations and Systems of Differential Equations*

Horatiu Barbulescu

Preface

The purpose of this handbook is to present the reader with a teachable text that includes theory and examples. Useful analytical techniques provide the student and the practitioner with powerful tools for mechanical design. This book may also serve as a reference for the designer and as a source book for the researcher.

This handbook is comprehensive, convenient, detailed, and is a guide for the mechanical engineer. It covers a broad spectrum of critical engineering topics and helps the reader understand the fundamentals.

This handbook contains the fundamental laws and theories of science basic to mechanical engineering including controls and mathematics. It provides readers with a basic understanding of the subject, together with suggestions for more specific literature. The general approach of this book involves the presentation of a systematic explanation of the basic concepts of mechanical systems.

This handbook's special features include authoritative contributions, chapters on mechanical design, useful formulas, charts, tables, and illustrations. With this handbook the reader can study and compare the available methods of analysis. The reader can also become familiar with the methods of solution and with their implementation.

Dan B. Marghitu

Contributors

Numbers in parentheses indicate the pages on which the authors' contributions begin.

Horatiu Barbulescu, (715) Department of Mechanical Engineering, Auburn University, Auburn, Alabama 36849

Bogdan O. Ciocirlan, (1, 51, 119, 559) Department of Mechanical Engineering, Auburn University, Auburn, Alabama 36849

Nicolae Craciunoiu, (243, 559) Department of Mechanical Engineering, Auburn University, Auburn, Alabama 36849

Cristian I. Diaconescu, (1, 51, 119, 243) Department of Mechanical Engineering, Auburn University, Auburn, Alabama 36849

Mircea Ivanescu, (611) Department of Electrical Engineering, University of Craiova, Craiova 1100, Romania

Dan B. Marghitu, (1, 51, 119, 189, 243, 339) Department of Mechanical Engineering, Auburn University, Auburn, Alabama 36849

Dumitru Mazilu, (339) Department of Mechanical Engineering, Auburn University, Auburn, Alabama 36849

Alexandru Morega, (445) Department of Electrical Engineering, "Politehnica" University of Bucharest, Bucharest 6-77206, Romania

P. K. Raju, (339) Department of Mechanical Engineering, Auburn University, Auburn, Alabama 36849

1

Statics

DAN B. MARGHITU, CRISTIAN I. DIACONESCU, AND
BOGDAN O. CIOCIRLAN

*Department of Mechanical Engineering,
Auburn University, Auburn, Alabama 36849*

Inside

1. Vector Algebra

1.1 Terminology and Notation

The *characteristics* of a vector are the magnitude, the orientation, and the sense. The *magnitude* of a vector is specified by a positive number and a unit having appropriate dimensions. No unit is stated if the dimensions are those of a pure number. The *orientation* of a vector is specified by the relationship between the vector and given reference lines and/or planes. The *sense* of a vector is specified by the order of two points on a line parallel to the vector. Orientation and sense together determine the *direction* of a vector. The *line of action* of a vector is a hypothetical infinite straight line collinear with the vector. Vectors are denoted by boldface letters, for example, **a**, **b**, **A**, **B**, **CD**. The symbol $|\mathbf{v}|$ represents the magnitude (or module, or absolute value) of the vector **v**. The vectors are depicted by either straight or curved arrows. A vector represented by a straight arrow has the direction indicated by the arrow. The direction of a vector represented by a curved arrow is the same as the direction in which a right-handed screw moves when the screw's axis is normal to the plane in which the arrow is drawn and the screw is rotated as indicated by the arrow.

Figure 1.1 shows representations of vectors. Sometimes vectors are represented by means of a straight or curved arrow together with a measure number. In this case the vector is regarded as having the direction indicated by the arrow if the measure number is positive, and the opposite direction if it is negative.

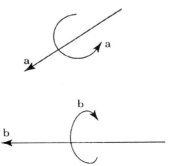

Figure 1.1

A *bound* vector is a vector associated with a particular point P in space (Fig. 1.2). The point P is the *point of application* of the vector, and the line passing through P and parallel to the vector is the line of action of the vector. The point of application may be represented as the tail, Fig. 1.2a, or the head of the vector arrow, Fig. 1.2b. A *free* vector is not associated with a particular point P in space. A *transmissible vector* is a vector that can be moved along its line of action without change of meaning.

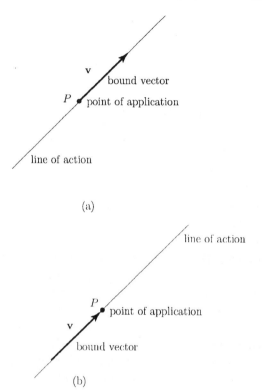

Figure 1.2

To move the body in Fig. 1.3 the force vector $\mathbf{F}$ can be applied anywhere along the line Δ or may be applied at specific points A, B, C. The force vector $\mathbf{F}$ is a transmissible vector because the resulting motion is the same in all cases.

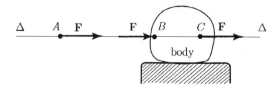

Figure 1.3

The force $\mathbf{F}$ applied at B will cause a different deformation of the body than the same force $\mathbf{F}$ applied at a different point C. The points B and C are on the body. If we are interested in the deformation of the body, the force $\mathbf{F}$ positioned at C is a bound vector.

The operations of vector analysis deal only with the characteristics of vectors and apply, therefore, to both bound and free vectors.

1.2 Equality

Two vectors **a** and **b** are said to be equal to each other when they have the same characteristics. One then writes

$$\mathbf{a} = \mathbf{b}.$$

Equality does not imply physical equivalence. For instance, two forces represented by equal vectors do not necessarily cause identical motions of a body on which they act.

1.3 Product of a Vector and a Scalar

DEFINITION The product of a vector **v** and a scalar s, $s\mathbf{v}$ or $\mathbf{v}s$, is a vector having the following characteristics:

1. *Magnitude.*
$$|s\mathbf{v}| \equiv |\mathbf{v}s| = |s||\mathbf{v}|,$$

where $|s|$ denotes the absolute value (or magnitude, or module) of the scalar s.

2. *Orientation.* $s\mathbf{v}$ is parallel to **v**. If $s = 0$, no definite orientation is attributed to $s\mathbf{v}$.

3. *Sense.* If $s > 0$, the sense of $s\mathbf{v}$ is the same as that of **v**. If $s < 0$, the sense of $s\mathbf{v}$ is opposite to that of **v**. If $s = 0$, no definite sense is attributed to $s\mathbf{v}$. ▲

1.4 Zero Vectors

DEFINITION A *zero vector* is a vector that does not have a definite direction and whose magnitude is equal to zero. The symbol used to denote a zero vector is **0**. ▲

1.5 Unit Vectors

DEFINITION A *unit vector* (versor) is a vector with the magnitude equal to 1. ▲

Given a vector **v**, a unit vector **u** having the same direction as **v** is obtained by forming the quotient of **v** and $|\mathbf{v}|$:

$$\mathbf{u} = \frac{\mathbf{v}}{|\mathbf{v}|}.$$

on

$_1$ and a vector $\mathbf{v}_2$: $\mathbf{v}_1 + \mathbf{v}_2$ or $\mathbf{v}_2 + \mathbf{v}_1$ is a vector whose
und by either graphical or analytical processes. The
according to the parallelogram law: $\mathbf{v}_1 + \mathbf{v}_2$ is equal to
llelogram formed by the graphical representation of the
e vectors $\mathbf{v}_1 + \mathbf{v}_2$ is called the *resultant* of $\mathbf{v}_1$ and $\mathbf{v}_2$.
lded by moving them successively to parallel positions
ne vector connects to the tail of the next vector. The
whose tail connects to the tail of the first vector, and
to the head of the last vector (Fig. 1.4b).
$\mathbf{v}_2$) is called the *difference* of $\mathbf{v}_1$ and $\mathbf{v}_2$ and is denoted
and 1.4d).

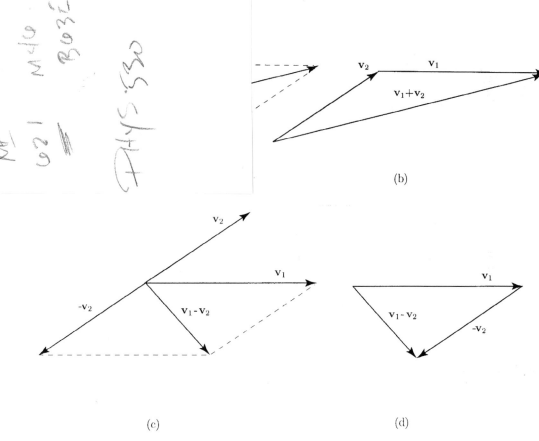

Figure 1.4

The sum of n vectors $\mathbf{v}_i$, $i = 1, \ldots, n$,

$$\sum_{i=1}^{n} \mathbf{v}_i \text{ or } \mathbf{v}_1 + \mathbf{v}_2 + \cdots + \mathbf{v}_n,$$

is called the *resultant* of the vectors $\mathbf{v}_i$, $i = 1, \ldots, n$.

The vector addition is:

1. Commutative, that is, the characteristics of the resultant are independent of the order in which the vectors are added (commutativity):

$$\mathbf{v}_1 + \mathbf{v}_2 = \mathbf{v}_2 + \mathbf{v}_1.$$

2. Associative, that is, the characteristics of the resultant are not affected by the manner in which the vectors are grouped (associativity):

$$\mathbf{v}_1 + (\mathbf{v}_2 + \mathbf{v}_3) = (\mathbf{v}_1 + \mathbf{v}_2) + \mathbf{v}_3.$$

3. Distributive, that is, the vector addition obeys the following laws of distributivity:

$$\mathbf{v} \sum_{i=1}^{n} s_i = \sum_{i=1}^{n} (\mathbf{v} s_i), \quad \text{for } s_i \neq 0, s_i \in \mathcal{R}$$

$$s \sum_{i=1}^{n} \mathbf{v}_i = \sum_{i=1}^{n} (s \mathbf{v}_i), \quad \text{for } s \neq 0, s \in \mathcal{R}.$$

Here $\mathcal{R}$ is the set of real numbers.

Every vector can be regarded as the sum of n vectors ($n = 2, 3, \ldots$) of which all but one can be selected arbitrarily.

1.7 Resolution of Vectors and Components

Let $\mathbf{1}_1$, $\mathbf{1}_2$, $\mathbf{1}_3$ be any three unit vectors not parallel to the same plane

$$|\mathbf{1}_1| = |\mathbf{1}_2| = |\mathbf{1}_3| = 1.$$

For a given vector $\mathbf{v}$ (Fig. 1.5), there exists three unique scalars v_1, v_1, v_3, such that $\mathbf{v}$ can be expressed as

$$\mathbf{v} = v_1 \mathbf{1}_1 + v_2 \mathbf{1}_2 + v_3 \mathbf{1}_3.$$

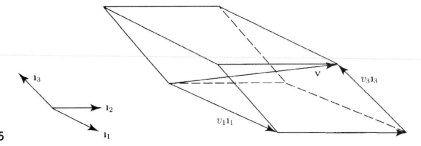

Figure 1.5

The opposite action of addition of vectors is the *resolution* of vectors. Thus, for the given vector $\mathbf{v}$ the vectors $v_1 \mathbf{1}_1$, $v_2 \mathbf{1}_2$, and $v_3 \mathbf{1}_3$ sum to the original vector. The vector $v_k \mathbf{1}_k$ is called the $\mathbf{1}_k$ *component* of $\mathbf{v}$, and v_k is called the $\mathbf{1}_k$ *scalar component* of $\mathbf{v}$, where $k = 1, 2, 3$. A vector is often replaced by its components since the components are equivalent to the original vector.

Every vector equation $\mathbf{v} = \mathbf{0}$, where $\mathbf{v} = v_1\mathbf{1}_1 + v_2\mathbf{1}_2 + v_3\mathbf{1}_3$, is equivalent to three scalar equations $v_1 = 0$, $v_2 = 0$, $v_3 = 0$.

If the unit vectors $\mathbf{1}_1$, $\mathbf{1}_2$, $\mathbf{1}_3$ are mutually perpendicular they form a *cartesian reference frame*. For a cartesian reference frame the following notation is used (Fig. 1.6):

$$\mathbf{1}_1 \equiv \mathbf{1}, \quad \mathbf{1}_2 \equiv \mathbf{J}, \quad \mathbf{1}_3 \equiv \mathbf{k}$$

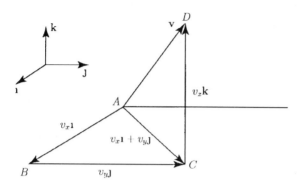

Figure 1.6

and

$$\mathbf{1} \perp \mathbf{J}, \quad \mathbf{1} \perp \mathbf{k}, \quad \mathbf{J} \perp \mathbf{k}.$$

The symbol $\perp$ denotes perpendicular.

When a vector $\mathbf{v}$ is expressed in the form $\mathbf{v} = v_x\mathbf{1} + v_y\mathbf{J} + v_z\mathbf{k}$ where $\mathbf{1}$, $\mathbf{J}$, $\mathbf{k}$ are mutually perpendicular unit vectors (cartesian reference frame or orthogonal reference frame), the magnitude of $\mathbf{v}$ is given by

$$|\mathbf{v}| = \sqrt{v_x^2 + v_y^2 + v_z^2}.$$

The vectors $\mathbf{v}_x = v_x\mathbf{1}$, $\mathbf{v}_y = v_y\mathbf{J}$, and $\mathbf{v}_z = v_y\mathbf{k}$ are the *orthogonal* or *rectangular component vectors* of the vector $\mathbf{v}$. The measures v_x, v_y, v_z are the *orthogonal* or *rectangular scalar components* of the vector $\mathbf{v}$.

If $\mathbf{v}_1 = v_{1x}\mathbf{1} + v_{1y}\mathbf{J} + v_{1z}\mathbf{k}$ and $\mathbf{v}_2 = v_{2x}\mathbf{1} + v_{2y}\mathbf{J} + v_{2z}\mathbf{k}$, then the sum of the vectors is

$$\mathbf{v}_1 + \mathbf{v}_2 = (v_{1x} + v_{2x})\mathbf{1} + (v_{1y} + v_{2y})\mathbf{J} + (v_{1z} + v_{2z})v_{1z}\mathbf{k}.$$

1.8 Angle between Two Vectors

Let us consider any two vectors $\mathbf{a}$ and $\mathbf{b}$. One can move either vector parallel to itself (leaving its sense unaltered) until their initial points (tails) coincide. The *angle* between $\mathbf{a}$ and $\mathbf{b}$ is the angle θ in Figs. 1.7a and 1.7b. The angle between $\mathbf{a}$ and $\mathbf{b}$ is denoted by the symbols $(\mathbf{a}, \mathbf{b})$ or $(\mathbf{b}, \mathbf{a})$. Figure 1.7c represents the case $(\mathbf{a}, \mathbf{b}) = 0$, and Fig. 1.7d represents the case $(\mathbf{a}, \mathbf{b}) = 180°$.

The direction of a vector $\mathbf{v} = v_x\mathbf{1} + v_y\mathbf{J} + v_z\mathbf{k}$ and relative to a cartesian reference, $\mathbf{1}$, $\mathbf{J}$, $\mathbf{k}$, is given by the cosines of the angles formed by the vector

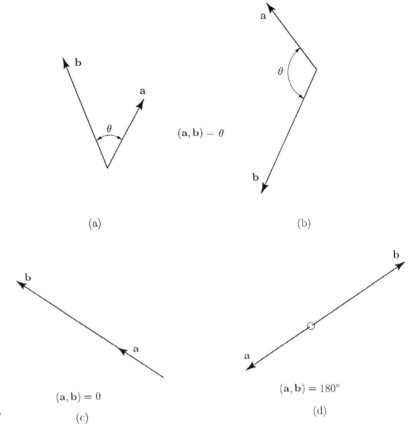

$$(\mathbf{a}, \mathbf{b}) = \theta$$

(a)

(b)

$$(\mathbf{a}, \mathbf{b}) = 0$$

$$(\mathbf{a}, \mathbf{b}) = 180°$$

Figure 1.7   (c)

(d)

and the representative unit vectors. These are called *direction cosines* and are denoted as (Fig. 1.8)

$$\cos(\mathbf{v}, \mathbf{1}) = \cos \alpha = l; \quad \cos(\mathbf{v}, \mathbf{j}) = \cos \beta = m; \quad \cos(\mathbf{v}, \mathbf{k}) = \cos \gamma = n.$$

The following relations exist:

$$v_x = |\mathbf{v}| \cos \alpha; \quad v_y = |\mathbf{v}| \cos \beta; \quad v_z = |\mathbf{v}| \cos \gamma.$$

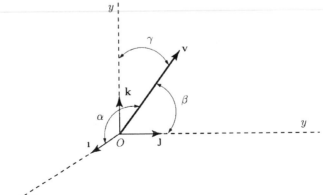

Figure 1.8

1.9 Scalar (Dot) Product of Vectors

DEFINITION The scalar (dot) product of a vector **a** and a vector **b** is

$$\mathbf{a} \cdot \mathbf{b} = \mathbf{b} \cdot \mathbf{a} = |\mathbf{a}||\mathbf{b}| \cos(\mathbf{a}, \mathbf{b}).$$

For any two vectors **a** and **b** and any scalar s

$$(s\mathbf{a}) \cdot \mathbf{b} = s(\mathbf{a} \cdot \mathbf{b}) = \mathbf{a} \cdot (s\mathbf{b}) = s\mathbf{a} \cdot \mathbf{b} \quad \blacktriangle$$

If

$$\mathbf{a} = a_x\mathbf{1} + a_y\mathbf{j} + a_z\mathbf{k}$$

and

$$\mathbf{b} = b_x\mathbf{1} + b_y\mathbf{j} + b_z\mathbf{k},$$

where **1**, **j**, **k** are mutually perpendicular unit vectors, then

$$\mathbf{a} \cdot \mathbf{b} = a_xb_x + a_yb_y + a_zb_z.$$

The following relationships exist:

$$\mathbf{1} \cdot \mathbf{1} = \mathbf{j} \cdot \mathbf{j} = \mathbf{k} \cdot \mathbf{k} = 1,$$

$$\mathbf{1} \cdot \mathbf{j} = \mathbf{j} \cdot \mathbf{k} = \mathbf{k} \cdot \mathbf{1} = 0.$$

Every vector **v** can be expressed in the form

$$\mathbf{v} = \mathbf{1} \cdot \mathbf{vi} + \mathbf{j} \cdot \mathbf{vj} + \mathbf{k} \cdot \mathbf{vk}.$$

The vector **v** can always be expressed as

$$\mathbf{v} = v_x\mathbf{1} + v_y\mathbf{j} + v_z\mathbf{k}.$$

Dot multiply both sides by **1**:

$$\mathbf{1} \cdot \mathbf{v} = v_x\mathbf{1} \cdot \mathbf{1} + v_y\mathbf{1} \cdot \mathbf{j} + v_z\mathbf{1} \cdot \mathbf{k}.$$

But,

$$\mathbf{1} \cdot \mathbf{1} = 1, \quad \text{and} \quad \mathbf{1} \cdot \mathbf{j} = \mathbf{1} \cdot \mathbf{k} = 0.$$

Hence,

$$\mathbf{1} \cdot \mathbf{v} = v_x.$$

Similarly,

$$\mathbf{j} \cdot \mathbf{v} = v_y \quad \text{and} \quad \mathbf{k} \cdot \mathbf{v} = v_z.$$

1.10 Vector (Cross) Product of Vectors

DEFINITION The vector (cross) product of a vector **a** and a vector **b** is the vector (Fig. 1.9)

$$\mathbf{a} \times \mathbf{b} = |\mathbf{a}||\mathbf{b}| \sin(\mathbf{a}, \mathbf{b})\mathbf{n}$$

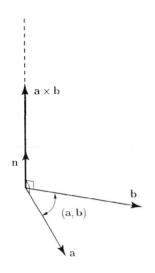

$a \times b \perp a$

$a \times b \perp b$

Figure 1.9

where **n** is a unit vector whose direction is the same as the direction of advance of a right-handed screw rotated from **a** toward **b**, through the angle (**a**, **b**), when the axis of the screw is perpendicular to both **a** and **b**. ▲

The magnitude of $\mathbf{a} \times \mathbf{b}$ is given by

$$|\mathbf{a} \times \mathbf{b}| = |\mathbf{a}||\mathbf{b}| \sin(\mathbf{a}, \mathbf{b}).$$

If **a** is parallel to **b**, **a**||**b**, then $\mathbf{a} \times \mathbf{b} = 0$. The symbol $\parallel$ denotes parallel. The relation $\mathbf{a} \times \mathbf{b} = \mathbf{0}$ implies only that the product $|\mathbf{a}||\mathbf{b}| \sin(\mathbf{a}, \mathbf{b})$ is equal to zero, and this is the case whenever $|\mathbf{a}| = 0$, or $|\mathbf{b}| = 0$, or $\sin(\mathbf{a}, \mathbf{b}) = 0$. For any two vectors **a** and **b** and any real scalar s,

$$(s\mathbf{a}) \times \mathbf{b} = s(\mathbf{a} \times \mathbf{b}) = \mathbf{a} \times (s\mathbf{b}) = s\mathbf{a} \times \mathbf{b}.$$

The sense of the unit vector **n** that appears in the definition of $\mathbf{a} \times \mathbf{b}$ depends on the order of the factors **a** and **b** in such a way that

$$\mathbf{b} \times \mathbf{a} = -\mathbf{a} \times \mathbf{b}.$$

Vector multiplication obeys the following law of distributivity (Varignon theorem):

$$\mathbf{a} \times \sum_{i=1}^{n} \mathbf{v}_i = \sum_{i=1}^{n} (\mathbf{a} \times \mathbf{v}_i).$$

A set of mutually perpendicular unit vectors $\mathbf{\imath}, \mathbf{\jmath}, \mathbf{k}$ is called *right-handed* if $\mathbf{\imath} \times \mathbf{\jmath} = \mathbf{k}$. A set of mutually perpendicular unit vectors $\mathbf{\imath}, \mathbf{\jmath}, \mathbf{k}$ is called *left-handed* if $\mathbf{\imath} \times \mathbf{\jmath} = -\mathbf{k}$.

If

$$\mathbf{a} = a_x\mathbf{\imath} + a_y\mathbf{\jmath} + a_z\mathbf{k},$$

and

$$\mathbf{b} = b_x\mathbf{\imath} + b_y\mathbf{\jmath} + b_z\mathbf{k},$$

where $\mathbf{i}, \mathbf{j}, \mathbf{k}$ are mutually perpendicular unit vectors, then $\mathbf{a} \times \mathbf{b}$ can be expressed in the following determinant form:

$$\mathbf{a} \times \mathbf{b} = \begin{vmatrix} \mathbf{i} & \mathbf{j} & \mathbf{k} \\ a_x & a_y & a_z \\ b_x & b_y & b_z \end{vmatrix}.$$

The determinant can be expanded by minors of the elements of the first row:

$$\begin{vmatrix} \mathbf{i} & \mathbf{j} & \mathbf{k} \\ a_x & a_y & a_z \\ b_x & b_y & b_z \end{vmatrix} = \mathbf{i} \begin{vmatrix} a_y & a_z \\ b_y & b_z \end{vmatrix} - \mathbf{j} \begin{vmatrix} a_x & a_z \\ b_x & b_z \end{vmatrix} + \mathbf{k} \begin{vmatrix} a_x & a_y \\ b_x & b_y \end{vmatrix}$$

$$= \mathbf{i}(a_y b_z - a_z b_y) - \mathbf{j}(a_x b_z - a_z b_x) + \mathbf{k}(a_x b_y - a_y b_x)$$

$$= (a_y b_z - a_z b_y)\mathbf{i} + (a_z b_x - a_x b_z)\mathbf{j} + (a_x b_y - a_y b_x)\mathbf{k}.$$

1.11 Scalar Triple Product of Three Vectors

DEFINITION The scalar triple product of three vectors $\mathbf{a}, \mathbf{b}, \mathbf{c}$ is

$$[\mathbf{a}, \mathbf{b}, \mathbf{c}] \equiv \mathbf{a} \cdot (\mathbf{b} \times \mathbf{c}) = \mathbf{a} \cdot \mathbf{b} \times \mathbf{c}. \quad \blacktriangle$$

It does not matter whether the dot is placed between $\mathbf{a}$ and $\mathbf{b}$, and the cross between $\mathbf{b}$ and $\mathbf{c}$, or vice versa, that is,

$$[\mathbf{a}, \mathbf{b}, \mathbf{c}] = \mathbf{a} \cdot \mathbf{b} \times \mathbf{c} = \mathbf{a} \times \mathbf{b} \cdot \mathbf{c}.$$

A change in the order of the factor appearing in a scalar triple product at most changes the sign of the product, that is,

$$[\mathbf{b}, \mathbf{a}, \mathbf{c}] = -[\mathbf{a}, \mathbf{b}, \mathbf{c}],$$

and

$$[\mathbf{b}, \mathbf{c}, \mathbf{a}] = [\mathbf{a}, \mathbf{b}, \mathbf{c}].$$

If $\mathbf{a}, \mathbf{b}, \mathbf{c}$ are parallel to the same plane, or if any two of the vectors $\mathbf{a}, \mathbf{b}, \mathbf{c}$ are parallel to each other, then $[\mathbf{a}, \mathbf{b}, \mathbf{c}] = 0$.

The scalar triple product $[\mathbf{a}, \mathbf{b}, \mathbf{c}]$ can be expressed in the following determinant form:

$$[\mathbf{a}, \mathbf{b}, \mathbf{c}] = \begin{vmatrix} a_x & a_y & a_z \\ b_x & b_y & b_z \\ c_x & c_y & c_z \end{vmatrix}.$$

1.12 Vector Triple Product of Three Vectors

DEFINITION The vector triple product of three vectors $\mathbf{a}, \mathbf{b}, \mathbf{c}$ is the vector $\mathbf{a} \times (\mathbf{b} \times \mathbf{c})$. $\quad \blacktriangle$

The parentheses are essential because $\mathbf{a} \times (\mathbf{b} \times \mathbf{c})$ is not, in general, equal to $(\mathbf{a} \times \mathbf{b}) \times \mathbf{c}$.

For any three vectors $\mathbf{a}, \mathbf{b},$ and $\mathbf{c}$,

$$\mathbf{a} \times (\mathbf{b} \times \mathbf{c}) = \mathbf{a} \cdot \mathbf{c}\,\mathbf{b} - \mathbf{a} \cdot \mathbf{b}\,\mathbf{c}.$$

1.13 *Derivative of a Vector*

The derivative of a vector is defined in exactly the same way as is the derivative of a scalar function. The derivative of a vector has some of the properties of the derivative of a scalar function.

The derivative of the sum of two vector functions **a** and **b** is

$$\frac{d}{dt}(\mathbf{a} + \mathbf{b}) = \frac{d\mathbf{a}}{dt} + \frac{d\mathbf{b}}{dt},$$

The time derivative of the product of a scalar function f and a vector function **u** is

$$\frac{d(f\mathbf{a})}{dt} = \frac{df}{dt}\mathbf{a} + f\frac{d\mathbf{a}}{dt}.$$

2. Centroids and Surface Properties

2.1 *Position Vector*

The position vector of a point P relative to a point O is a vector $\mathbf{r}_{OP} = \mathbf{OP}$ having the following characteristics:

- Magnitude the length of line OP
- Orientation parallel to line OP
- Sense OP (from point O to point P)

The vector $\mathbf{r}_{OP}$ is shown as an arrow connecting O to P (Fig. 2.1). The position of a point P relative to P is a zero vector.

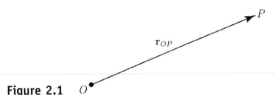

Figure 2.1

Let $\mathbf{1}, \mathbf{j}, \mathbf{k}$ be mutually perpendicular unit vectors (cartesian reference frame) with the origin at O (Fig. 2.2). The axes of the cartesian reference frame are x, y, z. The unit vectors $\mathbf{1}, \mathbf{j}, \mathbf{k}$ are parallel to x, y, z, and they have the senses of the positive x, y, z axes. The coordinates of the origin O are $x = y = z = 0$, that is, $O(0, 0, 0)$. The coordinates of a point P are $x = x_P$, $y = y_P, z = z_P$, that is, $P(x_P, y_P, z_P)$. The position vector of P relative to the origin O is

$$\mathbf{r}_{OP} = \mathbf{r}_P = \mathbf{OP} = x_P\mathbf{1} + y_P\mathbf{j} + z_P\mathbf{k}.$$

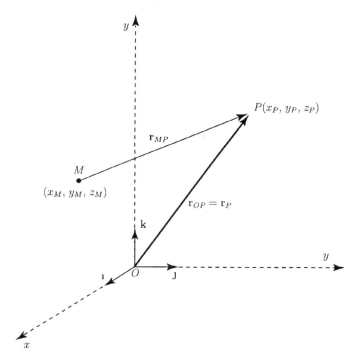

Figure 2.2

The position vector of the point P relative to a point M, $M \neq O$, of coordinates (x_M, y_M, z_M) is

$$\mathbf{r}_{MP} = \mathbf{MP} = (x_P - x_M)\mathbf{\imath} + (y_P - y_M)\mathbf{\jmath} + (z_P - z_M)\mathbf{k}.$$

The distance d between P and M is given by

$$d = |\mathbf{r}_P - \mathbf{r}_M| = |\mathbf{r}_{MP}| = |\mathbf{MP}| = \sqrt{(x_P - x_M)^2 + (y_P - y_M)^2 + (z_P - z_M)^2}.$$

2.2 First Moment

The position vector of a point P relative to a point O is $\mathbf{r}_P$ and a scalar associated with P is s, for example, the mass m of a particle situated at P. The *first moment* of a point P with respect to a point O is the vector $\mathbf{M} = s\mathbf{r}_P$. The scalar s is called the *strength* of P.

2.3 Centroid of a Set of Points

The set of n points P_i, $i = 1, 2, \ldots, n$, is $\{S\}$ (Fig. 2.3a)

$$\{S\} = \{P_1, P_2, \ldots, P_n\} = \{P_i\}_{i=1,2,\ldots,n}.$$

The strengths of the points P_i are s_i, $i = 1, 2, \ldots, n$, that is, n scalars, all having the same dimensions, and each associated with one of the points of $\{S\}$. The *centroid* of the set $\{S\}$ is the point C with respect to which the sum of the first moments of the points of $\{S\}$ is equal to zero.

The position vector of C relative to an arbitrarily selected reference point O is $\mathbf{r}_C$ (Fig. 2.3b). The position vector of P_i relative to O is $\mathbf{r}_i$. The position

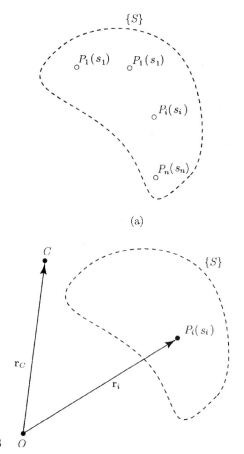

Figure 2.3

vector of P_i relative to C is $\mathbf{r}_i - \mathbf{r}_C$. The sum of the first moments of the points P_i with respect to C is $\sum_{i=1}^{n} s_i(\mathbf{r}_i - \mathbf{r}_C)$. If C is to be the centroid of $\{S\}$, this sum is equal to zero:

$$\sum_{i=1}^{n} s_i(\mathbf{r}_i - \mathbf{r}_C) = \sum_{i=1}^{n} s_i\mathbf{r}_i - \mathbf{r}_C \sum_{i=1}^{n} s_i = 0.$$

The position vector $\mathbf{r}_C$ of the centroid C, relative to an arbitrarily selected reference point O, is given by

$$\mathbf{r}_C = \frac{\sum_{i=1}^{n} s_i\mathbf{r}_i}{\sum_{i=1}^{n} s_i}.$$

If $\sum_{i=1}^{n} s_i = 0$ the centroid is not defined.

The centroid C of a set of points of given strength is a unique point, its location being independent of the choice of reference point O.

The cartesian coordinates of the centroid $C(x_C, y_C, z_C)$ of a set of points P_i, $i = 1, \ldots, n$, of strengths s_i, $i = 1, \ldots, n$, are given by the expressions

$$x_C = \frac{\sum\limits_{i=1}^{n} s_i x_i}{\sum\limits_{i=1}^{n} s_i}, \quad y_C = \frac{\sum\limits_{i=1}^{n} s_i y_i}{\sum\limits_{i=1}^{n} s_i}, \quad z_C = \frac{\sum\limits_{i=1}^{n} s_i z_i}{\sum\limits_{i=1}^{n} s_i}.$$

The *plane of symmetry* of a set is the plane where the centroid of the set lies, the points of the set being arranged in such a way that corresponding to every point on one side of the plane of symmetry there exists a point of equal strength on the other side, the two points being equidistant from the plane.

A set $\{S'\}$ of points is called a *subset* of a set $\{S\}$ if every point of $\{S'\}$ is a point of $\{S\}$. The centroid of a set $\{S\}$ may be located using the *method of decomposition*:

- Divide the system $\{S\}$ into subsets
- Find the centroid of each subset
- Assign to each centroid of a subset a strength proportional to the sum of the strengths of the points of the corresponding subset
- Determine the centroid of this set of centroids

2.4 Centroid of a Curve, Surface, or Solid

The position vector of the centroid C of a curve, surface, or solid relative to a point O is

$$\mathbf{r}_C = \frac{\int_D \mathbf{r} \, d\tau}{\int_D d\tau},$$

where D is a curve, surface, or solid; $\mathbf{r}$ denotes the position vector of a typical point of D, relative to O; and $d\tau$ is the length, area, or volume of a differential element of D. Each of the two limits in this expression is called an "integral over the domain D (curve, surface, or solid)."

The integral $\int_D d\tau$ gives the total length, area, or volume of D, that is,

$$\int_D d\tau = \tau.$$

The position vector of the centroid is

$$\mathbf{r}_C = \frac{1}{\tau} \int_D \mathbf{r} \, d\tau.$$

Let $\mathbf{i}, \mathbf{j}, \mathbf{k}$ be mutually perpendicular unit vectors (cartesian reference frame) with the origin at O. The coordinates of C are x_C, y_C, z_C and

$$\mathbf{r}_C = x_C \mathbf{i} + y_C \mathbf{j} + z_C \mathbf{k}.$$

It results that

$$x_C = \frac{1}{\tau} \int_D x \, d\tau, \quad y_C = \frac{1}{\tau} \int_D y \, d\tau, \quad z_C = \frac{1}{\tau} \int_D z \, d\tau.$$

2.5 Mass Center of a Set of Particles

The *mass center* of a set of particles $\{S\} = \{P_1, P_2, \ldots, P_n\} = \{P_i\}_{i=1,2,\ldots,n}$ is the centroid of the set of points at which the particles are situated with the strength of each point being taken equal to the mass of the corresponding particle, $s_i = m_i$, $i = 1, 2, \ldots, n$. For the system of n particles in Fig. 2.4, one can say

$$\left(\sum_{i=1}^{n} m_i \right) \mathbf{r}_C = \sum_{i=1}^{n} m_i \mathbf{r}_i.$$

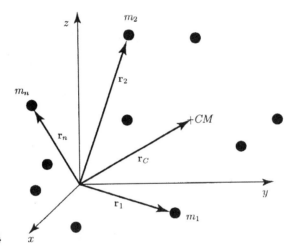

Figure 2.4

Therefore, the mass center position vector is

$$\mathbf{r}_C = \frac{\displaystyle\sum_{i=1}^{n} m_i \mathbf{r}_i}{M}, \tag{2.1}$$

where M is the total mass of the system.

2.6 Mass Center of a Curve, Surface, or Solid

The position vector of the mass center C of a continuous body D, curve, surface, or solid, relative to a point O is

$$\mathbf{r}_C = \frac{1}{m} \int_D \mathbf{r} \rho \, d\tau,$$

or using the orthogonal cartesian coordinates

$$x_C = \frac{1}{m}\int_D x\rho \, d\tau, \quad y_C = \frac{1}{m}\int_D y\rho \, d\tau, \quad z_C = \frac{1}{m}\int_D z\rho \, d\tau,$$

where ρ is the mass density of the body: mass per unit of length if D is a curve, mass per unit area if D is a surface, and mass per unit of volume if D is a solid; $\mathbf{r}$ is the position vector of a typical point of D, relative to O; $d\tau$ is the length, area, or volume of a differential element of D; $m = \int_D \rho \, d\tau$ is the total mass of the body; and x_C, y_C, z_C are the coordinates of C.

If the mass density ρ of a body is the same at all points of the body, ρ constant, the density, as well as the body, are said to be *uniform*. The mass center of a uniform body coincides with the centroid of the figure occupied by the body.

The *method of decomposition* may be used to locate the mass center of a continuous body B:

- Divide the body B into a number of bodies, which may be particles, curves, surfaces, or solids
- locate the mass center of each body
- assign to each mass center a strength proportional to the mass of the corresponding body (e.g., the weight of the body)
- locate the centroid of this set of mass centers

2.7 First Moment of an Area

A planar surface of area A and a reference frame xOy in the plane of the surface are shown in Fig. 2.5. The first moment of area A about the x axis is

$$M_x = \int_A y \, dA, \tag{2.2}$$

and the first moment about the y axis is

$$M_y = \int_A x \, dA. \tag{2.3}$$

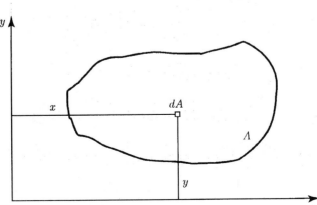

Figure 2.5 O

The first moment of area gives information of the shape, size, and orientation of the area.

The entire area A can be concentrated at a position $C(x_C, y_C)$, the centroid (Fig. 2.6). The coordinates x_C and y_C are the centroidal coordinates. To compute the centroidal coordinates one can equate the moments of the distributed area with that of the concentrated area about both axes:

$$Ay_C = \int_A y \, dA, \quad \Rightarrow y_C = \frac{\int_A y \, dA}{A} = \frac{M_x}{A} \tag{2.4}$$

$$Ax_C = \int_A x \, dA, \quad \Rightarrow x_C = \frac{\int_A x \, dA}{A} = \frac{M_y}{A}. \tag{2.5}$$

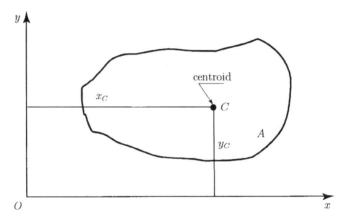

Figure 2.6

The location of the centroid of an area is independent of the reference axes employed, that is, the centroid is a property only of the area itself.

If the axes xy have their origin at the centroid, $O \equiv C$, then these axes are called *centroidal axes*. The first moments about centroidal axes are zero. All axes going through the centroid of an area are called centroidal axes for that area, and the first moments of an area about any of its centroidal axes are zero. The perpendicular distance from the centroid to the centroidal axis must be zero.

In Fig. 2.7 is shown a plane area with the axis of symmetry collinear with the axis y. The area A can be considered as composed of area elements in symmetric pairs such as that shown in Fig. 2.7. The first moment of such a pair about the axis of symmetry y is zero. The entire area can be considered as composed of such symmetric pairs and the coordinate x_C is zero:

$$x_C = \frac{1}{A} \int_A x \, dA = 0.$$

Thus, *the centroid of an area with one axis of symmetry must lie along the axis of symmetry*. The axis of symmetry then is a centroidal axis, which is another indication that the first moment of area must be zero about the axis

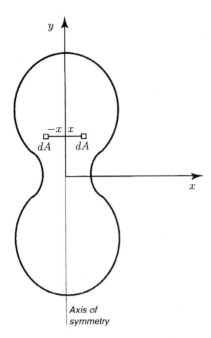

Figure 2.7

Axis of symmetry

of symmetry. With two orthogonal axes of symmetry, the centroid must lie at the intersection of these axes. For such areas as circles and rectangles, the centroid is easily determined by inspection.

In many problems, the area of interest can be considered formed by the addition or subtraction of simple areas. For simple areas the centroids are known by inspection. The areas made up of such simple areas are *composite* areas. For composite areas,

$$x_C = \frac{\sum_i A_i x_{Ci}}{A}$$

$$y_C = \frac{\sum_i A_i y_{Ci}}{A},$$

where x_{Ci} and y_{Ci} (with proper signs) are the centroidal coordinates to simple area A_i, and where A is the total area.

The centroid concept can be used to determine the simplest resultant of a distributed loading. In Fig. 2.8 the distributed load $w(x)$ is considered. The resultant force F_R of the distributed load $w(x)$ loading is given as

$$F_R = \int_0^L w(x)\, dx. \tag{2.6}$$

From this equation, the *resultant force equals the area under the loading curve*. The position, $\bar{x}$, of the *simplest* resultant load can be calculated from the relation

$$F_R \bar{x} = \int_0^L x w(x)\, dx \Rightarrow \bar{x} = \frac{\int_0^L x w(x)\, dx}{F_R}. \tag{2.7}$$

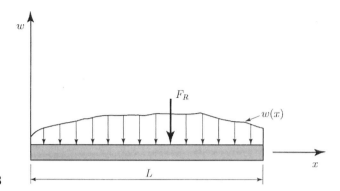

Figure 2.8

The position $\bar{x}$ is actually the centroid coordinate of the loading curve area. Thus, the *simplest resultant force of a distributed load acts at the centroid of the area under the loading curve.*

Example 1

For the triangular load shown in Fig. 2.9, one can replace the distributed loading by a force F equal to $(\frac{1}{2})(w_0)(b-a)$ at a position $\frac{1}{3}(b-a)$ from the right end of the distributed loading. ▲

Example 2

For the curved line shown in Fig. 2.10 the centroidal position is

$$x_C = \frac{\int x \, dl}{L}, \quad y_C = \frac{\int y \, dl}{L}, \tag{2.8}$$

where L is the length of the line. Note that the centroid C will not generally lie along the line. Next one can consider a curve made up of simple curves. For each simple curve the centroid is known. Figure 2.11 represents a curve

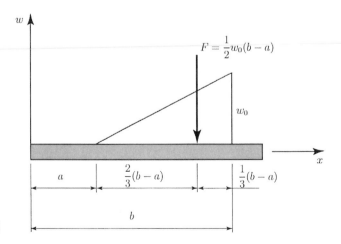

Figure 2.9

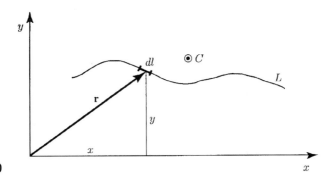

Figure 2.10

made up of straight lines. The line segment L_1 has the centroid C_1 with coordinates x_{C1}, y_{C1}, as shown in the diagram. For the entire curve

$$x_C = \frac{\sum_{i=1}^{4} x_{Ci} L_i}{L}, \quad y_C = \frac{\sum_{i=1}^{4} y_{Ci} L_i}{L}. \quad \blacktriangle$$

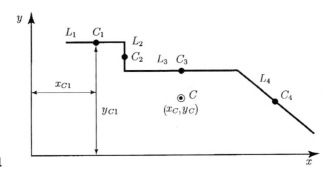

Figure 2.11

2.8 Theorems of Guldinus–Pappus

The theorems of Guldinus–Pappus are concerned with the relation of a surface of revolution to its generating curve, and the relation of a volume of revolution to its generating area.

THEOREM   Consider a coplanar generating curve and an axis of revolution in the plane of this curve (Fig. 2.12). The surface of revolution A developed by rotating the generating curve about the axis of revolution equals the product of the length of the generating L curve times the circumference of the circle formed by the centroid of the generating curve y_C in the process of generating a surface of revolution

$$A = 2\pi y_C L.$$

The generating curve can touch but must not cross the axis of revolution. $\blacktriangle$

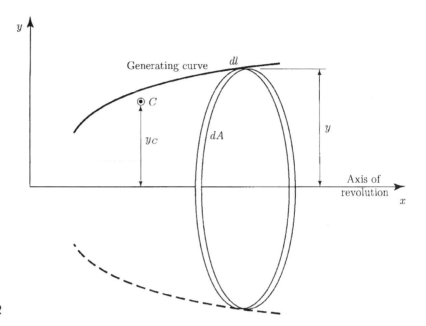

Figure 2.12

Proof

An element dl of the generating curve is considered in Fig. 2.12. For a single revolution of the generating curve about the x axis, the line segment dl traces an area

$$dA = 2\pi y \, dl.$$

For the entire curve this area, dA, becomes the surface of revolution, A, given as

$$A = 2\pi \int y \, dl = 2\pi y_C L, \tag{2.9}$$

where L is the length of the curve and y_C is the centroidal coordinate of the curve. The circumferential length of the circle formed by having the centroid of the curve rotate about the x axis is $2\pi y_C$. ▲

The surface of revolution A is equal to 2π times the first moment of the generating curve about the axis of revolution.

If the generating curve is composed of simple curves, L_i, whose centroids are known (Fig. 2.11), the surface of revolution developed by revolving the composed generating curve about the axis of revolution x is

$$A = 2\pi \left(\sum_{i=1}^{4} L_i y_{Ci} \right), \tag{2.10}$$

where y_{Ci} is the centroidal coordinate to the ith line segment L_i.

THEOREM Consider a generating plane surface A and an axis of revolution coplanar with the surface (Fig. 2.13). The volume of revolution V developed by rotating the generating plane surface about the axis of revolution equals the product of the area of the surface times the circumference of the circle formed by the centroid of the surface y_C in the process of generating the body of revolution

$$V = 2\pi y_C A.$$

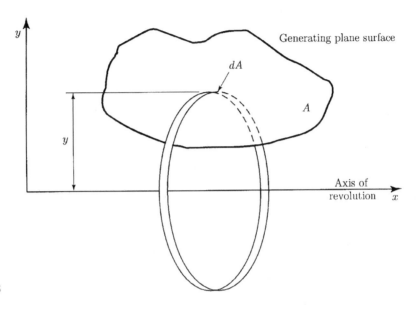

Figure 2.13

The axis of revolution can intersect the generating plane surface only as a tangent at the boundary or can have no intersection at all.   ▲

Proof

The plane surface A is shown in Fig. 2.13. The volume generated by rotating an element dA of this surface about the x axis is

$$dV = 2\pi y \, dA.$$

The volume of the body of revolution formed from A is then

$$V = 2\pi \int_A y \, dA = 2\pi y_C A. \tag{2.11}$$

Thus, the volume V equals the area of the generating surface A times the circumferential length of the circle of radius y_C.   ▲

The volume V equals 2π times the first moment of the generating area A about the axis of revolution.

2.9 Second Moments and the Product of Area

The *second moments* of the area A about x and y axes (Fig. 2.14), denoted as I_{xx} and I_{yy}, respectively, are

$$I_{xx} = \int_A y^2 \, dA \tag{2.12}$$

$$I_{yy} = \int_A x^2 \, dA. \tag{2.13}$$

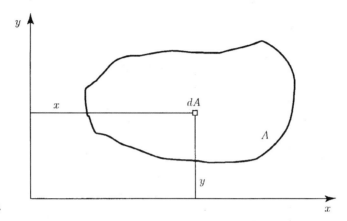

Figure 2.14

The second moment of area cannot be negative.

The entire area may be concentrated at a single point (k_x, k_y) to give the same moment of area for a given reference. The distance k_x and k_y are called the *radii of gyration*. Thus,

$$Ak_x^2 = I_{xx} = \int_A y^2 \, dA \Rightarrow k_x^2 = \frac{\int_A y^2 \, dA}{A}$$

$$Ak_y^2 = I_{yy} = \int_A x^2 \, dA \Rightarrow k_y^2 = \frac{\int_A x^2 \, dA}{A}. \tag{2.14}$$

This point (k_x, k_y) depends on the shape of the area and on the position of the reference. The centroid location is independent of the reference position.

DEFINITION The *product of area* is defined as

$$I_{xy} = \int_A xy \, dA. \tag{2.15}$$

This quantity may be negative and relates an area directly to a set of axes. ▲

If the area under consideration has an axis of symmetry, the product of area for this axis and any axis orthogonal to this axis is zero. Consider the area in Fig. 2.15, which is symmetrical about the vertical axis y. The planar cartesian frame is xOy. The centroid is located somewhere along the symmetrical axis y. Two elemental areas that are positioned as mirror images about the y axis are shown in Fig. 2.15. The contribution to the product of area of each elemental area is $xy\, dA$, but with opposite signs, and so the result is zero. The entire area is composed of such elemental area pairs, and the product of area is zero.

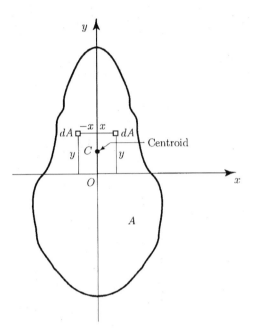

Figure 2.15

2.10 Transfer Theorems or Parallel-Axis Theorems

The x axis in Fig. 2.16 is parallel to an axis x' and it is at a distance b from the axis x'. The axis x' is going through the centroid C of the A area, and it is a *centroidal axis*. The second moment of area about the x axis is

$$I_{xx} = \int_A y^2\, dA = \int_A (y' + b)^2\, dA,$$

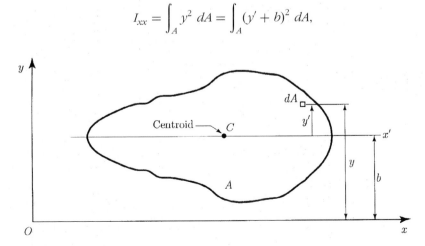

Figure 2.16

where the distance $y = y' + b$. Carrying out the operations,

$$I_{xx} = \int_A y'^2\,dA + 2b\int_A y'\,dA + Ab^2.$$

The first term of the right-hand side is by definition $I_{x'x'}$,

$$I_{x'x'} = \int_A y'^2\,dA.$$

The second term involves the first moment of area about the x' axis, and it is zero because the x' axis is a centroidal axis:

$$\int_A y'\,dA = 0.$$

THEOREM The second moment of the area A about any axis I_{xx} is equal to the second moment of the area A about a parallel axis at centroid $I_{x'x'}$ plus Ab^2, where b is the perpendicular distance between the axis for which the second moment is being computed and the parallel centroidal axis

$$I_{xx} = I_{x'x'} + Ab^2. \quad \blacktriangle$$

With the transfer theorem, one can find second moments or products of area about any axis in terms of second moments or products of area about a *parallel* set of axes going through the *centroid* of the area in question.

In handbooks the areas and second moments about various centroidal axes are listed for many of the practical configurations, and using the parallel-axis theorem second moments can be calculated for axes not at the centroid.

In Fig. 2.17 are shown two references, one x', y' at the centroid and the other x, y arbitrary but positioned parallel relative to x', y'. The coordinates of the centroid $C(x_C, y_C)$ of area A measured from the reference x, y are c and b, $x_C = c$, $y_C = b$. The centroid coordinates must have the proper signs. The product of area about the noncentroidal axes xy is

$$I_{xy} = \int_A xy\,dA = \int_A (x' + c)(y' + b)\,dA,$$

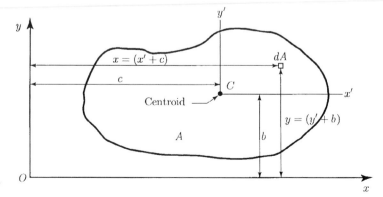

Figure 2.17

or

$$I_{xy} = \int_A x'y' \, dA + c \int_A y' \, dA + b \int_A x' \, dA + Abc.$$

The first term of the right-hand side is by definition $I_{x'y'}$,

$$I_{x'y'} = \int_A x'y' \, dA.$$

The next two terms of the right-hand side are zero since x' and y' are centroidal axes:

$$\int_A y' \, dA = 0 \quad \text{and} \quad \int_A x' \, dA = 0.$$

Thus, the parallel-axis theorem for products of area is as follows.

THEOREM The product of area for any set of axes I_{xy} is equal to the product of area for a parallel set of axes at centroid $I_{x'y'}$ plus Acb, where c and b are the coordinates of the centroid of area A,

$$I_{xy} = I_{x'y'} + Acb. \quad \blacktriangle$$

With the transfer theorem, one can find second moments or products of area about any axis in terms of second moments or products of area about a parallel set of axes going through the centroid of the area.

2.11 Polar Moment of Area

In Fig. 2.18, there is a reference xy associated with the origin O. Summing I_{xx} and I_{yy},

$$I_{xx} + I_{yy} = \int_A y^2 \, dA + \int_A x^2 \, dA$$

$$= \int_A (x^2 + y^2) \, dA = \int_A r^2 \, dA,$$

where $r^2 = x^2 + y^2$. The distance r^2 is independent of the orientation of the reference, and the sum $I_{xx} + I_{yy}$ is independent of the orientation of the

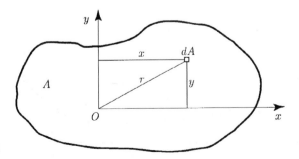

Figure 2.18

coordinate system. Therefore, the sum of second moments of area about orthogonal axes is a function only of the position of the origin O for the axes.

The *polar moment of area* about the origin O is

$$I_O = I_{xx} + I_{yy}. \tag{2.16}$$

The polar moment of area is an *invariant* of the system. The group of terms $I_{xx}I_{yy} - I_{xy}^2$ is also invariant under a rotation of axes.

2.12 Principal Axes

In Fig. 2.19, an area A is shown with a reference xy having its origin at O. Another reference $x'y'$ with the same origin O is rotated with angle α from xy (counterclockwise as positive). The relations between the coordinates of the area elements dA for the two references are

$$x' = x \cos \alpha + y \sin \alpha$$
$$y' = -x \sin \alpha + y \cos \alpha.$$

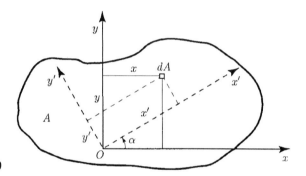

Figure 2.19

The second moment $I_{x'x'}$ can be expressed as

$$I_{x'x'} = \int_A (y')^2 dA = \int_A (-x \sin \alpha + y \cos \alpha)^2 dA$$
$$= \sin^2 \alpha \int_A x^2 dA - 2 \sin \alpha \cos \alpha \int_A xy \, dA + \cos^2 \alpha \int_A y^2 dA$$
$$= I_{yy} \sin^2 + I_{xx} \cos^2 - 2I_{xy} \sin \alpha \cos \alpha. \tag{2.17}$$

Using the trigonometric identities

$$\cos^2 \alpha = 0.5(1 + \cos 2\alpha)$$
$$\sin^2 \alpha = 0.5(1 - \cos 2\alpha)$$
$$2 \sin \alpha \cos \alpha = \sin 2\alpha,$$

Eq. (2.17) becomes

$$I_{x'x'} = \frac{I_{xx} + I_{yy}}{2} + \frac{I_{xx} - I_{yy}}{2} \cos 2\alpha - I_{xy} \sin 2\alpha. \tag{2.18}$$

If we replace α with $\alpha + \pi/2$ in Eq. (2.18) and use the trigonometric relations

$$\cos(2\alpha + \pi) = -\cos 2\alpha, \quad \sin(2\alpha + \pi) = -\cos 2\sin,$$

the second moment $I_{y'y'}$ can be computed:

$$I_{y'y'} = \frac{I_{xx} + I_{yy}}{2} - \frac{I_{xx} - I_{yy}}{2}\cos 2\alpha + I_{xy}\sin 2\alpha. \tag{2.19}$$

Next, the product of area $I_{x'y'}$ is computed in a similar manner:

$$I_{x'y'} = \int_A x'y'\,dA = \frac{I_{xx} - I_{yy}}{2}\sin 2\alpha + I_{xy}\cos 2\alpha. \tag{2.20}$$

If i_{xx}, I_{yy}, and I_{xy} are known for a reference xy with an origin O, then the second moments and products of area for every set of axes at O can be computed.

Next, it is assumed that I_{xx}, I_{yy}, and I_{xy} are known for a reference xy (Fig. 2.20). The sum of the second moments of area is constant for any reference with origin at O. The *minimum* second moment of area corresponds to an axis at *right angles* to the axis having the *maximum* second moment. The second moments of area can be expressed as functions of the angle variable α. The maximum second moment may be determined by setting the partial derivative of $I_{x'y'}$ with respect to α equals to zero. Thus,

$$\frac{\partial I_{x'x'}}{\partial \alpha} = (I_{xx} - I_{yy})(-\sin 2\alpha) - 2I_{xy}\cos 2\alpha = 0, \tag{2.21}$$

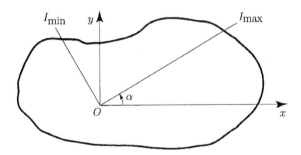

Figure 2.20

or

$$(I_{yy} - I_{xx})\sin 2\alpha_0 - 2I_{xy}\cos 2\alpha_0 = 0,$$

where α_0 is the value of α that satisfies Eq. (2.21). Hence,

$$\tan 2\alpha_0 = \frac{2I_{xy}}{I_{yy} - I_{xx}}. \tag{2.22}$$

The angle α_0 corresponds to an extreme value of $I_{x'x'}$ (i.e., to a maximum or minimum value). There are two possible values of $2\alpha_0$, which are π radians apart, that will satisfy the equation just shown. Thus,

$$2\alpha_{01} = \tan^{-1}\frac{2I_{xy}}{I_{yy} - I_{xx}} \Rightarrow \alpha_{01} = 0.5\tan^{-1}\frac{2I_{xy}}{I_{yy} - I_{xx}},$$

or

$$2\alpha_{02} = \tan^{-1}\frac{2I_{xy}}{I_{yy} - I_{xx}} + \pi \Rightarrow \alpha_{02} = 0.5\tan^{-1}\frac{2I_{xy}}{I_{yy} - I_{xx}} + 0.5\pi.$$

This means that there are two axes orthogonal to each other having extreme values for the second moment of area at 0. On one of the axes is the maximum second moment of area, and the minimum second moment of area is on the other axis. These axes are called the *principal axes*.

With $\alpha = \alpha_0$, the product of area $I_{x'y'}$ becomes

$$I_{x'y'} = \frac{I_{xx} - I_{yy}}{2}\sin 2\alpha_0 + I_{xy}\cos 2\alpha_0. \tag{2.23}$$

By Eq. (2.22), the sine and cosine expressions are

$$\sin 2\alpha_0 = \frac{2I_{xy}}{\sqrt{(I_{yy} - I_{xx})^2 + 4I_{xy}^2}}$$

$$\cos 2\alpha_0 = \frac{I_{yy} - I_{xx}}{\sqrt{(I_{yy} - I_{xx})^2 + 4I_{xy}^2}}.$$

Substituting these results into Eq. (2.23) gives

$$I_{x'y'} = -(I_{yy} - I_{xx})\frac{I_{xy}}{[(I_{yy} - I_{xx})^2 + 4I_{xy}^2]^{1/2}} + I_{xy}\frac{I_{yy} - I_{xx}}{[(I_{yy} - I_{xx})^2 + 4I_{xy}^2]^{1/2}}.$$

Thus,

$$I_{x'y'} = 0.$$

The *product of area corresponding to the principal axes is zero.*

3. Moments and Couples

3.1 Moment of a Bound Vector about a Point

DEFINITION The moment of a bound vector $\mathbf{v}$ about a point A is the vector

$$\mathbf{M}_A^{\mathbf{v}} = \mathbf{AB} \times \mathbf{v} = \mathbf{r}_{AB} \times \mathbf{v}, \tag{3.1}$$

where $\mathbf{r}_{AB} = \mathbf{AB}$ is the position vector of B relative to A, and B is any point of line of action, Δ, of the vector $\mathbf{v}$ (Fig. 3.1). ▲

The vector $\mathbf{M}_A^{\mathbf{v}} = \mathbf{0}$ if the line of action of $\mathbf{v}$ passes through A or $\mathbf{v} = \mathbf{0}$. The magnitude of $\mathbf{M}_A^{\mathbf{v}}$ is

$$|\mathbf{M}_A^{\mathbf{v}}| = M_A^{\mathbf{v}} = |\mathbf{r}_{AB}||\mathbf{v}|\sin\theta,$$

where θ is the angle between $\mathbf{r}_{AB}$ and $\mathbf{v}$ when they are placed tail to tail. The perpendicular distance from A to the line of action of $\mathbf{v}$ is

$$d = |\mathbf{r}_{AB}|\sin\theta,$$

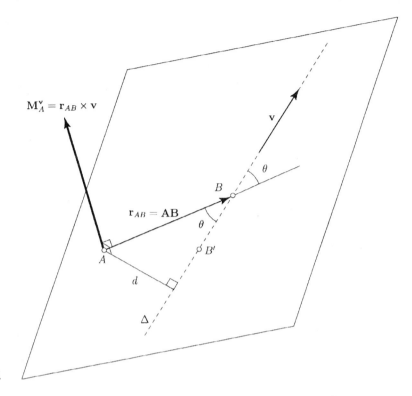

Figure 3.1

and the magnitude of $\mathbf{M}_A^{\mathbf{v}}$ is

$$|\mathbf{M}_A^{\mathbf{v}}| = M_A^{\mathbf{v}} = d|\mathbf{v}|.$$

The vector $\mathbf{M}_A^{\mathbf{v}}$ is perpendicular to both $\mathbf{r}_{AB}$ and $\mathbf{v}$: $\mathbf{M}_A^{\mathbf{v}} \perp \mathbf{r}_{AB}$ and $\mathbf{M}_A^{\mathbf{v}} \perp \mathbf{v}$. The vector $\mathbf{M}_A^{\mathbf{v}}$ being perpendicular to $\mathbf{r}_{AB}$ and $\mathbf{v}$ is perpendicular to the plane containing $\mathbf{r}_{AB}$ and $\mathbf{v}$.

The moment vector $\mathbf{M}_A^{\mathbf{v}}$ is a free vector, that is, a vector associated neither with a definite line nor with a definite point. The moment given by Eq. (3.1) does not depend on the point B of the line of action of $\mathbf{v}$, Δ, where $\mathbf{r}_{AB}$ intersects Δ. Instead of using the point B, one could use the point B' (Fig. 3.1). The vector $\mathbf{r}_{AB} = \mathbf{r}_{AB'} + \mathbf{r}_{B'B}$ where the vector $\mathbf{r}_{B'B}$ is parallel to $\mathbf{v}$, $\mathbf{r}_{B'B} \| \mathbf{v}$. Therefore,

$$\mathbf{M}_A^{\mathbf{v}} = \mathbf{r}_{AB} \times \mathbf{v} = (\mathbf{r}_{AB'} + \mathbf{r}_{B'B}) \times \mathbf{v} = \mathbf{r}_{AB'} \times \mathbf{v} + \mathbf{r}_{B'B} \times \mathbf{v} = \mathbf{r}_{AB'} \times \mathbf{v},$$

because $\mathbf{r}_{B'B} \times \mathbf{v} = \mathbf{0}$.

3.2 Moment of a Bound Vector about a Line

DEFINITION The moment $\mathbf{M}_\Omega^{\mathbf{v}}$ of a bound vector $\mathbf{v}$ about a line Ω is the Ω resolute (Ω component) of the moment $\mathbf{v}$ about any point on Ω (Fig. 3.2). ▲

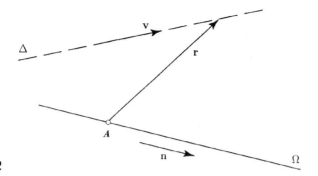

Figure 3.2

The $\mathbf{M}_\Omega^\mathbf{v}$ is the Ω resolute of $\mathbf{M}_A^\mathbf{v}$,

$$\mathbf{M}_\Omega^\mathbf{v} = \mathbf{n} \cdot \mathbf{M}_A^\mathbf{v}\mathbf{n}$$
$$= \mathbf{n} \cdot (\mathbf{r} \times \mathbf{v})\mathbf{n}$$
$$= [\mathbf{n}, \mathbf{r}, \mathbf{v}]\mathbf{n},$$

where $\mathbf{n}$ is a unit vector parallel to Ω, and $\mathbf{r}$ is the position vector of a point on the line of action $\mathbf{v}$ relative to the point on Ω.

The magnitude of $\mathbf{M}_\Omega^\mathbf{v}$ is given by

$$|\mathbf{M}_\Omega^\mathbf{v}| = |[\mathbf{n}, \mathbf{r}, \mathbf{v}]|.$$

The moment of a vector about a line is a free vector.

If a line Ω is parallel to the line of action Δ of a vector $\mathbf{v}$, then $[\mathbf{n}, \mathbf{r}, \mathbf{v}]\mathbf{n} = \mathbf{0}$ and $\mathbf{M}_\Omega^\mathbf{v} = \mathbf{0}$.

If a line Ω intersects the line of action Δ of $\mathbf{v}$, then $\mathbf{r}$ can be chosen in such a way that $\mathbf{r} = \mathbf{0}$ and $\mathbf{M}_\Omega^\mathbf{v} = \mathbf{0}$.

If a line Ω is perpendicular to the line of action Δ of a vector $\mathbf{v}$, and d is the shortest distance between these two lines, then

$$|\mathbf{M}_\Omega^\mathbf{v}| = d|\mathbf{v}|.$$

3.3 Moments of a System of Bound Vectors

DEFINITION The moment of a system $\{S\}$ of bound vectors $\mathbf{v}_i$, $\{S\} = \{\mathbf{v}_1, \mathbf{v}_2, \ldots, \mathbf{v}_n\} = \{\mathbf{v}_i\}_{i=1,2,\ldots,n}$ about a point A is

$$\mathbf{M}_A^{\{S\}} = \sum_{i=1}^{n} \mathbf{M}_A^{\mathbf{v}_i}. \quad \blacktriangle$$

DEFINITION The moment of a system $\{S\}$ of bound vectors $\mathbf{v}_i$, $\{S\} = \{\mathbf{v}_1, \mathbf{v}_2, \ldots, \mathbf{v}_n\} = \{\mathbf{v}_i\}_{i=1,2,\ldots,n}$ about a line Ω is

$$\mathbf{M}_\Omega^{\{S\}} = \sum_{i=1}^{n} \mathbf{M}_\Omega^{\mathbf{v}_i}.$$

The moments of a system of vectors about points and lines are free vectors.

The moments $\mathbf{M}_A^{\{S\}}$ and $\mathbf{M}_{A'}^{\{S\}}$ of a system $\{S\}$, $\{S\} = \{\mathbf{v}_i\}_{i=1,2,\dots,n}$, of bound vectors, $\mathbf{v}_i$, about two points A and P are related to each other as

$$\mathbf{M}_A^{\{S\}} = \mathbf{M}_P^{\{S\}} + \mathbf{r}_{AP} \times \mathbf{R}, \tag{3.2}$$

where $\mathbf{r}_{AP}$ is the position vector of P relative to A, and $\mathbf{R}$ is the resultant of $\{S\}$. ▲

Proof

Let B_i be a point on the line of action of the vector $\mathbf{v}_i$, and let $\mathbf{r}_{ABi}$ and $\mathbf{r}_{PBi}$ be the position vectors of B_i relative to A and P (Fig. 3.3). Thus,

$$\mathbf{M}_A^{\{S\}} = \sum_{i=1}^n \mathbf{M}_A^{\mathbf{v}_i} = \sum_{i=1}^n \mathbf{r}_{ABi} \times \mathbf{v}_i$$

$$= \sum_{i=1}^n (\mathbf{r}_{AP} + \mathbf{r}_{PBi}) \times \mathbf{v}_i = \sum_{i=1}^n (\mathbf{r}_{AP} \times \mathbf{v}_i + \mathbf{r}_{PBi} \times \mathbf{v}_i)$$

$$= \sum_{i=1}^n \mathbf{r}_{AP} \times \mathbf{v}_i + \sum_{i=1}^n \mathbf{r}_{PBi} \times \mathbf{v}_i$$

$$= \mathbf{r}_{AP} \times \sum_{i=1}^n \mathbf{v}_i + \sum_{i=1}^n \mathbf{r}_{PBi} \times \mathbf{v}_i$$

$$= \mathbf{r}_{AP} \times \mathbf{R} + \sum_{i=1}^n \mathbf{M}_P^{\mathbf{v}_i}$$

$$= \mathbf{r}_{AP} \times \mathbf{R} + \mathbf{M}_P^{\{S\}}.$$

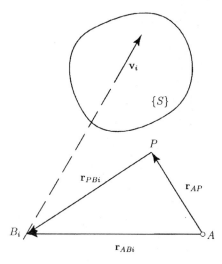

Figure 3.3

If the resultant $\mathbf{R}$ of a system $\{S\}$ of bound vectors is not equal to zero, $\mathbf{R} \neq \mathbf{0}$, the points about which $\{S\}$ has a minimum moment $\mathbf{M}_{min}$ lie on a line called the *central axis*, (CA), of $\{S\}$, which is parallel to $\mathbf{R}$ and passes through a point P whose position vector $\mathbf{r}$ relative to an arbitrarily selected reference point O is given by

$$\mathbf{r} = \frac{\mathbf{R} \times \mathbf{M}_O^{\{S\}}}{\mathbf{R}^2}.$$

The minimum moment $\mathbf{M}_{min}$ is given by

$$\mathbf{M}_{min} = \frac{\mathbf{R} \cdot \mathbf{M}_O^{\{S\}}}{\mathbf{R}^2} \mathbf{R}.$$

3.4 Couples

DEFINITION   A *couple* is a system of bound vectors whose resultant is equal to zero and whose moment about some point is not equal to zero.

A system of vectors is not a vector; therefore, couples are not vectors.

A couple consisting of only two vectors is called a *simple couple*. The vectors of a simple couple have equal magnitudes, parallel lines of action, and opposite senses. Writers use the word "couple" to denote a simple couple.

The moment of a couple about a point is called the *torque* of the couple, $\mathbf{M}$ or $\mathbf{T}$. The moment of a couple about one point is equal to the moment of the couple about any other point, that is, it is unnecessary to refer to a specific point.

The torques are vectors, and the magnitude of the torque of a simple couple is given by

$$|\mathbf{M}| = d|\mathbf{v}|,$$

where d is the distance between the lines of action of the two vectors comprising the couple, and $\mathbf{v}$ is one of these vectors.   ▲

Proof

In Fig. 3.4, the torque $\mathbf{M}$ is the sum of the moments of $\mathbf{v}$ and $-\mathbf{v}$ about any point. The moments about point A

$$\mathbf{M} = \mathbf{M}_A^{\mathbf{v}} + \mathbf{M}_A^{-\mathbf{v}} = \mathbf{r} \times \mathbf{v} + \mathbf{0}.$$

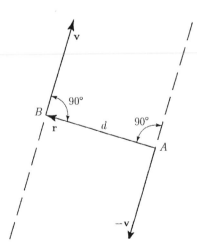

Figure 3.4

Hence,

$$|\mathbf{M}| = |\mathbf{r} \times \mathbf{v}| = |\mathbf{r}||\mathbf{v}| \sin(\mathbf{r}, \mathbf{v}) = d|\mathbf{v}|. \quad \blacktriangle$$

The direction of the torque of a simple couple can be determined by inspection: $\mathbf{M}$ is perpendicular to the plane determined by the lines of action of the two vectors comprising the couple, and the sense of $\mathbf{M}$ is the same as that of $\mathbf{r} \times \mathbf{v}$.

The moment of a couple about a line Ω is equal to the Ω resolute of the torque of the couple.

The moments of a couple about two parallel lines are equal to each other.

3.5 Equivalence

DEFINITION   Two systems $\{S\}$ and $\{S'\}$ of bound vectors are said to be *equivalent* when:

1. The resultant of $\{S\}$, $\mathbf{R}$, is equal to the resultant of $\{S'\}$, $\mathbf{R'}$:

$$\mathbf{R} = \mathbf{R'}.$$

2. There exists at least one point about which $\{S\}$ and $\{S'\}$ have equal moments:

$$\text{exists } P \colon \mathbf{M}_P^{\{S\}} = \mathbf{M}_P^{\{S'\}}. \quad \blacktriangle$$

Figures 3.5a and 3.5b each shown a rod subjected to the action of a pair of forces. The two pairs of forces are equivalent, but their effects on the rod are different from each other. The word "equivalence" is not to be regarded as implying physical equivalence.

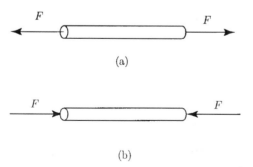

(a)

(b)

Figure 3.5

For given a line Ω and two equivalent systems $\{S\}$ and $\{S'\}$ of bound vectors, the sum of the Ω resolutes of the vectors in $\{S\}$ is equal to the sum of the Ω resolutes of the vectors in $\{S'\}$.

The moments of two equivalent systems of bound vectors, about a point, are equal to each other.

The moments of two equivalent systems $\{S\}$ and $\{S'\}$ of bound vectors, about *any* line Ω, are equal to each other.

3.5.1 TRANSIVITY OF THE EQUIVALENCE RELATION

If $\{S\}$ is equivalent to $\{S'\}$, and $\{S'\}$ is equivalent to $\{S''\}$, then $\{S\}$ is equivalent to $\{S''\}$.

Every system $\{S\}$ of bound vectors with the resultant $\mathbf{R}$ can be replaced with a system consisting of a couple C and a single bound vector $\mathbf{v}$ whose line of action passes through an arbitrarily selected *base point* O. The torque $\mathbf{M}$ of C depends on the choice of base point $\mathbf{M} = \mathbf{M}_O^{\{S\}}$. The vector $\mathbf{v}$ is independent of the choice of base point, $\mathbf{v} = \mathbf{R}$.

A couple C can be replaced with any system of couples the sum of whose torque is equal to the torque of C.

When a system of bound vectors consists of a couple of torque $\mathbf{M}$ and a single vector parallel to $\mathbf{M}$, it is called a *wrench*.

3.6 Representing Systems by Equivalent Systems

To simplify the analysis of the forces and moments acting on a given system, one can represent the system by an equivalent, less complicated one. The actual forces and couples can be replaced with a total force and a total moment.

In Fig. 3.6 is shown an arbitrary system of forces and moments, {system 1}, and a point P. This system can be represented by a system, {system 2}, consisting of a single force $\mathbf{F}$ acting at P and a single couple of torque $\mathbf{M}$. The conditions for equivalence are

$$\sum \mathbf{F}^{\{\text{system 2}\}} = \sum \mathbf{F}^{\{\text{system 1}\}} \Rightarrow \mathbf{F} = \sum \mathbf{F}^{\{\text{system 1}\}}$$

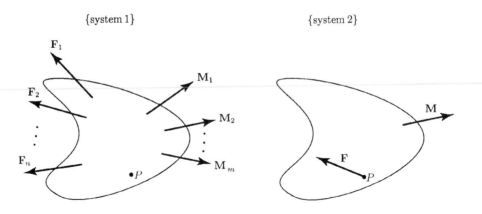

{system 1}   {system 2}

$$\mathbf{F} = \sum \mathbf{F}^{\{\text{system 1}\}}$$

$$\mathbf{M} = \sum \mathbf{M}_P^{\{\text{system 1}\}}$$

Figure 3.6

and

$$\sum \mathbf{M}_P^{\{\text{system 2}\}} = \sum \mathbf{M}_P^{\{\text{system 1}\}} \Rightarrow \mathbf{M} = \sum \mathbf{M}_P^{\{\text{system 1}\}}.$$

These conditions are satisfied if $\mathbf{F}$ equals the sum of the forces in {system 1}, and $\mathbf{M}$ equals the sum of the moments about P in {system 1}. Thus, no matter how complicated a system of forces and moments may be, it can be represented by a single force acting at a given point and a single couple. Three particular cases occur frequently in practice.

3.6.1 FORCE REPRESENTED BY A FORCE AND A COUPLE

A force $\mathbf{F}_P$ acting at a point P {system 1} in Fig. 3.7 can be represented by a force $\mathbf{F}$ acting at a different point Q and a couple of torque $\mathbf{M}$ {system 2}. The moment of {system 1} about point Q is $\mathbf{r}_{QP} \times \mathbf{F}_P$, where $\mathbf{r}_{QP}$ is the vector from Q to P. The conditions for equivalence are

$$\sum \mathbf{M}_P^{\{\text{system 2}\}} = \sum \mathbf{M}_P^{\{\text{system 1}\}} \Rightarrow \mathbf{F} = \mathbf{F}_P$$

{system 1} {system 2}

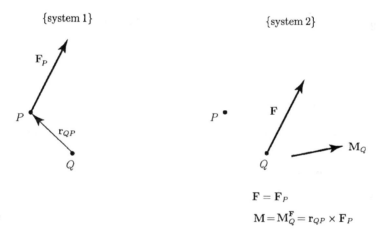

$$\mathbf{F} = \mathbf{F}_P$$
$$\mathbf{M} = \mathbf{M}_Q^{\mathbf{F}} = \mathbf{r}_{QP} \times \mathbf{F}_P$$

Figure 3.7

and

$$\sum \mathbf{M}_Q^{\{\text{system 2}\}} = \sum \mathbf{M}_Q^{\{\text{system 1}\}} \Rightarrow \mathbf{M} = \mathbf{M}_Q^{\mathbf{F}_P} = \mathbf{r}_{QP} \times \mathbf{F}_P.$$

The systems are equivalent if the force $\mathbf{F}$ equals the force $\mathbf{F}_P$ and the couple of torque $\mathbf{M}_Q^{\mathbf{F}_P}$ equals the moment of $\mathbf{F}_P$ about Q.

3.6.2 CONCURRENT FORCES REPRESENTED BY A FORCE

A system of concurrent forces whose lines of action intersect at a point P {system 1} (Fig. 3.8) can be represented by a single force whose line of action intersects P, {system 2}. The sums of the forces in the two systems are equal if

$$\mathbf{F} = \mathbf{F}_1 + \mathbf{F}_2 + \cdots + \mathbf{F}_n.$$

The sum of the moments about P equals zero for each system, so the systems are equivalent if the force $\mathbf{F}$ equals the sum of the forces in {system 1}.

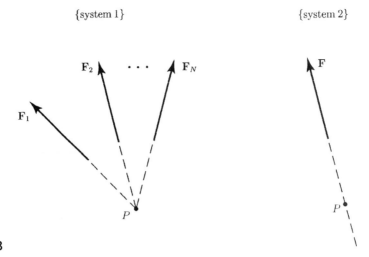

Figure 3.8

3.6.3 PARALLEL FORCES REPRESENTED BY A FORCE

A system of parallel forces whose sum is not zero can be represented by a single force **F** (Fig. 3.9).

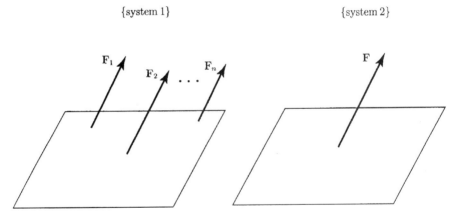

Figure 3.9

3.6.4 SYSTEM REPRESENTED BY A WRENCH

In general any system of forces and moments can be represented by a single force acting at a given point and a single couple.

Figure 3.10 shows an arbitrary force **F** acting at a point P and an arbitrary couple of torque **M**, {system 1}. This system can be represented by a simpler one, that is, one may represent the force **F** acting at a different point Q and the component of **M** that is parallel to **F**. A coordinate system is chosen so that **F** is along the y axis,

$$\mathbf{F} = F\mathbf{j},$$

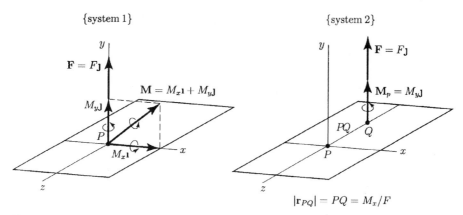

Figure 3.10

and **M** is contained in the xy plane,

$$\mathbf{M} = M_x\mathbf{i} + M_y\mathbf{j}.$$

The equivalent system, {system 2}, consists of the force **F** acting at a point Q on the z axis,

$$\mathbf{F} = F\mathbf{j},$$

and the component of **M** parallel to **F**,

$$\mathbf{M}_p = M_y\mathbf{j}.$$

The distance PQ is chosen so that $|\mathbf{r}_{PQ}| = PQ = M_x/F$. The {system 1} is equivalent to {system 2}. The sum of the forces in each system is the same **F**.

The sum of the moments about P in {system 1} is **M**, and the sum of the moments about P in {system 2} is

$$\sum \mathbf{M}_P^{\{\text{system 2}\}} = \mathbf{r}_{PQ} \times \mathbf{F} + M_y\mathbf{j} = (-PQ\mathbf{k}) \times (F\mathbf{j}) + M_y\mathbf{j} = M_x\mathbf{i} + M_y\mathbf{j} = \mathbf{M}.$$

The system of the force $\mathbf{F} = F\mathbf{j}$ and the couple $\mathbf{M}_p = M_y\mathbf{j}$ that is parallel to **F** is a wrench. A wrench is the simplest system that can be equivalent to an arbitrary system of forces and moments.

The representation of a given system of forces and moments by a wrench requires the following steps:

1. Choose a convenient point P and represent the system by a force **F** acting at P and a couple **M** (Fig. 3.11a).

2. Determine the components of **M** parallel and normal to **F** (Fig. 3.11b):

$$\mathbf{M} = \mathbf{M}_p + \mathbf{M}_n, \quad \text{where } \mathbf{M}_p \| \mathbf{F}.$$

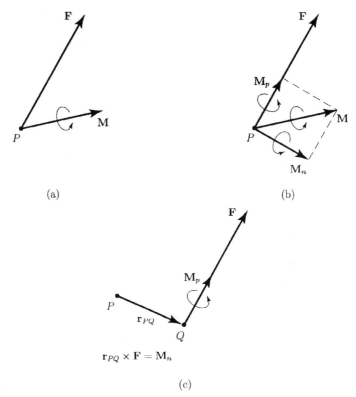

(a)

(b)

$$\mathbf{r}_{PQ} \times \mathbf{F} = \mathbf{M}_n$$

Figure 3.11

(c)

3. The wrench consists of the force **F** acting at a point Q and the parallel component $\mathbf{M}_p$ (Fig. 3.11c). For equivalence, the following condition must be satisfied:

$$\mathbf{r}_{PQ} \times \mathbf{F} = \mathbf{M}_n.$$

$\mathbf{M}_n$ is the normal component of **M**.

In general, {system 1} cannot be represented by a force **F** alone.

4. Equilibrium

4.1 Equilibrium Equations

A body is in *equilibrium* when it is stationary or in steady translation relative to an inertial reference frame. The following conditions are satisfied when a body, acted upon by a system of forces and moments, is in equilibrium:

1. The sum of the forces is zero:

$$\sum \mathbf{F} = 0. \tag{4.1}$$

2. The sum of the moments about any point is zero:

$$\sum \mathbf{M}_P = 0, \forall P. \tag{4.2}$$

If the sum of the forces acting on a body is zero and the sum of the moments about one point is zero, then the sum of the moments about every point is zero.

Proof

The body shown in Fig. 4.1 is subjected to forces $\mathbf{F}_{Ai}$, $i = 1, \ldots, n$, and couples $\mathbf{M}_j$, $j = 1, \ldots, m$. The sum of the forces is zero,

$$\sum \mathbf{F} = \sum_{i=1}^{n} \mathbf{F}_{Ai} = \mathbf{0},$$

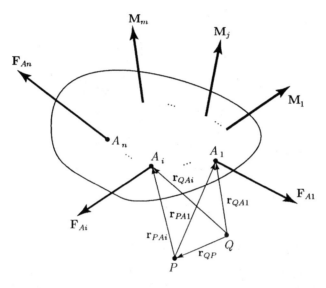

Figure 4.1

$$\mathbf{r}_{QAi} = \mathbf{r}_{QP} + \mathbf{r}_{PAi}$$

and the sum of the moments about a point P is zero,

$$\sum \mathbf{M}_P = \sum_{i=1}^{n} \mathbf{r}_{PAi} \times \mathbf{F}_{Ai} + \sum_{j=1}^{m} \mathbf{M}_j = \mathbf{0},$$

where $\mathbf{r}_{PAi} = \mathbf{PA}_i$, $i = 1, \ldots, n$. The sum of the moments about any other point Q is

$$
\begin{aligned}
\sum \mathbf{M}_Q &= \sum_{i=1}^{n} \mathbf{r}_{QAi} \times \mathbf{F}_{Ai} + \sum_{j=1}^{m} \mathbf{M}_j \\
&= \sum_{i=1}^{n} (\mathbf{r}_{QP} + \mathbf{r}_{PAi}) \times \mathbf{F}_{Ai} + \sum_{j=1}^{m} \mathbf{M}_j \\
&= \mathbf{r}_{QP} \times \sum_{i=1}^{n} \mathbf{F}_{Ai} + \sum_{i=1}^{n} \mathbf{r}_{PAi} \times \mathbf{F}_{Ai} + \sum_{j=1}^{m} \mathbf{M}_j \\
&= \mathbf{r}_{QP} \times \mathbf{0} + \sum_{i=1}^{n} \mathbf{r}_{PAi} \times \mathbf{F}_{Ai} + \sum_{j=1}^{m} \mathbf{M}_j \\
&= \sum_{i=1}^{n} \mathbf{r}_{PAi} \times \mathbf{F}_{Ai} + \sum_{j=1}^{m} \mathbf{M}_j = \sum \mathbf{M}_P = \mathbf{0}.
\end{aligned}
$$

A body is subjected to concurrent forces $\mathbf{F}_1, \mathbf{F}_2, \ldots, \mathbf{F}_n$ and no couples. If the sum of the concurrent forces is zero,

$$\mathbf{F}_1 + \mathbf{F}_2 + \cdots + \mathbf{F}_n = \mathbf{0},$$

the sum of the moments of the forces about the concurrent point is zero, so the sum of the moments about every point is zero. The only condition imposed by equilibrium on a set of concurrent forces is that their sum is zero.

4.2 Supports

4.2.1 PLANAR SUPPORTS

The *reactions* are forces and couples exerted on a body by its supports.

Pin Support

Figure 4.2 shows a pin support. A beam is attached by a smooth pin to a bracket. The pin passes through the bracket and the beam. The beam can rotate about the axis of the pin. The beam cannot translate relative to the bracket because the support exerts a reactive force that prevents this movement. Thus a pin support can exert a force on a body in any direction. The force (Fig. 4.3) is expressed in terms of its components in plane,

$$\mathbf{F}_A = A_x\mathbf{i} + A_y\mathbf{j}.$$

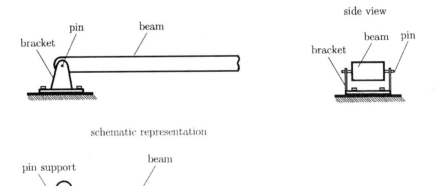

Figure 4.2

The directions of the reactions A_x and A_y are positive. If one determines A_x or A_y to be negative, the reaction is in the direction opposite to that of the arrow. The pin support is not capable of exerting a couple.

Roller Support

Figure 4.4 represents a roller support, which is a pin support mounted on rollers. The roller support can only exert a force normal (perpendicular) to the surface on which the roller support moves freely,

$$\mathbf{F}_A = A_y\mathbf{j}.$$

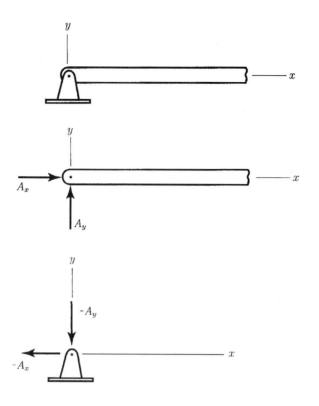

Figure 4.3

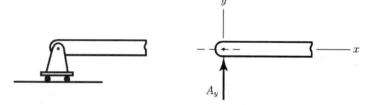

Figure 4.4

The roller support cannot exert a couple about the axis of the pin, and it cannot exert a force parallel to the surface on which it translates.

Fixed Support

Figure 4.5 shows a fixed support or built-in support. The body is literally built into a wall. A fixed support can exert two components of force and a couple:

$$\mathbf{F}_A = A_x\mathbf{i} + A_y\mathbf{j}, \quad \mathbf{M}_A = M_{Az}\mathbf{k}.$$

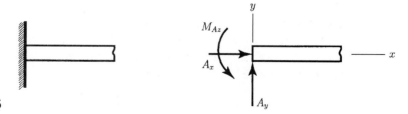

Figure 4.5

4.2.2 THREE-DIMENSIONAL SUPPORTS

Ball and Socket Support

Figure 4.6 shows a ball and socket support, where the supported body is attached to a ball enclosed within a spherical socket. The socket permits the body only to rotate freely. The ball and socket support cannot exert a couple to prevent rotation. The ball and socket support can exert three components of force:

$$\mathbf{F}_A = A_x\mathbf{i} + A_y\mathbf{j} + A_z\mathbf{k}.$$

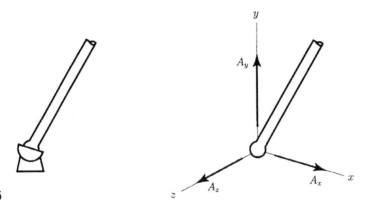

Figure 4.6

4.3 *Free-Body Diagrams*

Free-body diagrams are used to determine forces and moments acting on simple bodies in equilibrium.

The beam in Fig. 4.7a has a pin support at the left end A and a roller support at the right end B. The beam is loaded by a force F and a moment M at C. To obtain the free-body diagram, first the beam is isolated from its supports. Next, the reactions exerted on the beam by the supports are shown on the free-body diagram (Fig. 4.7b). Once the free-body diagram is obtained one can apply the equilibrium equations.

The steps required to determine the reactions on bodies are:

1. Draw the free-body diagram, isolating the body from its supports and showing the forces and the reactions

2. Apply the equilibrium equations to determine the reactions

For two-dimensional systems, the forces and moments are related by three scalar equilibrium equations:

$$\sum F_x = 0 \qquad (4.3)$$

$$\sum F_y = 0 \qquad (4.4)$$

$$\sum M_P = 0, \forall P. \qquad (4.5)$$

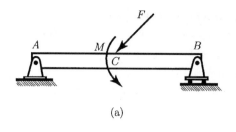

(a)

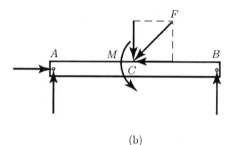

Figure 4.7

(b)

One can obtain more than one equation from Eq. (4.5) by evaluating the sum of the moments about more than one point. The additional equations will not be independent of Eqs. (4.3)–(4.5). One cannot obtain more than three independent equilibrium equations from a two-dimensional free-body diagram, which means one can solve for at most three unknown forces or couples.

For three-dimensional systems, the forces and moments are related by six scalar equilibrium equations:

$$\sum F_x = 0 \tag{4.6}$$

$$\sum F_y = 0 \tag{4.7}$$

$$\sum F_z = 0 \tag{4.8}$$

$$\sum M_x = 0 \tag{4.9}$$

$$\sum M_y = 0 \tag{4.10}$$

$$\sum M_x = 0 \tag{4.11}$$

One can evaluate the sums of the moments about any point. Although one can obtain other equations by summing the moments about additional points, they will not be independent of these equations. For a three-dimensional free-body diagram one can obtain six independent equilibrium equations and one can solve for at most six unknown forces or couples.

A body has *redundant supports* when the body has more supports than the minimum number necessary to maintain it in equilibrium. Redundant supports are used whenever possible for strength and safety. Each support added to a body results in additional reactions. The difference between the number of reactions and the number of independent equilibrium equations is called the *degree of redundancy*.

5. Dry Friction

If a body rests on an inclined plane, the friction force exerted on it by the surface prevents it from sliding down the incline. The question is: what is the steepest incline on which the body can rest?

A body is placed on a horizontal surface. The body is pushed with a small horizontal force F. If the force F is sufficiently small, the body does not move. Figure 5.1 shows the free-body diagram of the body, where the force W is the weight of the body, and N is the normal force exerted by the surface. The force F is the horizontal force, and F_f is the friction force exerted by the surface. Friction force arises in part from the interactions of the roughness, or asperities, of the contacting surfaces. The body is in equilibrium and $F_f = F$.

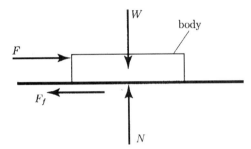

Figure 5.1

The force F is slowly increased. As long as the body remains in equilibrium, the friction force F_f must increase correspondingly, since it equals the force F. The body slips on the surface. The friction force, after reaching the maximum value, cannot maintain the body in equilibrium. The force applied to keep the body moving on the surface is smaller than the force required to cause it to slip. The fact that more force is required to start the body sliding on a surface than to keep it sliding is explained in part by the necessity to break the asperities of the contacting surfaces before sliding can begin.

The theory of dry friction, or *Coloumb friction*, predicts:

- The maximum friction forces that can be exerted by dry, contacting surfaces that are stationary relative to each other
- The friction forces exerted by the surfaces when they are in relative motion, or sliding

5.1 Static Coefficient of Friction

The magnitude of the maximum friction force, F_f, that can be exerted between two plane dry surfaces in contact is

$$F_f = \mu_s N, \tag{5.1}$$

where μ_s is a constant, the *static coefficient of friction*, and N is the normal component of the contact force between the surfaces. The value of the static coefficient of friction, μ_s, depends on:

- The materials of the contacting surfaces
- The conditions of the contacting surfaces: smoothness and degree of contamination

Typical values of μ_s for various materials are shown in Table 5.1.

Table 5.1 *Typical Values of the Static Coefficient of Friction*

Materials	μ_s
Metal on metal	0.15–0.20
Metal on wood	0.20–0.60
Metal on masonry	0.30–0.70
Wood on wood	0.25–0.50
Masonry on masonry	0.60–0.70
Rubber on concrete	0.50–0.90

Equation (5.1) gives the maximum friction force that the two surfaces can exert without causing it to slip. If the static coefficient of friction μ_s between the body and the surface is known, the largest value of F one can apply to the body without causing it to slip is $F = F_f = \mu_s W$. Equation (5.1) determines the magnitude of the maximum friction force but not its direction. The friction force resists the impending motion.

5.2 Kinetic Coefficient of Friction

The magnitude of the friction force between two plane dry contacting surfaces that are in motion relative to each other is

$$F_f = \mu_k N, \tag{5.2}$$

where μ_k is the *kinetic coefficient of friction* and N is the normal force between the surfaces. The value of the kinetic coefficient of friction is generally smaller than the value of the static coefficient of friction, μ_s.

To keep the body in Fig. 5.1 in uniform motion (sliding on the surface) the force exerted must be $F = F_f = \mu_k W$. The friction force resists the relative motion, when two surfaces are sliding relative to each other.

The body RB shown in Fig. 5.2a is moving on the fixed surface 0. The direction of motion of RB is the positive axis x. The friction force on the body RB acts in the direction opposite to its motion, and the friction force on the fixed surface is in the opposite direction (Fig. 5.2b).

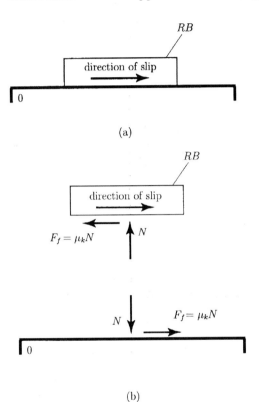

(a)

(b)

Figure 5.2

5.3 Angles of Friction

The *angle of friction*, θ, is the angle between the friction force, $F_f = |\mathbf{F}_f|$, and the normal force, $N = |\mathbf{N}|$, to the surface (Fig. 5.3). The magnitudes of the normal force and friction force and that of θ are related by

$$F_f = R \sin \theta$$
$$N = R \cos \theta,$$

where $R = |\mathbf{R}| = |\mathbf{N} + \mathbf{F}_f|$.

The value of the angle of friction when slip is impending is called the *static angle of friction*, θ_s:

$$\tan \theta_s = \mu_s.$$

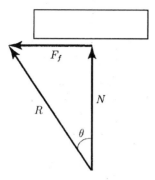

Figure 5.3

The value of the angle of friction when the surfaces are sliding relative to each other is called the *kinetic angle of friction*, θ_k:

$$\tan \theta_k = \mu_k.$$

References

1. A. Bedford and W. Fowler, *Dynamics*. Addison Wesley, Menlo Park, CA, 1999.

2. A. Bedford and W. Fowler, *Statics*. Addison Wesley, Menlo Park, CA, 1999.

3. F. P. Beer and E. R. Johnston, Jr., *Vector Mechanics for Engineers: Statics and Dynamics*. McGraw-Hill, New York, 1996.

4. R. C. Hibbeler, *Engineering Mechanics: Statics and Dynamics*. Prentice-Hall, Upper Saddle River, NJ, 1995.

5. T. R. Kane, *Analytical Elements of Mechanics*, Vol. 1. Academic Press, New York, 1959.

6. T. R. Kane, *Analytical Elements of Mechanics*, Vol. 2. Academic Press, New York, 1961.

7. T. R. Kane and D. A. Levinson, *Dynamics*. McGraw-Hill, New York, 1985.

8. D. J. McGill and W. W. King, *Engineering Mechanics: Statics and an Introduction to Dynamics*. PWS Publishing Company, Boston, 1995.

9. R. L. Norton, *Machine Design*. Prentice-Hall, Upper Saddle River, NJ, 1996.

10. R. L. Norton, *Design of Machinery*. McGraw-Hill, New York, 1999.

11. W. F. Riley and L. D. Sturges, *Engineering Mechanics: Statics*. John Wiley & Sons, New York, 1993.

12. I. H. Shames, *Engineering Mechanics: Statics and Dynamics*. Prentice-Hall, Upper Saddle River, NJ, 1997.

13. R. W. Soutas-Little and D. J. Inman, *Engineering Mechanics: Statics*. Prentice-Hall, Upper Saddle River, NJ, 1999.

2

Dynamics

**DAN B. MARGHITU, BOGDAN O. CIOCIRLAN, AND
CRISTIAN I. DIACONESCU**

*Department of Mechanical Engineering,
Auburn University, Auburn, Alabama 36849*

Inside

1. Fundamentals

1.1 Space and Time

S*pace* is the three-dimensional universe. The distance between two points in space is the length of the straight line joining them. The unit of length in the International System of units, or SI units, is the meter (m). In U.S. customary units, the unit of length is the foot (ft). The U.S. customary units use the mile (mi) (1 mi = 5280 ft) and the inch (in) (1 ft = 12 in).

The *time* is a scalar and is measured by the intervals between repeatable events. The unit of time is the second (s) in both SI units and U.S. customary units. The minute (min), hour (hr), and day are also used.

The *velocity* of a point in space relative to some reference is the rate of change of its position with time. The velocity is expressed in meters per second (m/s) in SI units, and is expressed in feet per second (ft/s) in U.S. customary units.

The *acceleration* of a point in space relative to some reference is the rate of change of its velocity with time. The acceleration is expressed in meters per second squared (m/s^2) in SI units, and is expressed in feet per second squared (ft/s^2) in U.S. customary units.

1.2 Numbers

Engineering measurements, calculations, and results are expressed in numbers. Significant digits are the number of meaningful digits in a number, counting to the right starting with the first nonzero digit. Numbers can be rounded off to a certain number of significant digits. The value of π can be expressed to three significant digits, 3.14, or can be expressed to six significant digits, 3.14159.

The multiples of units are indicated by prefixes. The common prefixes, their abbreviations, and the multiples they represent are shown in Table 1.1. For example, 5 km is 5 kilometers, which is 5000 m.

Table 1.1 *Prefixes Used in SI Units*

Prefix	Abbreviation	Multiple
nano-	n	10^{-9}
micro-	μ	10^{-6}
mili-	m	10^{-3}
kilo-	k	10^{3}
mega-	M	10^{6}
giga-	G	10^{9}

Dynamics

Table 1.2 *Unit Conversions*

Time	1 minute	=	60 seconds
	1 hour	=	60 minutes
	1 day	=	24 hours
Length	1 foot	=	12 inches
	1 mile	=	5280 feet
	1 inch	=	25.4 millimeters
	1 foot	=	0.3048 meter
Angle	2π radians	=	360 degrees

Some useful unit conversions are presented in Table 1.2. For example, 1 mi/hr in terms of ft/s is (1 mi equals 5280 ft and 1 hr equals 3600 s)

$$1\frac{\text{mi}}{\text{hr}} = 1\frac{1\,\text{mi} = 5280\,\text{ft}}{1\,\text{hr} = 3600\,\text{s}} = 1\frac{5280\,\text{ft}}{3600\,\text{s}} = 1.47\frac{\text{ft}}{\text{s}}.$$

1.3 Angular Units

Angles are expressed in radians (rad) in both SI and U.S. customary units. The value of an angle θ in radians (Fig. 1.1) is the ratio of the part of the

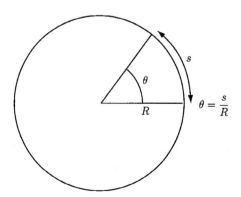

Figure 1.1

circumference s subtended by θ to the radius R of the circle,

$$\theta = \frac{s}{R}.$$

Angles are also expressed in degrees. There are 360 degrees (360°) in a complete circle. The complete circumference of the circle is $2\pi R$. Therefore,

$$360° = 2\pi \text{ rad.}$$

2. Kinematics of a Point

2.1 Position, Velocity, and Acceleration of a Point

One may observe students and objects in a classroom, and their positions relative to the room. Some students may be in the front of the classroom, some in the middle of the classroom, and some in the back of the classroom. The classroom is the "frame of reference." One can introduce a cartesian coordinate system x, y, z with its axes aligned with the walls of the classroom. A reference frame is a coordinate system used for specifying the positions of points and objects.

The position of a point P relative to a given reference frame with origin O is given by the position vector $\mathbf{r}$ from point O to point P (Fig. 2.1). If the point P is in motion relative to the reference frame, the position vector $\mathbf{r}$ is a function of time t (Fig. 2.1) and can be expressed as

$$\mathbf{r} = \mathbf{r}(t).$$

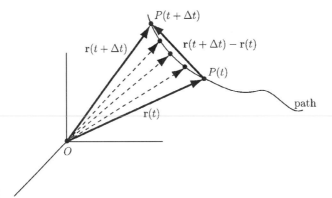

Figure 2.1

The velocity of the point P relative to the reference frame at time t is defined by

$$\mathbf{v} = \frac{d\mathbf{r}}{dt} = \dot{\mathbf{r}} = \lim_{\Delta t \to 0} \frac{\mathbf{r}(t + \Delta t) - \mathbf{r}(t)}{\Delta t}, \qquad (2.1)$$

where the vector $\mathbf{r}(t + \Delta t) - \mathbf{r}(t)$ is the change in position, or displacement of P, during the interval of time Δt (Fig. 2.1). The velocity is the rate of change of the position of the point P. The magnitude of the velocity $\mathbf{v}$ is the speed $v = |\mathbf{v}|$. The dimensions of $\mathbf{v}$ are (distance)/(time). The position and velocity of a point can be specified only relative to a reference frame.

The acceleration of the point P relative to the given reference frame at time t is defined by

$$\mathbf{a} = \frac{d\mathbf{v}}{dt} = \dot{\mathbf{v}} = \lim_{\Delta t \to 0} \frac{\mathbf{v}(t + \Delta t) - \mathbf{v}(t)}{\Delta t}, \qquad (2.2)$$

where $\mathbf{v}(t + \Delta t) - \mathbf{v}(t)$ is the change in the velocity of P during the interval of time Δt (Fig. 2.1). The acceleration is the rate of change of the velocity of P at time t (the second time derivative of the displacement), and its dimensions are (distance)/(time)^2.

2.2 Angular Motion of a Line

The angular motion of the line L, in a plane, relative to a reference line L_0, in the plane, is given by an angle θ (Fig. 2.2). The angular velocity of L relative to L_0 is defined by

$$\omega = \frac{d\theta}{dt} = \dot{\theta}, \qquad (2.3)$$

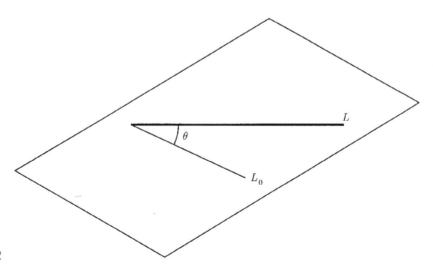

Figure 2.2

and the angular acceleration of L relative to L_0 is defined by

$$\alpha = \frac{d\omega}{dt} = \frac{d^2\theta}{dt^2} = \dot{\omega} = \ddot{\theta}. \qquad (2.4)$$

The dimensions of the angular position, angular velocity, and angular acceleration are (rad), (rad/s), and (rad/s^2), respectively. The scalar coordinate θ can be positive or negative. The counterclockwise (ccw) direction is considered positive.

2.3 Rotating Unit Vector

The angular motion of a unit vector **u** in a plane can be described as the angular motion of a line. The direction of **u** relative to a reference line L_0 is specified by the angle θ in Fig. 2.3a, and the rate of rotation of **u** relative to L_0 is defined by the angular velocity

$$\omega = \frac{d\theta}{dt}.$$

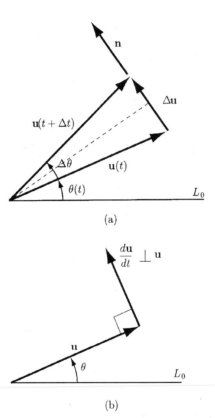

(a)

(b)

Figure 2.3

The time derivative of **u** is specified by

$$\frac{d\mathbf{u}}{dt} = \lim_{\Delta t \to 0} \frac{\mathbf{u}(t + \Delta t) - \mathbf{u}(t)}{\Delta t}.$$

Figure 2.3a shows the vector **u** at time t and at time $t + \Delta t$. The change in **u** during this interval is $\Delta\mathbf{u} = \mathbf{u}(t + \Delta t)\mathbf{u}(t)$, and the angle through which **u** rotates is $\Delta\theta = \theta(t + \Delta t) - \theta(t)$. The triangle in Fig. 2.3a is isosceles, so the magnitude of $\Delta\mathbf{u}$ is

$$|\Delta\mathbf{u}| = 2|\mathbf{u}|\sin(\Delta\theta/2) = 2\sin(\Delta\theta/2).$$

The vector $\Delta\mathbf{u}$ is

$$\Delta\mathbf{u} = |\Delta\mathbf{u}|\mathbf{n} = 2\sin(\Delta\theta/2)\mathbf{n},$$

where $\mathbf{n}$ is a unit vector that points in the direction of $\Delta\mathbf{u}$ (Fig. 2.3a). The time derivative of $\mathbf{u}$ is

$$\frac{d\mathbf{u}}{dt} = \lim_{\Delta t \to 0}\frac{\Delta\mathbf{u}}{\Delta t} = \lim_{\Delta t \to 0}\frac{2\sin(\Delta\theta/2)\mathbf{n}}{\Delta t} = \lim_{\Delta t \to 0}\frac{\sin(\Delta\theta/2)}{\Delta\theta/2}\frac{\Delta\theta}{\Delta t}\mathbf{n}$$

$$= \lim_{\Delta t \to 0}\frac{\sin(\Delta\theta/2)}{\Delta\theta/2}\frac{\Delta\theta}{\Delta t}\mathbf{n} = \lim_{\Delta t \to 0}\frac{\Delta\theta}{\Delta t}\mathbf{n} = \frac{d\theta}{dt}\mathbf{n},$$

where $\displaystyle\lim_{\Delta t \to 0}\frac{\sin(\Delta\theta/2)}{\Delta\theta/2} = 1$ and $\displaystyle\lim_{\Delta t \to 0}\frac{\Delta\theta}{\Delta t} = \frac{d\theta}{dt}$. So the time derivative of the unit vector $\mathbf{u}$ is

$$\frac{d\mathbf{u}}{dt} = \frac{d\theta}{dt}\mathbf{n} = \omega\mathbf{n},$$

where $\mathbf{n}$ is a unit vector that is perpendicular to $\mathbf{u}$, $\mathbf{n} \perp \mathbf{u}$, and points in the positive θ direction (Fig. 2.3b).

2.4 Straight Line Motion

The position of a point P on a straight line relative to a reference point O can be indicated by the coordinate s measured along the line from O to P (Fig. 2.4). In this case the reference frame is the straight line and the origin of the reference frame is the point O. The reference frame and its origin are used to describe the position of point P. The coordinate s is considered to be positive to the right of the origin O and is considered to be negative to the left

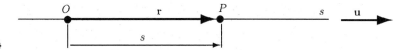

Figure 2.4

of the origin.

Let $\mathbf{u}$ be a unit vector parallel to the straight line and pointing in the positive s direction (Fig. 2.4). The position vector of the point P relative to the origin O is

$$\mathbf{r} = s\mathbf{u}.$$

The velocity of the point P relative to the origin O is

$$\mathbf{v} = \frac{d\mathbf{r}}{dt} = \frac{ds}{dt}\mathbf{u} = \dot{s}\mathbf{u}.$$

Dynamics

The magnitude v of the velocity vector $\mathbf{v} = v\mathbf{u}$ is the speed (velocity scalar)

$$v = \frac{ds}{dt} = \dot{s}.$$

The speed v of the point P is equal to the slope at time t of the line tangent to the graph of s as a function of time.

The acceleration of the point P relative to O is

$$\mathbf{a} = \frac{d\mathbf{v}}{dt} = \frac{d}{dt}(v\mathbf{u}) = \frac{dv}{dt}\mathbf{u} = \dot{v}\mathbf{u} = \ddot{s}\mathbf{u}.$$

The magnitude a of the acceleration vector $\mathbf{a} = a\mathbf{u}$ is the acceleration scalar

$$a = \frac{dv}{dt} = \frac{d^2 s}{dt^2}.$$

The acceleration a is equal to the slope at time t of the line tangent to the graph of v as a function of time.

2.5 Curvilinear Motion

The motion of the point P along a curvilinear path, relative to a reference frame, can be specified in terms of its position, velocity, and acceleration vectors. The directions and magnitudes of the position, velocity, and acceleration vectors do not depend on the particular coordinate system used to express them. The representations of the position, velocity, and acceleration vectors are different in different coordinate systems.

2.5.1 CARTESIAN COORDINATES

Let $\mathbf{r}$ be the position vector of a point P relative to the origin O of a cartesian reference frame (Fig. 2.5). The components of $\mathbf{r}$ are the x, y, z coordinates of the point P,

$$\mathbf{r} = x\mathbf{1} + y\mathbf{J} + z\mathbf{k}.$$

The velocity of the point P relative to the reference frame is

$$\mathbf{v} = \frac{d\mathbf{r}}{dt} = \dot{\mathbf{r}} = \frac{dx}{dt}\mathbf{1} + \frac{dy}{dt}\mathbf{J} + \frac{dz}{dt}\mathbf{k} = \dot{x}\mathbf{1} + \dot{y}\mathbf{J} + \dot{z}\mathbf{k}. \tag{2.5}$$

The velocity in terms of scalar components is

$$\mathbf{v} = v_x\mathbf{1} + v_y\mathbf{J} + v_z\mathbf{k}. \tag{2.6}$$

Three scalar equations can be obtained:

$$v_x = \frac{dx}{dt} = \dot{x}, \quad v_y = \frac{dy}{dt} = \dot{y}, \quad v_z = \frac{dz}{dt} = \dot{z}. \tag{2.7}$$

The acceleration of the point P relative to the reference frame is

$$\mathbf{a} = \frac{d\mathbf{v}}{dt} = \dot{\mathbf{v}} = \ddot{\mathbf{r}} = \frac{dv_x}{dt}\mathbf{1} + \frac{dv_y}{dt}\mathbf{J} + \frac{dv_z}{dt}\mathbf{k} = \dot{v}_x\mathbf{1} + \dot{v}_y\mathbf{J} + \dot{v}_z\mathbf{k} = \ddot{x}\mathbf{1} + \ddot{y}\mathbf{J} + \ddot{z}\mathbf{k}.$$

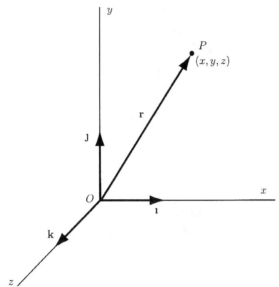

Figure 2.5

If we express the acceleration in terms of scalar components,

$$\mathbf{a} = a_x\mathbf{1} + a_y\mathbf{J} + a_z\mathbf{k}, \tag{2.8}$$

three scalar equations can be obtained:

$$a_x = \frac{dv_x}{dt} = \dot{v}_x = \ddot{x}, \quad a_y = \frac{dv_y}{dt} = \dot{v}_y = \ddot{y}, \quad a_z = \frac{dv_z}{dt} = \dot{v}_z = \ddot{z}. \tag{2.9}$$

Equations (2.7) and (2.9) describe the motion of a point relative to a cartesian coordinate system.

2.6 Normal and Tangential Components

The position, velocity, and acceleration of a point will be specified in terms of their components tangential and normal (perpendicular) to the path.

2.6.1 PLANAR MOTION

The point P is moving along a plane curvilinear path relative to a reference frame (Fig. 2.6). The position vector $\mathbf{r}$ specifies the position of the point P relative to the reference point O. The coordinate s measures the position of the point P along the path relative to a point O' on the path. The velocity of P relative to O is

$$\Delta\mathbf{v} = \frac{d\mathbf{r}}{dt} = \lim_{\Delta t \to 0} \frac{\mathbf{r}(t+\Delta t) - \mathbf{r}(t)}{\Delta t} = \lim_{\Delta t \to 0} \frac{\Delta\mathbf{r}}{\Delta t}, \tag{2.10}$$

where $\Delta\mathbf{r} = \mathbf{r}(t+\Delta t) - \mathbf{r}(t)$ (Fig. 2.6). The distance travelled along the path from t to $t+\Delta t$ is Δs. One can write Eq. (2.10) as

$$\mathbf{v} = \lim_{\Delta t \to 0} \frac{\Delta s}{\Delta t}\mathbf{u},$$

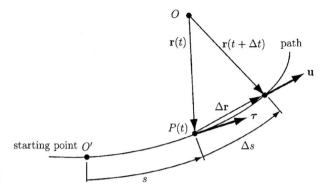

Figure 2.6

where **u** is a unit vector in the direction of $\Delta\mathbf{r}$. In the limit as Δt approaches zero, the magnitude of $\Delta\mathbf{r}$ equals ds because a chord progressively approaches the curve. For the same reason, the direction of $\Delta\mathbf{r}$ approaches tangency to the curve, and **u** becomes a unit vector, $\boldsymbol{\tau}$, tangent to the path at the position of P (Fig. 2.6):

$$\mathbf{v} = v\boldsymbol{\tau} = \frac{ds}{dt}\boldsymbol{\tau}. \tag{2.11}$$

The *tangent direction* is defined by the unit tangent vector $\boldsymbol{\tau}$, which is a path variable parameter

$$\frac{d\mathbf{r}}{dt} = \frac{ds}{dt}\boldsymbol{\tau}$$

or

$$\boldsymbol{\tau} = \frac{d\mathbf{r}}{ds}. \tag{2.12}$$

The velocity of a point in curvilinear motion is a vector whose magnitude equals the rate of change of distance traveled along the path and whose direction is tangent to the path.

To determine the acceleration of P, the time derivative of Eq. (2.11) is taken:

$$\mathbf{a} = \frac{d\mathbf{v}}{dt} = \frac{dv}{dt}\boldsymbol{\tau} + v\frac{d\boldsymbol{\tau}}{dt}. \tag{2.13}$$

If the path is not a straight line, the unit vector $\boldsymbol{\tau}$ rotates as P moves on the path, and the time derivative of $\boldsymbol{\tau}$ is not zero. The path angle θ defines the direction of $\boldsymbol{\tau}$ relative to a reference line shown in Fig. 2.7. The time derivative of the rotating tangent unit vector $\boldsymbol{\tau}$ is

$$\frac{d\boldsymbol{\tau}}{dt} = \frac{d\theta}{dt}\boldsymbol{\nu},$$

where $\boldsymbol{\nu}$ is a unit vector that is normal to $\boldsymbol{\tau}$ and points in the positive θ direction if $d\theta/dt$ is positive. The normal unit vector $\boldsymbol{\nu}$ defines the normal

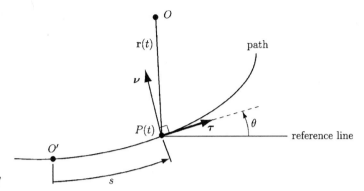

Figure 2.7

direction to the path. If we substitute this expression into Eq. (2.13), the acceleration of P is obtained:

$$\mathbf{a} = \frac{dv}{dt}\boldsymbol{\tau} + v\frac{d\theta}{dt}\boldsymbol{\nu}. \tag{2.14}$$

If the path is a straight line at time t, the normal component of the acceleration equals zero, because in that case $d\theta/dt$ is zero.

The tangential component of the acceleration arises from the rate of change of the magnitude of the velocity. The normal component of the acceleration arises from the rate of change in the direction of the velocity vector.

Figure 2.8 shows the positions on the path reached by P at time t, $P(t)$, and at time $t + dt$, $P(t + dt)$. If the path is curved, straight lines extended from these points $P(t)$ and $P(t + dt)$ perpendicular to the path will intersect at C as shown in Fig. 2.8. The distance ρ from the path to the point where these two lines intersect is called the *instantaneous radius of curvature* of the path.

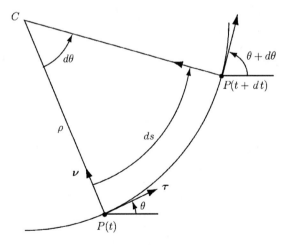

Figure 2.8

If the path is circular with radius a, then the radius of curvature equals the radius of the path, $\rho = a$. The angle $d\theta$ is the change in the path angle, and ds is the distance traveled, from t to $t + dt$. The radius of curvature ρ is related to ds by (Fig. 2.8)

$$ds = \rho \, d\theta.$$

Dividing by dt, one can obtain

$$\frac{ds}{dt} = v = \rho \frac{d\theta}{dt}.$$

Using this relation, one can write Eq. (2.14) as

$$\mathbf{a} = \frac{dv}{dt}\boldsymbol{\tau} + \frac{v^2}{\rho}\boldsymbol{\nu}.$$

For a given value of v, the normal component of the acceleration depends on the instantaneous radius of curvature. The greater the curvature of the path, the greater the normal component of the acceleration. When the acceleration is expressed in this way, the normal unit vector $\boldsymbol{\nu}$ must be defined to point toward the concave side of the path (Fig. 2.9). The velocity and acceleration in terms of normal and tangential components are (Fig. 2.10)

$$\mathbf{v} = v\boldsymbol{\tau} = \frac{ds}{dt}\boldsymbol{\tau}, \tag{2.15}$$

$$\mathbf{a} = a_t\boldsymbol{\tau} + a_n\boldsymbol{\nu}, \tag{2.16}$$

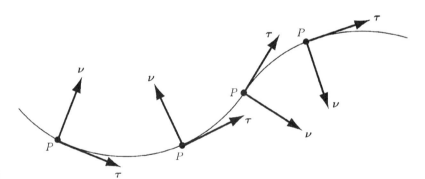

Figure 2.9

where

$$a_t = \frac{dv}{dt}, \quad a_n = v\frac{d\theta}{dt} = \frac{v^2}{\rho}. \tag{2.17}$$

If the motion occurs in the x–y plane of a cartesian reference frame (Fig. 2.11), and θ is the angle between the x axis and the unit vector $\boldsymbol{\tau}$, the unit vectors $\boldsymbol{\tau}$ and $\boldsymbol{\nu}$ are related to the cartesian unit vectors by

$$\boldsymbol{\tau} = \cos\theta\mathbf{i} + \sin\theta\mathbf{j}$$
$$\boldsymbol{\nu} = -\sin\theta\mathbf{i} + \cos\theta\mathbf{j}. \tag{2.18}$$

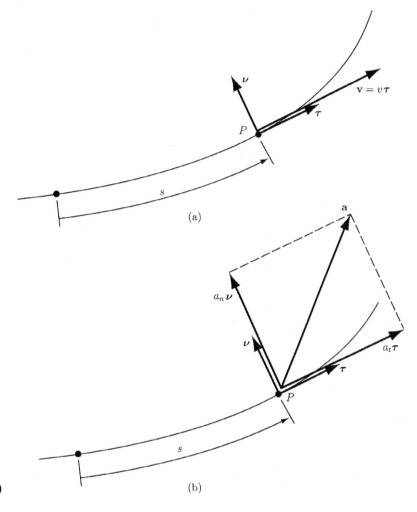

Figure 2.10

(a)

(b)

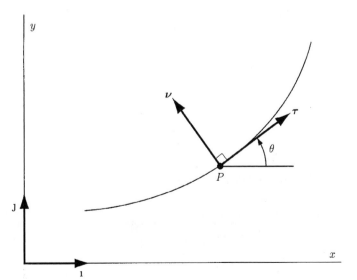

Figure 2.11

If the path in the x–y plane is described by a function $y = y(x)$, it can be shown that the instantaneous radius of curvature is given by

$$\rho = \frac{\left[1 + \left(\dfrac{dy}{dx}\right)^2\right]^{3/2}}{\left|\dfrac{d^2 y}{dx^2}\right|}.$$ (2.19)

2.6.2 CIRCULAR MOTION

The point P moves in a plane circular path of radius R as shown in Fig. 2.12. The distance s is

$$s = R\theta,$$ (2.20)

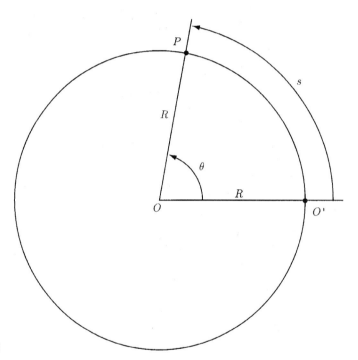

Figure 2.12

where the angle θ specifies the position of the point P along the circular path. The velocity is obtained taking the time derivative of Eq. (2.20),

$$v = \dot{s} = R\dot{\theta} = R\omega,$$ (2.21)

where $\omega = \dot{\theta}$ is the angular velocity of the line from the center of the path O to the point P. The tangential component of the acceleration is $a_t = dv/dt$, and

$$a_t = \dot{v} = R\dot{\omega} = R\alpha,$$ (2.22)

where $\alpha = \dot{\omega}$ is the angular acceleration. The normal component of the acceleration is

$$a_n = \frac{v^2}{R} = R\omega^2. \qquad (2.23)$$

For the circular path the instantaneous radius of curvature is $\rho = R$.

2.6.3 SPATIAL MOTION (FRENET'S FORMULAS)

The motion of a point P along a three-dimensional path is considered (Fig. 2.13a). The tangent direction is defined by the unit tangent vector $\boldsymbol{\tau}(|\boldsymbol{\tau}| = 1)$

$$\boldsymbol{\tau} = \frac{d\mathbf{r}}{ds}. \qquad (2.24)$$

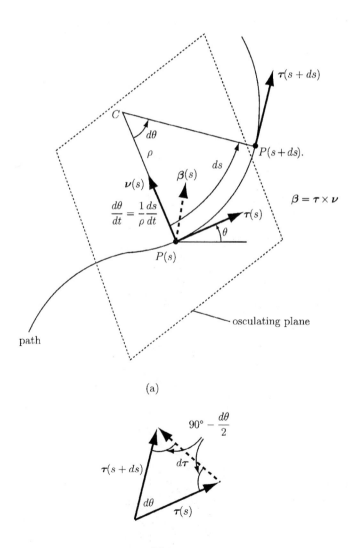

(a)

(b)

Figure 2.13

The second unit vector is derived by considering the dependence of $\boldsymbol{\tau}$ on s, $\boldsymbol{\tau} = \boldsymbol{\tau}(s)$. The dot product $\boldsymbol{\tau} \cdot \boldsymbol{\tau}$ gives the magnitude of the unit vector $\boldsymbol{\tau}$, that is,

$$\boldsymbol{\tau} \cdot \boldsymbol{\tau} = 1. \tag{2.25}$$

Equation (2.25) can be differentiated with respect to the path variable s:

$$\frac{d\boldsymbol{\tau}}{ds} \cdot \boldsymbol{\tau} + \boldsymbol{\tau} \cdot \frac{d\boldsymbol{\tau}}{ds} = 0 \Rightarrow \boldsymbol{\tau} \cdot \frac{d\boldsymbol{\tau}}{ds} = 0. \tag{2.26}$$

Equation (2.26) means that the vector $d\boldsymbol{\tau}/ds$ is always perpendicular to the vector $\boldsymbol{\tau}$. The normal direction, with the unit vector is $\boldsymbol{\nu}$, is defined to be parallel to the derivative $d\boldsymbol{\tau}/ds$. Because parallelism of two vectors corresponds to their proportionality, the normal unit vector may be written as

$$\boldsymbol{\nu} = \rho \frac{d\boldsymbol{\tau}}{ds}, \tag{2.27}$$

or

$$\frac{d\boldsymbol{\tau}}{ds} = \frac{1}{\rho} \boldsymbol{\nu}, \tag{2.28}$$

where ρ is the radius of curvature.

Figure 2.13a depicts the tangent and normal vectors associated with two points, $P(s)$ and $P(s + ds)$. The two points are separated by an infinitesimal distance ds measured along an arbitrary planar path. The point C is the intersection of the normal vectors at the two positions along the curve, and it is the center of curvature. Because ds is infinitesimal, the arc $P(s)P(s + ds)$ seems to be circular. The radius ρ of this arc is the radius of curvature. The formula for the arc of a circle is

$$d\theta = ds/\rho.$$

The angle $d\theta$ between the normal vectors in Fig. 2.13a is also the angle between the tangent vectors $\boldsymbol{\tau}(s + ds)$ and $\boldsymbol{\tau}(s)$. The vector triangle $\boldsymbol{\tau}(s + ds)$, $\boldsymbol{\tau}(s)$, $d\boldsymbol{\tau} = \boldsymbol{\tau}(s + ds) - \boldsymbol{\tau}(s)$ in Fig. 2.13b is isosceles because $|\boldsymbol{\tau}(s + ds)| = |\boldsymbol{\tau}(s)| = 1$. Hence, the angle between $d\boldsymbol{\tau}$ and either tangent vector is $90° - d\theta/2$. Since $d\theta$ is infinitesimal, the vector $d\boldsymbol{\tau}$ is perpendicular to the vector $\boldsymbol{\tau}$ in the direction of $\boldsymbol{\nu}$. A unit vector has a length of 1, so

$$|d\boldsymbol{\tau}| = d\theta|\boldsymbol{\tau}| = \frac{ds}{\rho}.$$

Any vector may be expressed as the product of its magnitude and a unit vector defining the sense of the vector

$$d\boldsymbol{\tau} = |d\boldsymbol{\tau}|\boldsymbol{\nu} = \frac{ds}{\rho} \boldsymbol{\nu}. \tag{2.29}$$

Note that the radius of curvature ρ is generally not a constant.

The tangent ($\boldsymbol{\tau}$) and normal ($\boldsymbol{\nu}$) unit vectors at a selected position form a plane, the *osculating plane*, that is tangent to the curve. Any plane containing $\boldsymbol{\tau}$ is tangent to the curve. When the path is not planar, the orientation of the oscillating plane containing the $\boldsymbol{\tau}$, $\boldsymbol{v}$ pair will depend on the position along the curve. The direction perpendicular to the osculating plane is called the *binormal*, and the corresponding unit vector is $\boldsymbol{\beta}$. The cross product of two unit vectors is a unit vector perpendicular to the original two, so the binormal direction may be defined such that

$$\boldsymbol{\beta} + \boldsymbol{\tau} \times \boldsymbol{v}.$$

Next the derivative of the $\boldsymbol{v}$ unit vector with respect to s in terms of its tangent, normal, and binormal components will be calculated. The component of any vector in a specific direction may be obtained from a dot product with a unit vector in that direction:

$$\frac{d\boldsymbol{v}}{ds} = \left(\boldsymbol{\tau} \cdot \frac{d\boldsymbol{v}}{ds}\right)\boldsymbol{\tau} + \left(\boldsymbol{v} \cdot \frac{d\boldsymbol{v}}{ds}\right) + \left(\boldsymbol{\beta} \cdot \frac{d\boldsymbol{v}}{ds}\right)\boldsymbol{\beta}. \tag{2.31}$$

The orthogonality of the unit vectors $\boldsymbol{\tau}$ and $\boldsymbol{v}$, $\boldsymbol{\tau} \perp \boldsymbol{v}$, requires that

$$\boldsymbol{\tau} \cdot \boldsymbol{v} = 0. \tag{2.32}$$

Equation (2.32) can be differentiated with respect to the path variable s:

$$\boldsymbol{\tau} \cdot \frac{d\boldsymbol{v}}{ds} + \boldsymbol{v} \cdot \frac{d\boldsymbol{\tau}}{ds} = 0$$

or

$$\boldsymbol{\tau} \cdot \frac{d\boldsymbol{v}}{ds} = -\boldsymbol{v} \cdot \frac{d\boldsymbol{\tau}}{ds} = -\boldsymbol{v} \cdot \left(\frac{1}{\rho}\boldsymbol{v}\right) = -\frac{1}{\rho}. \tag{2.33}$$

Because $\boldsymbol{v} \cdot \boldsymbol{v} = 1$, one may find that

$$\boldsymbol{v} \cdot \frac{d\boldsymbol{v}}{ds} = 0. \tag{2.34}$$

The derivative of the binormal component is

$$\frac{1}{T} = \boldsymbol{\beta} \cdot \frac{d\boldsymbol{v}}{ds}, \tag{2.35}$$

or

$$\frac{d\boldsymbol{v}}{ds} = \frac{1}{T}\boldsymbol{\beta}, \tag{2.36}$$

where T is the *torsion*. The reciprocal is used for consistency with Eq. (2.28). The torsion T has the dimension of length.

Substitution of Eqs. (2.33), (2.34), and (2.35) into Eq. (2.31) results in

$$\frac{d\boldsymbol{v}}{ds} = -\frac{1}{\rho}\boldsymbol{\tau} + \frac{1}{T}\boldsymbol{\beta}. \tag{2.37}$$

The derivative of $\boldsymbol{\beta}$,

$$\frac{d\boldsymbol{\beta}}{ds} = \left(\boldsymbol{\tau} \cdot \frac{d\boldsymbol{\beta}}{ds}\right)\boldsymbol{\tau} + \left(\boldsymbol{v} \cdot \frac{d\boldsymbol{v}}{ds}\right)\boldsymbol{v} + \left(\boldsymbol{\beta} \cdot \frac{d\boldsymbol{\beta}}{ds}\right)\boldsymbol{\beta}, \tag{2.38}$$

may be obtained by a similar approach.

Using the fact that $\boldsymbol{\tau}$, $\boldsymbol{\nu}$, and $\boldsymbol{\beta}$ are mutually orthogonal, and Eqs. (2.28) and (2.37), yields

$$\boldsymbol{\tau} \cdot \boldsymbol{\beta} = 0 \Rightarrow \boldsymbol{\tau} \cdot \frac{d\boldsymbol{\beta}}{ds} = -\frac{d\boldsymbol{\tau}}{ds} \cdot \boldsymbol{\beta} = -\frac{1}{\rho}\boldsymbol{\nu} \cdot \boldsymbol{\beta} = 0$$

$$\boldsymbol{\nu} \cdot \boldsymbol{\beta} = 0 \Rightarrow \boldsymbol{\nu} \cdot \frac{d\boldsymbol{\beta}}{ds} = -\frac{d\boldsymbol{\nu}}{ds} \cdot \boldsymbol{\beta} = -\frac{1}{T} \qquad (2.39)$$

$$\boldsymbol{\beta} \cdot \boldsymbol{\beta} = 1 \Rightarrow \boldsymbol{\beta} \cdot \frac{d\boldsymbol{\beta}}{ds} = 0.$$

The result is

$$\frac{d\boldsymbol{\beta}}{ds} = -\frac{1}{T}\boldsymbol{\nu}. \qquad (2.40)$$

Because $\boldsymbol{\nu}$ is a unit vector, this relation provides an alternative to Eq. (2.36) for the torsion:

$$\frac{1}{T} = -\left|\frac{d\boldsymbol{\beta}}{ds}\right|. \qquad (2.41)$$

Equations (2.28), (2.37), and (2.40) are the Frenet's formulas for a spatial curve.

Next the path is given in parametric form, the x, y, and z coordinates are given in terms of a parameter α. The position vector may be written as

$$\mathbf{r} = x(\alpha)\mathbf{1} + y(\alpha)\mathbf{j} + z(\alpha)\mathbf{k}. \qquad (2.42)$$

The unit tangent vector is

$$\boldsymbol{\tau} = \frac{d\mathbf{r}}{d\alpha}\frac{d\alpha}{ds} = \frac{\mathbf{r}'(\alpha)}{s'(\alpha)}, \qquad (2.43)$$

where a prime denotes differentiation with respect to α and

$$\mathbf{r}' = x'\mathbf{1} + y'\mathbf{j} + z'\mathbf{k}.$$

Using the fact that $|\boldsymbol{\tau}| = 1$, one may write

$$s' = (\mathbf{r}' \cdot \mathbf{r}')^{1/2} = [(x')^2 + (y')^2 + (z')^2]^{1/2}. \qquad (2.44)$$

The arc length s may be computed with the relation

$$s = \int_{\alpha_o}^{\alpha} [(x')^2 + (y')^2 + (z')^2]^{1/2} \, d\alpha, \qquad (2.45)$$

where α_o is the value at the starting position. The value of s' found from Eq. (2.44) may be substituted into Eq. (2.43) to calculate the tangent vector

$$\boldsymbol{\tau} = \frac{x'\mathbf{1} + y'\mathbf{j} + z'\mathbf{k}}{\sqrt{(x')^2 + (y')^2 + (z')^2}}. \qquad (2.46)$$

From Eqs. (2.43) and (2.28), the normal vector is

$$\boldsymbol{\nu} = \rho\frac{d\boldsymbol{\tau}}{ds} = \rho\frac{d\boldsymbol{\tau}}{d\alpha}\frac{d\alpha}{ds} = \frac{\rho}{s'}\left(\frac{\mathbf{r}''}{s'} - \frac{\mathbf{r}'s''}{(s')^2}\right) = \frac{\rho}{(s')^3}(\mathbf{r}''s' - \mathbf{r}'s''). \qquad (2.47)$$

2. **Kinematics of a Point**

The value of s' is given by Eq. (2.44) and the value of s'' is obtained differentiating Eq. (2.44):

$$s'' = \frac{\mathbf{r}' \cdot \mathbf{r}''}{(\mathbf{r}' \cdot \mathbf{r}')^{1/2}} = \frac{\mathbf{r}' \cdot \mathbf{r}''}{s'}. \tag{2.48}$$

The expression for the normal vector is obtained by substituting Eq. (2.48) into Eq. (2.47):

$$\boldsymbol{\nu} = \frac{\rho}{(s')^4}[\mathbf{r}''(s')^2 - \mathbf{r}'(\mathbf{r}' \cdot \mathbf{r}'')]. \tag{2.49}$$

Because $\boldsymbol{\nu} \cdot \boldsymbol{\nu} = 1$, the radius of curvature is

$$\frac{1}{\rho} = \frac{\rho}{(s')^4}\|[\mathbf{r}''(s')^2 - \mathbf{r}'(\mathbf{r}' \cdot \mathbf{r}'')]\|$$

$$= \frac{\rho}{(s')^4}[\mathbf{r}'' \cdot \mathbf{r}''(s')^4 - 2(\mathbf{r}' \cdot \mathbf{r}'')^2(s')^2 + \mathbf{r}'(\mathbf{r}' \cdot \mathbf{r}'')]^{1/2},$$

which simplifies to

$$\frac{1}{\rho} = \frac{1}{(s')^3}[\mathbf{r}'' \cdot \mathbf{r}''(s')^2 - (\mathbf{r}' \cdot \mathbf{r}'')^2]^{1/2}. \tag{2.50}$$

In the case of a planar curve $y = y(x)$ $(\alpha = x)$, Eq. (2.50) reduces to Eq. (2.19).

The binomial vector may be calculated with the relation

$$\boldsymbol{\beta} = \boldsymbol{\tau} \times \boldsymbol{\nu} = \frac{\mathbf{r}'}{s'} \times \frac{\rho}{(s')^4}[\mathbf{r}''(s')^2 - \mathbf{r}'(\mathbf{r}' \cdot \mathbf{r}'')]$$

$$= \frac{\rho}{(s')^3}\mathbf{r}' \times \mathbf{r}''. \tag{2.51}$$

The result of differentiating Eq. (2.51) may be written as

$$\frac{d\boldsymbol{\beta}}{ds} = \frac{1}{s'}\frac{d\boldsymbol{\beta}}{d\alpha} = \frac{1}{s'}\frac{d}{d\alpha}\left[\frac{\rho}{(s')^3}\right](\mathbf{r}' \times \mathbf{r}'') + \frac{\rho}{(s')^4}(\mathbf{r}' \times \mathbf{r}'''). \tag{2.52}$$

The torsion T may be obtained by applying the formula

$$\frac{1}{T} = -\boldsymbol{\nu} \cdot \frac{d\boldsymbol{\beta}}{ds}$$

$$= -\frac{\rho}{(s')^4}[\mathbf{r}''(s')^2 - \mathbf{r}'(\mathbf{r}' \cdot \mathbf{r}'')] \cdot \left[\frac{1}{s'}\frac{d}{d\alpha}\left(\frac{\rho}{(s')^3}\right)(\mathbf{r}' \times \mathbf{r}'') + \frac{\rho}{(s')^4}(\mathbf{r}' \times \mathbf{r}''')\right].$$

The preceding equation may be simplified and T can be calculated from

$$\frac{1}{T} = -\frac{\rho^2}{(s')^6}[\mathbf{r}'' \cdot (\mathbf{r}' \times \mathbf{r}''')]. \tag{2.53}$$

The expressions for the velocity and acceleration in normal and tangential directions for three-dimensional motions are identical in form to the expressions for planar motion. The velocity is a vector whose magnitude equals the rate of change of distance, and whose direction is tangent to the path. The acceleration has a component tangential to the path equal to the rate of change of the magnitude of the velocity, and a component perpendicular to the path that depends on the magnitude of the velocity and the instantaneous radius of curvature of the path. In planar motion, the normal unit vector $\boldsymbol{\nu}$ is parallel to the plane of motion. In three-dimensional motion, $\boldsymbol{\nu}$ is parallel to the osculating plane, whose orientation depends on the nature of the path. The binomial vector $\boldsymbol{\beta}$ is a unit vector that is perpendicular to the osculating plane and therefore defines its orientation.

2.6.4 POLAR COORDINATES

A point P is considered in the x–y plane of a cartesian coordinate system. The position of the point P relative to the origin O may be specified either by its cartesian coordinates x, y or by its polar coordinates r, θ (Fig. 2.14). The polar coordinates are defined by:

- The unit vector $\mathbf{u}_r$, which points in the direction of the radial line from the origin O to the point P
- The unit vector $\mathbf{u}_\theta$, which is perpendicular to $\mathbf{u}_r$ and points in the direction of increasing the angle θ

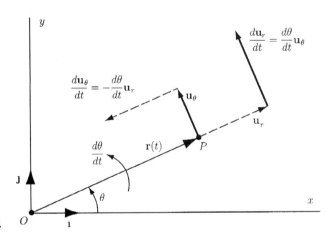

Figure 2.14

The unit vectors $\mathbf{u}_r$ and $\mathbf{u}_\theta$ are related to the cartesian unit vectors $\mathbf{1}$ and $\mathbf{j}$ by

$$\mathbf{u}_r = \cos\theta\mathbf{1} + \sin\theta\mathbf{j},$$
$$\mathbf{u}_\theta = -\sin\theta\mathbf{1} + \cos\theta\mathbf{j}. \tag{2.54}$$

The position vector $\mathbf{r}$ from O to P is

$$\mathbf{r} = r\mathbf{u}_r, \tag{2.55}$$

where r is the magnitude of the vector $\mathbf{r}$, $r = |\mathbf{r}|$.

The velocity of the point P in terms of polar coordinates is obtained by taking the time derivative of Eq. (2.55):

$$\mathbf{v} = \frac{d\mathbf{r}}{dt} = \frac{dr}{dt}\mathbf{u}_r + r\frac{d\mathbf{u}_r}{dt}. \tag{2.56}$$

The time derivative of the rotating unit vector $\mathbf{u}_r$ is

$$\frac{d\mathbf{u}_r}{dt} = \frac{d\theta}{dt}\mathbf{u}_\theta = \omega\mathbf{u}_\theta,\tag{2.57}$$

where $\omega = d\theta/dt$ is the angular velocity.

If we substitute Eq. (2.57) into Eq. (2.56), the velocity of P is

$$\mathbf{v} = \frac{dr}{dt}\mathbf{u}_r + r\frac{d\theta}{dt}\mathbf{u}_\theta = \frac{dr}{dt}\mathbf{u}_r + r\omega\mathbf{u}_\theta = \dot{r}\mathbf{u}_r + r\omega\mathbf{u}_\theta,\tag{2.58}$$

or

$$\mathbf{v} = v_r\mathbf{u}_r + v_\theta\mathbf{u}_\theta,\tag{2.59}$$

where

$$v_r = \frac{dr}{dt} = \dot{r} \quad \text{and} \quad v_\theta = r\omega.\tag{2.60}$$

The acceleration of the point P is obtained by taking the time derivative of Eq. (2.58):

$$\begin{aligned}\mathbf{a} = \frac{d\mathbf{v}}{dt} &= \frac{d^2r}{dt^2}\mathbf{u}_r + \frac{dr}{dt}\frac{d\mathbf{u}_r}{dt} + \frac{dr}{dt}\frac{d\theta}{dt}\mathbf{u}_\theta \\ &+ r\frac{d^2\theta}{dt^2}\mathbf{u}_\theta + r\frac{d\theta}{dt}\frac{d\mathbf{u}_\theta}{dt}.\end{aligned}\tag{2.61}$$

As P moves, $\mathbf{u}_\theta$ also rotates with angular velocity $d\theta/dt$. The time derivative of the unit vector $\mathbf{u}_\theta$ is in the $-\mathbf{u}_r$ direction if $d\theta/dt$ is positive:

$$\frac{d\mathbf{u}_\theta}{dt} = -\frac{d\theta}{dt}\mathbf{u}_r.\tag{2.62}$$

If Eq. (2.62) and Eq. (2.57) are substituted into Eq. (2.61), the acceleration of the point P is

$$\mathbf{a} = \left[\frac{d^2r}{dt^2} - r\left(\frac{d\theta}{dt}\right)^2\right]\mathbf{u}_r + \left[r\frac{d^2\theta}{dt^2} + 2\frac{dr}{dt}\frac{d\theta}{dt}\right]\mathbf{u}_\theta.$$

Thus, the acceleration of P is

$$\mathbf{a} = a_r\mathbf{u}_r + a_\theta\mathbf{u}_\theta,\tag{2.63}$$

where

$$\begin{aligned}a_r &= \frac{d^2r}{dt^2} - r\left(\frac{d\theta}{dt}\right)^2 = \frac{d^2r}{dt^2} - r\omega^2 = \ddot{r} - r\omega^2 \\ a_\theta &= r\frac{d^2\theta}{dt^2} + 2\frac{dr}{dt}\frac{d\theta}{dt} = r\alpha + 2\omega\frac{dr}{dt} = r\alpha + 2\dot{r}\omega.\end{aligned}\tag{2.64}$$

The term

$$\alpha = \frac{d^2\theta}{dt^2} = \ddot{\theta}$$

is called the angular acceleration.

Dynamics

The radial component of the acceleration $-r\omega^2$ is called *the centripetal* acceleration. The transverse component of the acceleration $2\omega(dr/dt)$ is called the *Coriolis* acceleration.

2.6.5 CYLINDRICAL COORDINATES

The cylindrical coordinates r, θ, and z describe the motion of a point P in the xyz space as shown in Fig. 2.15. The cylindrical coordinates r and θ are the polar coordinates of P measured in the plane parallel to the $x-y$ plane, and the unit vectors $\mathbf{u}_r$, and $\mathbf{u}_\theta$ are the same. The coordinate z measures the position of the point P perpendicular to the $x-y$ plane. The unit vector $\mathbf{k}$ attached to the coordinate z points in the positive z axis direction. The position vector $\mathbf{r}$ of the point P in terms of cylindrical coordinates is

$$\mathbf{r} = r\mathbf{u}_r + z\mathbf{k}. \tag{2.65}$$

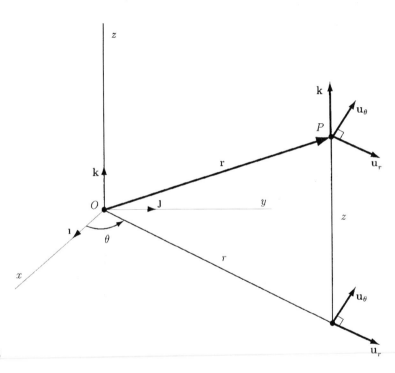

Figure 2.15

The coordinate r in Eq. (2.65) is not equal to the magnitude of $\mathbf{r}$ except when the point P moves along a path in the $x-y$ plane.

The velocity of the point P is

$$
\begin{aligned}
\mathbf{v} = \frac{d\mathbf{r}}{dt} &= v_r\mathbf{u}_r + v_\theta\mathbf{u}_\theta + v_z\mathbf{k} \\
&= \frac{dr}{dt}\mathbf{u}_r + r\omega\mathbf{u}_\theta + \frac{dz}{dt}\mathbf{k} \\
&= \dot{r}\mathbf{u}_r + r\omega\mathbf{u}_\theta + \dot{z}\mathbf{k},
\end{aligned}
\tag{2.66}
$$

and the acceleration of the point P is

$$\mathbf{a} = \frac{d\mathbf{v}}{dt} = a_r\mathbf{u}_r + a_\theta\mathbf{u}_\theta + a_z\mathbf{k}, \tag{2.67}$$

where

$$a_r = \frac{d^2r}{dt^2} - r\omega^2 = \ddot{r} - r\omega^2$$

$$a_\theta = r\alpha + 2\frac{dr}{dt}\omega = r\alpha + 2\dot{r}\omega \tag{2.68}$$

$$a_z = \frac{d^2z}{dt^2} = \ddot{z}.$$

2.7 Relative Motion

Suppose that A and B are two points that move relative to a reference frame with origin at point O (Fig. 2.16). Let $\mathbf{r}_A$ and $\mathbf{r}_B$ be the position vectors of points A and B relative to O. The vector $\mathbf{r}_{BA}$ is the position vector of point A relative to point B. These vectors are related by

$$\mathbf{r}_A = \mathbf{r}_B + \mathbf{r}_{BA}. \tag{2.69}$$

The time derivative of Eq. (2.69) is

$$\mathbf{v}_A = \mathbf{v}_B + \mathbf{v}_{AB}, \tag{2.70}$$

where $\mathbf{v}_A$ is the velocity of A relative to O, $\mathbf{v}_B$ is the velocity of B relative to O, and $\mathbf{v}_{AB} = d\mathbf{r}_{AB}/dt = \dot{r}_{AB}$ is the velocity of A relative to B. The time derivative of Eq. (2.70) is

$$\mathbf{a}_A = \mathbf{a}_B + \mathbf{a}_{AB}, \tag{2.71}$$

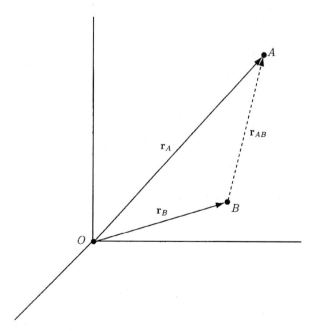

Figure 2.16

Dynamics

where $\mathbf{a}_A$ and $\mathbf{a}_B$ are the accelerations of A and B relative to O and $\mathbf{a}_{AB} = d\mathbf{v}_{AB}/dt = \ddot{\mathbf{r}}_{AB}$ is the acceleration of A relative to B.

3. Dynamics of a Particle

3.1 Newton's Second Law

Classical mechanics was established by Isaac Newton with the publication of *Philosophiae naturalis principia mathematica*, in 1687. Newton stated three "laws" of motion, which may be expressed in modern terms:

1. When the sum of the forces acting on a particle is zero, its velocity is constant. In particular, if the particle is initially stationary, it will remain stationary.
2. When the sum of the forces acting on a particle is not zero, the sum of the forces is equal to the rate of change of the *linear momentum* of the particle.
3. The forces exerted by two particles on each other are equal in magnitude and opposite in direction.

The linear momentum of a particle is the product of the mass of the particle, m, and the velocity of the particle, $\mathbf{v}$. Newton's second law may be written as

$$\mathbf{F} = \frac{d}{dt}(m\mathbf{v}), \qquad (3.1)$$

where $\mathbf{F}$ is the total force on the particle. If the mass of the particle is constant, $m = $ constant, the total force equals the product of its mass and acceleration, $\mathbf{a}$:

$$\mathbf{F} = m\frac{d\mathbf{v}}{dt} = m\mathbf{a}. \qquad (3.2)$$

Newton's second law gives precise meanings to the terms *mass* and *force*. In SI units, the unit of mass is the kilogram (kg). The unit of force is the newton (N), which is the force required to give a mass of 1 kilogram an acceleration of 1 meter per second squared:

$$1\text{ N} = (1\text{ kg})(1\text{ m/s}^2) = 1\text{ kg m/s}^2.$$

In U.S. customary units, the unit of force is the pound (lb). The unit of mass is the slug, which is the amount of mass accelerated at 1 foot per second squared by a force of 1 pound:

$$1\text{ lb} = (1\text{ slug})(1\text{ ft/s}^2),\text{ or }1\text{ slug} = 1\text{ lb s}^2/\text{lb}.$$

3.2 Newtonian Gravitation

Newton's postulate for the magnitude of gravitational force F between two particles in terms of their masses m_1 and m_2 and the distance r between them (Fig. 3.1) may be expressed as

$$F = \frac{Gm_1 m_2}{r^2},$$ (3.3)

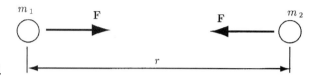

Figure 3.1

where G is called the *universal gravitational constant.* Equation (3.3) may be used to approximate the weight of a particle of mass m due to the gravitational attraction of the earth,

$$W = \frac{Gmm_E}{r^2},$$ (3.4)

where m_E is the mass of the earth and r is the distance from the center of the earth to the particle. When the weight of the particle is the only force acting on it, the resulting acceleration is called the acceleration due to gravity. In this case, Newton's second law states that $W = ma$, and from Eq. (3.4) the acceleration due to gravity is

$$a = \frac{Gm_E}{r^2}.$$ (3.5)

The acceleration due to gravity at sea level is denoted by g. From Eq. (3.5) one may write $Gm_E = gR_E^2$, where R_E is the radius of the earth. The expression for the acceleration due to gravity at a distance r from the center of the earth in terms of the acceleration due to gravity at sea level is

$$a = g\frac{R_E^2}{r^2}.$$ (3.6)

At sea level, the weight of a particle is given by

$$W = mg.$$ (3.7)

The value of g varies on the surface of the earth from one location to another. The values of g used in examples and problems are $g = 9.81$ m/s^2 in SI units and $g = 32.2$ ft/s^2 in U.S. customary units.

3.3 Inertial Reference Frames

Newton's laws do not give accurate results if a problem involves velocities that are not small compared to the velocity of light (3×10^8 m/s). Einstein's theory of relativity may be applied to such problems. Newtonian mechanics also fails in problems involving atomic dimensions. Quantum mechanics may be used to describe phenomena on the atomic scale.

The position, velocity, and acceleration of a point are specified, in general, relative to an arbitrary reference frame. Newton's second law cannot be expressed in terms of just any reference frame. Newton stated that the second law should be expressed in terms of a reference frame at rest with respect to the "fixed stars." Newton's second law may be applied with good results using reference frames that accelerate and rotate by properly accounting for the acceleration and rotation. Newton's second law, Eq. (3.2), may be expressed in terms of a reference frame that is fixed relative to the earth. Equation (3.2) may be applied using a reference that translates at constant velocity relative to the earth.

If a reference frame may be used to apply Eq. (3.2), it is said to be a *Newtonian* or *inertial* reference frame.

3.4 Cartesian Coordinates

To apply Newton's second law in a particular situation, one may choose a coordinate system. Newton's second law in a cartesian reference frame (Fig. 3.2) may be expressed as

$$\sum \mathbf{F} = m\mathbf{a}, \tag{3.8}$$

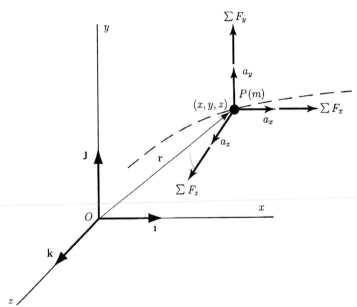

Figure 3.2

where $\sum \mathbf{F} = \sum F_x \mathbf{i} + \sum F_y \mathbf{j} + \sum F_z \mathbf{k}$ is the sum of the forces acting on a particle P of mass m, and

$$\mathbf{a} = a_x \mathbf{i} + a_y \mathbf{j} + a_z \mathbf{k} = \ddot{x}\mathbf{i} + \ddot{y}\mathbf{j} + \ddot{z}\mathbf{k}$$

is the acceleration of the particle. When x, y, and z components are equated, three scalar equations of motion are obtained,

$$\sum F_x = ma_x = m\ddot{x}, \quad \sum F_y = ma_y = m\ddot{y}, \quad \sum F_z = ma_z = m\ddot{z}, \quad (3.9)$$

or the total force in each coordinate direction equals the product of the mass and component of the acceleration in that direction.

Projectile Problem

An object P, of mass m, is launched through the air (Fig. 3.3). The force on the object is just the weight of the object (the aerodynamic forces are neglected). The sum of the forces is $\sum \mathbf{F} = -mg\mathbf{j}$. From Eq. (3.9) one may obtain

$$a_x = \ddot{x} = 0, \quad a_y = z\ddot{y} = -g, \quad a_z = \ddot{z} = 0.$$

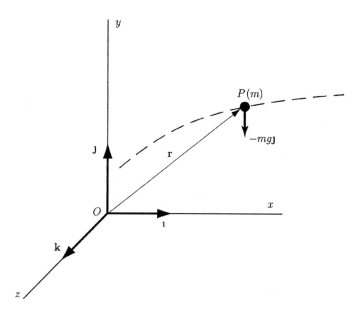

Figure 3.3

The projectile accelerates downward with the acceleration due to gravity.

Straight Line Motion

For straight line motion along the x axis, Eqs. (3.9) are

$$\sum F_x = m\ddot{x}, \quad \sum F_y = 0, \quad \sum F_z = 0.$$

3.5 Normal and Tangential Components

A particle P of mass m moves on a curved path (Fig. 3.4). One may resolve the sum of the forces $\sum \mathbf{F}$ acting on the particle into normal F_n and tangential F_t components:

$$\sum \mathbf{F} = F_t \boldsymbol{\tau} + F_n \boldsymbol{\nu}.$$

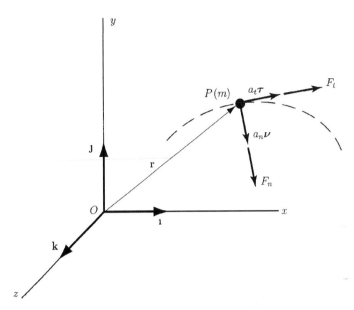

Figure 3.4

The acceleration of the particle in terms of normal and tangential components is

$$\mathbf{a} = a_t \boldsymbol{\tau} + a_n \boldsymbol{\nu}.$$

Newton's second law is

$$\sum \mathbf{F} = m\mathbf{a}$$

$$F_t \boldsymbol{\tau} + F_n \boldsymbol{\nu} = m(a_t \boldsymbol{\tau} + a_n \boldsymbol{\nu}), \qquad (3.10)$$

where

$$a_t = \frac{dv}{dt} = \dot{v}, \quad a_n = \frac{v^2}{\rho}.$$

When the normal and tangential components in Eq. (3.10) are equated, two scalar equations of motion are obtained:

$$F_t = m\dot{v}, \quad F_n = m\frac{v^2}{\rho}. \qquad (3.11)$$

The sum of the forces in the tangential direction equals the product of the mass and the rate of change of the magnitude of the velocity, and the sum of the forces in the normal direction equals the product of the mass and the normal component of acceleration. If the path of the particle lies in a plane, the acceleration of the particle perpendicular to the plane is zero, and so the sum of the forces perpendicular to the plane is zero.

3.6 Polar and Cylindrical Coordinates

A particle P with mass m moves in a plane curved path (Fig. 3.5). The motion of the particle may be described in terms of polar coordinates. Resolving the

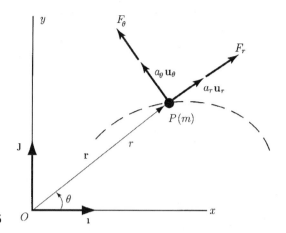

Figure 3.5

sum of the forces parallel to the plane into radial and transverse components gives

$$\sum \mathbf{F} = F_r \mathbf{u}_r + F_\theta \mathbf{u}_\theta,$$

and if the acceleration of the particle is expressed in terms of radial and transverse components, Newton's second law may be written in the form

$$F_r \mathbf{u}_r + F_\theta \mathbf{u}_\theta = m(a_r \mathbf{u}_r + a_\theta \mathbf{u}_\theta), \qquad (3.12)$$

where

$$a_r = \frac{d^2 r}{dt^2} - r\left(\frac{d\theta}{dt}\right)^2 = \ddot{r} - r\omega^2$$

$$a_\theta = r\frac{d^2\theta}{dt^2} + 2\frac{dr}{dt}\frac{d\theta}{dt} = r\alpha + 2\dot{r}\omega.$$

Two scalar equations are obtained:

$$\begin{aligned} F_r &= m(\ddot{r} - r\omega^2) \\ F_\theta &= m(r\alpha + 2\dot{r}\omega). \end{aligned} \qquad (3.13)$$

The sum of the forces in the radial direction equals the product of the mass and the radial component of the acceleration, and the sum of the forces in the transverse direction equals the product of the mass and the transverse component of the acceleration.

The three-dimensional motion of the particle P may be obtained using cylindrical coordinates (Fig. 3.6). The position of P perpendicular to the x–y plane is measured by the coordinate z and the unit vector $\mathbf{k}$. The sum of the forces is resolved into radial, transverse, and z components:

$$\sum \mathbf{F} = F_r \mathbf{u}_r + F_\theta \mathbf{u}_\theta + F_z \mathbf{k}.$$

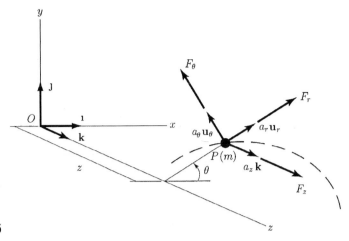

Figure 3.6

The three scalar equations of motion are the radial and transverse relations, Eq. (3.13) and the equation of motion in the z direction,

$$F_r = m(\ddot{r} - r\omega^2)$$
$$F_\theta = m(r\alpha + 2\dot{r}\omega) \tag{3.14}$$
$$F_z = m\ddot{z}.$$

3.7 Principle of Work and Energy

Newton's second law for a particle of mass m can be written in the form

$$\mathbf{F} = m\frac{d\mathbf{v}}{dt} = m\dot{\mathbf{v}}. \tag{3.15}$$

The dot product of both sides of Eq. (3.15) with the velocity $\mathbf{v} = d\mathbf{r}/dt$ gives

$$\mathbf{F} \cdot \mathbf{v} = m\dot{\mathbf{v}} \cdot \mathbf{v}, \tag{3.16}$$

or

$$\mathbf{F} \cdot \frac{d\mathbf{r}}{dt} = m\dot{\mathbf{v}} \cdot \mathbf{v}. \tag{3.17}$$

But

$$\frac{d}{dt}(\mathbf{v} \cdot \mathbf{v}) = \dot{\mathbf{v}} \cdot \mathbf{v} + \mathbf{v} \cdot \dot{\mathbf{v}} = 2\dot{\mathbf{v}} \cdot \mathbf{v},$$

and

$$\dot{\mathbf{v}} \cdot \mathbf{v} = \frac{1}{2}\frac{d}{dt}(\mathbf{v} \cdot \mathbf{v}) = \frac{1}{2}\frac{d}{dt}(v^2), \tag{3.18}$$

where $v^2 = \mathbf{v} \cdot \mathbf{v}$ is the square of the magnitude of $\mathbf{v}$. Using Eq. (3.18) one may write Eq. (3.17) as

$$\mathbf{F} \cdot d\mathbf{r} = \frac{1}{2}m\,d(\mathbf{v} \cdot \mathbf{v}) = \frac{1}{2}m\,d(v^2). \tag{3.19}$$

The term

$$dU = \mathbf{F} \cdot d\mathbf{r}$$

is the *work* where $\mathbf{F}$ is the total external force acting on the particle of mass m and $d\mathbf{r}$ is the infinitesimal displacement of the particle. Integrating Eq. (3.19), one may obtain

$$\int_{\mathbf{r}_1}^{\mathbf{r}_2} \mathbf{F} \cdot d\mathbf{r} = \int_{v_1^2}^{v_2^2} \frac{1}{2} m \, d(v^2) = \frac{1}{2} m v_2^2 - \frac{1}{2} m v_1^2, \tag{3.20}$$

where v_1 and v_2 are the magnitudes of the velocity at the positions $\mathbf{r}_1$ and $\mathbf{r}_2$.

The *kinetic energy* of a particle of mass m with the velocity $\mathbf{v}$ is the term

$$T = \frac{1}{2} m \mathbf{v} \cdot \mathbf{v} = \frac{1}{2} m v^2, \tag{3.21}$$

where $|\mathbf{v}| = v$. The work done as the particle moves from position $\mathbf{r}_1$ to position $\mathbf{r}_2$ is

$$U_{12} = \int_{\mathbf{r}_1}^{\mathbf{r}_2} \mathbf{F} \cdot d\mathbf{r}. \tag{3.22}$$

The *principle of work and energy* may be expressed as

$$U_{12} = \frac{1}{2} m v_2^2 - \frac{1}{2} m v_1^2. \tag{3.23}$$

The work done on a particle as it moves between two positions equals the change in its kinetic energy.

The dimensions of work, and therefore the dimensions of kinetic energy, are (force) × (length). In U.S. customary units, work is expressed in ft lb. In SI units, work is expressed in N m, or joules (J).

One may use the principle of work and energy on a system if no net work is done by internal forces. The internal friction forces may do net work on a system.

3.8 Work and Power

The position of a particle P of mass m in curvilinear motion is specified by the coordinate s measured along its path from a reference point O (Fig. 3.7a). The velocity of the particle is

$$\mathbf{v} = \frac{ds}{dt} \boldsymbol{\tau} = \dot{s}\boldsymbol{\tau},$$

where $\boldsymbol{\tau}$ is the tangential unit vector.

Using the relation $\dot{\mathbf{v}} = d\mathbf{r}/dt$, the infinitesimal displacement $d\mathbf{r}$ along the path is

$$d\mathbf{r} = \mathbf{v} \, dt = \frac{ds}{dt} \boldsymbol{\tau} \, dt = ds \, \boldsymbol{\tau}.$$

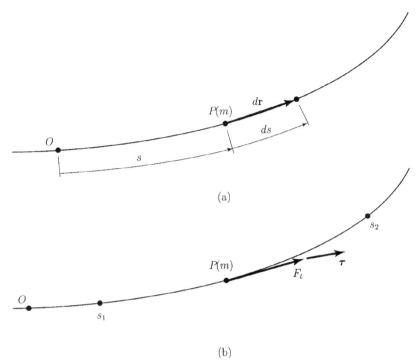

Figure 3.7   (b)

The work done by the external forces acting on the particle as result of the displacement $d\mathbf{r}$ is

$$\mathbf{F} \cdot d\mathbf{r} = \mathbf{F} \cdot ds\,\boldsymbol{\tau} = \mathbf{F} \cdot \boldsymbol{\tau}\, ds = F_t ds,$$

where $F_t = \mathbf{F} \cdot \boldsymbol{\tau}$ is the tangential component of the total force.

The work as the particle moves from a position s_1 to a position s_2 is (Fig. 3.7b)

$$U_{12} = \int_{s_1}^{s_2} F_t\, ds. \tag{3.24}$$

The work is equal to the integral of the tangential component of the total force with respect to distance along the path. Components of force perpendicular to the path do not do any work.

The work done by the external forces acting on a particle during an infinitesimal displacement $d\mathbf{r}$ is

$$dU = \mathbf{F} \cdot d\mathbf{r}.$$

The *power*, P, is the rate at which work is done. The power P is obtained by dividing the expression of the work by the interval of time dt during which the displacement takes place:

$$P = \mathbf{F} \cdot \frac{d\mathbf{r}}{dt} = \mathbf{F} \cdot \mathbf{v}.$$

In SI units, the power is expressed in newton meters per second, which is joules per second (J/s) or watts (W). In U.S. customary units, power is expressed in foot pounds per second or in horsepower (hp), which is 746 W or approximately 550 ft lb/s.

The power is also the rate of change of the kinetic energy of the object,

$$P = \frac{d}{dt}\left(\frac{1}{2}mv^2\right).$$

3.8.1 WORK DONE ON A PARTICLE BY A LINEAR SPRING

A linear spring connects a particle P of mass m to a fixed support (Fig. 3.8). The force exerted on the particle is

$$\mathbf{F} = -k(r - r_0)\mathbf{u}_r,$$

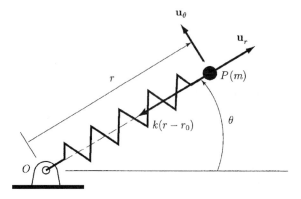

Figure 3.8

where k is the spring constant, r_0 is the unstretched length of the spring, and $\mathbf{u}_r$ is the polar unit vector. If we use the expression for the velocity in polar coordinates, the vector $d\mathbf{r} = \mathbf{v}dt$ is

$$d\mathbf{r} = \left(\frac{dr}{dt}\mathbf{u}_r + r\frac{d\theta}{dt}\mathbf{u}_\theta\right)dt = dr\,\mathbf{u}_r + rd\theta\,\mathbf{u}_\theta \tag{3.25}$$
$$\mathbf{F} \cdot d\mathbf{r} = [-k(r - r_0)\mathbf{u}_r] \cdot (dr\,\mathbf{u}_r + rd\theta\,\mathbf{u}_\theta) = -k(r - r_0)\,dr.$$

One may express the work done by a spring in terms of its stretch, defined by $\delta = r - r_0$. In terms of this variable, $\mathbf{F} \cdot d\mathbf{r} = -k\delta\,d\delta$, and the work is

$$U_{12} = \int_{\mathbf{r}_1}^{\mathbf{r}_2} \mathbf{F} \cdot d\mathbf{r} = \int_{\delta_1}^{\delta_2} -k\delta\,d\delta = -\frac{1}{2}k(\delta_2^2 - \delta_1^2),$$

where δ_1 and δ_2 are the values of the stretch at the initial and final positions.

3.8.2 WORK DONE ON A PARTICLE BY WEIGHT

A particle P of mass m (Fig. 3.9) moves from position 1 with coordinates (x_1, y_1, z_1) to position 2 with coordinates (x_2, y_2, z_2) in a cartesian reference frame with the y axis upward. The force exerted by the weight is

$$\mathbf{F} = -mg\mathbf{j}.$$

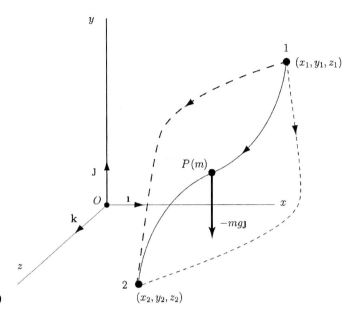

Figure 3.9

Because $\mathbf{v} = d\mathbf{r}/dt$, the expression for the vector $d\mathbf{r}$ is

$$d\mathbf{r} = \left(\frac{dx}{dt}\mathbf{i} + \frac{dy}{dt}\mathbf{j} + \frac{dz}{dt}\mathbf{k}\right) dt = dx\mathbf{i} + dy\mathbf{j} + dz\mathbf{k}.$$

The dot product of $\mathbf{F}$ and $d\mathbf{r}$ is

$$\mathbf{F} \cdot d\mathbf{r} = (-mg\mathbf{j}) \cdot (dx\mathbf{i} + dy\mathbf{j} + dz\mathbf{k}) = -mg \, dy.$$

The work done as P moves from position 1 to position 2 is

$$U_{12} = \int_{\mathbf{r}_1}^{\mathbf{r}_2} \mathbf{F} \cdot d\mathbf{r} = \int_{y_1}^{y_2} -mg \, dy = -mg(y_2 - y_1).$$

The work is the product of the weight and the change in the height of the particle. The work done is negative if the height increases and positive if it decreases. The work done is the same no matter what path the particle follows from position 1 to position 2. To determine the work done by the weight of the particle, only the relative heights of the initial and final positions must be known.

3.9 Conservation of Energy

The change in the kinetic energy is

$$U_{12} = \int_{r_1}^{r_2} \mathbf{F} \cdot d\mathbf{r} = \frac{1}{2} mv_2^2 - \frac{1}{2} mv_1^2. \tag{3.26}$$

A scalar function of position V called *potential energy* may be determined as

$$dV = -\mathbf{F} \cdot d\mathbf{r}. \tag{3.27}$$

If we use the function V, the integral defining the work is

$$U_{12} = \int_{r_1}^{r_2} \mathbf{F} \cdot d\mathbf{r} = \int_{V_1}^{V_2} -dV = -(V_2 - V_1), \tag{3.28}$$

where V_1 and V_2 are the values of V at the positions r_1 and r_2. The principle of work and energy would then have the form

$$\frac{1}{2} mv_1^2 + V_1 = \frac{1}{2} mv_2^2 + V_2, \tag{3.29}$$

which means that the sum of the kinetic energy and the potential energy V is constant:

$$\frac{1}{2} mv^2 + V = \text{constant} \tag{3.30}$$

or

$$E = T + V = \text{constant}. \tag{3.31}$$

If a potential energy V exists for a given force $\mathbf{F}$, i.e., a function of position V exists such that $dV = -\mathbf{F} \cdot d\mathbf{r}$, then $\mathbf{F}$ is said to be *conservative*.

If all the forces that do work on a system are conservative, the total energy — the sum of the kinetic energy and the potential energies of the forces — is constant, or conseved. The system is said to be conservative.

3.10 Conservative Forces

A particle moves from position 1 to position 2. Equation (3.28) states that the work depends only on the values of the potential energy at positions 1 and 2. The work done by a conservative force as a particle moves from position 1 to position 2 is independent of the path of the particle.

A particle P of mass m slides with friction along a path of length L. The magnitude of the friction force is μmg and is opposite to the direction of the motion of the particle. The coefficient of friction is μ. The work done by the friction force is

$$U_{12} = \int_0^L -\mu mg \ ds = -\mu mgL.$$

The work is proportional to the length L of the path and therefore is not independent of the path of the particle. Friction forces are not conservative.

3.10.1 POTENTIAL ENERGY OF A FORCE EXERTED BY A SPRING

The force exerted by a linear spring attached to a fixed support is a conservative force.

In terms of polar coordinates, the force exerted on a particle (Fig. 3.8) by a linear spring is $\mathbf{F} = -k(r - r_0)\mathbf{u}_r$. The potential energy must satisfy

$$dV = -\mathbf{F} \cdot d\mathbf{r} = k(r - r_0) \ dr,$$

or

$$dV = k\delta \, d\delta,$$

where $\delta = r - r_0$ is the stretch of the spring. Integrating this equation, the potential energy of a linear spring is

$$V = \frac{1}{2} k\delta^2. \tag{3.32}$$

3.10.2 POTENTIAL ENERGY OF WEIGHT

The weight of a particle is a conservative force. The weight of the particle P of mass m (Fig. 3.9) is $F = -mg\mathbf{j}$. The potential energy V must satisfy the relation

$$dV = -\mathbf{F} \cdot d\mathbf{r} = (mg\mathbf{j}) \cdot (dx\mathbf{i} + dy\mathbf{j} + dz\mathbf{k}) = mg \, dy, \tag{3.33}$$

or

$$\frac{dV}{dy} = mg.$$

After integration of this equation, the potential energy is

$$V = mgy + C,$$

where C is an integration constant. The constant C is arbitrary, because this expression satisfies Eq. (3.33) for any value of C. For $C = 0$ the potential energy of the weight of a particle is

$$V = mgy. \tag{3.34}$$

The potential energy V is a function of position and may be expressed in terms of a cartesian reference frame as $V = V(x, y, z)$. The differential of dV is

$$dV = \frac{\partial V}{\partial x} dx + \frac{\partial V}{\partial y} dy + \frac{\partial V}{\partial z} dz. \tag{3.35}$$

The potential energy V satisfies the relation

$$\begin{aligned} dV = -\mathbf{F} \cdot d\mathbf{r} &= -(F_x\mathbf{i} + F_y\mathbf{j} + F_z\mathbf{k}) \cdot (dx\mathbf{i} + dy\mathbf{j} + dz\mathbf{k}) \\ &= -(F_x \, dx + F_y \, dy + F_z \, dz), \end{aligned} \tag{3.36}$$

where $\mathbf{F} = F_x\mathbf{i} + F_y\mathbf{j} + F_z\mathbf{k}$. Using Eqs. (3.35) and (3.36), one may obtain

$$\frac{\partial V}{\partial x} dx + \frac{\partial V}{\partial y} dy + \frac{\partial V}{\partial z} dz = -(F_x \, dx + F_y \, dy + F_z \, dz),$$

which implies that

$$F_x = -\frac{\partial V}{\partial x}, \quad F_y = -\frac{\partial V}{\partial y}, \quad F_z = -\frac{\partial V}{\partial z}. \tag{3.37}$$

Given the potential energy $V = V(x, y, z)$ expressed in cartesian coordinates, the force $\mathbf{F}$ is

$$\mathbf{F} = -\left(\frac{\partial V}{\partial x}\mathbf{1} + \frac{\partial V}{\partial y}\mathbf{J} + \frac{\partial V}{\partial z}\mathbf{k}\right) = -\nabla V, \tag{3.38}$$

where ∇V is the *gradient* of V. The gradient expressed in cartesian coordinates is

$$\nabla = \frac{\partial}{\partial x}\mathbf{1} + \frac{\partial}{\partial y}\mathbf{J} + \frac{\partial}{\partial z}\mathbf{k}. \tag{3.39}$$

The *curl* of a vector force $\mathbf{F}$ in cartesian coordinates is

$$\nabla \times \mathbf{F} = \begin{vmatrix} \mathbf{1} & \mathbf{J} & \mathbf{k} \\ \dfrac{\partial}{\partial x} & \dfrac{\partial}{\partial y} & \dfrac{\partial}{\partial z} \\ F_x & F_y & F_z \end{vmatrix}. \tag{3.40}$$

If a force F is conservative, its curl $\nabla \times \mathbf{F}$ is zero. The converse is also true: A force $\mathbf{F}$ is conservative if its curl is zero.

In terms of cylindrical coordinates the force $\mathbf{F}$ is

$$\mathbf{F} = -\nabla V = -\left(\frac{\partial V}{\partial r}\mathbf{u}_r + \frac{1}{r}\frac{\partial V}{\partial \theta}\mathbf{u}_\theta + \frac{\partial V}{\partial z}\mathbf{k}\right). \tag{3.41}$$

In terms of cylindrical coordinates, the curl of the force $\mathbf{F}$ is

$$\nabla \times \mathbf{F} = \frac{1}{r}\begin{vmatrix} \mathbf{u}_r & r\mathbf{u}_\theta & \mathbf{k} \\ \dfrac{\partial}{\partial r} & \dfrac{\partial}{\partial \theta} & \dfrac{\partial}{\partial z} \\ F_r & rF_\theta & F_z \end{vmatrix}. \tag{3.42}$$

3.11 Principle of Impulse and Momentum

Newton's second law,

$$\mathbf{F} = m\frac{d\mathbf{v}}{dt},$$

is integrated with respect to time to give

$$\int_{t_1}^{t_2} \mathbf{F}\, dt = m\mathbf{v}_2 - m\mathbf{v}_1, \tag{3.43}$$

where $\mathbf{v}_1$ and $\mathbf{v}_2$ are the velocities of the particle P at the times t_1 and t_2.

The term $\int_{t_1}^{t_2} \mathbf{F}\, dt$ is called the *linear impulse*, and the term $m\mathbf{v}$ is called the *linear momentum*.

The principle of impulse and momentum: The impulse applied to a particle during an interval of time is equal to the change in its linear momentum (Fig. 3.10).

The dimensions of the linear impulse and linear momentum are (mass) times (length)/(time).

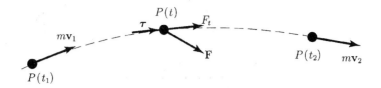

$$mv_1 + \int_{t_1}^{t_2} \mathbf{F}\, dt = mv_2$$

Figure 3.10

The average with respect to time of the total force acting on a particle from t_1 to t_2 is

$$\mathbf{F}_{av} = \frac{1}{t_2 - t_1} \int_{t_1}^{t_2} \mathbf{F}\, dt,$$

so one may write Eq. (3.43) as

$$\mathbf{F}_{av}(t_2 - t_1) = m\mathbf{v}_2 - m\mathbf{v}_1. \qquad (3.44)$$

An *impulsive force* is a force of relatively large magnitude that acts over a small interval of time (Fig. 3.11).

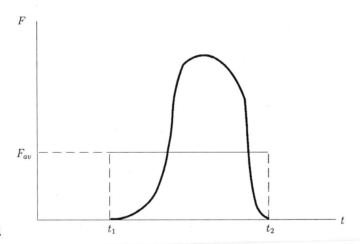

Figure 3.11

Equations (3.43) and (3.44) may be expressed in scalar forms. The sum of the forces in the tangent direction τ to the path of the particle equals the product of its mass m and the rate of change of its velocity along the path:

$$F_t = ma_t = m\frac{dv}{dt}.$$

Integrating this equation with respect to time, one may obtain

$$\int_{t_1}^{t_2} F_t\, dt = mv_2 - mv_1, \qquad (3.45)$$

where v_1 and v_2 are the velocities along the path at the times t_1 and t_2. The impulse applied to an object by the sum of the forces tangent to its path during an interval of time is equal to the change in its linear momentum along the path.

3.12 Conservation of Linear Momentum

Consider the two particles P_1 of mass m_1 and P_2 of mass m_2 shown in Fig. 3.12. The vector $\mathbf{F}_{12}$ is the force exerted by P_1 on P_2, and $\mathbf{F}_{21}$ is the force exerted by P_2 on P_1. These forces could be contact forces or could be exerted by a spring connecting the particles. As a consequence of Newton's third law, these forces are equal and opposite:

$$\mathbf{F}_{12} + \mathbf{F}_{21} = \mathbf{0}. \tag{3.46}$$

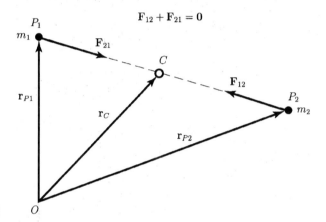

Figure 3.12

Consider that no external forces act on P_1 and P_2, or the external forces are negligible. The principle of impulse and momentum to each particle for arbitrary times t_1 and t_2 gives

$$\int_{t_1}^{t_2} \mathbf{F}_{21}\, dt = m_1 \mathbf{v}_{P1}(t_2) - m_1 \mathbf{v}_{P1}(t_1)$$

$$\int_{t_1}^{t_2} \mathbf{F}_{12}\, dt = m_2 \mathbf{v}_{P2}(t_2) - m_1 \mathbf{v}_{P2}(t_1),$$

where $\mathbf{v}_{P1}(t_1)$, $\mathbf{v}_{P1}(t_2)$ are the velocities of P_1 at the times t_1, t_2, and $\mathbf{v}_{P2}(t_1)$, $\mathbf{v}_{P2}(t_2)$ are the velocities of P_2 at the times t_1, t_2. The sum of these equations is

$$m_1 \mathbf{v}_{P1}(t_1) + m_2 \mathbf{v}_{P2}(t_1) = m_1 \mathbf{v}_{P1}(t_2) + m_2 \mathbf{v}_{P2}(t_2),$$

or the total linear momentum of P_1 and P_2 is conserved:

$$m_1 \mathbf{v}_{P1} + m_2 \mathbf{v}_{P2} = \text{constant}. \tag{3.47}$$

The position of the center of mass of P_1 and P_2 is (Fig. 3.12)

$$\mathbf{r}_C = \frac{m_1\mathbf{r}_{P1} + m_2\mathbf{r}_{P2}}{m_1 + m_2},$$

where $\mathbf{r}_{P1}$ and $\mathbf{r}_{P2}$ are the position vectors of P_1 and P_2. Taking the time derivative of this equation and using Eq. (3.47) one may obtain

$$(m_1 + m_2)\mathbf{v}_C = m_1\mathbf{v}_{P1} + m_2\mathbf{v}_{P2} = \text{constant,} \tag{3.48}$$

where $\mathbf{v}_C = d\mathbf{r}_C/dt$ is the velocity of the combined center of mass. The total linear momentum of the particles is conserved and the velocity of the combined center of mass of the particles P_1 and P_2 is constant.

3.13 Impact

Two particles A and B with the velocities $\mathbf{v}_A$ and $\mathbf{v}_B$ collide. The velocities of A and B after the impact are $\mathbf{v}'_A$ and $\mathbf{v}'_B$. The effects of external forces are negligible and the total linear momentum of the particles is conserved (Fig. 3.13):

$$m_A\mathbf{v}_A + m_B\mathbf{v}_B = m_A\mathbf{v}'_A + m_B\mathbf{v}'_B. \tag{3.49}$$

Furthermore, the velocity $\mathbf{v}$ of the center of mass of the particles is the same before and after the impact:

$$\mathbf{v} = \frac{m_A\mathbf{v}_A + m_B\mathbf{v}_B}{m_A + m_B}. \tag{3.50}$$

If A and B remain together after the impact, they are said to undergo a *perfectly plastic impact*. Equation (3.50) gives the velocity of the center of mass of the object they form after the impact (Fig. 3.13b).

If A and B rebound, linear momentum conservation alone does not provide enough equations to determine the velocities after the impact.

3.13.1 DIRECT CENTRAL IMPACTS

The particles A and B move along a straight line with velocities v_A and v_B before their impact (Fig. 3.14a). The magnitude of the force the particles exert on each other during the impact is R (Fig. 3.14b). The impact force is parallel to the line along which the particles travel (direct central impact). The particles continue to move along the same straight line after their impact (Fig. 3.14c). The effects of external forces during the impact are negligible, and the total linear momentum is conserved:

$$m_A v_A + m_B v_B = m_A v'_A + m_B v'_B. \tag{3.51}$$

Another equation is needed to determine the velocities v'_A and v'_B.

Let t_1 be the time at which A and B first come into contact (Fig. 3.14a). As a result of the impact, the objects will deform. At the time t_m the particles will have reached the maximum compression (period of compression, $t_1 < t < t_m$; Fig. 3.14b). At this time the relative velocity of the particles is zero, so they have the same velocity, v_m. The particles then begin to move apart and separate at a time t_2 (Fig. 3.14c). The second period, from the

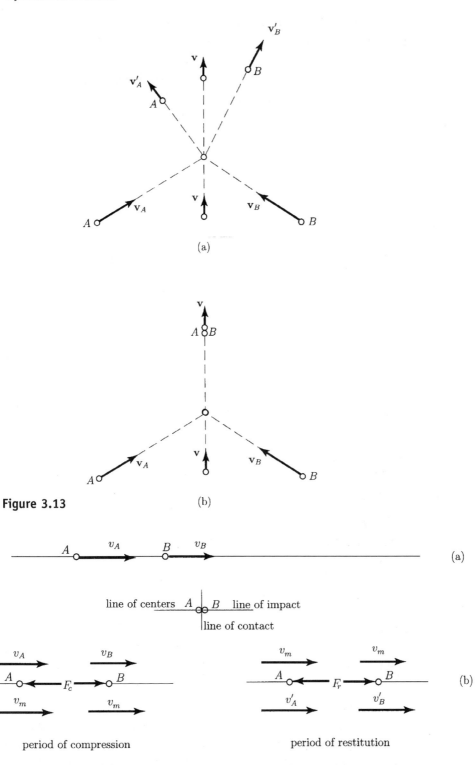

Figure 3.13

(a)

(b)

Figure 3.14

(a)

(b)

(c)

maximum compression to the instant at which the particles separate, is termed the period of restitution, $t_m < t < t_2$.

The principle of impulse and momentum is applied to A during the intervals of time from t_1 to the time of closest approach t_m and also from t_m to t_2,

$$\int_{t_1}^{t_m} -F_c \, dt = m_A v_m - m_A v_A \tag{3.52}$$

$$\int_{t_m}^{t_2} -F_r \, dt = m_A v_A' - m_A v_m, \tag{3.53}$$

where F_c is the magnitude of the contact force during the compression phase and F_r is the magnitude of the contact force during the restitution phase.

Then the principle of impulse and momentum is applied to B for the same intervals of time:

$$\int_{t_1}^{t_m} F_c + dt = m_B v_m - m_B v_B \tag{3.54}$$

$$\int_{t_m}^{t_2} F_r \, dt = m_B v_B' - m_B v_m, \tag{3.55}$$

As a result of the impact, part of the kinetic energy of the particles can be lost (because of a permanent deformation, generation of heat and sound, etc.). The impulse during the restitution phase of the impact from t_m to t_2 is in general smaller than the impulse during the compression phase t_1 to t_m.

The ratio of these impulses is called the *coefficient of restitution* (this definition was introduced by Poisson):

$$e = \frac{\int_{t_m}^{t_2} F_r \, dt}{\int_{t_1}^{t_m} F_c \, dt}. \tag{3.56}$$

The value of the coefficient of restitution depends on the properties of the objects as well as their velocities and orientations when they collide, and it can be determined by experiment or by a detailed analysis of the deformations of the objects during the impact.

If Eq. (3.53) is divided by Eq. (3.52) and Eq. (3.55) is divided by Eq. (3.54), the resulting equations are

$$(v_m - v_A)e = v_A' - v_m$$
$$(v_m - v_B)e = v_B' - v_m.$$

If the first equation is subtracted from the second one, the coefficient of restitution is

$$e = \frac{v_B' - v_A'}{v_A - v_B}. \tag{3.57}$$

Thus, the coefficient of restitution is related to the relative velocities of the objects before and after the impact (this is the kinematic definition of e introduced by Newton). If the coefficient of restitution e is known, Eq. (3.57) together with the equation of conservation of linear momentum, Eq. (3.51), may be used to determine v_A' and v_B'.

If $e = 0$ in Eq. (3.57), then $v'_B = v'_A$ and the objects remain together after the impact. The impact is perfectly plastic.

If $e = 1$, the total kinetic energy is the same before and after the impact:

$$\frac{1}{2} m_A v_A^2 + \frac{1}{2} m_B v_B^2 = \frac{1}{2} m_A (v'_A)^2 + \frac{1}{2} m_B (v'_B)^2.$$

An impact in which kinetic energy is conserved is called *perfectly elastic*.

3.13.2 OBLIQUE CENTRAL IMPACTS

Two small spheres A and B with the masses m_A and m_B approach with arbitrary velocities $\mathbf{v}_A$ and $\mathbf{v}_B$ (Fig. 3.15a). The initial velocities are not parallel, but they are in the same plane.

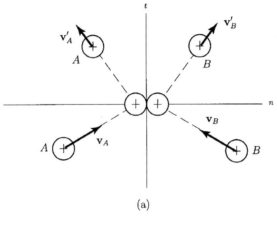

(a)

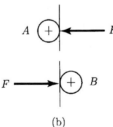

(b)

Figure 3.15

The forces they exert on each other during their impact are parallel to the n axis (center line axis) and point toward their centers of mass (Fig. 3.15b). There are no forces in the t direction at the contact point (tangent direction at the contact point). The velocities in the t direction are unchanged by the impact:

$$(\mathbf{v}'_A)_t = (\mathbf{v}_A)_t \quad \text{and} \quad (\mathbf{v}'_B)_t = (\mathbf{v}_B)_t. \tag{3.58}$$

In the n direction the linear momentum is conserved:

$$m_A (\mathbf{v}_A)_n + m_B (\mathbf{v}_B)_n = m_A (\mathbf{v}'_A)_n + m_B (\mathbf{v}'_B)_n. \tag{3.59}$$

The coefficient of restitution is defined as

$$e = \frac{(\mathbf{v}'_B)_n - (\mathbf{v}'_A)_n}{(\mathbf{v}_A)_n - (\mathbf{v}_B)_n}. \tag{3.60}$$

If B is a stationary object (fixed relative to the earth), then

$$(\mathbf{v}'_A)_n = -e(\mathbf{v}_A)_n.$$

3.14 Principle of Angular Impulse and Momentum

The position of a particle P of mass m relative to an inertial reference frame with origin O is given by the position vector $\mathbf{r} = \mathbf{OP}$ (Fig. 3.16). The cross product of Newton's second law with the position vector $\mathbf{r}$ is

$$\mathbf{r} \times \mathbf{F} = \mathbf{r} \times m\mathbf{a} = \mathbf{r} \times m\frac{d\mathbf{v}}{dt}. \tag{3.61}$$

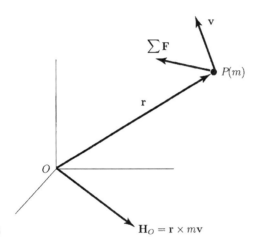

Figure 3.16

The time derivative of the quantity $\mathbf{r} \times m\mathbf{v}$ is

$$\frac{d}{dt}(\mathbf{r} \times m\mathbf{v}) = \left(\frac{d\mathbf{r}}{dt} \times m\mathbf{v}\right) + \left(\mathbf{r} \times m\frac{d\mathbf{v}}{dt}\right) = \mathbf{r} \times m\frac{d\mathbf{v}}{dt},$$

because $d\mathbf{r}/dt = \mathbf{v}$, and the cross product of parallel vectors is zero. Equation (3.61) may be written as

$$\mathbf{r} \times \mathbf{F} = \frac{d\mathbf{H}_O}{dt}, \tag{3.62}$$

where the vector

$$\mathbf{H}_O = \mathbf{r} \times m\mathbf{v} \tag{3.63}$$

is called the *angular momentum* about O (Fig. 3.16). The angular momentum may be interpreted as the moment of the linear momentum of the particle about point O. The moment $\mathbf{r} \times \mathbf{F}$ equals the rate of change of the moment of momentum about point O.

Integrating Eq. (3.62) with respect to time, one may obtain

$$\int_{t_1}^{t_2} (\mathbf{r} \times \mathbf{F}) \, dt = (\mathbf{H}_O)_2 - (\mathbf{H}_O)_1. \tag{3.64}$$

The integral on the left-hand side is called the *angular impulse.*

The principle of angular impulse and momentum: The angular impulse applied to a particle during an interval of time is equal to the change in its angular momentum.

The dimensions of the angular impulse and angular momentum are $(\text{mass}) \times (\text{length})^2/(\text{time})$.

4. Planar Kinematics of a Rigid Body

A rigid body is an idealized model of an object that does not deform, or change shape. A rigid body is by definition an object with the property that the distance between every pair of points of the rigid body is constant. Although any object does not deform as it moves, if its deformation is small one may approximate its motion by modeling it as a rigid body.

4.1 Types of Motion

The rigid body motion is described with respect to a reference frame (coordinate system) relative to which the motions of the points of the rigid body and its angular motion are measured. In many situations it is convenient to use a reference frame that is fixed with respect to the earth.

Rotation about a fixed axis. Each point of the rigid body on the axis is stationary, and each point not on the axis moves in a circular path about the axis as the rigid body rotates (Fig. 4.1a).

Translation. Each point of the rigid body describes parallel paths (Fig. 4.1b). Every point of a rigid body in translation has the same velocity and acceleration. The motion of the rigid body may be described the motion of a single point.

Planar motion. Consider a rigid body intersected by a plane fixed relative to a given reference frame (Fig. 4.1c). The points of the rigid body intersected by the plane remain in the plane for two-dimensional, or planar, motion. The fixed plane is the plane of the motion. Planar motion or complex motion exhibits a simultaneous combination of rotation and translation. Points on the rigid body will travel nonparallel paths, and there will be at every instant a center of rotation, which will continuously change location.

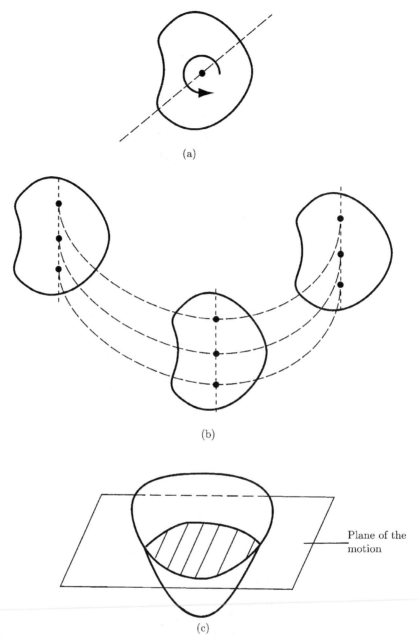

Figure 4.1

        The rotation of a rigid body about a fixed axis is a special case of planar motion.

4.2 Rotation about a Fixed Axis

Figure 4.2 shows a rigid body rotating about a fixed axis a. The reference line b is fixed and is perpendicular to the fixed axis a, $b \perp a$. The body-fixed line

c rotates with the rigid body and is perpendicular to the fixed axis a, $c \perp a$. The angle θ between the reference line and the body-fixed line describes the position, or orientation, of the rigid body about the fixed axis. The angular velocity (rate of rotation) of the rigid body is

$$\omega = \frac{d\theta}{dt} = \dot{\theta},$$ (4.1)

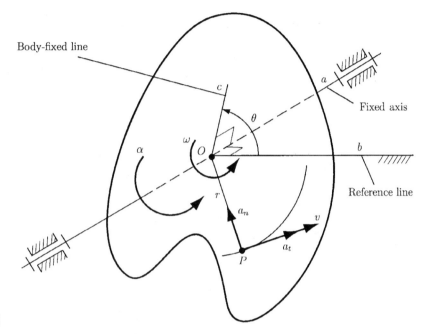

Figure 4.2

and the angular acceleration of the rigid body is

$$\alpha = \frac{d\omega}{dt} = \frac{d^2\theta}{dt^2} = \ddot{\theta}.$$ (4.2)

The velocity of a point P, of the rigid body, at a distance r from the fixed axis is tangent to its circular path (Fig. 4.2) and is given by

$$v = r\omega.$$ (4.3)

The normal and tangential accelerations of P are

$$a_t = r\alpha, \quad a_n = \frac{v^2}{r} = r\omega^2.$$ (4.4)

4.3 Relative Velocity of Two Points of the Rigid Body

Figure 4.3 shows a rigid body in planar translation and rotation. The position vector of the point A of the rigid body $\mathbf{r}_A = \mathbf{OA}$, and the position vector of the point B of the rigid body is $\mathbf{r}_B = \mathbf{OB}$. The point O is the origin of a given reference frame. The position of point A relative to point B is the vector $\mathbf{BA}$.

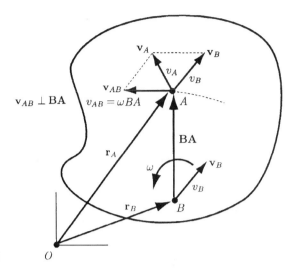

Figure 4.3

The position vector of point A relative to point B is related to the positions of A and B relative to O by

$$\mathbf{r}_A = \mathbf{r}_B + \mathbf{BA}. \tag{4.5}$$

The derivative of Eq. (4.5) with respect to time gives

$$\mathbf{v}_A = \mathbf{v}_B + \mathbf{v}_{AB}, \tag{4.6}$$

where $\mathbf{v}_A$ and $\mathbf{v}_B$ are the velocities of A and B relative to the reference frame. The velocity of point A relative to point B is

$$\mathbf{v}_{AB} = \frac{d\mathbf{BA}}{dt}.$$

Since A and B are points of the rigid body, the distance between them, $BA = |\mathbf{BA}|$, is constant. That means that relative to B, A moves in a circular path as the rigid body rotates. The velocity of A relative to B is therefore tangent to the circular path and equal to the product of the angular velocity ω of the rigid body and BA:

$$v_{AB} = |\mathbf{v}_{AB}| = \omega BA. \tag{4.7}$$

The velocity $\mathbf{v}_{AB}$ is perpendicular to the position vector $\mathbf{BA}$, $\mathbf{v}_{AB} \perp \mathbf{BA}$. The sense of $\mathbf{v}_{AB}$ is the sense of ω (Fig. 4.3). The velocity of A is the sum of the velocity of B and the velocity of A relative to B.

4.4 Angular Velocity Vector of a Rigid Body

EULER'S THEOREM

A rigid body constrained to rotate about a fixed point can move between any two positions by a single rotation about some axis through the fixed point. ▲

With Euler's theorem the change in position of a rigid body relative to a fixed point A during an interval of time from t to $t + dt$ may be expressed as a single rotation through an angle $d\theta$ about some axis. At the time t the rate of rotation of the rigid body about the axis is its angular velocity $\omega = d\theta/dt$, and the axis about which it rotates is called the *instantaneous axis of rotation*.

The angular velocity vector of the rigid body, denoted by $\boldsymbol{\omega}$, specifies both the direction of the instantaneous axis of rotation and the angular velocity. The vector $\boldsymbol{\omega}$ is defined to be parallel to the instantaneous axis of rotation (Fig. 4.4), and its magnitude is the rate of rotation, the absolute value of ω. The direction of $\boldsymbol{\omega}$ is related to the direction of the rotation of the rigid body through a right-hand rule: if one points the thumb of the right hand in the direction of $\boldsymbol{\omega}$, the fingers curl around $\boldsymbol{\omega}$ in the direction of the rotation.

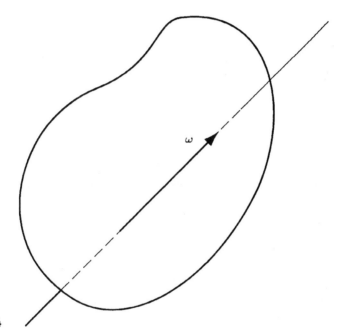

Figure 4.4

Figure 4.5 shows two points A and B of a rigid body. The rigid body has the angular velocity $\boldsymbol{\omega}$. The velocity of A relative to B is given by the equation

$$\mathbf{v}_{AB} = \frac{d\mathbf{BA}}{dt} = \boldsymbol{\omega} \times \mathbf{BA}. \qquad (4.8)$$

Proof

The point A is moving at the present instant in a circular path relative to the point B. The radius of the path is $|\mathbf{BA}| \sin \beta$, where β is the angle between the vectors $\mathbf{BA}$ and $\boldsymbol{\omega}$. The magnitude of the velocity of A relative to B is equal to the product of the radius of the circular path and the angular velocity of the

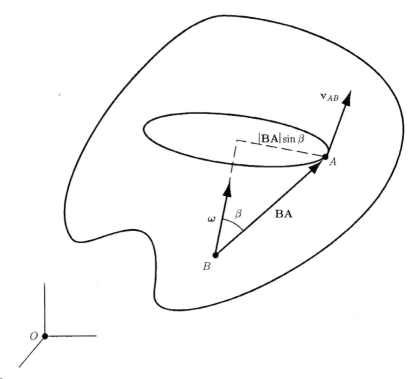

Figure 4.5

rigid body, $|\mathbf{v}_{AB}| = (|\mathbf{BA}|\sin\beta)|\omega|$, which is the magnitude of the cross product of **BA** and **ω** or

$$\mathbf{v}_{AB} = \boldsymbol{\omega} \times \mathbf{BA}.$$

The relative velocity $\mathbf{v}_{AB}$ is perpendicular to **ω** and perpendicular to **BA**.

When Eq. (4.8) is substituted into Eq. (4.6), the relation between the velocities of two points of a rigid body in terms of its angular velocity is obtained:

$$\mathbf{v}_A = \mathbf{v}_B + \mathbf{v}_{AB} = \mathbf{v}_B + \boldsymbol{\omega} \times \mathbf{BA}. \tag{4.9}$$

▲

4.5 Instantaneous Center

The *instantaneous center* of a rigid body is a point whose velocity is zero at the instant under consideration. Every point of the rigid body rotates about the instantaneous center at the instant under consideration.

The instantaneous center may be or may not be a point of the rigid body. When the instantaneous center is not a point of the rigid body, the rigid body is rotating about an external point at that instant.

Figure 4.6 shows two points A and B of a rigid body and their directions of motion Δ_A amd Δ_B,

$$\mathbf{v}_A \| \Delta_A \quad \text{and} \quad \mathbf{v}_B \| \Delta_B,$$

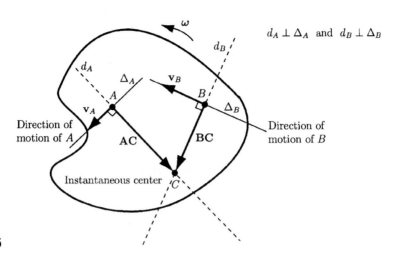

$$d_A \perp \Delta_A \quad \text{and} \quad d_B \perp \Delta_B$$

Figure 4.6

where $\mathbf{v}_A$ is the velocity of point A, and $\mathbf{v}_B$ is the velocity of point B.

Through the points A and B perpendicular lines are drawn to their directions of motion:

$$d_A \perp \Delta_A \quad \text{and} \quad d_B \perp \Delta_B.$$

The perpendicular lines intersect at point C:

$$d_A \cap d_B = C.$$

The velocity of point C in terms of the velocity of point A is

$$\mathbf{v}_C = \mathbf{v}_A + \boldsymbol{\omega} \times \mathbf{AC},$$

where $\boldsymbol{\omega}$ is the angular velocity vector of the rigid body. Since the vector $\boldsymbol{\omega} \times \mathbf{AC}$ is perpendicular to $\mathbf{AC}$,

$$(\boldsymbol{\omega} \times \mathbf{AC}) \perp \mathbf{AC},$$

this equation states that the direction of motion of C is parallel to the direction of motion of A:

$$\mathbf{v}_C \| \mathbf{v}_A. \tag{4.10}$$

The velocity of point C in terms of the velocity of point B is

$$\mathbf{v}_C = \mathbf{v}_B + \boldsymbol{\omega} \times \mathbf{BC}.$$

The vector $\boldsymbol{\omega} \times \mathbf{BC}$ is perpendicular to $\mathbf{BC}$,

$$(\boldsymbol{\omega} \times \mathbf{BC}) \perp \mathbf{BC},$$

so this equation states that the direction of motion of C is parallel to the direction of motion of B:

$$\mathbf{v}_C \| \mathbf{v}_B. \tag{4.11}$$

But C cannot be moving parallel to A and parallel to B, so Eqs. (4.10) and (4.11) are contradictory unless $\mathbf{v}_C = \mathbf{0}$. So the point C, where the perpendicular lines through A and B to their directions of motion intersect, is the instantaneous center. This is a simple method to locate the instantaneous center of a rigid body in planar motion.

If the rigid body is in translation (the angular velocity of the rigid body is zero) the instantaneous center of the rigid body C moves to infinity.

4.6 Relative Acceleration of Two Points of the Rigid Body

The velocities of two points A and B of a rigid body in planar motion relative to a given reference frame with the origin at point O are related by (Fig. 4.7)

$$\mathbf{v}_A = \mathbf{v}_B + \mathbf{v}_{AB}.$$

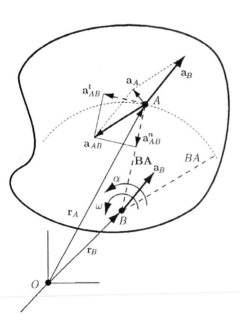

Figure 4.7

Taking the time derivative of this equation, one may obtain

$$\mathbf{a}_A = \mathbf{a}_B + \mathbf{a}_{AB},$$

where $\mathbf{a}_A$ and $\mathbf{a}_B$ are the accelerations of A and B relative to the origin O of the reference frame and $\mathbf{a}_{AB}$ is the acceleration of point A relative to point B. Because the point A moves in a circular path relative to the point B as the rigid body rotates, $\mathbf{a}_{AB}$ has a normal component and a tangential component (Fig. 4.7):

$$\mathbf{a}_{AB} = \mathbf{a}_{AB}^n + \mathbf{a}_{AB}^t.$$

The normal component points toward the center of the circular path (point B), and its magnitude is

$$|\mathbf{a}_{AB}^n| = |\mathbf{v}_{AB}|^2/|\mathbf{BA}| = \omega^2 BA.$$

The tangential component equals the product of the distance $BA = |\mathbf{BA}|$ and the angular acceleration α of the rigid body:

$$|\mathbf{a}_{AB}^t| = \alpha BA.$$

The velocity of the point A relative to the point B in terms of the angular velocity vector, $\boldsymbol{\omega}$, of the rigid body is given by Eq. (4.8):

$$\mathbf{v}_{AB} = \boldsymbol{\omega} \times \mathbf{BA}.$$

Taking the time derivative of this equation, one may obtain

$$\mathbf{a}_{AB} = \frac{d\boldsymbol{\omega}}{dt} \times \mathbf{BA} + \boldsymbol{\omega} \times \mathbf{v}_{AB}$$

$$= \frac{d\boldsymbol{\omega}}{dt} \times \mathbf{BA} + \boldsymbol{\omega} \times (\boldsymbol{\omega} \times \mathbf{BA}).$$

Defining the angular acceleration vector $\boldsymbol{\alpha}$ to be the rate of change of the angular velocity vector,

$$\boldsymbol{\alpha} = \frac{d\boldsymbol{\omega}}{dt}, \tag{4.12}$$

one finds that the acceleration of A relative to B is

$$\mathbf{a}_{AB} = \boldsymbol{\alpha} \times \mathbf{BA} + \boldsymbol{\omega} \times (\boldsymbol{\omega} \times \mathbf{BA}).$$

The velocities and accelerations of two points of a rigid body in terms of its angular velocity and angular acceleration are

$$\mathbf{v}_A = \mathbf{v}_B + \boldsymbol{\omega} \times \mathbf{BA} \tag{4.13}$$

$$\mathbf{a}_A = \mathbf{a}_B + \boldsymbol{\alpha} \times \mathbf{BA} + \boldsymbol{\omega} \times (\boldsymbol{\omega} \times \mathbf{BA}). \tag{4.14}$$

In the case of planar motion, the term $\boldsymbol{\alpha} \times \mathbf{BA}$ in Eq. (4.14) is the tangential component of the acceleration of A relative to B, and $\boldsymbol{\omega} \times (\boldsymbol{\omega} \times \mathbf{BA})$ is the normal component (Fig. 4.7). Equation (4.14) may be written for planar motion in the form

$$\mathbf{a}_A = \mathbf{a}_B + \boldsymbol{\alpha} \times \mathbf{BA} - \omega^2 \mathbf{BA}. \tag{4.15}$$

4.7 Motion of a Point That Moves Relative to a Rigid Body

A reference frame that moves with the rigid body is a *body fixed* reference frame. Figure 4.8 shows a rigid body RB, in motion relative to a primary reference frame with its origin at point O_0, XO_0YZ. The primary reference frame is a fixed reference frame or an earth fixed reference frame. The unit vectors $\mathbf{i}_0$, $\mathbf{j}_0$, and $\mathbf{k}_0$ of the primary reference frame are constant.

The body fixed reference frame, $xOyz$, has its origin at a point O of the rigid body ($O \in RB$) and is a moving reference frame relative to the primary

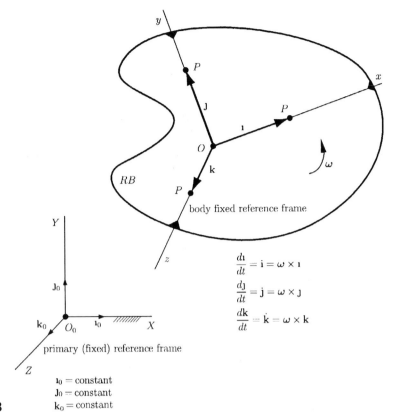

Figure 4.8

reference. The unit vectors $\mathbf{i}$, $\mathbf{j}$, and $\mathbf{k}$ of the body fixed reference frame are not constant, because they rotate with the body fixed reference frame.

The position vector of a point P of the rigid body ($P \in RB$) relative to the origin, O, of the body fixed reference frame is the vector $\mathbf{OP}$. The velocity of P relative to O is

$$\frac{d\mathbf{OP}}{dt} = \mathbf{v}_{PO} = \boldsymbol{\omega} \times \mathbf{OP},$$

where $\boldsymbol{\omega}$ is the angular velocity vector of the rigid body. The unit vector $\mathbf{i}$ may be regarded as the position vector of a point P of the rigid body (Fig. 4.8), and its time derivative may be written as $d\mathbf{i}/dt = \dot{\mathbf{i}} = \boldsymbol{\omega} \times \mathbf{i}$. In a similar way the time derivative of the unit vectors $\mathbf{j}$ and $\mathbf{k}$ may be obtained. The expressions

$$\frac{d\mathbf{i}}{dt} = \dot{\mathbf{i}} = \boldsymbol{\omega} \times \mathbf{i}$$

$$\frac{d\mathbf{j}}{dt} = \dot{\mathbf{j}} = \boldsymbol{\omega} \times \mathbf{j} \qquad (4.16)$$

$$\frac{d\mathbf{k}}{dt} = \dot{\mathbf{k}} = \boldsymbol{\omega} \times \mathbf{k}$$

are known as Poisson's relations.

The position vector of a point A (the point A is not assumed to be a point of the rigid body) relative to the origin O_0 of the primary reference frame is (Fig. 4.9)

$$\mathbf{r}_A = \mathbf{r}_O + \mathbf{r},$$

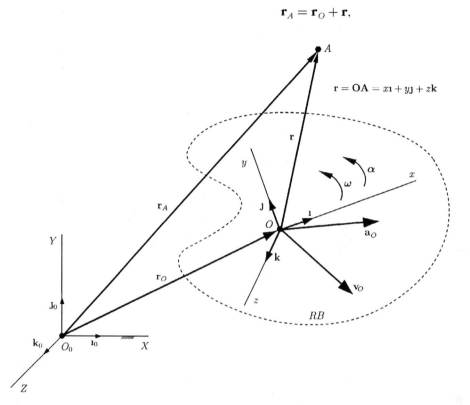

Figure 4.9

where

$$\mathbf{r} = \mathbf{OA} = x\mathbf{\imath} + y\mathbf{\jmath} + z\mathbf{k}$$

is the position vector of A relative to the origin O, of the body fixed reference frame, and x, y, and z are the coordinates of A in terms of the body fixed reference frame. The velocity of the point A is the time derivative of the position vector $\mathbf{r}_A$:

$$\mathbf{v}_A = \frac{d\mathbf{r}_O}{dt} + \frac{d\mathbf{r}}{dt} = \mathbf{v}_O + \mathbf{v}_{AO}$$

$$= \mathbf{v}_O + \frac{dx}{dt}\mathbf{\imath} + x\frac{d\mathbf{\imath}}{dt} + \frac{dy}{dt}\mathbf{\jmath} + y\frac{d\mathbf{\jmath}}{dt} + \frac{dz}{dt}\mathbf{k} + z\frac{d\mathbf{k}}{dt}.$$

Using Eqs. (4.16), one finds that the total derivative of the position vector $\mathbf{r}$ is

$$\frac{d\mathbf{r}}{dt} = \dot{\mathbf{r}} = \dot{x}\mathbf{\imath} + \dot{y}\mathbf{\jmath} + \dot{z}\mathbf{k} + \boldsymbol{\omega} \times \mathbf{r}.$$

The velocity of A relative to the body fixed reference frame is a local derivative:

$$\mathbf{v}_{Arel} = \frac{\partial \mathbf{r}}{\partial t} = \frac{dx}{dt}\mathbf{1} + \frac{dy}{dt}\mathbf{J} + \frac{dz}{dt}\mathbf{k} = \dot{x}\mathbf{1} + \dot{y}\mathbf{J} + \dot{z}\mathbf{k}, \tag{4.17}$$

A general formula for the total derivative of a moving vector $\mathbf{r}$ may be written as

$$\frac{d\mathbf{r}}{dt} = \frac{\partial \mathbf{r}}{\partial t} + \boldsymbol{\omega} \times \mathbf{r}. \tag{4.18}$$

This relation is known as the *transport theorem*. In operator notation the transport theorem is written as

$$\frac{d}{dt}(\ ) = \frac{\partial}{\partial t}(\ ) + \boldsymbol{\omega} \times (\ ). \tag{4.19}$$

The velocity of the point A relative to the primary reference frame is

$$\mathbf{v}_A = \mathbf{v}_O + \mathbf{v}_{Arel} + \boldsymbol{\omega} \times \mathbf{r}, \tag{4.20}$$

Equation (4.20) expresses the velocity of a point A as the sum of three terms:

- The velocity of a point O of the rigid body
- The velocity $\mathbf{v}_{Arel}$ of A relative to the rigid body
- The velocity $\boldsymbol{\omega} \times \mathbf{r}$ of A relative to O due to the rotation of the rigid body

The acceleration of the point A relative to the primary reference frame is obtained by taking the time derivative of Eq. (4.20):

$$\begin{aligned} \mathbf{a}_A &= \mathbf{a}_O + \mathbf{a}_{AO} \\ &= \mathbf{a}_O + \mathbf{a}_{Arel} + 2\boldsymbol{\omega} \times \mathbf{v}_{Arel} + \boldsymbol{\alpha} \times \mathbf{r} + \boldsymbol{\omega} \times (\boldsymbol{\omega} \times \mathbf{r}), \end{aligned} \tag{4.21}$$

where

$$\mathbf{a}_{Arel} = \frac{\partial^2 \mathbf{r}}{\partial t^2} = \frac{d^2 x}{dt^2}\mathbf{1} + \frac{d^2 y}{dt^2}\mathbf{J} + \frac{d^2 z}{dt^2}\mathbf{k} \tag{4.22}$$

is the acceleration of A relative to the body fixed reference frame or relative to the rigid body. The term

$$\mathbf{a}_{Cor} = 2\boldsymbol{\omega} \times \mathbf{v}_{Arel}$$

is called the Coriolis acceleration force.

In the case of planar motion, Eq. (4.21) becomes

$$\begin{aligned} \mathbf{a}_A &= \mathbf{a}_O + \mathbf{a}_{AO} \\ &= \mathbf{a}_O + \mathbf{a}_{Arel} + 2\boldsymbol{\omega} \times \mathbf{v}_{Arel} + \boldsymbol{\alpha} \times \mathbf{r} - \omega^2 \mathbf{r}, \end{aligned} \tag{4.23}$$

The motion of the rigid body (RB) is described relative to the primary reference frame. The velocity $\mathbf{v}_A$ and the acceleration $\mathbf{a}_A$ of point A are relative to the primary reference frame. The terms $\mathbf{v}_{Arel}$ and $\mathbf{a}_{Arel}$ are the velocity and acceleration of point A relative to the body fixed reference

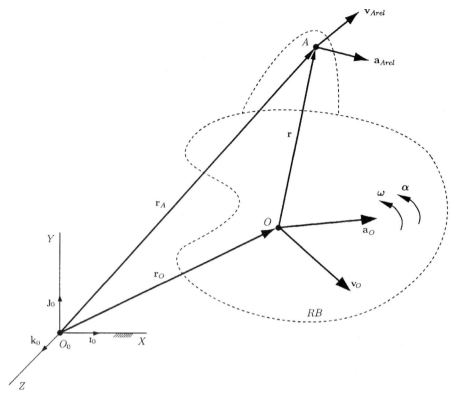

Figure 4.10

frame, i.e., they are the velocity and acceleration measured by an observer moving with the rigid body (Fig. 4.10).

If A is a point of the rigid body, $A \in RB$, $\mathbf{v}_{Arel} = \mathbf{0}$ and $\mathbf{a}_{Arel} = \mathbf{0}$.

4.7.1 MOTION OF A POINT RELATIVE TO A MOVING REFERENCE FRAME

The velocity and acceleration of an arbitrary point A relative to a point O of a rigid body, in terms of the body fixed reference frame, are given by Eqs. (4.20) and (4.21):

$$\mathbf{v}_A = \mathbf{v}_O + \mathbf{v}_{Arel} + \boldsymbol{\omega} \times \mathbf{OA} \tag{4.24}$$

$$\mathbf{a}_A = \mathbf{a}_O + \mathbf{a}_{Arel} + 2\boldsymbol{\omega} \times \mathbf{v}_{Arel} + \boldsymbol{\alpha} \times \mathbf{OA} + \boldsymbol{\omega} \times (\boldsymbol{\omega} \times \mathbf{OA}). \tag{4.25}$$

These results apply to any reference frame having a moving origin O and rotating with angular velocity $\boldsymbol{\omega}$ and angular acceleration $\boldsymbol{\alpha}$ relative to a primary reference frame (Fig. 4.11). The terms $\mathbf{v}_A$ and $\mathbf{a}_A$ are the velocity and acceleration of an arbitrary point A relative to the primary reference frame. The terms $\mathbf{v}_{Arel}$ and $\mathbf{a}_{Arel}$ are the velocity and acceleration of A relative to the secondary moving reference frame, i.e., they are the velocity and acceleration measured by an observer moving with the secondary reference frame.

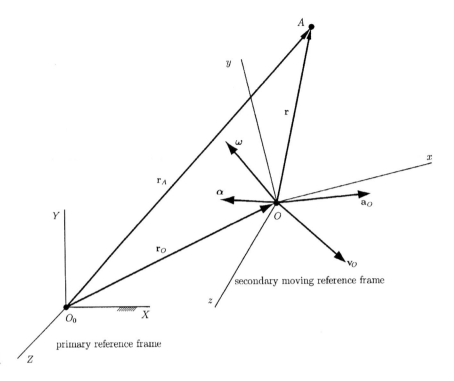

Figure 4.11

4.7.2 INERTIAL REFERENCE FRAMES

A reference frame is inertial if one may use it to apply Newton's second law in the form $\sum \mathbf{F} = m\mathbf{a}$.

Figure 4.12 shows a nonaccelerating, nonrotating reference frame with the origin at O_0, and a secondary nonrotating, earth centered reference frame with the origin at O. The nonaccelerating, nonrotating reference frame with the origin at O_0 is assumed to be an inertial reference. The acceleration of the earth, due to the gravitational attractions of the sun, moon, etc., is $\mathbf{g}_O$. The earth centered reference frame has the acceleration $\mathbf{g}_O$ as well.

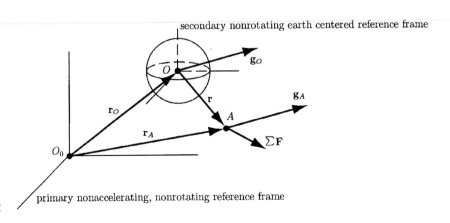

Figure 4.12

Newton's second law for an object A of mass m, using the hypothetical nonaccelerating, nonrotating reference frame with the origin at O_0, may be written as

$$m\mathbf{a}_A = m\mathbf{g}_A + \sum \mathbf{F}, \tag{4.26}$$

where $\mathbf{a}_A$ is the acceleration of A relative to O_0, $\mathbf{g}_A$ is the resulting gravitational acceleration, and $\sum \mathbf{F}$ is the sum of all other external forces acting on A.

By Eq. (4.25) the acceleration of A relative to O_0 is

$$\mathbf{a}_A = \mathbf{a}_O + \mathbf{a}_{Arel},$$

where $\mathbf{a}_{Arel}$ is the acceleration of A relative to the earth centered reference frame and the acceleration of the origin O is equal to the gravitational acceleration of the earth, $\mathbf{a}_O = \mathbf{g}_O$. The earth-centered reference frame does not rotate ($\boldsymbol{\omega} = \mathbf{0}$).

If the object A is on or near the earth, its gravitational acceleration $\mathbf{g}_A$ due to the attraction of the sun, etc., is nearly equal to the gravitational acceleration of the earth $\mathbf{g}_O$, and Eq. (4.26) becomes

$$\sum \mathbf{F} = m\mathbf{a}_{Arel}. \tag{4.27}$$

One may apply Newton's second law using a nonrotating, earth centered reference frame if the object is near the earth.

In most applications, Newton's second law may be applied using an earth fixed reference frame. Figure 4.13 shows a nonrotating reference frame with its origin at the center of the earth O and a secondary earth fixed reference frame with its origin at a point B. The earth fixed reference frame with the origin at B may be assumed to be an inertial reference and

$$\sum \mathbf{F} = m\mathbf{a}_{Arel}, \tag{4.28}$$

where $\mathbf{a}_{Arel}$ is the acceleration of A relative to the earth fixed reference frame.

The motion of an object A may be analysed using a primary inertial reference frame with its origin at the point O (Fig. 4.14). A secondary reference frame with its origin at B undergoes an arbitrary motion with angular velocity $\boldsymbol{\omega}$ and angular acceleration $\boldsymbol{\alpha}$. Newton's second law for the object A of mass m is

$$\sum \mathbf{F} = m\mathbf{a}_A, \tag{4.29}$$

where $\mathbf{a}_A$ is the acceleration of A acceleration relative to O. Equation (4.29) may be written in the form

$$\sum \mathbf{F} - m[\mathbf{a}_B + 2\boldsymbol{\omega} \times \mathbf{v}_{Arel} + \boldsymbol{\alpha} \times \mathbf{BA} \\ + \boldsymbol{\omega} \times (\boldsymbol{\omega} \times \mathbf{BA})] = m\mathbf{a}_{Arel}, \tag{4.30}$$

where $\mathbf{a}_{Arel}$ is the acceleration of A relative to the secondary reference frame. The term $\mathbf{a}_B$ is the acceleration of the origin B of the secondary reference frame relative to the primary inertial reference. The term $2\boldsymbol{\omega} \times \mathbf{v}_{Arel}$ is the Coriolis acceleration, and the term $-2m\boldsymbol{\omega} \times \mathbf{v}_{Arel}$ is called the Coriolis force.

Dynamics

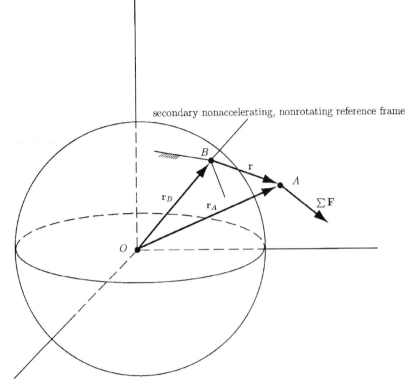

Figure 4.13   primary nonrotating earth centered reference

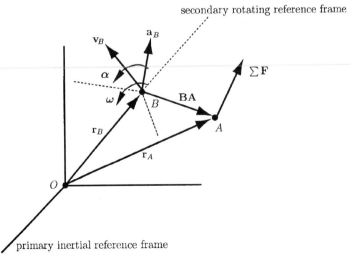

Figure 4.14   primary inertial reference frame

This is Newton's second law expressed in terms of a secondary reference frame undergoing an arbitrary motion relative to an inertial primary reference frame.

5. Dynamics of a Rigid Body

5.1 Equation of Motion for the Center of Mass

Newton stated that the total force on a particle is equal to the rate of change of its linear momentum, which is the product of its mass and velocity. Newton's second law is postulated for a particle, or small element of matter. One may show that the total external force on an arbitrary rigid body is equal to the product of its mass and the acceleration of its center of mass. An arbitrary rigid body with the mass m may be divided into N particles. The position vector of the i particle is $\mathbf{r}_i$ and the mass of the i particle is m_i (Fig. 5.1):

$$m = \sum_{i=1}^{N} m_i.$$

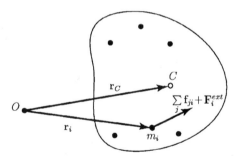

Figure 5.1

The position of the center of mass of the rigid body is

$$\mathbf{r}_C = \frac{\sum_{i=1}^{N} m_i \mathbf{r}_i}{m}. \tag{5.1}$$

Taking two time derivatives of Eq. (5.1), one may obtain

$$\sum_{i=1}^{N} m_i \frac{d^2 \mathbf{r}_i}{dt^2} = m \frac{d^2 \mathbf{r}_C}{dt^2} = m\mathbf{a}_C, \tag{5.2}$$

where $\mathbf{a}_C$ is the acceleration of the center of mass of the rigid body.

Let $\mathbf{f}_{ij}$ be the force exerted on the j particle by the i particle. Newton's third law states that the j particle exerts a force on the i particle of equal magnitude and opposite direction (Fig. 5.1):

$$\mathbf{f}_{ji} = -\mathbf{f}_{ij}.$$

Newton's second law for the i particle is

$$\sum_j \mathbf{f}_{ji} + \mathbf{F}_i^{ext} = m_i \frac{d^2 \mathbf{r}_i}{dt^2}, \qquad (5.3)$$

where $\mathbf{F}_i^{ext}$ is the external force on the i particle. Equation (5.3) may be written for each particle of the rigid body. Summing the resulting equations from $i = 1$ to N, one may obtain

$$\sum_i \sum_j \mathbf{f}_{ji} + \sum_i \mathbf{F}_i^{ext} = m\mathbf{a}_C, \qquad (5.4)$$

The sum of the internal forces on the rigid body is zero (Newton's third law):

$$\sum_i \sum_j \mathbf{f}_{ji} = 0.$$

The term $\sum_i \mathbf{F}_i^{ext}$ is the sum of the external forces on the rigid body:

$$\sum_i \mathbf{F}_i^{ext} = \sum \mathbf{F}.$$

One may conclude that the sum of the external forces equals the product of the mass and the acceleration of the center of mass:

$$\sum \mathbf{F} = m\mathbf{a}_C. \qquad (5.5)$$

If the rigid body rotates about a fixed axis O (Fig. 5.2), the sum of the moments about the axis due to external forces and couples acting on the body is

$$\sum M_O = I_O \alpha,$$

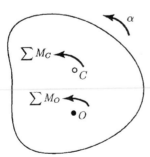

Figure 5.2

where I_O is the moment of inertia of the rigid body about O and α is the angular acceleration of the rigid body. In the case of general planar motion, the sum of the moments about the center of mass of a rigid body is related to its angular acceleration by

$$\sum M_C = I_C \alpha, \qquad (5.6)$$

where I_C is the moment of inertia of the rigid body about its center of mass C.

If the external forces and couples acting on a rigid body in planar motion are known, one may use Eqs. (5.5) and (5.6) to determine the acceleration of the center of mass of the rigid body and the angular acceleration of the rigid body.

5.2 Angular Momentum Principle for a System of Particles

An arbitrary system with mass m may be divided into N particles $P_1, P_2, \ldots, P_N$. The position vector of the i particle is $\mathbf{r}_i = \mathbf{OP}_i$ and the mass of the i particle is m_i (Fig. 5.3). The position of the center of mass, C, of the system is $\mathbf{r}_C = \sum_{i=1}^{N} m_i \mathbf{r}_i / m$. The position of the particle P_i of the system relative to O is

$$\mathbf{r}_i = \mathbf{r}_C + \mathbf{CP}_i. \tag{5.7}$$

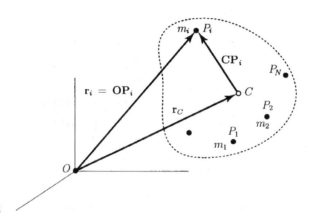

Figure 5.3

Multiplying Eq. (5.7) by m_i, summing from 1 to N, one may find that

$$\sum_{i=1}^{N} m_i \mathbf{CP}_i = \mathbf{0}. \tag{5.8}$$

The total angular momentum of the system about its center of mass C is the sum of the angular momenta of the particles about C,

$$\mathbf{H}_C = \sum_{i=1}^{N} \mathbf{CP}_i \times m_i \mathbf{v}_i, \tag{5.9}$$

where $\mathbf{v}_i = d\mathbf{r}_i/dt$ is the velocity of the particle P_i.

The total angular momentum of the system about O is the sum of the angular momenta of the particles,

$$\mathbf{H}_O = \sum_{i=1}^{N} \mathbf{r}_i \times m_i \mathbf{v}_i = \sum_{i=1}^{N} (\mathbf{r}_C + \mathbf{CP}_i) \times m_i \mathbf{v}_i = \mathbf{r}_C \times m\mathbf{v}_C + \mathbf{H}_C, \tag{5.10}$$

or the total angular momentum about O is the sum of the angular momentum about O due to the velocity $\mathbf{v}_C$ of the center of mass of the system and the total angular momentum about the center of mass (Fig. 5.4).

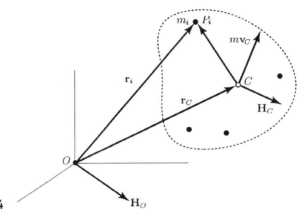

Figure 5.4

Newton's second law for the i particle is

$$\sum_j \mathbf{f}_{ji} + \mathbf{F}_i^{ext} = m_i \frac{d\mathbf{v}_i}{dt},$$

and the cross product with the position vector $\mathbf{r}_i$, and sum from $i = 1$ to N gives

$$\sum_i \sum_j \mathbf{r}_i \times \mathbf{f}_{ji} + \sum_i \mathbf{r}_i \times \mathbf{F}_i^{ext} = \sum_i \mathbf{r}_i \times \frac{d}{dt}(m_i\mathbf{v}_i). \qquad (5.11)$$

The first term on the left side of Eq. (5.11) is the sum of the moments about O due to internal forces, and

$$\mathbf{r}_i \times \mathbf{f}_{ji} + \mathbf{r}_i \times \mathbf{f}_{ij} = \mathbf{r}_i \times (\mathbf{f}_{ji} + \mathbf{f}_{ij}) = \mathbf{0}.$$

The term vanishes if the internal forces between each pair of particles are equal, opposite, and directed along the straight line between the two particles (Fig. 5.5).

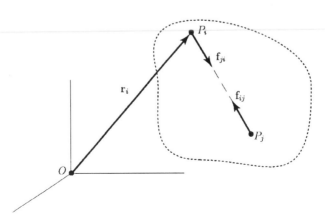

Figure 5.5

The second term on the left side of Eq. (5.11),

$$\sum_i \mathbf{r}_i \times \mathbf{F}_i^{ext} = \sum \mathbf{M}_O,$$

represents the sum of the moments about O due to the external forces and couples. The term on the right side of Eq. (5.11) is

$$\sum_i \mathbf{r}_i \times \frac{d}{dt}(m_i \mathbf{v}_i) = \sum_i \left[\frac{d}{dt}(\mathbf{r}_i \times m_i \mathbf{v}_i) - \mathbf{v}_i \times m_i \mathbf{v}_i \right] = \frac{d\mathbf{H}_O}{dt}, \qquad (5.12)$$

which represents the rate of change of the total angular momentum of the system about the point O.

Equation (5.11) may be rewritten as

$$\sum \mathbf{M}_O = \frac{d\mathbf{H}_O}{dt}. \qquad (5.13)$$

The sum of the moments about O due to external forces and couples equals the rate of change of the angular momentum about O.

Using Eqs. (5.10) and (5.13), one may obtain

$$\sum \mathbf{M}_O = \frac{d}{dt}(\mathbf{r}_C \times m\mathbf{v}_C + \mathbf{H}_C) = \mathbf{r}_C \times m\mathbf{a}_C + \frac{d\mathbf{H}_C}{dt}, \qquad (5.14)$$

where $\mathbf{a}_C$ is the acceleration of the center of mass.

If the point O is coincident with the center of mass at the present instant $C = O$, then $\mathbf{r}_C = \mathbf{0}$ and Eq. (5.14) becomes

$$\sum \mathbf{M}_C = \frac{d\mathbf{H}_C}{dt}. \qquad (5.15)$$

The sum of the moments about the center of mass equals the rate of change of the angular momentum about the center of mass.

5.3 Equations of Motion for General Planar Motion

An arbitrary rigid body with the mass m may be divided into N particles P_i, $i = 1, 2, \ldots, N$. The position vector of the P_i particle is $\mathbf{r}_i = \mathbf{OP}_i$ and the mass of the particle is m_i (Fig. 5.6).

Let d_O be the axis through the fixed origin point O that is perpendicular to the plane of the motion of a rigid body $x, y, d_O \perp (x, y)$. Let d_C be the parallel axis through the center of mass C, $d_C \| d_O$. The rigid body has a general planar motion, and one may express the angular velocity vector as $\boldsymbol{\omega} = \omega \mathbf{k}$.

The velocity of the P_i particle relative to the center of mass is

$$\frac{d\mathbf{R}_i}{dt} = \omega \mathbf{k} \times \mathbf{R}_i,$$

where $\mathbf{R}_i = \mathbf{CP}_i$. The sum of the moments about O due to external forces and couples is

$$\sum \mathbf{M}_O = \frac{d\mathbf{H}_O}{dt} = \frac{d}{dt}[(\mathbf{r}_C \times m\mathbf{v}_C) + \mathbf{H}_C], \qquad (5.16)$$

Dynamics

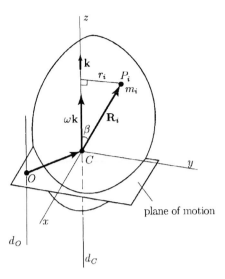

Figure 5.6

where

$$\mathbf{H}_C = \sum_i [\mathbf{R}_i \times m_i(\omega\mathbf{k} \times \mathbf{R}_i)]$$

is the angular momentum about d_C. The magnitude of the angular momentum about d_C is

$$H_C = \mathbf{H}_C \cdot \mathbf{k} = \sum_i [\mathbf{R}_i \times m_i(\omega\mathbf{k} \times \mathbf{R}_i)] \cdot \mathbf{k}$$

$$= \sum_i m_i[(\mathbf{R}_i \times \mathbf{k}) \times \mathbf{R}_i] \cdot \mathbf{k}\omega = \sum_i m_i[(\mathbf{R}_i \times \mathbf{k}) \cdot (\mathbf{R}_i \times \mathbf{k})]\omega \qquad (5.17)$$

$$= \sum_i m_i|\mathbf{R}_i \times \mathbf{k}|^2\omega = \sum_i m_i r_i\omega,$$

where the term $|\mathbf{k} \times \mathbf{R}_i| = r_i$ is the perpendicular distance from d_C to the P_i particle. The identity

$$(\mathbf{a} \times \mathbf{b}) \cdot \mathbf{c} = \mathbf{a} \cdot (\mathbf{b} \times \mathbf{c})$$

has been used.

The moment of inertia of the rigid body about d_C is

$$I = \sum_i m_i r_i^2,$$

Equation (5.17) defines the angular momentum of the rigid body about d_C:

$$H_C = I\omega \quad \text{or} \quad \mathbf{H}_C = I\omega\mathbf{k} = I\boldsymbol{\omega}.$$

Substituting this expression into Eq. (5.16), one may obtain

$$\sum \mathbf{M}_O = \frac{d}{dt}[(\mathbf{r}_C \times m\mathbf{v}_C) + I\boldsymbol{\omega}] = (\mathbf{r}_C \times m\mathbf{a}_C) + I\boldsymbol{\alpha}. \qquad (5.18)$$

If the fixed axis d_O is coincident with d_C at the present instant, $\mathbf{r} = \mathbf{0}$, and from Eq. (5.18) one may obtain

$$\sum \mathbf{M}_C = I\boldsymbol{\alpha}.$$

The sum of the moments about d_C equals the product of the moment of inertia about d_C and the angular acceleration.

5.4 D'Alembert's Principle

Newton's second law may be written as

$$\mathbf{F} + (-m\mathbf{a}_C) = \mathbf{0}, \quad \text{or} \quad \mathbf{F} + \mathbf{F}_{in} = \mathbf{0}, \tag{5.19}$$

where the term $\mathbf{F}_{in} = -m\mathbf{a}_C$ is the *inertial force*. Newton's second law may be regarded as an "equilibrium" equation.

Equation (5.18) relates the total moment about a fixed point O to the acceleration of the center of mass and the angular acceleration:

$$\sum \mathbf{M}_O = (\mathbf{r}_C \times m\mathbf{a}_C) + I\boldsymbol{\alpha}$$

or

$$\sum \mathbf{M}_O + [\mathbf{r}_C \times (-m\mathbf{a}_C)] + (-I\boldsymbol{\alpha}) = \mathbf{0}. \tag{5.20}$$

The term $\mathbf{M}_{in} = -I\boldsymbol{\alpha}$ is the *inertial couple*. The sum of the moments about any point, including the moment due to the inertial force $-m\mathbf{a}$ acting at the center of mass and the inertial couple, equals zero.

The equations of motion for a rigid body are analogous to the equations for static equilibrium: The sum of the forces equals zero and the sum of the moments about any point equals zero when the inertial forces and couples are taking into account. This is called *D'Alembert's principle*.

References

1. H. Baruh, *Analytical Dynamics*. McGraw-Hill, New York, 1999.

2. A. Bedford and W. Fowler, *Dynamics*. Addison Wesley, Menlo Park, CA, 1999.

3. F. P. Beer and E. R. Johnston, Jr., *Vector Mechanics for Engineers: Statics and Dynamics*. McGraw-Hill, New York, 1996.

4. J. H. Ginsberg, *Advanced Engineering Dynamics*, Cambridge University Press, Cambridge, 1995.

5. D. T. Greenwood, *Principles of Dynamics*. Prentice-Hall, Englewood Cliffs, NJ, 1998.

6. T. R. Kane, *Analytical Elements of Mechanics*, Vol. 1. Academic Press, New York, 1959.

7. T. R. Kane, *Analytical Elements of Mechanics*, Vol. 2. Academic Press, New York, 1961.

8. T. R. Kane and D. A. Levinson, *Dynamics*. McGraw-Hill, New York, 1985.

9. J. L. Meriam and L. G. Kraige, *Engineering Mechanics: Dynamics*. John Wiley & Sons, New York, 1997.

10. D. J. McGill and W. W. King, *Engineering Mechanics: Statics and an Introduction to Dynamics*. PWS Publishing Company, Boston, 1995.

11. L. A. Pars, *A treatise on Analytical Dynamics*. Wiley, New York, 1965.

12. I. H. Shames, *Engineering Mechanics: Statics and Dynamics*. Prentice-Hall, Upper Saddle River, NJ, 1997.

13. R. W. Soutas-Little and D. J. Inman, *Engineering Mechanics: Statics and Dynamics*. Prentice-Hall, Upper Saddle River, NJ, 1999.

3 Mechanics of Materials

DAN B. MARGHITU, CRISTIAN I. DIACONESCU, AND BOGDAN O. CIOCIRLAN

Department of Mechanical Engineering, Auburn University, Auburn, Alabama 36849

Inside

1. Stress

I n the design process, an important problem is to ensure that the strength of the mechanical element to be designed always exceeds the stress due to any load exerted on it.

1.1 Uniformly Distributed Stresses

Uniform distribution of stresses is an assumption that is frequently considered in the design process. Depending upon the way the force is applied to a mechanical element — for example, whether the force is an axial force or a shear one — the result is called *pure tension* (*compression*) or *pure shear*, respectively.

Let us consider a tension load F applied to the ends of a bar. If the bar is cut at a section remote from the ends and we remove one piece, the effect of the removed part can be replaced by applying a uniformly distributed force of magnitude σA to the cut end, where σ is the *normal stress* and A the cross-sectional area of the bar. The stress σ is given by

$$\sigma = \frac{F}{A}. \tag{1.1}$$

This uniform stress distribution requires that the bar be straight and made of a homogeneous material, that the line of action of the force contain the centroid of the section, and that the section be taken remote from the ends and from any discontinuity or abrupt change in cross-section. Equation (1.1) and the foregoing assumptions also hold for pure compression.

If a body is in shear, one can assume the uniform stress distribution and use

$$\tau = \frac{F}{A}, \tag{1.2}$$

where τ is the *shear stress*.

1.2 Stress Components

A general three-dimensional *stress element* is illustrated in Fig. 1.1a. Three normal positive stresses, σ_x, σ_y, and σ_z, and six positive shear stresses, τ_{xy}, $\tau_{yx}, \tau_{yz}, \tau_{zy}, \tau_{zx}$, and τ_{xz}, are shown. To ensure the static equilibrium, the following equations must hold:

$$\tau_{xy} = \tau_{yx}, \quad \tau_{yz} = \tau_{zy}, \quad \tau_{xz} = \tau_{zx}. \tag{1.3}$$

The normal stresses σ_x, σ_y, and σ_z are called *tension* or *tensile stresses* and considered positive if they are oriented in the direction shown in the figure. *Shear stresses* on a positive face of an element are positive if they act in the positive direction of the reference axis. The first subscript of any shear stress component denotes the axis to which it is perpendicular. The second subscript denotes the axis to which the shear stress component is parallel.

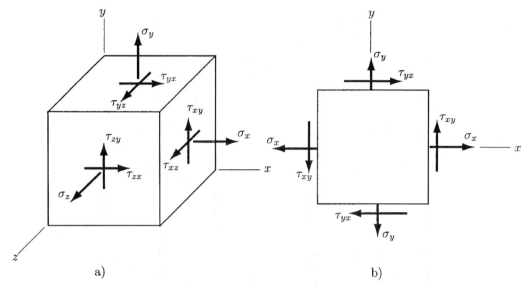

Figure 1.1   *Stress element. (a) Three-dimensional case; (b) planar case.*

A general two-dimensional stress element is illustrated in Fig. 1.1b. The two normal stresses σ_x and σ_y, respectively, are in the positive direction. Shear stresses are positive when they are in the clockwise (cw) and negative when they are in the counterclockwise (ccw) direction. Thus, τ_{yx} is positive (cw), and τ_{xy} is negative (ccw).

1.3 Mohr's Circle

Let us consider the element illustrated in Fig. 1.1b cut by an oblique plane at angle ϕ with respect to the x axis (Fig. 1.2). The stresses σ and τ act on this

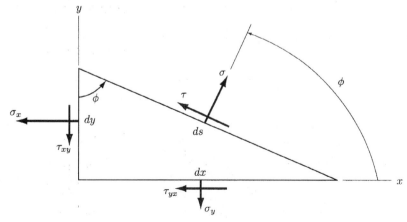

Figure 1.2
Normal and shear stresses on a planar surface.

oblique plane. The stresses σ and τ can be calculated by summing the forces caused by all stress components to zero, that is,

$$\sigma = \frac{\sigma_x + \sigma_y}{2} + \frac{\sigma_x - \sigma_y}{2} \cos 2\phi + \tau_{xy} \sin 2\phi, \qquad (1.4)$$

$$\tau = -\frac{\sigma_x - \sigma_y}{2} \sin 2\phi + \tau_{xy} \cos 2\phi. \qquad (1.5)$$

Differentiating Eq. (1.4) with respect to the angle ϕ and setting the result equal to zero yields

$$\tan 2\phi = \frac{2\tau_{xy}}{\sigma_x - \sigma_y}. \qquad (1.6)$$

The solution of Eq. (1.6) provides two values for the angle 2ϕ defining the maximum normal stress σ_1 and the minimum normal stress σ_2. These minimum and maximum normal stresses are called the *principal stresses*. The corresponding directions are called the *principal directions*. The angle between the principal directions is $\phi = 90°$.

Similarly, differentiating Eq. (1.5) and setting the result to zero we obtain

$$\tan 2\phi = -\frac{\sigma_x - \sigma_y}{2\tau_{xy}}. \qquad (1.7)$$

The solutions of Eq. (1.7) provides the angles 2ϕ at which the shear stress τ reaches an extreme value.

Equation (1.6) can be rewritten as

$$2\tau_{xy} \cos 2\phi = (\sigma_x - \sigma_y) \sin 2\phi,$$

or

$$\sin 2\phi = \frac{2\tau_{xy} \cos 2\phi}{\sigma_x - \sigma_y}. \qquad (1.8)$$

Substituting Eq. (1.8) into Eq. (1.5) gives

$$\tau = -\frac{\sigma_x - \sigma_y}{2} \frac{2\tau_{xy} \cos 2\phi}{\sigma_x - \sigma_y} + \tau_{xy} \cos 2\phi = 0. \qquad (1.9)$$

Hence, the shear stress associated with both principal directions is zero.

Substituting $\sin 2\phi$ from Eq. (1.7) into Eq. (1.4) yields

$$\sigma = \frac{\sigma_x + \sigma_y}{2}. \qquad (1.10)$$

The preceding equation states that the two normal stresses associated with the directions of the two maximum shear stresses are equal.

The analytical expressions for the two principal stresses can be obtained by manipulating Eqs. (1.6) and (1.4):

$$\sigma_1, \sigma_2 = \frac{\sigma_x + \sigma_y}{2} \pm \sqrt{\left(\frac{\sigma_x - \sigma_y}{2}\right)^2 + \tau_{xy}^2}. \qquad (1.11)$$

Similarly, the maximum and minimum values of the shear stresses are obtained using

$$\tau_1, \tau_2 = \pm\sqrt{\left(\frac{\sigma_x - \sigma_y}{2}\right)^2 + \tau_{xy}^2}. \tag{1.12}$$

Mohr's circle diagram (Fig. 1.3) is a graphical method to visualize the stress state. The normal stresses are plotted along the abscissa axis of the coordinate system and the shear stresses along the ordinate axis. Tensile normal stresses are considered positive (σ_x and σ_y are positive in Fig. 1.3) and compressive normal stresses negative. Clockwise (cw) shear stresses are considered positive, whereas counterclockwise (ccw) shear stresses are negative.

The following notation is used: OA as σ_x, AB as τ_{xy}, OC as σ_y, and CD as τ_{yx}. The center of the Mohr's circle is at point E on the σ axis. Point B has the stress coordinates σ_x, τ_{xy} on the x faces and point D the stress coordinates σ_y, τ_{yx} on the y faces. The angle 2ϕ between EB and ED is $180°$; hence the

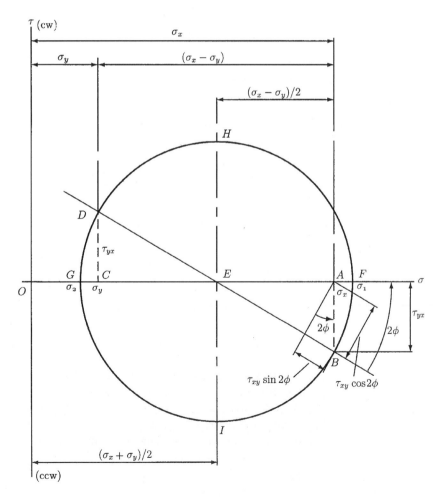

Figure 1.3
Mohr's circle.

angle between x and y on the stress element is $\phi = 90°$ (Fig. 1.1b). The maximum principal normal stress is σ_1 at point F, and the minimum principal normal stress is σ_2 at point G. The two extreme values of the shear stresses are plotted at points I and H, respectively. Thus, the Mohr's diagram is a circle of center E and diameter BD.

EXAMPLE 1.1   For a stress element having $\sigma_x = 100$ MPa and $\tau_{xy} = 60$ MPa (cw), find the principal stresses and plot the principal directions on a stress element correctly aligned with respect to the xy system. Also plot the maximum and minimum shear stresses τ_1 and τ_2, respectively, on another stress element and find the corresponding normal stresses. The stress components not given are zero.

Solution

First, we will construct the Mohr's circle diagram corresponding to the given data. Then, we will use the diagram to calculate the stress components. Finally, we will draw the stress components.

The first step to construct Mohr's diagram is to draw the σ and τ axes (Fig. 1.4a) and locate points A of $\sigma_x = 100$ MPa and C of $\sigma_y = 0$ MPa on the σ axis. Then, we represent $\tau_{xy} = 60$ MPa in the cw direction and $\tau_{yx} = 60$ MPa in the ccw direction. Hence, point B has the coordinates $\sigma_x = 100$ MPa, $\tau_{xy} = 60$ MPa and point D the coordinates $\sigma_x = 0$ MPa, $\tau_{yx} = 60$ MPa. The line BD is the diameter and point E the center of the Mohr's circle. The intersection of the circle with the σ axis gives the principal stresses σ_1 and σ_2 at points F and G, respectively.

The x axis of the stress elements is line EB and the y axis line ED. The segments BA and AE have the length of 60 and 50 MPa, respectively. The length of segment BE is

$$BE = HE = \tau_1 = \sqrt{(60)^2 + (50)^2} = 78.1 \text{ MPa}.$$

Since the intersection E is 50 MPa from the origin, the principal stresses are

$$\sigma_1 = 50 + 78.1 = 128.1 \text{ MPa}, \qquad \sigma_2 = 50 - 78.1 = -28.1 \text{ MPa}.$$

The angle 2ϕ with respect to the x axis cw to σ_1 is

$$2\phi = \tan^{-1}\frac{60}{50} = 50.2°.$$

To draw the principal stress element, we start with the x and y axes parallel to the original axes as shown in Fig. 1.4b. The angle ϕ is in the same direction as the angle 2ϕ in the Mohr's circle diagram. Thus, measuring 25.1° (half of 50.2°) clockwise from x axis, we can locate the σ_1 axis. The σ_2 axis will be at 90° with respect to the σ_1 axis, as shown in Fig. 1.4b.

To draw the second stress element, we note that the two extreme shear stresses occur at the points H and I in Fig. 1.4a. The two normal stresses corresponding to these shear stresses are each equal to 50 MPa. Point H is

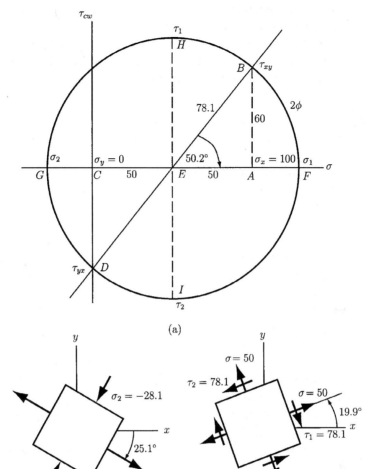

Figure 1.4
*Mohr's circle
application. (a)
Mohr's circle
diagram; (b)
principal stres-
ses; (c) extreme
value of the
shear stresses.*

39.8° ccw from point B in the Mohr's circle diagram. Therefore, we draw a stress element oriented 19.9° (half of 39.8°) ccw from x as shown in Fig. 1.4c. ▲

1.4 Triaxial Stress

For three-dimensional stress elements, a particular orientation occurs in space when all shear stress components are zero. As in the case of plane stress, the principal directions are the normals to the faces for this particular orientation. Since the stress element is three-dimensional, there are three principal directions and three principal stresses σ_1, σ_2, and σ_3, associated with the principal directions. In three dimensions, only six components of stress are required to specify the stress state, namely, σ_x, σ_y, σ_z, τ_{xy}, τ_{yz}, and τ_{zx}.

To plot Mohr's circles for triaxial stress, the principal normal stresses are ordered so that $\sigma_1 > \sigma_2 > \sigma_3$. The result is shown in Fig. 1.5a. The three *principal shear stresses* $\tau_{1/2}$, $\tau_{2/3}$, and $\tau_{1/3}$ are also shown in Fig. 1.5a. Each of these shear stresses occurs on two planes, one of the planes being shown in Fig. 1.5b. The principal shear stresses can be calculated by the following equations

$$\tau_{1/2} = \frac{\sigma_1 - \sigma_2}{2}; \quad \tau_{2/3} = \frac{\sigma_2 - \sigma_3}{2}; \quad \tau_{1/3} = \frac{\sigma_1 - \sigma_3}{2}. \tag{1.13}$$

If the normal principal stresses are ordered ($\sigma_1 > \sigma_2 > \sigma_3$), then $\tau_{max} = \tau_{1/3}$.

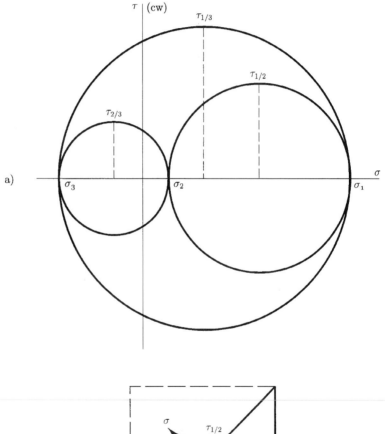

a)

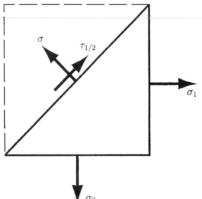

b)

Figure 1.5
Mohr's circle for triaxial stress element. (a) Mohr's circle diagram; (b) planar case.

Let us consider a principal stress element having the stresses σ_1, σ_2, and σ_3 as shown in Fig. 1.6. The stress element is cut by a plane ABC that forms equal angles with each of the three principal stresses. This plane is called an *octahedral plane*. Figure 1.6 can be interpreted as a free-body diagram when each of the stress components shown is multiplied by the area over which it acts. Summing the forces thus obtained to zero along each direction of the coordinate system, one can notice that a force called *octahedral force* exists on plane ABC. Dividing this force by the area of ABC, the result can be described by two components. One component, called *octahedral normal stress*, is normal to the plane ABC, and the other component, called *octahedral shear stress*, is located in the plane ABC.

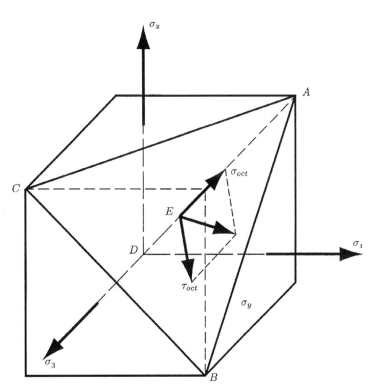

Figure 1.6
Octahedral stress element.

1.5 Elastic Strain

If a tensile load is applied to a straight bar, it becomes longer. The amount of elongation is called the *total strain*. The elongation per unit length of the bar ϵ is called *strain*. The expression for strain is given by

$$\epsilon = \frac{\delta}{l},\qquad (1.14)$$

where δ is the total elongation (total strain) of the bar of length l.

Shear strain γ is the change in a right angle of an element subjected to pure shear stresses.

Elasticity is a property of materials that allows them to regain their original geometry when the load is removed. The elasticity of a material can be expressed in terms of Hooke's law, which states that, within certain limits, the stress in a material is proportional to the strain which produced it. Hence, Hooke's law can be written as

$$\sigma = E\epsilon, \qquad \tau = G\gamma, \tag{1.15}$$

where E and G are constants of proportionality. The constant E is called the *modulus of elasticity* and the constant G is called the *shear modulus of elasticity* or the *modulus of rigidity*. A material that obeys Hooke's law is called *elastic*.

Substituting $\sigma = F/A$ and $\epsilon = \delta/l$ into Eq. (1.15) and manipulating, we obtain the expression for the total deformation δ of a bar loaded in axial tension or compression:

$$\delta = \frac{Fl}{AE}. \tag{1.16}$$

When a tension load is applied to a body, not only does an axial strain occur, but also a lateral one. If the material obeys Hooke's law, it has been demonstrated that the two strains are proportional to each other. This proportionality constant is called *Poisson's ratio*, given by

$$v = \frac{\text{lateral strain}}{\text{axial strain}}, \tag{1.17}$$

It can be proved that the elastic constants are related by

$$E = 2G(1 + v). \tag{1.18}$$

The stress state at a point can be determined if the relationship between stress and strain is known and the state of strain has already been measured. The *principal strains* are defined as the strains in the direction of the principal stresses. As is the case of shear stresses, the shear strains are zero on the faces of an element aligned along the principal directions. Table 1.1 lists the relationships for all types of stress. The values of Poisson's ratio v for various materials are listed in Table 1.2.

1.6 Equilibrium

Considering a particle of nonnegligible mass, any force $\mathbf{F}$ acting on it will produce an acceleration of the particle. The foregoing statement is derived from Newton's second law, which can be expressed as

$$\sum \mathbf{F} = m\mathbf{a},$$

where $\sum \mathbf{F}$ is the sum of the forces acting on the particle, m the mass of the particle, and $\mathbf{a}$ the acceleration. If all members of a system under investiga-

Table 1.1 *Elastic Stress–Strain Relations*

Type of stress	Principal strains	Principal stresses
Uniaxial	$\epsilon_1 = \dfrac{\sigma_1}{E}$	$\sigma_1 = E\epsilon_1$
	$\epsilon_2 = -v\epsilon_1$	$\sigma_2 = 0$
	$\epsilon_3 = -v\epsilon_1$	$\sigma_3 = 0$
Biaxial	$\epsilon_1 = \dfrac{\sigma_1}{E} - \dfrac{v\sigma_2}{E}$	$\sigma_1 = \dfrac{E(\epsilon_1 + v\epsilon_2)}{1 - v^2}$
	$\epsilon_2 = \dfrac{\sigma_2}{E} - \dfrac{v\sigma_1}{E}$	$\sigma_2 = \dfrac{E(\epsilon_2 + v\epsilon_1)}{1 - v^2}$
	$\epsilon_3 = -\dfrac{v\sigma_1}{E} - \dfrac{v\sigma_2}{E}$	$\sigma_3 = 0$
Triaxial	$\epsilon_1 = \dfrac{\sigma_1}{E} - \dfrac{v\sigma_2}{E} - \dfrac{v\sigma_3}{E}$	$\sigma_1 = \dfrac{E\epsilon_1(1 - v) + vE(\epsilon_2 + \epsilon_3)}{1 - v - 2v^2}$
	$\epsilon_2 = \dfrac{\sigma_2}{E} - \dfrac{v\sigma_1}{E} - \dfrac{v\sigma_3}{E}$	$\sigma_2 = \dfrac{E\epsilon_2(1 - v) + vE(\epsilon_1 + \epsilon_3)}{1 - v - 2v^2}$
	$\epsilon_3 = \dfrac{\sigma_3}{E} - \dfrac{v\sigma_1}{E} - \dfrac{v\sigma_2}{E}$	$\sigma_1 = \dfrac{E\epsilon_3(1 - v) + vE(\epsilon_1 + \epsilon_2)}{1 - v - 2v^2}$

Source: J. E. Shigley and C. R. Mischke, *Mechanical Engineering Design.* McGraw-Hill, New York, 1989. Used with permission.

Mechanics

tion are assumed motionless or in motion with a constant velocity, then every particle has zero acceleration $\mathbf{a} = \mathbf{0}$ and

$$\mathbf{F}_1 + \mathbf{F}_2 + \cdots + \mathbf{F}_i = \sum \mathbf{F} = \mathbf{0}, \qquad (1.19)$$

The forces acting on the particle are said to be *balanced* and the particle is said to be in *equilibrium* if Eq. (1.19) holds. If the velocity of the particle is zero, then the particle is said to be in *static equilibrium.*

A *system* may denote any part of a structure, that is, just one particle, several particles, a portion of a rigid body, an entire rigid body, or several rigid bodies. We can define the *internal forces* and the *internal moments* of a system as the action of one part of the system on another part of the same system. If forces and moments are applied to the considered system from the outside, then these forces and moments are called *external forces* and *external moments.*

The condition for the equilibrium of a single particle is expressed by Eq. (1.19). For a system containing many particles, Eq. (1.19) can be applied to each particle in the system. Let us select a particle, say the jth one, and let $\mathbf{F}_e$ be the sum of the external forces and $\mathbf{F}_i$ be the sum of the internal forces. Then, Eq. (1.19) becomes

$$\sum \mathbf{F}_j = \mathbf{F}_e + \mathbf{F}_i = \mathbf{0}. \qquad (1.20)$$

Table 1.2 *Physical Constants of Materials*

Material	Modulus of elasticity E		Modulus of rigidity G		Poisson's ratio ν	Unit weight w		
	Mpsi	GPa	Mpsi	GPa		lb/in^3	lb/ft^3	kN/m^3
Aluminum (all alloys)	10.3	71.0	3.80	26.2	0.334	0.098	169	26.6
Beryllium copper	18.0	124.0	7.0	48.3	0.285	0.297	513	80.6
Brass	15.4	106.0	5.82	40.1	0.324	0.309	534	83.8
Carbon steel	30.0	207.0	11.5	79.3	0.292	0.282	487	76.5
Cast iron, gray	14.5	100.0	6.0	41.4	0.211	0.260	450	70.6
Cooper	17.2	119.0	6.49	44.7	0.326	0.322	556	87.3
Douglas fir	1.6	11.0	0.6	4.1	0.33	0.016	28	4.3
Glass	6.7	46.2	2.7	18.6	0.245	0.094	162	25.4
Inconel	31.0	214.0	11.0	75.8	0.290	0.307	530	83.3
Lead	5.3	36.5	1.9	13.1	0.425	0.411	710	111.5
Magnesium	6.5	44.8	2.4	16.5	0.350	0.065	112	17.6
Molybdenum	48.0	331.0	17.0	117.0	0.307	0.368	636	100.0
Monel metal	26.0	179.0	9.5	65.5	0.320	0.319	551	86.6
Nickel silver	18.5	127.0	7.0	48.3	0.322	0.316	546	85.8
Nickel steel	30.0	207.0	11.5	79.3	0.291	0.280	484	76.0
Phosphor bronze	16.1	111.0	6.0	41.4	0.349	0.295	510	80.1
Stainless steel (18-8)	27.6	190.0	10.6	73.1	0.305	0.280	484	76.0

Source: J. E. Shigley and C. R. Mischke, *Mechanical Engineering Design.* McGraw-Hill, New York, 1989. Used with permission.

For *n* particles in the system we get

$$\sum_1^n \mathbf{F}_e + \sum_1^n \mathbf{F}_i = \mathbf{0}. \qquad (1.21)$$

The law of action and reaction (Newton's third law) states that when two particles react, a pair of interacting forces exist, these forces having the same magnitude and opposite senses, and acting along the line common to the two particles. Hence, the second term in Eq. (1.21) is zero and Eq. (1.21) becomes

$$\sum_1^n \mathbf{F}_e = \mathbf{0}. \qquad (1.22)$$

Equation (1.22) states that the sum of the forces exerted from the outside upon a system in equilibrium is zero. Similarly, we can prove that the sum of the external moments exerted upon a system in equilibrium is zero, that is,

$$\sum_1^n \mathbf{M}_e = \mathbf{0}. \qquad (1.23)$$

If Eqs. (1.22) and (1.23) are simultaneously satisfied, then the system is in *static equilibrium.*

To study the behavior of any part of a system, we can isolate that part and replace the original effects of the system on it by interface forces and moments. Figure 1.7 is a symbolic illustration of the process. The forces and moment are internal for the whole system, but they become external when applied to the isolated part. The interface forces, for example, are represented symbolically by the force vectors $\mathbf{F}_1, \mathbf{F}_2$, and $\mathbf{F}_i$ in Fig. 1.7. The isolated part, along with all forces and moments, is called the *free-body diagram.*

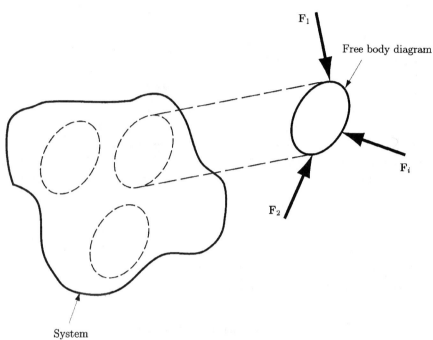

F₁

Free body diagram

Fᵢ

F₂

Figure 1.7
Free body diagram.

System

1.7 Shear and Moment

Let us consider a beam supported by the reactions R_1 and R_2 and loaded by the transversal forces F_1, F_2 as shown in Fig. 1.8a. The reactions R_1 and R_2 are considered positive since they act in the positive direction of the y axis. Similarly, the concentrated forces F_1 and F_2 are considered negative since they act in the negative y direction. Let us consider a cut at a section located at $x = a$ and take only the left-hand part of the beam with respect to the cut as a free body. To ensure equilibrium, an internal shear force V and an internal bending moment M must act on the cut surface (Fig. 1.8b). As we noted in the preceding section, the internal forces and moments become external when applied to an isolated part. Therefore, from Eq. (1.22), the shear force is the sum of the forces to the left of the cut section. Similarly, from Eq. (1.23), the bending moment is the sum of the moments of the forces to the left of the section. It can be proved that the shear force and the bending moment are related by

$$V = \frac{dM}{dx}.$$

(1.24)

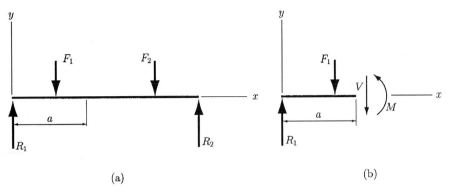

Figure 1.8 *Free body diagram of a simply supported beam.*

If bending is caused by a uniformly distributed load w, then the relation between shear force and bending moment is

$$\frac{dV}{dx} = \frac{d^2M}{dx^2} = -w, \tag{1.25}$$

A general force distribution called *load intensity* can be expressed as

$$q = \lim_{\Delta x \to 0} \frac{\Delta F}{\Delta x}.$$

Integrating Eqs. (1.24) and (1.25) between two points on the beam of coordinates x_A and x_B yields

$$\int_{V_A}^{V_B} dV = \int_{x_A}^{x_B} q\, dx = V_B - V_A. \tag{1.26}$$

The preceding equation states that the changes in shear force from A to B is equal to the area of the loading diagram between x_A and x_B. Similarly,

$$\int_{M_A}^{M_B} dM = \int_{x_A}^{x_B} V\, dx = M_B - M_A, \tag{1.27}$$

which states that the changes in moment from A to B is equal to the area of the shear force diagram between x_A and x_B.

1.8 Singularity Functions

Table 1.3 lists a set of five *singularity functions* that are useful in developing the general expressions for the shear force and the bending moment in a beam when it is loaded by concentrated forces or moments.

EXAMPLE 1.2 Develop the expressions for load intensity, shear force, and bending moment for the beam illustrated in Fig. 1.9.

Table 1.3 *Singularity Functions*

Function	Graph of $f_n(x)$	Meaning
Concentrated moment (unit doublet)	$\langle x - a \rangle^{-2}$	$\langle x - a \rangle^{-2} = 0 \quad x \neq a$ $\int_{-\infty}^{x} \langle x - a \rangle^{-2} dx = \langle x - a \rangle^{-1}$ $\langle x - a \rangle^{-2} = \pm\infty \quad x = a$
Concentrated force (unit impulse)	$\langle x - a \rangle^{-1}$	$\langle x - a \rangle^{-1} = 0 \quad x \neq a$ $\int_{-\infty}^{x} \langle x - a \rangle^{-1} dx = \langle x - a \rangle^{0}$ $\langle x - a \rangle^{-1} = +\infty \quad x = a$
Unit step	$\langle x - a \rangle^{0}$	$\langle x - a \rangle^{0} = \begin{cases} 0 & x < a \\ 1 & x \geq a \end{cases}$ $\int_{-\infty}^{x} \langle x - a \rangle^{0} dx = \langle x - a \rangle^{1}$
Ramp	$\langle x - a \rangle^{1}$	$\langle x - a \rangle^{1} = \begin{cases} 0 & x < a \\ x - a & x \geq a \end{cases}$ $\int_{-\infty}^{x} \langle x - a \rangle^{1} dx = \dfrac{\langle x - a \rangle^{2}}{2}$
Parabolic	$\langle x - a \rangle^{2}$	$\langle x - a \rangle^{2} = \begin{cases} 0 & x < a \\ (x - a)^{2} & x \geq a \end{cases}$ $\int_{-\infty}^{x} \langle x - a \rangle^{2} dx = \dfrac{\langle x - a \rangle^{3}}{3}$

Mechanics

Source: J. E. Shigley and C. R. Mischke, *Mechanical Engineering Design.* McGraw-Hill, New York, 1989. Used with permission.

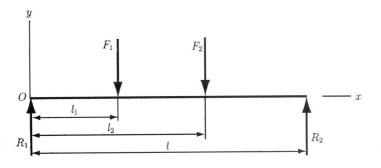

Figure 1.9
Free-body diagram of a simply supported beam loaded by concentrated forces.

Solution

We note that the beam shown in Fig. 1.9 is loaded by the concentrated forces F_1 and F_2. The reactions R_1 and R_2 are also concentrated loads. Thus, using Table 1.3, the load intensity has the following expression:

$$q(x) = R_1 \langle x \rangle^{-1} - F_1 \langle x - l_1 \rangle^{-1} - F_2 \langle x - l_2 \rangle^{-1} + R_2 \langle x - l \rangle^{-1}.$$

The shear force $V = 0$ at $x = -\infty$. Hence,

$$V(x) = \int_{-\infty}^{x} q(x)\, dx = R_1 \langle x \rangle^{0} - F_1 \langle x - l_1 \rangle^{0} - F_2 \langle x - l_2 \rangle^{0} + R_2 \langle x - l \rangle^{0}.$$

A second integration yields

$$M(x) = \int_{-\infty}^{x} V(x)\, dx = R_1 \langle x \rangle^{1} - F_1 \langle x - l_1 \rangle^{1} - F_2 \langle x - l_2 \rangle^{1} + R_2 \langle x - l \rangle^{1}.$$

To calculate the reactions R_1 and R_2, we will evaluate $V(x)$ and $M(x)$ at x slightly larger than l. At that point, both shear force and bending moment must be zero. Therefore, $V(x) = 0$ at x slightly larger than l, that is,

$$V = R_1 - F_1 - F_2 + R_2 = 0.$$

Similarly, the moment equation yields

$$M = R_1 l - F_1 (l - l_1) - F_2 (l - l_2) = 0.$$

The preceding two equations can be solved to obtain the reaction forces R_1 and R_2. ▲

EXAMPLE 1.3 A cantilever beam with a uniformly distributed load w is shown in Fig. 1.10. The load w acts on the portion $a \leq x \leq l$. Develop the shear force and bending moment expressions.

Solution

First, we note that M_1 and R_1 are the support reactions. Using Table 1.3, we find that the load intensity function is

$$q(x) = -M_1 \langle x \rangle^{-2} + R_1 \langle x \rangle^{-1} - w \langle x - a \rangle^{0}.$$

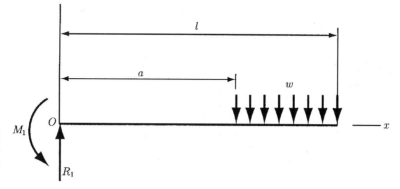

Figure 1.10
Free-body diagram for a cantilever beam.

Integrating successively two times gives

$$V(x) = \int_{-\infty}^{x} q(x)\, dx = -M_1\langle x\rangle^{-1} + R_1\langle x\rangle^0 - w\langle x - a\rangle^1$$

$$M(x) = \int_{-\infty}^{x} V(x)\, dx = -M_1\langle x\rangle^0 + R_1\langle x\rangle^1 - \frac{w}{2}\langle x - a\rangle^2.$$

The reactions can be calculated by evaluating $V(x)$ and $M(x)$ at x slightly larger than l and observing that both V and M are zero in this region. The shear force equation yields

$$-M_1 \cdot 0 + R_1 - w(l - a) = 0,$$

which can be solved to obtain the reaction R_1. The moment equation gives

$$-M_1 + R_1 l - \frac{w}{2}(l - a) = 0,$$

which can be solved to obtain the moment M_1. ▲

1.9 Normal Stress in Flexure

The relationships for the normal stresses in beams are derived considering that the beam is subjected to pure bending, that the material is isotropic and homogeneous and obeys Hooke's law, that the beam is initially straight with a constant cross-section throughout all its length, that the beam axis of symmetry is in the plane of bending, and that the beam cross-sections remain plane during bending.

A part of a beam on which a positive bending bending moment $\mathbf{M}_z = M\mathbf{k}$ ($\mathbf{k}$ being the unit vector associated with the z axis) is applied as shown in Fig. 1.11. A *neutral plane* is a plane that is coincident with the elements of the beam of zero strain. The xz plane is considered as the neutral plane. The x axis is coincident with the *neutral axis* of the section and the y axis is coincident with the axis of symmetry of the section.

Applying a positive moment on the beam, the upper surface will bend downward and, therefore, the neutral axis will also bend downward (Fig. 1.11). Because of this fact, the section PQ initially parallel to RS will

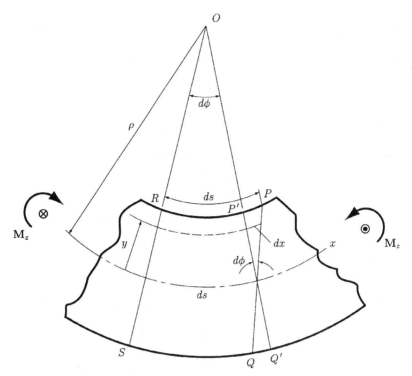

Figure 1.11
Normal stress in flexure.

twist through the angle $d\phi$ with respect to $P'Q'$. In Fig. 1.11, ρ is the radius of curvature of the neutral axis, ds is the length of a differential element of the neutral axis, and ϕ is the angle between the two adjacent sides RS and $P'Q'$. The definition of the curvature is

$$\frac{1}{\rho} = \frac{d\phi}{ds}. \tag{1.28}$$

The deformation of the beam at distance y from the neutral axis is

$$dx = yd\phi, \tag{1.29}$$

and the strain

$$\epsilon = -\frac{dx}{ds}, \tag{1.30}$$

where the negative sign suggests that the beam is in compression. Manipulating Eqs. (1.28), (1.29), and (1.30), we obtain

$$\epsilon = -\frac{y}{\rho}. \tag{1.31}$$

Since $\sigma = E\epsilon$, the expression for stress is

$$\sigma = -\frac{Ey}{\rho}. \tag{1.32}$$

Observing that the force acting on an element of area dA is σdA and integrating this force, we get

$$\int \sigma\, dA = -\frac{E}{\rho} \int y\, dA = 0. \tag{1.33}$$

Since the x axis is the neutral axis, the preceding equation states that the moment of the area about the neutral axis is zero. Thus, Eq. (1.33) defines the location of the neutral axis, that is, the neutral axis passes through the centroid of the cross-sectional area.

To ensure equilibrium, the internal bending moment created by the stress σ must be the same as the external moment $\mathbf{M}_z = M\mathbf{k}$, namely,

$$M = \int y\sigma\, dA = \frac{E}{\rho} \int y^2\, dA, \tag{1.34}$$

where the second integral in the foregoing equation is the second moment of area I about the z axis. It is given by

$$I = \int y^2\, dA. \tag{1.35}$$

Manipulating Eqs. (1.34) and (1.35), we obtain

$$\frac{1}{\rho} = \frac{M}{EI}. \tag{1.36}$$

Finally, eliminating ρ from Eqs. (1.32) and (1.36) yields

$$\sigma = -\frac{My}{I}. \tag{1.37}$$

Equation (1.37) states that the stress σ is directly proportional to the bending moment M and the distance y from the neutral axis (Fig. 1.12). The maximum stress is

$$\sigma = \frac{Mc}{I}, \tag{1.38}$$

where $c = y_{max}$. Equation (1.38) can also be written in the two forms

$$\sigma = \frac{M}{I/c}, \quad \sigma = \frac{M}{Z}, \tag{1.39}$$

where $Z = I/c$ is called the *section modulus*.

EXAMPLE 1.4 Determine the diameter of a solid round shaft OC in Fig. 1.13, 36 in long, such that the bending stress does not exceed 10 kpsi. The transversal loads of 800 lb and 300 lb act on the shaft.

Solution

The moment equation about C yields

$$\sum M_C = -36R_1 + 24(800) + 8(300) = 0.$$

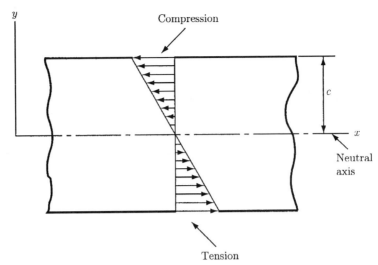

Figure 1.12
Bending stress in flexure.

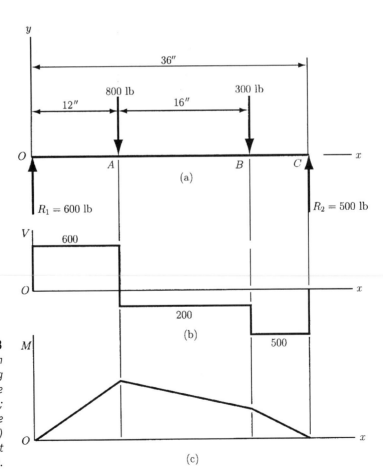

Figure 1.13
Loading diagram of a rotating shaft. (a) Free body diagram; (b) shear force diagram; (c) bending moment diagram.

The foregoing equation gives $R_1 = 600$ lb. The force equation with respect to the y axis is

$$R_1 - 400 - 150 + R_2 = 0,$$

yielding $R_2 = 500$ lb. The next step is to draw the shear force and the bending moment diagrams shown in Figs. 1.13b and 1.13c. From the bending moment diagram, we observe that the maximum bending moment is

$$M = 600(12) = 7200 \text{ lb in.}$$

The section modulus is

$$\frac{I}{c} = \frac{\pi d^3}{32} = 0.0982 d^3.$$

Then, the bending stress is

$$\sigma = \frac{M}{I/c} = \frac{7200}{0.0982 d^3}.$$

Considering $\sigma = 10,000$ psi and solving for d, we obtain

$$d = \sqrt[3]{\frac{7200}{0.0982(10000)}} = 1.94 \text{ in.} \quad \blacktriangle$$

1.10 Beams with Asymmetrical Sections

Considering the restriction of having the plane of bending coincident with one of the two principal axes of the section, the results of the preceding section can be applied to beams with asymmetrical sections.

From Eq. (1.33), the stress at a distance y from the neutral axis is

$$\sigma = -\frac{Ey}{\rho}. \tag{1.40}$$

and, thus, the force on the element of area dA in Fig. 1.14 is

$$dF = \sigma \, dA = \frac{Ey}{\rho} \, dA.$$

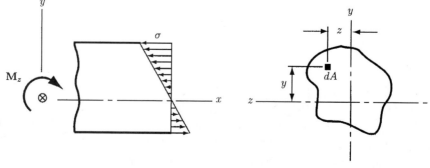

Figure 1.14
Normal stress of asymmetrical section beam.

The moment of this force about the y axis gives

$$M_y = \int z\, dF = \int \sigma z\, dA = -\frac{E}{\rho}\int yz\, dA. \qquad (1.41)$$

The last integral in Eq. (1.41) is the product of inertia I_{yz}. It can be shown that

$$I_{yz} = \int yz\, dA = 0 \qquad (1.42)$$

if the bending moment on the beam is in the plane of one of the principal axes. Hence, the relations developed in the preceding section can be applied to beams having asymmetrical sections only if one takes into account the restriction given by Eq. (1.42).

1.11 Shear Stresses in Beams

In the general case, beams have both shear forces and bending moments acting upon them. Let us consider a beam of constant cross-section subjected to a shear force $\mathbf{V} = V\mathbf{j}$ and a bending moment $\mathbf{M}_z = M\mathbf{k}$, $\mathbf{j}$ and $\mathbf{k}$ being the unit vectors corresponding to the y and z axes, respectively (Fig. 1.15). From Eq. (1.24), the relationship between V and M is

$$V = \frac{dM}{dx}. \qquad (1.43)$$

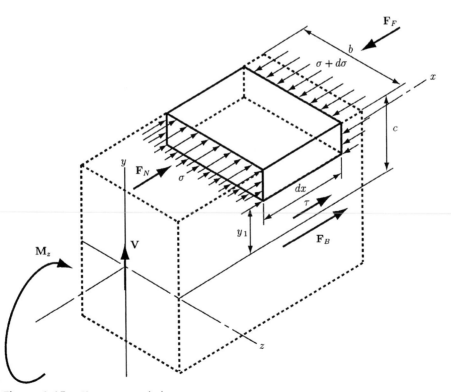

Figure 1.15   *Shear stresses in beams.*

Let us consider an element of length dx located at a distance y_1 above the neutral axis. Because of the shear force, the bending moment is not constant along the x axis. Let us denote by M the bending moment on the near side of the section and by $M + dM$ the bending moment on the far side. These two moments produce normal stresses σ and $\sigma + d\sigma$, respectively. The normal stresses generate forces perpendicular to the vertical faces of the element. Since the force acting on the far side is larger than that one acting on the near side, the resultant of these forces would cause the element to tend to slide in the $-x$ direction. To ensure equilibrium, the resultant must be balanced by a shear force acting in the $+x$ direction on the bottom of the section. A shear stress τ is generated by this shear force. Summarizing, three resultant forces are exerted on the element, that is, $\mathbf{F}_N = F_N\mathbf{i}$, $\mathbf{i}$ being the unit vector corresponding to the x axis, due to the normal stress σ acting on the near face; $\mathbf{F}_F = F_F\mathbf{i}$ due to the normal stress $\sigma + d\sigma$ acting on the far face; and $\mathbf{F}_B = F_B\mathbf{i}$ due to the shear stress τ acting on the bottom face.

Let us consider a small element of area dA on the near face. Since stress acting on this area is σ, the force exerted on the small area can be calculated as σdA. Integrating, we find that the force on the near face is

$$F_N = \int_{y_1}^{c} \sigma \, dA, \tag{1.44}$$

where the limits show that the integration is from the bottom of the element $y = y_1$ to the top $y = c$. If we use the expression $\sigma = My/I$, the preceding equation yields

$$F_N = \frac{M}{I} \int_{y_1}^{c} y \, dA. \tag{1.45}$$

The force acting on the far face can be calculated in a similar fashion, namely,

$$F_F = \int_{y_1}^{c} (\sigma + d\sigma) \, dA = \frac{M + dM}{I} \int_{y_1}^{c} y \, dA. \tag{1.46}$$

The force on the bottom face is

$$F_B = \tau b \, dx, \tag{1.47}$$

where b is the width of the element and $b \, dx$ is the area of the bottom face. Observing that all three forces act in the x direction, we can sum them algebraically:

$$\sum F_x = F_N - F_F + F_B = 0. \tag{1.48}$$

Substituting F_N and F_F and solving for F_B yields

$$F_B = F_F - F_N = \frac{M + dM}{I} \int_{y_1}^{c} y \, dA - \frac{M}{I} \int_{y_1}^{c} y \, dA = \frac{dM}{I} \int_{y_1}^{c} y \, dA. \tag{1.49}$$

Substituting Eq. (1.47) for F_B and solving for shear stress gives

$$\tau = \frac{dM}{dx} \frac{1}{Ib} \int_{y_1}^{c} y \, dA. \tag{1.50}$$

Using Eq. (1.43), we find that the shear stress formula becomes

$$\tau = \frac{V}{Ib}\int_{y_1}^{c} y\, dA. \tag{1.51}$$

In the preceding equation, the integral is the first moment of area of the vertical face about the neutral axis usually denoted as Q, and

$$Q = \int_{y_1}^{c} y\, dA. \tag{1.52}$$

Therefore, Eq. (1.51) can be rewritten as

$$\tau = \frac{VQ}{Ib}, \tag{1.53}$$

where I is the second moment of area of the section about the neutral axis.

1.12 Shear Stresses in Rectangular Section Beams

Figure 1.16 shows a part of a beam acted upon by a shear force $\mathbf{V} = V\mathbf{j}$ and a bending moment $\mathbf{M}_z = M\mathbf{k}$. Because of the bending moment, a normal stress σ is produced on a cross-section of the beam, such as $A-A$. The beam is in compression above the neutral axis and in tension below. Let us consider an element of area dA located at a distance y above the neutral axis. Observing that $dA = b\, dy$, we find that Eq. (1.52) becomes

$$Q = \int_{y_1}^{c} y\, dA = b\int_{y_1}^{c} y\, dy = \frac{by^2}{2}\Big|_{y_1}^{c} = \frac{b}{2}(c^2 - y_1^2). \tag{1.54}$$

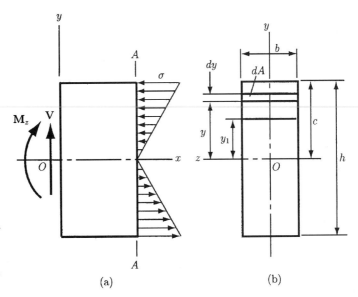

Figure 1.16
Stresses in rectangular section beams. (a) Side view; (b) cross-section.

(a)   (b)

Substituting Eq. (1.54) into Eq. (1.53) gives

$$\tau = \frac{V}{2I}(c^2 - y_1^2).$$

(1.55)

This equation represents the general equation for shear stress in a beam of rectangular cross-section. The expression for the second moment of area I for a rectangular section is

$$I = \frac{bh^3}{12},$$

and, if we substitute $h = 2c$ and $A = bh = 2bc$, the expression for I becomes

$$I = \frac{Ac^2}{3}.$$

(1.56)

Substituting Eq. (1.56) into Eq. (1.55) yields

$$\tau = \frac{3V}{2A}\left(1 - \frac{y_1^2}{c^2}\right) = C\frac{V}{A}.$$

(1.57)

The values C versus y_1 are listed in Table 1.4. The maximum shear stress is obtained for $y_1 = 0$, that is,

$$\tau_{max} = \frac{3V}{2A},$$

(1.58)

and the zero shear stress is obtained at the outer surface where $y_1 = c$. Formulas for the maximum flexural shear stress for the most commonly used shapes are listed in Table 1.5.

1.13 Torsion

A *torque vector* is a moment vector collinear with an axis of a mechanical element, causing the element to twist about that axis. A torque $\mathbf{T}_x = T\mathbf{i}$ applied to a solid round bar is shown in Fig. 1.17. The torque vectors are the arrows shown on the x axis. The angle of twist is given by the relationship

$$\theta = \frac{Tl}{GJ},$$

(1.59)

where T is the torque, l the length, G the modulus of rigidity, and J the polar second moment of area. Since the shear stress is zero at the center and

Table 1.4 *Variation of Shear Stress $\tau = C\dfrac{V}{A}$*

Distance y_1	0	0.2c	0.4c	0.6c	0.8c	c
Factor C	1.50	1.44	1.26	0.96	0.54	0

Source: J. E. Shigley and C. R. Mischke, *Mechanical Engineering Design.* McGraw-Hill, New York, 1989. Used with permission.

Table 1.5 *Formulas for Maximum Shear Stress Due to Bending*

Beam shape	Formula
Rectangular	$\tau_{max} = \dfrac{3V}{2A}$
Circular	$\tau_{max} = \dfrac{4V}{3A}$
Hollow round	$\tau_{max} = \dfrac{2V}{A}$
Structural (Web)	$\tau_{max} = \dfrac{V}{A_{web}}$

maximum at the surface for a solid round bar, the shear stress is proportional to the radius ρ, namely,

$$\tau = \frac{T\rho}{J}. \tag{1.60}$$

If r is the radius to the outer surface, then

$$\tau_{max} = \frac{Tr}{J}. \tag{1.61}$$

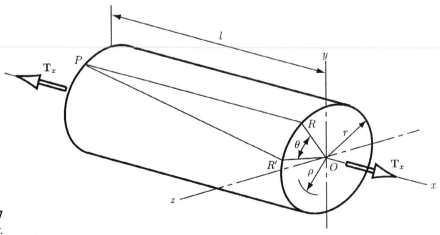

Figure 1.17
Torsion bar.

For a rotating shaft, the torque T can be expressed in terms of power and speed. One form of this relationship can be

$$H = \frac{2\pi Tn}{33,000(12)} = \frac{FV}{33,000} = \frac{Tn}{63,000}, \tag{1.62}$$

where H is the power (hp), T the torque (lb in), n the shaft speed (rev/min), F the force (lb), and V the velocity (ft/min). When SI units are used, the equation becomes

$$H = T\omega, \tag{1.63}$$

where H is the power (W), T the torque (N m), and ω the angular velocity (rad/s). Thus, the torque T can be approximated by

$$T = 9.55\frac{H}{n}, \tag{1.64}$$

where n is in rev/min.

For rectangular sections, the following approximate formula applies:

$$\tau_{max} = \frac{T}{wt^2}\left(3 + 1.8\frac{t}{w}\right). \tag{1.65}$$

Here w and t are the width and the thickness of the bar, respectively.

EXAMPLE 1.5 (*Source:* J. E. Shigley and C. R. Mischke, *Mechanical Engineering Design.* McGraw-Hill, New York, 1989.)

Figure 1.18 shows a crank loaded by a force $F = 1000$ lb that causes twisting and bending of a $\frac{3}{4}$-in diameter shaft fixed to a support at the origin of the reference system. Draw separate free body diagrams of the shaft AB and the arm BC, and compute the values of all exerted forces, moments, and torques. Compute the maximum torsional stress and the bending stress in the arm BC and indicate where they occur.

Solution

The two free body diagrams are shown in Fig. 1.19. The force and torque at point C are

$$\mathbf{F} = -1000\mathbf{j}\text{ lb}, \qquad \mathbf{T} = -1000\mathbf{k}\text{ lb in}.$$

At the end B of the arm BC,

$$\mathbf{F} = 1000\mathbf{j}\text{ lb}, \qquad \mathbf{M} = 4000\mathbf{i}\text{ lb in}, \qquad \mathbf{T} = 1000\mathbf{k}\text{ lb in},$$

whereas at the end B of the shaft AB,

$$\mathbf{F} = -1000\mathbf{j}\text{ lb}, \qquad \mathbf{T} = -4000\mathbf{i}\text{ lb in}, \qquad \mathbf{M} = -1000\mathbf{k}\text{ lb in}.$$

At point A,

$$\mathbf{F} = 1000\mathbf{j}\text{ lb}, \qquad \mathbf{M} = 6000\mathbf{k}\text{ lb in}, \qquad \mathbf{T} = 4000\mathbf{i}\text{ lb in}.$$

Mechanics

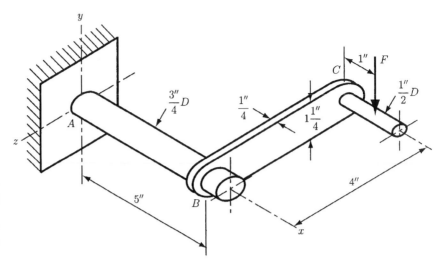

Figure 1.18
*Crank
mechanism.
Used with
permission from
Ref. 16.*

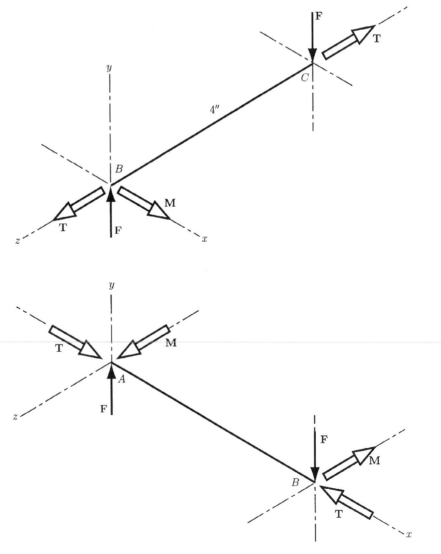

Figure 1.19
*Free body
diagrams of a
crank
mechanism.
Used with
permission from
Ref. 16.*

For the arm BC, the bending stress will reach a maximum near the shaft at B. The bending stress for the rectangular cross-section of the arm is

$$\sigma = \frac{M}{I/c} = \frac{6M}{bb^2} = \frac{6(4000)}{0.25(1.25)^2} = 61,440 \text{ psi.}$$

The torsional stress is

$$\tau_{max} = \frac{T}{wt^2}\left(3 + 1.8\frac{t}{w}\right) = \frac{1000}{1.25(0.25)^2}\left(3 + 1.8\frac{0.25}{1.25}\right) = 43,008 \text{ psi.}$$

The stress occurs at the middle of the $1\frac{1}{4}$-in side.   ▲

1.14 Contact Stresses

The theory presented in this section is based on the Hertzian stresses approach. A typical case of contact stresses occurs when the bodies in contact have a double radius of curvature. This means that the radius in the plane of rolling is different from the radius in a perpendicular plane, both planes taken through the axis of the contacting force. When such bodies are pressed together the produced stresses are three-dimensional. As the bodies are pressed, the initial point of contact between the bodies becomes an area of contact.

For example, if a force F is applied to two solid spheres of diameters d_1 and d_2, respectively, the spheres are pressed together and a circular area of contact of radius a is developed. The radius a is given by

$$a = \sqrt[3]{\frac{3F}{8}\frac{(1 - v_1^2)/E_1 + (1 - v_2^2)/E_2}{1/d_1 + 1/d_2}}, \qquad (1.66)$$

where E_1, v_1 and E_2, v_2 are the elastic constants of the two spheres, respectively. The pressure distribution within each sphere is semielliptical (Fig. 1.20). The maximum pressure P_{max} is obtained at the center of the contact area given by

$$P_{max} = \frac{3F}{2\pi a^2}. \qquad (1.67)$$

In the case of the contact between a sphere and a plane surface or a sphere and an internal spherical surface, Eqs. (1.66) and (1.67) can also be applied. If we observe that for a plane surface $d = \infty$ and for an internal surface the diameter is a negative quantity, the principal stresses are

$$\sigma_x = \sigma_y = -P_{max}\left[\left(1 - \frac{z}{a}\tan^{-1}\frac{1}{\frac{z}{a}}\right)(1 + \mu) - \frac{1}{2\left(1 + \frac{z^2}{a^2}\right)}\right], \qquad (1.68)$$

$$\sigma_z = \frac{-P_{max}}{1 + \frac{z^2}{a^2}}. \qquad (1.69)$$

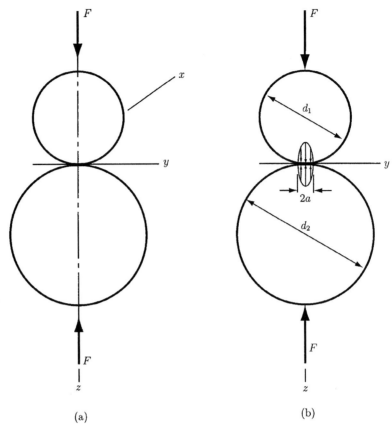

Figure 1.20
*Contact stress of
two spheres. (a)
Spheres in
contact; (b)
contact stress
distribution.
Used with
permission from
Ref. 16.*

In the case of the contact between two cylinders of length l and diameters d_1 and d_2 (Fig. 1.21), the area of contact is a narrow rectangle of width $2b$ and length l. The pressure distribution is elliptical. The half-width b is given by

$$b = \sqrt{\frac{2F}{\pi l} \frac{(1 - v_1^2)/E_1 + (1 - v_2^2)/E_2}{1/d_1 + 1/d_2}}, \tag{1.70}$$

and the maximum pressure by

$$P_{max} = \frac{2F}{\pi b l}. \tag{1.71}$$

As in the case of the sphere contact, setting $d = \infty$, the preceding equations can be applied for the contact of a cylinder and a plane surface or a cylinder

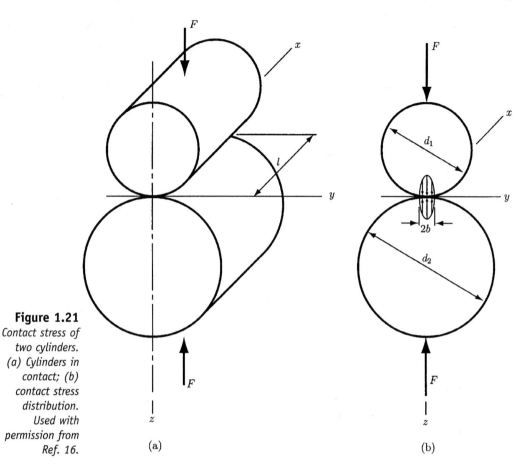

Figure 1.21
Contact stress of two cylinders. (a) Cylinders in contact; (b) contact stress distribution. Used with permission from Ref. 16.

(a)

(b)

and an internal cylindrical surface (for the last situation d is negative). The stresses on the z axis are given by the following equations, respectively:

$$\sigma_x = -2\nu P_{max}\left(\sqrt{1 + \frac{z^2}{b^2}} - \frac{z}{b}\right) \tag{1.72}$$

$$\sigma_y = -P_{max}\left[\left(2 - \frac{1}{1 + \frac{z^2}{b^2}}\right)\sqrt{1 + \frac{z^2}{b^2}} - 2\frac{z}{b}\right] \tag{1.73}$$

$$\sigma_z = \frac{-P_{max}}{\sqrt{1 + \frac{z^2}{b^2}}}. \tag{1.74}$$

2. Deflection and Stiffness

A *rigid* is a mechanical element that does not bend, deflect, or twist when an external action is exerted on it. Conversely, a *flexible* is a mechanical element

Mechanics

that changes its geometry when an external force, moment, or torque is applied. Therefore, *rigidity* and *flexibility* are terms that apply to particular situations. This chapter deals with deflection analysis that is frequently performed in the design of, for example, transmissions, springs, or automotive suspensions.

2.1 Springs

The property of a material that enables it to regain its original geometry after having been deformed is called *elasticity*. Let us consider a straight beam of length l which is simply supported at the ends and loaded by the transversal force F (Fig. 2.1a). If the elastic limit of the material is not exceeded (as indicated by the graph), the deflection y of the beam is linearly related to the force, and, therefore, the beam can be described as a *linear spring*.

The case of a straight beam supported by two cylinders is illustrated in Fig. 2.1b. As the force F is applied to the beam, the length between the supports decreases and, therefore, a larger force is needed to deflect a short beam than that required for a long one. Hence, the more this beam is deflected, the stiffer it becomes. The force is not linearly related to the deflection, and, therefore, the beam can be described as a *nonlinear stiffening spring*.

A dish-shaped round disk acted upon by the load F is shown in Fig. 2.1c. To flatten the disk, a larger force is needed, so the force increases first. Then, the force decreases as the disk approaches a flat configuration. A mechanical element having this behavior is called a *nonlinear softening spring*.

If we consider the relationship between force and deflection as

$$F = F(y), \tag{2.1}$$

then the *spring rate* is defined as

$$k(y) = \lim_{\Delta y \to 0} \frac{\Delta F}{\Delta y} = \frac{dF}{dy}, \tag{2.2}$$

where y is measured at the point of application of F in the direction of F. For a linear spring, k is a constant called the *spring constant*, and Eq. (2.2) becomes

$$k = \frac{F}{y}. \tag{2.3}$$

2.2 Spring Rates for Tension, Compression, and Torsion

The total extension or deformation of a uniform bar is

$$\delta = \frac{Fl}{AE}, \tag{2.4}$$

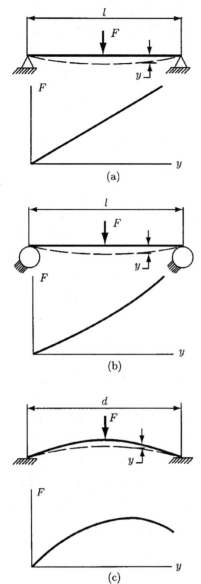

Figure 2.1
*Springs. (a)
Linear spring;
(b) stiffening
spring; (c) soft-
ening spring.
Used with
permission from
Ref. 16.*

where F is the force applied on the bar, l the length of the bar, A the cross-sectional area, and E the modulus of elasticity. From Eqs. (2.3) and (2.4), the spring constant of an axially loaded bar is obtained:

$$k = \frac{AE}{l}. \tag{2.5}$$

If a uniform round bar is subjected to a torque T, the angular deflection is

$$\theta = \frac{Tl}{GJ}, \tag{2.6}$$

where T is the torque, l the length of the bar, G the modulus of rigidity, and J the polar moment of area. If we multiply Eq. (2.6) by $180/\pi$ and substitute $J = \pi d^4/32$ (for a solid round bar), the expression for θ becomes

$$\theta = \frac{583.6\,Tl}{Gd^4}, \tag{2.7}$$

where θ is in degrees and d is the diameter of the round cross-section. If we rewrite Eq. (2.6) as a ratio between T and θ, we can define the spring rate:

$$k = \frac{T}{\theta} = \frac{GJ}{l}. \tag{2.8}$$

2.3 Deflection Analysis

If a beam is subjected to a positive bending moment M, the beam will deflect downward. The relationship between the curvature of the beam and the external moment M is

$$\frac{1}{\rho} = \frac{M}{EI}, \tag{2.9}$$

where ρ is the radius of curvature, E the modulus of elasticity, and I the second moment of area. It can be proved mathematically that the curvature of a plane curve can be described by

$$\frac{1}{\rho} = \frac{d^2y/dx^2}{[1 + (dy/dx)^2]^{3/2}}, \tag{2.10}$$

where y is the deflection of the beam at any point of coordinate x along its length. The slope of the beam at point x is

$$\theta = \frac{dy}{dx}. \tag{2.11}$$

If the slope is very small, that is, $\theta \approx 0$, then the denominator of Eq. (2.10)

$$\left[1 + \left(\frac{dy}{dx}\right)^2\right]^{3/2} = [1 + \theta^2]^{3/2} \approx 1.$$

Hence, Eq. (2.9) yields

$$\frac{M}{EI} = \frac{d^2y}{dx^2}. \tag{2.12}$$

Differentiating Eq. (2.12) two times successively gives

$$\frac{V}{EI} = \frac{d^3y}{dx^3} \tag{2.13}$$

$$\frac{q}{EI} = \frac{d^4y}{dx^4}, \tag{2.14}$$

where q is the load intensity and V the shear force,

$$V = \frac{dM}{dx} \quad \text{and} \quad \frac{dV}{dx} = \frac{d^2M}{dx^2} = q.$$

The preceding relations can be arranged as follows:

$$\frac{q}{EI} = \frac{d^4y}{dx^4} \tag{2.15}$$

$$\frac{V}{EI} = \frac{d^3y}{dx^3} \tag{2.16}$$

$$\frac{M}{EI} = \frac{d^2y}{dx^2} \tag{2.17}$$

$$\theta = \frac{dy}{dx} \tag{2.18}$$

$$y = f(x). \tag{2.19}$$

Figure 2.2 shows a beam of length $l = 10$ in loaded by the uniform load $w = 10\,\text{lb/in}$. All quantities are positive if upward, and negative if downward. Figure 2.2 also shows the shear force, bending moment, slope, and deflection diagrams. The values of these quantities at the ends of the beam, that is, at $x = 0$ and $x = l$, are called *boundary values*. For example, the bending moment and the deflection are zero at each end because the beam is simply supported.

2.4 Deflections Analysis Using Singularity Functions

Let us consider a simply supported beam acted upon by a concentrated load at the distance a from the origin of the xy coordinate system (Fig. 2.3). We want to develop an analytical expression for the deflection of the beam by using the singularity functions studied in Section 1.8. Since the beam is simply supported, we are interested in determining the deflection of the beam in between the supports, namely for $0 < x < l$. Thus, Eq. (2.15) yields

$$EI\frac{d^4y}{dx^4} = q = -F\langle x - a\rangle^{-1}. \tag{2.20}$$

Because of the range chosen for x, the reactions R_1 and R_2 do not appear in the preceding equation. Integrating from 0 to x Eq. (2.20) and using Eq. (2.16) gives

$$EI\frac{d^3y}{dx^3} = V = -F\langle x - a\rangle^0 + C_1, \tag{2.21}$$

where C_1 is an integration constant. Using Eq. (2.17) and integrating again, we obtain

$$EI\frac{d^2y}{dx^2} = M = -F\langle x - a\rangle^1 + C_1 x + C_2, \tag{2.22}$$

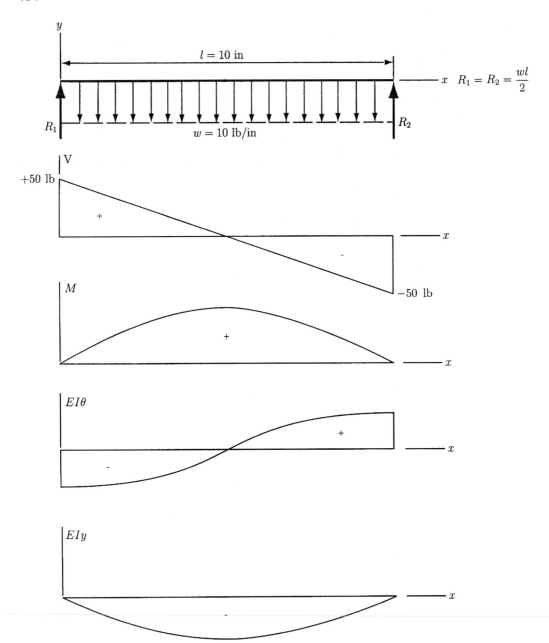

Figure 2.2   *Uniformly loaded beam.*

where C_2 is also an integration constant. We can determine the constants C_1 and C_2 by considering two boundary conditions. The boundary condition can be $M = 0$ at $x = 0$ applied to Eq. (2.22), which gives $C_2 = 0$ and $M = 0$ at $x = l$ also applied to Eq. (2.22), which gives

$$C_1 = \frac{F(l - a)}{l} = \frac{Fb}{l}.$$

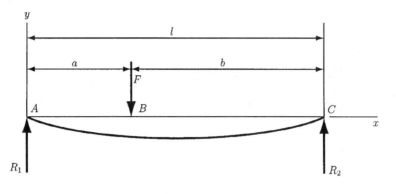

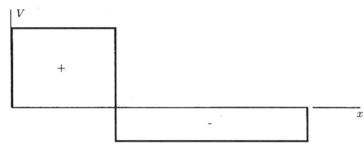

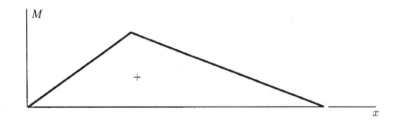

Figure 2.3
Simply supported beam loaded by a concentrated force.

Substituting C_1 and C_2 in Eq. (2.22) gives

$$EI\frac{d^2y}{dx^2} = M = \frac{Fbx}{l} - F\langle x - a\rangle^1. \tag{2.23}$$

Integrating Eq. (2.23) twice according to Eqs. (2.18) and (2.19) yields

$$EI\frac{dy}{dx} = EI\theta = \frac{Fbx^2}{2l} - \frac{F\langle x - a\rangle^2}{2} + C_3 \tag{2.24}$$

$$EIy = \frac{Fbx^3}{6l} - \frac{F\langle x - a\rangle^3}{6} + C_3 x + C_4. \tag{2.25}$$

The integration constants C_3 and C_4 in the preceding equations can be evaluated by considering the boundary conditions $y = 0$ at $x = 0$ and $y = 0$ at $x = l$. Substituting the first boundary condition in Eq. (2.25) yields $C_4 = 0$. The second condition substituted in Eq. (2.25) yields

$$0 = \frac{Fbl^2}{6} - \frac{Fb^3}{6} + C_3 l,$$

or

$$C_3 = -\frac{Fb}{6l}(l^2 - b^2).$$

If we substitute C_3 and C_4 in Eq. (2.25), the analytical expression for the deflection y is obtained:

$$y = \frac{F}{6EIl}[bx(x^2 + b^2 - l^2) - l\langle x - a\rangle^3].\qquad(2.26)$$

The shear force and bending moment diagrams are shown in Fig. 2.3.

As a second example, let us consider the beam shown in Fig. 2.4a. The loading diagram and the approximate deflection curve are shown in Fig. 2.4b. We will develop the analytical expression for the deflection y as

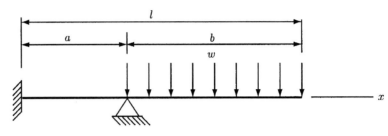

(a)

Figure 2.4
Cantilever beam loaded by a uniformly distributed force at the free end. (a) Loading diagram; (b) free-body diagram. Used with permission from Ref. 16.

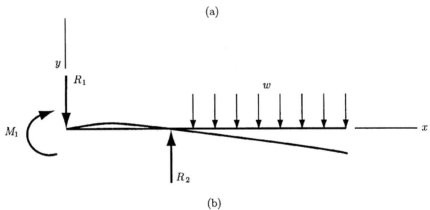

(b)

a function of x in a similar manner as for the preceding example. The loading equation for x in the range $0 < x < l$ is

$$q = R_2\langle x - a\rangle^{-1} - w\langle x - a\rangle^0.\qquad(2.27)$$

Integrating this equation four times according to Eqs. (2.15) to (2.19) yields

$$V = R_2\langle x - a\rangle^0 - w\langle x - a\rangle^1 + C_1\qquad(2.28)$$

$$M = R_2\langle x - a\rangle^1 - \frac{w}{2}\langle x - a\rangle^2 + C_1 x + C_2\qquad(2.29)$$

$$EI\theta = \frac{R_2}{2}\langle x - a\rangle^2 - \frac{w}{6}\langle x - a\rangle^3 + \frac{C_1}{2}x^2 + C_2 x + C_3\qquad(2.30)$$

$$EIy = \frac{R_2}{6}\langle x - a\rangle^3 - \frac{w}{24}\langle x - a\rangle^4 + \frac{C_1}{6}x^3 + \frac{C_2}{2}x^2 + C_3 x + C_4.\qquad(2.31)$$

The integration constants C_1 to C_4 can be evaluated by considering appropriate boundary conditions. Both $EI\theta = 0$ and $EIy = 0$ at $x = 0$. This gives $C_3 = 0$ and $C_4 = 0$. At $x = 0$ the shear force is equal to $-R_1$. Therefore, Eq. (2.28) gives $V(0) = R_1 = C_1$. The deflection must be zero at $x = a$. Thus, Eq. (2.31) yields

$$\frac{C_1}{6}a^3 + \frac{C_2}{2}a^2 = 0 \quad \text{or} \quad C_1\frac{a}{3} + C_2 = 0. \tag{2.32}$$

At the overhanging end, at $x = l$, the moment must be zero. For this boundary condition Eq. (2.29) gives

$$R_2(l - a) - \frac{w}{2}(l - a)^2 + C_1 l + C_2 = 0,$$

or, if we use the notation $l - a = b$ and the equation resulting from the sum of the forces in the y direction, namely $R_2 = R_1 + wb = -C_1 + wb$,

$$C_1 a + C_2 = -\frac{wb^2}{2}. \tag{2.33}$$

Solving Eqs. (2.32) and (2.33) simultaneously for C_1 and C_2 gives

$$C_1 = \frac{3wb^2}{4a}, \qquad C_2 = \frac{wb^2}{4}.$$

Therefore, the reaction R_2 is obtained:

$$R_2 = -C_1 + wb = \frac{wb}{4a}(4a + 3b).$$

Equation (2.29) for $x = 0$ gives

$$M(0) = M_1 = C_2 = \frac{wb^2}{4}.$$

The analytical expression for the deflection curve is obtained by substituting the expressions for R_2 and the constants C_1, C_2, C_3, and C_4 in Eq. (2.31), that is,

$$EIy = \frac{wb}{24a}(4a + 3b)\langle x - a \rangle^3 - \frac{w}{24}\langle x - a \rangle^4 - \frac{wb^2 x^3}{8a} + \frac{wb^2 x^2}{8}. \tag{2.34}$$

2.5 Impact Analysis

An impact situation is shown in Fig. 2.5, where a weight W moving with a constant velocity v on a frictionless surface strikes a spring of constant k. We are interested in finding the maximum force and the maximum deflection of the beam caused by the impact.

The following equation of motion for the weight can be developed after the impact:

$$\frac{W}{g}\ddot{y} = -ky. \tag{2.35}$$

Mechanics

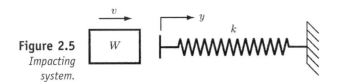

Figure 2.5
*Impacting
system.*

Here, the left-hand side term is actually mass times acceleration and the right-hand side term is the force due to the spring. It tends to retard the motion of the weight and, therefore, the spring force is negative. Written as

$$\ddot{y} + \frac{kg}{W} y = 0,$$

Eq. (2.35) is a homogeneous second-order differential equation having the solution

$$y = A\cos\left(\frac{kg}{W}\right)^{1/2} t + B\sin\left(\frac{kg}{W}\right)^{1/2} t, \qquad (2.36)$$

where A and B are two constants to be determined. Differentiating Eq. (2.36), we obtain the equation for the velocity of W after impact:

$$\dot{y} = A\left(\frac{kg}{W}\right)^{1/2} \sin\left(\frac{kg}{W}\right)^{1/2} t + B\left(\frac{kg}{W}\right)^{1/2} \cos\left(\frac{kg}{W}\right)^{1/2} t. \qquad (2.37)$$

Considering the initial conditions $y = 0$ and $\dot{y} = v$ at $t = 0$ in Eqs. (2.36) and (2.37), respectively, the constants A and B are obtained:

$$A = 0, \quad B = \frac{v}{(kg/W)^{1/2}}.$$

Substituting the preceding expressions in Eq. (2.36), we can write the solution of the equation of motion (2.35) as

$$y = \frac{v}{(kg/W)^{1/2}} \sin\left(\frac{kg}{W}\right)^{1/2} t. \qquad (2.38)$$

The maximum deflection is obtained when $\sin\left(\frac{kg}{W}\right)^{1/2} t = 1$. Thus,

$$y_{max} = \frac{v}{(kg/W)^{1/2}} = v\left(\frac{kg}{W}\right)^{1/2}. \qquad (2.39)$$

The maximum force acting on the spring is

$$F_{max} = ky_{max} = kv\left(\frac{kg}{W}\right)^{1/2}. \qquad (2.40)$$

Another impact situation is shown in Fig. 2.6a. A weight W falls a distance h and impacts a structure or member of spring rate k. The origin of the coordinate y corresponds to the position of the weight at $t = 0$. We want to find the maximum deflection and the maximum force acting on the spring caused by the impact.

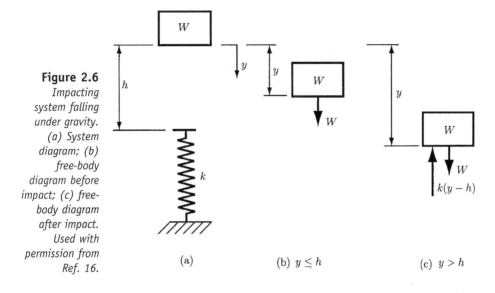

Figure 2.6
Impacting system falling under gravity. (a) System diagram; (b) free-body diagram before impact; (c) free-body diagram after impact. Used with permission from Ref. 16.

(a)   (b) $y \leq h$ (c) $y > h$

The free-body diagrams of the weight shown in Fig. 2.6b and 2.6c depict two different situation, namely $y \leq h$ and $y > h$, respectively. For each free-body diagram, Newton's second law yields, respectively,

$$\frac{W}{g}\ddot{y} = W, \qquad\qquad y \leq h$$

$$\frac{W}{g}\ddot{y} = k(y - h) + W, \quad y > h. \qquad (2.41)$$

The foregoing equations are linear, but each applies only for a certain range of y. The solution to the first equation is

$$y = \frac{gt^2}{2}, \qquad y \leq h. \qquad (2.42)$$

The preceding solution is no longer valid if $y \geq h$, which corresponds to the moment of time

$$t_1 = \left(\frac{2h}{g}\right)^{1/2}. \qquad (2.43)$$

Differentiating Eq. (2.42) with respect to time gives the equation for the weight velocity,

$$\dot{y} = gt, \qquad y \leq h. \qquad (2.44)$$

The velocity of the weight at $t = t_1$ is

$$\dot{y}_1 = gt_1 = (2gh)^{1/2}. \qquad (2.45)$$

To solve the second equation of the system (2.41), it is convenient to define another time variable $t' = t - t_1$. Thus, $t' = 0$ when the weight strikes the spring. The solution of the second equation of the system (2.41) in terms of the new time variable t' is

$$y = A\cos\left(\frac{kg}{W}\right)^{1/2}t' + B\sin\left(\frac{kg}{W}\right)^{1/2}t' + b + \frac{W}{k}. \qquad (2.46)$$

The constants A and B can be evaluated by considering the initial conditions in a similar fashion as for the preceding example. Therefore, the final expression for y will be

$$y = \left[\left(\frac{W}{k}\right)^2 + \frac{2Wb}{k}\right]^{1/2}\cos\left[\left(\frac{kg}{W}\right)^{1/2}t' - \phi\right] + b + \frac{W}{k}, \quad y > b, \quad (2.47)$$

where the phase angle ϕ is given by

$$\phi = \frac{\pi}{2} + \tan^{-1}\left(\frac{W}{2kb}\right)^{1/2}. \qquad (2.48)$$

The maximum deflection of the spring $\delta = y - b$ occurs when the cosine term in Eq. (2.47) is equal to unity. Therefore,

$$\delta = \frac{W}{k} + \frac{W}{k}\left[1 + \left(\frac{2kb}{W}\right)\right]^{1/2}. \qquad (2.49)$$

The maximum force acting on the spring is

$$F = k\delta = W + W\left[1 + \left(\frac{2bk}{W}\right)\right]^{1/2}. \qquad (2.50)$$

2.6 Strain Energy

The work done by the external forces on a deforming elastic member is transformed into *strain*, or *potential energy*. If y is the distance a member is deformed, then the strain energy is

$$U = \frac{F}{2}y = \frac{F^2}{2k}, \qquad (2.51)$$

where $y = F/k$. In the preceding equation, F can be a force, moment, or torque.

For tension (compression) and torsion, the potential energy is, respectively,

$$U = \frac{F^2 l}{2AE} \qquad (2.52)$$

$$U = \frac{T^2 l}{2GJ}. \qquad (2.53)$$

Let us consider now the element with one side fixed (Fig. 2.7a). The force F places the element in pure shear and the work done is $U = F\delta/2$.

The shear strain is $\gamma = \delta/l = \tau/G = F/AG$. Therefore, the strain energy due to shear is

$$U = \frac{F^2 l}{2AG}. \tag{2.54}$$

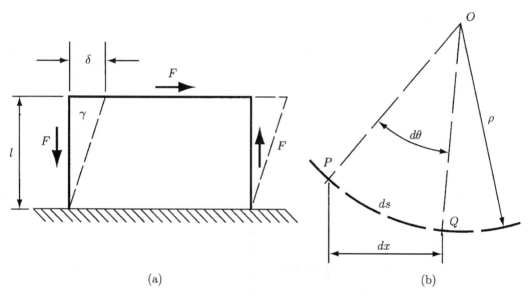

Figure 2.7   *Strain energy. (a) Strain energy due to direct shear; (b) strain energy due to bending.*

The expression for the strain energy due to bending can be developed by considering a section of a beam as shown in Fig. 2.7b. PQ is a section of the elastic curve of length ds and radius of curvature ρ. The strain energy is $dU = (M/2)d\theta$. Since $\rho\,d\theta = ds$, the strain energy becomes

$$dU = \frac{M\,ds}{2\rho}. \tag{2.55}$$

Considering Eq. (2.9), we can eliminate ρ in Eq. (2.55) and obtain

$$dU = \frac{M^2\,ds}{2EI}. \tag{2.56}$$

The strain energy due to bending for the whole beam can be obtained by integrating Eq. (2.56) and considering that $ds \approx dx$ for small deflections of the beam, that is,

$$U = \int \frac{M^2\,dx}{2EI}. \tag{2.57}$$

The strain energy stored in a unit volume u can be obtained by dividing Eqs. (2.52)–(2.54) by the total volume lA:

$$u = \frac{\sigma^2}{2E} \quad \text{(tension and compression)}$$

$$u = \frac{\tau^2}{2G} \quad \text{(direct shear)}$$

$$u = \frac{\tau_{max}^2}{4G} \quad \text{(torsion)}.$$

Even if shear is present and the beam is not very short, Eq. (2.57) still gives good results. The expression of the strain energy due to shear loading of a beam can be approximated by considering Eq. (2.54) multiplied by a correction factor C. The values of C depend upon the shape of the cross-section of the beam. Thus, the strain energy due to shear in bending is

$$U = \int \frac{CV^2 \, dx}{2AG}, \qquad (2.58)$$

where V is the shear force. Table 2.1 lists the values of the correction factor C for various cross-sections.

Table 2.1 *Strain Energy Correction Factors for Shear*

Beam cross-sectional shape	Factor C
Rectangular	1.50
Circular	1.33
Tubular, round	2.00
Box sections	1.00
Structural sections	1.00

Source: Arthur P. Boresi, Omar M. Sidebottom, Fred B. Seely, and James O. Smith, *Advanced Mechanics of Materials*, 3rd ed., p. 173. Wiley, New York, 1978.

EXAMPLE 2.1 Consider a simply supported beam shown of length l and rectangular cross-section (Fig. 2.8). A uniformly distributed load w is applied transversally. Find the strain energy due to shear.

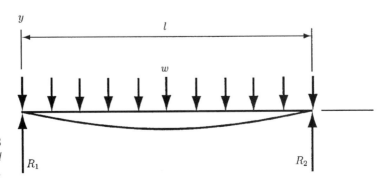

Figure 2.8
Uniformly loaded beam.

Solution

If we consider a cut at an arbitrary distance x from the origin and take only the left part of the beam as a free body, the shear force can be expressed as

$$V = R_1 - wx = \frac{wl}{2} - wx.$$

The strain energy given by Eq. (2.58) with $C = 1.5$ (see Table 2.1) is

$$U = \frac{1.5}{2AG} \int_0^l \left(\frac{wl}{2} - wx \right)^2 dx = \frac{3w^2 l^3}{48AG}. \quad \blacktriangle$$

EXAMPLE 2.2 A concentrated load F is applied to the end of a cantilever beam (Fig. 2.9). Find the strain energy by neglecting the shear.

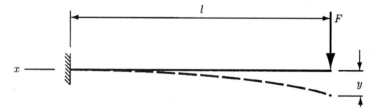

Figure 2.9
Cantilever beam loaded by a concentrated force at the free end.

Solution

The bending moment at any point x along the beam has the expression $M = -Fx$. If we substitute M into Eq. (2.57), the strain energy is

$$U = \int_0^l \frac{F^2 x^2}{2EI} dx = \frac{F^2 l^3}{6EI}. \quad \blacktriangle$$

2.7 Castigliano's Theorem

Castigliano's theorem provides a unique approach to deflection analysis. It states that *when forces act on elastic systems subject to small displacements, the displacement corresponding to any force, collinear with the force, is equal to the partial derivative of the total strain energy with respect to that force* (Ref. 16). Castigliano's theorem can be written as

$$\delta_i = \frac{\partial U}{\partial F_i}, \tag{2.59}$$

where δ_i is the displacement of the point of application of the force F_i in the direction of F_i, and U is the strain energy. For example, applying Castigliano's theorem for the cases of axial and torsional deflections and considering the expressions for the strain energy given by Eqs. (2.52) and (2.53), we obtain, respectively,

$$\delta = \frac{\partial}{\partial F}\left(\frac{F^2 l}{2AE}\right) = \frac{Fl}{AE} \tag{2.60}$$

$$\theta = \frac{\partial}{\partial T}\left(\frac{T^2 l}{2GJ}\right) = \frac{Tl}{GJ}. \tag{2.61}$$

Even if no force or moment acts at a point where we want to determine the deflection, Castigliano's theorem can still be used. In such a case, to apply Castigliano's theorem, first we need to consider a fictitious (dummy) force or moment Q_i at the point of interest and develop the expression for the strain energy including the energy due to that dummy force and moment. Then, we find the expression for the deflection by using Eq. (2.61) where the differentiation will be performed with respect to the dummy force or moment Q_i, that is, $\delta_i = \partial U / \partial Q_i$. Finally, since Q_i is a fictitious force or moment, we will set $Q_i = 0$ in the expression for δ_i.

EXAMPLE 2.3 A cantilever of length l is loaded by a transversal force F at a distance a as shown in Fig. 2.10. Find the maximum deflection of the cantilever if shear is neglected.

Figure 2.10
Castigliano's theorem applied for a cantilever beam.

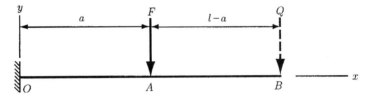

Solution

The maximum deflection of the cantilever will be at its free end. Therefore, to apply Castigliano's theorem, we consider a fictitious force Q at that point. Since we will apply Eq. (2.57) to develop the expression for the strain energy, we need to find the expression for the bending moment M. Hence, the bending moments corresponding to the segments OA and AB are, respectively,

$$M_{OA} = F(x - a) + Q(x - l)$$
$$M_{AB} = Q(x - l).$$

The total strain energy is obtained:

$$U = \int_0^a \frac{M_{OA}^2}{2EI}\, dx + \int_a^l \frac{M_{AB}^2}{2EI}\, dx.$$

When Castigliano's theorem is applied, the deflection is

$$y = \frac{\partial U}{\partial Q} = \frac{1}{2EI}\left[\int_0^a 2M_{OA}\frac{\partial M_{OA}}{\partial Q}\,dx + \int_a^l 2M_{AB}\frac{\partial M_{AB}}{\partial Q}\,dx\right].$$

Since

$$\frac{\partial M_{OA}}{\partial Q} = \frac{\partial M_{AB}}{\partial Q} = x - l,$$

the expression for the deflection becomes

$$y = \frac{F}{EI}\left\{\int_0^a [F(x-a) + Q(x-l)](x-l)\,dx + \int_a^l [Q(x-l)](x-l)\,dx\right\}.$$

Since Q is a dummy force, setting $Q = 0$ in the preceding equation gives

$$y = \frac{F}{EI}\int_0^a (x-a)(x-l)\,dx = \frac{a^2(3l-a)}{6EI}. \quad \blacktriangle$$

2.8 Compression

The analysis and design of *compression members* depends upon whether these members are loaded in tension or in torsion. The term *column* is applied to those members for which failure is not produced because of pure compression. It is convenient to classify the columns according to their length and to whether the loading is central or eccentric. Thus, for example, there are long columns with central loading, intermediate-length columns with central loading, columns with eccentric loading, or short columns with eccentric loading. The problem of compression members is to find the *critical load* that produces the failure of the member.

2.9 Long Columns with Central Loading

Figure 2.11 shows long columns of length l having applied an axial load P and various *end conditions*. The load P is applied along the vertical symmetry axis of the column. The end conditions shown in Fig. 2.11 are rounded–rounded ends (Fig. 2.11a), fixed–fixed ends (Fig. 2.11b), free–fixed ends (Fig. 2.11c), and rounded–fixed ends (Fig. 2.11d).

To develop the relationship between the critical load P_{cr} and the column material and geometry, first let us consider the situation shown in Fig. 2.11a. The figure shows that the bar is bent in the positive y direction and, thus, a negative moment is required:

$$M = -Py. \tag{2.62}$$

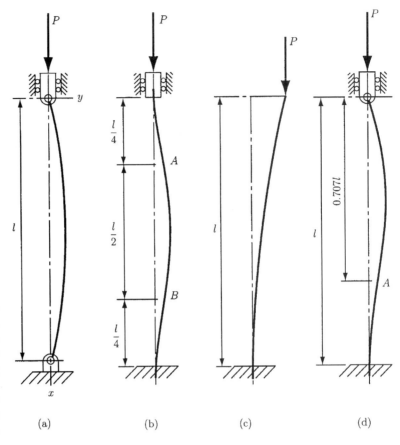

Figure 2.11
Loading columns. (a) Column with both ends pivoted; (b) column with both ends fixed; (c) column with one free end and one fixed end; (d) column with one pivoted and guided end and one fixed end. Used with permission from Ref. 16.

(a)   (b)   (c)   (d)

Conversely, if the bar is bent in the negative y direction, a positive moment would is needed, and also $M = -Py$ because of $-y$. Using Eqs. (2.17) and (2.62), we obtain

$$\frac{d^2y}{dx^2} = -\frac{P}{EI}y, \tag{2.63}$$

or

$$\frac{d^2y}{dx^2} + \frac{P}{EI}y = 0. \tag{2.64}$$

The solution of the preceding equation is

$$y = A \sin\sqrt{\frac{P}{EI}}x + B \cos\sqrt{\frac{P}{EI}}x, \tag{2.65}$$

where A and B are constants of integration that can be determined by considering the boundary conditions $y = 0$ at $x = 0$ and $y = 0$ at $x = l$. Substituting the two boundary conditions in Eq. (2.65), we obtain $B = 0$ and

$$0 = A \sin\sqrt{\frac{P}{EI}}l. \tag{2.66}$$

If we consider $A = 0$ into the foregoing equation, we obtain the trivial solution of no *buckling*. If $A \neq 0$, then

$$\sin \sqrt{\frac{P}{EI}} \, l = 0, \qquad (2.67)$$

which is satisfied if $\sqrt{P/EI} \, l = n\pi$, where $n = 1, 2, 3, \ldots$. The critical load associated with $n = 1$ is called the *first critical load* and is given by

$$P_{cr} = \frac{\pi^2 EI}{l^2}. \qquad (2.68)$$

This equation is called *Euler column formula* and is applied only for rounded-ends columns. Substituting Eq. (2.68) into Eq. (2.65), we find the equation of the deflection curve:

$$y = A \sin \frac{\pi x}{l}. \qquad (2.69)$$

This equation emphasizes that the deflection curve is a half-wave sine.

We observe that the minimum critical load occurs for $n = 1$. Values of n greater than 1 lead to deflection curves that cross the vertical axis at least once. The intersections of these curves with the vertical axis occur at the points of inflection of the curve, and the shape of the deflection curve is composed of several half-wave sines.

Consider the relation $I = Ak^2$ for the second moment of area I, where A is the cross-sectional area and k the radius of gyration. Equation (2.68) can be rewritten as

$$\frac{P_{cr}}{A} = \frac{\pi^2 E}{(l/k)^2}, \qquad (2.70)$$

where the ratio l/k is called the *slenderness ratio* and P_{cr}/A the *critical unit load*. The critical unit load is the load per unit area that can place the column in *unstable equilibrium*. Equation (2.70) shows that the critical unit load depends only upon the modulus of elasticity and the slenderness ratio.

Figure 2.11b depicts a column with both ends fixed. The inflection points are at A and B located at a distance $l/4$ from the ends. Comparing Figs. 2.11a and 2.11b, we can notice that AB is the same curve as for the column with rounded ends. Hence, we can substitute the length l by $l/2$ in Eq. (2.68) and obtain the expression for the first critical load:

$$P_{cr} = \frac{\pi^2 EI}{(l/2)^2} = \frac{4\pi^2 EI}{l^2}. \qquad (2.71)$$

Figure 2.11c shows a column with one end free and the other one fixed. Comparing Figs. 2.11a and 2.11c, we observe that the curve of the free–fixed ends column is equivalent to half of the curve for columns with rounded ends. Therefore, if $2l$ is substituted in Eq. (2.68) for l, the critical load for this case is obtained:

$$P_{cr} = \frac{\pi^2 EI}{(2l)^2} = \frac{\pi^2 EI}{4l^2}. \tag{2.72}$$

Figure 2.11d shows a column with one end fixed and the other one rounded. The inflection point is the point A located at a distance of $0.707l$ from the rounded end. Therefore,

$$P_{cr} = \frac{\pi^2 EI}{(0.707l)^2} = \frac{2\pi^2 EI}{l^2}. \tag{2.73}$$

The preceding situations can be summarized by writing the Euler equation in the forms

$$P_{cr} = \frac{C\pi^2 EI}{l^2}, \quad \frac{P_{cr}}{A} = \frac{C\pi^2 E}{(l/k)^2}, \tag{2.74}$$

where C is called the *end-condition constant*. It can have one of the values listed in Table 2.2.

Table 2.2 *End-Condition Constants for Euler Columns*

Column end conditions	End-condition constant C		
	Theoretical value	**Conservative value**	**Recommended value**[a]
Fixed–free	1/4	1/4	1/4
Rounded–rounded	1	1	1
Fixed–rounded	2	1	1.2
Fixed–fixed	4	1	1.2

[a]To be used only with liberal factors of safety when the column load is accurately known.
Source: Joseph E. Shigley and Charles R. Mischke, *Mechanical Engineering Design*, 5th ed., p. 123. McGraw-Hill, New York, 1989. Used with permission.

Figure 2.12 plots the unit load P_{cr}/A as a function of the slenderness ratio l/k. The curve PQR is obtained. In this figure, the quantity S_y that corresponds to point Q represents the yield strength of the material. Thus, one would consider that any compression member having an l/k value less than $(l/k)_Q$ should be treated as a pure compression member, whereas all others can be treated as Euler columns. In practice, this fact is not true. Several tests showed the failure of columns with a slenderness ratio below or very close to point Q. For this reason, neither simple compression methods nor the Euler column equation should be used when the slenderness ratio is near $(l/k)_Q$. The solution in this case is to consider a point T on the Euler curve of Fig. 2.12 such that, if the slenderness ratio corresponding to T is $(l/k)_1$, the Euler equation should be used only when the actual slenderness ratio of the

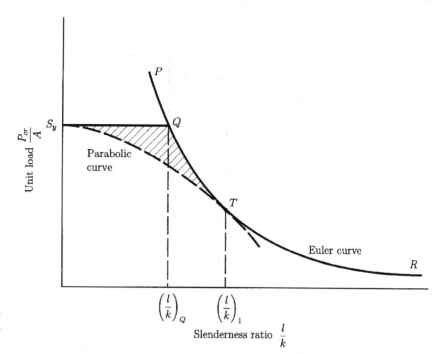

Mechanics

Figure 2.12
Euler's curve.
Used with
permission from
Ref. 16.

column is greater than $(l/k)_1$. Point T can be selected such that $P_{cr}/A = S_y/2$. From Eq. (2.74), the slenderness ratio $(l/k)_1$ is obtained:

$$\left(\frac{l}{k}\right)_1 = \left(\frac{2\pi^2 CE}{S_y}\right)^{1/2}. \tag{2.75}$$

2.10 Intermediate-Length Columns with Central Loading

When the actual slenderness ratio l/k is less than $(l/k)_1$, and so is in the region in Fig. 2.12 where the Euler formula is not suitable, one can use the *parabolic* or *J. B. Johnson formula* of the form

$$\frac{P_{cr}}{A} = a - b\left(\frac{l}{k}\right)^2, \tag{2.76}$$

where a and b are constants that can be obtained by fitting a parabola (the dashed line tangent at T) to the Euler curve in Fig. 2.12. Thus, we find

$$a = S_y \tag{2.77}$$

and

$$b = \left(\frac{S_y}{2\pi}\right)^2 \frac{1}{CE}. \tag{2.78}$$

Substituting Eqs. (2.77) and (2.78) into Eq. (2.76) yields

$$\frac{P_{cr}}{A} = S_y - \left(\frac{S_y}{2\pi}\frac{l}{k}\right)^2 \frac{1}{CE}, \tag{2.79}$$

which can be applied if $\dfrac{l}{k} \leq \left(\dfrac{l}{k}\right)_1$.

2.11 Columns with Eccentric Loading

Figure 2.13a shows a column acted upon by a force P that is applied at a distance e, also called eccentricity, from the centroidal axis of the column. To solve this problem, we consider the free-body diagram in Fig. 2.13b. Equating the sum of moments about the origin O to zero gives

$$\sum M_O = M + Pe + Py = 0. \tag{2.80}$$

Substituting M from Eq. (2.80) into Eq. (2.17) gives a nonhomogeneous second-order differential equation,

$$\frac{d^2y}{dx^2} + \frac{P}{EI}y = -\frac{Pe}{EI}. \tag{2.81}$$

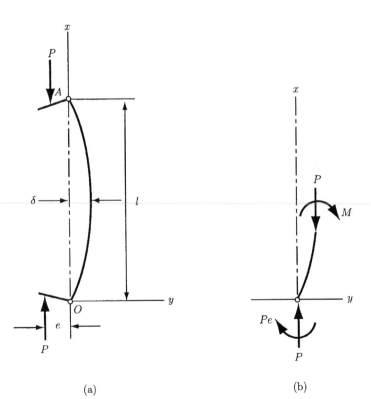

Figure 2.13
(a) Eccentric loaded column; (b) free-body diagram.

(a)

(b)

Considering the boundary conditions

$$x = 0, \quad y = 0$$

$$x = \frac{l}{2}, \quad \frac{dy}{dx} = 0,$$

and substituting $x = l/2$ in the resulting solution, we obtain the maximum deflection δ and the maximum bending moment M_{max}:

$$\delta = e\left[\sec\left(\frac{1}{2}\sqrt{\frac{P}{EI}}\right) - 1\right] \tag{2.82}$$

$$M_{max} = -P(e + \delta) = -Pe\sec\left(\frac{1}{2}\sqrt{\frac{P}{EI}}\right). \tag{2.83}$$

At $x = l/2$, the compressive stress σ_c is maximum and can be calculated by adding the axial component produced by the load P and the bending component produced by the bending moment M_{max}, that is,

$$\sigma_c = \frac{P}{A} - \frac{Mc}{I} = \frac{P}{A} - \frac{Mc}{Ak^2}. \tag{2.84}$$

Substituting Eq. (2.83) into the preceding equation yields

$$\sigma_c = \frac{P}{A}\left[1 + \frac{ec}{k^2}\sec\left(\frac{1}{2k}\sqrt{\frac{P}{EA}}\right)\right]. \tag{2.85}$$

Considering the yield strength S_y of the column material as σ_c and manipulating Eq. (2.85) gives

$$\frac{P}{A} = \frac{S_{yc}}{1 + (ec/k^2)\sec[(l/2k)\sqrt{P/AE}]}. \tag{2.86}$$

The preceding equation is called the *secant column formula*, and the term ec/k^2 the *eccentricity ratio*. Since Eq. (2.86) cannot be solved explicitly for the load P, root-finding techniques using numerical methods can be applied.

2.12 Short Compression Members

A short compression member is illustrated in Fig. 2.14. At point D, the compressive stress in the x direction has two component, namely, one due to the axial load P that is equal to P/A and another due to the bending moment that is equal to My/I. Therefore,

$$\sigma_c = \frac{P}{A} + \frac{My}{I} = \frac{P}{A} + \frac{PeyA}{IA} = \frac{P}{A}\left(1 + \frac{ey}{k^2}\right), \tag{2.87}$$

where $k = (I/A)^{1/2}$ is the radius of gyration, y the coordinate of point D, and e the eccentricity of loading. Setting the foregoing equation equal to zero and

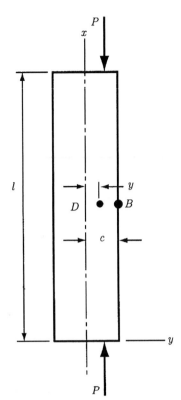

Figure 2.14
*Short compres-
sion member.*

solving, we obtain the y coordinate of a line parallel to the x axis along which the normal stress is zero:

$$y = -\frac{k^2}{e}.$$ (2.88)

If $y = c$, that is, at point B in Fig. 2.14, we obtain the largest compressive stress. Hence, substituting $y = c$ in Eq. (2.87) gives

$$\sigma_c = \frac{P}{A}\left(1 + \frac{ec}{k^2}\right).$$ (2.89)

For design or analysis, the preceding equation can be used only if the range of lengths for which the equation is valid is known. For a strut, it is desired that the effect of bending deflection be within a certain small percentage of eccentricity. If the limiting percentage is 1% of e, then the slenderness ratio is bounded by

$$\left(\frac{1}{k}\right)_2 = 0.282\left(\frac{AE}{P_{cr}}\right)^{1/2}.$$ (2.90)

Therefore, the limiting slenderness ratio for using Eq. (2.89) is given by Eq. (2.90).

3. Fatigue

A periodic stress oscillating between some limits applied to a machine member is called *repeated, alternating,* or *fluctuating.* The machine members often fail under the action of these stresses, and the failure is called *fatigue failure.* Generally, a small crack is enough to initiate fatigue failure. Since the stress concentration effect becomes greater around it, the crack progresses rapidly. We know that if the stressed area decreases in size, the stress increases in magnitude. Therefore, if the remaining area is small, the member can fail suddenly. A member failed because of fatigue shows two distinct regions. The first region is due to the progressive development of the crack; the other is due to the sudden fracture.

3.1 Endurance Limit

The strength of materials acted upon by fatigue loads can be determined by performing a fatigue test provided by R. R. Moore's high-speed rotating beam machine. During the test, the specimen is subjected to pure bending by using weights and rotated with constant velocity. For a particular magnitude of the weights, one records the number of revolutions at which the specimen fails. Then, a second test is performed for a specimen identical with the first one, but the magnitude of the weight is reduced. Again, the number of revolutions at which the fatigue failure occurs is recorded. The process is repeated several times. Finally, the fatigue strengths considered for each test are plotted against the corresponding number of revolutions. The resulting chart is called the *S–N diagram.*

Numerous tests have established that the ferrous materials have an *endurance limit* defined as the highest level of alternating stress that can be withstood indefinitely by a test specimen without failure. The symbol for endurance limit is S_e'. The endurance limit can be related to the tensile strength through some relationships. For example, for steel, Mischke[1] predicted the following relationships

$$S_e' = \begin{cases} 0.504 S_{ut}, & S_{ut} \le 200 \text{ kpsi } (1400 \text{ MPa}) \\ 100 \text{ kpsi}, & S_{ut} > 200 \text{ kpsi} \\ 700 \text{ MPa}, & S_{ut} > 1400 \text{ MPa}, \end{cases} \qquad (3.1)$$

where S_{ut} is the minimum tensile strength. Table 3.1 lists the values of the endurance limit for various classes of cast iron. The symbol S_e' refers to the endurance limit of the test specimen that can be significantly different from the endurance limit S_e of any machine element subjected to any kind of loads. The endurance limit S_e' can be affected by several factors called *modifying factors.* Some of these factors are the surface factor k_a, the size

[1] C. R. Mischke, "Prediction of stochastic endurance strength," *Trans. ASME, J. Vibration, Acoustics, Stress, and Reliability in Design* **109**(1), 113–122 (1987).

Table 3.1 *Typical Properties of Gray Cast Iron*

| ASTM number | Tensile strength S_{ut} (kpsi) | Compressive strength S_{uc} (kpsi) | Shear modulus of rupture S_{su} (kpsi) | Modulus of elasticity (Mpsi) | | Endurance limit S_e (kpsi) | Brinell hardness H_B | Fatigue stress concentration factor K_f |
				tension	torsion			
20	22	83	26	9.6–14	3.9–5.6	10	156	1.00
25	26	97	32	11.5–14.8	4.6–6.0	11.5	174	1.05
30	31	109	40	13–16.4	5.2–6.6	14	201	1.10
35	36.5	124	48.5	14.5–17.2	5.8–6.9	16	212	1.15
40	42.5	140	57	16–20	6.4–7.8	18.5	235	1.25
50	52.5	164	73	18.8–22.8	7.2–8.0	21.5	262	1.35
50	62.5	187.5	88.5	20.4–23.5	7.8–8.5	24.5	302	1.50

Source: Joseph E. Shigley and Charles R. Mischke, *Mechanical Engineering Design*, 5th ed., p. 123. McGraw-Hill, New York, 1989. Used with permission.

Table 3.2 *Generalized Fatigue Strength Factors for Ductile Materials*

	Bending	**Axial**	**Torsion**
a. Endurance limit			
$S_e = k_a k_b k_c S'_e$, where S'_e is the specimen endurance limit			
k_c (load factor)	1	1	0.58
k_b (gradient factor)			
diameter $<$ (0.4 in or 10 mm)	1	0.7–0.9	1
(0.4 in or 10 mm) $<$ diameter $<$ (2 in or 50 mm)	0.9	0.7–0.9	0.9
k_a (surface factor)		See Fig. 3.5	
b. 10^3-cycle strength	$0.9S_u$	$0.75S_u$	$0.9S_{us}$[a]

[a]$S_{us} \approx 0.8S_u$ for steel; $S_{us} \approx 0.7S_u$ for other ductile materials.
Source: R. C. Juvinall and K. M. Marshek, *Fundamentals of Machine Component Design.* John Wiley & Sons, New York, 1991. Used with permission.

factor k_b, or the load factor k_c. Thus, the endurance limit of a member can be related to the endurance limit of the test specimen by

$$S_e = k_a \, k_b \, k_c \, S'_e. \tag{3.2}$$

Some values of the foregoing factors for bending, axial loading, and torsion are listed in Table 3.2.

3.1.1 SURFACE FACTOR k_a

The influence of the surface of the specimen is described by the modification factor k_a, which depends upon the quality of the finishing. The following formula describes the surface factor:

$$k_a = a S_{ut}^b. \tag{3.3}$$

S_{ut} is the tensile strength. Some values for a and b are listed in Table 3.3.

Table 3.3 *Surface Finish Factor*

	Factor a		**Exponent b**
Surface finish	**kpsi**	**MPa**	
Ground	1.34	1.58	-0.085
Machined or cold-drawn	2.70	4.51	-0.256
Hot-rolled	14.4	57.7	-0.718
As forged	39.9	272.0	-0.995

Source: J. E. Shigley and C. R. Mischke, *Mechanical Engineering Design.* McGraw-Hill, New York, 1989. Used with permission.

Mechanics

3.1.2 SIZE FACTOR k_b

The results of the tests performed to evaluate the size factor in the case of bending and torsion loading of a bar, for example, can be expressed as

$$k_b = \begin{cases} \left(\dfrac{d}{0.3}\right)^{-0.1133} \text{in}, & 0.11 \leq d \leq 2 \text{ in} \\ \left(\dfrac{d}{7.62}\right)^{-0.1133} \text{mm}, & 2.79 \leq d \leq 51 \text{ mm}, \end{cases} \tag{3.4}$$

where d is the diameter of the test bar. For larger sizes, the size factor varies from 0.06 to 0.075. The tests also revealed that there is no size effect for axial loading; thus, $k_b = 1$.

To apply Eq. (3.4) for a nonrotating round bar in bending or for a noncircular cross section, we need to define the *effective dimension* d_e. This dimension is obtained by considering the volume of material stressed at and above 95% of the maximum stress and a similar volume in the rotating beam specimen. When these two volumes are equated, the lengths cancel and only the areas have to be considered. For example, if we consider a rotating round section (Fig. 3.1a) or a rotating hollow round, the 95% stress area is a ring having the outside diameter d and the inside diameter $0.95d$. Hence, the 95% stress area is

$$A_{0.95\sigma} = \frac{\pi}{4}[d^2 - (0.95d)^2] = 0.0766d^2. \tag{3.5}$$

If the solid or hollow rounds do not rotate, the 95% stress area is twice the area outside two parallel chords having a spacing of $0.95D$, where D is the diameter. Therefore, the 95% stress area in this case is

$$A_{0.95\sigma} = 0.0105D^2. \tag{3.6}$$

Setting Eq. (3.5) equal to Eq. (3.6) and solving for d, we obtain the effective diameter

$$d_e = 0.370D, \tag{3.7}$$

which is the effective size of the round corresponding to a nonrotating solid or hollow round.

A rectangular section shown in Fig. 3.1b has $A_{0.95\sigma} = 0.05hb$ and

$$d_e = 0.808(hb)^{1/2}. \tag{3.8}$$

For a channel section,

$$A_{0.95\sigma} = \begin{cases} 0.5ab, & \text{axis 1-1}, \\ 0.052xa + 0.1t_f(b - x), & \text{axis 2-2}, \end{cases} \tag{3.9}$$

where a, b, x, t_f are the dimensions of the channel section as depicted in Fig. 3.1c.

The 95% area for an I-beam section is (Fig. 3.1d)

$$A_{0.95\sigma} = \begin{cases} 0.10at_f, & \text{axis 1-1}, \\ 0.05ba, & t_f > 0.025a, & \text{axis 2-2}. \end{cases} \tag{3.10}$$

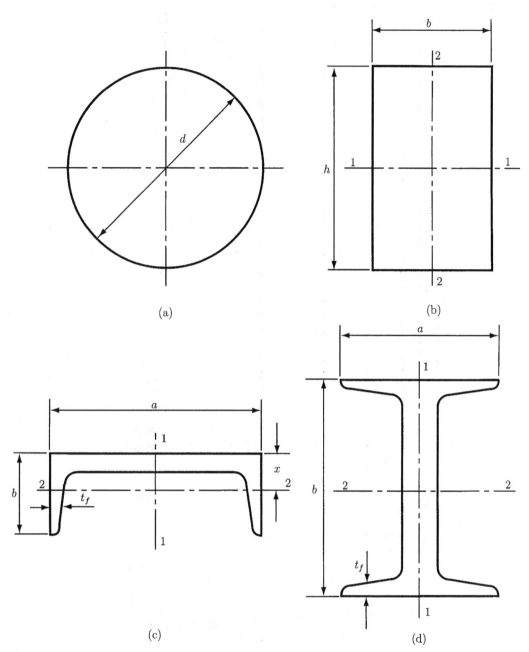

Figure 3.1 *Beam cross-sections. (a) Solid round; (b) rectangular section; (c) channel section; (d) web section. Used with permission from Ref. 16.*

3.1.3 LOAD FACTOR k_c

Tests revealed that the load factor has the following values:

$$
k_c = \begin{cases}
0.923, & \text{axial loading,} & S_{ut} \leq 220 \text{ kpsi (1520 MPa),} \\
1, & \text{axial loading,} & S_{ut} > 220 \text{ kpsi (1520 MPa),} \\
1, & \text{bending,} \\
0.577, & \text{torsion and shear.}
\end{cases}
\tag{3.11}
$$

3.2 Fluctuating Stresses

In design problems, it is frequently necessary to determine the stress of parts corresponding to the situation when the stress fluctuates without passing through zero (Fig. 3.2). A *fluctuating stress* is a combination of a static plus a completely reversed stress. The components of the stresses are depicted in Fig. 3.2, where σ_{min} is minimum stress, σ_{max} the maximum stress, σ_a the stress amplitude or the alternating stress, σ_m the midrange or the mean stress, σ_r the stress range, and σ_s the steady or static stress. The steady stress can have any value between σ_{min} and σ_{max} and exists because of a fixed load. It is usually independent of the varying portion of the load. The following relations between the stress components are useful:

$$\sigma_m = \frac{\sigma_{max} + \sigma_{min}}{2} \qquad (3.12)$$

$$\sigma_a = \frac{\sigma_{max} - \sigma_{min}}{2}. \qquad (3.13)$$

The stress ratios

$$R = \frac{\sigma_{min}}{\sigma_{max}}$$

$$A = \frac{\sigma_a}{\sigma_m} \qquad (3.14)$$

are also used to describe the fluctuating stresses.

3.3 Constant Life Fatigue Diagram

Figure 3.3 illustrates the graphical representation of various combinations of mean and alternating stress. This diagram is called the *constant life fatigue diagram* because it has lines corresponding to a constant 10^6-cycle or "infinite" life. The horizontal axis ($\sigma_a = 0$) corresponds to static loading. Yield and tensile strength are represented by points A and B, while the compressive yield strength $-S_y$ is at point A'. If $\sigma_m = 0$ and $\sigma_a = S_y$ (point A''), the stress fluctuates between $+S_y$ and $-S_y$. Line AA'' corresponds to fluctuations having a tensile peak of S_y, and line $A'A''$ corresponds to compressive peaks of $-S_y$. Points C, D, E, and F correspond to $\sigma_m = 0$ for various values of fatigue life, and lines CB, DB, EB, and FB are the estimated lines of constant life (from the S–N diagram). Since Goodman developed this empirical procedure to obtain constant life lines, these lines are called the *Goodman lines*.

From Fig. 3.3, we observe that area $A''HCGA$ corresponds to a life of at least 10^6 cycles and no yielding. Area $HCGA''H$ corresponds to less than 10^6 cycles of life and no yielding. Area AGB along with area $A'HCGA$ corresponds to 10^6 cycles of life when yielding is acceptable.

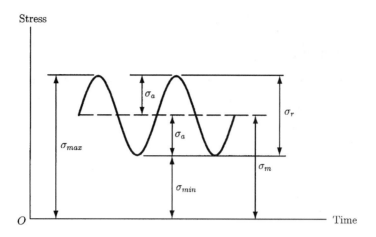

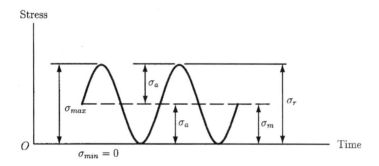

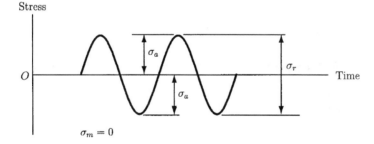

Figure 3.2
Sinusoidal fluctuating stress.

EXAMPLE 3.1 (*Source:* R. C. Juvinall and K. M. Marshek, *Fundamentals of Machine Component Design.* John Wiley & Sons, New York, 1991.)

Estimate the *S–N* curve and a family of constant life fatigue curves pertaining to the axial loading of precision steel parts having $S_u = 120$ ksi, $S_y = 100$ ksi (Fig. 3.4). All cross-sectional dimensions are under 2 in.

Solution

According to Table 3.2, the gradient factor $k_b = 0.9$. The 10^3-cycle peak alternating strength for axially loaded material is $S = 0.75 S_u = 0.75(120) = 90$ ksi. The 10^6-cycle peak alternating strength for axially loaded ductile material is $S_e = k_a k_b k_c S'_e$, where $S'_e = (0.5)(120) = 60$ ksi from Eq. (3.1),

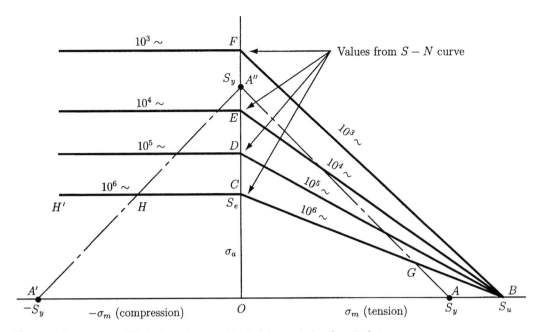

Figure 3.3   *Constant life fatigue diagram. Used with permission from Ref. 9.*

$k_c = 1$, and $k_a = 0.9$ from Fig. 3.5. The endurance limit is $S_e = 48.6$ ksi. The estimated S–N curve is plotted in Fig. 3.6. From the estimated S–N curve, the peak alternating strengths at 10^4 and 10^5 cycles are, respectively, 76.2 and 62.4 ksi. The σ_m–σ_a curves for 10^3, 10^4, 10^5, and 10^6 cycles of life are given in Fig. 3.6.   ▲

(a)

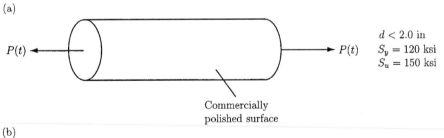

(b)

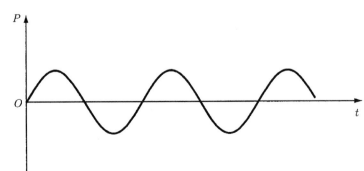

Figure 3.4   *Axial loading cylinder. (a) Loading diagram; (b) fluctuating load.*

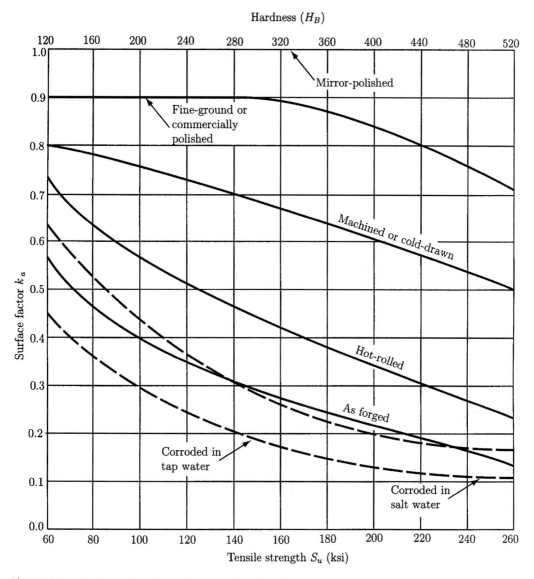

Figure 3.5 *Surface factor. Used with permission from Ref. 9.*

3.4 Fatigue Life for Randomly Varying Loads

For the most mechanical parts acted upon by randomly varying stresses, the prediction of fatigue life is not an easy task. The procedure for dealing with this situation is often called the *linear cumulative damage rule*. The idea is as follows: If a part is cyclically loaded at a stress level causing failure in 10^5 cycles, then each cycle of that loading consumes one part in 10^5 of the life of the part. If other stress cycles are interposed corresponding to a life of 10^4 cycles, each of these consumes one part in 10^4 of the life, and so on. Fatigue failure is predicted when 100% of the life has been consumed.

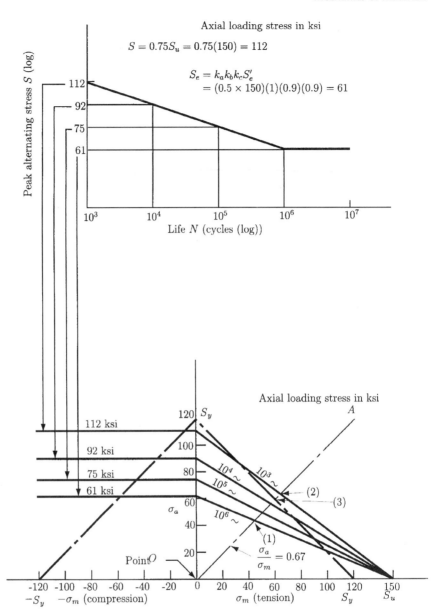

Figure 3.6
*Life diagram.
Used with
permission from
Ref. 9.*

The linear cumulative damage rule is expressed by

$$\frac{n_1}{N_1} + \frac{n_2}{N_2} + \cdots + \frac{n_k}{N_k} = 1 \quad \text{or} \quad \sum_{j=1}^{j=k} \frac{n_j}{N_j} = 1, \tag{3.15}$$

where $n_1, n_2, \ldots, n_k$ represent the number of cycles at specific overstress levels and $N_1, N_2, \ldots, N_k$ represent the life (in cycles) at these overstress levels, as taken from the appropriate S–N curve. Fatigue failure is predicted when the above equation holds.

EXAMPLE 3.2 (*Source:* R. C. Juvinall and K. M. Marshek, *Fundamentals of Machine Component Design*, John Wiley & Sons, New York, 1991.)

The stress fluctuation of a part during 6 s of operation is shown in Fig. 3.7a. The part has $S_u = 500$ MPa, and $S_y = 400$ MPa. The $S-N$ curve for bending is given in Fig. 3.7c. Estimate the life of the part.

Solution

The 6-s test period includes, in order, two cycles of fluctuation a, three cycles of fluctuation b, and two cycles of c. Each of these fluctuations corresponds to a point in Fig. 3.7b. For a the stresses are $\sigma_m = 50$ MPa, $\sigma_a = 100$ MPa.

Points (a), (b), (c) in Fig. 3.7b are connected to the point $\sigma_m = S_u$, which gives a family of four "Goodman lines" corresponding to some constant life.

The Goodman lines intercept the vertical axis at points a' through c'. Points a through d correspond to the same fatigue lives as points a' through d'. These lives are determined from the $S-N$ curve in Fig. 3.7c. The life for a and a' can be considered infinite.

Adding the portions of life cycles b and c gives

$$\frac{n_b}{N_b} + \frac{n_c}{N_c} = \frac{3}{3 \times 10^6} + \frac{2}{2 \times 10^4} = 0.000011.$$

This means that the estimated life corresponds to 1/0.000011, or 90,909 periods of 6-s duration. This is equivalent to 151.5 hr. ▲

3.5 Criteria of Failure

There are various techniques for plotting the results of the fatigue failure test of a member subjected to fluctuating stress. One of them is called the *modified Goodman diagram* and is shown in Fig. 3.8. For this diagram the mean stress is plotted on the abscissa and the other stress components on the ordinate. As shown in the figure, the mean stress line forms a 45° angle with the abscissa. The resulting line drawn to S_e above and below the origin are actually the modified Goodman diagram. The yield strength S_y is also plotted on both axes, since yielding can be considered as a criterion of failure if $\sigma_{max} > S_y$.

Four other criteria of failure are shown in the diagram in Fig. 3.9, that is, Soderberg, the modified Goodman, Gerber, and yielding. The fatigue limit S_e (or the finite life strength S_f) and the alternating stress S_a are plotted on the ordinate. The yield strength S_{yt} is plotted on both coordinate axes and the tensile strength S_{ut} and the mean stress S_m on the abscissa. As we can observe from Fig. 3.9, only the Soderberg criterion guards against yielding.

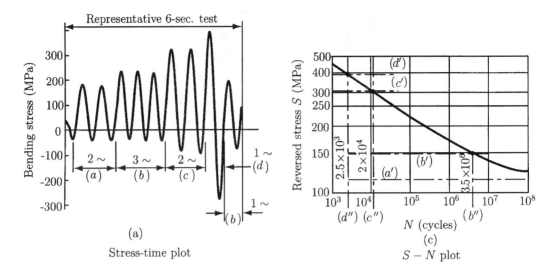

(a)

Stress-time plot

(c)

$S - N$ plot

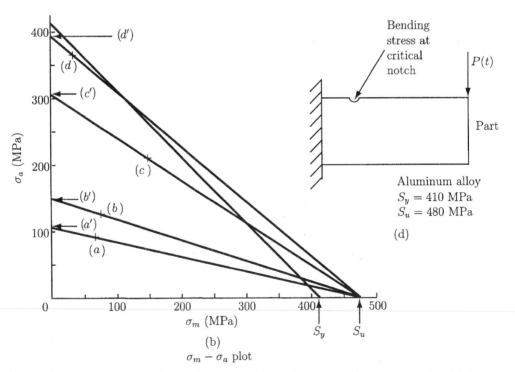

(b)

$\sigma_m - \sigma_a$ plot

(d)

Aluminum alloy
$S_y = 410$ MPa
$S_u = 480$ MPa

Figure 3.7 *Fatigue analysis of a cantilever beam. (a) Bending stress; (b) stress fluctuation; (c) life diagram; (d) loading diagram. Used with permission from Ref. 9.*

We can describe the linear criteria shown in Fig. 3.9, namely Soderberg, Goodman, and yield, by the equation of a straight line of general form

$$\frac{x}{a} + \frac{y}{b} = 1. \tag{3.16}$$

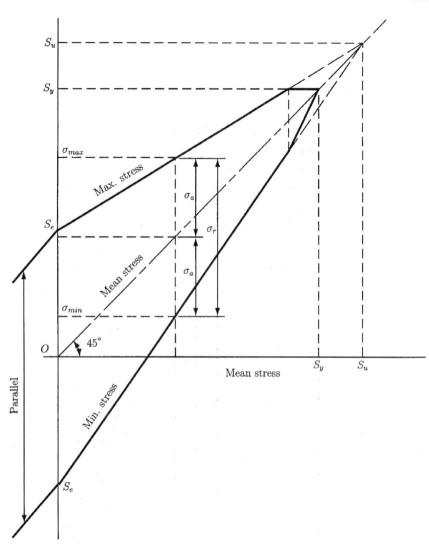

Figure 3.8
Goodman diagram. Used with permission from Ref. 16.

In this equation, a and b are the coordinates of the points of intersection of the straight line with the x and y axes, respectively. For example, the equation for the Soderberg line is

$$\frac{S_a}{S_e} + \frac{S_m}{S_{yt}} = 1. \tag{3.17}$$

Similarly, the modified Goodman relation is

$$\frac{S_a}{S_e} + \frac{S_m}{S_{ut}} = 1. \tag{3.18}$$

The yielding line is described by the equation

$$\frac{S_a}{S_y} + \frac{S_m}{S_{yt}} = 1. \tag{3.19}$$

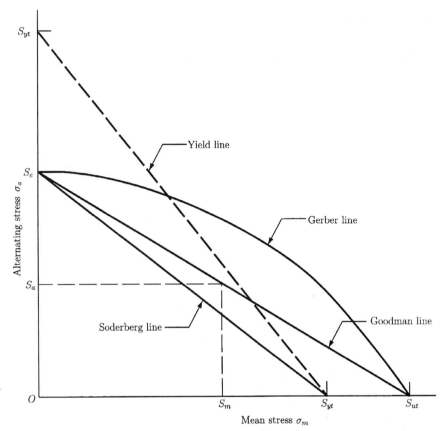

Figure 3.9
Various criteria of failure. Used with permission from Ref. 16.

The curve representing the Gerber theory is a better predictor since it passes through the central region of the failure points. The Gerber criterion is also called the *Gerber parabolic relation* because the curve can be modeled by a parabolic equation of the form

$$\frac{S_a}{S_e} + \left(\frac{S_m}{S_{yt}}\right)^2 = 1. \tag{3.20}$$

If each strength in Eqs. (3.17) to (3.20) is divided by a factor of safety n, the stresses σ_a and σ_m can replace S_a and S_m. Therefore, the Soderberg equation becomes

$$\frac{\sigma_a}{S_e} + \frac{\sigma_m}{S_y} = \frac{1}{n}, \tag{3.21}$$

the modified Goodman equation becomes

$$\frac{\sigma_a}{S_e} + \frac{\sigma_m}{S_{ut}} = \frac{1}{n}, \tag{3.22}$$

and the Gerber equation becomes

$$\frac{n\sigma_a}{S_e} + \left(\frac{n\sigma_m}{S_{ut}}\right)^2 = 1. \tag{3.23}$$

Figure 3.10 illustrates the Goodman line and the way in which the Goodman equation can be used in practice. Given an arbitrary point A of coordinates σ_m, σ_a as shown in the figure, we can draw a safe stress line through A parallel to the modified Goodman line. The safe stress line is the locus of all points of coordinates σ_M, σ_m for which the same factor of safety n is considered, that is, $S_m = n\sigma_m$ and $S_a = n\sigma_a$.

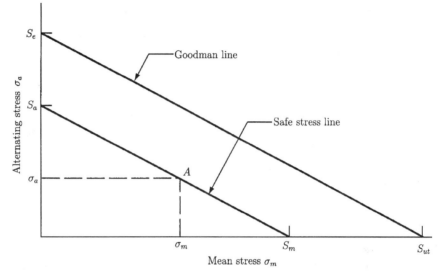

Figure 3.10
*Safe stress line.
Used with
permission from
Ref. 16.*

References

1. J. S. Arora, *Introduction to Optimum Design.* McGraw-Hill, New York, 1989.

2. F. P. Beer and E. R. Johnston, Jr., *Mechanics of Materials.* McGraw-Hill, New York, 1992.

3. K. S. Edwards, Jr., and R. B. McKee, *Fundamentals of Mechanical Component Design.* McGraw-Hill, New York, 1991.

4. A. Ertas and J. C. Jones, *The Engineering Design Process.* John Wiley & Sons, New York, 1996.

5. A. S. Hall, Jr., A. R. Holowenko, and H. G. Laughlin, *Theory and Problems of Machine Design.* McGraw-Hill, New York, 1961.

6. B. J. Hamrock, B. Jacobson, and S. R. Schmid, *Fundamentals of Machine Elements.* McGraw-Hill, New York, 1999.

7. R. C. Hibbeler, *Mechanics of Materials*. Prentice-Hall, Upper Saddle River, NJ, 2000.

8. R. C. Juvinall, *Engineering Considerations of Stress, Strain, and Strength*. McGraw-Hill, New York, 1967.

9. R. C. Juvinall and K. M. Marshek, *Fundamentals of Machine Component Design*. John Wiley & Sons, New York, 1991.

10. G. W. Krutz, J. K. Schueller, and P. W. Claar II, *Machine Design for Mobile and Industrial Applications*. Society of Automotive Engineers, Warrendale, PA, 1994.

11. W. H. Middendorf and R. H. Engelmann, *Design of Devices and Systems*. Marcel Dekker, New York, 1998.

12. R. L. Mott, *Machine Elements in Mechanical Design*. Prentice Hall, Upper Saddle River, NJ, 1999.

13. R. L. Norton, *Design of Machinery*. McGraw-Hill, New York, 1992.

14. R. L. Norton, *Machine Design*. Prentice Hall, Upper Saddle River, NJ, 2000.

15. W. C. Orthwein, *Machine Component Design*. West Publishing Company, St. Paul, MN, 1990.

16. J. E. Shigley and C. R. Mischke, *Mechanical Engineering Design*. McGraw-Hill, NY, 1989.

17. C. W. Wilson, *Computer Integrated Machine Design*. Prentice Hall, Upper Saddle River, NJ, 1997.

4

Theory of Mechanisms

DAN B. MARGHITU

Department of Mechanical Engineering,
Auburn University, Auburn, Alabama 36849

Inside

1. Fundamentals

1.1 Motions

For the planar case the following motions are defined (Fig. 1.1):

- *Pure rotation*: The body possesses one point (center of rotation) that has no motion with respect to a fixed reference frame (Fig. 1.1a). All other points on the body describe arcs about that center.
- *Pure translation*: All the points on the body describe parallel paths (Fig. 1.1b).
- *Complex (general) motion*: A simultaneous combination of rotation and translation (Fig. 1.1c).

1.2 Mobility

The *number of degrees of freedom* (DOF) or *mobility* of a system is equal to the number of independent parameters (measurements) that are needed to

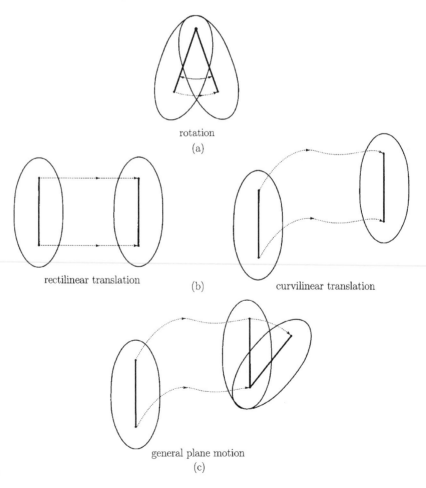

rotation
(a)

rectilinear translation   (b)   curvilinear translation

general plane motion

Figure 1.1   (c)

uniquely define its position in space at any instant of time. The number of DOF is defined with respect to a reference frame.

Figure 1.2 shows a free rigid body, RB, in planar motion. The rigid body is assumed to be incapable of deformation, and the distance between two particles on the rigid body is constant at any time. The rigid body always remains in the plane of motion xy. Three parameters (three DOF) are required to completely define the position of the rigid body: two linear coordinates (x, y) to define the position of any one point on the rigid body, and one angular coordinate θ to define the angle of the body with respect to the axes. The minimum number of measurements needed to define its position are shown in the figure as x, y, and θ. A free rigid body in a plane then has three degrees of freedom. The rigid body may translate along the x axis, v_x, may translate along the y axis, v_y, and may rotate about the z axis, ω_z.

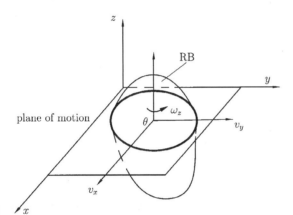

Figure 1.2

The particular parameters chosen to define the position of the rigid body are not unique. Any alternative set of three parameters could be used. There is an infinity of possible sets of parameters, but in this case there must always be three parameters per set, such as two lengths and an angle, to define the position because a rigid body in plane motion has three DOF.

Six parameters are needed to define the position of a free rigid body in a three-dimensional (3D) space. One possible set of parameters that could be used are three lengths (x, y, z) plus three angles $(\theta_x, \theta_y, \theta_z)$. Any free rigid body in three-dimensional space has six degrees of freedom.

1.3 Kinematic Pairs

Linkages are basic elements of all mechanisms. Linkages are made up of links and kinematic pairs (joints). A *link*, sometimes known as an *element* or a *member*, is an (assumed) rigid body that possesses nodes. *Nodes* are defined as points at which links can be attached. A link connected to its neighboring links by *s* nodes is a link of *degree s*. A link of degree 1 is also called unary (Fig. 1.3a), of degree 2, binary (Fig. 1.3b), of degree 3, ternary (Fig. 1.3c), etc.

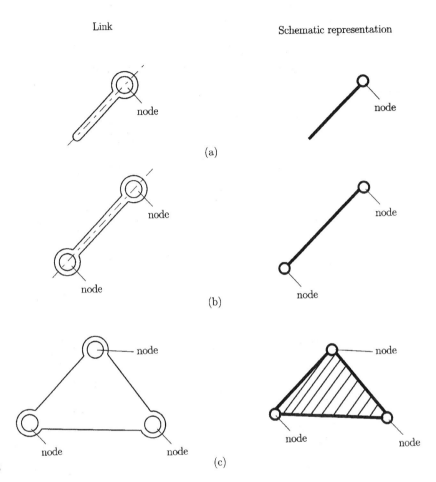

Link   Schematic representation

Figure 1.3

The first step in the motion analysis of a mechanism is to sketch the equivalent *kinematic* or *skeleton diagram*. The kinematic diagram is a stick diagram and display only the essential of the mechanism. The links are numbered (starting with the ground link as number 0) and the kinematic pairs are lettered. The input link is also labeled.

A *kinematic pair* or *joint* is a connection between two or more links (at their nodes). A kinematic pair allows some relative motion between the connected links.

The number of independent coordinates that uniquely determine the relative position of two constrained links is termed the *degree of freedom* of a given kinematic pair. Alternatively, the term *kinematic pair class* is introduced. A kinematic pair is of the *j*th class if it diminishes the relative motion of linked bodies by *j* degrees of freedom; that is, *j* scalar constraint conditions correspond to the given kinematic pair. It follows that such a kinematic pair has $6j$ independent coordinates. The number of degrees of freedom is the fundamental characteristic quantity of kinematic pairs. One of the links of a system is usually considered to be the reference link, and the position of other RBs is determined in relation to this reference body. If the reference link is stationary, the term *frame* or *ground* is used.

The coordinates in the definition of degree of freedom can be linear or angular. Also, the coordinates used can be absolute (measured with regard to the frame) or relative. Figures 1.4–1.9 show examples of kinematic pairs commonly found in mechanisms. Figures 1.4a and 1.4b show two forms of a planar kinematic pair with one degree of freedom, namely, a rotating pin kinematic pair and a translating slider kinematic pair. These are both typically referred to as *full kinematic pairs* and are of the fifth class. The pin kinematic pair allows one rotational (R) DOF, and the slider kinematic pair allows one translational (T) DOF between the joined links. These are both special cases of another common kinematic pair with one degree of freedom, the screw and nut (Fig. 1.5a). Motion of either the nut or the screw relative to the other results in helical motion. If the helix angle is made zero (Fig. 1.5b), the nut rotates without advancing and it becomes a pin kinematic pair. If the helix

Type of full joint   Schematic representation

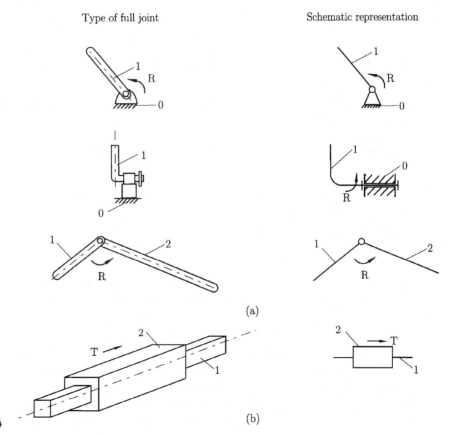

(a)

(b)

Figure 1.4

Type of full joint   Schematic representation

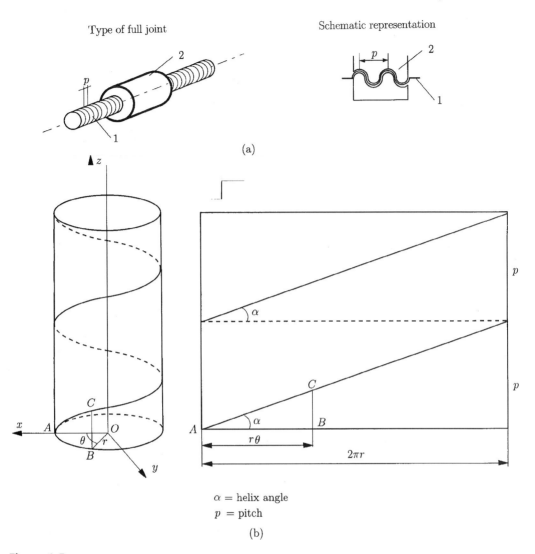

α = helix angle
p = pitch

(b)

Figure 1.5

angle is made 90°, the nut will translate along the axis of the screw, and it becomes a slider kinematic pair.

Figure 1.6 shows examples of kinematic pairs with two degrees of freedom, which simultaneously allow two independent, relative motions, namely translation (T) and rotation (R), between the joined links. A kinematic pair with two degrees of freedom is usually referred to as a *half kinematic pair* and is of the 4th class. A half kinematic pair is sometimes also called a roll–slide kinematic pair because it allows both rotation (rolling) and translation (sliding).

A joystick, ball-and-socket kinematic pair, or sphere kinematic pair (Fig. 1.7a), is an example of a kinematic pair with three degrees of freedom (third class), which allows three independent angular motions between the

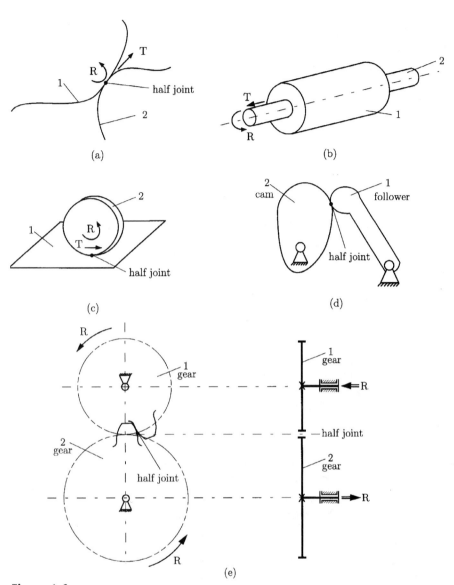

Figure 1.6

two links that are joined. This ball kinematic pair would typically be used in a three-dimensional mechanism, one example being the ball kinematic pairs used in automotive suspension systems. A plane kinematic pair (Fig. 1.7b) is also an example of a kinematic pair with three degrees of freedom, which allows two translations and one rotation.

Note that to visualize the degree of freedom of a kinematic pair in a mechanism, it is helpful to "mentally disconnect" the two links that create the kinematic pair from the rest of the mechanism. It is easier to see how many degrees of freedom the two joined links have with respect to one another.

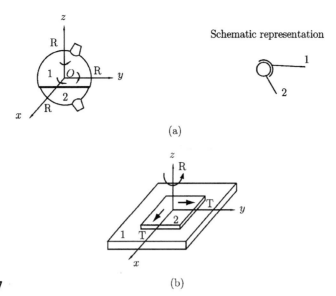

Schematic representation

(a)

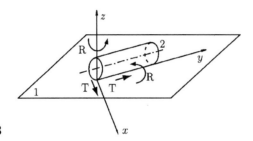

(b)

Figure 1.7

Figure 1.8

Figure 1.8 shows an example of a second-class kinematic pair (cylinder on plane), and Fig. 1.9 represents a first-class kinematic pair (sphere on plane).

The type of contact between the elements can be point (P), curve (C), or surface (S). The term *lower kinematic pair* was coined by Reuleaux to describe kinematic pairs with surface contact. He used the term *higher kinematic pair* to describe kinematic pairs with point or curve contact. The main practical advantage of lower kinematic pairs over higher kinematic pairs is their better ability to trap lubricant between their enveloping surfaces. This is especially true for the rotating pin kinematic pair.

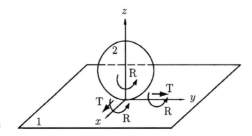

Figure 1.9

A *closed kinematic pair* is a kinematic pair that is kept together or closed by its geometry. A pin in a hole or a slider in a two-sided slot is a form of closed kinematic pair. A *force closed kinematic pair,* such as a pin in a half-bearing or a slider on a surface, requires some external force to keep it together or closed. This force could be supplied by gravity, by a spring, or by some external means. In linkages, closed kinematic pairs are usually preferred and are easy to accomplish. For cam–follower systems, force closure is often preferred.

The *order of a kinematic pair* is defined as the number of links joined minus one. The simplest kinematic pair combination of two links has order 1 and is a single kinematic pair (Fig. 1.10a). As additional links are placed on the same kinematic pair, the order is increased on a one-for-one basis (Fig. 1.10b. Kinematic pair order has significance in the proper determination of overall degrees of freedom for an assembly.

Bodies linked by kinematic pairs form a *kinematic chain.* Simple kinematic chains are shown in Fig. 1.11.

A *contour* or *loop* is a configuration described by a polygon (Fig. 1.11a).

The presence of loops in a mechanical structure can be used to define the following types of chains:

Mechanisms

joint of order one   joint of order two

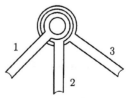

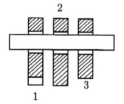

Schematic representation   Schematic representation

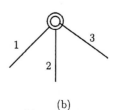

Figure 1.10
Used with permission from Ref. 15.

(a)   (b)

Schematic representation

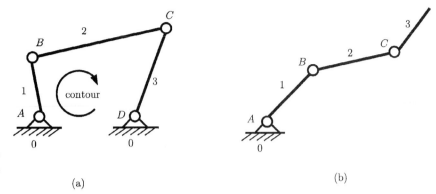

(a)   (b)

- *Closed kinematic chains* have one or more loops so that each link and each kinematic pair is contained in at least one of the loops (Fig. 1.11a). A closed kinematic chain has no open attachment point.
- *Open kinematic chains* contain no loops (Fig. 1.11b). A common example of an open kinematic chain is an industrial robot.
- *Mixed kinematic chains* are a combination of closed and open kinematic chains.

Another classification is also useful:

- *Simple chains* contain only binary elements.
- *Complex chains* contain at least one element of degree 3 or higher.

A *mechanism* is defined as a kinematic chain in which at least one link has been "grounded" or attached to the frame (Figs. 1.11a and 1.12). By Reuleaux's definition, a *machine* is a collection of mechanisms arranged to transmit forces and do work. He viewed all energy- or force-transmitting devices as machines that utilize mechanisms as their building blocks to provide the necessary motion constraints.

The following terms can be defined (Fig. 1.12):

- *A crank* is a link that makes a complete revolution about a fixed grounded pivot.
- *A rocker* is a link that has oscillatory (back and forth) rotation and is fixed to a grounded pivot.
- *A coupler* or connecting rod is a link that has complex motion and is not fixed to ground.

Ground is defined as any link or links that are fixed (nonmoving) with respect to the reference frame. Note that the reference frame may in fact itself be in motion.

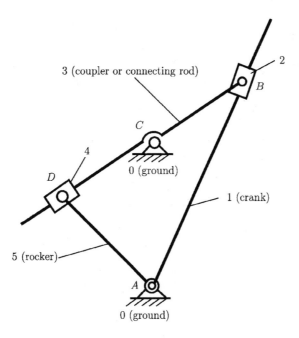

Figure 1.12
Used with permission from Ref. 15.

1.4 Number of Degrees of Freedom

The concept of *number of degrees of freedom* is fundamental to analysis of mechanisms. It is usually necessary to be able to determine quickly the number of DOF of any collection of links and kinematic pairs that may be suggested as a solution to a problem. The number of degrees of freedom or the *mobility* of a system can be defined as either of the following:

- The number of inputs that need to be provided in order to create a predictable system output
- The number of independent coordinates required to define the position of the system

The *family f* of a mechanism is the number of DOF eliminated from all the links of the system.

Every free body in space has six degrees of freedom. A system of family *f* consisting of *n* movable links has $(6 - f)n$ degrees of freedom. Each kinematic pair of class *j* diminishes the freedom of motion of the system by $j - f$ degrees of freedom. If we designate the number of kinematic pairs of class *k* as c_k, it follows that the number of degrees of freedom of the particular system is

$$M = (6 - f)n - \sum_{j=f+1}^{5} (j - f)c_j. \tag{1.1}$$

In the literature, this is referred to as the Dobrovolski formula.

A *driver* link is that part of a mechanism which causes motion, such as the crank. The number of driver links is equal to the number of DOF of the mechanism. A *driven* link or *follower* is that part of a mechanism whose motion is affected by the motion of the driver.

1.5 Planar Mechanisms

For the special case of planar mechanisms ($f = 3$), formula (1.1) has the form

$$M = 3n - 2c_5 - c_4, \tag{1.2}$$

where c_5 is the number of full kinematic pairs and c_4 is the number of half kinematic pairs.

The mechanism in Fig. 1.11a has three moving links ($n = 3$) and four rotational kinematic pairs at A, B, C, and D ($c_5 = 4$). The number of DOF for this mechanism is

$$M = 3n - 2c_5 - c_4 = 3(3) - 2(4) = 1.$$

For the mechanism shown in Fig. 1.12 there are seven kinematic pairs of class 5 ($c_5 = 7$) in the system:

- At A, one rotational kinematic pair between link 0 and link 1
- At B, one rotational kinematic pair between link 1 and link 2
- At B, one translational kinematic pair between link 2 and link 3
- At C, one rotational kinematic pair between link 0 and link 3
- At D, one rotational kinematic pair between link 3 and link 4
- At D, one translational kinematic pair between link 4 and link 5
- At A, one rotational kinematic pair between link 5 and link 0

The number of moving links is five ($n = 5$). The number of DOF for this mechanism is

$$M = 3n - 2c_5 - c_4 = 3(5) - 2(7) = 1,$$

and this mechanism has one driver link.

There is a special significance to kinematic chains that do not change their mobility after being connected to an arbitrary system. Kinematic chains defined in this way are called system groups. Connecting them to or disconnecting them from a given system enables given systems to be modified or structurally new systems to be created while maintaining the original freedom of motion. The term "system group" has been introduced for the classification of planar mechanisms used by Assur and further investigated by Artobolevskij. If we limit this to planar systems containing only kinematic pairs of class 2, from Eq. (1.2) we can obtain

$$3n - 2c_5 = 0, \tag{1.3}$$

according to which the number of system group links n is always even. The simplest system is the binary group with two links ($n = 2$) and three full kinematic pairs ($c_5 = 3$). The binary group is also called a *dyad*. The sets of links shown in Fig. 1.13 are dyads, and one can distinguish the following types:

- Rotation rotation rotation (dyad RRR) (Fig. 1.13a)
- Rotation rotation translation (dyad RRT) (Fig.1.13b)
- Rotation translation rotation (dyad RTR) (Fig. 1.13c)
- Translation rotation translation (dyad TRT) (Fig. 1.13d)
- Translation translation rotation (dyad TTR) (Fig. 1.13e)

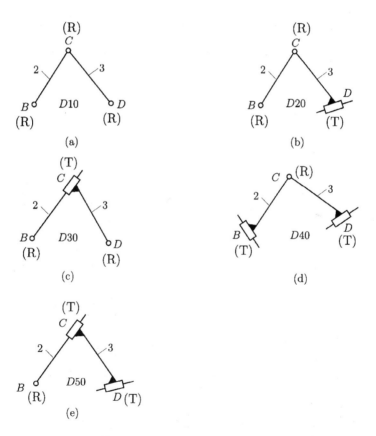

Figure 1.13

The advantage of the group classification of a system lies in its simplicity. The solution of the whole system can then be obtained by composing partial solutions.

The mechanism in Fig. 1. 11a as one driver, link 1, with rotational motion and one dyad RRR, link 2 and link 3.

The mechanism in Fig. 1.12 is obtained by composing the following:

■ The driver link 1 with rotational motion
■ The dyad RTR: links 2 and 3, and the kinematic pairs B rotation (B_R), B translation (B_T), C rotation (C_R)
■ The dyad RTR: links 4 and 5, and the kinematic pairs D rotation (D_R), D translation (D_T), A rotation (A_R).

2. Position Analysis

2.1 Cartesian Method

The position analysis of a kinematic chain requires the determination of the kinematic pair positions and/or the position of the mass center of the link. A planar link with the end nodes A and B is considered in Fig. 2.1. Let (x_A, y_A) be the coordinates of the kinematic pair A with respect to the reference frame xOy, and (x_B, y_B) be the coordinates of the kinematic pair B with the same reference frame. Using Pythagoras, one can write

$$(x_B - x_A)^2 + (y_B - y_A)^2 = AB^2 = L_{AB}^2, \qquad (2.1)$$

where L_{AB} is the length of the link AB.

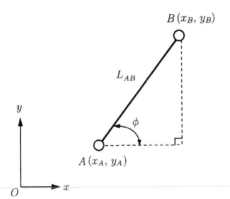

Figure 2.1

Let ϕ be the angle of the link AB with the horizontal axis Ox. Then, the slope m of the link AB is defined as

$$m = \tan \phi = \frac{y_B - y_A}{x_B - x_A}. \qquad (2.2)$$

Let n be the intercept of AB with the vertical axis Oy. Using the slope m and the intercept n, one finds that the equation of the straight link, in the plane, is

$$y = mx + n, \qquad (2.3)$$

where x and y are the coordinates of any point on this link.

For a link with a translational kinematic pair (Fig. 2.2), the sliding direction (Δ) is given by the equation

$$x \cos \alpha - y \sin \alpha - p = 0, \tag{2.4}$$

where p is the distance from the origin O to the sliding line (Δ). The position function for the kinematic pair $A(x_A, y_A)$ is

$$x_A \cos \alpha - y_A \sin \alpha - p = \pm d, \tag{2.5}$$

where d is the distance from A to the sliding line. The relation between the kinematic pair A and a point B on the sliding direction, $B \in (\Delta)$ (the symbol $\in$ means "belongs to"), is

$$(x_A - x_B) \sin \beta - (y_A - y_B) \cos \beta = \pm d, \tag{2.6}$$

where $\beta = \alpha + \dfrac{\pi}{2}$.

If $Ax + By + C = 0$ is the linear equation of the line (Δ), then the distance d is (Fig. 2.2)

$$d = \frac{|Ax_A + By_A + C|}{\sqrt{A^2 + B^2}}. \tag{2.7}$$

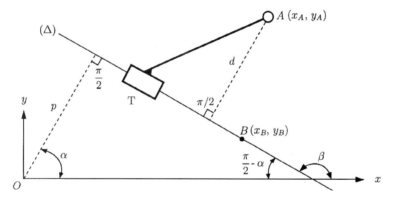

Figure 2.2

For a driver link in rotational motion (Fig. 2.3a), the following relations can be written:

$$\begin{aligned} x_B &= x_A + L_{AB} \cos \phi \\ y_B &= y_A + L_{AB} \sin \phi. \end{aligned} \tag{2.8}$$

From Fig. 2.3b, for a driver link in transitional motion one can have

$$\begin{aligned} x_B &= x_A + s \cos \phi + L_1 \cos(\phi + \alpha), \\ y_B &= y_A + s \sin \phi + L_1 \sin(\phi + \alpha). \end{aligned} \tag{2.9}$$

For the RRR dyad (Fig. 2.4), there are two quadratic equations of the form

$$\begin{aligned} (x_A - x_C)^2 + (y_A - y_C)^2 &= AC^2 = L_{AC}^2 = L_2^2 \\ (x_B - x_C)^2 + (y_B - y_C)^2 &= BC^2 = L_{BC}^2 = L_3^2, \end{aligned} \tag{2.10}$$

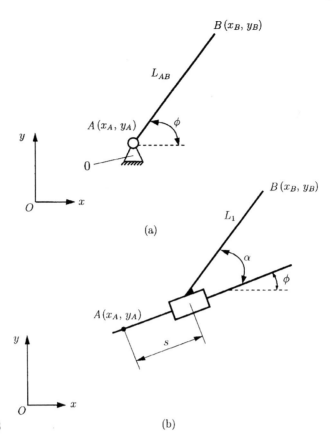

Figure 2.3

where the coordinates of the kinematic pair C, x_C and y_C, are the unknowns. With x_C and y_C determined, the angles ϕ_2 and ϕ_3 are computed from the relations

$$y_C - y_A - (x_C - x_A)\tan\phi_2 = 0$$
$$y_C - y_B - (x_C - x_B)\tan\phi_3 = 0. \qquad (2.11)$$

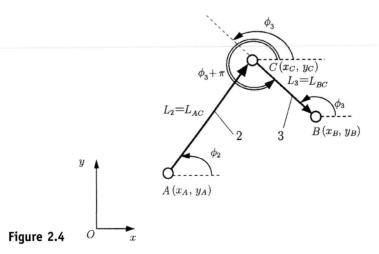

Figure 2.4

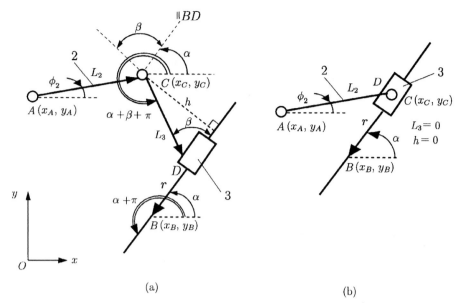

Figure 2.5

The following relations can be written for the RRT dyad (Fig. 2.5a):

$$(x_A - x_C)^2 + (y_A - y_C)^2 = AC^2 = L_{AC}^2 = L_2^2$$
$$(x_C - x_B) \sin \alpha - (y_C - y_B) \cos \alpha = \pm h. \tag{2.12}$$

From these two equations, the two unknowns x_C and y_C are computed. Figure 2.5b depicts the particular case for the RRT dyad when $L_3 = h = 0$.

For the RTR dyad (Fig. 2.6a), the known data are the positions of the kinematic pair A and B, x_A, y_A, x_B, y_B, the angle α, and the length L_2 ($h = L_2 \sin \alpha$). There are four unknowns in the position of $C(x_C, y_C)$ and in the equation for the sliding line (Δ): $y = mx + n$. The unknowns in the sliding line m and n are computed from the relations

$$L_2 \sin \alpha = \frac{|mx_A - y_A + n|}{\sqrt{m^2 + 1}}$$
$$y_B = mx_B + n. \tag{2.13}$$

The coordinates of the kinematic pair C can be obtained using

$$(x_A - x_C)^2 + (y_A - y_C)^2 = L_2^2$$
$$y_C = mx_C + n. \tag{2.14}$$

In Fig. 2.6b the particular case when $L_2 = h = 0$ is shown.

To compute the coordinates of the kinematic pair C for the TRT dyad (Fig. 2.7), two equations can be written:

$$(x_C - x_A) \sin \alpha - (y_C - y_A) \cos \alpha = \pm d$$
$$(x_C - x_B) \sin \beta - (y_C - y_B) \cos \beta = \pm h. \tag{2.15}$$

The input data are x_A, y_A, x_B, y_B, α, β, d, h, and the output data are x_C, y_C.

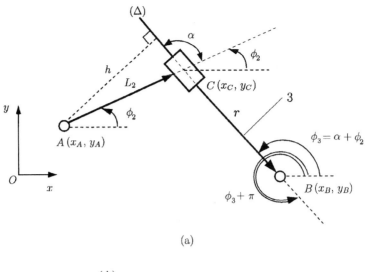

(a)

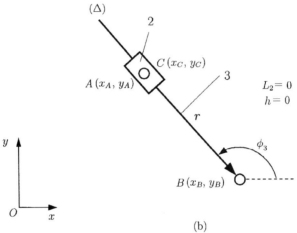

(b)

Figure 2.6

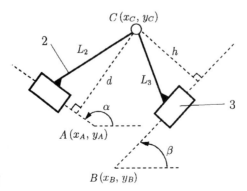

Figure 2.7

EXAMPLE Determine the positions of the kinematic pairs of the mechanism shown in Fig. 2.8. The known elements are: $AB = l_1$, $CD = l_3$, $CE = l_4$, $AD = d_1$, and h is the distance from the slider 5 to the horizontal axis Ax.

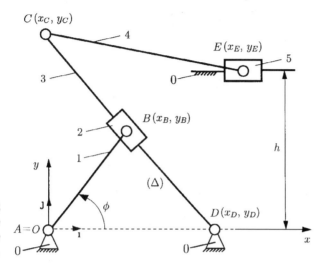

Figure 2.8
Used with permission from Ref. 15.

Solution

The origin of the system is at A, $A \equiv O$, that is, $x_A = y_A = 0$. The coordinates of the R kinematic pair at B are

$$x_B = l_1 \cos \phi, \quad y_B = l_1 \sin \phi.$$

For the dyad DBB (RTR) the following equations can be written with respect to the sliding line CD:

$$m x_B - y_B + n = 0, \quad y_D = m x_D + n.$$

With $x_D = d_1$ and $y_D = 0$ from the preceding system, the slope m of the link CD and the intercept n can be calculated:

$$m = \frac{l_1 \sin \phi}{l_1 \cos \phi - d_1}, \quad n = \frac{d_1 l_1 \sin \phi}{d_1 - l_1 \cos \phi}.$$

The coordinates x_C and y_C of the center of the R kinematic pair C result from the system of two equations

$$y_C = \frac{l_1 \sin \phi}{l_1 \cos \phi - d_1} x_C + \frac{d_1 l_1 \sin \phi}{d_1 - l_1 \cos \phi}$$

$$(x_C - x_D)^2 + (y_C - y_D)^2 = l_3^2.$$

Because of the quadratic equation, two solutions are obtained for x_C and y_C. For continuous motion of the mechanism, there are constraint relations for choosing the correct solution, that is, $x_C < x_B < x_D$ and $y_C > 0$.

For the last dyad CEE (RRT), a position function can be written for the kinematic pair E:

$$(x_C - x_E)^2 + (y_C - b)^2 = l_4^2.$$

It results the values x_{E1} and x_{E2}, and the solution $x_E > x_C$ will be selected for continuous motion of the mechanism.   ▲

2.2 Vector Loop Method

First the independent closed loops are identified.

A vector equation corresponding to each independent loop is established. The vector equation gives rise to two scalar equations, one for the horizontal axis x, and one for the vertical axis y.

For an open kinematic chain (Fig. 2.9), with general kinematic pairs (pin kinematic pairs, slider kinematic pairs, etc.), a vector loop equation can be considered:

$$\mathbf{r}_A + \mathbf{r}_1 + \cdots \mathbf{r}_n = \mathbf{r}_B \tag{2.16}$$

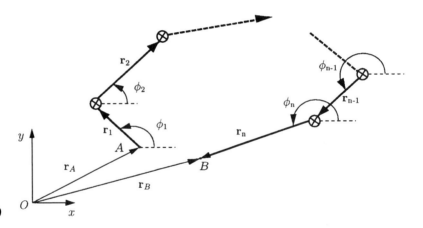

Figure 2.9

or

$$\sum_{k=1}^{n} \mathbf{r}_k = \mathbf{r}_B - \mathbf{r}_A. \tag{2.17}$$

The vectorial Eq. (2.17) can be projected on the reference frame xOy:

$$\sum_{k=1}^{n} r_k \cos \phi_k = x_B - x_A$$

$$\sum_{k=1}^{n} r_k \sin \phi_k = y_B - y_A. \tag{2.18}$$

2.2.1 RRR DYAD

The input data are that the position of A is (x_A, y_A), the position of B is (x_B, y_B), the length of AC is $L_{AC} = L_2$, and the length of BC is $L_{BC} = L_3$ (Fig. 2.4). The unknown data are the position of $C(x_C, y_C)$ and the angles ϕ_2 and ϕ_3.

The position equation for the RRR dyad is $\mathbf{AC} + \mathbf{CB} = \mathbf{r}_B - \mathbf{r}_A$, or

$$L_2 \cos \phi^2 + L_3 \cos(\phi_3 + \pi) = x_B - x_A,$$
$$L_2 \sin \phi_2 + L_3 \sin(\phi_3 + \pi) = y_B - y_A. \tag{2.19}$$

The angles ϕ_2 and ϕ_3 can be computed from Eq. (2.19). The position of C can be computed using the known angle ϕ_2:

$$x_C = x_A + L_2 \cos \phi_2$$
$$y_C = y_A + L_2 \sin \phi_2. \tag{2.20}$$

2.2.2 RRT DYAD

The input data are that the position of A is (x_A, y_A), the position of B is (x_B, y_B), the length of AC is L_2, the length of CB is L_3, and the angles α and β are constants (Fig. 2.5a).

The unknown data are the position of $C(x_C, y_C)$, the angle ϕ_2, and the distance $r = DB$.

The vectorial equation for this kinematic chain is $\mathbf{AC} + \mathbf{CD} + \mathbf{DB} = \mathbf{r}_B - \mathbf{r}_A$, or

$$L_2 \cos \phi_2 + L_3 \cos(\alpha + \beta + \pi) + r \cos(\alpha + \pi) = x_B - x_A$$
$$L_2 \sin \phi_2 + L_3 \sin(\alpha + \beta + \pi) + r \sin(\alpha + \pi) = y_B - y_A. \tag{2.21}$$

One can compute r and ϕ_2 from Eq. (2.21). The position of C can be found with Eq. (2.20).

Particular case $L_3 = 0$ (Fig. 2.5b): In this case Eq. (2.21) can be written as

$$L_2 \cos \phi_2 + r \cos(\alpha + \pi) = x_B - x_A$$
$$L_2 \sin \phi_2 + r \sin(\alpha + \pi) = y_B - y_A. \tag{2.22}$$

2.2.3 RTR DYAD

The input data are that the position of A is (x_A, y_A), the position of B is (x_B, y_B), the length of AC is L_2, and the angle α is constant (Fig. 2.6a).

The unknown data are the distance $r = CB$ and the angles ϕ_2 and ϕ_3.

The vectorial loop equation can be written as $\mathbf{AC} + \mathbf{CB} = \mathbf{r}_B - \mathbf{r}_A$, or

$$L_2 \cos \phi_2 + r \cos(\alpha + \phi_2 + \pi) = x_B - x_A$$
$$L_2 \sin \phi_2 + r \sin(\alpha + \phi_2 + \pi) = y_B - y_A. \tag{2.23}$$

One can compute r and ϕ_2 from Eq. (2.23). The angle ϕ_3 can be written

$$\phi_3 = \phi_2 + \alpha. \tag{2.24}$$

Particular case $L_2 = 0$ (Fig. 2.6b): In this case, from Eqs. (2.23) and (2.24) one can obtain

$$r\cos(\phi_3) = x_B - x_A,$$
$$r\sin(\phi_3) = y_B - y_A. \qquad (2.25)$$

EXAMPLE Figure 2.10a shows a quick-return shaper mechanism. Given the lengths $AB = 0.20$ m, $AD = 0.40$ m, $CD = 0.70$ m, $CE = 0.30$ m, and the input angle $\phi = \phi_1 = 45°$, obtain the positions of all the other kinematic pairs. The distance from the slider 5 to the horizontal axis Ax is $y_E = 0.35$ m.

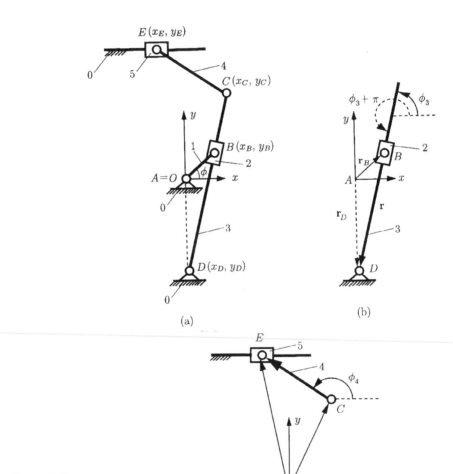

Figure 2.10
Used with permission from Ref. 15.

Solution

The coordinates of the kinematic pair B are

$$x_B = AB \cos \phi = 0.20 \cos 45° = 0.141 \text{ m}$$
$$y_B = AB \sin \phi = 0.20 \sin 45° = 0.141 \text{ m}.$$

The vector diagram (Fig. 2.10b) is drawn by representing the RTR (BBD) dyad. The vector equation corresponding to this loop is written as

$$\mathbf{r}_B + \mathbf{r} - \mathbf{r}_D = \mathbf{0},$$

or

$$\mathbf{r} = \mathbf{r}_D - \mathbf{r}_B,$$

where $\mathbf{r} = \mathbf{BD}$ and $|\mathbf{r}| = r$. Projecting the preceding vectorial equation on x and y axis, we obtain two scalar equations:

$$r \cos(\pi + \phi_3) = x_D - x_B = -0.141 \text{ m}$$
$$r \sin(\pi + \phi_3) = y_D - y_B = -0.541 \text{ m}.$$

The angle ϕ_3 is obtained by solving the equation

$$\tan \phi_3 = \frac{0.541}{0.141} \Longrightarrow \phi_3 = 75.36°.$$

The distance r is

$$r = \frac{x_D - x_B}{\cos(\pi + \phi_3)} - 0.56 \text{ m}.$$

The coordinates of the kinematic pair C are

$$x_C = CD \cos \phi_3 = 0.17 \text{ m}$$
$$y_C = CD \sin \phi_3 = -AD = 0.27 \text{ m}.$$

For the next dyad RRT (CEE) (Fig. 2.10c), one can write

$$CE \cos(\pi - \phi_4) = x_E - x_C$$
$$CE \sin(\pi - \phi_4) = y_E - y_C.$$

Solving this system, we obtain the unknowns ϕ_4 and x_E:

$$\phi_4 = 165.9° \quad \text{and} \quad x_E = -0.114 \text{ m}. \quad \blacktriangle$$

3. Velocity and Acceleration Analysis

The classical method for obtaining the velocities and/or accelerations of links and kinematic pairs is to compute the derivatives of the positions and/or velocities with respect to time.

3.1 Driver Link

For a driver link in rotational motion (Fig. 2.3a), the following position relation can be written:

$$x_B(t) = x_A + L_{AB} \cos \phi(t),$$
$$y_B(t) = y_A + L_{AB} \sin \phi(t). \tag{3.1}$$

Differentiating Eq. (3.1) with respect to time, we obtain the following expressions:

$$v_{Bx} = \dot{x}_B = \frac{dx_B(t)}{dt} = -L_{AB}\dot{\phi} \sin \phi$$
$$v_{By} = \dot{y}_B = \frac{dy_B(t)}{dt} = L_{AB}\dot{\phi} \cos \phi. \tag{3.2}$$

The angular velocity of the driver link is $\omega = \dot{\phi}$.

The time derivative of Eq. (3.2) yields

$$a_{Bx} = \ddot{x}_B = \frac{dv_B(t)}{dt} = -L_{AB}\dot{\phi}^2 \cos \phi - L_{AB}\ddot{\phi} \sin \phi$$
$$a_{By} = \ddot{y}_B = \frac{dv_B(t)}{dt} = -L_{AB}\dot{\phi}^2 \sin \phi + L_{AB}\ddot{\phi} \cos \phi, \tag{3.3}$$

where $\alpha = \ddot{\phi}$ is the angular acceleration of the driver link AB.

3.2 RRR Dyad

For the RRR dyad (Fig. 2.4), there are two quadratic equations of the form

$$[x_C(t) - x_A]^2 + [y_C(t) - y_A]^2 = L_{AC} = L_2^2$$
$$[x_C(t) - x_B]^2 + [y_C(t) - y_B]^2 = L_{BC} = L_3^2. \tag{3.4}$$

Solving this system of quadratic equations, we obtain the coordinates $x_C(t)$ and $y_C(t)$.

The derivative of Eq. (3.4) with respect to time yields

$$(x_C - x_A)(\dot{x}_C - \dot{x}_A) + (y_C - y_A)(\dot{y}_C - \dot{y}_A) = 0$$
$$(x_C - x_B)(\dot{x}_C - \dot{x}_B) + (y_C - y_B)(\dot{y}_C - \dot{y}_B) = 0, \tag{3.5}$$

The velocity vector $\mathbf{v}_C = [\dot{x}_C, \dot{y}_C]^T$ of the preceding system of equations can be written in matrix form as

$$\mathbf{v}_C = \mathbf{M}_1 \cdot \mathbf{v}, \tag{3.6}$$

where

$$\mathbf{v} = [\dot{x}_A, \dot{y}_A, \dot{x}_B, \dot{y}_B]^T$$
$$\mathbf{M}_1 = \mathbf{A}_1^{-1} \cdot \mathbf{A}_2$$
$$\mathbf{A}_1 = \begin{bmatrix} x_C - x_A & y_C - y_A \\ x_C - x_B & y_C - y_B \end{bmatrix}$$
$$\mathbf{A}_2 = \begin{bmatrix} x_C - x_A & y_C - y_A & 0 & 0 \\ 0 & 0 & x_C - x_B & y_C - y_B \end{bmatrix}.$$

Similarly, by differentiating Eq. (3.5), the following acceleration equations are obtained:

$$(\dot{x}_C - \dot{x}_A)^2 + (x_C - x_A)(\ddot{x}_C - \ddot{x}_A) + (\dot{y}_C - \dot{y}_A)^2 + (y_C - y_A)(\ddot{y}_C - \ddot{y}_A) = 0$$
$$(\dot{x}_C - \dot{x}_B)^2 + (x_C - x_B)(\ddot{x}_C - \ddot{x}_B) + (\dot{y}_C - \dot{y}_B)^2 + (y_C - y_B)(\ddot{y}_C - \ddot{y}_B) = 0.$$

$$(3.7)$$

The acceleration vector is obtained from the preceding system of equations:

$$\mathbf{a}_C = [\ddot{x}_C, \ddot{y}_C]^T = \mathbf{M}_1 \cdot \mathbf{a} + \mathbf{M}_2. \tag{3.8}$$

Here,

$$\mathbf{a} = [\ddot{x}_A, \ddot{y}_A, \ddot{x}_B, \ddot{y}_B]^T$$
$$\mathbf{M}_2 = -\mathbf{A}_1^{-1} \cdot \mathbf{A}_3$$
$$\mathbf{A}_3 = \begin{bmatrix} (\dot{x}_C - \dot{x}_A)^2 + (\dot{y}_C - \dot{y}_A)^2 \\ (\dot{x}_C - \dot{x}_B)^2 + (\dot{y}_C - \dot{y}_B)^2 \end{bmatrix}.$$

To compute the angular velocity and acceleration of the RRR dyad, the following equations can be written:

$$y_C(t) - y_A + [x_C(t) - x_A]\tan\phi_2(t) = 0$$
$$y_C(t) - y_B + [x_C(t) - x_B]\tan\phi_3(t) = 0. \tag{3.9}$$

The derivative with respect to time of Eq. (3.9) yields

$$\dot{y}_C - \dot{y}_A - (\dot{x}_C - \dot{x}_A)\tan\phi_2 - (x_C - x_A)\frac{1}{\cos^2\phi_2}\dot{\phi}_2 = 0$$

$$\dot{y}_C - \dot{y}_B - (\dot{x}_C - \dot{x}_B)\tan\phi_3 - (x_C - x_B)\frac{1}{\cos^2\phi_3}\dot{\phi}_x = 0. \tag{3.10}$$

The angular velocity vector is computed as

$$\boldsymbol{\omega} = [\dot{\phi}_2, \dot{\phi}_3]^T = \boldsymbol{\Omega}_1 \cdot \mathbf{v} + \boldsymbol{\Omega}_2 \cdot \mathbf{v}_C, \tag{3.11}$$

where

$$\boldsymbol{\Omega}_1 = \begin{bmatrix} \dfrac{x_C - x_A}{L_2^2} & -\dfrac{x_C - x_A}{L_2^2} & 0 & 0 \\ 0 & 0 & \dfrac{x_C - x_B}{L_3^2} & -\dfrac{x_C - x_B}{L_3^2} \end{bmatrix}$$

$$\boldsymbol{\Omega}_2 = \begin{bmatrix} -\dfrac{x_C - x_A}{L_2^2} & \dfrac{x_C - x_A}{L_2^2} \\ -\dfrac{x_C - x_B}{L_3^2} & \dfrac{x_C - x_B}{L_3^2} \end{bmatrix}.$$

When Eq. (3.11) is differentiated, the angular vector $\boldsymbol{\alpha} = \dot{\boldsymbol{\omega}} = [\ddot{\phi}_2, \ddot{\phi}_3]^T$ is

$$\boldsymbol{\alpha} = \dot{\boldsymbol{\Omega}}_1 \cdot \mathbf{v} + \dot{\boldsymbol{\Omega}}_2 \cdot \mathbf{v}_C + \boldsymbol{\Omega}_1 \cdot \mathbf{a} + \boldsymbol{\Omega}_2 \cdot \mathbf{a}_C. \tag{3.12}$$

Mechanisms

3.3 RRT Dyad

For the RRT dyad (Fig. 2.5a), the following equations can be written for position analysis:

$$[x_C(t) - x_A]^2 + [y_C(t) - y_A]^2 = AC^2 = L_{AC}^2 = L_2^2$$
$$[x_C(t) - x_B]\sin\alpha - [y_C(t) - y_B]\cos\alpha = \pm h. \tag{3.13}$$

From this system of equations, $x_C(t)$ and $y_C(t)$ can be computed.

The time derivative of Eq. (3.13) yields

$$(x_C - x_A)(\dot{x}_C - \dot{x}_A) + (y_C - y_A)(\dot{y}_C - \dot{y}_A) = 0$$
$$(\dot{x}_C - \dot{x}_B)\sin\alpha - (\dot{y}_C - \dot{y}_B)\cos\alpha = 0. \tag{3.14}$$

The solution for the velocity vector from Eq. (3.14) is

$$\mathbf{v}_C = [\dot{x}_C, \dot{y}_C]^T = \mathbf{M}_3 \cdot \mathbf{v}, \tag{3.15}$$

where

$$\mathbf{M}_3 = \mathbf{A}_4^{-1} \cdot \mathbf{A}_5$$

$$\mathbf{A}_4 = \begin{bmatrix} x_C - x_A & y_C - y_A \\ \sin\alpha & -\cos\alpha \end{bmatrix}$$

$$\mathbf{A}_5 = \begin{bmatrix} x_C - x_A & y_C - y_A & 0 & 0 \\ 0 & 0 & \sin\alpha & -\cos\alpha \end{bmatrix}.$$

Differentiating Eq. (3.14) with respect to time,

$$(\dot{x}_C - \dot{x}_A)^2 + (x_C - x_A)(\ddot{x}_C - \ddot{x}_A) + (\dot{y}_C - \dot{y}_A)^2 + (y_C - y_A)(\ddot{y}_C - \ddot{y}_A) = 0$$
$$(\ddot{x}_C - \ddot{x}_B)\sin\alpha - (\ddot{y}_C - \ddot{y}_B)\cos\alpha = 0, \tag{3.16}$$

gives the acceleration vector

$$\mathbf{a}_C = [\ddot{x}_C, \ddot{y}_C]^T = \mathbf{M}_3 \cdot \mathbf{a} + \mathbf{M}_4, \tag{3.17}$$

where

$$\mathbf{M}_4 = -\mathbf{A}_4^{-1} \cdot \mathbf{A}_6$$

$$\mathbf{A}_6 = \begin{bmatrix} (\dot{x})_C - \dot{x}_A)^2 + (\dot{y}_C - \dot{y}_A)^2 \\ 0 \end{bmatrix}. \tag{3.18}$$

The angular position of the element 2 is described by

$$y_C(t) - y_A - [x_C(t) - x_A]\tan\phi_2(t) = 0. \tag{3.19}$$

The time derivative of Eq. (3.19) yields

$$\dot{y}_C - \dot{y}_A - (\dot{x}_C - \dot{x}_A)\tan\phi_2 - (x_C - x_A)\frac{1}{\cos^2\phi_2}\dot{\phi}_2 = 0, \tag{3.20}$$

and the angular velocity of the element 2 is

$$\omega_2 = \frac{x_C - x_A}{L_2^2}[(\dot{y}_C - \dot{y}_A) - (\dot{x}_C - \dot{x}_A)\tan\phi_2]. \tag{3.21}$$

The angular acceleration of the element 2 is $\alpha_2 = \dot{\omega}_2$.

3.4 RTR Dyad

For the RTR dyad (Fig. 2.6a), the following relations can be written:

$$[x_C(t) - x_A]^2 + [y_C(t) - y_A]^2 = L_2^2$$

$$\tan \alpha = \frac{\left(\dfrac{y_C - y_B}{x_C - x_B}\right) - \left(\dfrac{y_C - y_A}{x_C - x_A}\right)}{1 + \left(\dfrac{y_C - y_B}{x_C - x_B}\right) \cdot \left(\dfrac{y_C - y_A}{x_C - x_A}\right)}$$

$$= \frac{(y_C - y_B)(x_C - x_A) - (y_C - y_A)(x_C - x_B)}{(x_C - x_B)(x_C - x_A) + (y_C - y_B)(y_C - y_A)}. \qquad (3.22)$$

The time derivative of Eq. (3.22) yields

$$(x_C - x_A)(\dot{x}_C - \dot{x}_A) + (y_C - y_A)(\dot{y}_C - \dot{y}_A) = 0$$
$$\tan \alpha[(\dot{x}_C - \dot{x}_B)(x_C - x_A) + (x_C - x_B)(\dot{x}_C - \dot{x}_A)]$$
$$+ \tan \alpha[(\dot{y}_C - \dot{y}_A)(y_C - y_B) + (y_C - y_A)(\dot{y}_C - \dot{y}_B)]$$
$$+ (\dot{y}_C - \dot{y}_A)(x_C - x_B) + (y_C - y_A)(\dot{x}_C - \dot{x}_B)$$
$$- (\dot{y}_C - \dot{y}_B)(x_C - x_A) - (y_C - y_B)(\dot{x}_C - \dot{x}_A) = 0, \qquad (3.23)$$

or in matrix form,

$$\mathbf{A}_7 \cdot \mathbf{v}_C = \mathbf{A}_8 \cdot \mathbf{v}, \qquad (3.24)$$

where

$$\mathbf{A}_7 = \begin{bmatrix} x_C - x_A & y_C - y_A \\ \gamma_1 & \gamma_2 \end{bmatrix}$$

$$\mathbf{A}_8 = \begin{bmatrix} x_C - x_A & y_C - y_A & 0 & 0 \\ \gamma_3 & \gamma_4 & \gamma_5 & \gamma_6 \end{bmatrix},$$

and

$$\gamma_1 = [(x_C - x_B) + (x_C - x_A)] \tan \alpha - (y_C - y_B) + (y_C - y_A)$$
$$\gamma_2 = [(y_C - y_A) + (y_C - y_B)] \tan \alpha - (x_C - x_A) + (x_C - x_B)$$
$$\gamma_3 = (x_C - x_B) \tan \alpha + (y_C - y_B)$$
$$\gamma_4 = (x_C - x_A) \tan \alpha + (y_C - y_A)$$
$$\gamma_5 = (y_C - y_B) \tan \alpha + (x_C - x_B)$$
$$\gamma_6 = (y_C - y_A) \tan \alpha - (x_C - x_A).$$

The solution for the velocity vector from Eq. (3.24) is

$$\mathbf{v}_C = \mathbf{M}_5 \cdot \mathbf{v}, \qquad (3.25)$$

where

$$\mathbf{M}_5 = \mathbf{A}_7^{-1} \cdot \mathbf{A}_8.$$

Differentiating Eq. (3.24), we obtain

$$\mathbf{A}_7 \cdot \mathbf{a}_C = \mathbf{A}_8 \cdot \mathbf{a} - \mathbf{A}_9, \tag{3.26}$$

where

$$\mathbf{A}_9 = \begin{bmatrix} (\dot{x}_C - \dot{x}_A)^2 + (\dot{y}_C - \dot{y}_A)^2 \\ \gamma_7 \end{bmatrix}$$

$$\gamma_7 = 2(\dot{x}_C - \dot{x}_B)(\dot{x}_C - \dot{x}_A)\tan\alpha + 2(\dot{y}_C - \dot{y}_B)(\dot{y}_C - \dot{y}_A)\tan\alpha$$
$$- 2(\dot{y}_C - \dot{y}_B)(\dot{x}_C - \dot{x}_A) + 2(\dot{y}_C - \dot{y}_A)(\dot{x}_C - \dot{x}_B).$$

The acceleration vector is

$$\mathbf{a}_C = \mathbf{M}_5 \cdot \mathbf{a} - \mathbf{M}_6, \tag{3.27}$$

where

$$\mathbf{M}_6 = \mathbf{A}_7^{-1} \cdot \mathbf{A}_9.$$

To compute the angular velocities for the RTR dyad, the following equations can be written:

$$y_C(t) - y_A = [x_C(t) - x_A]\tan\phi_2$$
$$\phi_3 = \phi_2 + \alpha. \tag{3.28}$$

The time derivative of Eq. (3.28) yields

$$(\dot{y}_C - \dot{y}_A) = (\dot{x}_C - \dot{x}_A)\tan\phi_2 + (x_C - x_A)\frac{1}{\cos^2\phi_2}\dot{\phi}_2$$
$$\dot{\phi}_3 = \dot{\phi}_2. \tag{3.29}$$

The angular velocities are

$$\omega_2 = \omega_3 = \frac{\cos^2\phi_2}{x_C - x_A}[(\dot{y}_C - \dot{y}_A) - (\dot{x}_C - \dot{x}_A)\tan\phi_2]. \tag{3.30}$$

The angular accelerations are found to be

$$\alpha_2 = \alpha_3 = \dot{\omega}_2 = \dot{\omega}_3. \tag{3.31}$$

3.5 TRT Dyad

For the TRT dyad (Fig. 2.7), two equations can be written:

$$[x_C(t) - x_A]\sin\alpha - [y_C(t) - y_A]\cos\alpha = \pm d$$
$$[x_C(t) - x_B]\sin\beta - [y_C(t) - y_B]\cos\beta = \pm b. \tag{3.32}$$

The derivative with respect to time of Eq. (3.32) yields

$$(\dot{x}_C - \dot{x}_A)\sin\alpha - (\dot{y}_C - \dot{y}_A)\cos\alpha + (x_C - x_A)\dot{\alpha}\cos\alpha + (y_C - y_A)\dot{\alpha}\sin\alpha = 0$$
$$(\dot{x}_C - \dot{x}_B)\sin\beta - (\dot{y}_C - \dot{y}_B)\cos\beta + (x_C - x_B)\dot{\beta}\cos\beta + (y_C - y_B)\dot{\beta}\sin\beta = 0,$$
$$\tag{3.33}$$

which can be rewritten as

$$\mathbf{A}_{10} \cdot \mathbf{v}_C = \mathbf{A}_{11} \cdot \mathbf{v}_1, \tag{3.34}$$

where

$$\mathbf{v}_1 = [\dot{x}_A, \ \dot{y}_A, \ \dot{\alpha}, \ \dot{x}_B, \ \dot{y}_B, \ \dot{\beta}]^T$$

$$\mathbf{A}_{10} = \begin{bmatrix} -\sin\alpha & -\cos\alpha \\ \sin\beta & -\cos\beta \end{bmatrix}$$

$$\mathbf{A}_{11} = \begin{bmatrix} \sin\alpha & -\cos\alpha & 0 & \xi_1 & 0 & 0 \\ 0 & 0 & 0 & \sin\beta & -\cos\beta & \xi_2 \end{bmatrix}$$

$$\xi_1 = (x_A - x_C)\cos\alpha + (y_A - y_C)\sin\alpha$$

$$\xi_2 = (x_B - x_C)\cos\beta + (y_B - y_C)\sin\beta.$$

The solution of Eq. (3.34) gives the velocity of kinematic pair C,

$$\mathbf{v}_C = \mathbf{M}_7 \cdot \mathbf{v}_1, \tag{3.35}$$

where

$$\mathbf{M}_7 = \mathbf{A}_{10}^{-1} \cdot \mathbf{A}_{11}.$$

Differentiating Eq. (3.34) with respect to time, we obtain

$$\mathbf{A}_{10} \cdot \mathbf{a}_C = \mathbf{A}_{11} \cdot \mathbf{a}_1 - \mathbf{A}_{12}, \tag{3.36}$$

where

$$\mathbf{a}_1 = [\ddot{x}_A, \ \ddot{y}_A, \ \ddot{\alpha}, \ \ddot{x}_B, \ \ddot{y}_B, \ \ddot{\beta}]^T$$

$$\mathbf{A}_{12} = \begin{bmatrix} \xi_3 \\ \xi_4 \end{bmatrix}$$

$$\xi_3 = 2(\dot{x}_C - \dot{x}_A)\dot{\alpha}\cos\alpha + 2(\dot{y}_C - \dot{y}_A)\dot{\beta}\sin\alpha$$
$$- (x_C - x_A)\dot{\alpha}^2\sin\alpha + (y_C - y_A)\dot{\alpha}^2\cos\alpha,$$

$$\xi_4 = 2(\dot{x}_C - \dot{x}_B)\dot{\beta}\cos\beta + 2(\dot{y}_C - \dot{y}_B)\dot{\beta}\sin\beta$$
$$- (x_C - x_B)\dot{\beta}^2\sin\beta + (y_C - y_B)\dot{\beta}^2\cos\beta.$$

The solution of the preceding equations give the acceleration of kinematic pair C,

$$\mathbf{a}_C = \mathbf{M}_7 \cdot \mathbf{a} + \mathbf{M}_8, \tag{3.37}$$

where

$$\mathbf{M}_8 = \mathbf{A}_{10}^{-1} \cdot \mathbf{A}_{12}.$$

EXAMPLE The mechanism considered is shown in Fig. 3.1. Find the motion of the mechanism for the following data: $AB = 0.100$ m, $CD = 0.075$ m, $DE = 0.200$ m, $AC = 0.150$ m, driver link angle $\phi = \phi_1 = 45°$, and angular speed of the driver link 1 $\omega = \omega_1 = 4.712$ rad/s. ▲

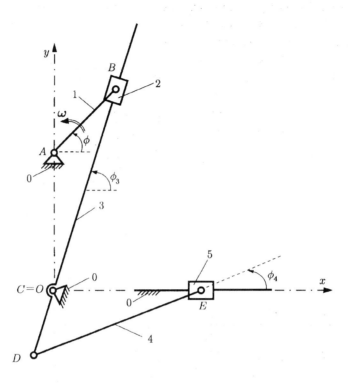

Figure 3.1
*Used with
permission from
Ref. 15.*

3.5.1 VELOCITY AND ACCELERATION ANALYSIS OF THE MECHANISM

The origin of the fixed reference frame is at $C \equiv 0$. The position of the fixed kinematic pair A is

$$x_A = 0, \quad y_A = AC. \tag{3.38}$$

Kinematic Pair B

The position of kinematic pair B is

$$x_B(t) = x_A + AB \cos \phi(t), \quad y_B(t) = y_A + AB \sin \phi(t), \tag{3.39}$$

and for $\phi = 45°$,

$$x_B = 0.000 + 0.100 \cos 45° = 0.070 \text{ m},$$
$$y_B = 0.150 + 0.100 \sin 45° = 0.220 \text{ m}.$$

Velocity Analysis: The linear velocity vector of B is

$$\mathbf{v}_B = \dot{x}_B \mathbf{i} + \dot{y}_B \mathbf{j},$$

where

$$\dot{x}_B = \frac{dx_B}{dt} = -AB\dot{\phi}\sin\phi, \quad \dot{y}_B = \frac{dy_B}{dt} = AB\dot{\phi}\cos\phi. \tag{3.40}$$

With $\phi = 45°$ and $\dot{\phi} = \omega = 4.712\,\text{rad/s}$,

$$\dot{x}_B = -0.100(4.712)\sin 45° = -0.333\,\text{m/s}$$
$$\dot{y}_B = 0.100(4.712)\cos 45° = 0.333\,\text{m/s}$$

$$v_B = |\mathbf{v}_B| = \sqrt{\dot{x}_B^2 + \dot{y}_B^2} = \sqrt{(-0.333)^2 + 0.333^2} = 0.471\,\text{m/s}. \tag{3.41}$$

Acceleration analysis: The linear acceleration vector of B is

$$\mathbf{a}_B = \ddot{x}_B\mathbf{\imath} + \ddot{y}_B\mathbf{\jmath},$$

where

$$\ddot{x}_B = \frac{d\dot{x}_B}{dt} = -AB\dot{\phi}^2\cos\phi - AB\ddot{\phi}\sin\phi$$
$$\ddot{y}_B = \frac{d\dot{y}_B}{dt} = -AB\dot{\phi}^2\sin\phi + AB\ddot{\phi}\cos\phi. \tag{3.42}$$

The angular acceleration of link 1 is $\ddot{\phi} = \dot{\omega} = 0$. The numerical values are

$$\ddot{x}_B = -0.100(4.712)^2\cos 45° = -1.569\,\text{m/s}^2$$
$$\ddot{y}_B = -0.100(4.712)^2\sin 45° = -1.569\,\text{m/s}^2$$

$$a_B = |\mathbf{a}_B| = \sqrt{\ddot{x}_B^2 + \ddot{y}_B^2} = \sqrt{(-1.569)^2 + (-1.569)^2} = 2.220\,\text{m/s}^2. \tag{3.43}$$

Link 3

The kinematic pairs B, C, and D are located on the same link BD. The following equation can be written:

$$y_B(t) - y_C - [x_B(t) - x_C]\tan\phi_3(t) = 0. \tag{3.44}$$

The angle $\phi_3 = \phi_2$ is computed using

$$\phi_3 = \phi_2 = \arctan\frac{y_B - y_C}{x_B - x_C}, \tag{3.45}$$

and for $\phi = 45°$ one can obtain

$$\phi_3 = \arctan\frac{0.22}{0.07} = 72.235°.$$

Velocity analysis: The derivative of Eq. (3.44) yields

$$\dot{y}_B - \dot{y}_C - (\dot{x}_B - \dot{x}_C)\tan\phi_3 - (x_B - x_C)\frac{1}{\cos^2\phi_3}\dot{\phi}_3 = 0. \tag{3.46}$$

The angular velocity of link 3, $\omega_3 = \omega_2 = \dot{\phi}_3$, can be computed as

$$\omega_3 = \omega_2 = \frac{\cos^2\phi_3[\dot{y}_B - \dot{y}_C - (\dot{x}_B - \dot{x}_C)\tan\phi_3]}{x_B - x_C}, \tag{3.47}$$

220 Theory of Mechanisms

and

$$\omega_3 = \frac{\cos^2 72°(0.333 + 0.333 \tan 72.235°)}{0.07} = 1.807 \text{ rad/s.}$$

Acceleration analysis: The angular acceleration of link 3, $\alpha_3 = \alpha_2 = \ddot{\phi}_3$, can be computed from the time derivative of Eq. (3.46):

$$\ddot{y}_B - \ddot{y}_C - (\ddot{x}_B - \ddot{x}_C)\tan\phi_3 - 2(\dot{x}_B - \dot{x}_C)\frac{1}{\cos^2\phi_3}\dot{\phi}_3$$

$$- 2(x_B - x_C)\frac{\sin\phi_3}{\cos^3\phi_3}\dot{\phi}_3^2 - (x_B - x_C)\frac{1}{\cos^2\phi_3}\ddot{\phi}_3 = 0. \qquad (3.48)$$

The solution of the previous equation is

$$\alpha_3 = \alpha_2 = \left[\ddot{y}_B - \ddot{y}_C - (\ddot{x}_B - \ddot{x}_C)\tan\phi_3 - 2(\dot{x}_B - \dot{x}_C)\frac{1}{\cos^2\phi_3}\dot{\phi}_3\right.$$

$$\left. - 2(x_B - x_C)\frac{\sin\phi_3}{\cos^3\phi_3}\dot{\phi}_3^2\right]\frac{\cos^2\phi_3}{x_B - x_C}, \qquad (3.49)$$

and for the given numerical data

$$\alpha_3 = \alpha_2 = \left[-1.569 + 1.569\tan 72.235° + 2(0.333)\frac{1}{\cos^2 72.235°}1.807\right.$$

$$\left. - 2(0.07)\frac{\sin 72.235°}{\cos^2 72.235°}(1.807)^2\right]\frac{\cos^2 72.235°}{0.07} = 1.020 \text{ rad/s}^2. \qquad (3.50)$$

Kinematic Pair D

For the position analysis of kinematic pair D, the following quadratic equations can be written:

$$[x_D(t) - x_C]^2 + [y_D(t) - y_C]^2 = BC^2 \qquad (3.51)$$

$$[x_D(t) - x_C]\sin\phi_3(t) - [y_D(t) - y_C]\cos\phi_3(t) = 0. \qquad (3.52)$$

The previous equations can be rewritten as follows:

$$x_D^2(t) + y_D^2(t) = CD^2$$

$$x_D(t)\sin\phi_3(t) - y_D(t)\cos\phi_3(t) = 0. \qquad (3.53)$$

The solution of the preceding system of equations is

$$x_D = \pm\frac{CD}{\sqrt{1 + \tan^2\phi_3}} = \pm\frac{0.075}{\sqrt{1 + \tan^2 72.235°}} = -0.023 \text{ m}$$

$$y_D = x_D\tan\phi_3 = -0.023\tan 72.235° = -0.071 \text{ m.} \qquad (3.54)$$

The negative value for x_D was selected for this position of the mechanism.

Velocity analysis: The velocity analysis is carried out by differentiating Eq. (3.53),

$$x_D\dot{x}_D + y_D\dot{y}_D = 0$$

$$\dot{x}_D\sin\phi_3 + x_D\cos\phi_3\dot{\phi}_3 - \dot{y}_D\cos\phi_3 + y_D\sin\phi_3\dot{\phi}_3 = 0, \qquad (3.55)$$

which can be rewritten as

$$-0.023\dot{x}_D - 0.07\dot{y}_D = 0$$

$$0.95\dot{x}_D - 0.023(0.3)(1.807) - 0.3\dot{y}_D - 0.07(0.95)(1.807) = 0. \qquad (3.56)$$

The solution is

$$\dot{x}_D = 0.129 \text{ m/s}, \quad \dot{y}_D = -0.041 \text{ m/s}.$$

The magnitude of the velocity of kinematic pair D is

$$v_D = |\mathbf{v}_D| = \sqrt{\dot{x}_D^2 + \dot{y}_D^2} = \sqrt{0.129^2 + (-0.041)^2} = 0.135 \text{ m/s}.$$

Acceleration analysis: The acceleration analysis is obtained using the derivative of the velocity given by Eq. (3.55):

$$\dot{x}_D^2 + x_D\ddot{x}_D + \dot{y}_D^2 + y_D\ddot{y}_D = 0$$

$$\ddot{x}_D \sin\phi_3 + 2\dot{x}_D\dot{\phi}_3\cos\phi_3 - x_D\dot{\phi}u_3^2\sin\phi_3 + x_D\ddot{\phi}_3\cos\phi_3 - \ddot{y}_D\cos\phi_3$$

$$+ 2\dot{y}_D\dot{\phi}_3\sin\phi_3 + y_D\dot{\phi}_3^2\cos\phi_3 + y_D\ddot{\phi}_3^2\sin\phi_3 = 0. \qquad (3.57)$$

The solution of this system is

$$\ddot{x}_D = 0.147 \text{ m/s}^2, \quad \ddot{y}_D = 0.210 \text{ m/s}^2.$$

The absolute acceleration of kinematic pair D is

$$a_D = |\mathbf{a}_D| = \sqrt{\ddot{x}_d^2 + \ddot{y}_D^2} = \sqrt{(0.150)^2 + (0.212)^2} = 0.256 \text{ m/s}^2.$$

Kinematic Pair E

The position of kinematic pair E is determined from

$$[x_E(t) - x_D(t)]^2 + [y_E(t) - y_D(t)]^2 = DE^2. \qquad (3.58)$$

For kinematic pair E, with the coordinate $y_E = 0$, Eq. (3.58) becomes

$$[x_E(t) - x_D(t)]^2 + y_D^2(t) = DE^2, \qquad (3.59)$$

or

$$(x_E + 0.023)^2 + (0.071)^2 = 0.2^2,$$

with the correct solution $x_E = 0.164$ m.

Velocity analysis: The velocity of kinematic pair E is determined by differentiating Eq. (3.59) as follows:

$$2(\dot{x}_E - \dot{x}_D)(x_E - x_D) + 2y_D\dot{y}_D = 0. \qquad (3.60)$$

This can be rewritten as

$$\dot{x}_E - \dot{x}_D = \frac{y_D\dot{y}_D}{x_E - x_D}.$$

The solution of the foregoing equation is

$$\dot{x}_E = 0.129 - \frac{(-0.071)(-0.041)}{0.164 + 0.023} = 0.113 \text{ m/s}.$$

Acceleration analysis: The derivative of Eq. (3.60) yields

$$(\ddot{x}_E - \ddot{x}_D)(x_E - x_D) + (\dot{x}_E - \dot{x}_D)^2 + \dot{y}_D^2 + y_D\ddot{y}_D = 0, \qquad (3.61)$$

with the solution

$$\ddot{x}_E = \ddot{x}_D - \frac{\dot{y}_D^2 + y_D\ddot{y}_D + (\dot{x}_E - \dot{x}_D)^2}{x_E - x_D},$$

or with numerical values

$$\ddot{x}_E = 0.150 - \frac{(-0.041)^2 + (-0.07)(0.21) + (0.112 - 0.129)^2}{0.164 + 0.023} = 0.271 \text{ m/s}^2.$$

Link 4

The angle ϕ_4 is determined from

$$y_E - y_D(t) - [x_E(t) - x_D(t)]\tan\phi_4(t) = 0, \qquad (3.62)$$

where $y_E = 0$. The preceding equation can be rewritten as

$$-y_D(t) - [x_E(t) - x_D(t)]\tan\phi_4(t) = 0, \qquad (3.63)$$

and the solution is

$$\phi_4 = \arctan\left(\frac{-y_D}{x_E - x_D}\right) = \arctan\left(\frac{0.071}{0.164 + 0.023}\right) = 20.923°.$$

Velocity analysis: The derivative of Eq. (3.63) yields

$$-\dot{y}_D - (\dot{x}_E - \dot{x}_D)\tan\phi_4 - (x_E - x_D)\frac{1}{\cos^2\phi_4}\dot{\phi}_4 = 0. \qquad (3.64)$$

Hence,

$$\omega_4 = \dot{\phi}_4 = -\frac{\cos^2\phi_4[\dot{y}_d + (\dot{x}_e - \dot{x}_d)\tan\phi_4]}{x_e - x_d} = 0.221 \text{ rad/s.}$$

Acceleration analysis: The angular acceleration of link 4 is determined by differentiating Eq. (3.64) as follows:

$$-\ddot{y}_D - (\ddot{x}_E - \ddot{x}_D)\tan\phi_4 - 2(\dot{x}_E - \dot{x}_D)\frac{1}{\cos^2\phi_4}\dot{\phi}_4$$

$$- 2(x_E - x_D)\frac{\sin\phi_4}{\cos^3\phi_4}\dot{\phi}_4^2 - (x_E - x_D)\frac{1}{\cos^2\phi_4}\ddot{\phi}_4 = 0.$$

The solution of this equation is

$$\alpha_4 = \ddot{\phi}_4 = -1.105 \text{ rad/s}^2.$$

4. Kinetostatics

For a kinematic chain it is important to know how forces and torques are transmitted from the input to the output, so that the links can be properly designated. The friction effects are assumed to be negligible in the dynamic force analysis or kinetostatics analysis presented here. The first part of this chapter is a review of general force analysis principles and conventions.

4.1 Moment of a Force about a Point

A *force vector* $\mathbf{F}$ has a magnitude, an orientation, and a sense. The magnitude of a vector is specified by a positive number and a unit having appropriate dimensions. The orientation of a vector is specified by the relationship between the vector and given reference lines and/or planes. The sense of a vector is specified by the order of two points on a line parallel to the vector. Orientation and sense together determine the direction of a vector. The line of action of a vector is a hypothetical infinite straight line collinear with the vector.

The force vector $\mathbf{F}$ can be expressed in terms of a cartesian reference frame, with the unit vectors $\mathbf{1}$, $\mathbf{\jmath}$, and $\mathbf{k}$ (Fig. 4.1):

$$\mathbf{F} = F_x\mathbf{1} + F_y\mathbf{\jmath} + F_z\mathbf{k}. \tag{4.1}$$

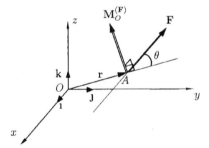

Figure 4.1

The components of the force in the x, y, and z directions are F_x, F_y, and F_z. The resultant of two forces $\mathbf{F}_1 = F_{1x}\mathbf{1} + F_{1y}\mathbf{\jmath} + F_{1z}\mathbf{k}$ and $\mathbf{F}_2 = F_{2x}\mathbf{1} + F_{2y}\mathbf{\jmath} + F_{2z}\mathbf{k}$ is the vector sum of those forces:

$$\mathbf{R} = \mathbf{F}_1 + \mathbf{F}_2 = (F_{1x} + F_{2x})\mathbf{1} + (F_{1y} + F_{2y})\mathbf{\jmath} + (F_{1z} + F_{2z})\mathbf{k}. \tag{4.2}$$

A *moment* (*torque*) is defined as the moment of a force about (with respect to) a point. The moment of the force $\mathbf{F}$ about the point O is the cross product vector

$$\mathbf{M}_O^{(\mathbf{F})} = \mathbf{r} \times \mathbf{F}$$

$$= \begin{vmatrix} \mathbf{1} & \mathbf{\jmath} & \mathbf{k} \\ r_x & r_y & r_z \\ F_x & F_y & F_z \end{vmatrix}$$

$$= (r_y F_z - r_z f_y)\mathbf{1} + (r_z F_x - r_x F_z)\mathbf{\jmath} + (r_x F_y - r_y F_x)\mathbf{k}, \tag{4.3}$$

where $\mathbf{r} = \mathbf{OA} = r_x\mathbf{1} + r_y\mathbf{J} + r_z\mathbf{k}$ is a position vector directed from the point about which the moment is taken (O in this case) to any point on the line of action of the force (Fig. 4.1).

The magnitude of $\mathbf{M}_O^{(F)}$ is

$$|\mathbf{M}_O^{(F)}| = M_O^{(F)} = rF|\sin\theta|,$$

where $\theta = \angle(\mathbf{r}, \mathbf{F})$ is the angle between vectors $\mathbf{r}$ and $\mathbf{F}$, and $r = |\mathbf{r}|$ and $F = |\mathbf{F}|$ are the magnitudes of the vectors.

The line of action of $\mathbf{M}_O^{(F)}$ is perpendicular to the plane containing $\mathbf{r}$ and $\mathbf{F}$ ($\mathbf{M}_O^{(F)} \perp \mathbf{r}$ and $\mathbf{M}_O^{(F)} \perp \mathbf{F}$), and the sense is given by the right-hand rule.

Two forces, $\mathbf{F}_1$ and $\mathbf{F}_2$, that have equal magnitudes $|\mathbf{F}_1| = |\mathbf{F}_2|$, opposite senses $\mathbf{F} = -\mathbf{F}_2$, and parallel directions ($\mathbf{F}_1 \| \mathbf{F}_2$) are called a *couple*. The resultant force of a couple is zero $\mathbf{R} = \mathbf{F}_1 + \mathbf{F}_2 = \mathbf{0}$. The resultant moment $\mathbf{M} \neq \mathbf{0}$ about an arbitrary point is

$$\mathbf{M} = \mathbf{r}_1 \times \mathbf{F}_1 + \mathbf{r}_2 \times \mathbf{F}_2,$$

or

$$\mathbf{M} = \mathbf{r}_1 \times (-\mathbf{F}_2) + \mathbf{r}_2 \times \mathbf{F}_2 = (\mathbf{r}_2 - \mathbf{r}_1) \times \mathbf{F}_2 = \mathbf{r} \times \mathbf{F}_2, \qquad (4.4)$$

where $\mathbf{r} = \mathbf{r}_2 - \mathbf{r}_1$ is a vector from any point on the line of action of $\mathbf{F}_1$ to any point of the line of action of $\mathbf{F}_2$. The direction of the torque is perpendicular to the plane of the couple and the magnitude is given by (Fig. 4.2)

$$|\mathbf{M}| = M = rF_2|\sin\theta| = hF_2, \qquad (4.5)$$

where $h = r|\sin\theta|$ is the perpendicular distance between the line of action. The resultant moment of a couple is independent of the point with respect to which moments are taken.

4.2 Inertia Force and Inertia Moment

Newton's second law of motion states that *a particle acted on by forces whose resultant is not zero will move in such a way that the time rate of*

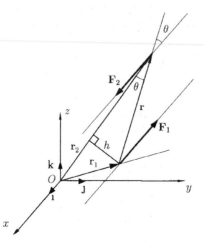

Figure 4.2

change of its momentum will at any instant be proportional to the resultant force.

In the case of a particle with constant mass (m = constant), Newton's second law is expressed as

$$\mathbf{F} = m\mathbf{a}, \tag{4.6}$$

where the vector $\mathbf{F}$ is the resultant of the external forces on the particle and $\mathbf{a}$ is the acceleration vector of the particle.

In the case of a rigid body, RB, with constant mass m, Newton's second law is (Fig. 4.3)

$$\mathbf{F} = m\mathbf{a}_C \tag{4.7}$$

$$\mathbf{M}_C = I_C\boldsymbol{\alpha}, \tag{4.8}$$

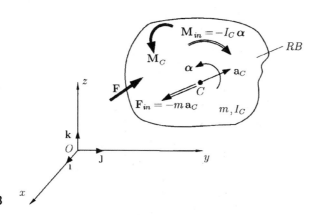

Figure 4.3

where

- $\mathbf{F}$ is the resultant of external force on the rigid body
- $\mathbf{a}_C$ is the acceleration of the center of mass, C, of the rigid body
- $\mathbf{M}_C$ is the resultant external moment on the rigid body about the center of mass C
- I_C is the mass moment of inertia of the rigid body with respect to an axis passing through the center of mass C and perpendicular to the plane of rotation of the rigid body
- $\boldsymbol{\alpha}$ is the angular acceleration of the rigid body

Equations (4.7) and (4.8) can be interpreted in two ways:

1. The forces and moments are known and the equations can be solved for the motion of the rigid body (direct dynamics).
2. The motion of the RB is known and the equations can be solved for the force and moments (inverse dynamics).

The dynamic force analysis in this chapter is based on the known motion of the mechanism.

D'Alembert's principle is derived from Newton's second law and is expressed as

$$\mathbf{F} + (-m\mathbf{a}_C) = \mathbf{0} \tag{4.9}$$

$$\mathbf{M}_C + (-I_C\boldsymbol{\alpha}) = \mathbf{0}. \tag{4.10}$$

The terms in parentheses in Eqs. (4.9) and (4.10) are called the *inertia force* and the *inertia moments*, respectively. The inertia force $\mathbf{F}_{in}$ is

$$\mathbf{F}_{in} = -m\mathbf{a}_C, \tag{4.11}$$

and the inertia moment is

$$\mathbf{M}_{in} = -I_C\boldsymbol{\alpha}. \tag{4.12}$$

The dynamic force analysis can be expressed in a form similar to static force analysis,

$$\sum \mathbf{R} = \sum \mathbf{F} + \mathbf{F}_{in} = \mathbf{0} \tag{4.13}$$

$$\sum \mathbf{T}_C = \sum \mathbf{M}_C + \mathbf{M}_{in} = \mathbf{0}, \tag{4.14}$$

where $\sum \mathbf{F}$ is the vector sum of all external forces (resultant of external force), and $\sum \mathbf{M}_C$ is the sum of all external moments about the center of mass C (resultant external moment).

For a rigid body in plane motion in the xy plane,

$$\mathbf{a}_C = \ddot{x}_C\mathbf{1} + \ddot{y}_C\mathbf{J}, \quad \boldsymbol{\alpha} = \alpha\mathbf{k},$$

with all external forces in that plane, Eqs. (4.13) and (4.14) become

$$\sum R_x = \sum F_x + F_{in\,x} = \sum F_x + (-m\ddot{x}_C) = 0 \tag{4.15}$$

$$\sum R_y = \sum F_y + F_{in\,y} = \sum F_y + (-m\ddot{y}_C) = 0 \tag{4.16}$$

$$\sum T_C = \sum M_C + M_{in} = \sum M_C + (-I_C\alpha) = 0. \tag{4.17}$$

With d'Alembert's principle the moment summation can be about any arbitrary point P,

$$\sum \mathbf{T}_P = \sum \mathbf{M}_P + \mathbf{M}_{in} + \mathbf{r}_{PC} \times \mathbf{F}_{in} = \mathbf{0}, \tag{4.18}$$

where

- $\sum \mathbf{M}_P$ is the sum of all external moments about P
- $\mathbf{M}_{in} = -I_C\boldsymbol{\alpha}$ is the inertia moment
- $\mathbf{F}_{in} = -m\mathbf{a}_C$ is the inertia force
- $\mathbf{r}_{PC} = \mathbf{PC}$ is a vector from P to C

The dynamic analysis problem is reduced to a static force and moment balance problem where the inertia forces and moments are treated in the same way as external forces and torques.

4.3 Free-Body Diagrams

A free-body diagram is a drawing of a part of a complete system, isolated in order to determine the forces acting on that rigid body. The following force convention is defined: $\mathbf{F}_{ij}$ represents the force exerted by link i on link j.

Figure 4.4 shows various free-body diagrams that can be considered in the analysis of a crank slider mechanism (Fig. 4.4a).

In Fig. 4.4b, the free body consists of the three moving links isolated from the frame 0. The forces acting on the system include a driving torque $\mathbf{M}$, an external driven force $\mathbf{F}$, and the forces transmitted from the frame at kinematic pair A, $\mathbf{F}_{01}$, and at kinematic pair C, $\mathbf{F}_{03}$. Figure 4.4c is a free-body diagram of the two links 1 and 2. Figure 4.4d is a free-body diagram of a single link.

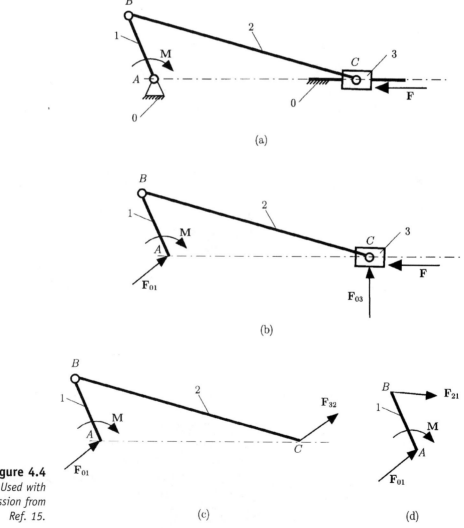

Figure 4.4
Used with permission from Ref. 15.

Mechanisms

The force analysis can be accomplished by examining individual links or subsystems of links. In this way the reaction forces between links as well as the required input force or moment for a given output load are computed.

4.4 Reaction Forces

Figure 4.5a is a schematic diagram of a crank slider mechanism comprising of a crank 1, a connecting rod 2, and a slider 3. The center of mass of link 1 is C_1, the center of mass of link 2 is C_2, and the center of mass of slider 3 is C. The mass of the crank is m_1, the mass of the connecting road is m_2, and the mass of the slider is m_3. The moment of inertia of link i is I_{Ci}, $i = 1, 2, 3$.

The gravitational force is $G_i = -m_i g\mathbf{j}$, $i = 1, 2, 3$, where $g = 9.81$ m/s^2 is the acceleration of gravity.

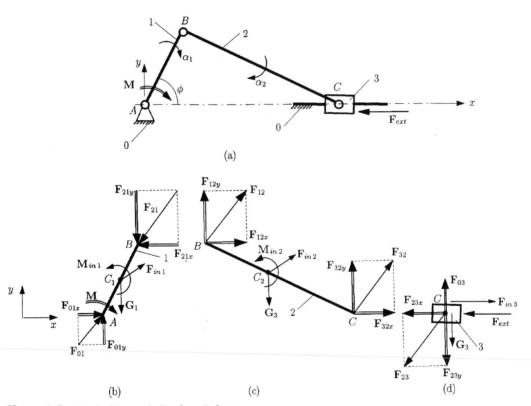

Figure 4.5   *Used with permission from Ref. 15.*

For a given value of the crank angle ϕ and a known driven force $\mathbf{F}_{ext}$, the kinematic pair reactions and the drive moment $\mathbf{M}$ on the crank can be computed using free-body diagrams of the individual links.

Figures 4.5b, 4.5c, and 4.5d show free-body diagrams of the crank 1, the connecting rod 2, and the slider 3. For each moving link the dynamic equilibrium equations are applied.

For the slider 3 the vector sum of the all the forces (external forces $\mathbf{F}_{ext}$, gravitational force $\mathbf{G}_3$, inertia forces $\mathbf{F}_{in3}$, reaction forces $\mathbf{F}_{23}$, $\mathbf{F}_{03}$) is zero (Fig. 4.5d):

$$\sum \mathbf{F}^{(3)} = \mathbf{F}_{23} + \mathbf{F}_{in3} + \mathbf{G}_3 + \mathbf{F}_{ext} + \mathbf{F}_{03} = 0.$$

Projecting this force onto the x and y axes gives

$$\sum \mathbf{F}^{(3)} \cdot \mathbf{1} = F_{23x} + (-m_3\ddot{x}_C) + F_{ext} = 0 \tag{4.19}$$

$$\sum \mathbf{F}^{(3)} \cdot \mathbf{J} = F_{23y} - m_3 g + F_{03y} = 0. \tag{4.20}$$

For the connecting rod 2 (Fig. 4.5c), two vertical equations can be written:

$$\sum \mathbf{F}^{(2)} = \mathbf{F}_{32} + \mathbf{F}_{in2} + \mathbf{G}_2 + \mathbf{F}_{12} = 0$$

$$\sum \mathbf{M}_B^{(2)} = (\mathbf{r}_C - \mathbf{r}_B) \times \mathbf{F}_{32} + (\mathbf{r}_{C2} - \mathbf{r}_B) \times (\mathbf{F}_{in2} + \mathbf{G}_2) + \mathbf{M}_{in2} = 0,$$

or

$$\sum \mathbf{F}^{(2)} \cdot \mathbf{1} = F_{32x} + (-m_2\ddot{x}_{C2}) + F_{12x} = 0 \tag{4.21}$$

$$\sum \mathbf{F}^{(2)} \cdot \mathbf{J} = F_{32y} + (-m_2\ddot{y}_{C2}) - m_2 g + F_{12y} = 0 \tag{4.22}$$

$$\begin{vmatrix} \mathbf{1} & \mathbf{J} & \mathbf{k} \\ x_C - x_B & y_C - y_B & 0 \\ F_{32x} & F_{32y} & 0 \end{vmatrix} + \begin{vmatrix} \mathbf{1} & \mathbf{J} & \mathbf{k} \\ x_{C2} - x_B & y_{C2} - y_B & 0 \\ -m_2\ddot{x}_{C2} & -m_2\ddot{y}_{C2} - m_2 g & 0 \end{vmatrix}$$

$$- I_{C2}\alpha_2\mathbf{k} = 0. \tag{4.23}$$

For the crank 1 (Fig. 4.5b), there are two vectorial equations,

$$\sum \mathbf{F}^{(1)} = \mathbf{F}_{21} + \mathbf{F}_{in1} + \mathbf{G}_1 + \mathbf{F}_{01} = 0$$

$$\sum \mathbf{M}_A^{(1)} = \mathbf{r}_B \times \mathbf{F}_{21} + \mathbf{r}_{C1} \times (\mathbf{F}_{in1} + \mathbf{G}_1) + \mathbf{M}_{in1} + \mathbf{M} = 0$$

or

$$\sum \mathbf{F}^{(1)} \cdot \mathbf{1} = F_{21x} + (-m_1\ddot{x}_{C1}) + F_{01x} = 0 \tag{4.24}$$

$$\sum \mathbf{F}^{(1)} \cdot \mathbf{J} = F_{21y} + (-m_1\ddot{y}_{C1}) - m_1 g + F_{01y} = 0 \tag{4.25}$$

$$\begin{vmatrix} \mathbf{1} & \mathbf{J} & \mathbf{k} \\ x_B & y_B & 0 \\ F_{21x} & F_{21y} & 0 \end{vmatrix} + \begin{vmatrix} \mathbf{1} & \mathbf{J} & \mathbf{k} \\ x_{C1} & y_{C1} & 0 \\ -m_1\ddot{x}_{C1} & -m_1\ddot{y}_{C1} - m_1 g & 0 \end{vmatrix} - I_{C1}\alpha_1\mathbf{k} + M\mathbf{k} = 0,$$

$$\tag{4.26}$$

where $M = |\mathbf{M}|$ is the magnitude of the input torque on the crank.

The eight scalar unknowns F_{03y}, $F_{23x} = -F_{32x}$, $F_{23y} = -F_{32y}$, $F_{12x} = -F_{21x}$, $F_{12y} = -F_{21y}$, F_{01x}, F_{01y}, and M are computed from the set of eight equations (4.19), (4.20), (4.21), (4.22), (4.23), (4.24), (4.25), and (4.26).

4.5 Contour Method

An analytical method to compute reaction forces that can be applied for both planar and spatial mechanisms will be presented. The method is based on the decoupling of a closed kinematic chain and writing the dynamic equilibrium equations. The kinematic links are loaded with external forces and inertia forces and moments.

A general monocontour closed kinematic chain is considered in Fig. 4.6. The reaction force between the links $i-1$ and i (kinematic pair A_i) will be determined. When these two links $i-1$ and i are separated (Fig. 4.6b), the reaction forces $\mathbf{F}_{i-1,i}$ and $\mathbf{F}_{i,i-1}$ are introduced and

$$\mathbf{F}_{i-1,i} + \mathbf{F}_{i,i-1} = \mathbf{0}. \tag{4.27}$$

Table 4.1 shows the reaction forces for several kinematic pairs. The following notations have been used: $\mathbf{M}_\Delta$ is the moment with respect to the axis Δ, and F_Δ is the projection of the force vector $\mathbf{F}$ onto the axis Δ.

It is helpful to "mentally disconnect" the two links $(i-1)$ and i, which create the kinematic pair A_i, from the rest of the mechanism. The kinematic pair at A_i will be replaced by the reaction forces $\mathbf{F}_{i-1,i}$, and $\mathbf{F}_{i,i-1}$. The closed kinematic chain has been transformed into two open kinematic chains, and two paths I and II can be associated. The two paths start from A_i.

For the path I (counterclockwise), starting at A_i and following I the first kinematic pair encountered is A_{i-1}. For the link $i-1$ left behind, dynamic equilibrium equations can be written according to the type of kinematic pair

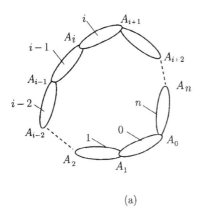

(a)

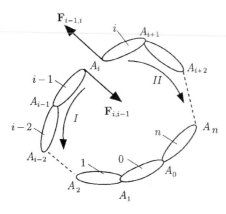

Figure 4.6   (b)

Table 4.1 *Reaction Forces for Several Kinematic Pairs*

Type of joint	Joint force or moment	Unknowns	Equilibrium condition
 Rotational joint	$F_x + F_y = F$ $F \perp \Delta\Delta$	$|F_x| = F_x$ $|F_y| = F_y$	$M_\Delta = 0$
 Translational joint	$F \perp \Delta\Delta$	$|F| = F$ x	$F_\Delta = 0$
 Cylindrical joint	$F_x + F_y = F$ $F \perp \Delta\Delta$	$|F_x| = F_x$ $|F_y| = F_y$ x	$F_\Delta = 0$ $M_\Delta = 0$
 Roll-slide joint	$F \perp \Delta\Delta$ $F \parallel n$	$|F| = F$ x	$F_\Delta = 0$ $M_\Delta = 0$
 Sphere joint	$F_x + F_y + F_z = F$	$|F_x| = F_x$ $|F_y| = F_y$ $|F_z| = F_z$	$M_{\Delta_1} = 0$ $M_{\Delta_2} = 0$ $M_{\Delta_3} = 0$

Mechanisms

at A_{i-1}. Following the same path I, the next kinematic pair encountered is A_{i-2}. For the subsystem ($i - 1$ and $i - 2$), equilibrium conditions corresponding to the type of the kinematic pair at A_{i-2} can be specified, and so on. A similar analysis can be performed for the path II of the open kinematic chain. The number of equilibrium equations written is equal to the number of unknown scalars introduced by the kinematic pair A_i (reaction forces at this kinematic pair). For a kinematic pair, the number of equilibrium conditions is equal to the number of relative mobilities of the kinematic pair.

The five-link ($j = 1, 2, 3, 4, 5$) mechanism shown in Fig. 4.7a has the center of mass locations designated by $C_j(x_{C_j}, y_{C_j}, 0)$. The following analysis will consider the relationships of the inertia forces $\mathbf{F}_{inj}$, the inertia moments $\mathbf{M}_{inj}$, the gravitational force $\mathbf{G}_j$, the driven force, $\mathbf{F}_{ext}$, to the joint reactions $\mathbf{F}_{ij}$, and the drive torque $\mathbf{M}$ on the crank 1 [15].

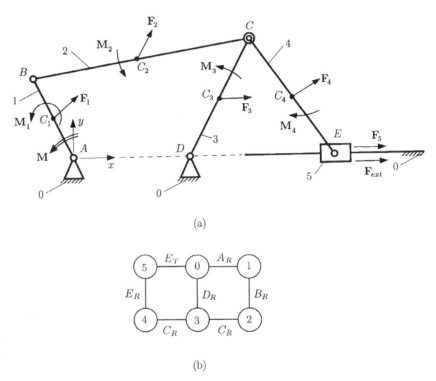

(a)

(b)

Figure 4.7
Used with permission from Ref. 15.

To simplify the notation, the total vector force at C_j is written as $\mathbf{F}_j = \mathbf{F}_{inj} + \mathbf{G}_j$ and the inertia torque of link j is written as $\mathbf{M}_j = \mathbf{M}_{inj}$. The diagram representing the mechanism is depicted in Fig. 4.7b and has two contours 0-1-2-3-0 and 0-3-4-5-0.

Remark

The kinematic pair at C represents a ramification point for the mechanism and the diagram, and the dynamic force analysis will start with this kinematic pair. The force computation starts with the contour 0-3-4-5-0 because the driven load $\mathbf{F}_{ext}$ on link 5 is given.

4.5.1 (I) CONTOUR 0-3-4-5-0

Reaction $\mathbf{F}_{34}$

The rotation kinematic pair at C (or C_R, where the subscript R means rotation), between 3 and 4, is replaced with the unknown reaction (Fig. 4.8)

$$\mathbf{F}_{34} = -\mathbf{F}_{43} = F_{34x}\mathbf{1} + F_{34y}\mathbf{J}.$$

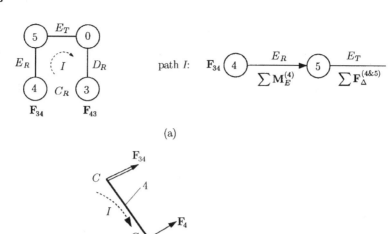

Figure 4.8
*Used with
permission from
Ref. 15.*

If the path I is followed (Fig. 4.8a), for the rotation kinematic pair at $E(E_R)$ a moment equation can be written as

$$\sum \mathbf{M}_E^{(4)} = (\mathbf{r}_C - \mathbf{r}_E) \times \mathbf{F}_{32} + (\mathbf{r}_{C4} - \mathbf{r}_D) \times \mathbf{F}_4 + \mathbf{M}_4 = 0,$$

or

$$\begin{vmatrix} \mathbf{1} & \mathbf{j} & \mathbf{k} \\ x_C - x_E & y_C - y_E & 0 \\ F_{34x} & F_{34y} & 0 \end{vmatrix} + \begin{vmatrix} \mathbf{1} & \mathbf{j} & \mathbf{k} \\ x_{C4} - x_E & y_{C4} - y_E & 0 \\ F_{4x} & F_{4y} & 0 \end{vmatrix} + M_4 \mathbf{k} = \mathbf{0}. \quad (4.28)$$

Continuing on path I, the next kinematic pair is the translational kinematic pair at $D (D_T)$. The projection of all the forces that act on 4 and 5 onto the sliding direction Δ (x axis) should be zero:

$$\sum \mathbf{F}_\Delta^{(4\,\&\,5)} = \sum \mathbf{F}^{(4\,\&\,5)} \cdot \mathbf{1} = (\mathbf{F}_{34} + \mathbf{F}_4 + \mathbf{F}_5 + \mathbf{F}_{ext}) \cdot \mathbf{1}$$
$$= F_{34x} + F_{4x} + F_{5x} + F_{ext} = 0. \quad (4.29)$$

After the system of Eqs. (4.28) and (4.29) are solved, the two unknowns F_{34x} and F_{34y} are obtained.

Reaction $\mathbf{F}_{45}$

The rotation kinematic pair at $E (E_R)$, between 4 and 5, is replaced with the unknown reaction (Fig. 4.9)

$$\mathbf{F}_{45} = -\mathbf{F}_{54} = F_{45x}\mathbf{1} + F_{45y}\mathbf{j}.$$

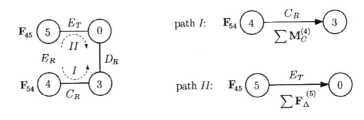

(a)

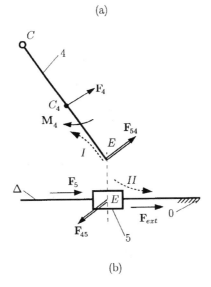

Figure 4.9
*Used with
permission from
Ref. 15.*

(b)

If the path I is traced (Fig. 4.9a), for the pin kinematic pair at C (C_R) a moment equation can be written,

$$\sum \mathbf{M}_C^{(4)} = (\mathbf{r}_E - \mathbf{r}_C) \times \mathbf{F}_{54} + (\mathbf{r}_{C4} - \mathbf{r}_C) \times \mathbf{F}_4 + \mathbf{M}_4 = \mathbf{0},$$

or

$$\begin{vmatrix} \mathbf{1} & \mathbf{J} & \mathbf{k} \\ x_E - x_C & y_E - y_C & 0 \\ -F_{45x} & -F_{45y} & 0 \end{vmatrix} + \begin{vmatrix} \mathbf{1} & \mathbf{J} & \mathbf{k} \\ x_{C4} - x_C & y_{C4} - y_C & 0 \\ F_{4x} & F_{4y} & 0 \end{vmatrix} + M_4 \mathbf{k} = \mathbf{0}. \quad (4.30)$$

For the path II the slider kinematic pair at E (E_T) is encountered. The projection of all forces that act on 5 onto the sliding direction Δ (x axis) should be zero:

$$\sum \mathbf{F}_\Delta^{(5)} = \sum \mathbf{F}^{(5)} \cdot \mathbf{1} = (\mathbf{F}_{45} + \mathbf{F}_5 + \mathbf{F}_{ext}) \cdot \mathbf{1}$$
$$= F_{45x} + F_{5x} + F_{ext} = 0. \quad (4.31)$$

The unknown force components F_{45x} and F_{45y} are calculated from Eqs. (4.30) and (4.31).

Reaction $\mathbf{F}_{05}$

The slider kinematic pair at E (E_T), between 0 and 5, is replaced with the unknown reaction (Fig. 4.10)

$$\mathbf{F}_{05} = F_{05y}\mathbf{J}.$$

The reaction kinematic pair introduced by the translational kinematic pair is perpendicular to the sliding direction, $\mathbf{F}_{05} \perp \Delta$. The application point P of the force $\mathbf{F}_{05}$ is unknown.

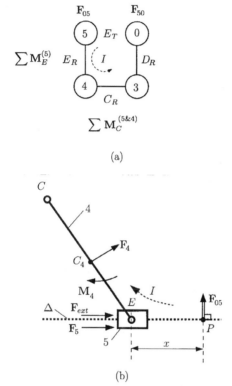

(a)

(b)

Figure 4.10
Used with permission from Ref. 15.

If the path I is followed, as in Fig. 4.10a, for the pin kinematic pair at E (E_R) a moment equation can be written for link 5,

$$\sum \mathbf{M}_E^{(5)} = (\mathbf{r}_P - \mathbf{r}_E) \times \mathbf{F}_{05} = 0,$$

or

$$xF_{05y} = 0 \Longrightarrow x = 0. \tag{4.32}$$

The application point is at E ($P \equiv E$).

Continuing on path I, the next kinematic pair is the pin kinematic pair C (C_R):

$$\sum \mathbf{M}_C^{(4\,\&\,5)} = (\mathbf{r}_E - \mathbf{r}_C) \times (\mathbf{F}_{05} + \mathbf{F}_5 + \mathbf{F}_{ext}) + (\mathbf{r}_{C4} - \mathbf{r}_C) \times \mathbf{F}_4 + \mathbf{M}_4 = 0,$$

Mechanisms

or

$$\begin{vmatrix} \mathbf{1} & \mathbf{J} & \mathbf{k} \\ x_E - x_C & y_E - y_C & 0 \\ F_{5x} + F_{ext} & F_{05y} & 0 \end{vmatrix} + \begin{vmatrix} \mathbf{1} & \mathbf{J} & \mathbf{k} \\ x_{C4} - x_C & y_{C4} - y_C & 0 \\ F_{4x} & F_{4y} & 0 \end{vmatrix} + M_4\mathbf{k} = \mathbf{0}. \quad (4.33)$$

The kinematic pair reaction force F_{05y} can be computed from Eq. (4.33).

4.5.2 (II) CONTOUR 0-1-2-3-0

For this contour the kinematic pair force $\mathbf{F}_{43} = -\mathbf{F}_{34}$ at the ramification point C is considered as a known external force.

Reaction $\mathbf{F}_{03}$

The pin kinematic pair D_R, between 0 and 3, is replaced with unknown reaction force (Fig. 4.11)

$$\mathbf{F}_{03} = F_{03x}\mathbf{1} + F_{03y}\mathbf{J}.$$

If the path I is followed (Fig. 4.11a), a moment equation can be written for the pin kinematic pair C_R for the link 3,

$$\sum \mathbf{M}_C^{(3)} = (\mathbf{r}_D - \mathbf{r}_C) \times \mathbf{F}_{03} + (\mathbf{r}_{C3} - \mathbf{r}_C) \times \mathbf{F}_3 + \mathbf{M}_3 = \mathbf{0},$$

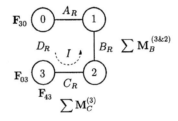

(a)

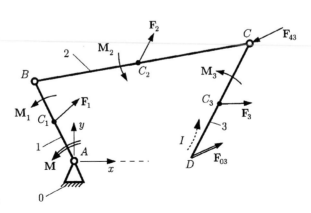

Figure 4.11
*Used with
permission from
Ref. 15.*

(b)

or

$$\begin{vmatrix} \mathbf{1} & \mathbf{j} & \mathbf{k} \\ x_D - x_C & y_D - y_C & 0 \\ F_{03x} & F_{03y} & 0 \end{vmatrix} + \begin{vmatrix} \mathbf{1} & \mathbf{j} & \mathbf{k} \\ x_{C3} - x_C & y_{C3} - y_C & 0 \\ F_{3x} & F_{3y} & 0 \end{vmatrix} + M_3\mathbf{k} = \mathbf{0}. \quad (4.34)$$

Continuing on path I, the next kinematic pair is the pin kinematic pair B_R, and a moment equation can be written for links 3 and 2,

$$\sum \mathbf{M}_B^{(3\,\&\,2)} = (\mathbf{r}_D - \mathbf{r}_B) \times \mathbf{F}_{03} + (\mathbf{r}_{C3} - \mathbf{r}_B) \times \mathbf{F}_3 + M_3 + (\mathbf{r}_C - \mathbf{r}_B) \times \mathbf{F}_{43}$$
$$+ (\mathbf{r}_{C2} - \mathbf{r}_B) \times \mathbf{F}_2 + M_2 = \mathbf{0},$$

or

$$\begin{vmatrix} \mathbf{1} & \mathbf{j} & \mathbf{k} \\ x_D - x_B & y_D - y_B & 0 \\ F_{03x} & F_{03y} & 0 \end{vmatrix} + \begin{vmatrix} \mathbf{1} & \mathbf{j} & \mathbf{k} \\ x_{C3} - x_B & y_{C3} - y_B & 0 \\ F_{3x} & F_{3y} & 0 \end{vmatrix}$$

$$+ M_3\mathbf{k} + \begin{vmatrix} \mathbf{1} & \mathbf{j} & \mathbf{k} \\ x_C - x_B & y_C - y_B & 0 \\ F_{43x} & F_{43y} & 0 \end{vmatrix}$$

$$+ \begin{vmatrix} \mathbf{1} & \mathbf{j} & \mathbf{k} \\ x_{C2} - x_B & y_{C2} - y_B & 0 \\ F_{2x} & F_{2y} & 0 \end{vmatrix} + M_2\mathbf{k} = \mathbf{0}. \quad (4.35)$$

The two components F_{03x} and F_{03y} of the reaction force are obtained from Eqs. (4.34) and (4.36).

Reaction $\mathbf{F}_{23}$

The pin kinematic pair C_R, between 2 and 3, is replaced with the unknown reaction force (Fig. 4.12)

$$\mathbf{F}_{23} = F_{23x}\mathbf{1} + F_{23y}\mathbf{j}.$$

If the path I is followed, as in Fig. 4.12a, a moment equation can be written for the pin kinematic pair D_R for the link 3,

$$\sum \mathbf{M}_D'^{(3)} = (\mathbf{r}_C - \mathbf{r}_D) \times (\mathbf{F}_{23} + \mathbf{F}_{43})(\mathbf{r}_{C3} - \mathbf{r}_D) \times \mathbf{F}_3 + M_3 = \mathbf{0},$$

or

$$\begin{vmatrix} \mathbf{1} & \mathbf{j} & \mathbf{k} \\ x_C - x_D & y_C - y_D & 0 \\ F_{23x} + F_{43x} & F_{23y} + F_{43y} & 0 \end{vmatrix} + \begin{vmatrix} \mathbf{1} & \mathbf{j} & \mathbf{k} \\ x_{C3} - x_D & y_{C3} - y_D & 0 \\ F_{3x} & F_{3y} & 0 \end{vmatrix} + M_3\mathbf{k} = \mathbf{0}.$$

$$(4.36)$$

For the path II the first kinematic pair encountered is the pin kinematic pair B_R, and a moment equation can be written for link 2,

$$\sum \mathbf{M}_B^{(2)} = (\mathbf{r}_C - \mathbf{r}_B) \times (-\mathbf{F}_{23}) + (\mathbf{r}_{C2} - \mathbf{r}_B) \times \mathbf{F}_2 + M_2 = \mathbf{0},$$

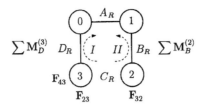

(a)

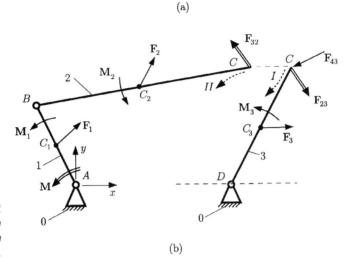

Figure 4.12
*Used with
permission from
Ref. 15.*

(b)

or

$$\begin{vmatrix} \mathbf{1} & \mathbf{J} & \mathbf{k} \\ x_C - x_B & y_C - y_B & 0 \\ -F_{23x} & -F_{23y} & 0 \end{vmatrix} + \begin{vmatrix} \mathbf{1} & \mathbf{J} & \mathbf{k} \\ x_{C2} - x_B & y_{C2} - y_B & 0 \\ F_{2x} & F_{2y} & 0 \end{vmatrix} + M_2\mathbf{k} = \mathbf{0}. \quad (4.37)$$

The two force components F_{23x} and F_{23y} of the reaction force are obtained from Eqs. (4.36) and (4.37).

Reaction $\mathbf{F}_{12}$

The pin kinematic pair B_R, between 1 and 2, is replaced with the unknown reaction force (Fig. 4.13)

$$\mathbf{F}_{12} = F_{12x}\mathbf{1} + F_{12y}\mathbf{J}.$$

If the path I is followed, as in Fig. 4.13a, a moment equation can be written for the pin kinematic pair C_R for the link 2,

$$\sum \mathbf{M}_C^{(2)} = (\mathbf{r}_B - \mathbf{r}_C) \times \mathbf{F}_{12} + (\mathbf{r}_{C2} - \mathbf{r}_C) \times \mathbf{F}_2 + \mathbf{M}_2 = \mathbf{0},$$

or

$$\begin{vmatrix} \mathbf{1} & \mathbf{J} & \mathbf{k} \\ x_B - x_C & y_B - y_C & 0 \\ F_{12x} & F_{12y} & 0 \end{vmatrix} + \begin{vmatrix} \mathbf{1} & \mathbf{J} & \mathbf{k} \\ x_{C2} - x_C & y_{C2} - y_C & 0 \\ F_{2x} & F_{2y} & 0 \end{vmatrix} + M_2\mathbf{k} = \mathbf{0}. \quad (4.38)$$

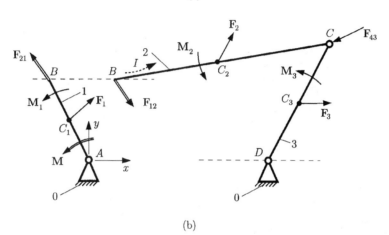

Figure 4.13
*Used with
permission from
Ref. 15.*

Continuing on path I, the next kinematic pair encountered is the pin kinematic pair D_R, and a moment equation can be written for links 2 and 3

$$\sum M_D^{(2\&3)} = (\mathbf{r}_B - \mathbf{r}_D) \times \mathbf{F}_{12} + (\mathbf{r}_{C2} - \mathbf{r}_D) \times \mathbf{F}_2 + \mathbf{M}_2$$
$$+ (\mathbf{r}_C - \mathbf{r}_D) \times \mathbf{F}_{43} + (\mathbf{r}_{C3} - \mathbf{r}_D) \times \mathbf{F}_3 + \mathbf{M}_3 = \mathbf{0},$$

or

$$\begin{vmatrix} \mathbf{1} & \mathbf{J} & \mathbf{k} \\ x_B - x_D & y_B - y_D & 0 \\ F_{12x} & F_{12y} & 0 \end{vmatrix} + \begin{vmatrix} \mathbf{1} & \mathbf{J} & \mathbf{k} \\ x_{C2} - x_D & y_{C2} - y_D & 0 \\ F_{2x} & F_{2y} & 0 \end{vmatrix} + M_2\mathbf{k}$$

$$+ \begin{vmatrix} \mathbf{1} & \mathbf{J} & \mathbf{k} \\ x_C - x_D & y_C - y_D & 0 \\ F_{43x} & F_{43y} & 0 \end{vmatrix} + \begin{vmatrix} \mathbf{1} & \mathbf{J} & \mathbf{k} \\ x_{C3} - x_D & y_{C3} - y_D & 0 \\ F_{3x} & F_{3y} & 0 \end{vmatrix} + M_3\mathbf{k} = \mathbf{0}. \quad (4.39)$$

The two components F_{12x} and F_{12y} of the kinematic pair force are computed from Eqs. (4.38) and (4.39).

Reaction $\mathbf{F}_{01}$ and Driver Torque $\mathbf{M}$

The pin kinematic pair A_R, between 0 and 1, is replaced with the unknown reaction force (Fig. 4.14)

$$\mathbf{F}_{01} = F_{01x}\mathbf{1} + F_{01y}\mathbf{J}.$$

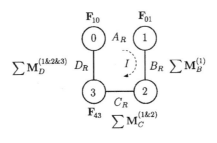

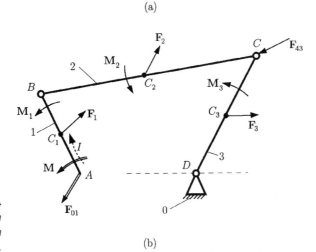

Figure 4.14
*Used with
permission from
Ref. 15.*

The unknown driver torque is $\mathbf{M} = M\mathbf{k}$. If the path I is followed (Fig. 4.14a), a moment equation can be written for the pin kinematic pair B_R for the link 1,

$$\sum \mathbf{M}_B^{(1)} = (\mathbf{r}_A - \mathbf{r}_B) \times \mathbf{F}_{01} + (\mathbf{r}_{C1} - \mathbf{r}_B) \times \mathbf{F}_1 + \mathbf{M}_1 + \mathbf{M} = 0,$$

or

$$\begin{vmatrix} \mathbf{i} & \mathbf{j} & \mathbf{k} \\ x_A - x_B & y_A - y_B & 0 \\ -F_{01x} & -F_{01y} & 0 \end{vmatrix} + \begin{vmatrix} \mathbf{i} & \mathbf{j} & \mathbf{k} \\ x_{C1} - x_B & y_{C1} - y_B & 0 \\ F_{1x} & F_{1y} & 0 \end{vmatrix} + M_1\mathbf{k} + M\mathbf{k} = \mathbf{0}.$$

$$(4.40)$$

Continuing on path I, the next kinematic pair encountered is the pin kinematic pair C_R, and a moment equation can be written for links 1 and 2,

$$\sum \mathbf{M}_C^{(1\,\&\,2)} = (\mathbf{r}_A - \mathbf{r}_C) \times \mathbf{F}_{01} + (\mathbf{r}_{C1} - \mathbf{r}_C) \times \mathbf{F}_1 + \mathbf{M}_1 + \mathbf{M}$$
$$+ (\mathbf{r}_{C2} - \mathbf{r}_C) \times \mathbf{F}_2 + \mathbf{M}_2 = \mathbf{0}. \qquad (4.41)$$

Equation (4.41) is the vector sum of the moments about D_R of all forces and torques that act on links 1, 2, and 3:

$$\sum \mathbf{M}_D^{(1\,\&\,2\,\&\,3)} = (\mathbf{r}_A - \mathbf{r}_D) \times \mathbf{F}_{01} + (\mathbf{r}_{C1} - \mathbf{r}_D) \times \mathbf{F}_1 + \mathbf{M}_1 + \mathbf{M}$$
$$+ (\mathbf{r}_{C2} - \mathbf{r}_D) \times \mathbf{F}_2 + \mathbf{M}_2 + (\mathbf{r}_C - \mathbf{r}_D) \times \mathbf{F}_{43} + (\mathbf{r}_{C3} - \mathbf{r}_D) \times \mathbf{F}_3$$
$$+ \mathbf{M}_3 = \mathbf{0}. \qquad (4.42)$$

The components F_{01x}, F_{01y}, and M are computed from Eqs. (4.40), (4.41), and (4.42).

References

1. P. Appell, *Traité de Mécanique Rationnelle*, Gautier-Villars, Paris, 1941.

2. A. Bedford and W. Fowler, *Dynamics.* Addison-Wesley, Menlo Park, CA, 1999.

3. A. Bedford and W. Fowler, *Statics.* Addison-Wesley, Menlo Park, CA, 1999.

4. M. Atanasiu, *Mecanica.* EDP, Bucharest, 1973.

5. I. I. Artobolevski, *Mechanisms in Modern Engineering Design.* MIR, Moscow, 1977.

6. M. I. Buculei, *Mechanisms-* University of Craiova Press, Craiova, Romania, 1976.

7. M. I. Buculei, D. Bagnaru, G. Nanu and D. B. Marghitu, *Analysis of Mechanisms with Bars*, Scrisul romanesc, Craiova, Romania, 1986.

8. A. G. Erdman and G. N. Sandor, *Mechanisms Design.* Prentice-Hall, Upper Saddle River, NJ, 1984.

9. R. C. Juvinall and K. M. Marshek, *Fundamentals of Machine Component Design.* John Wiley & Sons, New York, 1983.

10. T. R. Kane, *Analytical Elements of Mechanics*, Vol. 1. Academic Press, New York, 1959.

11. T. R. Kane, *Analytical Elements of Mechanics*, Vol. 2. Academic Press, New York, 1961.

12. T. R. Kane and D. A. Levinson, *Dynamics.* McGraw-Hill, New York, 1985.

13. J. T. Kimbrell, *Kinematics Analysis and Synthesis.* McGraw-Hill, New York, 1991.

14. N. I. Manolescu, F. Kovacs, and A. Oranescu, *The Theory of Mechanisms and Machines.* EDP, Bucharest, 1972.

15. D. B. Marghitu and M. J. Crocker, *Analytical Elements of Mechanism.* Cambridge University Press, 2001.

16. D. H. Myszka, *Machines and Mechanisms.* Prentice-Hall, Upper Saddle River, NJ, 1999.

17. R. L. Norton, *Machine Design.* Prentice-Hall, Upper Saddle River, NJ, 1996.

18. R. L. Norton, *Design of Machinery.* McGraw-Hill, New York, 1999.

19. R. M. Pehan, *Dynamics of Machinery.* McGraw-Hill, New York, 1967.

20. I. Popescu, *Planar Mechanisms.* Scrisul romanesc, Craiova, Romania, 1977.

21. I. Popescu, *Mechanisms.* University of Craiova Press, Romania, 1990.

22. F. Reuleaux, *The Kinematics of Machinery.* Dover, New York, 1963.

23. J. E. Shigley and C. R. Mischke, *Mechanical Engineering Design.* McGraw-Hill, New York, 1989.

Mechanisms

24. J. E. Shigley and J. J. Uicker, *Theory of Machines and Mechanisms.* McGraw-Hill, New York, 1995.

25. R. Voinea, D. Voiculescu, and V. Ceausu, *Mecanica.* EDP, Bucharest, 1983.

26. K. J. Waldron and G. L. Kinzel, *Kinematics, Dynamics, and Design of Machinery.* John Wiley & Sons, New York, 1999.

27. C. E. Wilson and J. P. Sadler, *Kinematics and Dynamics of Machinery.* Harper Collins College Publishers, 1991.

28. *The Theory of Mechanisms and Machines (Teoria mehanizmov i masin).* Vassaia scola, Minsc, Russia, 1970.

5

Machine Components

DAN B. MARGHITU, CRISTIAN I. DIACONESCU,
AND NICOLAE CRACIUNOIU

*Department of Mechanical Engineering,
Auburn University, Auburn, Alabama 36849*

Inside

1. Screws

T hreaded fasteners such as screws, nuts, and bolts are important components of mechanical structures and machines. Screws may be used as removable fasteners or as devices for moving loads.

1.1 Screw Thread

The basic arrangement of a helical thread wound around a cylinder is illustrated in Fig. 1.1. The terminology of an external screw threads is (Fig. 1.1):

- *Pitch*, denoted by p, is the distance, parallel to the screw axis, between corresponding points on adjacent thread forms having uniform spacing.
- *Major diameter*, denoted by d, is the largest (outside) diameter of a screw thread.
- *Minor diameter*, denoted by d_r or d_1, is the smallest diameter of a screw thread.
- *Pitch diameter*, denoted by d_m or d_2, is the imaginary diameter for which the widths of the threads and the grooves are equal.

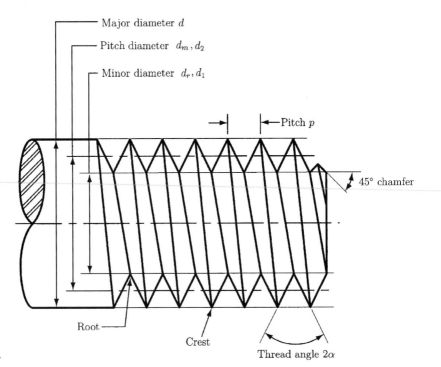

Major diameter d

Pitch diameter d_m, d_2

Minor diameter d_r, d_1

Pitch p

45° chamfer

Root

Crest

Thread angle 2α

Figure 1.1

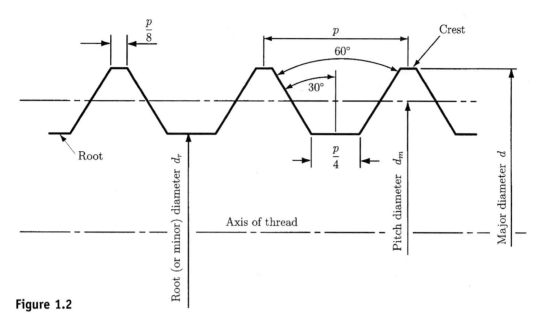

Figure 1.2

The standard geometry of a basic profile of an external thread is shown in Fig. 1.2, and it is basically the same for both *Unified* (inch series) and *ISO* (International Standards Organization, metric) threads.

The *lead*, denoted by *l*, is the distance the nut moves parallel to the screw axis when the nut is given one turn. A screw with two or more threads cut beside each other is called a *multiple-threaded* screw. The lead is equal to twice the pitch for a double-threaded screw, and to three times the pitch for a triple-threaded screw. The pitch *p*, lead *l*, and lead angle λ are represented in Fig. 1.3. Figure 1.3a shows a single-threaded, right-hand screw, and Fig. 1.3b shows a double-threaded left-hand screw. All threads are assumed to be right-hand, unless otherwise specified.

A standard geometry of an ISO profile, M (metric) profile, with 60° symmetric threads is shown in Fig. 1.4. In Fig. 1.4 $D(d)$ is the basic major diameter of an internal (external) thread, $D_1(d_1)$ is the basic minor diameter of an internal (external) thread, $D_2(d_2)$ is the basic pitch diameter, and $H = 0.5(3)^{1/2}p$.

Metric threads are specified by the letter M preceding the nominal major diameter in millimeters and the pitch in millimeters per thread. For example:

$$M\ 14 \times 2$$

M is the SI thread designation, 10 mm is the outside (major) diameter, and the pitch is 2 mm per thread.

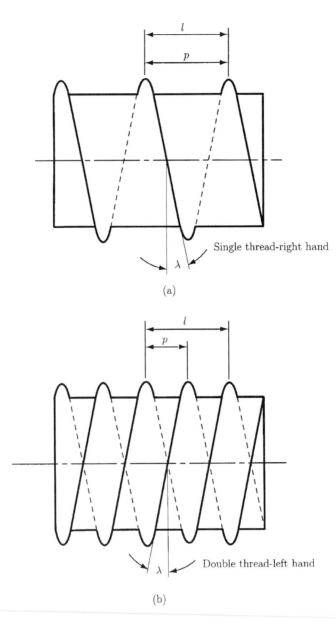

Figure 1.3   (b)

Screw size in the Unified system is designated by the size number for major diameter, the number of threads per inch, and the thread series, like this:

$$\frac{5''}{8} - 18 \text{ UNF}$$

$\frac{5''}{8}$ is the outside (major) diameter, where the double tick marks mean inches, and 18 threads per inch. Some Unified thread series are

UNC, Unified National Coarse
UNEF, Unified National Extra Fine

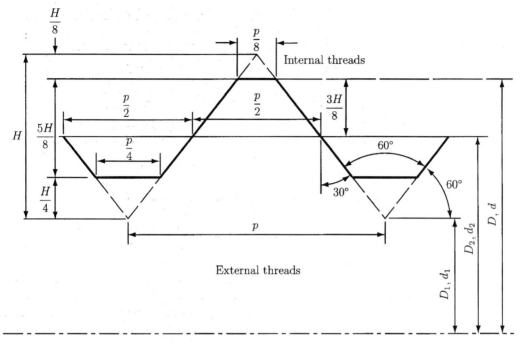

Figure 1.4

UNF, Unified National Fine
UNS, Unified National Special
UNR, Unified National Round (round root)

The UNR series threads have improved fatigue strengths.

1.2 Power Screws

For application that require power transmission, the Acme (Fig. 1.5) and square threads (Fig. 1.6) are used.

Power screws are used to convert rotary motion to linear motion of the meshing member along the screw axis. These screws are used to lift weights

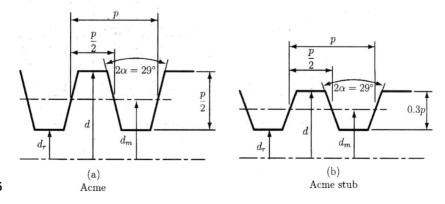

Figure 1.5   (a)   (b)

                               Acme   Acme stub

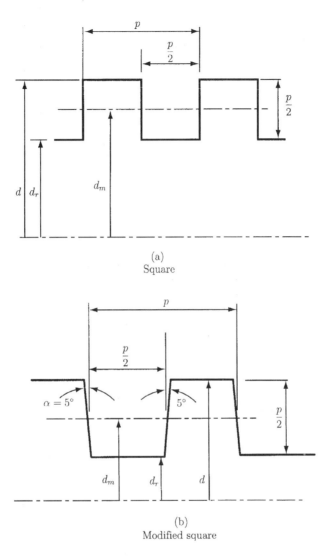

Figure 1.6

(a)
Square

(b)
Modified square

(screw-type jacks) or exert large forces (presses, tensile testing machines). The power screws can also be used to obtain precise positioning of the axial movement.

A square-threaded power screw with a single thread having the pitch diameter d_m, the pitch p, and the helix angle λ is considered in Fig. 1.7. Consider that a single thread of the screw is unrolled for exactly one turn. The edge of the thread is the hypotenuse of a right triangle and the height is the lead. The hypotenuse is the circumference of the pitch diameter circle (Fig. 1.8). The angle λ is the helix angle of the thread.

The screw is loaded by an axial compressive force F (Figs. 1.7 and 1.8).

The force diagram for lifting the load is shown in Fig. 1.8a (the force P_r acts to the right). The force diagram for lowering the load is shown in Fig. 1.8b (the force P_l acts to the left). The friction force is

$$F_f = \mu N,$$

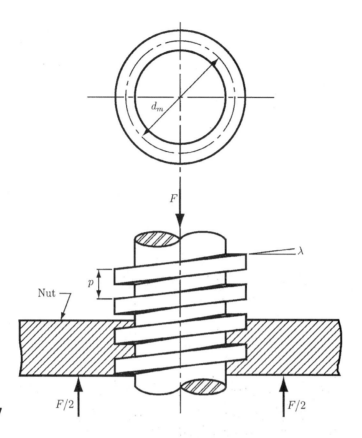

Figure 1.7

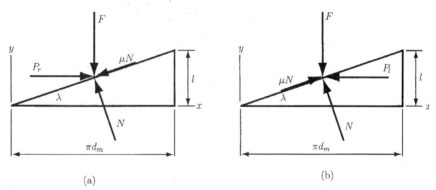

Figure 1.8

(a) (b)

where μ is the coefficient of dry friction and N is the normal force. The friction force is acting opposite to the motion.

The equilibrium of forces for raising the load gives

$$\sum F_x = P_r - N \sin \lambda - \mu N \cos \lambda = 0 \qquad (1.1)$$

$$\sum F_y = F + \mu N \sin \lambda - N \cos \lambda = 0. \qquad (1.2)$$

Similarly, for lowering the load one may write the equations

$$\sum F_x = -P_l - N \sin \lambda + \mu N \cos \lambda = 0 \qquad (1.3)$$

$$\sum F_y = F - \mu N \sin \lambda - N \cos \lambda = 0. \qquad (1.4)$$

Eliminating N and solving for P_r gives

$$P_r = \frac{F(\sin \lambda + \mu \cos \lambda)}{\cos \lambda - \mu \sin \lambda}, \qquad (1.5)$$

and for lowering the load,

$$P_l = \frac{F(\mu \cos \lambda - \sin \lambda)}{\cos \lambda + \mu \sin \lambda}. \qquad (1.6)$$

Using the relation

$$\tan \lambda = l/\pi d_m$$

and dividing the equations by $\cos \lambda$, one may obtain

$$P_r = \frac{F[(l/\pi d_m) + \mu]}{1 - (\mu l/\tau d_m)} \qquad (1.7)$$

$$P_l = \frac{F[\mu - (l/\pi d_m)]}{1 + (\mu l/\pi d_m)}. \qquad (1.8)$$

The torque required to overcome the thread friction and to raise the load is

$$T_r = P_r \frac{d_m}{2} = \frac{F d_m}{2} \left(\frac{l + \pi \mu d_m}{\pi d_m - \mu l} \right). \qquad (1.9)$$

The torque required to lower the load (and to overcome a part of the friction) is

$$T_l = \frac{F d_m}{2} \left(\frac{\pi \mu d_m - l}{\pi d_m + \mu l} \right). \qquad (1.10)$$

When the lead l is large or the friction μ is low, the load will lower itself. In this case the screw will spin without any external effort, and the torque T_l in Eq. (1.10) will be negative or zero. When the torque is positive, $T_l > 0$, the screw is said to be *self-locking*.

The condition for self-locking is

$$\pi \mu d_m > l.$$

Dividing both sides of this inequality by πd_m and using $l/\pi d_m = \tan \lambda$ yields

$$\mu > \tan \lambda. \qquad (1.11)$$

The self-locking is obtained whenever the coefficient of friction is equal to or greater than the tangent of the thread lead angle.

The torque, T_0, required only to raise the load when the friction is zero, $\mu = 0$, is obtained from Eq. (1.9):

$$T_0 = \frac{Fl}{2\pi}. \qquad (1.12)$$

The screw efficiency e can be defined as

$$e = \frac{T_0}{T_r} = \frac{Fl}{2\pi T_r}. \qquad (1.13)$$

For square threads the normal thread load, F, is parallel to the axis of the screw (Figs 1.6 and 1.7). The preceding equations can be applied for square threads.

For Acme threads (Figs 1.5) or other threads, the normal thread load is inclined to the axis because of the thread angle 2α and the lead angle λ. The lead angle can be neglected (is small), and only the effect of the thread angle is considered (Fig. 1.9). The angle α increases the frictional force by the wedging action of the threads. The torque required for raising the load is obtained from Eq. (1.9) where the frictional terms must be divided by $\cos\alpha$:

$$T_r = \frac{Fd_m}{2}\left(\frac{l + \pi\mu d_m \sec\alpha}{\pi d_m - \mu l \sec\alpha}\right). \tag{1.14}$$

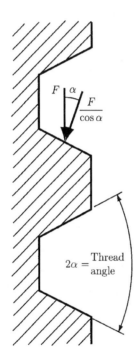

Figure 1.9

Equation (1.14) is an approximation because the effect of the lead angle has been neglected. For power screws the square thread is more efficient than the Acme thread. The Acme thread adds an additional friction due to the wedging action. It is easier to machine an Acme thread than a square thread.

In general, when the screw is loaded axially, a thrust bearing or thrust collar may be used between the rotating and stationary links to carry the axial component (Fig. 1.10). The load is concentrated at the mean collar diameter d_c. The torque required is

$$T_c = \frac{F\mu_c d_c}{2}, \tag{1.15}$$

where μ_c is the coefficient of collar friction.

Machine Components

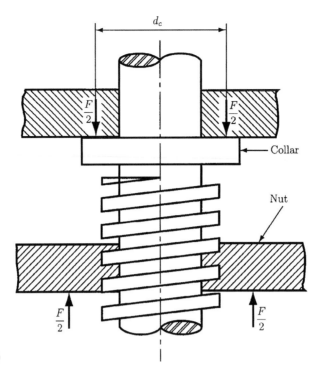

Figure 1.10

Machine Components

EXAMPLE A double square-thread power screw has the major diameter $d = 64$ mm and the pitch $p = 8$ mm. The coefficient of friction μ is 0.08, and the coefficient of collar friction μ_c is 0.08. The mean collar diameter d_c is 80 mm. The external load on the screw F is 10 kN.

Find:

1. The lead, the pitch (mean) diameter and the minor diameter
2. The torque required to raise the load
3. The torque required to lower the load
4. The efficiency

Solution

1. From Fig. 1.6a:
The minor diameter is

$$d_r = d - p = 64 - 8 = 56 \text{ mm};$$

the pitch (mean) diameter is

$$d_m = d - p/2 = 64 - 4 = 60 \text{ mm}.$$

The lead is

$$l = 2p = 2(8) = 16 \text{ mm}.$$

2. The torque required to raise the load is

$$
\begin{aligned}
T_r &= \frac{Fd_m}{2}\left(\frac{l+\pi\mu d_m}{\pi d_m - \mu l}\right) + \frac{F\mu_c d_c}{2} \\
&= \frac{10^4(60)(10^{-3})}{2}\left[\frac{16+\pi 0.08(60)}{\pi 60 - 0.08(16)}\right] + \frac{10^4(0.08)(80)(10^{-3})}{2} \\
&= 49.8 + 32 = 81.8 \text{ N m.}
\end{aligned}
$$

3. The torque required to lower the load is

$$
\begin{aligned}
T_l &= \frac{Fd_m}{2}\left(\frac{\pi\mu d_m - l}{\pi d_m + \mu l}\right) + \frac{F\mu_c d_c}{2} \\
&= \frac{10^4(60)(10^{-3})}{2}\left[\frac{\pi 0.08(60) - 16}{\pi 60 + 0.08(16)}\right] + \frac{10^4(0.08)(80)(10^{-3})}{2} \\
&= -1.54 + 32 = 30.45 \text{ N m.}
\end{aligned}
$$

The screw is not self-locking (the first term in the foregoing expression is negative).

4. The overall efficiency is

$$
e = \frac{Fl}{2\pi T_r} = \frac{10^4(16)(10^{-3})}{2\pi(81.8)} = 0.31. \quad \blacktriangle
$$

2. Gears

2.1 Introduction

Gears are toothed elements that transmit rotary motion from one shaft to another. Gears are generally rugged and durable and their power transmission efficiency is as high as 98%. Gears are usually more costly than chains and belts. The American Gear Manufacturers Association (AGMA) has established standard tolerances for various degrees of gear manufacturing precision. *Spurs gears* are the simplest and most common type of gears. They are used to transfer motion between parallel shafts, and they have teeth that are parallel to the shaft axes.

2.2 Geometry and Nomenclature

The basic requirement of gear-tooth geometry is the condition of angular velocity ratios that are exactly constant, that is, the angular velocity ratio between a 30-tooth and a 90-tooth gear must be precisely 3 in every position. The action of a pair of gear teeth satisfying this criteria is named conjugate gear-tooth action.

Machine Components

LAW OF CONJUGATE GEAR-TOOTH ACTION

The common normal to the surfaces at the point of contact of two gears in rotation must always intersect the line of centers at the same point P, called the pitch point.

The law of conjugate gear-tooth action can be satisfied by various tooth shapes, but the one of current importance is the involute of the circle. An *involute* (of the circle) is the curve generated by any point on a taut thread as it unwinds from a circle, called the base circle (Fig. 2.1a). The involute can

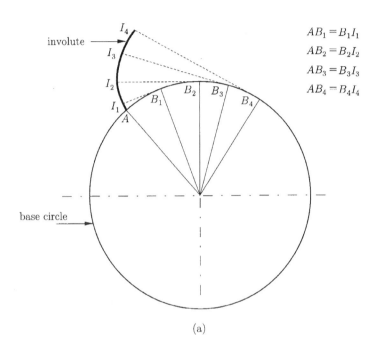

$$AB_1 = B_1I_1$$
$$AB_2 = B_2I_2$$
$$AB_3 = B_3I_3$$
$$AB_4 = B_4I_4$$

(a)

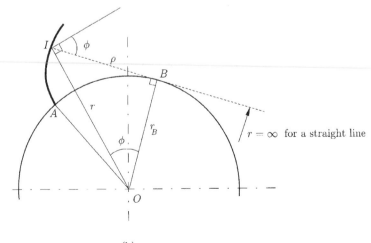

$r = \infty$ for a straight line

Figure 2.1   (b)

also be defined as the locus of a point on a taut string that is unwrapped from a cylinder. The circle that represents the cylinder is the *base circle*. Figure 2.1b represents an involute generated from a base circle of radius r_b starting at the point A. The radius of curvature of the involute at any point I is given by

$$\rho = \sqrt{r^2 - r_b^2},\qquad(2.1)$$

where $r = OI$. The involute pressure angle at I is defined as the angle between the normal to the involute IB and the normal to OI, $\phi = \angle IOB$.

In any pair of mating gears, the smaller of the two is called the pinion and the larger one the gear. The term "gear" is used in a general sense to indicate either of the members and also in a specific sense to indicate the larger of the two. The angular velocity ratio between a pinion and a gear is (Fig. 2.2).

$$i = \omega_p/\omega_g = -d_g/d_p,\qquad(2.2)$$

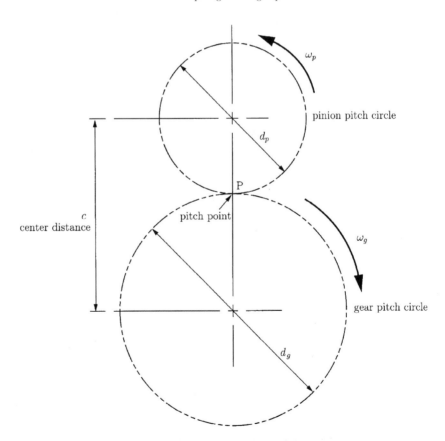

Figure 2.2

where ω is the angular velocity and d is the *pitch diameter*, the minus sign indicates that the two gears rotate in opposite directions. The *pitch circles* are the two circles, one for each gear, that remain tangent throughout the engagement cycle. The point of tangency is the pitch point. The diameter of the pitch circle is the pitch diameter. If the angular speed is expressed in rpm, then the symbol n is preferred instead of ω. The diameter (without a qualifying adjective) of a gear always refers to its pitch diameter. If other diameters (base, root, outside, etc.) are intended, they are always specified. Similarly, d, without subscripts, refers to pitch diameter. The pitch diameters of a pinion and gear are distinguished by subscripts p and g (d_p and d_g are their symbols; Fig. 2.2). The *center distance* is

$$c = (d_p + d_g)/2 = r_p + r_g, \qquad (2.3)$$

where $r = d/2$ is the *pitch circle radius*.

In Fig. 2.3 line *tt* is the common tangent to the pitch circles at the pitch point, and *AB* is the common normal to the surfaces at *C*, the point of contact of two gears. The inclination of *AB* with the line *tt* is called the *pressure angle*, ϕ. The pressure angle most commonly used, with both English and SI units, is 20°. In the United States 25° is also standard, and 14.5° was formerly an alternative standard value. Pressure angle affects the force that tends to separate mating gears.

The involute profiles are augmented outward beyond the pitch circle by a distance called the *addendum, a* (Fig. 2.4). The outer circle is usually

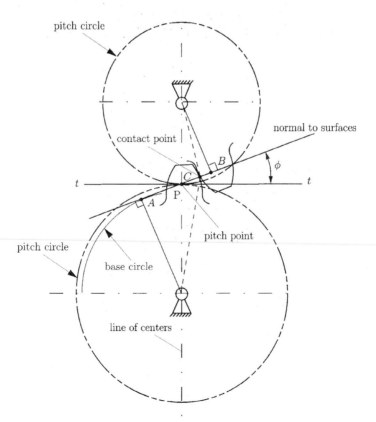

Figure 2.3

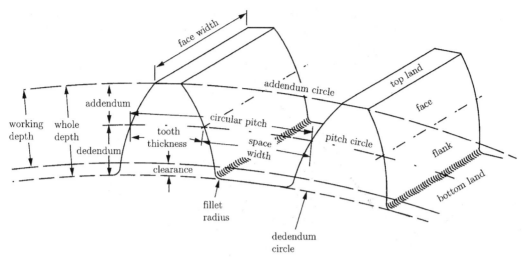

Figure 2.4

termed the *addendum circle*, $r_a = r + a$. Similarly, the tooth profiles are extended inward from the pitch circle a distance called the *dedendum, b*. The involute portion can extend inward only to the base circle. A fillet at the base of the tooth merges the profile into the dedendum circle. The fillet decreases the bending stress concentration. The *clearance* is the amount by which the dedendum in a given gear exceeds the addendum of its mating gear.

The *circular pitch* is designated as p and is measured in inches (English units) or millimeters (SI units). If N is the number of teeth in the gear (or pinion), then

$$p = \pi d/N, \quad p = \pi d_p/N_p, \quad p = \pi d_g/N_g. \tag{2.4}$$

More commonly used indices of gear-tooth size are *diametral pitch*, P_d (used only with English units) and *module, m* (used only with SI). Diametral pitch is defined as the number of teeth per inch of pitch diameter:

$$P_d = N/d, \quad P_d = N_p/d_p, \quad P_d = N_g/d_g. \tag{2.5}$$

Module m, which is essentially the complementary of P_d, is defined as the pitch diameter in millimeters divided by the number of teeth (number of millimeters of pitch diameter per tooth):

$$m = d/N, \quad m = d_p/N_p, \quad m = d_g/N_g. \tag{2.6}$$

One can easily verify that

$$pP_d = \pi \ (p \text{ in inches;} \quad P_d \text{ in teeth per inch})$$
$$p/m = \pi \ (p \text{ in millimeters;} \quad m \text{ in millimeters per tooth})$$
$$m = 25.4/P_d.$$

With English units the word "pitch," without a qualifying adjective, denotes diametral pitch (a "12-pitch gear" refers to a gear with $P_d = 12$ teeth per inch of pitch diameter). With SI units "pitch" means circular pitch (a "gear of pitch 3.14 mm" refers to a gear having a circular pitch p of 3.14 mm).

Standard diametral pitches P_d (English units) in common use are as follows:

> 1 to 2 by increments of 0.25
> 2 to 4 by increments of 0.5
> 4 to 10 by increments of 1
> 10 to 20 by increments of 2
> 20 to 40 by increments of 4

With SI units, commonly used standard values of module m are as follows:

> 0.2 to 1.0 by increments of 0.1
> 1.0 to 4.0 by increments of 0.25
> 4.0 to 5.0 by increments of 0.5

Addendum, minimum dedendum, and clearance for standard full-depth involute teeth (pressure angle is 20°) with English units in common use are

$$\text{Addendum } a = 1/P_d$$
$$\text{Minimum dedendum } b = 1.157/P_d.$$

For stub involute teeth with a pressure angle equal to 20°, the standard values are (English units)

$$\text{Addendum } a = 0.8/P_d$$
$$\text{Minimum dedendum } b = 1/P_d.$$

For SI units the standard values for full-depth involute teeth with a pressure angle of 20° are

$$\text{Addendum } a = m,$$
$$\text{Minimum dedendum } b = 1.25 \, m.$$

2.3 Interference and Contact Ratio

The contact of segments of tooth profiles that are not conjugate is called *interference*. The involute tooth form is only defined outside the base circle. In some cases, the dedendum will extend below the base circle. Then, the portion of tooth below the base circle will not be an involute and will interfere with the tip of the tooth on the mating gear, which is an involute. Interference will occur, preventing rotation of the mating gears, if either of the addendum circles extends beyond tangent points A and B (Fig. 2.5), which are called interference points. In Fig. 2.5 both addendum circles extend beyond the interference points.

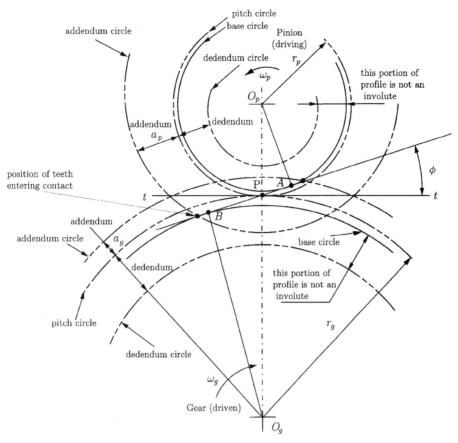

Figure 2.5

The maximum possible addendum circle radius of a pinion or gear without interference is

$$r_{a(max)} = \sqrt{r_b^2 + c^2 \sin^2 \phi},\qquad(2.7)$$

where $r_b = r \cos \phi$ is the base circle radius of the pinion or gear. The base circle diameter is

$$d_b = d \cos \phi.\qquad(2.8)$$

The average number of teeth in contact as the gears rotate together is the contact ratio CR, which is calculated (for external gears) from

$$CR = \frac{\sqrt{r_{ap}^2 - r_{bp}^2} + \sqrt{r_{ag}^2 - r_{bg}^2} - c \sin \phi}{p_b},\qquad(2.9)$$

where r_{ap}, r_{ag} are addendum radii of the mating pinion and gear, and r_{bp}, r_{bg} are base circle radii of the mating pinion and gear. The base pitch p_b is computed with

$$p_b = \pi d_b / N = p \cos \phi.\qquad(2.10)$$

The base pitch is like the circular pitch, except that it represents an arc of the base circle rather than an arc of the pitch circle. The acceptable values for contact ratio are $CR > 1.2$.

For internal gears the contact ratio is

$$CR = \frac{\sqrt{r_{ap}^2 - r_{bp}^2} - \sqrt{r_{ag}^2 - ru_{bg}^2} + c\sin\phi}{p_b}. \tag{2.11}$$

Gears are commonly specified according to AGMA Class Number, a code that denotes important quality characteristics. Quality numbers denote tooth-element tolerances. The higher the number, the tighter the tolerance. Gears are heat treated by case hardening, nitriding, precipitation hardening, or through hardening. In general, harder gears are stronger and last longer than soft ones.

EXAMPLE   Two involute spur gears of module 5, with 19 and 28 teeth, operate at a pressure angle of $20°$. Determine whether there will be interference when standard full-depth teeth are used. Find the contact ratio.

Solution

A standard full-depth tooth has on addendum of $a = m = 5\,\text{mm}$.

The gears will mesh at their pitch circles, and the pitch circle radii of pinion and gear are

$$r_p = mN_p/2 = 5(19)/2 = 47.5 \text{ mm}$$
$$r_g = mN_g/2 = 5(28)/2 = 70 \text{ mm}.$$

The theoretical center distance is

$$c = (d_p + d_g)/2 = r_p + r_g = 47.5 + 70 = 117.5 \text{ mm}.$$

The base circle radii of pinion and gear are

$$r_{bp} = r_p\cos\phi = 47.5\cos 20° = 44.635 \text{ mm}$$
$$r_{bg} = r_g\cos\phi = 70\cos 20° = 65.778 \text{ mm}.$$

The addendum circle radii of pinion and gear are

$$r_{ap} = r_p + a = m(N_p + 2)/2 = 52.5 \text{ mm}$$
$$r_{ag} = r_g + a = m(N_g + 2)/2 = 75 \text{ mm}.$$

The maximum possible addendum circle radii of pinion and gear, without interference, are

$$r_{a(max)p} = \sqrt{r_{bp}^2 + c^2\sin^2\phi} = 60.061 \text{ mm} > r_{ap} = 52.5 \text{ mm}$$
$$r_{a(max)g} = \sqrt{r_{bg}^2 + c^2\sin^2\phi} = 77.083 \text{ mm} > r_{ag} = 75 \text{ mm}.$$

Clearly, the use of standard teeth would not cause interference.

The contact ratio is

$$CR = \frac{\sqrt{r_{ap}^2 - r_{bp}^2} + \sqrt{r_{ag}^2 - r_{bg}^2} - c\sin\phi}{\pi m \cos\phi} = 1.590,$$

which should be a suitable value ($CR > 1.2$). ▲

2.4 Ordinary Gear Trains

A gear train is any collection of two or more meshing gears. Figure 2.6a shows a simple gear train with three gears in series. The train ratio is computed with the relation

$$i_{13} = \frac{\omega_1}{\omega_3} = \frac{\omega_1}{\omega_2}\frac{\omega_2}{\omega_3} = \left(-\frac{N_2}{N_1}\right)\left(-\frac{N_3}{N_2}\right) = \frac{N_3}{N_1}. \qquad (2.12)$$

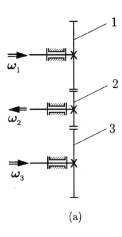

(a)

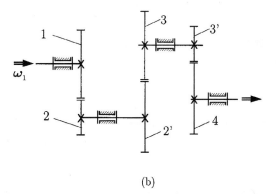

Figure 2.6 (b)

Only the sign of the overall ratio is affected by the intermediate gear 2, which is called an *idler*.

Figure 2.6b shows a compound gear train, without idler gears, with the train ratio

$$i_{14} = \frac{\omega_1}{\omega_2}\frac{\omega_{2'}}{\omega_3}\frac{\omega_{3'}}{\omega_4} = \left(-\frac{N_2}{N_1}\right)\left(-\frac{N_3}{N_{2'}}\right)\left(-\frac{N_4}{N_{3'}}\right) = -\frac{N_2 N_3 N_4}{N_1 N_{2'} N_{3'}}. \qquad (2.13)$$

2.5 Epicyclic Gear Trains

When at least one of the gear axes rotates relative to the frame in addition to the gear's own rotation about its own axes, the train is called a *planetary gear train* or *epicyclic gear train*. The term "epicyclic" comes from the fact that points on gears with moving axes of rotation describe epicyclic paths. When a generating circle (planet gear) rolls on the outside of another circle, called a directing circle (sun gear), each point on the generating circle describes an epicycloid, as shown in Fig. 2.7.

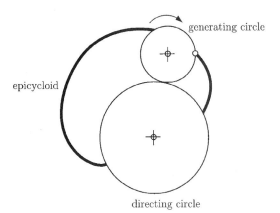

Figure 2.7   directing circle

Generally, the more planet gears there are, the greater is the torque capacity of the system. For better load balancing, new designs have two sun gears and up to 12 planetary assemblies in one casing.

In the case of simple and compound gears, it is not difficult to visualize the motion of the gears, and the determination of the speed ratio is relatively easy. In the case of epicyclic gear trains, it is often difficult to visualize the motion of the gears. A systematic procedure using the contour method is presented in what follows. The contour method is applied to determine the distribution of velocities for an epicyclic gear train.

The velocity equations for a simple closed kinematic chain are

$$\sum_{(i)} \boldsymbol{\omega}_{i,i-1} = \mathbf{0}$$

$$\sum_{(i)} \mathbf{AA}_i \times \boldsymbol{\omega}_{i,i-1} = \mathbf{0}, \qquad (2.14)$$

where $\boldsymbol{\omega}_{i,i-1}$ is the relative angular velocity of the rigid body (i) with respect to the rigid body ($i-1$), and $\mathbf{AA}_i$ is the position vector of the kinematic pair, A_i, between the rigid body (i) and the rigid body ($i-1$) with respect to a "fixed" reference frame.

EXAMPLE   The second planetary gear train considered is shown in Fig. 2.8a. The system consists of an input sun gear 1 and a planet gear 2 in mesh with 1 at B. Gear 2 is carried by the arm S fixed on the shaft of gear 3, as shown. Gear 3 meshes with the output gear 4 at F. The fixed ring gear 4 meshes with the planet gear 2 at D.

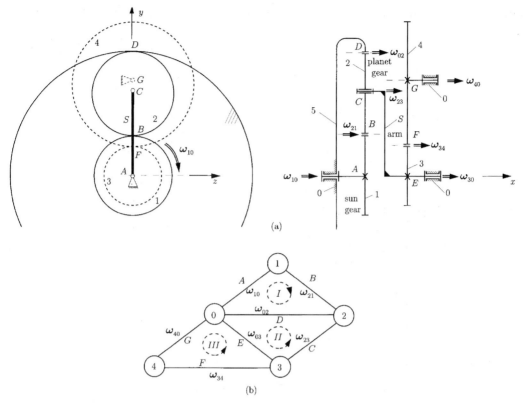

Figure 2.8 *Used with permission from Ref. 18.*

There are four moving gears (1, 2, 3, and 4) connected by the following:

- Four full joints ($c_5 = 4$): one at A, between the frame 0 and the sun gear 1, one at C, between the arm S and the planet gear 2, one at E, between the frame 0 and the gear 3, and one at G, between the frame 0 and the gear 3.
- three half joints ($c_4 = 3$): one at B, between the sun gear 1 and the planet gear 2, one at D, between the planet gear 2 and the ring gear, and one at F, between the gear 3 and the output gear 4. The module of the gears m is 5 mm.

The system possesses one DOF,

$$m = 3n - 2c_5 - c_4 = 3 \cdot 4 - 2 \cdot 4 - 3 = 1. \tag{2.15}$$

The sun gear has $N_1 = 19$-tooth external gear, the planet gear has $N_2 = 28$-tooth external gear, and the ring gear has $N_5 = 75$-tooth internal gear. The gear 3 has $N_3 = 18$-tooth external gear, and the output gear has $N_4 = 36$-tooth external gear. The sun gear rotates with input angular speed $n_1 = 2970$ rpm ($\omega_1 = \omega_{10} = \pi n_1/30 = 311.018$ rad/s). Find the absolute output angular velocity of the gear 4, the velocities of the pitch points B and F, and the velocity of joint C.

Solution

The velocity analysis is carried out using the contour equation method. The system shown in Fig. 2.8a has five elements (0, 1, 2, 3, 4) and seven joints. The number of independent loops is given by

$$n_c = l - p + 1 = 7 - 5 + 1 = 3.$$

This gear system has three independent contours. The graph of the kinematic chain and the independent contours are represented in Fig. 2.8b.

The position vectors **AB**, **AC**, **AD**, **AF**, and **AG** are defined as follows:

$$\mathbf{AB} = [x_B, y_B, 0] = [x_B, r_1, 0] = [x_B, mN_1/2, 0],$$
$$\mathbf{AC} = [x_C, y_C, 0] = [x_C, r_1 + r_2, 0] = [x_C, m(N_1 + N_2)/2, 0],$$
$$\mathbf{AD} = [x_D, y_D, 0] = [x_D, r_1 + 2r_2, 0] = [x_D, m(N_1 + 2N_2)/2, 0],$$
$$\mathbf{AF} = [x_F, y_F, 0] = [x_F, r_3, 0] = [x_F, mN_3/2, 0],$$
$$\mathbf{AG} = [x_G, y_G, 0] = [x_C, r_3 + r_4, 0] = [x_G, m(N_3 + N_4)/2, 0].$$

First Contour

The first closed contour contains the elements 0, 1, 2, and 0 (clockwise path). For the velocity analysis, the following vectorial equations can be written:

$$\boldsymbol{\omega}_{10} + \boldsymbol{\omega}_{21} + \boldsymbol{\omega}_{02} = \mathbf{0}$$
$$\mathbf{AB} \times \boldsymbol{\omega}_{21} + \mathbf{AD} \times \boldsymbol{\omega}_{02} = \mathbf{0}. \tag{2.16}$$

Here the input angular velocity is

$$\boldsymbol{\omega}_{10} = [\omega_{10}, 0, 0] = [\omega_1, 0, 0],$$

and the unknown angular velocities are

$$\boldsymbol{\omega}_{21} = [\omega_{21}, 0, 0]$$
$$\boldsymbol{\omega}_{02} = [\omega_{02}, 0, 0].$$

The sing of the relative angular velocities is selected positive, and then the numerical results will give the real orientation of the vectors.

Equation (2.16) becomes

$$\omega_1 \mathbf{1} + \omega_{21} \mathbf{1} + \omega_{02} \mathbf{1} = \mathbf{0}$$

$$\begin{vmatrix} \mathbf{1} & \mathbf{j} & \mathbf{k} \\ x_B & y_B & 0 \\ \omega_{21} & 0 & 0 \end{vmatrix} + \begin{vmatrix} \mathbf{1} & \mathbf{j} & \mathbf{k} \\ x_D & y_D & 0 \\ \omega_{02} & 0 & 0 \end{vmatrix} = \mathbf{0}. \tag{2.17}$$

Equation (2.17) projected on a "fixed" reference frame $xOyz$ is

$$\omega_1 + \omega_{21} + \omega_{02} = 0,$$
$$y_B \omega_{21} + y_D \omega_{02} = 0. \tag{2.18}$$

Equation (2.18) represents a system of two equations with two unknowns ω_{21} and ω_{02}. Solving the algebraic equations, we obtain the following value for the absolute angular velocity of planet gear 2:

$$\omega_{20} = -\omega_{02} = -\frac{N_1 \omega_1}{2 N_2} = -105.524 \text{ rad/s.} \tag{2.19}$$

Second Contour

The second closed contour contains the elements 0, 3, 2, and 0 (counterclockwise path). For the velocity analysis, the following vectorial equations can be written:

$$\boldsymbol{\omega}_{30} + \boldsymbol{\omega}_{23} + \boldsymbol{\omega}_{02} = \mathbf{0}$$
$$\mathbf{AE} \times \boldsymbol{\omega}_{30} + \mathbf{AC} \times \boldsymbol{\omega}_{23} + \mathbf{AD} \times \boldsymbol{\omega}_{02} = \mathbf{0}. \tag{2.20}$$

The unknown angular velocities are

$$\boldsymbol{\omega}_{30} = [\omega_{21},\ 0,\ 0]$$
$$\boldsymbol{\omega}_{23} = [\omega_{23},\ 0,\ 0].$$

When Eq. (2.20) is solved, the following value is obtained for the absolute angular velocity of the gear 3 and the arm S:

$$\omega_{30} = \frac{N_1 \omega_1}{2(N_1 + N_2)} = 62.865 \text{ rad/s.} \tag{2.21}$$

Third Contour

The third closed contour contains the links 0, 4, 3, and 0 (counterclockwise path). The velocity vectorial equations are

$$\boldsymbol{\omega}_{40} + \boldsymbol{\omega}_{34} + \boldsymbol{\omega}_{03} = \mathbf{0}$$
$$\mathbf{AG} \times \boldsymbol{\omega}_{40} + \mathbf{AF} \times \boldsymbol{\omega}_{34} + \mathbf{AE} \times \boldsymbol{\omega}_{03} = \mathbf{0} \tag{2.22}$$

or

$$\omega_{40}\mathbf{1} + \omega_{34}\mathbf{1} - \omega_{30}\mathbf{1} = \mathbf{0}$$
$$\begin{vmatrix} \mathbf{1} & \mathbf{j} & \mathbf{k} \\ x_G & y_G & 0 \\ \omega_{40} & 0 & 0 \end{vmatrix} + \begin{vmatrix} \mathbf{1} & \mathbf{j} & \mathbf{k} \\ x_F & y_F & 0 \\ \omega_{34} & 0 & 0 \end{vmatrix} = \mathbf{0}. \tag{2.23}$$

The unknown angular velocities are

$$\boldsymbol{\omega}_{40} = [\omega_{40},\ 0,\ 0]$$
$$\boldsymbol{\omega}_{34} = [\omega_{34},\ 0,\ 0].$$

The absolute angular velocity of the output gear 4 is

$$\omega_{40} = -\frac{N_1 N_3 \omega_1}{2(N_1 + N_2)N_4} = -31.432 \text{ rad/s.} \tag{2.24}$$

Linear Velocities of Pitch Points

The velocity of the pitch point B is

$$v_F = \omega_{40} r_4 = 2.828 \text{ m/s}.$$

The velocity of the joint C is

$$v_B = \omega_{10} r_1 = 14.773 \text{ m/s},$$

and the velocity of the pitch point F is

$$v_C = \omega_{30}(r_1 + r_2) = 7.386 \text{ m/s}.$$

Gear Geometrical Dimensions

For standard external gear teeth the addendum is $a = m$.

    Gear 1:

        Pitch circle diameter $d_1 = mN_1 = 95.0$ mm

      Addendum circle diameter $d_{a1} = m(N_1 + 2) - 105.0$ mm

      Dedendum circle diameter $d_{d1} = m(N_1 - 2.5) = 82.5$ mm.

    Gear 2:

        Pitch circle diameter $d_2 = mN_2 = 140.0$ mm

      Addendum circle diameter $d_{a2} = m(N_2 + 2) = 150.0$ mm

      Dedendum circle diameter $d_{d2} = m(N_2 - 2.5) = 127.5$ mm.

    Gear 3:

        Pitch circle diameter $d_3 = mN_3 = 90.0$ mm

      Addendum circle diameter $d_{a3} = m(N_3 + 2) = 100.0$ mm

      Dedendum circle diameter $d_{d3} = m(N_3 - 2.5) = 77.5$ mm.

    Gear 4:

        Pitch circle diameter $d_4 = mN_4 = 180.0$ mm

      Addendum circle diameter $d_{a4} = m(N_4 + 2) = 190.0$ mm

      Dedendum circle diameter $d_{d4} = m(N_4 - 2.5) = 167.5$ mm.

    Gear 5 (internal gear):

        Pitch circle diameter $d_5 = mN_5 = 375.0$ mm

      Addendum circle diameter $d_{a5} = m(N_5 - 2) = 365.0$ mm

      Dedendum circle diameter $d_{d5} = m(N_5 + 2.5) = 387.5$ mm.

Number of Planet Gears

The number of necessary planet gears k is given by the assembly condition

$$(N_1 + N_5)/k = \text{integer},$$

and for the planetary gear train $k = 2$ planet gears. The vicinity condition between the sun gear and the planet gear

$$m(N_1 + N_2) \sin(\pi/k) > d_{a2}$$

is verified.

The group drawings for this mechanism with planetary gears are given in Fig. 2.9. ▲

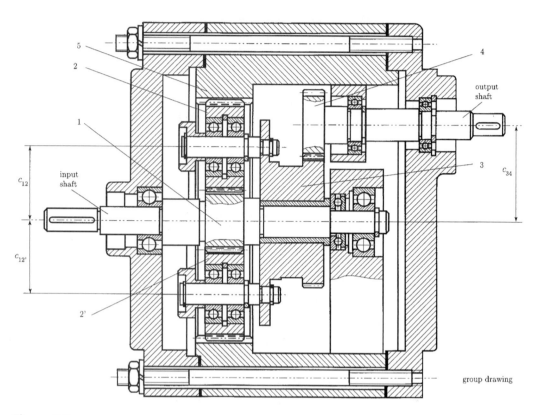

Figure 2.9

2.6 Differential

Figure 2.10 is a schematic drawing of the ordinary bevel-gear automotive differential. The drive shaft pinion 1 and the ring gear 2 are normally hypoid gears. The ring gear 2 acts as the planet carrier for the planet gear 3, and its speed can be calculated as for a simple gear train when the speed of the drive shaft is given. Sun gears 4 and 5 are connected, respectively, to each rear wheel.

When the car is traveling in a straight line, the two sun gears rotate in the same direction with exactly the same speed. Thus, for straight-line motion of the car, there is no relative motion between the planet gear 3 and ring 2. The planet gear 3, in effect, serves only as a key to transmit motion from the planet carrier to both wheels.

When the vehicle is making a turn, the wheel on the inside of the turn makes fewer revolutions than the wheel with a larger turning radius. Unless this difference in speed is accommodated in some manner, one or both of the tires would have to slide in order to make the turn. The differential permits

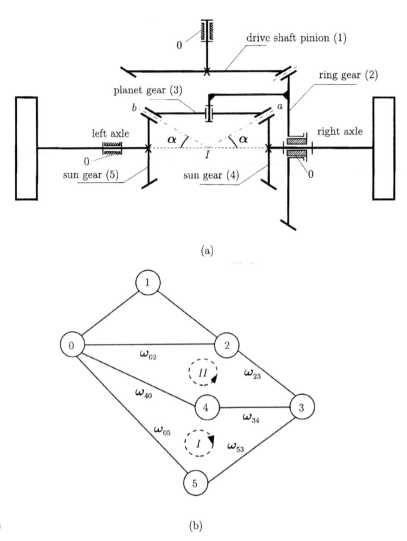

Figure 2.10   (b)

the two wheels to rotate at different velocities, while delivering power to
both. During a turn, the planet gear 3 rotates about its own axis, thus
permitting gears 4 and 5 to revolve at different velocities. The purpose of a
differential is to differentiate between the speeds of the two wheels. In the
usual passenger-car differential, the torque is divided equally whether the car
is traveling in a straight line or on a curve. Sometimes the road conditions are
such that the tractive effort developed by the two wheels is unequal. In this
case the total tractive effort available will be only twice that at the wheel
having the least traction, because the differential divides the torque equally.
If one wheel should happen to be resting on snow or ice, the total effort
available is very small and only a small torque will be required to cause the
wheel to spin. Thus, the car sits there with one wheel spinning and the other
at rest with no tractive effort. And, if the car is in motion and encounters
slippery surfaces, then all traction as well as control is lost.

It is possible to overcome the disadvantages of the simple bevel-gear differential by adding a coupling unit that is sensitive to wheel speeds. The object of such a unit is to cause most of the torque to be directed to the slow-moving wheel. Such a combination is then called a limited-slip differential.

2.6.1 ANGULAR VELOCITIES DIAGRAM

The velocity analysis is carried out using the contour equation method and the graphical angular velocities diagram.

There are five moving elements (1, 2, 3, 4, and 5) connected by the following:

- Five full joints ($c_5 = 5$): one between the frame 0 and the drive shaft pinion gear 1, one between the frame 0 and the ring gear 2, one between the planet carrier arm 2 and the planet gear 3, one between the frame 0 and the sun gear 4, and one between the frame 0 and the sun gear 5
- Three half joints ($c_4 = 3$): one between the drive shaft pinion gear 1 and the ring gear 2, one between the planet gear 3 and the sun gear 4, and one between the planet gear 3 and the sun gear 5

The system possesses two DOF:

$$M = 3n - 2c_5 - c_4 = 3 \cdot 5 - 2 \cdot 5 - 3 = 2.$$

The input data are the absolute angular velocities of the two wheels ω_{40} and ω_{50}.

The system shown in Fig. 2.10a has six elements (0, 1, 2, 3, 4, and 5) and eight joints ($c_4 + c_5$). The number of independent loops is given by

$$n_c = 8 - p + 1 = 8 - 6 + 1 = 3.$$

This gear system has three independent contours. The graph of the kinematic chain and the independent contours are represented in Fig. 2.10b.

The first closed contour contains the elements 0, 4, 3, 5, and 0 (clockwise path). For the velocity analysis, the following vectorial equations can be written:

$$\omega_{40} + \omega_{34} + \omega_{53} + \omega_{05} = \mathbf{0},$$

or

$$\omega_{40} + \omega_{34} = \omega_{50} + \omega_{35}. \tag{2.25}$$

The unknown angular velocities are ω_{34} and ω_{35}. The relative angular velocity of the planet gear 3 with respect to the sun gear 4 is parallel to the line Ia, and the relative angular velocity of the planet gear 3 with respect to the sun gear 5 is parallel to Ib. Equation (2.25) can be solved graphically (Fig. 2.11). The vectors OA and OB represent the velocities ω_{50} and ω_{40}. At A and B two parallels at Ib and Ia are drawn. The intersection between the two lines is the point C. The vector BC represents the relative angular velocity of

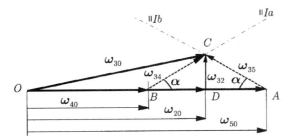

Figure 2.11

the planet gear 3 with respect to the sun gear 4, and the vector AC represents the relative angular velocity of the planet gear 3 with respect to the sun gear 5.

The absolute angular velocity of planet gear 3 is

$$\boldsymbol{\omega}_{30} = \boldsymbol{\omega}_{40} + \boldsymbol{\omega}_{34}.$$

The vector OC represents the absolute angular velocity of the planet gear.

The second closed contour contains the elements 0, 4, 3, 2, and 0 (counterclockwise path). For the velocity analysis, the following vectorial equation can be written:

$$\boldsymbol{\omega}_{40} + \boldsymbol{\omega}_{34} + \boldsymbol{\omega}_{23} + \boldsymbol{\omega}_{02} = \mathbf{0}. \tag{2.26}$$

If we use the velocities diagram (Fig. 2.11), the vector DC represents the relative angular velocity of the planet gear 3 with respect to the ring gear 2, $\boldsymbol{\omega}_{23}$, and the OD represents the absolute angular velocity of the ring gear 2, $\boldsymbol{\omega}_{20}$.

From Fig. 2.11 one can write

$$\omega_{20} = |OD| = \frac{1}{2}(\omega_{40} + \omega_{50})$$

$$\omega_{32} = |DC| = \frac{1}{2}(\omega_{50} - \omega_{40})\tan\alpha. \tag{2.27}$$

When the car is traveling in a straight line, the two sun gears rotate in the same direction with exactly the same speed, $\omega_{50} = \omega_{40}$, and there is no relative motion between the planet gear and the ring gear, $\omega_{32} = 0$. When the wheels are jacked up, $\omega_{50} = -\omega_{40}$ and the absolute angular velocity of the ring gear 2 is zero.

2.7 Gear Force Analysis

The force between mating teeth (neglecting the sliding friction) can be resolved at the pith point (P in Fig. 2.12) into two components:

- A tangential component F_t, which accounts for the power transmitted
- A radial component F_r, which does no work but tends to push the gears apart

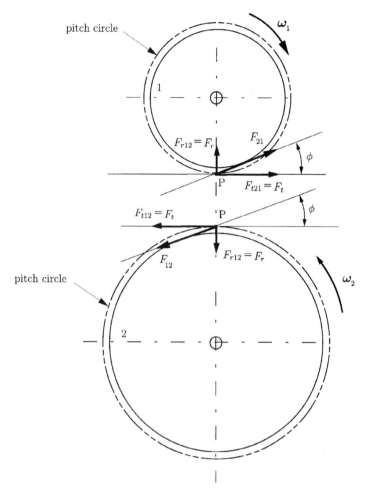

Figure 2.12

The relationship between these components is

$$F_r = F_t \tan \phi, \tag{2.28}$$

where ϕ is the pressure angle.

The pitch line velocity in feet per minute is equal to

$$V = \pi dn/12 \text{ (ft/min)}, \tag{2.29}$$

where d is the pitch diameter in inches of the gear rotating n rpm.

In SI units,

$$V = \pi dn/60{,}000 \text{ (m/s)}, \tag{2.30}$$

where d is the pitch diameter in millimeters of the gear rotating n rpm.

The transmitted power in horsepower is

$$H = F_t V/33{,}000 \text{ (hp)}, \tag{2.31}$$

where F_t is in pounds and V is in feet per minute.

Machine Components

In SI units the transmitted power in watts is

$$H = F_t V \text{ (W)},\tag{2.32}$$

where F_t is in newtons and V is in meters per second.

The transmitted torque can be expressed as

$$M_t = 63{,}000 \, H/n \text{ (lb in)},\tag{2.33}$$

where H is in horsepower and n in rpm.

In SI units,

$$M_t = 9549 \, H/n(\text{N m}),\tag{2.34}$$

where the power H is in kW and n in rpm.

EXAMPLE The planetary gear train considered is shown in Fig. 2.13. The sun gear has $N_1 = 19$-tooth external gear, the planet gear has $N_2 = N_{2'} = 28$-tooth external gear, and the ring gear has $N_5 = 75$-tooth internal gear. The gear 3 has $N_3 = 18$-tooth external gear, and the output gear has $N_4 = 36$-tooth external gear. The module of the gears is $m = 5\,\text{mm}$, and the pressure angle is $\phi = 20°$. The resistant or technological torque is $\mathbf{M}_4 = M_1\mathbf{1}$, where $M_4 = 500\,\text{Nm}$, and is opposed to the angular velocity of the output gear, $\boldsymbol{\omega}_{40} = \omega_4\mathbf{1}$, $\omega_4 < 0$ (Fig. 2.14). The joints are frictionless.

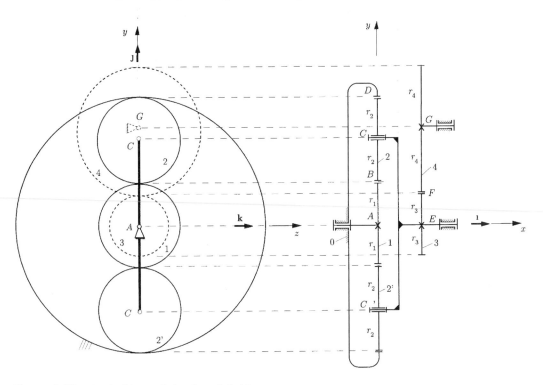

Figure 2.13 *Used with permission from Ref. 18.*

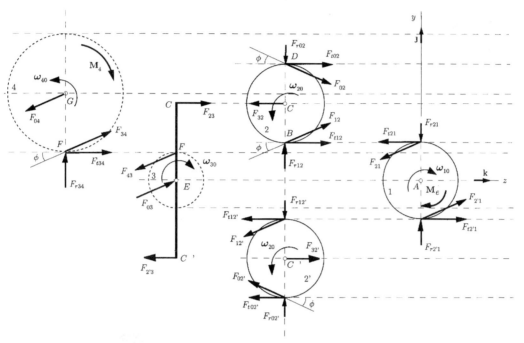

Figure 2.14 *Used with permission from Ref. 18.*

The position vectors of the joints are defined as follows (Fig. 2.13):

$$\mathbf{r}_A = [0, 0, 0]$$
$$\mathbf{r}_B = \mathbf{AB} = [0, r_1, 0] = [0, mN_1/2, 0]$$
$$\mathbf{r}_C = \mathbf{AB} = [0, r_1 + r_2, 0] = [0, m(N_1 + N_2)/2, 0]$$
$$\mathbf{r}_{C'} = \mathbf{AC'} = [0, -r_1 - r_2, 0] = [0, -m(N_1 + N_2)/2, 0]$$
$$\mathbf{r}_D = \mathbf{AD} = [0, r_1 + 2r_2, 0] = [0, m(N_1 + 2N_2)/2, 0]$$
$$\mathbf{r}_E = [\#, 0, 0]$$
$$\mathbf{r}_F = \mathbf{AF} = [\#, r_3, 0] = [\#, mN_3/2, 0]$$
$$\mathbf{r}_G = \mathbf{AG} = [\#, r_3 + r_4, 0] = [\#, m(N_3 + N_4)/2, 0]. \qquad (2.35)$$

The x parameter $\#$ is not important for the calculation.   ▲

Gear 4

The force of gear 3 that acts on gear 4 at the pitch point F is denoted by F_{34}. The force between mating teeth can be resolved at the pith point into two components, a tangential component $F_{t34} = F_{34} \cos \phi$, and a radial component $F_{r34} = F_{34} \sin \phi$, or

$$\mathbf{F}_{34} = [0, F_{r34}, F_{t34}] = F_{34} \sin \phi \mathbf{j} + F_{34} \cos \phi \mathbf{k}. \qquad (2.36)$$

The equilibrium of moments for the gear 4 with respect to its center G can be written as

$$\sum M_G^{(\text{gear }4)} = \mathbf{M}_4 + \mathbf{GF} \times \mathbf{F}_{34} = \mathbf{0}, \tag{2.37}$$

where $\mathbf{GF} = \mathbf{r}_F - \mathbf{r}_G = -r_4\mathbf{J}$. Equation (2.37) can be written as

$$M_4\mathbf{1} + \begin{vmatrix} \mathbf{1} & \mathbf{J} & \mathbf{k} \\ 0 & -r_4 & 0 \\ 0 & F_{34}\sin\phi & F_{34}\cos\phi \end{vmatrix} = \mathbf{0}. \tag{2.38}$$

Solving Eq. (2.38) gives the reaction F_{34},

$$F_{34} = \frac{2M_4}{mN_4\cos\phi} = \frac{2\cdot 500}{0.005\cdot 36\cdot\cos 20^\circ} = 5912.1 \text{ N.} \tag{2.39}$$

The reaction of the ground 0 on gear 4 at G is

$$\mathbf{F}_{04} = -\mathbf{F}_{34}.$$

Link 3

The link 3 is composed of the gear 3 and the planetary arm. The reaction of the gear 4 on gear 3 at F is known:

$$\mathbf{F}_{34} = -\mathbf{F}_{34} = -F_{34}\sin\phi\mathbf{J} - F_{34}\cos\phi\mathbf{k}.$$

The unknowns are the reactions of the planets gears 2 and 2′ on the planet arm at C and C':

$$\begin{aligned}\mathbf{F}_{23} &= F_{23r}\mathbf{J} + F_{23t}\mathbf{k} \\ \mathbf{F}_{2'3} &= -F_{23r}\mathbf{J} - F_{23t}\mathbf{k}.\end{aligned} \tag{2.40}$$

The reaction of the ground 0 on gear 3 at E is

$$\mathbf{F}_{03} = -\mathbf{F}_{43}.$$

From the free-body diagram of link 3 (Fig. 2.14), the tangential component of the force F_{23t} can be computed writing a moment equation with respect to the center of gear 3, E:

$$\sum \mathbf{M}_E^{(\text{link }3)} = (\mathbf{r}_F - \mathbf{r}_E) \times \mathbf{F}_{43} + (\mathbf{r}_C - \mathbf{r}_E) \times \mathbf{F}_{23} + (\mathbf{r}_{C'} - \mathbf{r}_E) \times \mathbf{F}_{2'3}$$

$$= \begin{vmatrix} \mathbf{1} & \mathbf{J} & \mathbf{k} \\ 0 & r_3 & 0 \\ 0 & -F_{34}\sin\phi & -F_{34}\cos\phi \end{vmatrix} + \begin{vmatrix} \mathbf{1} & \mathbf{J} & \mathbf{k} \\ \# & r_1+r_2 & 0 \\ 0 & F_{23r} & F_{23t} \end{vmatrix}$$

$$+ \begin{vmatrix} \mathbf{1} & \mathbf{J} & \mathbf{k} \\ \# & -r_1-r_2 & 0 \\ 0 & -F_{23r} & -F_{23t} \end{vmatrix} = -F_{34}r_3\cos\phi\mathbf{1} + 2F_{23t}(r_1+r_2)\mathbf{1} = \mathbf{0}.$$

$$\tag{2.41}$$

The force F_{23t} is

$$F_{23t} = \frac{M_4 r_3}{2(r_1+r_2)r_4} = 1063.83 \text{ N.} \tag{2.42}$$

Gear 2

The forces that act on gear 2 are

$\mathbf{F}_{32} = -F_{23r}\mathbf{j} - F_{23t}\mathbf{k}$, the reaction of the arm on the planet 2 at C, the tangential component $F_{32t} = -F_{23t}$ is known

$\mathbf{F}_{12} = F_{12}\sin\phi\mathbf{j} + F_{12}\cos\phi\mathbf{k}$, the reaction of the sun gear 1 on the planet 2 at B; unknown

$\mathbf{F}_{02} = -F_{02}\sin\phi\mathbf{j} + F_{02}\cos\phi\mathbf{k}$, the reaction of the ring gear 0 on the planet 2 at D; unknown

Two vectorial equilibrium equations can be written. The sum of moments that act on gear 2 with respect to the center C is zero,

$$\sum\mathbf{M}_C^{(\text{gear 2})} = (\mathbf{r}_D - \mathbf{r}_C) \times \mathbf{F}_{02} + (\mathbf{r}_B - \mathbf{r}_C) \times \mathbf{F}_{12}$$

$$\begin{vmatrix} \mathbf{1} & \mathbf{j} & \mathbf{k} \\ 0 & r_2 & 0 \\ 0 & -F_{02}\sin\phi & F_{02}\cos\phi \end{vmatrix} + \begin{vmatrix} \mathbf{1} & \mathbf{j} & \mathbf{k} \\ 0 & -r_2 & 0 \\ 0 & F_{12}\sin\phi & F_{12}\cos\phi \end{vmatrix} = \mathbf{0}, \qquad (2.43)$$

and the sum of all the forces that act on gear 2 is zero,

$$\sum\mathbf{F}^{(\text{gear 2})} = \mathbf{F}_{02} + \mathbf{F}_{12} + \mathbf{F}_{32}$$
$$= (-F_{02}\sin\phi\mathbf{j} + F_{02}\cos\phi\mathbf{k}) + (F_{12}\sin\phi\mathbf{j} + F_{12}\cos\phi\mathbf{k})$$
$$+ (F_{23r}\mathbf{j} + F_{23t}\mathbf{k}) = \mathbf{0}. \qquad (2.44)$$

Solving the system of Eqs. (2.43) and (2.44) results in

$$F_{32r} = 0, \quad F_{12} = F_{02} = \frac{M_4 r_3 \sec\phi}{4(r_1 + r_2)r_4} = 566.052 \text{ N}. \qquad (2.45)$$

Gear 1

The equilibrium torque $\mathbf{M}_e = M_e\mathbf{1}$ that acts on the input sun gear 1 is computed from the moment equation with respect to the center A,

$$\sum\mathbf{M}_A^{(\text{gear 1})} = \mathbf{M}_e + 2\mathbf{r}_B \times \mathbf{F}_{21}, \qquad (2.46)$$

and

$$M_e = \frac{M_4 r_1 r_3}{2(r_1 r_4 + r_2 r_4)} = 50.531 \text{ Nm}. \qquad (2.47)$$

The equilibrium torque $\mathbf{M}_e$ has the same direction and orientation as the angular velocity $\boldsymbol{\omega}_{10}$.

2.8 Strength of Gear Teeth

The flank of the driver tooth makes contact with the tip of the driven tooth at the beginning of action between a pair of gear teeth. The total load F is assumed to be carried by one tooth, and is normal to the tooth profile (see Fig. 2.15). The bending stress at the base of the tooth is produced by the tangential load component F_t which is perpendicular to the centerline of the tooth. The friction and the radial component F_r are neglected. The parabola in Fig. 2.15 outlines a beam of uniform strength. The weakest section of the gear tooth is at section A–A, where the parabola is tangent to the tooth outline.

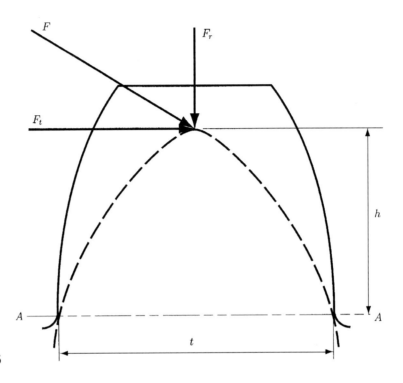

Figure 2.15

The bending stress σ is

$$\sigma = \frac{6M}{Bt^2} = \frac{6F_t h}{Bt^2}, \tag{2.48}$$

and

$$F_t = \sigma B(t^2/6h) = \sigma B(t^2/6hp)p, \tag{2.49}$$

where $M = F_t h$ is the bending moment, h is the distance between the section A–A and the point where the load is applied, and t is the tooth thickness. In the preceding equations, B is the face width and is limited to a maximum of 4 times the circular pitch, that is, $B = kp$, where $k \leq 4$.

The form factor $\gamma = t^2/6hp$ is a dimensionless quantity tabulated in Table 2.1.

If γ is substituted in the preceding equation, the usual form of the Lewis equation is

$$F_t = \sigma B p \gamma, \tag{2.50}$$

Table 2.1 *Form Factors γ for Use in Lewis Strength Equation*

Number of teeth	$14\frac{1}{2}°$ full-depth involute or composite	20° full-depth involute	20° stub involute
12	0.067	0.078	0.099
13	0.071	0.083	0.103
14	0.075	0.088	0.108
15	0.078	0.092	0.111
16	0.081	0.094	0.115
17	0.084	0.096	0.117
18	0.086	0.098	0.120
19	0.088	0.100	0.123
20	0.090	0.102	0.125
21	0.092	0.104	0.127
23	0.094	0.106	0.130
25	0.097	0.108	0.133
27	0.099	0.111	0.136
30	0.101	0.114	0.139
34	0.104	0.118	0.142
38	0.106	0.122	0.145
43	0.108	0.126	0.147
50	0.110	0.130	0.151
60	0.113	0.134	0.154
75	0.115	0.138	0.158
100	0.117	0.142	0.161
150	0.119	0.146	0.165
300	0.122	0.150	0.170
Rack	0.124	0.154	0.175

Source: A. S. Hall, A. R. Holowenko, and H. G. Laughlin, *Theory and Problems of Machine Design*, Schaum's Outline Series. McGraw-Hill, New York, 1961. Used with permission.

or

$$F_t = \sigma p^2 k\gamma = \sigma \pi^2 k\gamma / P_d^2. \qquad (2.51)$$

If the pitch diameter P_d is known, then the following form of the Lewis equation may be used:

$$P_d^2/\gamma = \sigma k\pi^2 / F_t. \qquad (2.52)$$

Here σ is the allowable stress, k is 4 (upper limit), $F_t = 2M_t/d$ is the transmitted force, and M_t is the torque on the weaker gear.

If the pitch diameter is unknown, the following form of the Lewis equation may be used:

$$\sigma = \frac{2M_t P_d^3}{k\pi^2\gamma N}. \qquad (2.53)$$

Here σ is stress less than or equal to the allowable stress, and N is the number of teeth on the weaker gear. The minimum number of teeth, N, is usually limited to 15.

2.8.1 ALLOWABLE TOOTH STRESSES

The allowable stress for gear tooth design is

$$\text{Allowable } \sigma = \sigma_0 \left(\frac{600}{600 + V} \right) \text{ for } V \text{ less than 2000 ft/min}$$

$$= \sigma_0 \left(\frac{1200}{1200 + V} \right) \text{ for } V \text{ 2000 to 4000 ft/min}$$

$$= \sigma \left(\frac{78}{78 + \sqrt{V}} \right) \text{ for } V \text{ greater than 4000 ft/min,}$$

where σ_0 is the endurance strength for released loading corrected for average stress concentration values of the gear material, measured in psi, and V is the pitch line velocity, measured in ft/min. The endurance strength is $\sigma_0 = 8000\,\text{psi}$ for cast iron, and $\sigma_0 = 12{,}000\,\text{psi}$ for bronze. For carbon steels the endurance strength range is from 10,000 psi to 50,000 psi.

2.8.2 DYNAMIC TOOTH LOADS

The dynamic forces on the teeth are produced by the transmitted force, and also by the velocity changes due to inaccuracies of the tooth profiles, spacing, misalignments in mounting, and tooth deflection under load.

The dynamic load F_d proposed by Buckingham is

$$F_d = \frac{0.05\, V(BC + F_t)}{0.05\, V + \sqrt{BC + F_t}} + F_t,$$

where F_d is the dynamic load (lb), F_t is the transmitted force (tangential load), and C is a constant that depends on the tooth material and form, and on the accuracy of the tooth cutting process. The constant C is tabulated in Table 2.2. The dynamic force F_d must be less than the allowable endurance

Table 2.2 *Values of Deformation Factor C for Dynamic Load Check*

Materials		Involute tooth form	Tooth error (inches)			
Pinion	Gear		0.0005	0.001	0.002	0.003
Cast iron	Cast iron	$14\frac{1}{2}°$	400	800	1600	2400
Steel	Cast iron	$14\frac{1}{2}°$	550	1100	2200	3300
Steel	Steel	$14\frac{1}{2}°$	800	1600	3200	4800
Cast iron	Cast iron	20° full depth	415	830	1660	2490
Steel	Cast iron	20° full depth	570	1140	2280	4320
Steel	Steel	20° full depth	830	1660	3320	4980
Cast iron	Cast iron	20° stub	430	860	1720	2580
Steel	Cast iron	20° stub	590	1180	2360	3540
Steel	Steel	20° stub	860	1720	3440	5160

Source: A. S. Hall, A. R. Holowenko, and H. G. Laughlin, *Theory and Problems of Machine Design*, Schaum's Outline Series, McGraw-Hill, New York, 1961. Used with permission.

load F_0. The allowable endurance load is $F_0 = \sigma_0 B \gamma p$, where σ_0 is based on average stress concentration values.

2.8.3 WEAR TOOTH LOADS

The wear load F_w is

$$F_w = d_p BKQ, \tag{2.54}$$

where d_p is the pitch diameter of the smaller gear (pinion), K is the stress factor for fatigue, $Q = 2N_g/(N_p + N_g)$, N_g is the number of teeth on the gear, and N_p is the number of teeth on the pinion.

The stress factor for fatigue has the expression

$$K = \frac{s_{es}^2(\sin\phi)(1/Ep + 1/Eg)}{1.4},$$

where s_{es} is the surface endurance limit of a gear pair (psi), E_p is the modulus of elasticity of the pinion material (psi), E_g is the modulus of elasticity of the gear material (psi), and ϕ is the pressure angle. An estimated value for surface endurance is

$$s_{es} = (400)(\text{BHN}) - 10,000 \text{ psi},$$

where BHN may be approximated by the average Brinell hardness number of the gear and pinion. The wear load F_w is an allowable load and must be greater than the dynamic load F_d. Table 2.3 presents several values of K for various materials and tooth forms.

Machine Components

EXAMPLE 1 A driver steel pinion with $\sigma_0 = 20,000$ rotates at $n_1 = 1500$ rpm and transmits 13.6 hp. The transmission ratio is $i = -4$ (external gearing). The gear is made of mild steel with $\sigma_0 = 15,000$ psi. Both gears have a 20° pressure angle and are full-depth involute gears. Design a gear with the smallest diameter that can be used. No fewer than 15 teeth are to be used on either gear.

Solution

In order to determine the smallest diameter gears that can be used, the minimum number of teeth for the pinion will be selected, $N_p = 15$. Then $N_g = N_p i = 15(4) = 60$. It is first necessary to determine which is weaker, the gear or the pinion. The load carrying capacity of the tooth is a function of the $\sigma_0\gamma$ product. For the pinion $\sigma_0\gamma = 20,000(0.092) = 1840$ psi, where $\gamma = 0.092$ was selected from Table 2.1 for a 20° full-depth involute gear with 15 teeth. For the gear $\sigma_0\gamma = 15,000(0.134) = 2010$ psi, where $\gamma = 0.134$ correspond to a 20° full-depth involute gear with 60 teeth. Hence, the pinion is weaker. The torque transmitted by the pinion is

$$M_t = 63,000\,H/n_1 = 63,000(13.6)/1500 = 571.2\text{lb in.} \tag{2.55}$$

Table 2.3 *Values for Surface Endurance Limit s_{es} and Stress Fatigue Factor K*

Average Brinell hardness number of steel pinion and steel gear	Surface endurance limit s_{es}	Stress fatigue factor K	
		$14\frac{1}{2}^{\circ}$	20°
150	50,000	30	41
200	70,000	58	79
250	90,000	96	131
300	110,000	114	196
400	150,000	268	366

Brinell hardness number, BHN				
Steel pinion	Gear			
150	C.I.	50,000	44	60
200	C.I.	70,000	87	119
250	C.I.	90,000	144	196
150	Phosphor bronze	50,000	46	62
200	Phosphor bronze	65,000	73	100
C.I. pinion	C.I. gear	80,000	152	208
C.I. pinion	C.I. gear	90,000	193	284

Source: A. S. Hall, A. R. Holowenko, and H. G. Laughlin, *Theory and Problems of Machine Design*, Schaum's Outline Series, McGraw-Hill, New York, 1961. Used with permission.

Since the diameter is unknown, the induced stress is

$$\sigma = \frac{2M_t P_d^3}{k\pi^2 \gamma N_p} = \frac{2(571.2)P_d^3}{4\pi^2(0.092)(15)} = 20.97 P_d^3, \tag{2.56}$$

where a maximum value of $k = 4$ was considered. Assume allowable stress $\sigma \approx \sigma_0/2 = 20,000/2 = 10,000$ psi. This assumption permits the determination of an approximate P_d. Equation (2.56) yields $P_d^3 \approx 10,000/20.97 = 476.87$. Hence, $P_d \approx 8$. Try $P_d = 8$. Then $d_p = 15/8 = 1.875$ in. The pitch line velocity is $V = d_p \pi n_1/12 = 1.875\pi(1500)/12 = 736.31$ ft/min. Because the pitch line velocity is less than 2000 ft/min, the allowable stress will be

$$\sigma = 20,000\left(\frac{600}{600 + 736.31}\right) = 8979.95 \text{ psi.}$$

Using Eq. (2.56) the induced stress will be $\sigma = 20.97(8^3) = 10736.64$ psi. The pinion is weak because the induced stress is larger than the allowable stress (10736×8979.95). Try a stronger tooth, $P_d = 7$. Then $d_p = 15/7 = 2.14$ in. The pitch line velocity is $V = d_p \pi n_1/12 = 2.14\pi(1500)/12 = 841.5$ ft/min. Because the pitch line velocity is less than 2000 ft/min, the allowable stress is

$$\sigma = 20,000\left(\frac{600}{600 + 736.31}\right) = 8324.66 \text{ psi.}$$

Using Eq. (2.56) the induced stress will be $\sigma = 20.97(7^3) = 7192.71$ psi. Now the pinion is stronger because the induced stress is smaller than the allowable stress. Then the parameter k can be reduced from the maximum value of $k = 4$ to $k = 4(7192.71/8324.66) = 3.45$. Hence, the face width $B = kp = 3.45(\pi/7) = 1.55$ in. Then $P_d = 7$, $B = 1.55$ in, $d_p = 2.14$ in, and $d_g = d_p(4) = 2.14(4) = 8.57$ in. The circular pitch for gears is $p = \pi d_p/N_p = \pi d_g/N_g = 0.448$ in, and the center distance is $c = (d_p + d_g)/2 = 5.35$ in. The addendum of the gears is $a = 1/P_d = 1/7 = 0.14$ in, while the minimum dedendum for $20°$ full-depth involute gears is $b = 1.157/P_d = 1.157/7 = 0.165$ in. The base circle diameter for pinion and gear are $d_{bp} = d_p \cos \phi = 2.14 \cos 20° = 2.01$ in, and $d_{bg} = d_g \cos \phi = 8.56 \cos 20° = 8.05$ in, respectively. The maximum possible addendum circle radius of pinion or gear without interference can be computed as

$$r_{a(max)} = \sqrt{r_b^2 + c^2 \sin^2 \phi},$$

where $r_b = d_b/2$. Hence, for pinion $r_{a(max)} = \sqrt{1 + 5.35^2 \sin 20°} = 3.29$ in, while for the gear $r_{a(max)} = \sqrt{4^2 + 5.35^2 \sin 20°} = 5.1$ in. The contact ratio CR is calculated from the equation

$$CR = \frac{\sqrt{r_{ap}^2 - r_{bp}^2} + \sqrt{r_{ag}^2 + r_{bg}^2} - c \sin \phi}{p_b},$$

where r_{ap}, r_{ag} are the addendum radii of the mating pinion and gear, and r_{bp}, r_{bg} are the base circle radii of the mating pinion and gear. Here, $r_{ap} = r_p + a = d_p/2 + a = 1.21$ in, $r_{ag} = r_g + a = 4.42$ in, $r_{bp}/2 = 1.0$ in and $r_{bg} = d_{bg}/2 = 4.02$ in. The base pitch is computed as $p_b = \pi d_b/N = p \cos 20° = 0.42$ in. Finally, the contact ratio will be $CR = 1.63$, which should be a suitable value (> 1.2). ▲

EXAMPLE 2 A steel pinion $(\sigma_0 = 137.9 \times 10^6 \text{ N/m}^2)$ rotates an iron gear $(\sigma_0 = 102.88 \times 10^6 \text{ N/m}^2)$ and transmits a power of 20 kW. The pinion operates at $n_1 = 2000$ rpm, and the transmission ratio is 4 to 1 (external gearing). Both gears are full-depth involute gears and have a pressure angle of $20°$. Design a gear with the smallest diameter that can be used. No less than 15 teeth are to be used on either gear.

Solution

To find the smallest diameter gears that can be used, the number of teeth for the pinion will be $N_p = 15$. Hence, $N_g = N_p 4 = 15(4) = 60$.

It is first necessary to determine which is weaker, the gear or the pinion. For the pinion, the product $\sigma_0 \gamma = 137.9(0.092) = 12.686 \times 10^6 \text{ N/m}^2$, where $\gamma = 0.092$ was selected from Table 2.1 for a $20°$ full-depth involute gear with 15 teeth. For gear $\sigma_0 \gamma = 102.88(0.134) = 13.785 \times 10^6 \text{ N/m}^2$, where $\gamma = 0.134$ corresponds to a $20°$ full-depth involute gear with 60 teeth. Hence, the pinion is weaker.

Machine Components

The torque transmitted by the pinion is

$$M_t = 9549 \, H/n_1 = 9549(20)/2000 = 95.49 \text{ Nm}. \qquad (2.57)$$

Since the diameter is unknown, the induced stress is

$$\sigma = \frac{2M_t}{k\pi^2 \gamma N_p m^3} = \frac{2(95.49)}{4\pi^2(0.092)(15)m^3} = \frac{3.5}{m^3}, \qquad (2.58)$$

where P_d was replaced by $1/m$, and a maximum value of $k = 4$ was considered. Assume allowable stress $\sigma \approx \sigma_0/2 = 137.9/2 = 68.95 \times 10^6$ N/m². This assumption permits the determination of an approximate m. Equation (2.58) yields $m^3 \approx 3.5/68.95 = 3.7$ mm. Try $m = 3$ mm. Then $d_p = N_p m = 15(3) = 45$ mm. The pitch line velocity is $V = d_p \pi n_1 / 60,000 = 45\pi(2000)/60,000 = 4.71$ m/s. The allowable stress will be

$$\sigma = 137.9 \left(\frac{600}{600 + 4.71} \right) = 136.85 \times 10^6 \text{ N/m}^2.$$

According to Eq. (2.58), the induced stress will be $\sigma = 3.5(3 \times 10^{-3})^3 = 129.83 \times 10^6$ N/m². The pinion is stronger. Because the smallest diameter is required, we will determine the smallest m such that the induced stress remains lower than the allowable stress. Try $m = 2.75$ mm. Then $d_p = N_p m = 15(2.75) = 41.25$ mm. The pitch line velocity is $V = d_p \pi n_1 / 60,000 = 41.25\pi(2000)/60,000 = 4.32$ m/s. The allowable stress will be

$$\sigma = 137.9 \left(\frac{600}{600 + 4.32} \right) = 136.91 \times 10^6 \text{ N/m}^2.$$

The induced stress will be $\sigma = 3.5/(2.75 \times 10^{-3})^3 = 168.56 \times 10^6$ N/m². Now the pinion is weak. Hence, the minimum m that satisfies the stress constraints is $m = 3$ mm. Then the parameter k can be reduced from the maximum value of $k = 4$ to $k = 4(129.83/136.85) = 3.79$. Hence, the face width $B = kp = 3.79(\pi m) = 35.77$ mm, and $d_p = 45$ mm. Then $d_g = d_p(4) = 45(4) = 180$ mm. The circular pitch for gears is $p = \pi d_p/N_p = \pi d_g/N_g = 9.42$ mm, and the center distance is $c = (d_p + d_g)/2 = 112.5$ mm. The addendum of the gears is $a = m = 3$ mm, while the minimum dedendum for 20° full-depth involute gears is $b = 1.26m = 3.75$ mm. The base circle diameters for pinion and gear are $d_{bp} = d_p \cos\phi = 45 \cos 20° = 42.28$ mm, and $d_{bg} = d_g \cos\phi = 180 \cos 20° = 169.14$ mm, respectively. The maximum possible addendum circle radius without interference for the pinion is $r_{a(max)} = \sqrt{21.14^2 + 112.5^2} \sin 20° = 69.1$ mm, and for the gear is $r_{a(max)} = \sqrt{84.57^2 + 112.5^2} \sin 20° = 107.15$ mm. The contact ratio CR is

$$CR = \frac{\sqrt{r_{ap}^2 - r_{bp}^2} + \sqrt{r_{ap}^2 - r_{bg}^2} - c\sin\phi}{p_b},$$

Here, r_{ap}, r_{ag} are the addendum radii of the pinion and the gear, and r_{bp}, r_{bg} are the base circle radii of the pinion and the gear. Here, $r_{ap} = r_p + a = d_p/2 + a = 25.5$ mm, $r_{ag} = r_g + a = 93$ mm, $r_{bp} = d_{bp}/2 = 21.14$ mm, and $r_{bg} = d_{bg}/2 = 84.57$ mm. The base pitch is computed as $p_b = \pi d_b/N = p\cos 20° = 8.85$ mm. Hence, $CR = 1.63 > 1.2$ should be a suitable value. ▲

3. Springs

3.1 Introduction

Springs are mechanical elements that exert forces or torques and absorb energy. The absorbed energy is usually stored and later released. Springs are made of metal. For light loads the metal can be replaced by plastics. Some applications that require minimum spring mass use structural composite materials. Blocks of rubber can be used as springs, in bumpers and vibration isolation mountings of electric or combustion motors.

3.2 Materials for Springs

The hot and cold working processes are used for spring manufacturing. Plain carbon steels, alloy steels, corrosion-resisting steels, or nonferrous materials can be used for spring manufacturing.

Spring materials are compared by an examination of their tensile strengths, which requires the wire size to be known. The material and its processing also have an effect on tensile strength. The tensile strength S_{ut} is a linear function of the wire diameter d, which is estimated by

$$S_{ut} = \frac{A}{d^m},$$ (3.1)

where the constant A and the exponent m are presented in Table 3.1.

Table 3.1 *Constants of Tensile Strength Expression*

Material	m	A	
		kpsi	MPa
Music wire	0.163	186	2060
Oil-tempered wire	0.193	146	1610
Hard-drawn wire	0.201	137	1510
Chrome vanadium	0.155	173	1790
Chrome silicon	0.091	218	1960

Source: J. E. Shigley and C. R. Mischke, *Mechanical Engineering Design.* McGraw-Hill, New York, 1989. Used with permission.

The torsional yield strength can be obtained by assuming that the tensile yield strength is between 60% and 90% of the tensile strength. According to the distortion-energy theory, the torsional yield strength for steels is

$$0.35 S_{ut} \leq S_y \leq 0.52 S_{ut}.$$ (3.2)

for static application, the *maximum allowable torsional stress* τ_{all} may be used instead of S_{sy}:

$$S_{sy} = \tau_{all} \approx \begin{cases} 0.45 S_{ut}, \text{ cold-drawn carbon steel} \\ 0.50 S_{ut}, \text{ hardened and tempered carbon and low-alloy steel} \\ 0.35 S_{ut}, \text{ austenitic stainless steel and nonferrous alloys.} \end{cases}$$

(3.3)

Machine Components

3.3 Helical Extension Springs

Extension springs (Fig. 3.1) are used for maintaining the torsional stress in the wire. The initial tension is the external force, F, applies to the spring. Spring manufacturers recommended that the initial tension be

$$\tau_{initial} = (0.4 - 0.8)\frac{S_{ut}}{C},\tag{3.4}$$

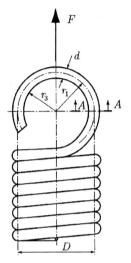 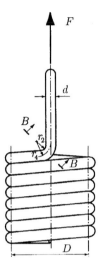

Figure 3.1

where S_{ut} is the tensile strength in psi. The constant C is the spring index, defined by $C = D/d$, where D is the mean diameter of the coil and d is the diameter of the wire (Fig. 3.1).

The bending stress, which occurs in section A–A, is

$$\sigma = \frac{16FD}{\pi d^3}\left(\frac{r_1}{r_3}\right),\tag{3.5}$$

and torsional stress in section B–B is

$$\tau = \frac{FD}{\pi d^3}\left(\frac{r_4}{r_2}\right).\tag{3.6}$$

In practical application the radius r_4 is greater than twice the wire diameter. Hook stresses can be further reduced by winding the last few coils with a decreasing diameter D (Fig. 3.2). This lower the nominal stress by reducing the bending and torsional moment arms.

3.4 Helical Compression Springs

The helical springs are usually made of circular cross-section wire or rod (Fig. 3.3). These springs are subjected to a torsional shear stress and to a transverse shear stress. There is also an additional stress effect due to the curvature of the helix.

Figure 3.2

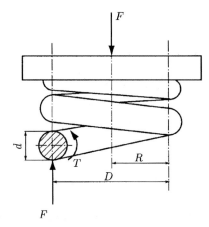

Figure 3.3 F

3.4.1 SHEAR STRESS, τ

The total shear stress, τ (psi), induced in a helical spring is

$$\tau = \frac{Tr}{J} + \frac{F}{A} = \frac{16T}{\pi d^3} + \frac{4F}{\pi d^2},$$ (3.7)

where

$T = FD/2$, is the torque (lb in)

$r = d/2$, is the wire radius (in)

F is the axial load (lb)

$A = \pi d^2/4$ is the cross-section area (in^2)

$J = \pi d^4/32$ is the polar second moment of inertia (in in^4).

The shear stress expressed in Eq. (3.7) can be rewritten as

$$\tau = K_s P \frac{FD}{\pi d^3}, \tag{3.8}$$

where K_s is the shear stress multiplication factor,

$$K_s = \frac{2C + 1}{2C}. \tag{3.9}$$

The spring index $C = D/C$ is in the range 6 to 12.

3.4.2 CURVATURE EFFECT

The curvature of the wire increases the stress on the inside of the spring and decreases it on the outside. One may write the stress equation as

$$\tau = K_w \frac{8FD}{\pi d^3}, \tag{3.10}$$

where K_w is called the Wahl factor and is given by

$$K_w = \frac{4C - 1}{4C - 4} + \frac{0.615}{C}. \tag{3.11}$$

The Wahl factor corrects both curvature and direct shear effects. The effect of the curvature alone is defined by the curvature correction factor K_c, which can be obtained as

$$K_c = \frac{K_w}{K_s}. \tag{3.12}$$

3.4.3 DEFLECTION, δ

The deflection–force relations are obtained using Castigliano's theorem. The total strain energy for a helical spring is

$$U = U_t + U_s = \frac{T^2 l}{2GJ} + \frac{F^2 l}{2AG}, \tag{3.13}$$

where

$$U_t = \frac{T^2 l}{2GJ} \tag{3.14}$$

is the torsional component of the energy, and

$$U_s = \frac{F^2 l}{2AG} \tag{3.15}$$

is the shear component of the energy. The spring load is F, the torsion torque is T, the length of the wire is l, the second moment of inertia is J, the cross-sectional area of the wire is A, and the modulus of rigidity is G.

Substituting $T = FD/2$, $l = \pi DN$, $J = \pi d^4/32$, and $A = \pi d^2/4$ in Eq. (3.13), one may obtain the total strain energy as

$$U = \frac{4F^2 D^3 N}{d^4 G} + \frac{2F^2 DN}{d^2 G}, \tag{3.16}$$

where $N = N_a$ is the number of active coils.

When Castigliano's theorem is applied, the deflection of the helical spring is

$$\delta = \frac{\partial U}{\partial F} = \frac{8FD^3N}{d^4G} + \frac{4FDN}{d^2G}. \tag{3.17}$$

If we use the spring index $C = D/d$, the deflection becomes

$$\delta = \frac{8FD^3N}{d^4G}\left(1 + \frac{1}{2C^2}\right) \approx \frac{8FD^3N}{d^4G}. \tag{3.18}$$

3.4.4 SPRING RATE

The general relationship between force and deflection can be written as

$$F = F(\delta). \tag{3.19}$$

Then the *spring rate* is defined as

$$k(\delta) = \lim_{\Delta\delta\to0}\frac{\Delta F}{\Delta\delta} = \frac{dF}{d\delta}, \tag{3.20}$$

where δ must be measured in the direction of the load F and at the point of application of F. Because most of the force–deflection equations that treat the springs are linear, k is constant and is named the *spring constant*. For this reason Eq. (3.20) may be written as

$$k = \frac{F}{\delta}. \tag{3.21}$$

From Eq. (3.18), with the substitution $C = D/d$, the spring rate for a helical spring under an axial load is

$$k = \frac{Gd}{8C^3N}. \tag{3.22}$$

For springs in parallel having individual spring rates k_i (Fig. 3.4a), the spring rate k is

$$k = k_1 + k_2 + k_3. \tag{3.23}$$

For springs in series, with individual spring rates k_i (Fig. 3.4b), the spring rate k is

$$k = \frac{1}{\dfrac{1}{k_1} + \dfrac{1}{k_2} + \dfrac{1}{k_3}}. \tag{3.24}$$

Machine Components

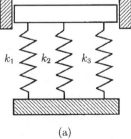
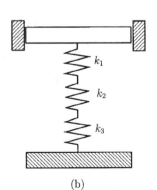

Figure 3.4   (a)   (b)

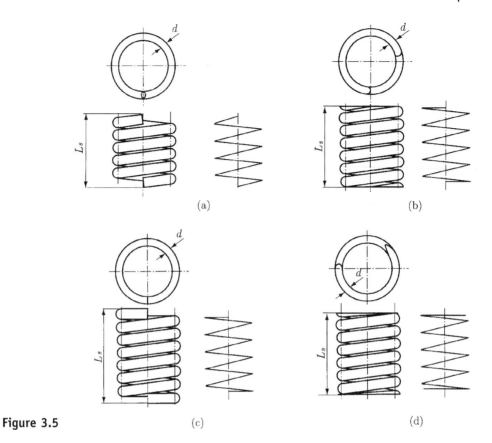

Figure 3.5

3.4.5 SPRING ENDS

For helical springs the ends can be specified as (Fig. 3.5) (a) plain ends; (b) plain and ground ends; (c) squared ends; (d) squared and ground ends. A spring with plain ends (Fig. 3.5a) has a noninterrupted helicoid, and the ends are the same as if a long spring had been cut into sections. A spring with plain and ground ends (Fig. 3.5b) or squared ends (Fig. 3.5c) is obtained by deforming the ends to a zero-degree helix angle. Springs should always be both squared and ground (Fig. 3.5d) because a better transfer of the load is obtained. Table 3.2 presents the type of ends and how that affects the number of coils and the spring length. In Table 3.2, N_a is the number of active coils, and d is the wire diameter.

Table 3.2 *Types of Spring Ends*

Term	End coils, N_e	Total coil, N_t	Free length, L_0	Solid length, L_s	Pitch, p
Plain	0	N_a	$pN_a + d$	$d(N_t + 1)$	$(L_0 - d)/N_a$
Plain and ground	1	$N_a + 1$	$p(N_a + 1)$	dN_t	$L_0/(N_a + 1)$
Squared or closed	2	$N_a + 2$	$pN_a + 3d$	$d(N_t + 1)$	$(L_0 - 3d)/N_a$
Squared and ground	2	$N_a + 2$	$pN_a + 2d$	dN_t	$(L_0 - 2d)/N_a$

Source: J. E. Shigley and C. R. Mischke, *Mechanical Engineering Design.* McGraw-Hill, New York, 1989. Used with permission.

EXAMPLE An oil-tempered wire is used for a helical compression spring. The wire diameter is $d = 0.025$ in, and the outside diameter of the spring is $D_0 = 0.375$ in. The ends are plain and the number of total turns is 10.5. Find:

The torsional yield strength
The static load corresponding to the yield strength
The rate of the spring
The deflection that would be caused by the static load found
The solid length of the spring
The length of the spring so that no permanent change of the free length occurs when the spring is compressed solid and then released
The pitch of the spring for the free length

Solution

From Eq. (3.3) the torsional yield strength, for hardened and tempered carbon and low-alloy steel, is

$$S_{sy} = 0.50 S_{ut}.$$

The minimum tensile strength given from Eq. (3.1) is

$$S_{ut} = \frac{A}{d^m},$$

where, from Table 3.1, the constant $A = 146$ kpsi and the exponent $m = 0.193$.

The minimum tensile strength is

$$S_{ut} = \frac{A}{d^m} = \frac{146}{(0.025)^{0.193}} = 297.543 \text{ kpsi.}$$

The torsional yield strength is

$$S_{sy} - 0.50 S_{ut} = 0.50(297.543) = 148.772 \text{ kpsi.}$$

To calculate the static load F corresponding to the yields strength, it is necessary to find the spring index, C, and the shear stress correction factor, K_s. The mean diameter D is the difference between the outside diameter and the wire diameter d,

$$D = D_0 - d = 0.375 - 0.025 = 0.350 \text{ in.}$$

The spring index is

$$C = \frac{D}{d} = \frac{0.350}{0.025} = 14.$$

From Eq. (3.9), the shear stress correction factor is

$$K_s = \frac{2C + 1}{2C} = \frac{2(14) + 1}{2(14)} = 1.035.$$

With the use of Eq. (3.7), the static load is calculated with

$$F = \frac{\pi d^3 S_{sy}}{8 K_s D} = \frac{\pi (0.025^3)(148.772)(10^3)}{8(1.035)(0.350)} = 2.520 \text{ lb.}$$

From Table 3.2, the number of active coils is $N_a = N_t = 10.5$.

Machine Components

The spring rate, Eq. (3.22) for $N = N_a$ is

$$k = \frac{Gd}{8C^3 N_a} = \frac{(11.5)(10^6)(0.025)}{8(14^3)(10.5)} = 1.24 \text{ lb/in.}$$

The deflection of the spring is

$$\delta = \frac{F}{k} = \frac{2.520}{1.24} = 2.019 \text{ in.}$$

The solid length, L_s, is calculated using Table 3.2:

$$L_s = d(N_t + 1) = 0.025(10.5 + 1) = 0.287 \text{ in.}$$

To avoid yielding, the spring can be no longer than the sold length plus the defection caused by the load. The free length is

$$L_0 = \delta + L_s = 2.019 + 0.287 = 2.306 \text{ in.}$$

From Table 3.2, the pitch p is calculated using the relation

$$p = \frac{L_0 - d}{N_a} = \frac{2.306 - 0.025}{10.5} 0.217 \text{ in.}$$

3.5 Torsion Springs

Helical torsion springs (Fig. 3.6) are used in door hinges, in automobile starters, and for any application where torque is required. Torsion springs are of two general types: helical (Fig. 3.7) and spiral (Fig. 3.8). The primary stress in torsion springs is bending. The bending moment Fa is applied to each end of the wire. The highest stress acting inside of the wire is

$$\sigma_i = \frac{K_i Mc}{I}, \tag{3.25}$$

where the factor for inner surface stress concentration K_i is given in Fig. 3.9, and I is the moment of inertia. The distance from the neutral axis to the extreme fiber for round solid bar is $c = d/2$, and $c = h/2$ for rectangular bar.

For a solid round bar section, $I = \pi d^4/64$, and for a rectangular bar, $I = bh^3/12$. Substituting the product Fa for bending moment and the equations for section properties of round and rectangular wire, one may write for round wire

$$\frac{I}{c} = \frac{\pi d^3}{32}, \quad \sigma_i = \frac{32 Fa}{\pi d^3} K_{i,round}, \tag{3.26}$$

and for rectangular wire

$$\frac{I}{c} = \frac{bh^2}{6}, \quad \sigma_i = \frac{6 Fa}{bh^2} K_{i,rectangular}. \tag{3.27}$$

The angular deflection of a beam subjected to bending is

$$\theta = \frac{ML}{EI}, \tag{3.28}$$

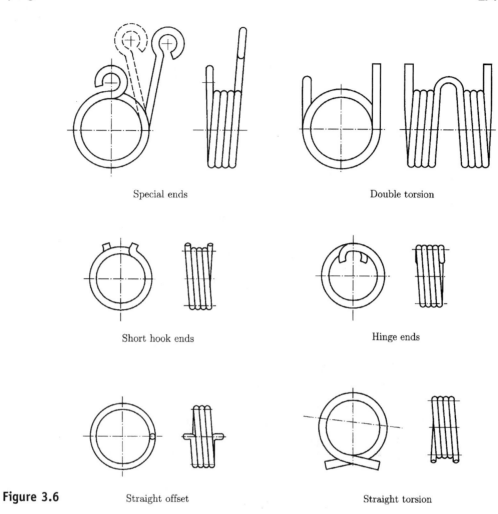

Special ends · Double torsion

Short hook ends · Hinge ends

Figure 3.6 Straight offset · Straight torsion

Figure 3.7

where M is the bending moment, L is the beam length, E is the modulus of elasticity, and I is the momentum of inertia.

Equation (3.28) can be used for helical and spiral torsion springs. Helical torsion springs and spiral springs can be made from thin rectangular wire. Round wire is often used in noncritical applications.

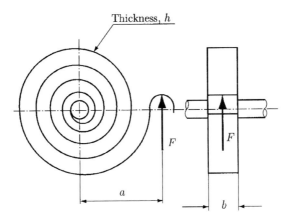

Figure 3.8

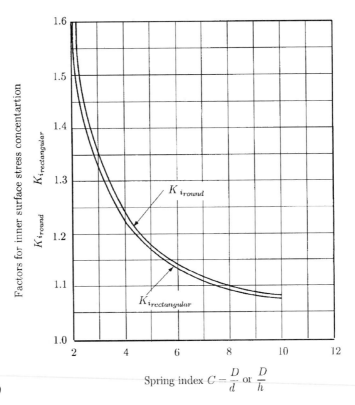

Figure 3.9

3.6 Torsion Bar Springs

The torsion bar spring, shown in Fig. 3.10, is used in automotive suspension. The stress, angular deflection, and spring rate equation are

$$\tau = \frac{Tr}{J} \qquad (3.29)$$

$$\theta = \frac{Tl}{JG} \qquad (3.30)$$

$$k = \frac{JG}{l}, \qquad (3.31)$$

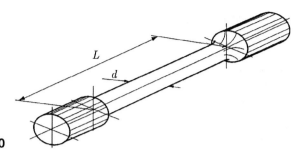

Figure 3.10

where T is the torque, $r = d/2$ is the bar radius, l is the length of the spring, G is the modulus of rigidity, and J is the second polar moment of area.

For a solid round section, J is

$$J = \frac{\pi d^4}{32}. \qquad (3.32)$$

For a solid rectangular section,

$$J = \frac{bh^3}{12}. \qquad (3.33)$$

For a solid round rod of diameter d, Eqs. (3.29), (3.30), and (3.31) become

$$\tau = \frac{16T}{\pi d^3} \qquad (3.34)$$

$$\theta = \frac{32Tl}{\pi d^4 G} \qquad (3.35)$$

$$k = \frac{\pi d^4 G}{32l}. \qquad (3.36)$$

3.7 Multileaf Springs

The multileaf spring can be a simple cantilever (Fig. 3.11a) or a semielliptic leaf (Fig. 3.11b). The design of multileaf springs is based on force F, length L, deflection, and stress relationships. The multileaf spring may be considered as a triangular plate (Fig. 3.12a) cut into n strips of width b or stacked in a graduated manner (Fig. 3.12b).

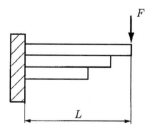

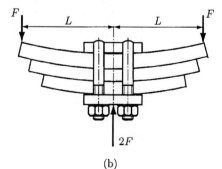

Figure 3.11   (a)   (b)

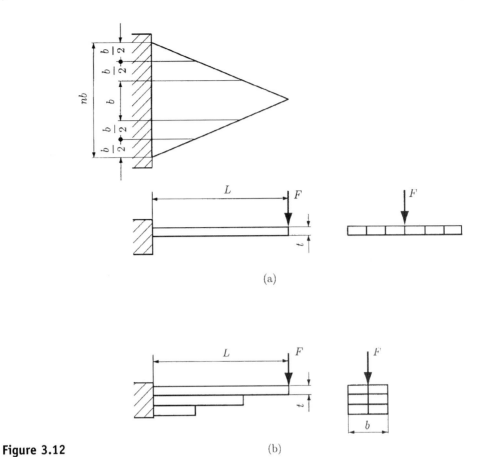

(a)

Figure 3.12   (b)

To support transverse shear N_e, extra full-length leaves are added on the graduated stack, as shown in Fig. 3.13. The number N_e is always one less than the total number of full length leaves N.

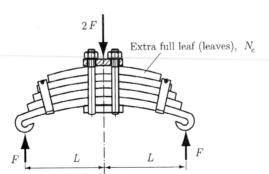

Figure 3.13

The prestressed leaves have a different radius of curvature than the graduated leaves. This will leave a gap h between the extra full-length leaves and the graduated leaves before assembly (Fig. 3.14).

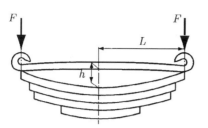

Figure 3.14

3.7.1 BENDING STRESS σ_e

The bending stress in the extra full-length leaves installed without initial prestress is

$$\sigma_e = \frac{18FL}{bt^2(3N_e + 2N_g)},\tag{3.37}$$

where F is the total applied load at the end of the spring (lb), L is the length of the cantilever or half the length of the semielliptic spring (in), b is the width of each spring leaf (in), t is the thickness of each spring leaf (in), N_e is the number of extra full-length leaves, and N_g is the number of graduated leaves.

3.7.2 BENDING STRESS σ_g

For graduated leaves assembled with extra full-length leaves without initial prestress, the bending stress is

$$\sigma_g = \frac{12FL}{bt^2(3N_e + 2N_g)} = \frac{2\sigma_e}{3}.\tag{3.38}$$

3.7.3 DEFLECTION OF A MULTILEAF SPRING, δ

The deflection of a multileaf spring with graduated and extra full-length leaves is

$$\delta = \frac{12Fl^3}{bt^2E(3N_e + 2N_g)},\tag{3.39}$$

where E is the modulus of elasticity (psi).

3.7.4 BENDING STRESS, σ

The bending stress of multileaf springs without extra leaves or with extra full length prestressed leaves that have the same stress after the full load has been applied is

$$\sigma = \frac{6Fl}{Nbt^2},\tag{3.40}$$

where N is the total number of leaves.

Machine Components

3.7.5 GAP

The gap between preassembled graduated leaves and extra full-length leaves (Fig. 3.14) is

$$b = \frac{2FL^3}{Nbt^3E},$$ (3.41)

3.8 Belleville Springs

Belleville springs are made from tapered washers (Fig. 3.15a) stacked in series, parallel, or a combination of parallel–series, as shown in Fig. 3.15b. The load–deflection and stress–deflection are

$$F = \frac{E\delta}{(1-\mu^2)(d_o/2)^2 M}[(b-\delta/2)(b-\delta)t + t^3]$$ (3.42)

$$\sigma = \frac{E\delta}{(1-\mu^2)(d_o/2)^2 M}[C_1(b-\delta/2) = C_2 t],$$ (3.43)

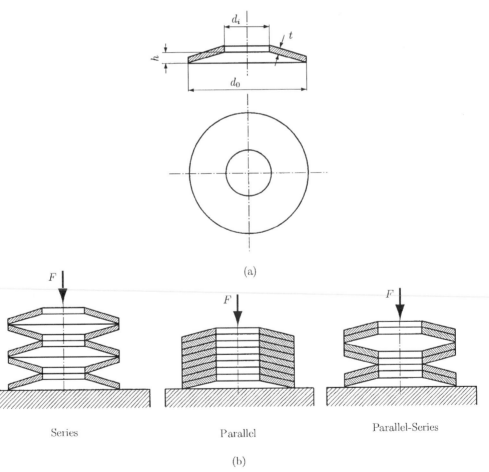

(a)

(b)

Series   Parallel   Parallel-Series

Figure 3.15

where F is the axial load (lb), δ is the deflection (in), t is the thickness of the washer (in), h, is the free height minus thickness (in), E is the modulus of elasticity (psi), σ is the stress at inside circumference (psi), d_o is the outside diameter of the washer (in), d_i is the inside diameter of the washer (in), and μ is the Poisson's ratio. The constant M, C_1, and C_2 are given by the equations

$$M = \frac{6}{\pi \log_e(d_o/d_i)} \left(\frac{d_o/d_i - 1}{d_o/d_i} \right)^2$$

$$C_1 = \frac{6}{\pi \log_e(d_o/d_i)} \left[\frac{d_o/d_i - 1}{\log_e(d_o/d_i)} - 1 \right]$$

$$C_2 = \frac{6}{\pi \log_e(d_o/d_i)} \left[\frac{d_o/d_i - 1}{2} \right].$$

4. Rolling Bearings

4.1 Generalities

A bearing is a connector that permits the connected members to rotate or to move relative to one another. Often one of the members is fixed, and the bearing acts as a support for the moving member. Most bearings support rotating shafts against either transverse (radial) or thrust (axial) forces. To minimize friction, the contacting surfaces in a bearing may be partially or completely separated by a film of liquid (usually oil) or gas. These are *sliding bearings*, and the part of the shaft that turns in the bearing is the journal. Under certain combinations of force, speed, fluid viscosity, and bearing geometry, a fluid film forms and separates the contacting surfaces in a sliding bearing, and this is known as a *hydrodynamic* film. An oil film can also be developed with a separate pumping unit that supplies pressurized oil to the bearing, and this is known as a *hydrostatic* film.

The surfaces in a bearing may also be separated by balls, rollers, or needles; these are known as *rolling bearings*. Because shaft speed is required for the development of a hydrodynamic film, the starting friction in hydrodynamic bearings is higher than in rolling bearings. To minimize friction when metal-to-metal contact occurs, low-friction bearing materials have been developed, such as bronze alloys and babbitt metal.

The principal advantage of these bearings is the ability to operate at friction levels considerably lower at startup, the friction coefficient having the values $\mu = 0.001$–0.003. Also, they have the following advantages over bearings with sliding contact: they maintain accurate shaft alignment for long periods of time; they can carry heavy momentary overloads without failure; their lubrication is very easy and requires little attention; and they are easily replaced in case of failure.

Machine Components

Rolling bearings have the following disadvantages: The shaft and house design and processing are more complicated; there is more noise for the higher speeds; the resistance to shock forces is lower; the cost is higher.

4.2 Classification

The important parts of rolling bearings are illustrated in Fig. 4.1: outer ring, inner ring, rolling element and separator (retainer). The role of the separator is to maintain an equal distance between the rolling elements. The races are the outer ring or the inner ring of a bearing. The raceway is the path of the rolling element on either ring of the bearing.

Rolling bearings may be classified using the following criteria (Fig. 4.2):

- The rolling element share: ball bearings (Fig. 4.2a–f), roller bearings (cylinder, Figs. 4.2g,h; cone, Fig. 4.2i; barrel, Fig. 4.2j), and needle bearings (Fig. 4.2k)
- The direction of the principal force: radial bearings (Figs. 4.2a,b,g,h), thrust bearings (Figs. 4.2d,e), radial-thrust bearings (Figs. 4.2c,i), or thrust-radial bearings (Fig. 4.2f)
- The number of rolling bearing rows: rolling bearings with one row (Figs. 4.2a,d,g,k), with two rows (Figs. 4.2b,e,h), etc.

The radial bearing is primarily designed to support a force perpendicular to the shaft axis. The thrust bearing is primarily designed to support a force parallel to the shaft axis.

Single row rolling bearings are manufactured to take radial forces and some thrust forces. The angular contact bearings provide a greater thrust capacity. Double row bearings are made to carry heavier radial and thrust forces. The single row bearings will withstand a small misalignment or deflection of the shaft. The self-aligning bearings (Fig. 4.2f), are used for severe misalignments and deflections of the shaft.

Cylinder roller bearings provide a greater force than ball bearings of the same size because of the greater contact area. This type of bearing will not take thrust forces. Tapered (cone) roller bearings combined the advantages of ball and cylinder roller bearings, because they can take either radial or thrust forces, and they have high force capacity.

Needle bearings are used where the radial space is limited, and when the separators are used they have high force capacity. In many practical cases they are used without the separators.

4.3 Geometry

In Fig. 4.3 is shown a ball bearing with the *pitch diameter* given by

$$d_m \approx \frac{d_0 + d}{2}, \qquad (4.1)$$

where d_0 is the outer diameter of the ball bearing and d is the bore.

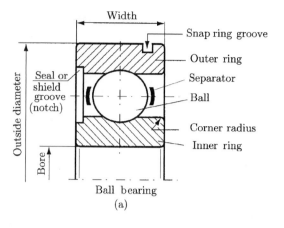

Ball bearing
(a)

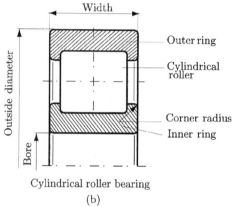

Cylindrical roller bearing
(b)

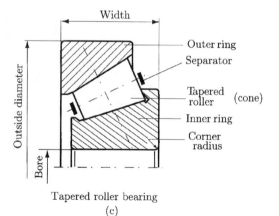

Tapered roller bearing
(c)

Figure 4.1

The *pitch diameter* can be calculated exactly as

$$d_m = \frac{D_i + D_e}{2},$$ (4.2)

where D_i is the race diameter of the inner ring and D_e is the race diameter of the outer ring.

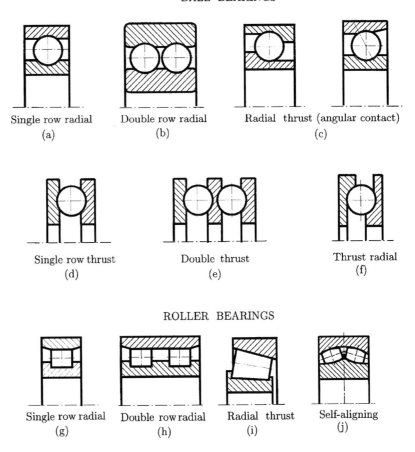

BALL BEARINGS

Single row radial
(a)

Double row radial
(b)

Radial thrust (angular contact)
(c)

Single row thrust
(d)

Double thrust
(e)

Thrust radial
(f)

ROLLER BEARINGS

Single row radial
(g)

Double row radial
(h)

Radial thrust
(i)

Self-aligning
(j)

NEEDLE BEARINGS

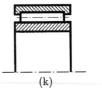

(k)

Figure 4.2

In general the ball bearings are manufactured with a clearance between the balls and the raceways. The clearance measured in the radial plane is the *diametral clearance*, s_d, and is computed with the relation (Fig. 4.3)

$$s_d = D_e - D_i - 2D, \qquad (4.3)$$

where D is the ball diameter.

Because a radial ball bearing has a diametral clearance in the no-load state, the bearing also has an axial clearance. Removal of this axial clearance

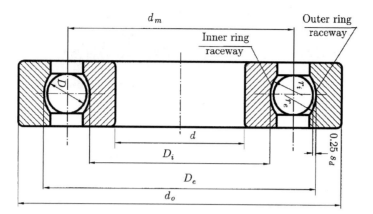

Figure 4.3

causes the ball–raceway contact to assume an oblique angle with the radial plane. Angular-contact ball bearings are designed to operate under thrust force, and the clearance built into the unloaded bearing along with the raceway groove curvatures determines the bearing free contact angle. Because of the diametral clearance for radial ball bearing there is a *free endplay*, s_a, as shown in Fig. 4.4. In Fig. 4.4 the center of the outer ring raceway circle is O_e, the center of the inner ring raceway circle is O_i, and the center of the ball is O.

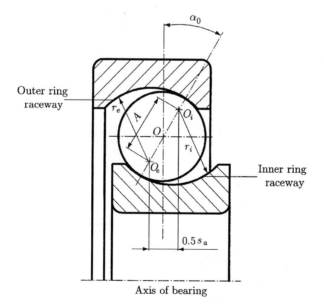

Figure 4.4

The distance between the centers O_e and O_i is

$$A = r_e + r_i - D,$$ (4.4)

where r_e is the radius of the outer ring raceway and r_i is the radius of the inner ring raceway.

If the raceway groove curvature radius is $r = fD$, where f is a dimensionless coefficient, then

$$A = (f_e = f_i - 1)D = BD, \tag{4.5}$$

where $B = f_e + f_i - 1$ is defined, as the *total curvature of the bearing*. In the preceding formula, $r_e = f_e D$ and $r_i = f_i D$, where f_e and f_i are adimensional coefficients.

The *free contact angle* α_0 is the angle made by the line passing through the points of contact of the ball and both raceways and a plane perpendicular to the bearing axis of rotation (Fig. 4.4). The magnitude of the free contact angle can be written

$$\sin \alpha_0 = 0.5 s_a / A. \tag{4.6}$$

The diametral clearance can allow the ball bearing to misalign slightly under no load. The *free angle of misalignment* θ is defined as the maximum angle through which the axis of the inner ring can be rotated with respect to the axis of the outer ring before stressing bearing components,

$$\theta = \theta_i + \theta_e, \tag{4.7}$$

where θ_i (Fig. 4.5a) is the misalignment angle for the inner ring,

$$\cos \theta_i = 1 - \frac{s_d[(2f_i - 1)D - s_d/4]}{2d_m[d_m + (2f_i - 1)D + s_d/2]}, \tag{4.8}$$

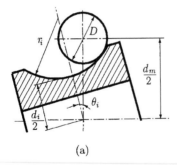

(a)

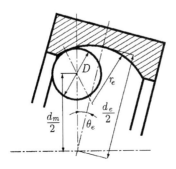

Figure 4.5   (b)

and θ_e (Fig. 4.5b) is the misalignment angle for the outer ring,

$$\cos\theta_e = 1 - \frac{s_d[(2f_e - 1)D - s_d/4]}{2d_m[d_m - (2f_e - 1)D + s_d/2]}. \tag{4.9}$$

With the trigonometric identity

$$\cos\theta_i + \cos\theta_e = 2\cos[(\theta_i + \theta_e)/2]\cos(\theta_i - \theta_e)/2] \tag{4.10}$$

and with the approximation $\theta_i - \theta_e \approx 0$, the free angle of misalignment becomes

$$\theta = 2\arccos[(\cos\theta_i + \cos\theta_e)/2], \tag{4.11}$$

or

$$\theta = 2\arccos\left\{1 - \frac{s_d}{4d_m}\left[\frac{(2f_i - 1)D - s_d/4}{d_m + (2f_i - 1)D + s_d/2} + \frac{(2f_e - 1)D - s_d/4}{d_m - (2f_e - 1)D + s_d/2}\right]\right\}. \tag{4.12}$$

4.4 Static Loading

In Fig. 4.6a is shown a single row radial thrust (angular contact) ball bearing. The *contact angle* α is the angle of the axis of contact between balls and

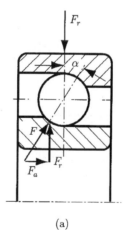

(a)

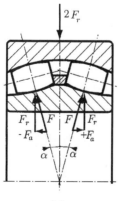

(b)

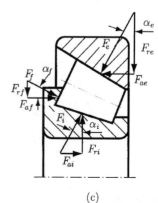

(c)

Figure 4.6

races. For a single row radial ball bearing, the angle α is zero. If F_r is the radial force applied to the ball, then the normal force to be supported by the ball is

$$F = \frac{F_r}{\cos \alpha}, \tag{4.13}$$

and the axial force, F_a (or F_t), is

$$F_a = F \sin \alpha. \tag{4.14}$$

For self-aligning roller bearings (Fig. 4.6b), the preceding relations are valid for each roller, and the total axial force is zero.

For tapered roller bearings (Fig. 4.6c), there are three contact angles: α_i, the contact angle for the inner ring; α_e, the contact angle for the outer ring; and α_f, the contact angle for the frontal face.

The normal and axial forces for the inner ring are

$$F_i = \frac{F_{ri}}{\cos \alpha_i} \quad \text{and} \quad F_{ai} = D_{ri} \tan \alpha_i, \tag{4.15}$$

where F_{ri} is the radial force acting on the inner ring. The normal and axial forces for the outer ring are

$$F_e = \frac{F_{re}}{\cos \alpha_e} \quad \text{and} \quad F_{ae} = F_{re} \tan \alpha_e, \tag{4.16}$$

where F_{re} is the radial force acting on the outer ring. The normal and axial forces for the frontal face are

$$F_f = \frac{F_{rf}}{\cos \alpha_f} \quad \text{and} \quad F_{af} = F_{rf} \tan \alpha_f, \tag{4.17}$$

where F_{rf} is the radial force acting on the frontal face.

The equilibrium equations for the radial and axial directions are

$$F_{ri} - F_{rf} - F_{re} = 0 \quad \text{or} \quad F_{ri} - F_f \cos \alpha_f - F_e \cos \alpha_e = 0 \tag{4.18}$$

$$F_{ai} + F_{af} - F_{ae} = 0 \quad \text{or} \quad F_{ri} \tan \alpha_i + F_f \sin \alpha_f - F_e \sin \alpha_e = 0. \tag{4.19}$$

From Eqs. (4.18) and (4.19), the forces F_e and F_f are obtained:

$$F_e = \frac{F_{ri}(\sin \alpha_f + \tan \alpha_i \cos \alpha_f)}{\sin(\alpha_e + \alpha_f)} \tag{4.20}$$

$$F_f = \frac{F_{ri}(\sin \alpha_e - \tan \alpha_i \cos \alpha_e)}{\sin(\alpha_e + \alpha_f)}. \tag{4.21}$$

4.5 *Standard Dimensions*

The Annular Bearing Engineers' Committee (ABEC) of the Anti-Friction Bearing Manufacturers Association (AFBMA) has established four primary grades of precision, designated ABEC 1, 5, 7, and 9. ABEC 1 is the standard grade and is adequate for most normal applications. The other grades have progressively finer tolerances. For example, tolerances on bearing bores between 35 and 50 mm range from +0.0000 to −0.0005 in for ABEC grade **I**, and from +0.0000 to −0.0001 in for ABEC grade 9. Tolerances on other dimensions are comparable. Similarly, the AFBMA Roller Bearing Engineers Committee has established RBEC standards I and 5 for cylindrical roller bearings.

The bearing manufacturers have established standard dimensions (Fig. 4.7 and Table 4.1) for ball and straight roller bearings, in metric sizes, which define the bearing bore d, the outside diameter d_0, the width w, the fillet sizes on the shaft and housing shoulders r; the shaft diameter d_S, and the housing diameter d_H.

For a given bore, there is an assortment of widths and outside diameters. Furthermore, the outside diameters selected are such that, for a particular outside diameter, one can usually find a variety of bearings having different bores. That is why the bearings are made in various proportions for different series (Fig. 4.8): extra-extra-light series (LL00), extra-light series (L00), light series (200), and medium series (300).

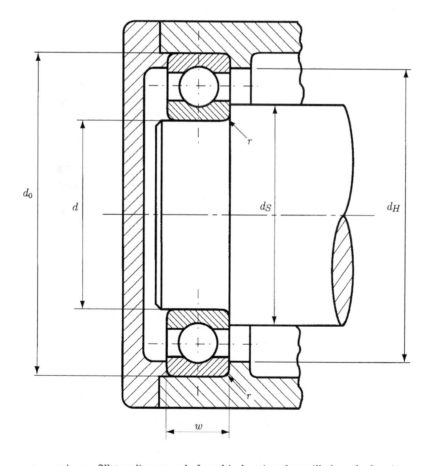

Figure 4.7 r - maximum fillet radius on a shaft and in housing that will clear the bearing corner radius

Table 4.1 *Bearing Dimensions*

Bearing Basic Number	d (mm)	Ball bearings					Roller bearings				
		d_0 (mm)	w (mm)	r (mm)	d_S (mm)	d_H (mm)	d_0 (mm)	w (mm)	r (mm)	d_S (mm)	d_H (mm)
L00	10	26	8	0.30	12.7	23.4					
200	10	30	9	0.64	13.8	26.7					
300	10	35	11	0.64	14.8	31.2					
L01	12	28	8	0.30	14.5	25.4					
201	12	32	10	0.64	16.2	28.4					
301	12	37	12	1.02	17.7	32.0					
L02	15	32	9	0.30	17.5	29.2					
202	15	35	11	0.64	19.0	31.2					
302	15	42	13	1.02	21.2	36.6					
L03	17	35	10	0.30	19.8	32.3	35	10	0.64	20.8	32.0
203	17	40	12	0.64	22.4	34.8	40	12	0.64	20.8	36.3
303	17	47	14	1.02	23.6	41.1	47	14	1.02	22.9	41.4
L04	20	42	12	0.64	23.9	38.1	42	12	0.64	24.4	36.8
204	20	47	14	1.02	25.9	41.7	47	14	1.02	25.9	42.7
304	20	52	15	1.02	27.7	27.7	52	15	1.02	25.9	46.2
L05	25	47	12	0.64	29.0	42.9	47	12	0.64	29.2	43.4
205	25	52	15	1.02	30.5	46.7	52	15	1.02	30.5	47.0
305	25	62	17	1.02	33.0	54.9	62	17	1.02	31.5	55.9
L06	30	55	13	1.02	34.8	49.3	47	9	0.38	3.33	43.9
206	30	62	16	1.02	36.8	55.4	62	16	1.02	36.1	56.4
306	30	72	19	1.02	38.4	64.8	72	19	1.52	37.8	64.0
L07	35	62	14	1.02	40.1	56.1	55	10	0.64	39.4	50.8
207	35	72	17	1.02	42.4	65.0	72	17	1.02	41.7	65.3
307	35	80	21	1.52	45.2	70.4	80	21	1.52	43.7	71.4
L08	40	68	15	1.02	45.2	62.0	68	15	1.02	45.7	62.7
208	40	80	18	1.02	48.0	72.4	80	18	1.52	47.2	72.9
308	40	90	23	1.52	50.8	80.0	90	23	1.52	49.0	81.3
L09	45	75	16	1.02	50.8	52.8	75	16	1.02	50.8	69.3
209	45	85	19	1.02	52.8	77.5	85	19	1.52	78.2	
309	45	100	25	1.52	57.2	88.9	100	25	2.03	55.9	90.4
L10	50	80	16	1.02	55.6	73.7	72	12	0.64	54.1	68.1
210	50	90	20	1.02	57.7	82.3	90	20	1.52	57.7	82.8
310	50	110	27	2.03	64.3	96.5	110	27	2.03	61.0	99.1
L11	55	90	18	1.02	61.7	83.1	90	18	1.52	62.0	83.6
211	55	100	21	1.52	65.0	90.2	100	21	2.03	64.0	91.4
311	55	120	29	2.03	69.8	106.2	120	29	2.03	66.5	108.7

(continued)

Table 4.1 *Continued*

Bearing Basic Number	d (mm)	Ball bearings						Roller bearings				
		d_0 (mm)	w (mm)	r (mm)	d_S (mm)	d_H (mm)	d_0 (mm)	w (mm)	r (mm)	d_S (mm)	d_H (mm)
L12	60	95	18	1.02	66.8	87.9	95	18	1.52	67.1	88.6
212	60	110	22	1.52	70.6	99.3	110	22	2.03	69.3	101.3
312	60	130	31	2.03	75.4	115.6	130	31	2.54	72.9	117.9
L13	65	100	18	1.02	71.9	92.7	100	18	1.52	72.1	93.7
213	65	120	23	1.52	76.5	108.7	120	23	2.54	77.0	110.0
313	65	140	33	2.03	81.3	125.0	140	33	2.54	78.7	127.0
L14	70	110	20	1.02	77.7	102.1	110	20			
214	70	125	24	1.52	81.0	114.0	125	24	2.54	81.8	115.6
314	70	150	35	2.03	86.9	134.4	150	35	3.18	84.3	135.6
L15	75	115	20	1.02	82.3	107.2	115	20			
215	75	130	25	1.52	86.1	118.9	130	25	2.54	85.6	120.1
215	75	160	37	2.03	92.7	143.8	160	37	3.18	90.4	145.8
L16	80	125	22	1.02	88.1	116.3	125	22	2.03	88.4	117.6
216	80	140	26	2.03	93.2	126.7	140	26	2.54	91.2	129.3
316	80	170	39	2.03	98.6	152.9	170	39	3.18	96.0	154.4
L17	85	130	22	1.02	93.2	121.4	130	22	2.03	93.5	122.7
217	85	150	28	2.03	99.1	135.6	150	28	3.18	98.	139.2
317	85	180	41	2.54	105.7	160.8	180	41	3.96	102.9	164.3
L18	85	140	24	1.02	99.6	129.0	140	24			
218	85	160	30	2.03	104.4	145.5	160	30	3.18	103.1	147.6
318	85	190	43	2.54	111.3	170.2	190	43	3.96	108.2	172.7
L19	90	145	24	1.52	104.4	134.1	145	24			
219	90	170	32	2.03	110.2	154.9	170	32	3.18	109.0	157.0
319	90	200	45	2.54	117.3	179.3	200	45	3.96	115.1	181.9
L20	95	150	24	1.52	109.5	139.2	150	24	2.54	109.5	141.7
220	95	180	34	2.03	116.1	164.1	180	34	3.96	116.1	167.1
320	95	215	47	2.54	122.9	194.1	215	47	4.75	122.4	194.6
L21	100	160	26	2.03	116.1	146.8	160	26			
221	100	190	36	2.03	121.9	173.5	190	36	3.96	121.4	175.3
321	100	225	49	2.54	128.8	203.5	225	49	4.75	128.0	203.5
L22	105	170	28	2.03	122.7	156.5	170	28	2.54	121.9	159.3
222	105	200	38	2.03	127.8	182.6	200	38	3.96	127.3	183.9
322	105	240	50	2.54	134.4	218.2	240	50	4.75	135.9	217.2
L24	120	180	28	2.03	132.6	166.6	180	28			
224	1220	215	40	2.03	138.2	197.1	215	40	4.75	139.2	198.9
324	120						260	55	6.35	147.8	235.2

(continued)

Table 4.1 *Continued*

Bearing Basic Number	d (mm)	Ball bearings					Roller bearings				
		d_0 (mm)	w (mm)	r (mm)	d_S (mm)	d_H (mm)	d_0 (mm)	w (mm)	r (mm)	d_S (mm)	d_H (mm)
L26	130	200	33	2.063	143.8	185.4	200	33	3.18	143.0	188.2
226	130	230	40	2.54	149.9	210.1	230	40	4.75	149.1	213.9
326	130	280	58	3.05	160.0	253.0	280	58	6.35	160.3	254.5
L28	140	210	33	2.03	153.7	195.3	210	33			
228	140	250	42	2.54	161.5	228.6	250	42	4.75	161.5	232.4
328	140						300	62	7.92	172.0	271.3
L30	150	225	35	2.03	164.3	209.8	225	35	3.96	164.3	212.3
230	150	270	45	2.54	173.0	247.6	270	45	6.35	174.2	251.0
L32	160	240	38	2.03	175.8	223.0	240	38			
232	160						290	48	6.35	185.7	269.5
L36	180	280	46	2.03	196.8	261.6	280	46	4.75	199.6	262.9
236	180						320	52	6.35	207.5	298.2
L40	200						310	51			
240	200						360	58	7.92	232.4	334.5
L44	220						340	56			
244	220						400	65	9.52	256.0	372.1
L48	240						360	56			
248	240						440	72	9.52	279.4	408.4

Source: R. C. Juvinall and K. M. Marshek, *Fundamentals of Machine Component Design.* John Wiley & Sons, New York, 1991. Used with permission.

4.6 Bearing Selection

Bearing manufacturers' catalogs identify bearings by number, give complete dimensional information, list rated load capacities, and furnish details concerning mounting, lubrication, and operation.

Often, special circumstances must be taken into account. Lubrication is especially important in high-speed bearing applications, the best being a fine oil mist or spray. This provides the necessary lubricant film and carries away friction heat with a minimum "churning loss" within the lubricant itself. For ball bearings, nonmetallic separators permit highest speeds.

The size of bearing selected for an application is usually influenced by the size of shaft required (for strength and rigidity considerations) and by the available space. In addition, the bearing must have a high enough load rating to provide an acceptable combination of life and reliability.

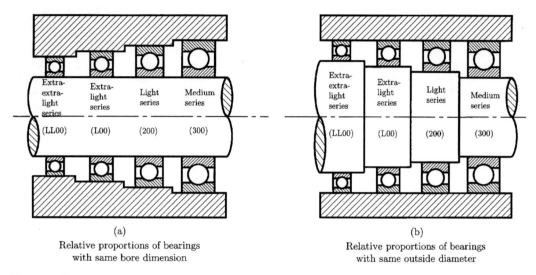

(a)
Relative proportions of bearings
with same bore dimension

(b)
Relative proportions of bearings
with same outside diameter

Figure 4.8

4.6.1 LIFE REQUIREMENT

The *life* of an individual ball or roller bearing is the number of revolutions (or hours at some constant speed) for which the bearings run before the first evidence of fatigue develops in the material of either the rings or of any of the rolling elements. Bearing applications usually require lives different from that used for the catalog rating. Palmegran determined that ball bearing life varies inversely with approximately the third power of the force. Later studies have indicated that this exponent ranges between 3 and 4 for various rolling-element bearings. Many manufacturers retain Palmegren's exponent of 3 for ball bearings and use 10/3 for roller bearings. Following the recommendation of other manufacturers, the exponent 10/3 will be used for both bearing types. Thus the *life required* by the application is

$$L = L_R (C/F_r)^{3.33}, \tag{4.22}$$

where C is the *rated capacity* (from Table 4.2), L_R is the life corresponding to the rated capacity (i.e., $L_r = 9 \times 10^7$ revolutions), and F_r is the radial force involved in the application. The required value of the rated capacity for the application is

$$C_{req} = F_r (L/L_R)^{0.3}. \tag{4.23}$$

For a group of apparently identical bearings the *rating life*, L_R, is the life in revolutions (at a given constant speed and force) that 90% of the group tested bearings will exceed before the first evidence of fatigue develops.

Different manufacturers' catalogs use different values of L_R (some use $L_R = 10^6$ revolutions).

Table 4.2 *Bearing Rated Capacities, C, for 90 × 10^6 Revolution Life (L_R) with 90% Reliability*

	Radial ball, $\alpha = 0°$			Angular ball, $\alpha = 25°$			Roller		
d (mm)	L00 xlt (kN)	200 lt (kN)	300 med (kN)	100 xlt (kN)	200 lt (kN)	300 med (kN)	1000 xlt (kN)	1200 lt (kN)	1300 med (kN)
10	1.02	1.42	1.90	1.02	1.10	1.88			
12	1.12	1.42	2.46	1.10	1.54	2.05			
15	1.22	1.56	3.05	1.28	1.66	2.85			
17	1.32	2.70	3.75	1.36	2.20	3.55	2.12	3.80	4.90
20	2.25	3.35	5.30	2.20	3.05	5.8	3.30	4.40	6.20
25	2.45	3.65	5.90	2.65	3.25	7.20	3.70	5.50	8.5
30	3.35	5.40	8.80	3.60	6.00	8.80		8.30	10.0
35	4.20	8.50	10.6	4.75	8.20	11.0		9.30	13.1
40	4.50	9.40	12.6	4.95	9.90	13.2	7.20	11.1	16.5
45	5.80	9.10	14.8	6.30	10.4	16.4	7.40	12.2	20.9
50	6.10	9.70	15.8	6.60	11.0	19.2		12.5	24.5
55	8.20	12.0	18.0	9.00	13.6	21.5	11.3	14.9	27.1
60	8.70	13.6	20.0	9.70	16.4	24.0	12.0	18.9	32.5
65	9.10	16.0	22.0	10.2	19.2	26.5	12.2	21.1	38.3
70	11.6	17.0	24.5	13.4	19.2	29.5		23.6	44.0
75	12.2	17.0	25.5	13.8	20.0	32.5		23.6	45.4
80	14.2	18.4	28.0	16.6	22.5	35.5	17.3	26.2	51.6
85	15.0	22.5	30.0	17.2	26.5	38.5	18.0	30.7	55.2
90	17.2	25.0	32.5	20.0	28.0	41.5		37.4	65.8
95	18.0	27.5	38.0	21.0	31.0	45.5		44.0	65.8
100	18.0	30.5	40.5	21.5	34.5		20.9	48.0	72.9
105	21.0	32.0	43.5	24.5	37.5			49.8	84.5
110	23.5	35.0	46.0	27.5	41.0	55.0	29.4	54.3	85.4
120	24.5	37.5		28.5	44.5			61.4	100.1
130	29.5	41.0		33.5	48.0	71.0	48.9	69.4	120.1
140	30.5	47.5		35.0	56.0			77.4	131.2
150	34.5			39.0	62.0		58.7	83.6	
160								113.4	
180	47.0			54.0			97.9	140.1	
200								162.4	
220								211.3	
240								258.0	

Source: New Departure-Hyatt Bearing Division, General Motors Corporation.

4.6.2 RELIABILITY REQUIREMENT

Tests show that the median life of rolling-element bearings is about five times the standard 10% failure fatigue life. The standard life is commonly designated as the L_{10} life (sometimes as the B_{10} life), and this life corresponds to 10% failures. It means that this is the life for which 90% have not failed, and it corresponds to 90% reliability ($r = 90\%$). Thus, the life for 50% reliability is about five times the life for 90% reliability. Using the general Weibull equation together with extensive experimental data, a life adjustment *reliability factor* K_r, is recommended. The life adjustment reliability factor, K_r, is plotted in Fig. 4.9. This factor is applicable to both ball and roller bearings. The rated bearing life for any given reliability (greater than 90%) is thus the product $K_r L_R$. Incorporating this factor into Eqs. (4.22) and (4.23) gives

$$L = K_r L_R (C/F_r)^{3.33}$$

$$C_{req} = F_r \left(\frac{L}{K_r L_R} \right)^{0.3}. \tag{4.24}$$

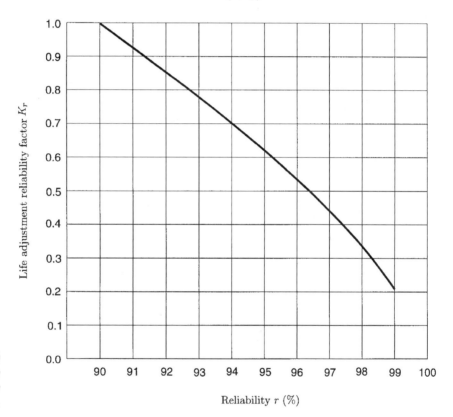

Figure 4.9
Used with permission from Ref. 8.

4.6.3 INFLUENCE OF AXIAL FORCE

For ball bearings (load angle $\alpha = 0°$), any combination of radial force (F_r) and thrust (F_a) results in approximately the same life as does a pure *radial equivalent force*, F_e, calculated as

- For $0.00 < F_a/F_r < 0.35 \Rightarrow F_e = F_r$
- For $0.35 < F_a/F_r < 10.0 \Rightarrow F_e = F_r[1 + 1.115(F_a/F_r - 0.35)]$
- For $F_a/F_r > 10.0 \Rightarrow F_e = 1.176 F_a$

Standard values of load angle α for angular ball bearings are $15°$, $25°$, and $35°$. Only the $25°$ angular ball bearings will be treated next. The radial equivalent force, F_e, for angular ball bearings with $\alpha = 25°$ is:

- For $0.00 < F_a/F_r < 0.68 \Rightarrow F_e = F_r$
- For $0.68 < F_a/F_r < 10.0 \Rightarrow F_e = F_r[1 + 0.87(F_a/F_r - 0.68)]$
- For $F_a/F_r > 10.0 \Rightarrow F_e = 0.911F_a$

4.6.4 SHOCK FORCE

The standard bearing rated capacity is for the condition of uniform force without shock, which is a desirable condition. In many applications there are various degrees of shock loading. This has the effect of increasing the nominal force by an *application factor*, K_a. In Table 4.3 some representative sample values of K_a are given. The force application factor in Table 4.3 serves the same purpose as factors of safety.

Table 4.3 *Application Factors K_a*

Type of application	Ball bearing	Roller bearing
Uniform force, no impact	1.0	1.0
Gearing	1.0–1.3	1.0
Light impact	1.2–1.5	1.0–1.1
Moderate impact	1.5–2.0	1.1–1.5
Heavy impact	2.0–3.0	1.5–2.0

Source: R. C. Juvinall and K. M. Marshek, *Fundamentals of Machine Component Design*. John Wiley & Sons, New York, 1991. Used with permission.

If we substitute F_e for F_r and adding K_a, Eq. (4.24) gives

$$L = K_r L_R \left(\frac{C}{K_a F_e}\right)^{3.33}$$
$$C_{req} = K_a F_e \left(\frac{L}{K_r L_R}\right)^{0.3}. \tag{4.25}$$

When more specific information is not available, Table 4.4 may be used as a guide for the life of a bearing in industrial applications. Table 4.4 contains recommendations on bearing life for some classes of machinery. The information has been accumulated by experience.

EXAMPLE 1 (FROM REF. 8) Select a ball bearing for a machine for continuous 24-hour service. The machine rotates at the angular speed of 900 rpm. The radial force is $F_r = 1\,kN$, and the thrust force is $F_a = 1.25\,kN$, with light impact.

Table 4.4 *Representative Bearing Design Lives*

Type of application	Design life (kh, thousands of hours)
Instruments and apparatus for infrequent use	0.1–0.5
Aircraft engines	0.5–2.0
Machines used intermittently, where service interruption is of minor importance	4–8
Machines intermittently used, where reliability is of great importance	8–14
Machines for 8-hour service, but not every day	14–20
Machines for 8-hour service, every working day	20–30
Machines for continuous 24-hour service	50–60
Machines for continuous 24-hour service where reliability is of extreme importance	100–200

Source: R. C. Juvinall and K. M. Marshek, *Fundamentals of Machine Component Design.* John Wiley & Sons, New York, 1991. Used with permission.

Solution

Both radial, $\alpha = 0°$, and angular, $\alpha = 25°$, ball bearings will be chosen. The equivalent radial force for radial and angular ball bearings is for $F_a/F_r = 1.25$,

- For $0.35 < F_a/F_r < 10.0 \Rightarrow F_e = F_r[1 + 1.115(F_a/F_r - 0.35)] = 2.4\,\text{kN}$ (radial bearing)
- For $0.68 < F_a/F_r < 10.0 \Rightarrow F_e = F_r[1 + 0.87(F_a/F_r - 0.68)] = 1.8\,\text{kN}$ (angular bearing)

From Table 4.3 choose (conservatively) $K_a = 1.5$ for light impact. From Table 4.4 choose (conservatively) 60,000 hour life. The life in revolutions is

$$l = 800 \text{ rpm} \times 60{,}000 \text{ h} \times 60 \text{ min/h} = 3240 \times 10^6 \text{ rev.}$$

For standard 90% reliability ($K_r = 1$, Fig. 4.9), and for $L_R = 90 \times 10^6$ rev (for use with Table 4.2), Eq. (4.25) gives

$$C_{req} = K_a F_e \left(\frac{L}{K_r L_R} \right)^{0.3}$$

$$= (2.4)(1.5)(3240/90)^{0.3} = 10.55 \text{ kN (radial bearing)}$$

$$= (1.8)(1.5)(3240/90)^{0.3} = 7.91 \text{ kN (angular bearing).}$$

Radial Bearings

- From Table 4.2 with 10.55 kN for L00 series $\Rightarrow C = 11.6\,\text{kN}$ and $d = 70\,\text{mm}$ bore. From Table 4.1 with 70 mm bore and L00 series, the bearing number is L14.

- From Table 4.2 with 10.55 kN for 200 series $\Rightarrow C = 12.0$ kN and $d = 55$ mm bore. From Table 4.1 with 55 mm bore and 200 series, the bearing number is 211.
- From Table 4.2 with 10.55 kN for 300 series $\Rightarrow C = 10.6$ kN and $d = 35$ mm bore. From Table 4.1 with 35 mm bore and 300 series, the bearing number is 307.

For angular contact bearings the appropriate choices would be L11, 207, and 306.

The final selection would be made on the basis of cost of the total installation, including shaft and housing.   ▲

EXAMPLE 2   Figure 4.10 shows a two-stage gear reducer with identical pairs of gears. An electric motor with power $H = 2$ kW and $n_1 = 900$ rpm is coupled to the shaft a. On this shaft there is rigidly connected the input driver gear 1 with the number of teeth $N_1 = N_p = 17$. The speed reducer uses a countershaft b with two rigidly connected gears 2 and 2′, having $N_2 = N_g = 51$ teeth and $N_{2'} = N_p = 17$ teeth. The output gear 3 has $N_3 = N_g = 51$ teeth and is rigidly fixed to the shaft c coupled to the driven machine. The input shaft a and output shaft c are collinear. The countershaft b turns freely in bearings A and B. The gears mesh along the pitch diameter and the shaft are parallel. The diametral pitch for each stage is $P_d = 5$, and the pressure angle is $\phi = 20°$. The distance between the bearings is $s = 100$ mm, and the distance $l = 25$ mm (Fig. 4.10). The gear reducer is a part of an industrial machine intended for continuous one-shift (8 hours per day). Select identical extra-light series (L00) ball bearings for A and b.

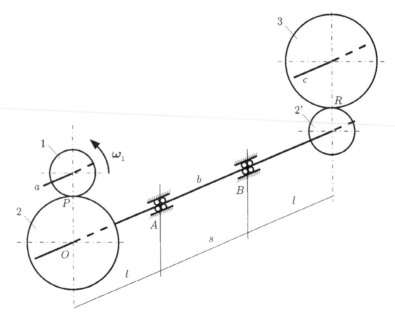

Figure 4.10

Solution

The pitch diameters of pinions 1 and 2' are $d_1 = d_{2'} = d_p = N_p/P_d = 17/5 = 3.4$ in. The pitch diameters of gears 2 and 3 are $d_2 = d_3 = d_g = N_g/P_d = 51/5 = 10.2$ in. The circular pitch is $p = \pi/P_d = 3.14/5 = 0.63$ in.

Angular Speeds

The following relation exists for the first stage:

$$\frac{n_1}{n_2} = \frac{N_2}{N_1} \Rightarrow n_2 = n_1 \frac{N_1}{N_2} = 900 \frac{17}{51} = 300 \text{ rpm.} \tag{4.26}$$

For the second stage

$$\frac{n_2}{n_3} = \frac{N_3}{N_{2'}} \Rightarrow n_3 = n_2 \frac{N_{2'}}{N_3} = 300 \frac{17}{51} = 100 \text{ rpm.} \tag{4.27}$$

The angular speed of the countershaft b is $n_b = n_2 = 300$ rpm, and the angular speed of the driven shaft c is $n_c = n_3 = 100$ rpm.

Torque Carried by Each Shaft

The relation between the power H_a of the motor and the torque M_a in shaft a is

$$H_a = \frac{M_a n_a}{9549}, \tag{4.28}$$

and the torque M_a in shaft a is

$$M_a = \frac{9549 H_a}{n_a} = \frac{9549(2 \text{ kW})}{900 \text{ rpm}} = 21.22 \text{ Nm.}$$

The torque in shaft b is

$$M_b = \frac{9549 H_a}{n_b} = M_a \frac{N_2}{N_1} = 21.22 \frac{51}{17} = 63.66 \text{ Nm,}$$

and the torque in shaft c is

$$M_c = \frac{9549 H_a}{n_c} = M_b \frac{N_3}{N_{2'}} = 63.66 \frac{51}{17} = 190.98 \text{ Nm.}$$

Bearing Reactions

All the gear radial and tangential force is transferred at the pitch point P. The tangential force on the motor pinion is

$$F_t = \frac{M_a}{r_p} = \frac{21.22}{0.0431} = 492.34 \text{ N,}$$

where $r_p = d_p/2 = 1.7$ in $= 0.0431$ m. The radial force on the motor pinion is

$$F_r = F_t \tan \phi = 492.34 \tan 20° = 179.2 \text{ N.}$$

The force on the motor pinion 1 at P (Fig. 4.11) is

$$\mathbf{F}_{21} = F_{r21}\mathbf{j} + F_{r21}\mathbf{k} = 179.2\mathbf{j} - 492.34\mathbf{k} \text{ N.} \tag{4.29}$$

Machine Components

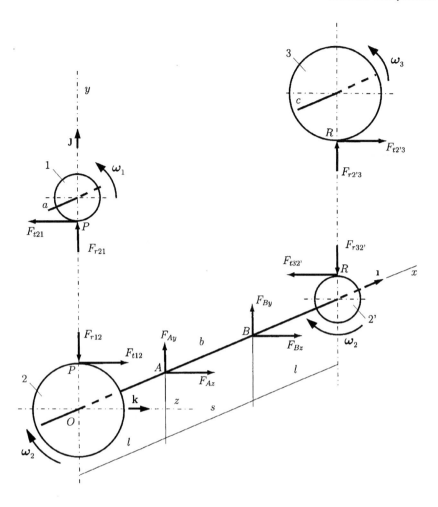

Figure 4.11

The force on the countershaft gear 2 at P is

$$\mathbf{F}_{12} = -\mathbf{F}_{21} = F_{r12}\mathbf{j} + F_{t12}\mathbf{k} = -179.2\mathbf{j} + 492.34\mathbf{k} \text{ N}. \qquad (4.30)$$

The forces on the countershaft pinion $2'$ at R are three times as large, that is,

$$F_{t'} = \frac{M_b}{r_p} = \frac{63.66}{0.0431} = 1477 \text{ N}$$

$$F_{r'} = F_{t'} \tan\phi = 1477 \tan 20° = 537.6 \text{ N}$$

and

$$\mathbf{F}_{32'} = F_{r32'}\mathbf{j} + F_{t32'}\mathbf{k} = -537.6\mathbf{j} - 1477\mathbf{k} \text{ N}. \qquad (4.31)$$

The unknown forces applied to bearings A and B can be written as

$$\mathbf{F}_A = F_{Ay}\mathbf{j} + F_{Az}\mathbf{k}$$

$$\mathbf{F}_B = F_{By}\mathbf{j} + F_{Bz}\mathbf{k}.$$

A sketch of the countershaft as a free body in equilibrium is shown in Fig. 4.11. To determine these forces, two vectorial equations are used. The sum of moments of all forces that act on the countershaft with respect to A is zero:

$$\sum \mathbf{M}_A = \mathbf{AP} \times \mathbf{F}_{12} + \mathbf{AR} \times \mathbf{F}_{32'} + \mathbf{AB} \times \mathbf{F}_B$$

$$= \begin{vmatrix} \mathbf{1} & \mathbf{j} & \mathbf{k} \\ -l & r_2 & 0 \\ 0 & F_{r12} & F_{t12} \end{vmatrix} + \begin{vmatrix} \mathbf{1} & \mathbf{j} & \mathbf{k} \\ s+l & r_{2'} & 0 \\ 0 & F_{r32'} & F_{t32'} \end{vmatrix} + \begin{vmatrix} \mathbf{1} & \mathbf{j} & \mathbf{k} \\ s & 0 & 0 \\ 0 & F_{By} & F_{Bz} \end{vmatrix} = \mathbf{0},$$

(4.32)

or

$$\sum \mathbf{M}_A \cdot \mathbf{j} = lF_{t12} - (s+l)F_{t32'} - sF_{Bz} = 0$$
$$\sum \mathbf{M}_A \cdot \mathbf{k} = -lF_{r12} + (s+l)F_{r32'} + sF_{By} = 0.$$

(4.33)

From the preceding equations, $F_{By} = 627.2\,\text{N}$ and $F_{Bz} = 1969.33\,\text{N}$. The radial force at B is

$$F_B = \sqrt{F_{By}^2 + F_{Bz}^2} = 2066.8 \text{ N}.$$

The sum of all forces that act on the countershaft is zero,

$$\sum \mathbf{F} = \mathbf{F}_{12} + \mathbf{F}_A + \mathbf{F}_B + \mathbf{F}_{32'} = \mathbf{0},$$

(4.34)

or

$$-F_{r12} + F_{Ay} + F_{By} - F_{r32'} = 0$$
$$F_{t12} + F_{Az} + F_{Bz} - F_{t32'} = 0.$$

(4.35)

From Eq. (4.35), $F_{Ay} = 89.6\,\text{N}$ and $F_{Az} = -984.67\,\text{N}$. The radial force at A is

$$F_A\sqrt{F_{Ay}^2 + F_{Az}^2} = 988.73 \text{ N}.$$

Ball Bearing Selection

Since the radial force at B is greater than the radial force at A, $F_B > F_A$, the bearing selection will be based on bearing B. The equivalent radial force for radial ball bearings is $F_e = F_B = 2066.8\,\text{N}$.

The ball bearings operate 8 hours per day, 5 days per week. From Table 4.3 choose $K_a = 1.1$ for gearing. From Table 4.4, choose (conservatively) 30,000 hour life.

The life in revolutions is

$$L = 300 \text{ rpm} \times 30,000 \text{ h} \times 60 \text{ min/h} = 540 \times 10^6 \text{ rev.}$$

For standard 90% reliability ($K_r = 1$, Fig. 4.9), and for $L_R = 90 \times 10^6$ rev (for use with Table 4.2), Eq. (4.25) gives

$$C_{req} = K_a F_e \left(\frac{L}{K_r L_R} \right)^{0.3}$$

$$= (1.1)(2066.8) \left[\frac{540 \times 10^6}{(1)90 \times 10^6} \right]^{0.3} = 3891.67 \text{ N}$$

$$\approx 3.9 \text{ kN}.$$

From Table 4.2, with 3.9 kN for L00 series $\Rightarrow C = 4.2$ kN and $d = 35$ mm bore. From Table 4.1, with 35 mm bore and L00 series, the bearing number is L07. The shaft size requirement may necessitate the use of a larger bore bearing.

5. Lubrication and Sliding Bearings

Lubrication reduces the friction, wear, and heating of machine parts in relative motion. A lubricant is a substance that is inserted between the moving parts.

5.1 Viscosity

5.1.1 NEWTON'S LAW OF VISCOUS FLOW

A surface of area A is moving with the linear velocity V on a film of lubricant as shown in Fig. 5.1a. The thickness of the lubricant is s and the deforming

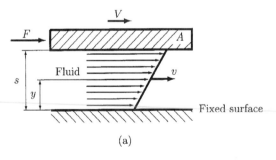

(a)

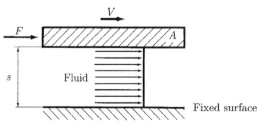

Figure 5.1   (b)

force acting on the film is F. The layers of the fluid in contact with the moving surface have the velocity $v = V$ and the layers of the fluid in contact with the fixed surface have the velocity $v = 0$.

Newton's law of viscous flow states that the shear stress τ in a fluid is proportional to the rate of change of the velocity v with respect to the distance y from the fixed surface,

$$\tau = \frac{F}{A} = \mu \frac{\partial v}{\partial y}, \tag{5.1}$$

where μ is a constant, the *absolute viscosity*, or the *dynamic viscosity*. The derivative $\partial v / \partial y$ is the rate of change of velocity with distance and represents the rate of shear, or the velocity gradient. For a constant velocity gradient Eq. (5.1) can be written as (Fig. 5.1b).

$$\tau = \mu \frac{V}{2}. \tag{5.2}$$

The unit of viscosity μ, U.S. customary units, is pound-second per square inch, lb s/in^2, or reyn (from Osborne Reynolds). In SI units, the viscosity is expressed as newton-seconds per square meter, N s/m^2, or pascal-seconds, Pa s. The conversion factor between the two is the same as for stress,

$$1 \text{ reyn} = 6890 \text{ Pa s.}$$

The reyn and pascal-second are such large units that microreyn (μreyn) and millipascal-second (mPa s) are more commonly used. The former standard metric unit of viscosity was the poise. One centipoise, cp, is equal to one millipascal-second,

$$1 cp = 1 \text{ mPa s.}$$

Liquid viscosities are determined by measuring the time required for a given quantity of the liquid to flow by gravity through a precision opening. For lubricating oils, the Saybolt universal viscometer, shown in Fig. 5.2, is an instrument used to measure the viscosity. The viscosity measurements are *Saybolt seconds*, or *SUS* (Saybolt universal seconds), *SSU* (Saybolt seconds universal), and *SUV* (Saybolt universal viscosity).

With a Saybolt universal viscometer one can measure the *kinematic viscosity*, v. The kinematic viscosity is defined as absolute viscosity, μ, divided by mass density, ρ,

$$v = \frac{\mu}{\rho}. \tag{5.3}$$

The units for kinematic viscosity are length^2/time, as cm^2/s, which is the stoke (St). Using SI units, $1 \text{ m}^2/\text{s} = 10^4$ St.

Absolute viscosities can be obtained from Saybolt viscometer measurements (time s, in seconds) by the equations

$$\mu(\text{mPa s or cp}) = \left(0.22 \text{ s} - \frac{180}{s} \right) \rho \tag{5.4}$$

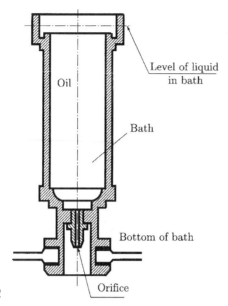

Figure 5.2

and

$$\mu(\mu\mathrm{reyn}) = 0.145\left(0.22\ \mathrm{s} - \frac{180}{\mathrm{s}}\right)\rho, \qquad (5.5)$$

where ρ is the mass density in grams per cubic centimeter, g/cm^3 (which is also called specific gravity). For petroleum oils the mass density at different temperatures is

$$\rho = 0.89 - 0.00063(^\circ\mathrm{C} - 15.6)\ \mathrm{g/cm}^3, \qquad (5.6)$$

or

$$\rho = 0.89 - 0.00035(^\circ\mathrm{F} - 60)\ \mathrm{g/cm}^3. \qquad (5.7)$$

The Society of Automotive Engineers (SAE) classifies oils according to viscosity.

The viscosity of SAE single viscosity motor oils is measured at 100°C (212°F):

SAE viscosity grade	Viscosity range μ at 100°C (cSt)
20	$5.6 \le \mu < 9.3$
30	$9.3 \le \mu < 12.5$
40	$12.5 \le \mu < 16.3$
50	$16.3 \le \mu < 21.9$

The viscosity of motor oils with a W suffix is measured at temperatures that depend upon the grade:

SAE viscosity grade	Viscosity range μ (cp)	Temperature (°C)	Viscosity range at 100°C μ (cSt)
0 W	$\mu \leq 3250$	-30	$3.8 \leq \mu$
5 W	$\mu \leq 3500$	-25	$3.8 \leq \mu$
10 W	$\mu \leq 3500$	-20	$4.1 \leq \mu$
15 W	$\mu \leq 3500$	-15	$5.6 \leq \mu$
20 W	$\mu \leq 4500$	-10	$5.6 \leq \mu$
25 W	$\mu \leq 6\,000$	-5	$9.3 \leq \mu$

Multigrade motor oils, such SAE 10W-30, must meet the viscosity standard at the W temperature and the SAE 30 viscosity requirement at 100°C.

In Figs. 5.3, 5.4, and 5.5 are shown graphics representing the absolute viscosity function of temperature for typical SAE numbered oils. Grease is a non-Newtonian material that does not begin to flow until a shear stress exceeding a yield point is applied.

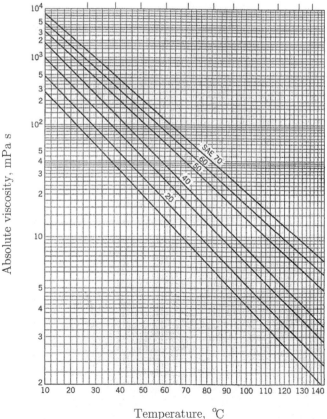

Figure 5.3
Used with permission from Ref. 8.

Absolute viscosity, mPa s

Temperature, °C

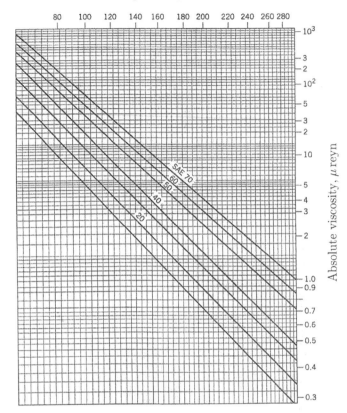

Temperature, °F

Figure 5.4
Used with permission from Ref. 8.

Multigrade oils, such as SAE 10W-40, have less variation of viscosity with temperature than a single-grade motor oil such as SAE 40 or SAE 10W. The *viscosity index* (VI) measures the variation in viscosity with temperature. The viscosity index, on the Dean and Davis scale, of Pennsylvania oils is VI = 100. The viscosity index, on the same scale, of Gulf Coast oils is VI = 0. Other oils are rated intermediately. Nonpetroleum-base lubricants have widely varying viscosity indices. Silicone oils have relatively little variation of viscosity with temperature. The viscosity index improvers (additives) can increase viscosity index of petroleum oils.

EXAMPLE   The Saybolt kinematic viscosity of an oil corresponds to 60 s at 90°C (Fig. 5.2). What is the corresponding absolute viscosity in millipascal-seconds (or centipoises) and in microreyns?

Solution

From Eq. (5.4), the absolute viscosity in centipoise is

$$\mu = \left(0.22s - \frac{180}{s}\right)\rho = \left[(0.22)(60) - \frac{180}{60}\right]0.843 = 8.598 \text{ cp (or } 8.598 \text{ mPa s).}$$

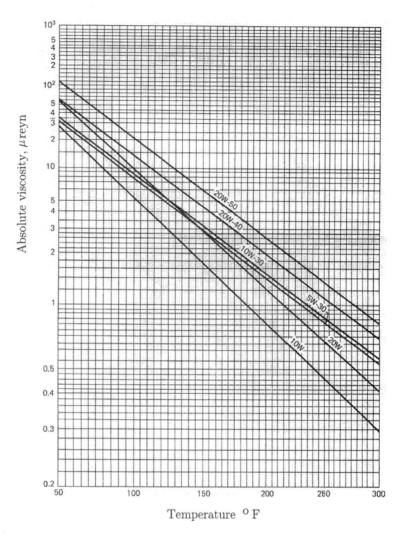

Figure 5.5
Used with permission from Ref. 15.

From Eq. (5.5), the absolute viscosity in microreyns, μreyn, is

$$\mu = 0.145\left(0.22s - \frac{180}{s}\right)\rho = 0.145\left[(0.22)(60) - \frac{180}{60}\right]0.843 = 1.246 \ \mu\text{reyn}.$$

From Eq. (5.6), the mass density of the oil is

$$\rho = 0.89 - 0.00063(^\circ\text{C} - 15.6) = 0.89 - 0.00063(90 - 15.6) = 0.843 \text{ g/cm}^3.$$

▲

5.2 Petroff's Equation

Hydrodynamic lubrication is defined when the surfaces of the bearing are separated by a film of lubricant, and does not depend upon the introduction of the lubricant under pressure. The pressure is created by the motion of the moving surface. Hydrostatic lubrication is defined when the lubricant is

Machine Components

introduced at a pressure sufficiently high to separate the surfaces of the bearing.

A hydrodynamic bearing (hydrodynamic lubrication) is considered in Fig. 5.6. There is no lubricant flow in the axial direction and the bearing carries a very small load. The radius of the shaft is R, the clearance is c, and the length of the bearing is L (Fig. 5.6). The shaft rotates with the angular speed n rev/s and its surface velocity is $V = 2\pi R n$.

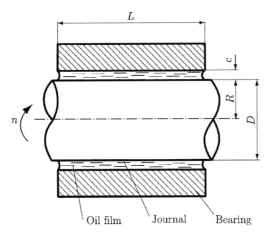

Figure 5.6

From Eq. (5.2) the shearing stress is

$$\tau = \mu \frac{V}{S} = \frac{2\pi R \mu n}{c}. \tag{5.8}$$

The force required to shear the film is the stress times the area,

$$F = \tau A,$$

where $A = 2\pi R L$.

The torque is the force times the lever arm,

$$T = FR = (\tau A)R = \left(\frac{2\pi R \mu n}{c} 2\pi R L \right) R = \frac{4\pi^2 \mu n L R^3}{c}. \tag{5.9}$$

If a small radial load W is applied on the bearing, the pressure P (the radial load per unit of projected bearing area) is

$$P = \frac{W}{2RL}.$$

The frictional force is fW, where f is the coefficient of friction, and the frictional torque is

$$T_f = -f(W)R = f(2RLP)R = 2R^2 fLP. \tag{5.10}$$

From Eqs. (5.9) and (5.10), the coefficient of friction is

$$f = 2\pi^2 \left(\frac{\mu n}{P} \right) \left(\frac{R}{c} \right). \tag{5.11}$$

This is called Petroff's law or Petroff's equation. In Petroff's equation there are two important bearing parameters: the dimensionless variable $(\mu n/P)$, and the clearance ratio (R/c) with the order between 500 and 1000.

The bearing *characteristic number*, or Sommerfeld number S, is given by

$$D = \frac{\mu n}{P}\left(\frac{R}{c}\right)^2,\tag{5.12}$$

where R is the journal radius (in), c is the radial clearance (in), μ is the absolute viscosity (reyn), n is the speed (rev/s), and P is the pressure (psi).

The power loss is calculated with the relation

$$H = 2\pi T_f n.\tag{5.13}$$

EXAMPLE 5.1 A shaft with a 120 mm diameter (Fig. 5.7) is supported by a bearing of 100 mm length with a diametral clearance of 0.2 mm and is lubricated by oil having a viscosity of 60 mPa s. The shaft rotates at 720 rpm. The radial load is 6000 N. Find the bearing coefficient of friction and the power loss.

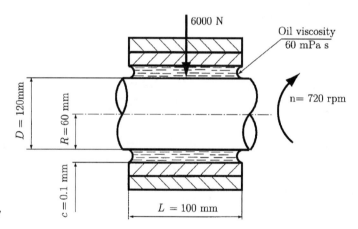

Figure 5.7

Solution

1. From Eq. (5.11), the coefficient of friction is

$$f = 2\pi^2\left(\frac{\mu n}{P}\right)\left(\frac{R}{c}\right).$$

With $R = 60\,\text{mm} = 60\,10^{-3}\,\text{m}$, $c = 0.1\,\text{mm}$, $n = 720\,\text{rev/min} = 12\,\text{rev/s}$, and oil viscosity $\mu = 60\,\text{mPa s} = 0.06\,\text{Pa s}$, the pressure is

$$P = \frac{W}{2RL} = \frac{6000\,\text{N}}{2(0.06\,\text{m})(0.1\,\text{m})} = 500{,}000\,\text{N/m}^2 = 500{,}000\,\text{Pa}.$$

Using the numerical data, we find that the value of the coefficient of friction is

$$f = 2\pi^2 \frac{(0.06 \text{ Pa s})(12 \text{ rev/s})}{500,000 \text{ Pa}} \frac{60 \text{ mm}}{0.1 \text{ mm}} = 0.017.$$

2. The friction torque is calculated with

$$T_f = fWR = (0.017)(6000 \text{ N})(0.06 \text{ m}) = 6.139 \text{ Nm}$$

3. The power loss is

$$H = 2\pi T_f n = 2\pi(6.139 \text{ Nm})(12 \text{ rev/s}) = 462.921 \text{ Nm/s} = 462.921 \text{ W}.$$

▲

5.3 Hydrodynamic Lubrication Theory

In Fig. 5.8 is shown a small element of lubricant film of dimensions dx, dy, and dz. The normal forces, due the pressure, act upon the right and left sides of the element. The shear forces, due to the viscosity and to the velocity, act upon the top and bottom of the element. The equilibrium of forces gives

$$p \, dx \, dz + \tau \, dx \, dz - \left(p + \frac{dp}{dx} dx\right) dy \, dz - \left(\tau + \frac{\partial\tau}{\partial y} dy\right) dx \, dz = 0, \quad (5.14)$$

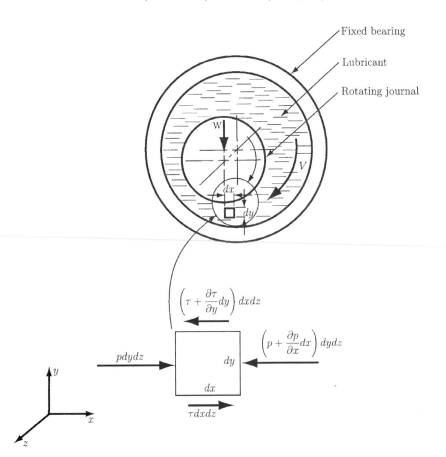

Figure 5.8

which reduces to

$$\frac{dp}{dx} = \frac{\partial \tau}{\partial y}. \tag{5.15}$$

In Eq. (5.15) the pressure of the film p is constant in the y direction and depends only on the coordinate x, $p = p(x)$. The shear stress τ is calculated from Eq. (5.1),

$$\tau = \mu \frac{\partial v(x, y)}{\partial y}. \tag{5.16}$$

The velocity v of any particle of lubricant depends on both coordinates x and y, $v = v(x, y)$.

From Eqs. (5.15) and (5.16) results

$$\frac{dp}{dx} = \mu \frac{\partial^2 v}{\partial^2 y}, \tag{5.17}$$

or

$$\frac{\partial^2 v}{\partial^2 y} = \frac{1}{\mu} \frac{dp}{dx}. \tag{5.18}$$

Holding x constant and integrating twice with respect to y gives

$$\frac{\partial v}{\partial y} = \frac{1}{\mu} \left(\frac{dp}{dx} y + C_1 \right), \tag{5.19}$$

and

$$v = \frac{1}{\mu} \left(\frac{dp}{dx} \frac{y^2}{2} + C_1 x + C_2 \right). \tag{5.20}$$

The constants C_1 and C_2 are calculated using the boundary conditions for $y = 0 \Rightarrow v = 0$, and for $y = s \Rightarrow v = V$.

With C_1 and C_2 values computed, Eq. (5.16) gives the equation for the velocity distribution of the lubricant film across any yz plane,

$$v = \frac{1}{2\mu} \frac{dp}{dx} (y^2 - sy) + \frac{V}{s} y. \tag{5.21}$$

The volume of lubricant Q flowing across the section for width of unity in the z direction is

$$Q = \int_0^s v(x, y) dy = \frac{Vs}{2} - \frac{s^3}{12\mu} \frac{dp}{dx}. \tag{5.22}$$

For incompressible lubricant the flow is the same for any section,

$$\frac{dQ}{dx} = 0.$$

By differentiating Eq. (5.22), one can write

$$\frac{dQ}{dx} = \frac{V}{2} \frac{ds}{dx} - \frac{d}{dx} \left(\frac{s^3}{12\mu} \frac{dp}{dx} \right),$$

or

$$\frac{d}{dx}\left(\frac{s^3}{\mu}\frac{dp}{dx}\right) = 6V\frac{ds}{dx}, \tag{5.23}$$

which is the classical Reynolds equation for one-dimensional flow.

The following assumptions were made:

The fluid is Newtonian, incompressible, of constant viscosity, and experiences no inertial or gravitational forces

The fluid has a laminar flow, with no slip at the boundary surfaces

The fluid experiences negligible pressure variation over its thickness

The journal radius can be considered infinite

The Reynolds equation for two-dimensional flow is (the z direction is included) is

$$\frac{\partial}{\partial x}\left(\frac{s^3}{\mu}\frac{\partial p}{\partial x}\right) + \frac{\partial}{\partial z}\left(\frac{s^3}{\mu}\frac{\partial p}{\partial z}\right) = 6V\frac{\partial s}{\partial x}. \tag{5.24}$$

For short bearings, one can neglect the x term in the Reynolds equation:

$$\frac{\partial}{\partial z}\left(\frac{s^3}{\mu}\frac{\partial p}{\partial z}\right) = 6V\frac{\partial b}{\partial x}. \tag{5.25}$$

Equation (5.25) can be used for analysis and design.

5.4 Design Charts

Raimondi and Boyd have transformed the solutions of the Reynolds Eq. (5.25) to chart form. The charts provide accurate solutions for bearings of all proportions. Some charts are shown in Figs. 5.9 to 5.15. The quantities given in the charts are shown in Fig. 5.16. The Raimondi and Boyd charts give plots of dimensionless bearing parameters as functions of the bearing characteristic number, or Sommerfeld variable, S.

EXAMPLE 5.2 A journal bearing has the diameter $D = 2.5$ in, the length $L = 0.625$ in, and the radial clearance $c = 0.002$ in, as shown in Fig. 5.17. The shaft rotates at $n = 3600$ rpm. The journal bearing supports a constant load $W = 1500$ lb. The lubricant film is SAE 40 oil at atmospheric pressure. The average temperature of the oil film is $T_{avg} = 140°F$.

Find the minimum oil film thickness b_0, the bearing coefficient of friction f, the maximum pressure p_{max}, the position angle of minimum film thickness ϕ, the angular position of the point of maximum pressure $\theta_{p_{max}}$, the terminating position of the oil film θ_{po}, the total oil flow rate Q, and the flow ratio (side flow/total flow) Q_s/Q.

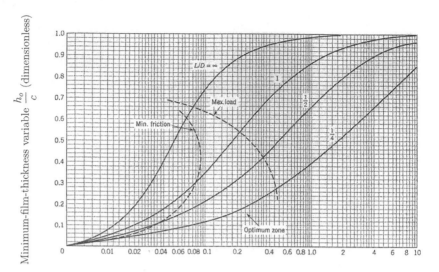

Figure 5.9
From Ref. 19.

Bearing characteristic number, $S = \left(\dfrac{R}{c}\right)^2 \dfrac{\mu n}{P}$

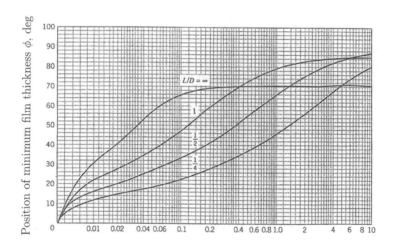

Figure 5.10
From Ref. 19.

Bearing characteristic number, $S = \left(\dfrac{R}{c}\right)^2 \dfrac{\mu n}{P}$

Solution

The pressure is

$$P = \frac{W}{LD} = \frac{1500}{(0.625)(2.5)} = 960 \text{ psi.}$$

The dynamic viscosity is $\mu = 5 \cdot 10^{-6}$ reyn (SAE 40, $T_{avg} = 140°$F), from Fig. 5.4. The Sommerfeld number is

$$S = \left(\frac{R}{c}\right)^2 \frac{\mu n}{P} = \left(\frac{1.25}{0.002}\right)^2 \frac{5 \times 10^{-6}(60)}{960} = 0.122.$$

For all charts, $S = 0.122$ and $L/D = 0.25$ are used.

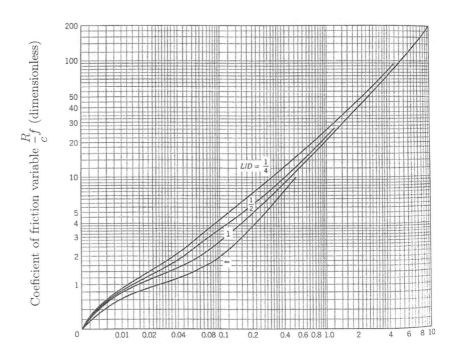

Figure 5.11
From Ref. 19.

Bearing characteristic number, $S = \left(\dfrac{R}{c}\right)^2 \dfrac{\mu n}{P}$

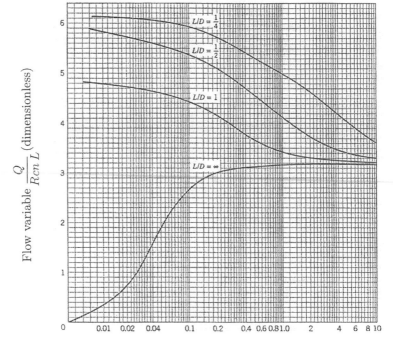

Figure 5.12
From Ref. 19.

Bearing characteristic number, $S = \left(\dfrac{R}{c}\right)^2 \dfrac{\mu n}{P}$

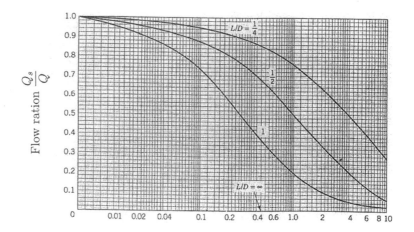

Figure 5.13
From Ref. 19.

Bearing characteristic number, $S = \left(\dfrac{R}{c}\right)^2 \dfrac{\mu n}{P}$

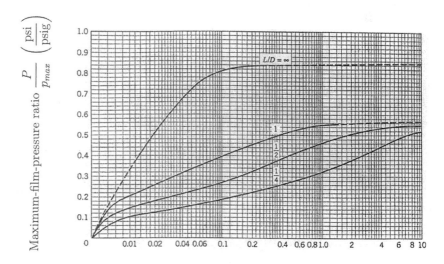

Figure 5.14
From Ref. 19.

Bearing characteristic number, $S = \left(\dfrac{R}{c}\right)^2 \dfrac{\mu n}{P}$

From Fig. 5.9, $h_o/c = 0.125$, that is, the minimum film thickness $h_0 = 0.00025$ in. From Fig. 5.11, $(R/c)f = 5.0$, that is, the coefficient of friction $f = 0.00832$. From Fig. 5.14, $P/p_{max} = 0.2$, that is, the maximum film pressure $p_{max} = 4800$ psi.

From Fig. 5.10, the position angle of minimum film thickness $\phi = 25°$. From Fig. 5.15, the terminating position of the oil film $\theta_{po} = 32°$, and the angular position of the point of maximum pressure $\theta_{p_{max}} = 9°$.

From Fig. 5.12, the flow variable $Q/RcnL = 5.9$, that is, the total flow $Q = 0.553$ in^3/s. From Fig. 5.13, the flow ratio (side flow/total flow) $Q_s/Q = 0.94$, that is, 6% of the flow is recirculated. ▲

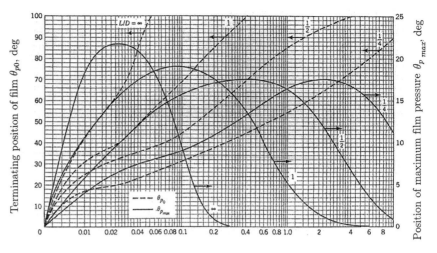

Figure 5.15
From Ref. 19.

$$\text{Bearing characteristic number, } S = \left(\frac{R}{c}\right)^2 \frac{\mu n}{P}$$

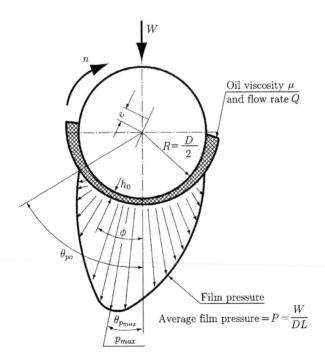

Figure 5.16
Used with permission from Ref. 16.

h_0 the minimum oil film thickness
$\theta_{p_{max}}$ the angular position of the point of maximum pressure
θ_{po} the terminating position of the oil film
p_{max} the maximum pressure
ϕ the position angle of minimum film thickness

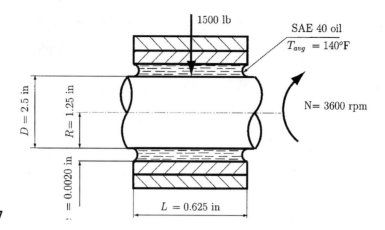

Figure 5.17

EXAMPLE 5.3 The oil lubricated bearing of a steam turbine has the diameter $D = 160\,\text{mm}$ (Fig. 5.18). The angular velocity of the rotor shaft is $n = 2400\,\text{rpm}$. The radial load is $W = 18\,\text{kN}$. The lubricant is SAE 20, controlled to an average temperature of 78°C.

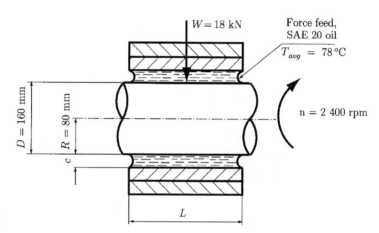

Figure 5.18

Find the bearing length, the radial clearance and the corresponding values of the minimum oil film thickness, the coefficient of friction, the friction power loss, and the oil flows.

Solution

From Table 5.1, for a steam turbine (1 to 2 MPa range) the unit load $P = 1.5\,\text{MPa}$ is arbitrarily selected. Based of this decision, the bearing length is

$$L = \frac{W}{PD} = \frac{18,000}{(1.5)(160)} = 75\text{mm}.$$

Table 5.1 *Representative Unit Sleeve Bearing Loads in Current Practice*

Application	Unit load $P = \dfrac{W_{max}}{LD}$	
Relatively steady load		
Electric motors	0.8–1.5 MPa	120–250 psi
Steam turbines	1.0–2.0 MPa	150–300 psi
Gear reducers	0.8–1.5 MPa	120–250 psi
Centrif. pumps	0.6–1.2 MPa	100–180 psi
Rapidly fluctuating load		
Diesel engines:		
Main bearings	6–12 MPa	900–1700 psi
Connecting rod bearings	8–15 MPa	1150–2399 psi
Automotive gasoline engines:		
Main bearings	4–5 MPa	600–750 psi
Connecting rod bearings	10–15 MPa	1700–2300 psi

Source: R. C. Juvinall and K. M. Marshek, *Fundamentals of Machine Component Design*. John Wiley & Sons, New York, 1991. Used with permission.

Arbitrarily round this up to $L = 80$ mm to give $L/D = \frac{1}{2}$ for convenient use of the Raimondi and Boyd charts.

With $L = 80$ mm, P is given by the relation

$$P = \frac{W}{LD} = \frac{18{,}000}{(80)(160)} = 1.406 \text{ MPa}.$$

From Fig. 5.9 for $L/D = \frac{1}{2}$, the optimum zone is between $S = 0.037$ and $S = 0.35$.

From Fig. 5.3 the viscosity of SAE 20 oil at 78°C is $\mu = 9.75$ mPa s. Substituting the known values into the equation for S,

$$S = \left(\frac{\mu n}{P}\right)\left(\frac{R}{c}\right)^2$$

gives

$$0.037 = \frac{(9.75 \cdot 10^{-3} \text{ Pa s})(40 \text{ rev/s})}{(1.406 \cdot 10^6 \text{ Pa})}\left(\frac{80}{c}\right)^2.$$

Hence, the clearance is $c = 0.2$ mm ($c/R = 0.00025$).

Similarly, for $S = 0.35$, $c = 0.0677$ mm ($c/R = 0.00083$).

The minimum oil film thickness h_o is calculated from the ratio h_o/c obtained from Fig. 5.9. The coefficient of friction f is calculated from the ratio Rf/c obtained from Fig. 5.11. The total oil flow rate Q is calculated from the ratio $Q/(RcnL)$ obtained from Fig. 5.12. The side leakage oil flow Q_s is determined from the ratio Q_s/Q obtained from Fig. 5.13. The values of h_o, f, Q, and Q_s, as functions of c ($0.03 \leq c \leq 0.23$), with c extending to either side of the optimum range, are listed in Table 5.2 and plotted in Fig. 5.19. Figure

Table 5.2 *Numerical Values for Example 5.3*

c (mm)	S	h_0 (mm)	f	Q (mm³/s)	Q_s (mm³/s)
0.0300	1.7453	0.0231	0.0147	26,730	9,620
0.0400	0.9817	0.0256	0.0111	39,980	19,990
0.0500	0.6283	0.0270	0.0093	52,980	31,790
0.0670	0.3499	0.0288	0.0075	77,450	54,220
0.0900	0.1939	0.0288	0.0068	110,540	88,430
0.1100	0.1298	0.0275	0.0062	140,400	117,940
0.1300	0.0929	0.0260	0.0060	169,060	145,390
0.1500	0.0698	0.0255	0.0055	198,680	176,830
0.1800	0.0485	0.0234	0.0050	242,750	220,910
0.2000	0.0393	0.0220	0.0047	272,140	253,090
0.2200	0.0325	0.0220	0.0047	300,940	281,380
0.2300	0.0297	0.2185	0.0046	315,730	296,780

5.19 indicates a good operation for the radial clearances between about 0.05 mm and 0.22 mm.

For the minimum acceptable oil film thickness, h_o, the following empirical relations are given (Trumpler empirical equation):

$$h_o \geq h_{omin} = 0.0002 + 0.00004D(h_o \text{ and } D \text{ in inches})$$
$$h_o \geq h_{omin} = 0.005 + 0.00004D(h_o \text{ and } D \text{ in millimeters}). \quad (5.26)$$

For $D = 160$ mm the minimum acceptable oil film thickness is

$$h_{omin} = 0.005 + 0.00004(160) = 0.0114 \text{ mm}.$$

The minimum film thickness using a safety factor of $C_s = 2$ applied to the load, and assuming an "extreme case" of $c = 0.22$ mm, is calculated as follows:

- The Sommerfeld number is

$$S = \left(\frac{\mu n}{C_s P}\right)\left(\frac{R}{c}\right)^2 = \frac{(9.75)(10^{-3})(40)}{(2)(1.406 \cdot 10^6)}\left(\frac{80}{0.22}\right)^2 = 0.01625.$$

- From Fig. 5.9, using $S = 0.01625$, we find $h_o/c = 0.06$, and the minimum film thickness h_o is 0.0132 mm.

This value satisfies the condition $h_o = 0.0132 \geq h_{omin} = 0.0114$.

Friction power loss can be computed using the value of the friction torque T_f. For the tightest bearing fit where $c = 0.05$ mm and $f = 0.0093$, the friction torque is

$$T_f = \frac{WfD}{2} = \frac{(18,000 \text{ N})(0.0093)(0.160 \text{ m})}{2} = 13.392 \text{ N m},$$

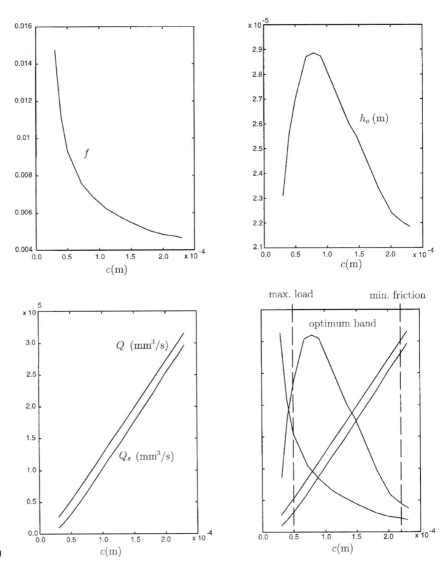

Figure 5.19

and the friction power is

$$\text{Friction power} = \frac{nT}{9549} = \frac{(2400)(13.392)}{9549} = 3.365 \text{ kW.} \quad \blacktriangle$$

References

1. J. S. Arora, *Introduction to Optimum Design.* McGraw-Hill, New York, 1989.

2. K. S. Edwards, Jr. and R. B. McKee, *Fundamentals of Mechanical Component Design.* McGraw-Hill, New York, 1991.

3. A. Ertas and J. C. Jones, *The Engineering Design Process.* John Wiley & Sons, New York, 1996.

4. A. S. Hall, Jr., A. R. Holowenko, and H. G. Laughlin, *Theory and Problems of Machine Design.* McGraw-Hill, New York, 1961.

5. B. J. Hancock, B. Jaconson, and S. R. Schmid, *Fundamentals of Machine Elements.* McGraw-Hill, New York, 1999.

6. T. A. Harris, *Rolling Bearing Analysis.* John Wiley & Sons, New York, 1984.

7. R. C. Juvinall, *Engineering Considerations of Stress, Strain, and Strength.* McGraw-Hill, New York, 1967.

8. R. C. Juvinall and K. M. Marshek, *Fundamentals of Machine Component Design.* John Wiley & Sons, New York, 1991.

9. G. W. Krutz, J. K. Schueller, and P. W. Claar, II, *Machine Design for Mobile and Industrial Applications.* Society of Automotive Engineers, Warrendale, PA, 1994.

10. W. H. Middendorf and R. H. Englemann, *Design of Devices and Systems.* Marcel Dekker, New York, 1998.

11. R. L. Mott, *Machine Elements in Mechanical Design.* Prentice-Hall, Upper Saddle River, NJ, 1999.

12. R. L. Norton, *Design of Machinery.* McGraw-Hill, New York, 1992.

13. R. L. Norton, *Machine Design.* Prentice-Hall, Upper Saddle River, NJ, 2000.

14. W. C. Orthwein, *Machine Component Design.* West Publishing Company, St. Paul, MN, 1990.

15. J. E. Shigley and C. R. Mitchell, *Mechanical Engineering Design.* McGraw-Hill, New York, 1983.

16. J. E. Shigley and C. R. Mischke, *Mechanical Engineering Design.* McGraw-Hill, New York, 1989.

17. C. W. Wilson, *Computer Integrated Machine Design.* Prentice Hall, Upper Saddle River, NJ, 1997.

18. D. B. Marghitu and M. J. Crocker, *Analytical Elements of Mechanisms.* Cambridge University Press, 2001.

19. A. A. Raimond and J. Boyd, *Trans. ASLE* **1**, 159–209. Pergamon Press, New York, 1958.

Machine Components

6 Theory of Vibration

DAN B. MARGHITU, P. K. RAJU, AND DUMITRU MAZILU

*Department of Mechanical Engineering,
Auburn University, Auburn, Alabama 36849*

Inside

1. Introduction

V ibration can be characterized by systems having mass and elasticity. Vibrations are categorized as free and forced. Free vibration occurs when external forces are absent and the system oscillates under the action of forces within the system itself. Natural frequencies are established by the mass of the system and the stiffness of the system. The calculation of the natural frequencies is important in the study of vibrations.

Forced vibration is when there are external forces (excitation forces) that act on the system. For the analysis of linear vibrating systems, the principle of superposition is valid.

Vibrating systems are subject to damping when energy is dissipated. The number of independent coordinates required to uniquely describe the motion of a system is called the degree of freedom of the system. A free particle in planar motion has two degrees of freedom (two translations), and a free particle in space has three degrees of freedom (three translations). A free rigid body in planar motion has three degrees of freedom (two translations and one rotation), and a free rigid body in space has six degrees of freedom (three translations and three rotations).

Consider a particle of mass m suspended from a spring with the elastic constant k, (Fig. 1.1). When the particle is displaced from its rest or

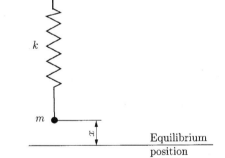

Figure 1.1
Oscillations of a mass suspended on a spring.

equilibrium position the mass m will oscillate about the equilibrium position with simple harmonic motion. The displacement x of the mass from the equilibrium position is a sine wave. This simple harmonic motion can be represented by a point P on the circumference of circle of radius r as shown in Fig. 1.2. The term "harmonic" was originally borrowed from music and means that the motion is exactly repeated after a certain period of time T. The linear tangential velocity v of the point P is

$$v = \omega r, \tag{1.1}$$

where ω is the angular velocity. The period T for one complete revolution of the point P is given by

$$T = \frac{2\pi r}{v} = \frac{2\pi r}{\omega r} = \frac{2\pi}{\omega}. \tag{1.2}$$

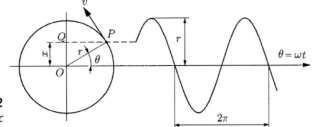

Figure 1.2
Simple harmonic motion.

The frequency f is the number of complete oscillations that take place in one second (the reciprocal of the period):

$$f = \frac{1}{T}.$$
(1.3)

The unit for frequency is hertz (Hz). The angular velocity ω is

$$\omega = 2\pi f$$
(1.4)

and is also known as the angular frequency, circular frequency, or radian frequency. The projection of P on the vertical diameter is the point Q, which moves up and down with simple harmonic motion

$$x = r\sin\theta = r\sin\omega t.$$
(1.5)

The velocity and acceleration of harmonic motion can be determined by differentiation of Eq. (1.5):

$$v = \dot{x} = \omega r\cos\omega t = \omega r\sin\left(\omega t + \frac{\pi}{2}\right)$$
$$a = \ddot{x} = -\omega^2 r\sin\omega t = \omega^2 r\sin(wt + \pi).$$
(1.6)

The velocity and acceleration are also harmonic functions with the same frequency of oscillation.

For trigonometric functions, Euler's equation is

$$e^{i\theta} = \cos\theta + i\sin\theta,$$
(1.7)

where $i = \sqrt{-1}$.

2. Linear Systems with One Degree of Freedom

The linear systems with one degree of freedom (DOF) are mechanical systems whose geometric configuration at a particular moment in time can be determined using a single scalar parameter (distance or angle).

2.1 Equation of Motion

The differential equation of motion for a body of mass m with one degree of freedom will be developed. Consider the mass attached to the end of a spring and a damper (Fig. 2.1). The mass m is in translational motion, and x is the linear displacement. The forces acting on the body are:

- The *elastic (spring) force* in spring $F_e = -kx$, where k is the *spring stiffness*. The minus sign is due to the fact that the elastic force tends to bring the body to the equilibrium position.
- The *damping force* in damper $F_r = -c\dot{x}$, which is opposite to the motion. The force F_r is also called *the viscous resistance force* and c is called *the coefficient of viscous damping*.
- The *exciting force* $F_p = F_0 \sin pt$, which disturbs the body from the equilibrium position. The exciting circular frequency is p (rad/s).

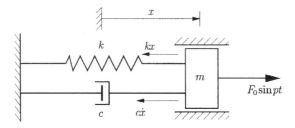

Figure 2.1
Mechanical model of a linear system with one DOF.

Newton's second law projected on the direction of motion gives

$$m\ddot{x} = -kx - c\dot{x} + F_0 \sin pt. \qquad (2.1)$$

The general differential equation of translational linear vibrations with one degree of freedom is

$$m\ddot{x} + c\dot{x} + kx = F_0 \sin pt. \qquad (2.2)$$

Another example of a linear system with one degree of freedom is a shaft with torsional spring stiffness k (the torque moment needed to perform a rotation of 1 radian) carrying at one end a mass of moment of inertia J (Fig. 2.2). A damper with the coefficient c is hooked on the shaft.

The following torques act on the shaft:

- The *elastic torque* $M_e = -k\theta$, which acts on the shaft and is opposed to the angular deformation θ.
- The *damping torque* $M_r = -c\dot{\theta}$, which is produced by the damper and is also opposed to the motion
- The *exciting torque* $M_p = M_0 \sin pt$.

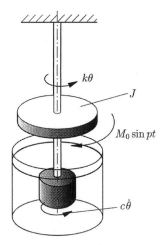

Figure 2.2
Shaft with torsional spring and damper.

The equation of motion with respect to the shaft axis gives

$$J\ddot{\theta} = -k\theta - c\dot{\theta} + M_0 \sin pt,\tag{2.3}$$

and the general differential equation of torsional vibrations of the shaft is

$$J\ddot{\theta} + c\dot{\theta} + k\theta = M_0 \sin pt.\tag{2.4}$$

2.2 Free Undamped Vibrations

The mechanical model of a free undamped vibration is given in Fig. 2.3 and consists of a particle of mass m in a rectilinear motion. The particle is attached to one end of a massless spring of elastic constant k.

The equation of motion on the $0x$ axis is

$$m\ddot{x} + kx = 0,\tag{2.5}$$

Figure 2.3
Mechanical model of a free undamped vibration.

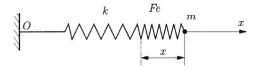

or

$$\ddot{x} + \frac{k}{m}x = 0.\tag{2.6}$$

If we denote $k/m = \omega^2 > 0$, where ω is the *undamped circular (angular) frequency*, the differential equation of motion becomes

$$\ddot{x} + \omega^2 x = 0.\tag{2.7}$$

Vibration

The characteristic equation in r is

$$r^2 + \omega^2 = 0, \tag{2.8}$$

and the general solution is

$$x = C_1 \sin \omega t + C_2 \cos \omega t, \tag{2.9}$$

where C_1 and C_2 are constants.

Equation (2.9) can be written as

$$x = A \sin(\omega t + \varphi), \tag{2.10}$$

with

$$A = \sqrt{C_1^2 + C_2^2}, \quad \tan \varphi = \frac{C_2}{C_1}, \tag{2.11}$$

where A is the *amplitude*, and φ is the *phase angle*.

Using the initial condition

$$t = 0 \Rightarrow \begin{cases} x = x_0 \\ \dot{x} = v_0 \end{cases} \tag{2.12}$$

and the derivative of Eq. (2.9) with respect to time,

$$\dot{x} = C_1 \omega \cos \omega t - C_2 \omega \sin \omega t, \tag{2.13}$$

the constants are

$$C_1 = \frac{v_0}{\omega}, \quad C_2 = x_0$$

$$A = \sqrt{x_0^2 + \frac{v_0^2}{\omega^2}}, \quad \varphi = \arctan \frac{x_0 \omega}{v_0}. \tag{2.14}$$

The equation of motion is now

$$x = \frac{v_0}{\omega} \sin \omega t + x_0 \cos \omega t, \tag{2.15}$$

or

$$x = \sqrt{x_0^2 + \frac{v_0^2}{\omega^2}} \sin \left(\omega t + \arctan \frac{x_0 \omega}{v_0} \right). \tag{2.16}$$

The motion in this case is a harmonic vibration. The displacement x, velocity $\dot{x}$, and acceleration $\ddot{x}$ are represented in Fig. 2.4, for the values $x_0 = 0\,\text{m}$, $v_0 = 1\,\text{m/s}$, and $\omega = 0.5\,\text{rad/s}$.

The *period* of vibration is given by

$$T = \frac{2\pi}{\omega} = 2\pi \sqrt{\frac{m}{k}}, \tag{2.17}$$

and the *frequency* of motion is

$$f = \frac{1}{T} = \frac{\omega}{2\pi} = \frac{1}{2\pi} \sqrt{\frac{k}{m}}. \tag{2.18}$$

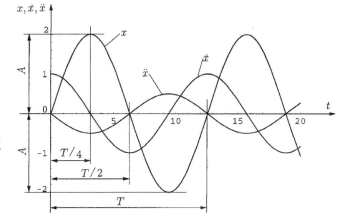

Figure 2.4

Harmonic vibration,
$x_0 = 0\,\mathrm{m}$,
$v_0 = 1\,\mathrm{m}$,
$\omega = 0.5\,\mathrm{rad/s}$.

The *undamped circular frequency* of the motion is called the *natural (circular or angular) frequency* of the system and is given by

$$\omega = \sqrt{\frac{k}{m}} = \sqrt{\frac{g}{f_{st}}}, \tag{2.19}$$

where f_{st} is the static deflection

$$f_{st} = \frac{mg}{k}, \tag{2.20}$$

where g is the acceleration of gravity.

2.3 Free Damped Vibrations

2.3.1 FUNDAMENTALS

All real systems dissipate energy when they vibrate, and the damping must be included in analysis, particularly when the amplitude of vibration is required. Energy is dissipated by frictional effects. The general expression of the viscous damping force is given by

$$F_r = -(\text{sign } \dot{x})R, \tag{2.21}$$

where

$$\text{sign } \dot{x} = \begin{cases} 1, & \dot{x} > 0 \\ 0, & \dot{x} = 0 \\ -1, & \dot{x} < 0. \end{cases} \tag{2.22}$$

The most common types of damping are as follows:

- *Dry damping*, when

$$R = \text{const} \tag{2.23}$$

- *Viscous damping*, when

$$R = cv = c\dot{x}, \tag{2.24}$$

where c is the coefficient of viscous damping and $v = \dot{x}$.

■ *Arbitrary damping,* where

$$R = cv^n (n > 1) \tag{2.25}$$

This damping leads to a nonlinear differential equation of motion.

2.3.2 FREE DAMPED VIBRATIONS WITH DRY DAMPING

In this case, the viscous damping force R is given by Coulomb's law of friction,

$$R = \mu N, \tag{2.26}$$

where N is the *normal force* and μ is the *coefficient of friction.*

The resistance force has a constant value and it is opposed to the direction of motion. A simple example is shown in Fig. 2.5. The differential equation of motion is

$$m\ddot{x} + kx = \pm \mu m g. \tag{2.27}$$

Figure 2.5
Mechanical model of a free damped vibration with dry damping.

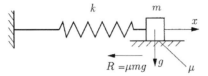

The minus sign is for the case of positive motion along the $0x$ axis.

The general solutions of the above equation are

$$\begin{cases} x = C_1 \sin \omega t + C_2 \cos \omega t - \dfrac{\mu m g}{k}, \\[2mm] x = C_2 \sin \omega t + C_2 \cos \omega t + \dfrac{\mu m g}{k}, \end{cases} \tag{2.28}$$

where the natural angular frequency is $\omega = \sqrt{k/m}$.

Assume the body of mass m at rest and the spring is compressed (or stretched) such that its initial displacement is $|x_0| > |x_s = \mu m g / k|$. With the initial conditions

$$t = 0 \Rightarrow \begin{cases} x = \pm x_0, \\ \dot{x} = 0 \end{cases}, \tag{2.29}$$

the equations of motion are

$$\begin{cases} x = -(x_0 - x_s) \cos \omega t - x_s \\ x = (x_0 - 3x_s) \cos \omega t + x_s \end{cases}. \tag{2.30}$$

Figure 2.6 represents the displacement x for $x_0 = 5\,\text{m}$, $x_s = 0.25\,\text{m}$, $\omega = 1\,\text{rad/s}$. The maximum distances $x_1, x_2, \ldots, x_n$ for the body at each step are determined from the condition $\dot{x} = 0$, and they are $x_1 = x_0 - 2x_s$, $|x_2| = x_0 - 4x_s, \ldots, |x_n| \le x_0 - 2nx_s$.

Figure 2.6
Oscillation decay for free damped vibration with dry damping:
$x_0 = 5$ m,
$x_s = 2.5$ m,
$\omega = 1$ rad/s.

The motion stops when the distance is $x_n \leq x_s$. The number of steps n for which the body vibrates back and forth is determined from

$$x_n = x_s \Rightarrow n = \frac{x_0 - x_s}{2x_s} = \frac{kx_0 - \mu mg}{2\mu mg}. \tag{2.31}$$

The manner of oscillation decay is shown in Fig. 2.6. The amplitudes of motion decrease linearly in time. The zone $x = \pm x_s$ is called the *dead zone*.

2.3.3 FREE DAMPED VIBRATIONS WITH VISCOUS DAMPING

Viscous damping is a common form of damping that is found in many engineering systems such as instruments and shock absorbers. In such cases the viscous damping force is proportional to the first power of the speed $R = cv$, and it always opposes the motion. This type of vibration appears for the case of motion in a liquid environment with low viscosity or in the case of motion in air with speed under 1 m/s. Figure 2.7a shows a mechanical model

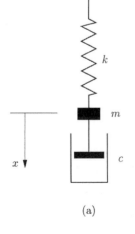

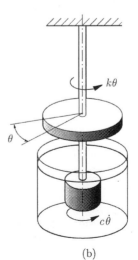

Figure 2.7
Free damped vibration with viscous damping.
(a) Linear damping;
(b) torsional damping.

(a)

(b)

for a free vibration with linear damping, and Fig. 2.7b represents a model for a free vibration with torsional damping.

The equation of motion for the linear displacement, x, is

$$m\ddot{x} = -kx - c\dot{x}, \qquad (2.32)$$

or

$$m\ddot{x} + c\dot{x} + kx = 0. \qquad (2.33)$$

The equation of motion for the angular displacement, θ, is

$$m\ddot{\theta} = -k\theta - c\dot{\theta}, \qquad (2.34)$$

or

$$m\ddot{\theta} + c\dot{\theta} + k\dot{\theta} = 0. \qquad (2.35)$$

Consider the single degree of freedom model with viscous damping shown in Fig. 2.8. The differential equation of motion is

$$m\ddot{x} + c\dot{x} + kx = 0, \qquad (2.36)$$

Figure 2.8
Mechanical model of free damped vibration with viscous damping.

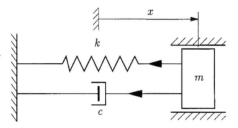

or

$$\ddot{x} + \frac{c}{m}\dot{x} + \frac{k}{m} = 0, \qquad (2.37)$$

or

$$\ddot{x} + 2\alpha\dot{x} + \omega^2 x = 0, \qquad (2.38)$$

where

$$\omega = \sqrt{k/m}$$

is the undamped natural (angular) frequency, and

$$\alpha = c/2m$$

is the damping ratio.

The characteristic equation in r for Eq. (2.38) is

$$r^2 + 2\alpha r + \omega^2 = 0, \qquad (2.39)$$

with the solutions

$$r_{1,2} = -\alpha \pm \sqrt{\alpha^2 - \omega^2}. \qquad (2.40)$$

Case 1. $\alpha^2 - \omega^2 < 0 \Rightarrow$ the roots r_1 and r_2 are complex conjugate ($r_1, r_2 \in \mathcal{C}$, where $\mathcal{C}$ is the set of complex numbers).

Case 2. $\alpha^2 - \omega^2 > 0 \Rightarrow$ the roots r_1 and r_2 are real and distinct ($r_1, r_2 \in \mathcal{R}$, $r_1 \neq r_2$, where $\mathcal{R}$ is the set of real numbers).

Case 3. $\alpha^2 - \omega^2 = 0 \Rightarrow$ the roots r_1 and r_2 are real and identical, $r_1 = r_2$. In this case, the damping coefficient is called the *critical damping coefficient*, $c = c_{cr}$, and

$$\alpha = \omega \Rightarrow \frac{c_{cr}}{2m} = \sqrt{\frac{k}{m}}. \tag{2.41}$$

The expression of the critical damping coefficient is

$$c_{cr} = 2\sqrt{km} = 2m\omega. \tag{2.42}$$

One can classify the vibrations with respect to critical damping coefficient as follows:

Case 1. $c < c_{cr} \Rightarrow$ complex conjugate roots (low damping and oscillatory motion).

Case 2. $c > c_{cr} \Rightarrow$ real and distinct roots (great damping and aperiodic motion).

Case 3. $c = c_{cr} \Rightarrow$ real and identical roots (great damping and aperiodic motion).

Case 1: Complex Conjugate Roots $\alpha^2 - \omega^2 < 0$

The term

$$\beta = \sqrt{\omega^2 - \alpha^2} \tag{2.43}$$

is the *quasicircular frequency*.

The roots of Eq. (2.39) are

$$r_{1,2} = \alpha \pm i\beta. \tag{2.44}$$

The solution of the differential equation is

$$x = e^{-\alpha t}(C_1 \sin \beta t + C_2 \cos \beta t), \tag{2.45}$$

or

$$x = Ae^{-\alpha t} \sin(\beta t + \varphi), \tag{2.46}$$

where C_1, C_2 (A and φ) are constants.

If we use the initial condition

$$t = 0 \Rightarrow \begin{cases} x = x_0 \\ \dot{x} = v_0 \end{cases} \tag{2.47}$$

and the derivative of Eq. (2.45) with respect to time,

$$\dot{x} = -\alpha e^{-\alpha t}(C_1 \sin \beta t + C_2 \cos \beta t) + \beta e^{-\alpha t}(C_1 \cos \beta t - C_2 \sin \beta t), \tag{2.48}$$

Vibration

the constants are

$$C_2 = x_0, \quad C_1 = \frac{v_0 + \alpha x_0}{\beta}, \tag{2.49}$$

or

$$A = \sqrt{C_1^2 + C_2^2} = \sqrt{x_0^2 + \left(\frac{v_0 + \alpha x_0}{\beta}\right)^2} \tag{2.50}$$

$$\tan \varphi = \frac{C_2}{C_2} = \frac{\beta x_0}{v_0 + \alpha x_0}. \tag{2.51}$$

The solution is

$$x = e^{-\alpha t}\left(\frac{v_0 + \alpha x_0}{\beta} \sin \beta t + x_0 \cos \beta t\right), \tag{2.52}$$

or

$$x = \sqrt{x_0^2 + \left(\frac{v_0 + \alpha x_0}{\beta}\right)^2}\, e^{-\alpha t} \sin\left(\beta t + \arctan\frac{\beta x_0}{v_0 + \alpha x_0}\right). \tag{2.53}$$

The exponential decay, $Ae^{-\alpha t}$, decreases in time, so one can obtain a motion that is modulated in amplitude. In Eq. (2.43) the quasicircular (or quasiangular) frequency β is

$$\beta = \sqrt{\omega^2 - \alpha^2} = \sqrt{\frac{k}{m} - \left(\frac{c}{2m}\right)^2} = \omega\sqrt{1 - \left(\frac{c}{c_{cr}}\right)^2}. \tag{2.54}$$

The *quasiperiod* of motion is

$$T_\beta = \frac{2\pi}{\beta} = \frac{2\pi}{\sqrt{\omega^2 - \alpha^2}}. \tag{2.55}$$

The diagram of motion for x is shown in Fig. 2.9, for $x_0 = 0$ m, $v_0 = 0.2$ m/s, $\alpha = 0.1\,\text{s}^{-1}$, and $\omega = 0.8\,\text{rad/s}$.

The rate of decay of oscillation is

$$\frac{x(t)}{x(t + T_\beta)} = \frac{Ae^{-\alpha t}\sin(\beta t + \varphi)}{Ae^{-\alpha(t_T \beta)}\sin[\beta(t + T_\beta) + \varphi]} = \frac{1}{e^{-2\pi\alpha/\beta}} = e^{2\pi\alpha/\beta} = \text{const.} \tag{2.56}$$

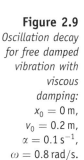

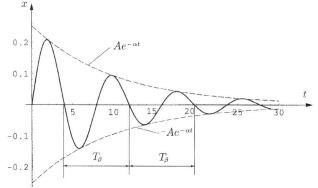

Figure 2.9
Oscillation decay for free damped vibration with viscous damping:
$x_0 = 0$ m,
$v_0 = 0.2$ m,
$\alpha = 0.1\,\text{s}^{-1}$,
$\omega = 0.8\,\text{rad/s}$.

Equation (2.56) shows that the displacement measured at equal time intervals of one quasiperiod decreases in geometric progression. To characterize this decay, the *logarithmic decrement* δ is introduced

$$\delta = \ln \frac{x(t)}{x(t + T_\beta)} = \ln e^{2\pi\alpha/\beta} = \frac{2\pi\alpha}{\sqrt{\omega^2 - \alpha^2}} = \frac{\pi c}{m\beta}. \qquad (2.57)$$

The manner of oscillation decay can be represented using x and ω as axes, (Fig. 2.10). The oscillation continues until the amplitude of motion is so small that the maximum spring force is unable to overcome the friction force.

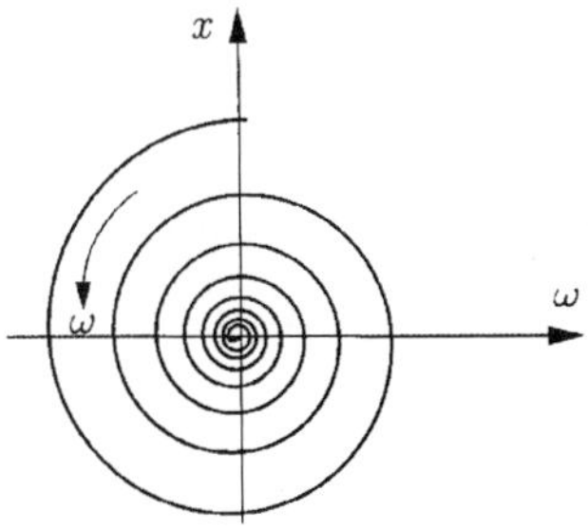

Figure 2.10
Oscillation decay represented using x and ω axes.

Case 2: Real and Distinct Roots $\alpha^2 - \omega^2 > 0$

The roots of the characteristic equation are negative

$$r_1 = -\lambda_1, \ r_2 = -\lambda_2; \quad \lambda_1 \text{ and } \lambda_2 > 0. \qquad (2.58)$$

The solution of Eq. (2.38) in this case is

$$x = C_1 e^{-\lambda_1 t} + C_2 e^{-\lambda_2 t}. \qquad (2.59)$$

The motion is aperiodic and tends asymptotically to the rest position ($x \to 0$ when $t \to \infty$). Figure 2.11 shows the diagrams of motion for different initial conditions.

Case 3: Real Identical Roots $\alpha^2 - \omega^2 = 0$

The roots of the characteristic equation are

$$r_1 = r_2 = -\alpha. \qquad (2.60)$$

Figure 2.11
Diagram of damped motion for different initial conditions: (a) $x_0 > 0$, $v_0 > 0$; (b) $x_0 > 0$, $v_0 < 0$ (low); (c) $x_0 > 0$, $v_0 < 0$ (great).

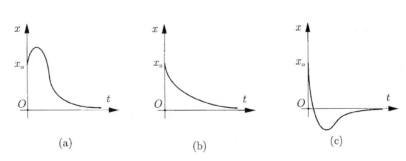

Vibration

The solution of the differential equation, namely the law of motion, in this case becomes

$$x = e^{-\alpha t}(C_1 t + C_2) = \frac{C_1 t + C_2}{e^{\alpha t}} \tag{2.61}$$

When $t \to \infty$, one obtains the undeterminate $x = \infty/\infty$. The l'Hospital rule is applied in this case:

$$x = \lim_{t \to \infty} \frac{C_1 t + C_2}{e^{\alpha t}} = \lim_{t \to \infty} \frac{C_1}{\alpha e^{\alpha t}} = \frac{C_1}{\infty} = 0, \tag{2.62}$$

namely, the motion stops aperiodically.

The diagrams of motion are similar to the previous case, with the same boundary conditions. Critical damping represents the limit of periodic motion; hence, the displaced body is restored to equilibrium in the shortest possible time, and without oscillation. Many devices, particularly electrical instruments, are critically damped to take advantage of this property.

2.4 Forced Undamped Vibrations

2.4.1 RESPONSE OF AN UNDAMPED SYSTEM TO A SIMPLE HARMONIC EXCITING FORCE WITH CONSTANT AMPLITUDE

Common sources of harmonic excitation imbalance in rotating machines, the motion of the machine itself, or forces produced by reciprocating machines. These excitations may be undesirable for equipment whose operation may be disturbed or for the safety of the structure if large vibration amplitudes develop. Resonance is to be avoided in most cases, and to prevent large amplitudes from developing, dampers and absorbers are often used.

General Case

An elastic system is excited by a harmonic force of the form

$$F_p = F_0 \sin pt, \tag{2.63}$$

where F_0 is the amplitude of the forced vibration and p is the forced angular frequencies. The differential equation of motion for the mechanical model in Fig. 2.12 is

$$m\ddot{x} + kx = F_0 \sin pt, \tag{2.64}$$

or

$$\ddot{x} + \frac{k}{m}x = \frac{F_0}{m}\sin pt. \tag{2.65}$$

Figure 2.12
Mechanical model of forced undamped vibrations.

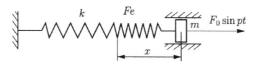

The following notation is used:

$$\frac{k}{m} = \omega^2; \quad \frac{F_0}{m} = q. \tag{2.66}$$

Equation (2.65) can be written as

$$\ddot{x} + \omega^2 x = q \sin pt, \tag{2.67}$$

with the solution

$$x = C_1 \sin \omega t + C_2 \cos \omega t + x_p. \tag{2.68}$$

The particular solution x_p of the nonhomogeneous differential equation is

$$x_p = C \sin pt, \tag{2.69}$$

and the second derivative with respect to time is

$$\ddot{x}_p = -Cp^2 \sin pt. \tag{2.70}$$

The constant C is determined using Eqs. (2.67), (2.69), and (2.70) for $\omega \neq p$:

$$-Cp^2 \sin pt + \omega^2 C \sin pt = q \sin pt, \tag{2.71}$$

or

$$C(\omega^2 - p^2) = q \Rightarrow C = \frac{q}{(\omega^2 - p^2)}. \tag{2.72}$$

Introducing in Eq. (2.68) the obtained value of C, one can get

$$x = C_1 \sin \omega t + C_2 \cos \omega t + \frac{q}{\omega^2 - p^2} \sin pt, \tag{2.73}$$

which may be written as

$$x = A \sin(\omega t + \varphi) + \frac{q}{\omega^2 - p^2} \sin pt. \tag{2.74}$$

The constants C_1 and C_2 (A and φ) are determined from the initial conditions

$$t = 0 \Rightarrow \begin{cases} x = 0 \\ \dot{x} = 0 \end{cases} \tag{2.75}$$

The derivative with respect to time of Eq. (2.73) is

$$\dot{x} = C_1 \omega \cos \omega t - C_2 \omega \sin \omega t + \frac{qp}{\omega^2 - p^2} \cos pt. \tag{2.76}$$

With the help of Eqs. (2.75), (2.73), and (2.76) the constants are

$$C_1 = -\frac{p}{\omega} \frac{q}{\omega^2 - p^2}$$

$$C_2 = 0.$$

The vibration equation is

$$x = \frac{q}{\omega^2 - p^2} \left[\sin pt - \frac{p}{\omega} \sin \omega t \right]. \tag{2.77}$$

Equation (2.77) is a combined motion of two vibrations: one with the natural frequency ω and one with the forced angular frequency p. The resultant is a nonharmonic vibration (Fig. 2.13), here for $\omega = 1\,\text{rad/s}$, $q = 1\,\text{N/kg}$, $p = 0.1\,\text{rad/s}$. The amplitude is

$$A = \frac{q}{\omega^2 - p^2} = \frac{\dfrac{F_0}{m}}{\omega^2 - p^2} = \frac{\dfrac{F_0}{m}\dfrac{1}{\omega^2}}{1 - \left(\dfrac{p^2}{\omega}\right)^2}$$

$$= \frac{\dfrac{F_0}{m}\dfrac{m}{k}}{1 - \left(\dfrac{p}{\omega}\right)^2} = \frac{F_0}{k}\frac{1}{1 - \left(\dfrac{p}{\omega}\right)^2} = \frac{x_{st}}{1 - \left(\dfrac{p}{\omega}\right)^2} = x_{st}A_0, \qquad (2.78)$$

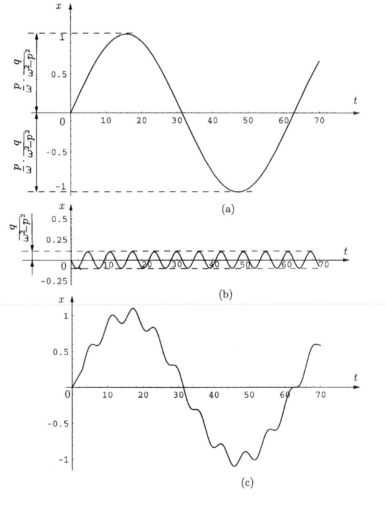

Figure 2.13
Combined motion of two vibrations:
$\omega = 1\,\text{rad/s}$,
$q = 1\,\text{N/kg}$,
$p = 0.1\,\text{rad/s}$.

where $x_{st} = F_0/k$ is the static deformation of the elastic system, under the maximum value F_0, and

$$|A_0| = \frac{1}{\left|1 - \left(\dfrac{p}{\omega}\right)^2\right|} \qquad (2.79)$$

is a magnification factor. In Fig. 2.14 is shown the magnification factor function of p/ω.

From Fig. 2.14 one can notice:

- Point a ($p = 0$, $A_0 = 1$) and $x = x_{st}$. The system vibrates in phase with force.
- Point b ($p \to \infty$, $A_0 \to 0$), which corresponds to great values of angular frequency p. The influence of forced force is practically null.
- Point c ($p = \omega$, $A_0 \to \infty$). This phenomenon called *resonance* and is very important in engineering applications.

The curve in Fig. 2.14 is called a *curve of resonance*.

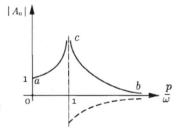

Figure 2.14
Curve of resonance.

Resonance

When the frequency of perturbation p is equal to the natural angular frequency ω, the resonance phenomenon appears. The resonance is characteristic through increasing amplitude to infinity. In Eq. (2.77) for $\omega = p$, the limit case is obtained, $\lim_{p \to \omega} x = \infty \cdot 0$. Using l'Hospital's rule one can calculate the limit:

$$\lim_{p \to \omega} x = q \lim_{p \to \omega} \frac{\sin pt - \dfrac{p}{\omega}\sin \omega t}{\omega^2 - p^2} = q \lim_{p \to \omega} \frac{t\cos pt}{-2p} = \left(\frac{-q}{2\omega}\right)\cos \omega t. \qquad (2.80)$$

A diagram of the motion is shown in Fig. 2.15 for $q = 1\,\text{N/kg}$ and $\omega = 0.2\,\text{rad/s}$. The amplitude increases linearly according to Eq. (2.80).

The Beat Phenomenon

The beat phenomenon appears in the case of two combined parallel vibrations with similar angular frequency ($\omega \approx p$). One can introduce the

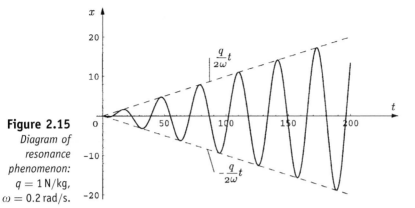

Figure 2.15
Diagram of resonance phenomenon:
$q = 1\,\text{N/kg}$,
$\omega = 0.2\,\text{rad/s}$.

factor $\varepsilon = \omega - p$, and in this case $p/\omega \approx 1$ and $p + \omega \approx 2\omega$. The vibration becomes

$$x = \frac{q}{\omega^2 - p^2}[\sin pt - \sin \omega t]$$

$$= \frac{q}{(\omega + p)(\omega - p)}2\sin\frac{(p-\omega)t}{2}\cos\frac{(p+\omega)t}{2}$$

$$= \frac{2q}{2\omega\varepsilon}\sin\frac{(-\varepsilon t)}{2}\cos\frac{2\omega t}{2} = -\frac{q}{\omega\varepsilon}\sin\left(\frac{\varepsilon}{2}t\right)\cos\omega t. \qquad (2.81)$$

The amplitude is in this case is

$$\phi(t) = -\frac{q}{\omega\varepsilon}\sin\left(\frac{\varepsilon}{2}t\right). \qquad (2.82)$$

The vibration diagram is shown in Fig. 2.16 for $q = 2\,\text{N/kg}$, $\varepsilon = 0.12\,\text{rad/s}$ and $\omega = 0.8\,\text{rad/s}$.

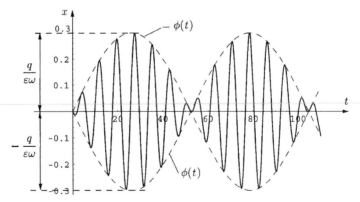

Figure 2.16
Diagram of beat phenomenon:
$q = 2\,\text{N/kg}$,
$\epsilon = 0.12 : \text{rad/s}$,
$\omega = 0.8\,\text{rad/s}$.

2.4.2 RESPONSE OF AN UNDAMPED SYSTEM TO A CENTRIFUGAL EXCITING FORCE

Unbalance in rotating machines is a common source of vibration excitation. Frequently, the excited (perturbation) harmonic force came from an unbalanced mass that is in a rotating motion that generates a centrifugal force. For this case the model is depicted in Fig. 2.17. The unbalanced mass m_0 is connected to the mass m_1 with a massless crank of lengths r. The crank and the mass m_0 rotate with a constant angular frequency p. The centrifugal force is

$$F_0 = m_0 r p^2, \tag{2.83}$$

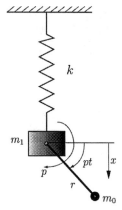

Figure 2.17
Mechanical model of undamped system with centrifugal exciting force.

and represents the amplitude of the forced vibration ($F_p = F_0 \sin pt$). The amplitude of the combined vibration is

$$A' = \frac{q}{\omega^2 - p^2} = \frac{F_0}{k} \frac{1}{1 - \left(\dfrac{p}{\omega}\right)^2} = \frac{m_0 r p^2}{k} \frac{1}{1 - \left(\dfrac{p}{\omega}\right)^2}$$

$$= \frac{\dfrac{m_0}{m} r p^2}{\dfrac{k}{m}} \frac{1}{1 - \left(\dfrac{p}{\omega}\right)^2} = \frac{m_0}{m} r \frac{\left(\dfrac{p}{\omega}\right)^2}{1 - \left(\dfrac{p}{\omega}\right)^2} = \frac{m_0 r}{m} A'_0, \tag{2.84}$$

where A'_0 is a magnification factor, and $m = m_1 + m_0$. The magnification factor A'_0 is

$$A'_0 = \frac{\left(\dfrac{p}{\omega}\right)^2}{1 - \left(\dfrac{p}{\omega}\right)^2} = \left(\dfrac{p}{\omega}\right)^2 A_0. \tag{2.85}$$

The variation of the magnification factor is shown in Fig. 2.18 where the representative points are a, b, and c.

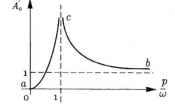

Figure 2.18
Variation of the magnification factor.

2.4.3 RESPONSE OF AN UNDAMPED SYSTEM TO A NONHARMONIC PERIODIC EXCITING FORCE

One may consider the nonharmonic periodic exciting force $F(t)$. The function $F(t)$ obeys the Dirichlet conditions. The exciting force $F(t)$ can be developed in Fourier series:

$$F_p = F(t) = a_0 + a_1 \cos pt + a_2 \cos 2pt + \cdots + b_1 \sin pt + b_2 \sin 2pt + \cdots$$

(2.86)

For n terms,

$$F(x) = F_0 + F_1 \sin(pt + \varphi_1) + F_2 \sin(2pt + \varphi_2) + \cdots$$
$$\approx \sum_{i=0}^{n} F_i \sin(ipt + \varphi_i).$$

(2.87)

The equation of motion is

$$m\ddot{x} + kx = F(t).$$

(2.88)

For the linear differential equation, the superposition principle is applied and the general solution is

$$x = C_1 \sin \omega t + C_2 \cos \omega t + \frac{F_0}{m\omega^2} + \frac{F_1}{m(\omega^2 - p^2)} \sin(pt + \varphi_1)$$
$$+ \frac{F_2}{m[\omega^2 - (2p)^2]} \sin(2pt + \varphi_2) + \frac{F_3}{m[\omega^2 - (3p)^2]} \sin(3pt + \varphi_3) + \cdots,$$

(2.89)

or

$$x = C_1 \sin \omega t + C_2 \cos \omega t + \sum_{i=0}^{n} \frac{F_i}{m[\omega^2 - (ip)^2]} \sin(ipt + \varphi_i).$$

(2.90)

The resonance appears for the first harmonic (the fundamental harmonic) and for superior harmonics ($\omega = p$, $\omega = 2p$, $\omega = 3p, \ldots, \omega = np$).

2.4.4 RESPONSE OF AN UNDAMPED SYSTEM TO AN ARBITRARY EXCITING FORCE

For this general case the exciting force is an arbitrary force Fig. (2.19). The differential equation of motion is

$$m\ddot{x} + kx = F(t).$$

(2.91)

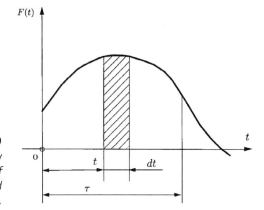

Figure 2.19
Arbitrary exciting force of an undamped system.

The vibration in this case is

$$x = C_1 \sin \omega t + C_2 \cos \omega t + \frac{1}{m\omega} \int_0^\tau F(t) \sin[\omega(\tau - t)]dt, \tag{2.92}$$

where τ is presented in Fig. 2.19 and C_1, C_2 are constants. The integral in Eq. (2.92) is called the Duhamel integral.

2.5 Forced Damped Vibrations

2.5.1 RESPONSE OF A DAMPED SYSTEM TO A SIMPLE HARMONIC EXCITING FORCE WITH CONSTANT AMPLITUDE

The mechanical model is shown in Fig. 2.20. The differential equation of motion is

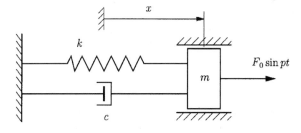

Figure 2.20
Mechanical model of forced damped vibration.

$$m\ddot{x} + c\dot{x} + kx = F_0 \sin pt. \tag{2.93}$$

The following notation is used:

$$\frac{c}{2m} = 2\alpha; \quad \frac{k}{m} = \omega; \quad \frac{F_0}{m} = q. \tag{2.94}$$

The equation of motion becomes

$$\ddot{x} + 2\alpha\dot{x} + \omega^2 x = q \sin pt. \tag{2.95}$$

Case 1: (c < c$_{cr}$) or $\alpha < \omega$.

The characteristic equation is

$$r^2 + 2\alpha r + \omega^2 = 0, \tag{2.96}$$

with the roots

$$r_{1,2} = -\alpha \pm \sqrt{\alpha^2 - \omega^2} = -\alpha \pm i\beta, \tag{2.97}$$

where

$$-\beta^2 = \alpha^2 - \omega^2. \tag{2.98}$$

The general solution of differential equation (2.95) is

$$x = x_1 + x_2, \tag{2.99}$$

where x_1 represent the solution of the differential homogenous equation, and x_2 is a particular solution of the differential nonhomogeneous equation. In this case x_1 represents the natural vibration of the system and x_2 is the forced vibration of the system. The solution of the free damped system is

$$x_1 = e^{-\alpha t}(C_1 \sin \omega t + C_2 \cos \omega t), \tag{2.100}$$

or

$$x_1 = B_1 e^{-\alpha t} \sin(\omega t + \varphi), \tag{2.011}$$

where

$$B_1 = \sqrt{C_1^2 + C_2^2} \quad \text{and} \quad \tan \varphi = \frac{C_2}{C_1}. \tag{2.102}$$

The solution of the forced (excited) vibration is

$$x_2 = D_1 \sin pt + D_2 \cos pt. \tag{2.103}$$

The constants D_1 and D_2 are determined using the identification method. One can calculate

$$\begin{cases} \dot{x}_2 = D_1 p \cos pt - D_2 p \sin pt \\ \ddot{x}_2 = -D_1 p^2 \sin pt - D_2 p^2 \cos pt. \end{cases} \tag{2.104}$$

Introducing Eq. (2.103) and Eq. (2.104) in Eq. (2.95) one can obtain

$$\begin{aligned} & \left[-D_1 p^2 \sin pt - D_2 p^2 \cos pt\right] + 2\alpha[D_1 p \cos pt - D_2 p \sin pt] \\ & + \omega^2 \left[D_1 \sin pt + D_2 \cos pt\right] = q \sin pt. \end{aligned} \tag{2.105}$$

Using the identification method, the linear system is obtained

$$\begin{cases} D_1(\omega^2 - p^2) - 2D_2 \alpha p = q \\ 2D_1 \alpha p + D_2(\omega^2 - p^2) = 0. \end{cases} \tag{2.106}$$

Solving this system, one finds

$$\begin{aligned} D_1 &= \frac{(\omega^2 - p^2)q}{(\omega^2 - p^2)^2 + 4\alpha^2 p^2} \\ D_2 &= -\frac{2\alpha pq}{(\omega^2 - p^2)^2 + 4\alpha^2 p^2}. \end{aligned} \tag{2.107}$$

Therefore the forced vibration x_2 has the expression

$$x_2 = \frac{(\omega^2 - p^2)q}{(\omega^2 - p^2)^2 + 4\alpha^2 p^2} \sin pt - \frac{2\alpha pq}{(\omega^2 - p^2)^2 + 4\alpha^2 p^2} \cos pt \tag{2.108}$$

or

$$x_2 = B_2 \sin(pt - \phi), \tag{2.109}$$

where

$$B_2 = \sqrt{D_1^2 + D_2^2} \quad \text{and} \quad \tan \phi = -\frac{D_2}{D_1}. \tag{2.110}$$

In conclusion, the motion of the system is represented by the equation

$$x = B_1 e^{-\alpha t} \sin(\omega t + \varphi) + B_2 \sin(pt - \phi). \tag{2.111}$$

The graphic of the vibration is shown in Fig. 2.21 for $x_0 = 0\,\text{m}$, $v_0 = 0.2\,\text{m/s}$, $\omega = 5\,\text{rad/s}$, $q = 1\,\text{N/kg}$, $p = 0.3\,\text{rad/s}$, $\alpha = 0.1\,\text{s}^{-1}$.

According to Eq. (2.110) the amplitude of the vibration is

$$B_2 = \sqrt{D_1^2 + D_2^2} = \frac{q}{(\omega^2 - p^2)^2 + 4\alpha^2 p^2} \sqrt{(\omega^2 - p^2)^2 + 4\alpha^2 p^2}$$

$$= \frac{q}{\sqrt{(\omega^2 - p^2)^2 + 4\alpha^2 p^2}}. \tag{2.112}$$

One can denote

$$x_{st} = \frac{F_0}{k} = \frac{F_0/m}{k/m} = \frac{q}{\omega^2}, \tag{2.113}$$

where x_{st} represents the static deformation.

The amplitude of forced vibration can be written as

$$B_2 = \frac{\dfrac{q}{\omega^2}}{\sqrt{\left(1 - \dfrac{p^2}{\omega^2}\right)^2 + 4\dfrac{\alpha^2 p^2}{\omega^2 \omega^2}}} = x_{st} A_1, \tag{2.114}$$

where A_1 is the magnification factor. The expression of the magnification factor is

$$A_1 = \frac{1}{\sqrt{\left(1 - \dfrac{p^2}{\omega^2}\right)^2 + 4\dfrac{\alpha^2 p^2}{\omega^2 \omega^2}}} = \frac{1}{\sqrt{\left(1 - \dfrac{p^2}{\omega^2}\right)^2 + 4\left(\dfrac{c}{c_{cr}}\right)^2 \left(\dfrac{p}{\omega}\right)^2}}, \tag{2.115}$$

where

$$\frac{\alpha}{\omega} = \frac{\dfrac{c}{2m}}{\sqrt{\dfrac{k}{m}}} = \frac{c}{2\sqrt{km}} = \frac{c}{c_{cr}}. \tag{2.116}$$

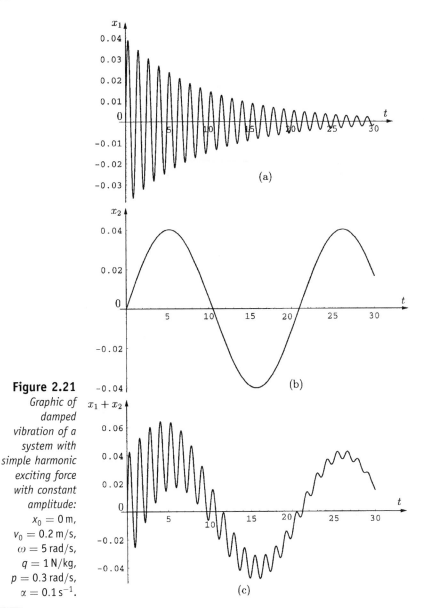

Figure 2.21
Graphic of damped vibration of a system with simple harmonic exciting force with constant amplitude:
$x_0 = 0\,\text{m}$,
$v_0 = 0.2\,\text{m/s}$,
$\omega = 5\,\text{rad/s}$,
$q = 1\,\text{N/kg}$,
$p = 0.3\,\text{rad/s}$,
$\alpha = 0.1\,\text{s}^{-1}$.

In Fig. 2.22 the magnification factor A_1 is plotted versus the ratio p/ω. Different curves for different values of the ratio c/c_{cr} are obtained. In the case of $c/c_{cr} = 0$ (system without damping),

$$A_1 = \frac{1}{1 - \dfrac{p^2}{\omega^2}} = A_0. \tag{2.117}$$

The damping factor c/c_{cr} has a large influence on the amplitude near resonance. It is recommended to avoid the domain $0.7 < p/\omega < 1.3$.

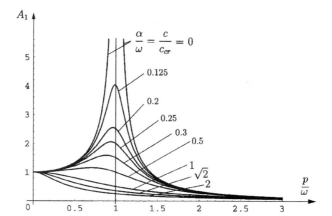

Figure 2.22
Diagram of the magnification factor A_1.

The maximum of A_1 is obtained from the relation

$$\frac{dA_1}{d\left(\frac{p^2}{\omega^2}\right)} = 0 \Rightarrow \frac{p}{\omega} = \sqrt{1 - 2\left(\frac{\alpha}{\omega}\right)^2} = \sqrt{1 - 2\left(\frac{c}{c_{cr}}\right)^2}.$$ (2.118)

The delay is compute with relation

$$\tan\varphi = -\frac{D_2}{D_1} = \frac{2\alpha p}{\omega^2 - p^2} = \frac{2\dfrac{\alpha}{\omega}\dfrac{p}{\omega}}{1 - \left(\dfrac{p}{\omega}\right)^2} = \frac{2\dfrac{c}{c_{cr}}\dfrac{p}{\omega}}{1 - \left(\dfrac{p}{\omega}\right)^2},$$ (2.119)

and

$$\phi = \arctan\frac{2\dfrac{c}{c_{cr}}\dfrac{p}{\omega}}{1 - \left(\dfrac{p}{\omega}\right)^2}.$$ (2.120)

The variation of delay versus frequency ratio is plotted in Fig. 2.23. From Fig. 2.23, for $p/\omega < 1$ is obtained $\phi = 0$, namely, the vibration is in phase with the perturbation force. For $p/\omega > 1$ is obtained $\phi = \pi$, that is, the vibration

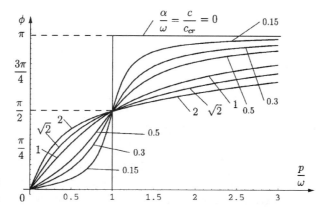

Figure 2.23
Variation delay.

Vibration

and perturbation forces are in opposition of phase. For the case $c/c_{cr} \neq 0$ (systems with damping), for $p/\omega < 1$ the phase angle between the vibration and perturbation forces is $0 < \phi < \pi/2$, and for $p/\omega > 1$ the phase angle is $\pi/2 < \phi < \pi$. For $p = \omega$ and for any parameter c/c_{cr}, the phase angle is the same.

Using d'Alambert's principle for a general system with one DOF, we obtain the differential equation

$$-m\ddot{x} - c\dot{x} - kx + F_0 \sin pt = 0, \tag{2.121}$$

where $m\ddot{x}$ is the inertia force, $c\dot{x}$ is the damping force, kx is the spring force, and $F_0 \sin pt$ is the perturbation force. Equation (2.121) is another form of Eq. (2.93).

To represent Eq. (2.121) with rotating vectors (Fig. 2.24), the equation of motion is considered as

$$x = A \sin(pt - \phi). \tag{2.122}$$

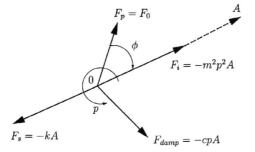

Figure 2.24
Damped vibration representation with rotating vectors.

In Fig. 2.24 the vector A represents the amplitude of the excited vibration. The angular frequency of all rotative vectors is p. A difference of phase ϕ between the vector A and perturbation force is noticed. The four vectors illustrated in Fig. 2.24 are in equilibrium and the projection on the vector A and its normal directions yields

$$\begin{cases} mp^2 A + F_0 \cos \phi - kA = 0 \\ F_0 \sin \phi - cpA = 0. \end{cases} \tag{2.123}$$

2.5.2 RESPONSE OF A DAMPED SYSTEM TO A CENTRIFUGAL EXCITING FORCE

Unbalance in rotating machines is a common source of vibration excitation. One can consider the system shown in Fig. 2.25 with the total mass

$$m = m_1 + m_0. \tag{2.124}$$

The mass is connected to the base through an elastic spring with the elastic constant k and through a damper with the coefficient of viscous resistance c. The unbalance is represented by an eccentric mass m_0 with eccentricity r that

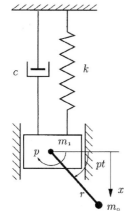

Figure 2.25
Mechanical model of a damped system with a centrifugal exciting force.

is rotating with angular velocity $p = \text{const}$ (Fig. 2.25). The mass m_0 produces the perturbation centrifugal force

$$F_0 = m_0 r p^2. \qquad (2.125)$$

In this case

$$q = \frac{F_0}{m} = \frac{m_0 r p^2}{m_1 + m_0}. \qquad (2.126)$$

With Eq. (2.114) one can obtain the expression of the amplitude to maintain excited vibration,

$$A'' = \frac{\dfrac{q}{\omega^2}}{\sqrt{\left(1 - \dfrac{p^2}{\omega^2}\right)^2 + 4\left(\dfrac{c}{c_{cr}}\right)^2 \dfrac{p^2}{\omega^2}}}$$

$$= \frac{\dfrac{m_0 r}{m_1 + m_0} \dfrac{p^2}{\omega^2}}{\sqrt{\left(1 - \dfrac{p^2}{\omega^2}\right)^2 + 4\left(\dfrac{c}{c_{cr}}\right)^2 \dfrac{p^2}{\omega^2}}} = \frac{m_0 r}{m_1 + m_0} A_2, \qquad (2.127)$$

where A_2 is a nondimensional magnification factor and has the expression

$$A_2 = \frac{\dfrac{p^2}{\omega^2}}{\sqrt{\left(1 - \dfrac{p^2}{\omega^2}\right)^2 + 4\left(\dfrac{c}{c_{cr}}\right)^2 \dfrac{p^2}{\omega^2}}}. \qquad (2.128)$$

Using Eqs. (2.115) and (2.128) one can write

$$A_2 = A_1 \left(\frac{p}{\omega}\right)^2. \qquad (2.129)$$

In Fig. 2.26 is shown the influence of c/c_{cr} on the magnification factor, A_2.

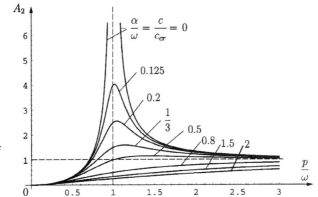

Figure 2.26
Influence of c/c_{cr} on the magnification factor A_2.

2.5.3 RESPONSE OF A DAMPED SYSTEM TO AN ARBITRARY EXCITING FORCE

In a general case, the perturbation force is a arbitrary function of time $F(t)$. The differential equation of motion is (Fig. 2.27)

$$m\ddot{x} + c\dot{x} + kx = F(t). \tag{2.130}$$

Figure 2.27
Mechanical model of a damped system with arbitrary exciting force.

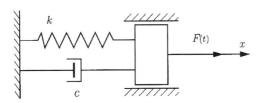

In the case of less friction ($c < c_{cr}$), with notation from section 2.5.1 and with the notation

$$\frac{F(t)}{m} = q_1, \tag{2.131}$$

the differential equation of motion becomes

$$\ddot{x} + 2\alpha\dot{x} + \omega^2 x = q_1. \tag{2.132}$$

Because Eq. (2.132) is a linear differential equation, the superposition principle can be applied:

$$x = x_1 + x_2. \tag{2.133}$$

The general solution of the homogenous differential equation x_1 corresponding to Eq. (2.132) is calculated with Eqs. (2.45) or (2.46), and it is the natural vibration of the system. The particular solution x_2 of Eq. (2.132) is determined with the help of the conservation of momentum theorem. The variation of perturbation force on mass unit q_1 is shown in Fig. 2.28. For an arbitrary time one can consider an elementary impulse $q_1 dt$, as represented in Fig. 2.28. For the direction of motion one can write

$$Fdt = mdv, \qquad (2.134)$$

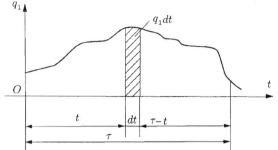

Figure 2.28
The variation of an arbitrary exciting force on mass unit, q_1.

and

$$dv = \frac{Fdt}{m} = q_1 dt. \qquad (2.135)$$

For the time τ, $\tau > t$, the variation of the speed dv can be considered as the initial speed of vibration (which begins at time $\tau - t$).

The case $c < c_{cr}$ is considered with the initial condition

$$t = 0 \Rightarrow \begin{cases} x = 0 \\ \dot{x} = v_0. \end{cases} \qquad (2.136)$$

Using the general solution given by Eq. (2.45), with notation from Eq. (2.43) and with initial condition Eq. (2.136), the values of the constants C_1 and C_2 are obtained:

$$C_1 = \frac{v_0}{\beta}; \quad C_2 = 0. \qquad (2.137)$$

The solution of motion in this case is

$$x = \frac{v_0}{\beta} e^{-\alpha t} \sin \beta t. \qquad (2.138)$$

Substituting in Eq. (2.138) the speed given by Eq. (2.135), one can write the elementary displacement

$$dx = \frac{q_1 dt}{\beta} e^{-\alpha(\tau - t)} \sin \beta(\tau - t). \qquad (2.139)$$

In Eq. (2.139) one can consider $(\tau - t)$ the time interval. For each elementary impulse $q_1 dt$ produced between time $t = 0$ and $t = \tau$ results an elementary displacement dx. The effect of continuous action of perturbation force in time interval $(0, \tau)$ is obtained through integration:

$$x_2 = \frac{1}{\beta} \int_0^\tau q_1 e^{-\alpha(\tau-t)} \sin \beta(\tau - t) dt$$

$$= \frac{1}{m\beta} \int_0^\tau F(t) e^{-\alpha(\tau-t)} \sin \beta(\tau - t) dt. \qquad (2.140)$$

The preceding expression is the Duhamel integral and represents the effect of an arbitrary perturbation force $F(t)$ in the time interval $(0,\tau)$. Applying the superposition principle, one can obtain the law of motion for the general case, which represents the complete solution of the differential equation (2.132) on the interval $(0, \tau)$:

$$x = e^{-\alpha\tau} \left(\frac{v_0 + \alpha x_0}{\beta} \sin \beta\tau + x_0 \cos \beta\tau \right)$$

$$+ \frac{1}{m\beta} \int_0^\tau F(t) e^{-\alpha(\tau-t)} \sin \beta(\tau - t) dt. \qquad (2.141)$$

In Eq. (2.141) the first term represents the effect of the initial displacement x_0 and of the initial speed v_0, and the second term is the effect of the perturbation force $F(t)$. If the damping is neglected, $\alpha = 0$ and $\beta = \omega$, the equation of motion is

$$x = \frac{v_0}{\omega} \sin \omega\tau + x_0 \cos \omega\tau + \frac{1}{m\omega} \int_0^\tau F(t) \sin \omega(\tau - t) dt. \qquad (2.142)$$

If we use the identity

$$\sin(\omega\tau - \omega t) = \sin \omega\tau \cos \omega t - \cos \omega\tau \sin \omega t, \qquad (2.143)$$

Eq. (2.142) becomes

$$x_0 = \frac{1}{m\omega} \left[\sin \omega\tau \int_0^\tau F(t) \cos \omega t dt - \cos \omega\tau \int_0^\tau F(t) \sin \omega t dt \right]. \qquad (2.144)$$

One can denote

$$A = \frac{1}{m\omega} \int_0^\tau F(t) \cos \omega t dt,$$

$$\qquad (2.145)$$

$$B = -\frac{1}{m\omega} \int_0^\tau F(t) \sin \omega dt,$$

and Eq. (2.142) can be rewritten as

$$x = \frac{v_0}{\omega} \sin \omega\tau + x_0 \cos \omega\tau + A \sin \omega\tau + B \cos \omega\tau. \qquad (2.146)$$

If

$$F(t) = F_0 \sin pt, \qquad (2.147)$$

one can obtain the results from section 2.4.1.

Equation (2.142) can be used in case of a mass system acted on by a series of discontinuous impulses, which produce jumps in speed Δv_0, Δv_1,

$\Delta v_2, \ldots,$ at time moments $t = 0, t = t', t = t''.$ For $x_0 = 0$ one can obtain the equation of motion on the $(0, \tau)$ interval:

$$x = \frac{1}{\omega}[\Delta v_0 \sin \omega\tau + \Delta v_1 \sin \omega(\tau - t') + \Delta v_2 \sin \omega(\tau - t'') + \cdots]. \quad (2.148)$$

2.6 Mechanical Impedance

For the mechanical system shown in Fig. 2.29a, the spring force is

$$F = kx, \quad (2.149)$$

Figure 2.29
Forced vibration:
(a) undamped
forced vibration;
(b) forced
vibration with
viscous.

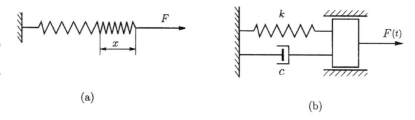

(a)

(b)

where x is the displacement and k is the spring constant. The case of a forced vibration with viscous damping is considered in Fig. 2.29b. The differential equation of motion is

$$m\ddot{x} + c\dot{x} + kx = F(t) = F_0 \cos pt. \quad (2.150)$$

This equation can be written in complex form:

$$m\ddot{z} + c\dot{z} + kz = F_0 e^{ipt}. \quad (2.151)$$

Equation (2.151) can be obtained from

$$m\ddot{y} + c\dot{y} + ky = F_0 \sin pt, \quad (2.152)$$

by multiplying by i ($i^2 = -1$) and adding term by term with Eq. (2.150). If in this equation $\dot{z} = ipz$ and $\ddot{z} = -p^2 z$ are replaced, one may obtain

$$F_0 e^{ipt} = (-mp^2 + icp + k)z. \quad (2.153)$$

If the notation

$$Z = -mp^2 + icp + k, \quad (2.154)$$

is used, Eq. (2.153) becomes

$$F_0 e^{ipt} = Zz, \quad (2.155)$$

where Z is mechanical *impedance.*

In this way, the study of the forced vibration is reduced to a static problem, and

$$z = \frac{e^{ipt}}{Z}F_0 = \frac{e^{ipt}}{-mp^2 + k + icp}F_0. \quad (2.156)$$

Vibration

The amplitude A is

$$A = |z| = \frac{|e^{ipt}|F_0}{|-mp^2 + k + icp|} = \frac{|e^{ipt}|F_0}{m|\omega^2 - p^2 + 2i\alpha p|}, \tag{2.157}$$

or

$$A = \frac{q}{\sqrt{(\omega^2 - p^2)^2 + 4\alpha^2 p^2}}. \tag{2.158}$$

Equation (2.158) is the same as Eq. (2.112).

In the case of n mechanical impedances $Z_1, Z_2, \ldots, Z_n$ in parallel, one can write

$$Z = Z_1 + Z_2 + \cdots + Z_n, \tag{2.159}$$

or

$$Z = \sum_{i=1}^{n} Z_i. \tag{2.160}$$

For mechanical impedances in series the total impedance is

$$Z = \frac{1}{\dfrac{1}{Z_1} + \dfrac{1}{Z_2} + \cdots + \dfrac{1}{Z_n}}, \tag{2.161}$$

or

$$Z = \frac{1}{\displaystyle\sum_{i=1}^{n} \dfrac{1}{Zi}}. \tag{2.162}$$

2.7 Vibration Isolation: Transmissibility

2.7.1 FUNDAMENTALS

A machine with mass m excited by a perturbation force $F_0 \sin pt$ is considered. The machine is on a foundation. The foundation is rigid, and the machine has only translational motions. *Vibration isolation* consists in diminishing the force that is transmitted to the foundation. The coefficient of transmissibility is

$$\tau = \frac{F_{trmax}}{F_{0max}}, \tag{2.163}$$

where F_{trmax} is the maximum transmitted force and F_{0max} is the maximum perturbation force.

Machine Directly on a Foundation

In this case (Fig. 2.30), the perturbation force is transmitted to the foundation. The transmissibility coefficient is $\tau = 1$ and the machine is not isolated.

Figure 2.30
Mechanical model of transmissibility in the case of a machine directly on a foundation.

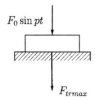

$F_0 \sin pt$

F_{trmax}

Machine on a Foundation with Elastic Elements

The machine is connected to a foundation through elastic elements with the equivalent elastic constant k (Fig. 2.31). The force is transmitted to the foundation through elastic elements; therefore, the transmitted force is the elastic force kx. In Fig. 2.32 is depicted the variation of magnification factor versus frequency ratio.

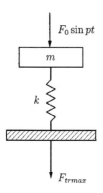

$F_0 \sin pt$

m

k

F_{trmax}

Figure 2.31
Mechanical model of transmissibility in the case of a machine on a foundation with elastic elements.

In the case of undamped forced vibration, the particular solution is

$$x_p = A \sin pt = \frac{q}{\omega^2 - p^2} \sin pt, \tag{2.164}$$

where

$$A = \frac{q}{\omega^2 - p^2} = \frac{\dfrac{F_0}{m}}{\omega^2 - p^2} = \frac{\dfrac{F_0}{m}\dfrac{1}{\omega^2}}{1 - \dfrac{p^2}{\omega^2}} \tag{2.165}$$

$$= \frac{\dfrac{F_0}{m}\dfrac{m}{k}}{1 - \dfrac{p^2}{\omega^2}} = \frac{F_0}{k}\frac{1}{1 - \dfrac{p^2}{\omega^2}} = x_{st}A_0. \tag{2.166}$$

The magnification factor is

$$A_0 = \frac{A}{x_{st}} = \frac{1}{\left|1 - \dfrac{p^2}{\omega^2}\right|}. \tag{2.167}$$

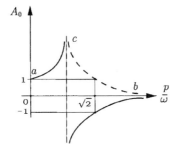

Figure 2.32
Variation of transmissibility coefficient.

The maximum transmitted force is kA, and the transmissibility coefficient is

$$\tau = \frac{F_{trmax}}{F_{0max}} = \frac{kA}{F_0} = \frac{A}{\frac{F_0}{k}} = \frac{A}{x_{st}} = A_0. \qquad (2.168)$$

The diagram in Fig. 2.32 represents the variation of the transmissibility coefficient. The case of a machine directly on a foundation (rigid joint) gives a particular case, $\omega \to \infty$, point a on the diagram in Fig. 2.32. Also, if $p < \sqrt{2}\omega$ one can obtain $|\tau| > 1$, i.e., the force transmitted to the foundation is greater than the perturbation force. For good operation it is necessary that $|\tau| > 1$. As a result in calculus one will take the negative values; thus,

$$\tau = \frac{1}{1 - \left(\frac{p}{\omega}\right)^2} < -1, \qquad (2.169)$$

where $(p/\omega)^2 > 2$.

In conclusion, in the case of a machine on a foundation with elastic elements, it is recommended that $p/\omega > \sqrt{2}$. The dangerous situation is when $p/\omega = 1$.

Machine on a Foundation with an Elastic Element and a Damper

The machine is settled on a foundation with the help of an elastic element with the elastic constant k, and a damper with the viscous damping coefficient c (Fig. 2.33). The transmitted force is not in the same phase with

Figure 2.33
Mechanical model of transmissibility in the case of a machine on a foundation with an elastic element and a damper.

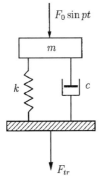

the perturbation force. In this case, the transmitted force is

$$F_{trmax} = [kx + c\dot{x}]_{max}. \tag{2.170}$$

The exciting vibration is

$$x = A\sin(pt - \phi), \tag{2.171}$$

and

$$\dot{x} = Ap\cos(pt - \phi). \tag{2.172}$$

Therefore, the force transmitted to the foundation is

$$\begin{aligned} F_{tr} &= kx + c\dot{x} \\ &= kA\sin(pt - \phi) + cAp\cos(pt - \phi) \\ &= M\sin(pt - \phi), \end{aligned} \tag{2.173}$$

The amplitude of resultant vibration is

$$M = \sqrt{k^2 A^2 + c^2 A^2 p^2} = F_{trmax}, \tag{2.174}$$

which represents the maximum transmitted force. Elastic force and damping force are delayed by $\pi/2$. By Eq. (2.110) the transmissibility coefficient is

$$\tau = \frac{A\sqrt{k^2 + c^2 p^2}}{F_0} = \frac{\dfrac{q}{\sqrt{(w^2 - p^2)^2 + 4\alpha^2 p^2}}\sqrt{k^2 + c^2 p^2}}{\dfrac{F_0}{m}m}$$

$$= \frac{\sqrt{\dfrac{k^2}{m^2} + \dfrac{c^2}{m^2}p^2}}{\sqrt{(\omega^2 - p^2)^2 + 4\alpha^2 p^2}} = \sqrt{\frac{\omega^4 + 4\alpha^2 p^2}{(\omega^2 - p^2)^2 + 4\alpha^2 p^2}}, \tag{2.175}$$

or

$$\sqrt{\frac{1 + 4\dfrac{\alpha^2}{\omega^2}\dfrac{p^2}{\omega^2}}{\left(1 - \dfrac{p^2}{\omega^2}\right)^2 + 4\dfrac{\alpha^2}{\omega^2}\dfrac{p^2}{\omega^2}}} = \sqrt{\frac{1 + 4\left(\dfrac{c}{c_{cr}}\right)^2\dfrac{p^2}{\omega^2}}{\left(1 - \dfrac{p^2}{\omega^2}\right)^2 + 4\left(\dfrac{c}{c_{cr}}\right)^2\dfrac{p^2}{\omega^2}}}. \tag{2.176}$$

From Eq. (2.175) one can observe that the transmissibility coefficient τ does not depend on the amplitude of perturbation force. Equation (2.175) is plotted in Fig. 2.34, which shows the variation of transmissibility coefficient as a function of the ratio c/c_{cr}. If $\tau = 1$, the perturbation force is transmitted integral to the foundation. For the case $\alpha/\omega = 0$ one may obtain

$$|\tau| = 1 = \frac{1}{\left|1 - \dfrac{p^2}{\omega^2}\right|}. \tag{2.177}$$

There are two cases:

Case a: $1 = 1 - p^2/\omega^2 \Rightarrow p/\omega = 0$, which corresponds to the point a on the diagram shown in Fig. 2.34.

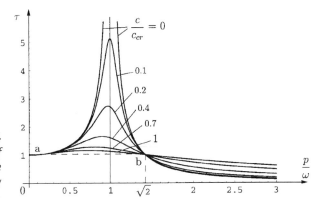

Figure 2.34
The influence of c/c_{cr} on the transmissibility coefficient τ.

Case b: $1 = -(1 - p^2/\omega^2) \Rightarrow p/\omega = \sqrt{2}$, which corresponds to the point b on the diagram of Fig. 2.34. All the curves of variation of transmissibility coefficient in Fig. 2.34 cross point $a(p/\omega = 0, \tau = 1)$ and point $b(p/\omega = \sqrt{2}, \tau = 1)$.

To demonstrate that all the curves cross at point b, one can put $p^2/\omega^2 = 2$. From Eq. (2.175) results

$$\sqrt{\frac{1 + 4\dfrac{\alpha^2}{\omega^2}2}{(1-2)^2 + 4\dfrac{\alpha^2}{\omega^2}2}} = 1. \tag{2.178}$$

Hence, for any value of ratio α/ω it results $\tau = 1$. The choice of the ratio $\alpha/\omega = c/c_{cr}$ is made from case to case, taking into account the transitory regime.

2.8 Energetic Aspect of Vibration with One DOF

2.8.1 MECHANICAL WORK AND POTENTIAL ENERGY FOR A SPRING WITH A LINEAR CHARACTERISTIC

The elastic force of a linear spring is is

$$F_e = -kx, \tag{2.179}$$

where x is the linear displacement. The elementary mechanical work of an elastic force is

$$dL = F_e dx = -kx dx, \tag{2.180}$$

and the mechanical work for a displacement from 0 to x is

$$L = \int_0^x dL = -k\int_0^x x dx = -k\frac{x^2}{2}. \tag{2.181}$$

The elastic force is a conservative force and

$$U = -\frac{1}{2}kx^2 + C. \tag{2.182}$$

One can write

$$dL = dU = -dV, \tag{2.183}$$

and the potential energy is

$$V = \frac{1}{2}kx^2. \tag{2.184}$$

2.8.2 SIMPLE HARMONIC VIBRATION

The mechanical model for this vibration was shown in Fig. 2.3. The differential equation of the model is

$$m\ddot{x} + kx = 0. \tag{2.185}$$

Multiplying Eq. (2.185) by x gives

$$mx\ddot{x} + kx\dot{x} = 0. \tag{2.186}$$

This can be written as

$$\frac{d}{dt}\left(\frac{1}{2}m\dot{x}^2 + \frac{1}{2}kx^2\right) = 0. \tag{2.187}$$

Using the kinetic energy $E = \frac{1}{2}m\dot{x}^2$ and potential energy $V = \frac{1}{2}kx^2$ results in

$$\frac{d}{dt}(E + V) = 0, \tag{2.188}$$

or

$$E_{max} = E + V = \text{const}, \tag{2.189}$$

that is, the total mechanical energy of a system remains constant, and the system is a *conservative system*. Figure 2.35 shows the variation of the kinetic energy and the potential energy during one period (the total energy is constant at any time).

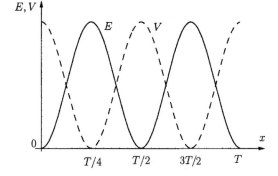

Figure 2.35
Variation of the kinetic energy and the potential energy during one period.

2.8.3 FREE DAMPED VIBRATION

The mechanical model was shown in Fig. 2.8, and the differential equation of motion is

$$m\ddot{x} + c\dot{x} + kx = 0, \tag{2.190}$$

or

$$m\ddot{x} + kx = -c\dot{x}. \tag{2.191}$$

Multiplying by x gives

$$\frac{d}{dt}\left(\frac{1}{2}m\dot{x}^2 + \frac{1}{2}kx^2\right) = -c\dot{x}^2 \tag{2.192}$$

or

$$\frac{d}{dt}(E + V) = -c\dot{x}\frac{dx}{dt}. \tag{2.193}$$

Therefore,

$$dE_{mec} = -c\dot{x}dx = dL_{damp}, \tag{2.194}$$

where $-c\dot{x}dx = dL_{damp}$ is the elementary mechanical work of the viscous damped force, which is the resistant mechanical work. The solution of the equation of motion is

$$x = Ae^{-\alpha t}\sin(\beta t + \varphi). \tag{2.195}$$

Hence,

$$\dot{x} = Ae^{-\alpha t}[-\alpha\sin(\beta t + \varphi) + \cos(\beta t + \varphi)]. \tag{2.196}$$

The quasiperiod is given by $T_\beta = 2\pi/\beta$.

In considering instances when the displacement $x = 0$, the mechanical energy of system is equal to the maximum kinetic energy,

$$E_{mech} = E_{max} = \tfrac{1}{2}m\dot{x}^2. \tag{2.197}$$

With Eq. (2.197), this gives the following results:

For $t = 0 \Rightarrow$

$$E_0 = \tfrac{1}{2}mA^2[-\alpha\sin\varphi + \beta\cos\varphi]^2, \tag{2.198}$$

For $t = T_\beta = 2\pi/\beta \Rightarrow$

$$\begin{aligned}
E_{T_\beta} &= \frac{1}{2}mA^2 e^{-2\alpha(2\pi/\beta)}\left[-\alpha\sin\left(\beta\frac{2\pi}{\beta} + \varphi\right) + \beta\cos\left(\beta\frac{2\pi}{\beta} + \varphi\right)\right]^2 \\
&= \frac{1}{2}mA^2 e^{-4\alpha\pi/\beta}[-\alpha\sin\varphi + \beta\cos\varphi]^2.
\end{aligned} \tag{2.199}$$

For $t = 2T_\beta = 4\pi/\beta \Rightarrow$

$$\begin{aligned}
E_{2T_\beta} &= \frac{1}{2}mA^2 e^{-2\alpha(4\pi/\beta)}\left[-\alpha\sin\left(\beta\frac{4\pi}{\beta} + \varphi\right) + \beta\cos\left(\beta\frac{4\pi}{\beta} + \varphi\right)\right]^2 \\
&= \frac{1}{2}mA^2 e^{-8\alpha\pi/\beta}[-\alpha\sin\varphi + \beta\cos\varphi]^2.
\end{aligned} \tag{2.200}$$

For comparison one can use the ratio $E_{T_\beta}/E_0 = e^{-4\alpha\pi/\beta}$ and $E_{2T_\beta}/E_{T_\beta} = e^{-4\alpha\pi/\beta}$ and the same values are obtained. With the logarithmic decrement one can get

$$E_{t+T_\beta} = E_t e^{-2\delta}. \tag{2.201}$$

2.8.4 UNDAMPED FORCED VIBRATION

The mechanical model was shown in Fig. 2.12.

The differential equation of motion is

$$m\ddot{x} + kx = F_0 \sin pt. \tag{2.202}$$

Multiplying by x, one can obtain

$$\frac{d}{dt}\left(\frac{1}{2}m\dot{x}^2 + \frac{1}{2}kx^2\right) = F_0\dot{x}\sin pt \tag{2.203}$$

or

$$\frac{d}{dt}(E + V) = F_0\frac{dx}{dt}\sin pt. \tag{2.204}$$

Therefore,

$$dE_{mec} = F_0 \sin pt dx = dL_{pert}, \tag{2.205}$$

where the elementary mechanical work of the perturbation force is denoted by

$$F_0 \sin pt dx = dL_{pert}. \tag{2.206}$$

Because the mechanical work of the perturbation force is an active mechanical work, it results in $dE_{mec} > 0$, that is, the mechanical energy of the system increases because of the perturbation force. The equation of motion is

$$x = \frac{q}{\omega^2 - p^2}\left(\sin pt - \frac{p}{\omega}\sin \omega t\right). \tag{2.207}$$

2.8.5 DAMPED FORCED VIBRATION

The mechanical model was shown in Fig. 2.20. The equation of motion is

$$m\ddot{x} + c\dot{x} + kx = F_0 \sin pt, \tag{2.208}$$

or

$$m\ddot{x} + kx = -c\dot{x} + F_0 \sin pt. \tag{2.209}$$

Multiplying by $\dot{x}$, one may obtain

$$\frac{d}{dt}\left(\frac{1}{2}m\dot{x}^2 + \frac{1}{2}kx^2\right) = c\dot{x}^2 + F_0\dot{x}\sin pt, \tag{2.210}$$

or

$$\frac{d}{dt}(E + V) = -c\dot{x}dx + F_0 \sin pt dx. \tag{2.211}$$

With Eq. (2.194) and Eq. (2.206) one can write

$$dE_{max} = dL_{damp} + dL_{pert}. \tag{2.212}$$

In a permanent regime the law of motion is

$$x = A\sin(pt - \phi), \tag{2.213}$$

and

$$\dot{x} = Ap\cos(pt - \phi), \tag{2.214}$$
$$dx = Ap\cos(pt - \varphi)dt. \tag{2.215}$$

In Eq. (2.112) the amplitude is

$$A = \frac{q}{\sqrt{(\omega^2 - p^2)^2 + 4\alpha^2 p^2}}, \tag{2.216}$$

the delay is

$$\tan\phi = \frac{2\alpha p}{\omega^2 - p^2}, \tag{2.217}$$

and

$$\sin\phi = \frac{\tan\phi}{\sqrt{1 + \tan^2\phi}} = \frac{2\alpha p}{\sqrt{(\omega^2 - p^2)^2 - 4\alpha^2 p^2}}. \tag{2.218}$$

For a period, when $\omega = p$, the mechanical work produced by the perturbation force and the damped force is

$$
\begin{aligned}
L_{pert} &= \int_0^{2\pi/\omega} F_0\sin pt\,dx = \int_0^{2\pi/p} F_0\sin pt Ap(\cos(pt - \phi)dt \\
&= F_0 Ap\int_0^{2\pi/p} \sin pt \cos(pt - \phi)dt \\
&= F_0 Ap\left[\int_0^{2\pi/p} \sin pt \cos pt \cos\phi\,dt + \int_0^{2\pi/p} \sin^2 pt \sin\phi\,dt\right] \\
&= F_0 Ap\sin\phi\int_0^{2\pi/p} \frac{1 - \cos 2pt}{2}\,dt \\
&= \frac{F_0 Ap\sin\phi}{2}\left(\frac{2\pi}{p}\right) + \frac{F_0 Ap\sin\phi}{2}\int_0^{2\pi/p} \cos 2pt\,dt. \tag{2.219}
\end{aligned}
$$

Therefore,

$$L_{pert} = \pi F_0 A\sin\phi. \tag{2.220}$$

Using Eqs. (2.220), (2.216), and (2.218) gives

$$L_{pert} = \pi F_0 \frac{q}{\sqrt{(\omega^2 - p^2)^2 + 4\alpha^2 p^2}} \frac{2\alpha p}{\sqrt{(\omega^2 - p^2)^2 + 4\alpha^2 p^2}}$$

$$= \frac{\pi F_0 \dfrac{F_0}{m} \dfrac{c}{m} p}{(\omega^2 - p^2)^2 + 4\alpha^2 p^2} = \frac{\pi \left(\dfrac{F_0}{m}\right)^2 cp}{(\omega^2 - p^2)^2 + 4\alpha^2 p^2}$$

$$= \frac{\pi c p q^2}{(\omega^2 - p^2)^2 + 4\alpha^2 p^2}. \tag{2.221}$$

One may obtain

$$L_{pert} = \pi c p A^2. \tag{2.222}$$

In a similar way

$$L_{damp} = \int_0^{2\pi/\omega} -c\dot{x}\,dx = -c\int_0^{2\pi/p} Ap\cos(pt - \phi)Ap\cos(pt - \phi)$$

$$= -cA^2 p^2 \int_0^{2\pi/p} \cos^2(pt - \phi)\,dt$$

$$= -cA^2 p^2 \int_0^{2\pi/p} \frac{1 + \cos 2(pt - \phi)}{2}\,dt$$

$$= -\frac{cA^2 p^2}{2}\left(\frac{2\pi}{p}\right) - \frac{cA^2 p^2}{2}\int_0^{2\pi/p} \cos 2(pt - \phi)\,dt. \tag{2.223}$$

Therefore,

$$L_{damp} = -\pi c A^2 p. \tag{2.224}$$

From Eqs. (2.222) and (2.224) results

$$L_{pert} + L_{damp} = 0, \tag{2.225}$$

that is,

$$E_{mech} = E + V = \text{const.} \tag{2.226}$$

One can find conservative systems in the case of damped forced vibration.

2.8.6 RAYLEIGH METHOD

The Rayleigh method is an approximative method used to compute the circular frequency of conservative mechanical systems with one or more DOF. One may consider a conservative mechanical system with one DOF. The kinetic energy T and the potential energy V were shown in Fig. 2.35. Hence,

$$E_{max} = V_{max}. \tag{2.227}$$

With Eq. (2.227) one can compute the approximative circular frequency.

2.9 Critical Speed of Rotating Shafts

Rotating shafts tend to bow out at certain speeds and whirl in a complicated manner. Whirling is defined as the rotation of the plane made by the bent shaft and the line of centers of the bearings. The phenomenon results from such various causes as mass unbalance, hysteresis damping in the shaft, gyroscopic forces, and fluid friction in bearings. Figure 2.36a shows a shaft with a wheel of mass m. The center of mass, G, of the wheel is at the distance e (eccentricity) from the center of the wheel, A. The shaft rotates with a constant (angular) speed $\Omega = p = $ const. The centrifugal force is $F_c = mep^2$ and acts at the center of mass, G. By projecting the centrifugal force on the horizontal (Ox) and the vertical axis (Oy) (radial and tangential directions, Fig. 2.36b), the following relations are obtained:

$$\begin{cases} F_{cH} = mep^2 \cos pt \\ F_{cV} = mep^2 \sin pt. \end{cases} \tag{2.228}$$

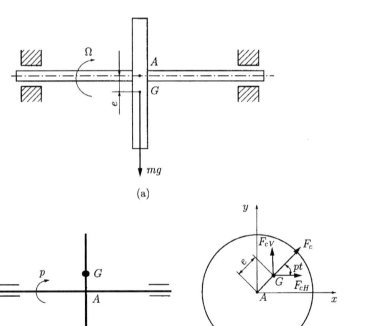

Figure 2.36
Shaft with a wheel: (a) shaft with a wheel of mass m; (b) mechanical model.

(a)

(b)

This perturbation periodic forces generate transversal vibrations in two planes. The shaft is loaded with a bending moment. If the circular frequency of transversal vibration (bending) is equal to the rotation angular speed, then the resonance will take place. In this case rotation angular speed is called *critical angular speed* Ω_{cr}, and the critical rpm is n_{cr}, given by

$$n_{cr} = \frac{30}{\pi}\Omega_{cr}. \tag{2.229}$$

Since the shaft has distributed mass and elasticity along its length, the system has more than one degree of freedom. One can assume that the mass of the shaft is negligible and its lateral stiffness is k. In this case the natural frequency is

$$\omega = \sqrt{\frac{k}{m}} = \sqrt{\frac{g}{f_{st}}},\qquad(2.230)$$

where f_{st} is the static deflection.

The center line of the support bearings intersects the plane of the wheel at O, (Fig. 2.37a), and the shaft center is deflected with $r = OA$. The lateral view of a general position of the rotating wheel of mass m is shown in Fig. 2.37b. A particular case is shown in Fig. 2.37b: OA and AG are in extension. The elastic force and the centrifugal force are in relative equilibrium,

$$kr = m(r+e)p^2 = mrp^2 + mep^2,\qquad(2.31)$$

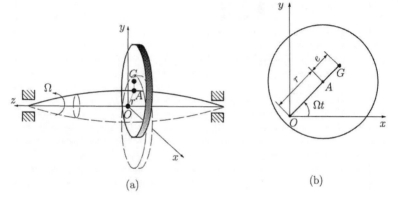

Figure 2.37
*(a) Shaft with a wheel in rotational motion;
(b) Lateral view of a general position of the rotating wheel.*

(a) (b)

and

$$r = e\frac{p^2}{\dfrac{k}{m}-p^2} = e\frac{\left(\dfrac{p}{\omega}\right)^2}{1-\left(\dfrac{p}{\omega}\right)^2}.\qquad(2.232)$$

This equation is plotted in Fig. 2.38. From this figure one can observe that there are two domains. One is undercritical ($p < \omega$) when $r > 0$ and the point G is outside of segment OA. The other one is overcritical ($p > \omega$) when $r < 0$ and the point G is inside of segment OA. If angular speed of the shaft increases, the point of mass G tends to point O (center line of bearings). This phenomenon is called *self-centering* or *self-aligning*.

In the situations when the wheel is not at middle of the shaft and it is at one extremity (Fig. 2.39) or at a distance a from the bearing (Fig. 2.40), the gyroscopic phenomenon appears. In this case the centrifugal force $P = myp^2$ and gyroscopic moment (moment of inertial forces) appear. To determine the gyroscopic moment M of the wheel, consider a plane made by the deflected

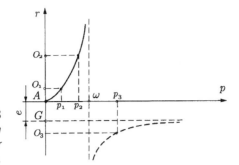

Figure 2.38
Variation of center deflection r.

shaft and the undeflected line of the shaft. The whirl ω_1 is defined as the speed of rotation of this plane about the undeflected line of the shaft. Resolving ω_1 into components perpendicular and parallel to the face of the wheel, one can obtain $\omega_1 \sin\theta$ and $\omega_1 \cos\theta$, as shown in Fig. 2.39:

$$M = J_z\left(1 + \frac{J_z - J_x}{J_z}\frac{\omega_1}{\omega}\cos\theta\right)\omega \times \omega_1. \tag{2.233}$$

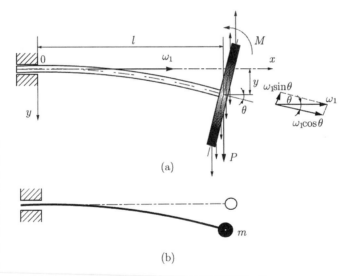

Figure 2.39
Shaft with a wheel at one extremity: (a) shaft with a wheel at one extremity; (b) mechanical model.

Here, $\omega_1 = \Omega = p$; θ is the deflection angle of the shaft at point A with respect to the axis of the bearings; $J_z = J_1$ and $J_x = J_y = J_2$ are the moments of inertia of the wheel; J_z is the moment of inertia of the wheel with respect to an axis perpendicular to the wheel at the point A (polar moment of inertia); and J_x, J_y are the moment of inertia with respect to two perpendicular diameters of the wheel in the plane (diametric moments of inertia). For the cases shown in Figs. 2.39 and 2.40, because angle θ is small one can approximate $\sin\theta = 0$ and $\cos\theta = 1$, and the gyroscopic moment is

$$M - (J_1 - J_2)\omega_1^2 \sin\theta\cos\theta \approx (J_1 - J_2)p^2\theta. \tag{2.234}$$

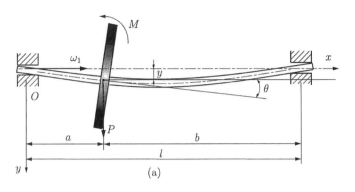

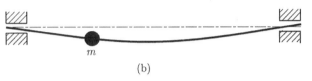

(a)

Figure 2.40
(a) Shaft with a wheel at distance a from the bearing; (b) mechanical model.

(b)

The gyroscopic torque has the effect of decreasing the deflection at bending. Therefore, the gyroscopic effect increases the stiffness of the shaft. In the case when the rotation speed of the wheel and the rotation speed of the elastic line of the shaft are different, the gyroscopic torque is changed. If Ω_0 is the angular rotation speed of the plane that contain the elastic line of the shaft, then

$$M = (J_1 p \Omega_0 - J_2 p^2)\theta. \tag{2.235}$$

If the speed of this plane is opposite to the angular speed of the shaft, the gyroscopic torque will change the sign. The deflection of bending increase and the critical rpm is low.

The elasticity of the supports of the shaft is another cause that produces modifications to the the critical rpm. If k_A and k_B (Fig. 2.41) are the elastic constants of the supports and k_i is the elastic constant of the shaft, the equivalent constant for the supports is

$$\frac{1}{k_r} = \frac{\dfrac{a^2}{k_B} + \dfrac{b^2}{k_A}}{(a+b)^2}, \tag{2.236}$$

and the equivalent constant for the supports and shaft is

$$\frac{1}{k} = \frac{21}{k_r} + \frac{1}{k_i}. \tag{2.237}$$

Figure 2.41
Elasticity of the supports of the shaft.

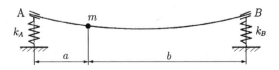

The natural frequency of the system, ω, is different from the circular frequency, ω_0, for the case of rigid supports:

$$\omega = \sqrt{\frac{k}{m}}, \quad \omega_0 = \sqrt{\frac{k_i}{m}}.$$

The ratio of the two circular frequencies is

$$\frac{\omega}{\omega_0} = \sqrt{\frac{k}{k_i}} = \sqrt{\frac{1}{\dfrac{k_i}{k_r} + 1}}. \tag{2.238}$$

If one can consider the weight of the wheel, the moment of the force is

$$M_0 = mge \sin pt, \tag{2.239}$$

The angular acceleration is

$$\varepsilon = \frac{M_0}{J_0} = \frac{mge \sin pt}{J_0}. \tag{2.240}$$

Because of this angular acceleration, the tangential force of inertia is (Fig. 2.42b)

$$F_t = me\varepsilon = \frac{m^2 ge^2}{J_0} \sin pt. \tag{2.241}$$

Figure 2.42
Tangential force of inertia: (a) weight of the wheel; (b) tangential force of inertia.

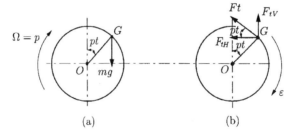

(a)   (b)

The two components are

$$F_{tH} = F_t \cos pt = \frac{m^2 ge^2}{J_0} \sin pt \cos pt = \frac{m^2 ge^2}{2J_0} \sin 2pt$$

$$F_{tV} = F_t \sin pt = \frac{m^2 ge^2}{J_0} \sin^2 pt = \frac{m^2 ge^2}{2J_0} (1 - \cos 2pt). \tag{2.242}$$

This perturbation force with circular frequency $2p$ presents the danger of resonance for the case $\omega = 2p \Rightarrow p = \omega/2$.

3. Linear Systems with Finite Numbers of Degrees of Freedom

Systems with n degrees of freedom are described by a set of n simultaneous ordinary differential equations of the second order with n generalized coordinates. The number of natural frequencies is equal to the number of degrees of freedom. A system with two degrees of freedom will have two natural frequencies. When free vibration takes place at one of these natural frequencies, a definite relationship exists between the amplitude of the two coordinates, and the configuration is referred to as the natural mode. The two degrees of freedom system will then have two normal mode vibrations corresponding to the two natural frequencies. Free vibration initiated under any condition will in general be the superposition of the two normal mode vibrations. However, forced harmonic vibration will take place at the frequency of the excitation, and the amplitude of the two coordinates will tend to a maximum at the two natural frequencies. A mode of vibration is associated with each natural frequency. Since the equations of motion are coupled, the motion of the masses is the combination of the motions of the individual modes.

The steps for solving a vibration problem are (Fig. 3.1)

- From the physical model, go to the mechanical model, involving the option to work with the continuous model or the discrete model (with a finite number of degrees of freedom).
- From the mechanical model, go to the mathematical model. The mathematical model consists of a set of differential equations.
- Perform qualitative analysis with or without determination of the dynamic response. The stability analysis, determination of natural frequency, etc., are parts of the qualitative analysis.

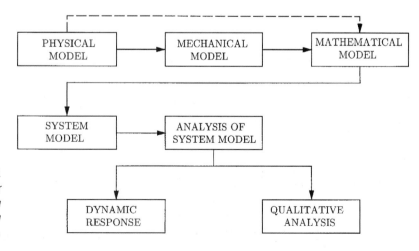

Figure 3.1
The steps for solving a vibration problem.

Vibration

3.1 Mechanical Models

3.1.1 ELASTIC CONSTANTS

The elastic constants are associated with a linear displacement q as the result of a force F, or are associated with an angular displacement θ as the result of a moment M:

$$k = \frac{F}{q} \quad \text{or} \quad k = \frac{M}{\theta}.$$

In both cases the elastic constant, denoted by k, is associated with a Hooke model (Fig. 3.2). For a series connection of Hooke models, with k_i elastic constant, the equivalent elastic constant k_e is

$$\frac{1}{k_e} = \sum_{i=1}^{n} \frac{1}{k_i}, \tag{3.1}$$

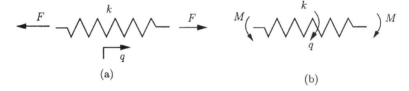

Figure 3.2
Hooke model.

(a)   (b)

and for parallel connection, the equivalent elastic constant k_e is

$$k_e = \sum_{i-1}^{n} k_i. \tag{3.2}$$

The elastic constants (spring stiffness) for useful cases are presented in Table 3.1.

EXAMPLE 3.1 The mechanical model for the physical model shown in Fig. 3.3 will be determined. The rods 1 and 2 are linear elastic. The R_1 spring has diameter of wire d_1, average (medium) radius r, and n_1 number of turns. Linear springs R_2, R_3, R_4, R_5 are identical with diameter of wire d_2, average radius $r/2$, and n_2 number of turns. The R_6 spring is identical with the R_1 spring.

Solution

Using Table 3.1, elastic constants (spring stiffness) are determined from rods 1 and 2. For the rod 1,

$$k_1 = \frac{3E_1 I_1}{l_1^3} \, [\text{N/m}].$$

For the rod 2,

$$k_2 = \frac{27 E_2 I_2}{8 l_2^3} \, [\text{N/m}].$$

Table 3.1

Mechanical model	Elastic constants
	$$k = \frac{3EI(a+b)}{a^2 b^2},$$
	$$k = \frac{3EI(a+b)^3}{a^3 b^3},$$
	$$k = \frac{3EI}{l^3},$$
	$$k = \frac{12EI(a+b)^3}{a^3 b^2 (3a+4b)},$$
	$$k_1 = \frac{12EI(a+c)^3}{a^3 c^3 (4a-3c)} \text{(for } m_1\text{)}, \qquad k_2 = \frac{3EI}{b^3} \text{(for } m_2\text{)},$$
	$$k_1 = \frac{3EI(a+c)}{a^2 c^2} \text{(for } m_1\text{)}, \qquad k_2 = \frac{3EI}{b^3} \text{(for } m_2\text{)}.$$

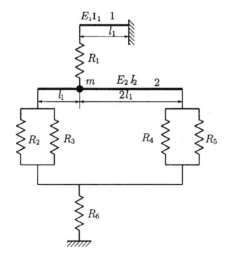

Figure 3.3
Physical model for Example 3.1.

The elastic constant for the spring R_1 and R_6 is

$$k_3 = \frac{Gd_1^4}{64r^3 n_1}\,[\text{N/m}].$$

For the springs R_2–R_5, the elastic constant is

$$k_4 = \frac{Gd_2^4}{8r^3 n_2}\,[\text{N/m}].$$

Using the mechanical model shown in Fig. 3.4, one may find

$$\frac{1}{k'} = \frac{1}{k_1} + \frac{1}{k_3} \Rightarrow k' = \frac{k_1 k_3}{k_1 + k_3}$$

$$\frac{1}{k''} = \frac{1}{k_2} + \frac{1}{k_3} + \frac{1}{4k_4} \Rightarrow k'' = \frac{4k_2 k_3 k_4}{k_2 + k_3 + 4k_4}.$$

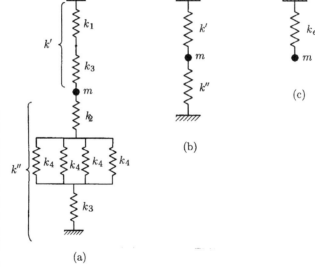

Figure 3.4
Mechanical model. (a) The equivalent mechanical model of the physical model; (b) two springs connected in parallel; (c) final model.

For the next model, shown in Fig. 3.4b, two springs of constants k' and k'' are connected in parallel. The final model is shown in Fig. 3.4c and has the elastic constant $k_e = k' + k''$. ▲

EXAMPLE 3.2 Determine the elastic constant k associated with the displacement q (parallel to the direction Δ–Δ) for the linear elastic curved beam in Fig. 3.5. The force **F** acts at A where the mass m is located. The direction of F is the direction of q.

Solution

First, one can evaluate the deformation at A on q direction (parallel with Δ–Δ).

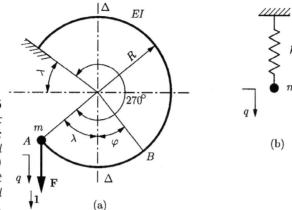

Figure 3.5
Linear elastic curved beam: (a) physical model; (b) equivalent mechanical model.

Next, one can compute the elastic constant k as the ratio between the force $\mathbf{F}$ and the computed deformation. To compute the deformation at A, the Mohr-Maxwell method is used.

The bending moment in section B of rod is $M(\varphi) = FR(\sin \lambda + \sin \varphi)$. If a unitary force **1** acts at A (this force is parallel to $\mathbf{F}$), the bending moment of the unitary force with respect to B is $m(\varphi) = R(\sin \lambda + \sin \varphi)$.

By the Mohr–Maxwell method, the deflection at A is

$$f = \frac{1}{EI} \int_0^{3\pi/2} M(\varphi) m(\varphi) ds,$$

where $ds = Rd\varphi$ is the elementary length of rod axis. Solving the integral gives for the deflection

$$f = \frac{FR^3}{EI} \left(\frac{3\pi}{2} \sin^2 \lambda + 2 \sin \lambda + \frac{3\pi}{4} \right).$$

The elastic constant is

$$k = \frac{F}{f} = \frac{EI}{\dfrac{3\pi}{2} \sin^2 \lambda + 2 \sin \lambda + \dfrac{3\pi}{4}},$$

and the mechanical model is shown in Fig. 3.5b. ▲

EXAMPLE 3.3 Determine the elastic constant k associated with the displacement q (parallel with direction $\Delta_2 \Delta_2$) for the system in Fig. 3.6a.

Solution

In section A–A (Fig. 3.6b), three static undetermined variables X_1, X_2, and X_3 are introduced. By symmetry, $X_2 = 0$ and $X_3 = F/2$. To compute X_1, the Mohr–Maxwell method or the Castigliano theorem can be used. The mechanical work of deformation is

$$L_d = \frac{4}{2EI_1} \int_0^{L_1/2} \left[X_1 - \frac{Fx}{2} \right] dx + \frac{4}{2EI_2} \int_0^{L_2/2} X_1^2 dx.$$

Figure 3.6
Linear elastic frame: (a) physical model; (b) three static undetermined variables X_1, X_2, and X_3 are introduced; (c) equivalent mechanical model.

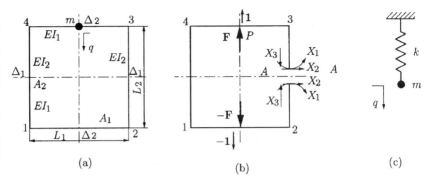

(a)   (b)   (c)

With $\partial L_d/\partial X_1 = 0$ (Castigliano theorem) results

$$X_1 = \frac{FL_1}{8}\frac{I_2 L_1}{(I_2 L_1 + I_1 L_2)}.$$

According to the Veresceaghin method (with force **1**) the bending deflection at point P is

$$f_b = \frac{FL_1^2}{96EI_1}\frac{L_1 I_2 + 4L_2 I_1}{L_2 I_2 + L_2 I_1}.$$

In this case the elastic constant is

$$k = \frac{F}{f_b} = \frac{96EI_1(L_1 I_2 + L_2 I_1)}{L_1^2(L_1 I_2 + 4L_2 I_1)}.$$

The mechanical model is shown in Fig. 3.6c. ▲

3.1.2 DAMPING

All real systems dissipate energy when they vibrate. The energy dissipated is often very small, that is, an undamped analysis is sometimes realistic. When the damping is significant, its effect must be included in the analysis, particularly when the amplitude of vibration is required. It is often difficult to model damping exactly because many mechanisms may be operating in a system. However, each type of damping can be analyzed, and since in many dynamical systems one form of damping predominates, a reasonably accurate analysis is usually possible.

Damping represents forces of friction and can be either external damping or internal damping.

External damping appears as the result of interaction between a mechanic system and the external environment. It can be of the following types:

1. *Columbian damping* is associated with dry friction, with the friction force given by

$$\mathbf{F}_f = -\mu \|\mathbf{F}_{kj}\| \cdot \frac{\mathbf{v}}{\|\mathbf{v}\|}, \tag{3.3}$$

where μ is the coefficient of friction, $\mathbf{F}_j$ is the joint force, and $\mathbf{v}$ is the relative velocity of sliding.

2. *Viscous damping* is associated with the dissipating force $\mathbf{F}_v = -c\mathbf{v}$, where $\mathbf{v}$ is the relatively velocity, and c is the coefficient of viscous resistance.

If $c = $ (const. the damping is called linear damping; if $c = c(t)$, the damping is called parametric damping; and if $c = c(q)$, $c = c(\dot{q})$, $c = c(q, \dot{q})$ or generally $c = c(q, \dot{q}, t)$, the damping is called complex damping.

Internal damping is associated with viscoelastic damping and hysteresis damping. The most common type of damping is linear viscous damping (Fig. 3.7). In case of series connection, the equivalent coefficient of viscous damping is

$$\frac{1}{c_e} = \sum_{i=1}^{n} \frac{1}{c_i}, \tag{3.4}$$

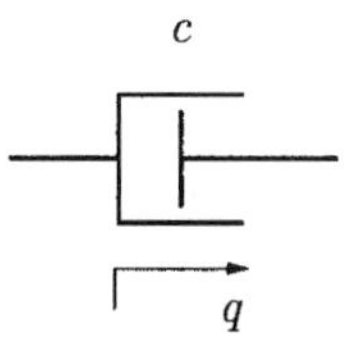

Figure 3.7
Linear viscous damping.

and in the case of a parallel connection, it is

$$c_e = \sum_{i=1}^{n} c_i, \tag{3.5}$$

The coefficient of viscous damping for one degree of freedom q is

$$c = \frac{\psi k}{\pi \omega}, \tag{3.6}$$

where ψ is relative dissipation of energy, k is the elastic constant, and ω is the natural frequency.

From Eq. (3.6) one can obtain

$$c = \begin{cases} \dfrac{q_1}{\pi} \sqrt{km} \\ \dfrac{q_2}{\pi} \sqrt{kJ} \end{cases}, \tag{3.7}$$

where q_1 is the linear displacement, m is the mass, q_2 is the angular displacement, and J is the mass moment of inertia.

3.1.3 MASS GEOMETRY

The reduction of a physical model to a mechanical model with a finite number of degrees of freedom implies the following:

- Elements of rigid body type (RB) or material point (MP) type
- Elements of continuous system (CS) type

For a discrete system with m_i, $i = 1, \ldots, n$ masses, the sum of all material point masses must be equal to the total mass M of the system:

$$\sum_{i=1}^{n} m_i = M. \tag{3.8}$$

3.2 Mathematical Models

3.2.1 NEWTON METHOD

The Newton method can be used for a system with material points. There are two situations.

Situation A

The material points (MP) are interconnected with linear springs k_i and linear dampers c_i. Application of the Newton equation for MP_i (Fig. 3.8) leads to a mathematical model for MP. For the system in Fig. 3.8, the equation of motion for the m_i mass is

$$m_i \ddot{q}_i = -k_{i-1}(q_i - q_{i-1}) - c_{i-1}(\dot{q}_i - \dot{q}_{i-1})$$
$$- k_i(q_i - q_{i+1}) - c_i(\dot{q}_i - \dot{q}_{i+1}) + F_i(t). \tag{3.9}$$

EXAMPLE 3.4 Determine the mathematical model associated with the mechanical model shown in Fig. 3.9.

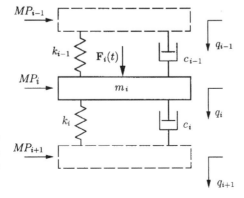

Figure 3.8
Material points (MP) interconnected with linear springs and linear dampers.

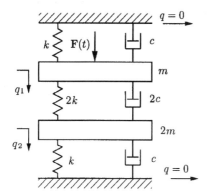

Figure 3.9
*Mechanical
model for
Example 3.4.*

Solution

The Newton method is applied, and

$$\begin{cases} m\ddot{q}_1 = -k(q_1 - 0) - c(\dot{q}_1 - 0) - 2k(q_1 - q_2) - 2c(\dot{q}_1 - \dot{q}_2) + F(t) \\ 2m\ddot{q}_2 = -2k(q_2 - q_1) - 2c(\dot{q}_2 - \dot{q}_1) - k(q_2 - 0) - c(\dot{q}_2 - 0), \end{cases}$$

or, finally, the mathematical model is

$$\begin{cases} m\ddot{q}_1 + 3c\dot{q}_1 - 2c\dot{q}_2 + 3kq_1 - 2kq_2 - F(t) = 0 \\ 2m\ddot{q}_2 - 2c\dot{q}_1 + 3x\dot{q}_2 - 2kq_1 + 3kq_2 = 0. \end{cases}$$

Situation B

The material points (MP) result from the lumped masses of rods (beams), plates, or elastic solids. For elastic displacements the coefficients of influence α_{ij} will be defined. For systems with one DOF, to move a mass m with a distance x a force F is required, that is, $F = kx$.

Similarly, if a force F is applied to the mass, the deflection will be $x = (1/k)F$. If α is defined as the inverse of k, then $x = \alpha F$, where α is the flexibility coefficient of influence. The stiffness matrix of the system is $[K]$. The elements of the matrix $[A] = [K]^{-1}$ are the flexibility coefficients of influence. For the system of masses shown in Fig. 3.10, the deflection due to the force F_i will be $\alpha_{ij}F_i$. Each time a new force acts on another mass the system will move to a new static configuration and the total displacement for the mass m_i will be

$$x_i = \alpha_{i1}F_1 + \alpha_{i2}F_2 + \cdots + \alpha_{in}F_n = \sum_{j=1}^{n} \alpha_{ij}F_j.$$

Figure 3.10
*Lumped masses
model for linear
rod in bending
vibration.*

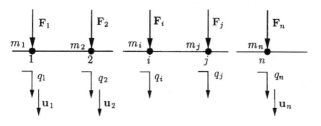

One can write

$$X = [A]F,$$

where

$$X = \{x_1\, x_2\, \ldots\, x_n\}^T, \quad F = \{F_1\, F_2\, \ldots\, F_n\}, \quad \text{and} \quad [A] = \begin{bmatrix} \alpha_{11} & \alpha_{12} & \cdots & \alpha_{1n} \\ \alpha_{21} & \alpha_{22} & \cdots & \alpha_{2n} \\ \vdots & \vdots & & \vdots \\ \alpha_{n1} & \alpha_{n2} & \cdots & \alpha_{nn} \end{bmatrix}.$$

▲

EXAMPLE 3.5 Using the coefficients of influence method, write the equation of motion for the bending vibration for a linear rod with n lumped masses (Fig. 3.10).

Solution

Let $\mathbf{u}_1, \ldots, \mathbf{u}_n$ be the unit vectors attached to the displacements of the sections $1, \ldots, n$. For these sections the inertial forces attached to the masses $m_1, \ldots, m_n$ are introduced as

$$\mathbf{F}_{in_1} = -m_1 \ddot{q}_1 \mathbf{u}_1, \ldots, \mathbf{F}_{in_n} = -m_n \ddot{q}_n \mathbf{u}_n.$$

One can write

$$\begin{cases} q_1 = \alpha_{11}(F_1 - m_1 \ddot{q}_1) + \alpha_{12}(F_2 - m_2 \ddot{q}_2) + \cdots + \alpha_{1n}(F_n - m_n \ddot{q}_n) \\ q_n = \alpha_{n1}(F_1 - m_q \ddot{q}_1) + \alpha_{n2}(F_2 - m_2 \ddot{q}_2) + \cdots + \alpha_{nn}(F_n - m_n \ddot{q}_n). \end{cases}$$

Finally, the equations of motions are

$$\begin{cases} \alpha_{11} m_1 \ddot{q}_1 + \cdots + \alpha_{1n} m_n \ddot{q}_n + q_1 = \alpha_{11} F_1 + \cdots + \alpha_{1n} F_n \\ \alpha_{21} m_1 \ddot{q}_1 + \cdots + \alpha_{2n} m_n \ddot{q}_n + q_2 = \alpha_{21} F_1 + \cdots + \alpha_{2n} F_n \\ \qquad\qquad\qquad \vdots \\ \alpha_{n1} m_1 \ddot{q}_1 + \cdots + \alpha_{nn} m_n \ddot{q}_n + q_n = \alpha_{n1} F_1 + \cdots + \alpha_{nn} F_n. \end{cases}$$

▲ (3.10)

EXAMPLE 3.6 Determine the mathematical model for the system shown in Fig. 3.11.

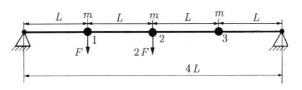

Figure 3.11
Linear beam system.

Solution

With the coefficients of influence given in Table 3.2, the result is

$$\alpha_{11} \frac{L^2(3L)^2}{3EI(4L)} = \frac{3L^3}{4EI}, \quad \alpha_{22} = \frac{4L^3}{3EI}, \quad \alpha_{33} = \alpha_{11}$$

$$= \frac{11L^3}{12EI}, \quad \alpha_{21} = \alpha_{12}, \quad \alpha_{13} = \alpha_{31} = \frac{7L^3}{12EI}$$

$$\alpha_{23} = \alpha_{32} = \alpha_{12} = \alpha_{21}.$$

Table 3.2

Mechanical model

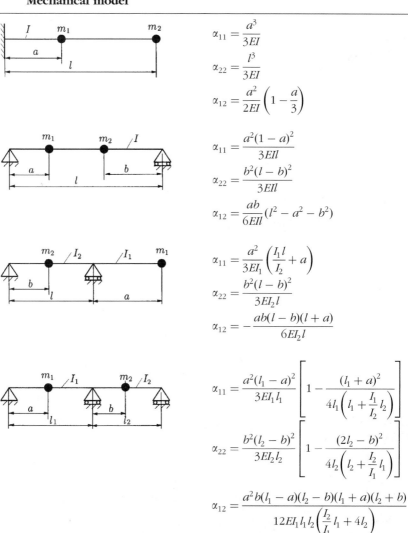

$$\alpha_{11} = \frac{a^3}{3EI}$$

$$\alpha_{22} = \frac{l^3}{3EI}$$

$$\alpha_{12} = \frac{a^2}{2EI}\left(1 - \frac{a}{3}\right)$$

$$\alpha_{11} = \frac{a^2(1-a)^2}{3EIl}$$

$$\alpha_{22} = \frac{b^2(l-b)^2}{3EIl}$$

$$\alpha_{12} = \frac{ab}{6EIl}(l^2 - a^2 - b^2)$$

$$\alpha_{11} = \frac{a^2}{3EI_1}\left(\frac{I_1 l}{I_2} + a\right)$$

$$\alpha_{22} = \frac{b^2(l-b)^2}{3EI_2 l}$$

$$\alpha_{12} = -\frac{ab(l-b)(l+a)}{6EI_2 l}$$

$$\alpha_{11} = \frac{a^2(l_1 - a)^2}{3EI_1 l_1}\left[1 - \frac{(l_1 + a)^2}{4l_1\left(l_1 + \frac{I_1}{I_2}l_2\right)}\right]$$

$$\alpha_{22} = \frac{b^2(l_2 - b)^2}{3EI_2 l_2}\left[1 - \frac{(2l_2 - b)^2}{4l_2\left(l_2 + \frac{I_2}{I_1}l_1\right)}\right]$$

$$\alpha_{12} = \frac{a^2 b(l_1 - a)(l_2 - b)(l_1 + a)(l_2 + b)}{12EI_1 l_1 l_2\left(\frac{I_2}{I_1}l_1 + 4l_2\right)}$$

Vibration

Using Eq. (3.10), one can obtain

$$\begin{cases} \alpha_{11}m\ddot{q}_1 + \alpha_{12}m\ddot{q}_2 + \alpha_{13}m\ddot{q}_3 + q_1 = \alpha_{11}F + \alpha_{12}(2F) \\ \alpha_{21}m\ddot{q}_1 + \alpha_{22}m\ddot{q}_2 + \alpha_{23}m\ddot{q}_3 + q_2 = \alpha_{21}F + \alpha_{22}(2F) \\ \alpha_{31}m\ddot{q}_1 + \alpha_{32}m\ddot{q}_2 + \alpha_{33}m\ddot{q}_3 + q_3 = \alpha_{31}F + \alpha_{32}(2F). \end{cases}$$

For complex geometry the coefficients of influence are determined using the Mohr–Maxwell method. ▲

EXAMPLE 3.7 Find the mathematical model for system shown in Fig. 3.12 (only flexural vibrations are considered). The stiffness of the rods is EI.

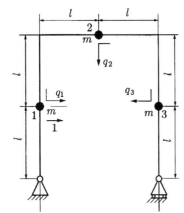

Figure 3.12
System in flexural vibrations.

Solution

Place the force **1** at section 2 ($\alpha_{12} = \alpha_{21}$) and at section 3 ($\alpha_{13} = \alpha_{31}$). Because of symmetry, $\alpha_{12} = \alpha_{23} = \alpha_{32}$. ▲

3.2.2 LAGRANGE METHOD

For a mechanical model with n degrees of freedom, with n generalized coordinates $q_1, \ldots, q_n$, the Lagrange equations are

$$\frac{d}{dt}\left(\frac{\partial T}{\partial \dot{q}_j}\right) = \frac{\partial T}{\partial q_j} = Q_j,$$

where t is the total kinetic energy of the mechanical model, and Q_j is the generalized force associated with the coordinate q_j.

Kinetic Energy

The kinetic energy is given by

$$T = T_{MP} + T_{RB}, \qquad (3.11)$$

where T_{RB} is the total kinetic energy of the rigid bodies, and T_{MP} is the total kinetic energy of the material points.

The kinetic energy of the material points is

$$T_{MP} = \sum_i \frac{m_i \mathbf{v}_i^2}{2},$$

(3.12)

where m_i are the masses of the MP and $\mathbf{v}_i$ are the velocities of the m_i. For a rigid body (RB), the kinetic energy is computed using the generalized relation

$$T = \frac{M \mathbf{v}_0^2}{2} + M(\mathbf{v}_0, \boldsymbol{\omega}, \mathbf{r}_c) + \frac{1}{2} \boldsymbol{\omega} [\vec{\mathbf{J}}_0 \cdot \boldsymbol{\omega}],$$

(3.13)

where M is the mass of the rigid body, $\mathbf{v}_0$ is the velocity of the origin O of a reference frame, J_0 is the tensor (matrix) of inertia of the rigid body with respect to O, $\boldsymbol{\omega}$ is the angular velocity vector of the rigid body, and $\mathbf{r}_c$ is the position vector of the mass center of the rigid body with respect to O. For k rigid bodies the kinetic energy is

$$T_{RB} = \sum_k \left\{ \frac{M_k \mathbf{v}_{0k}^2}{2} + M_k(\mathbf{v}_{0k}, \boldsymbol{\omega}_k, \mathbf{r}_{ck}) + \frac{1}{2} \boldsymbol{\omega}_k [\mathbf{J}_{0k} \cdot \boldsymbol{\omega}_k] \right\}.$$

(3.14)

Particular cases for rigid body motion:

■ Rotation motion:

$$T = \frac{J_\Delta \omega^2}{2},$$

(3.15)

where J_Δ is the moment of inertia of RB with respect to a fixed axis of revolution Δ

■ Planar motion:

$$T = \frac{M \mathbf{v}_c^2}{2} + \frac{J_c \omega^2}{2},$$

(3.16)

where $\mathbf{v}_c$ is the velocity of the mass center and J_c is the moment of inertia with respect to a perpendicular axis on the plane of motion at its center of mass.

Generalized Force Q_j

The generalized force Q_j is

$$Q_j = Q_j^{(E)} + Q_j^{(D)} + Q_j^{(F)}$$

(3.17)

where $Q_j^{(E)}$ is the component of generalized force due to the Hooke models, $Q_j^{(D)}$ is the component of generalized force due to the dissipating force, in particular linear damping, and $Q_j^{(F)}$ is the component of the generalized force due to the external forces (weight forces, technological forces etc.). If the Hooke model with the elastic constant k is between two points of position vectors $\mathbf{r}_1$ and $\mathbf{r}_2$, the elastic potential of the model is

$$V^{(E)} = \frac{k(\mathbf{r}_1 - \mathbf{r}_2)^2}{2}$$

(3.18)

and

$$Q_j^{(E)} = -\frac{\partial}{\partial q_j}\left(\sum_l V_l^{(E)}\right) = -\frac{\partial}{\partial q_j}\left[\sum_l \frac{k_l(\mathbf{r}_{1l} - \mathbf{r}_{2l})^2}{2}\right]. \qquad (3.19)$$

If linear damping c is between two point with velocities $\mathbf{v}_1$ and $\mathbf{v}_2$, then the Rayleigh function of dissipation is

$$D = \frac{c(v_1 - \mathbf{v}_2)^2}{2} \qquad (3.20)$$

and

$$Q_j^{(D)} = -\frac{\partial}{\partial \dot{q}_j}\left[\sum_m D_m\right] = \frac{\partial}{\partial \dot{q}_j}\left[\sum_m \frac{c_m(\mathbf{v}_{1m} - \mathbf{v}_{2m})^2}{2}\right]. \qquad (3.21)$$

The components $Q_j^{(F)}$ are computed using the general relation

$$Q_j^{(F)} = \sum_p \mathbf{F}_p \cdot \frac{\partial \mathbf{r}_p}{\partial q_j}, \qquad (3.22)$$

where $\mathbf{r}_p$ is the position vector of the force $\mathbf{F}_p$.

3.2.3 THE DERIVATIVE OF ANGULAR MOMENTUM METHOD

For the model with rotors in Fig. 3.13, the equations of motion are

$$J_i\ddot{q}_i = -k_{i-1}(q_i - q_{i-1}) - c_{i-1}(\dot{q}_i - \dot{q}_{i-1}) - k_i(q_i - q_{i+1})$$
$$- c_i(\dot{q}_i - \dot{q}_{i+1}) + M(t), \qquad (3.23)$$

where J_i is the inertia moment with respect to its axis.

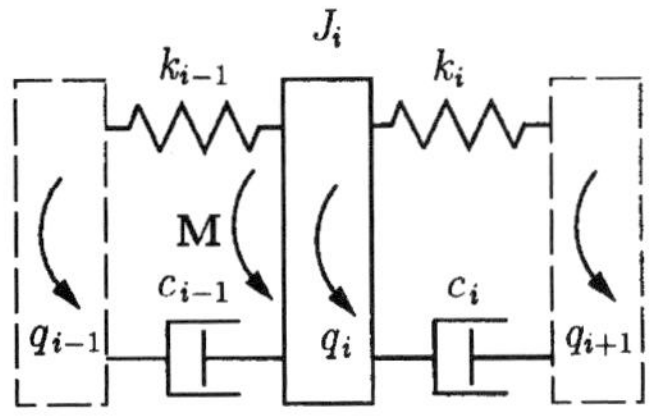

Figure 3.13
Model with rotors.

EXAMPLE 3.8   Determine the mathematical model of torsional vibrations for the physical model shown in Fig. 3.14. The friction torque and the viscous damping coefficients will be considered. The viscous damping coefficients are c_1, c_2, c_3, and c_4.

Solution

The elastic constants of the shaft sections are

$$k_i = \frac{GI_{pi}}{L_i}, \qquad i = 1, 2, 3, 4,$$

where $I_{pi} = \pi d_i^3/16$.

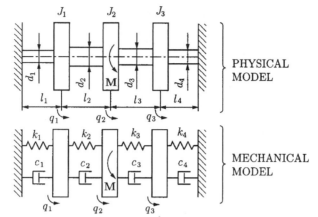

Figure 3.14
Physical and mechanical model of torsional vibrations.

With the help of Eq. (3.23) the following equations of motion are obtained:

$$\begin{cases} J_1\ddot{q}_1 = -k_1 q_1 - c_1\dot{q}_1 - k_2(q_1 - q_2) - c_2(\dot{q}_1 - \dot{q}_2) \\ J_2\ddot{q}_2 = -k_2(q_2 - q_1) - c_2(\dot{q}_2 - \dot{q}_1) - k_3(q_2 - q_3) - c_3(\dot{q}_2 - \dot{q}_3) + M(t) \quad \blacktriangle \\ J_3\ddot{q}_3 = -k_3(q_3 - q_2) - c_3(\dot{q}_3 - \dot{q}_2) - k_4 q_3 - c_4\dot{q}_3. \end{cases}$$

EXAMPLE 3.9 Write the mathematical model of the vibrations of the tool system in a shaping machine. The mechanical model is shown in Fig. 3.15.

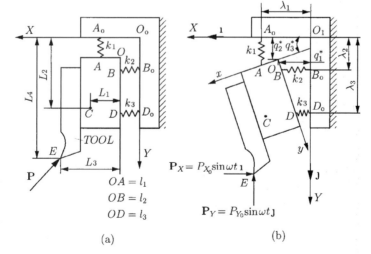

Figure 3.15
Tool assembly in a shaping machine. (a) Physical model; (b) mechanical model.

$OA = l_1$
$OB = l_2$
$OD = l_3$

$P_X = P_{X_0}\sin\omega t$
$P_Y = P_{Y_0}\sin\omega t$

(a) (b)

Solution

The Lagrange method is used. The generalized coordinates (Fig. 3.15b) are

$$q_1 = \Delta q_1^* = q_1^* - q_{10}^*, \qquad q_2 = \Delta q_2^* = q_2^* - q_{20}^*, \qquad q_3 = \Delta q_3^* = \Delta q_3^* - q_{30}^*,$$

Vibration

where q_{10}^*, q_{20}^*, q_{30}^* are the position of the system in static stable equilibrium, when the $Px_0\mathbf{1}$ and $Py_0\mathbf{J}$ act on the system. Therefore, we have a vibration around a stable equilibrium position. The relations of transformation from the reference xOy to the fixed reference frame XO_1YZ are

$$\begin{bmatrix} X_A \\ Y_A \end{bmatrix} = \begin{bmatrix} q_1^* \\ q_2^* \end{bmatrix} + [M] \begin{bmatrix} l_1 \\ 0 \end{bmatrix}, \qquad \begin{bmatrix} X_B \\ Y_B \end{bmatrix} = \begin{bmatrix} q_1^* \\ q_2^* \end{bmatrix} + [M] \begin{bmatrix} 0 \\ l_2 \end{bmatrix}$$

$$\begin{bmatrix} X_D \\ Y_D \end{bmatrix} = \begin{bmatrix} q_1^* \\ q_2^* \end{bmatrix} + [M] \begin{bmatrix} 0 \\ l_3 \end{bmatrix}, \qquad \begin{bmatrix} X_C \\ Y_C \end{bmatrix} = \begin{bmatrix} q_1^* \\ q_2^* \end{bmatrix} + [M] \begin{bmatrix} L_1 \\ L_2 \end{bmatrix}$$

$$\begin{bmatrix} X_E \\ Y_E \end{bmatrix} = \begin{bmatrix} q_1^* \\ q_2^* \end{bmatrix} + [M] \begin{bmatrix} L_3 \\ L_4 \end{bmatrix},$$

where

$$[M] = \begin{bmatrix} \cos q_3^* & -\sin q_3^* \\ \sin q_3^* & \cos q_3^* \end{bmatrix}$$

is the rotation matrix.

The following relations result:

$$\begin{cases} X_A = q_1^* + l_1 \cos q_3^* \\ Y_A = q_2^* + l_1 \sin q_3^* \end{cases}, \qquad \begin{cases} X_B = q_1^* - l_2 \sin q_3^* \\ Y_B = q_2^* + l_2 \cos q_3^* \end{cases}, \qquad \begin{cases} X_D = q_1^* - l_3 \sin q_3^* \\ Y_D = q_2^* + l_3 \cos q_3^* \end{cases}$$

$$\begin{cases} X_E = q_1^* + L_3 \cos q_3^* - L_4 \sin q_3^* \\ Y_E = q_2^* + L_3 \sin q_3^* + L_4 \cos q_3^* \end{cases}, \qquad \begin{cases} X_C = q_1^* + L_1 \cos q_3^* - L_2 \sin q_3^* \\ Y_C = q_2^* + L_1 \sin q_3^* + L_2 \cos q_3^* \end{cases}.$$

The kinetic energy is

$$T = T_{RB} = \frac{M\mathbf{v}_c^2}{2} + \frac{J_c\omega^2}{2},$$

where

$$\mathbf{v}_c = \dot{X}_C\mathbf{1} + \dot{Y}_C\mathbf{J} = [\dot{q}_1^* - L_1 \sin q_3^* \dot{q}_3^* - L_2 \cos q_3^* \dot{q}_3^*]\mathbf{1}$$
$$+ [\dot{q}_2^* + L_1 \cos q_3^* \dot{q}_3^* - L_2 \sin q_3^* \dot{q}_3^*]\mathbf{J}.$$

With $\dot{q}_3^* = \dot{q}_3$, $\dot{q}_2^* = \dot{q}_2$, $\dot{q}_1^* = \dot{q}_1$,

$$\mathbf{v}_c = [\dot{q}_1 - L_1 \sin(\dot{q}_{30}^* + q_3)\dot{q}_3 - L_2 \cos(\dot{q}_{30}^* + q_3)\dot{q}_3]\mathbf{1}$$
$$+ [\dot{q}_2 + L_1 \cos(\dot{q}_{30}^* + q_3)\dot{q}_3 - L_2 \sin(\dot{q}_{30}^* + q_3)\dot{q}_3]\mathbf{J}.$$

The angular velocity is

$$\boldsymbol{\omega} = \dot{q}_3^*\mathbf{k} = \dot{q}_3\mathbf{k},$$

where $\mathbf{k} = \mathbf{1} \times \mathbf{J}$. The following trigonometric relations are known:

$$\sin(q_{30}^* + q_3) = \sin q_{30}^* \cos q_3 + \cos q_{30}^* \sin q_3$$
$$\cos(q_{30}^* + q_3) = \cos q_{30}^* \cos q_3 - \sin q_{30}^* \sin q_3.$$

For small oscillations $q_3 \to 0$, on $\cos q_3 \cong 1$, $\sin q_3 \cong q_3$,

$$\sin(q_{30}^* + q_3) = \sin q_{30}^* + q_3 \cos q_{30}^*$$
$$\cos(q_{30}^* + q_3) = \cos q_{30}^* - q_3 \sin q_{30}^*.$$

The kinetic energy becomes

$$T = \frac{M}{2}[\dot{q}^2 + \dot{q}_2^2 + L_1^2 + L_2^2)\dot{q}_3^2 - 2(L_1 \sin \dot{q}_{30}^* + L_2 \cos \dot{q}_{30}^*)\dot{q}_1\dot{q}_3]$$

$$+ M[(L_1 \cos q_{30}^* - L_2 \sin q_{30}^*)\dot{q}_2\dot{q}_3] + \frac{J_C\dot{q}_3^2}{2}.$$

The symbol Δ is the variation of distance from the instant position (at time t) to the static equilibrium position, and one can write

$$\Delta(\sin q_3^*) = \sin q_3^* - \sin q_{30}^* \cong \frac{1}{1!}\cos q_{30}^*(q_3^* - q_{30}^*) = q_3 \cos q_{30}^*,$$

$$\Delta(\cos q_3^*) = -q_3 \sin q_{30}^*,$$

where the Taylor series has been used:

$$\mathbf{AA}_0 = (X_A - \lambda_1)\mathbf{1} + Y_A\mathbf{J} = (q_1^* - L_1 \cos q_3^* - \lambda_1)\mathbf{1} + (q_2^* + l_1 \sin q_3^*)\mathbf{J}$$

$$\mathbf{BB}_0 = (q_1^* - l_2 \sin q_3^*)\mathbf{1} + (q_2^* + l_2 \cos q_3^* - \lambda_2)\mathbf{J}$$

$$\mathbf{DD}_0 = (q_1^* - L_3 \sin q_3^*)\mathbf{1} + (q_2^* + l_2 \cos q_3^* - \lambda_3)\mathbf{J}$$

$$\Delta\mathbf{AA}_0 = (q_1 - l_1 q_3 \sin q_{30}^*)\mathbf{1} + (q_2 + l_1 q_3 \cos q_{30}^*)\mathbf{J}$$

$$\Delta\mathbf{BB}_0 = (q_1 - l_2 q_3 \sin q_{30}^*)\mathbf{1} + (q_2 - l_2 q_3 \cos q_{30}^*)\mathbf{J}$$

$$\Delta\mathbf{DD}_0 = (q_1 - l_3 q_3 \sin q_{30}^*)\mathbf{1} + (q_2 - l_3 q_3 \cos q_{30}^*)\mathbf{J}.$$

The potential of the elastic forces is

$$V^{(E)}\frac{k_1}{2}(\Delta\mathbf{AA}_0)^2 + \frac{k_2}{2}(\Delta\mathbf{BB}_0)^2 + \frac{k_3}{2}(\Delta\mathbf{DD}_0)^2.$$

After elementary calculation,

$$V^{(E)} = \frac{k_1}{2}[q_1^2 + q_2^2 + l_1^2 q_3^2 + 2l_1 q_3(q_2 \cos q_{30}^* - q_1 \sin q_{30}^*)]$$

$$+ \frac{k_2}{2}[q_1^2 + q_2^2 + l_2^2 q_3^2 - 2l_2 q_3(q_1 \sin q_{30}^* + q_2 \cos q_{30}^*)]$$

$$+ \frac{k_3}{2}[q_1^2 + q_2^2 + l_3^2 q_3^2 - 2l_3 q_3(q_1 \sin q_{30}^* + q_2 \cos q_{30}^*)].$$

The generalized forces corresponding to the cutting force $\mathbf{P} = -P_X\mathbf{1} - P_Y\mathbf{J}$, where $P_X = P_{X0} \sin(\omega t)$, $P_Y = P_{Y0} \sin(\omega t)$, are

$$Q_1 = \mathbf{P}\frac{\partial\mathbf{R}_E}{\partial q_1}, \quad Q_2 = \mathbf{P}\frac{\partial\mathbf{R}_E}{\partial q_2}, \quad Q_3 = \mathbf{P}\frac{\partial\mathbf{R}_E}{\partial q_3},$$

where

$$\mathbf{R}_E = (q_1^* + L_3 \cos q_3^* - L_4 \sin q_3^*)\mathbf{1} + (q_2^* + L_3 \sin q_3^* + L_4 \cos q_3^*)\mathbf{J}$$

$$q_1^* = q_{10}^* + q_1, \quad q_2^* = q_{20}^* + q_2, \quad q_3^* = q_{30}^* + q_3.$$

The generalized forces are

$$Q_1 = -P_{X0} \sin(\omega t), \quad Q_2 = -P_{Y0} \sin(\omega t)$$

$$Q_3 = [L_3(\sin q_{30}^* + q_3 \cos q_{30}^*) + L_4(\cos q_{30}^* - q_3 \sin q_{30}^*)]P_{X0} \sin(\omega t)$$

$$+ [L_4(\sin q_{30}^* + q_3 \cos q_{30}^*) - L_3(\cos q_{30}^* - q_3 \sin q_{30}^*)]P_{Y0} \sin(\omega t).$$

Vibration

The weight is negligible for this example.

The Lagrange equations are

$$\frac{d}{dt}\left(\frac{\partial T}{\partial \dot{q}_j}\right) - \frac{\partial T}{\partial q_j} = -\frac{\partial V^{(E)}}{\partial q_j} + Q_j, \qquad j = 1, 2, 3,$$

and the linear system with constant coefficients is

$$\begin{cases} M\ddot{q}_1 - M(L_1 \sin q_{30}^* + L_2 \cos q_{30}^*)\ddot{q}_3 + (k_1 + k_2 + k_3)q_1 \\ \quad -2(k_1 l_1 + k_2 l_2 + k_3 l_3)q_3 \sin q_{30}^* = -P_{X0}\sin(\omega t) \\ M\ddot{q}_2 - M(L_1 \cos q_{30}^* + L_2 \sin q_{30}^*)\ddot{q}_3 + (k_1 + k_2 + k_3)q_2 \\ \quad +2(k_1 l_1 - k_2 l_2 - k_3 l_3)q_3 \cos q_{30}^* = -P_{Y0}\sin(\omega t) \\ [M(L_1^2 + L_2^2) + J_c]\ddot{q}_3 - M(L_1 \sin q_{30}^* + L_2 \cos q_{30}^*)\ddot{q}_1 \\ \quad +M(L_1 \cos q_{30}^* - L_2 \sin q_{30}^*)\ddot{q}_2 + (k_1 l_1^2 + k_2 l_2^2 + k_3 l_3^2)q_3 \\ \quad -(k_1 l_1 + k_2 l_2 + k_3 l_3)q_1 \sin q_{30}^* + (k_1 l_1 - k_2 l_2 - k_3 l_3)q_2 \cos q_{30}^* \\ \quad = [L_3(\sin q_{30}^* + q_3 \cos q_{30}^*) + L_4(\cos q_{30}^* - q_3 \sin q_{30}^*)]P_{X0}\sin(\omega t) \\ \quad +[L_4(\sin q_{30}^* + q_3 \cos q_{30}^*) - L_3(\cos q_{30}^* - q_3 \sin q_{30}^*)]P_{Y0}\sin(\omega t), \end{cases}$$

which represents the mathematical model of motion. The values q_{10}^*, q_{20}^*, q_{30}^*, result from the static equilibrium position under the action of the force $\mathbf{P}_0 = -P_{X0}\mathbf{1} - P_{Y0}\mathbf{J}$. ▲

EXAMPLE 3.10 Determine the mathematical model of motion for the mass m in Fig. 3.16.

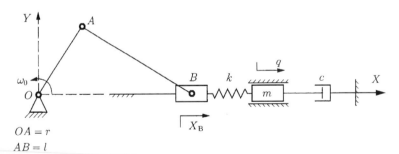

Figure 3.16
Model with one $OA = r$
DOF. $AB = l$

Solution

Using the Newton method for the mass m results in

$$m\ddot{q} = -k(q - X_B) - c\dot{q}.$$

With

$$X_B = r\cos(\omega_0 t) + l\sqrt{1 - \lambda^2 \sin^2(\omega_0 t)} \quad \text{and} \quad \lambda = \frac{r}{l},$$

the equation of motion becomes

$$m\ddot{q} + c\dot{q} + kq = k\left[r\cos(\omega_0 t) + l\sqrt{1 - \lambda^2 \sin^2(\omega_0 t)}\right]. \quad \blacktriangle$$

3.2.4 CONCLUSIONS ABOUT MATHEMATICAL MODELS OF LINEAR VIBRATIONS WITH A FINITE NUMBER OF DEGREES OF FREEDOM

Vibration with One Degree of Freedom q(t)

The mathematical model is a differential equation with constant coefficients,

$$\mu\ddot{q} + c\dot{q} + kq = F(t), \tag{3.24}$$

where μ can be a mass or inertia moment (according to q a linear displacement or angular displacement), c is the coefficient of viscous (damping) resistance, k is the elastic constant (spring stiffness), and F is the exciting (disturbing) force.

With the notations

$$\frac{c}{\mu} = 2n, \qquad \frac{k}{\mu} = \omega_n^2, \qquad \phi(t) = \frac{1}{1\mu}F(t),$$

the mathematical model becomes

$$\ddot{q} + 2n\dot{q} + \omega_n^2 q = \phi(t). \tag{3.25}$$

Vibration with a Finite Number of Degrees of Freedom

The mathematical model is a system of linear differential equation system with constant coefficients of the type

$$\sum_{j=1}^{n} [m_{ij}\ddot{q}_j + d_{ij}\dot{q}_j + r_{ij}q_j] = F_i(t), \qquad i = 1, \ldots, n. \tag{3.26}$$

The system given by Eq. (3.26) can be written in a matrix form

$$[M][\ddot{q}] + [D][\dot{q}] + [R][q] = [F], \tag{3.27}$$

where $[M] = [m_{ij}]$ is called the inertia matrix (its elements are masses or inertia moments), $[D] = [d_{ij}]$ is called the damping matrix (its elements are coefficients of damping), $[R] = [r_{ij}]$ is called the stiffness matrix (its elements are elastic constants), $[q] = [q_1, \ldots, q_n]^T$, $[F] = [F_1, \ldots, F_n]^T$.

Vibration

EXAMPLE 3.11 For the mathematical model from Example 3.8, the matrices are

$$[M] = \begin{bmatrix} J_1 & 0 & 0 \\ 0 & J_2 & 0 \\ 0 & 0 & J_3 \end{bmatrix}, \quad [D] = \begin{bmatrix} c_1 + c_2 & -c_2 & 0 \\ -c_2 & c_2 + c_3 & -c_3 \\ 0 & -C_3 & c_3 + c_4 \end{bmatrix}$$

$$[R] = \begin{bmatrix} k_1 + k_2 & -k_2 & 0 \\ -K_2 & k_2 + k_3 & -k_3 \\ 0 & -k_3 & k_3 + k_4 \end{bmatrix}, \quad [q] = \begin{bmatrix} q_1 \\ q_2 \\ q_3 \end{bmatrix},$$

$$[F] = \begin{bmatrix} 0 \\ M(t) \\ 0 \end{bmatrix}.$$

In conclusion, from Eq. (3.27) one can classify the vibrations:

$$[F] \equiv 0 \Rightarrow \text{free vibrations}$$
$$[F] \neq 0 \Rightarrow \text{force dvibrations}$$
$$[D] \equiv 0 \Rightarrow \text{undamped vibrations}$$
$$[D] \neq 0 \Rightarrow \text{damped vibrations.} \quad \blacktriangle$$

3.3 System Model

3.3.1 VIBRATIONS WITH ONE DEGREE OF FREEDOM

Mathematical models from Eqs. (3.24) and (3.25) are open monovariable systems (with one input and one output). The input value is $i(t) = F(t)$ or, respectively, $i(t) = \phi(t)$. The output value is $e(t) = q(t)$. The mathematical model can be written as

$$\dot{\mathbf{x}} = [A]\mathbf{x} + [B]\mathbf{1}$$
$$\mathbf{e} = [C]\mathbf{x},$$
(3.28)

where

$$[A] = \begin{bmatrix} 0 & 1 \\ -\omega_n^2 & -2n \end{bmatrix}, \ [B] = \begin{bmatrix} 0 \\ 1 \end{bmatrix}, \ [C] = \begin{bmatrix} 1 & 0 \end{bmatrix}, \ i = \phi(t), \ e = q(t).$$

3.3.2 VIBRATION WITH A FINITE NUMBER OF DEGREES OF FREEDOM

The mathematical model of Eqs. (3.26) and (3.27) are also open monovariable linear systems. The input vector is $\mathbf{1} = [F]$, the output vector is $\mathbf{e} = [q]$ and the state vector is

$$\mathbf{x} = \begin{bmatrix} [q] \\ [\dot{q}] \end{bmatrix}.$$

From the matrix form of Eq. (3.27), the canonical form results:

$$\dot{\mathbf{x}} = [A]\mathbf{x} + [B]\phi$$
$$[q] = [C]\mathbf{x}.$$
(3.29)

Here

$$[A]_{2n\times 2n} = \begin{bmatrix} [0]_{n\times n} & [I]_{n\times n} \\ -[M]^{-1}[R] & -[M]^{-1}[D] \end{bmatrix}, \qquad [B]_{2n\times 2n} = \begin{bmatrix} [0]_{n\times n} & [0]_{n\times n} \\ [I]_{n\times n} & [0]_{n\times n} \end{bmatrix}$$

$$\phi = [\phi] = \begin{bmatrix} [M]^{1}[F] \\ [0]_{n\times 1} \end{bmatrix}, \qquad [C] = \big[[I]_{n\times n} \ \ [0]_{n\times n}\big].$$

3.4 Analysis of System Model

3.4.1 DYNAMICAL RESPONSE

For a vibration with one degree of freedom, the dynamic response [solution of Eq. (3.27)] is

$$\mathbf{x}(t) = e^{[A]t}\mathbf{x}_0 + \int_0^t e^{[A](t-\tau)}[B]\phi(\tau)d\tau$$

$$q(t) = [C]\mathbf{x}, \qquad (3.30)$$

where $[A]$, $[B]$, $[C]$ are given in Eq. (3.27), and $\mathbf{x}_0 = \big[q(0)/\dot{q}(0)\big]$.

For vibrations with more than one degree of freedom, the dynamic response [solution of Eq. (3.29)] is given by

$$\mathbf{x(t)} = E^{[A]t}\mathbf{x}_0 + \int_0^t e^{[A](t-\tau)}[B]\cdot\phi(\tau)d\tau$$

$$[q] = [C]\mathbf{x}, \qquad (3.31)$$

where the matrices $[A]$, $[B]$, $[C]$, $\phi = [\phi]$ are given by Eqs. (3.29) and $\mathbf{x}_0 = [q_1(0), \ldots, q_n(0), \dot{q}_1(0), \ldots, \dot{q}_n(0)]^T$.

3.4.2 QUALITATIVE ANALYSIS OF THE SYSTEM MODEL

Natural Frequency of Vibrant System

Consider one free undamped system with n degrees of freedom. The mathematical model, Eq. (3.26) with $[D] = 0$ and $[F] = 0$, is

$$[M][\ddot{q}] + [R][q] = 0. \qquad (3.32)$$

In this model the motion of the system is "disconnected" from the environment. This is natural motion. The coincidence of exciting angular frequency with natural frequency leads to a phenomenon called resonance. For a system with n degrees of freedom, the dynamic response given by Eqs. (3.31) and (3.31) becomes

$$\mathbf{x}(t) = e^{[A]t}\mathbf{x}_0$$

$$[q] = [C]\mathbf{x}(t). \qquad (3.33)$$

In this particular case,

$$[A] = \begin{bmatrix} [0]_{n\times n} & [I]_{n\times n} \\ -[M]^{-1}[R] & [0]_{n\times n} \end{bmatrix}. \qquad (3.34)$$

But

$$e^{[A]t} = \mathcal{L}^{-1}([[I]s - [A]]^{-1}) = \mathcal{L}^{-1}\left(\frac{adj[[I]s - [A]]}{P(s)}\right), \qquad (3.35)$$

with $P(s) = \det[[I]s - [A]]$ (characteristic polynomial). For the roots of equation $P(j\omega) = \det[[I]s - [A]]$ we have $e^{[A]t} \rightarrow \infty$ and

$$P(s) = \det\begin{bmatrix} s[I]_{n\times n} & -[I]_{n\times n} \\ [M]^{-1}[R] & s[I]_{n\times n} \end{bmatrix}. \qquad (3.36)$$

With $s \rightarrow j\omega$ and $P(j\omega) = 0$, the equation for the natural frequency of a system is

$$\det[[R] - \omega^2[M]] = 0, \qquad (3.37)$$

where the solutions are the natural frequencies $\omega_1, \omega_2, \ldots, \omega_n$ of the vibrant system.

EXAMPLE 3.12 Determine the equation of the natural frequency for the vibrant system presented in Example 3.4.

Solution

In this case the matrices are

$$[M] = \begin{bmatrix} m & 0 \\ 0 & 2m \end{bmatrix}, \qquad [R] = \begin{bmatrix} 3k & -2k \\ -2k & 3k \end{bmatrix}.$$

Equation (3.38) becomes

$$\det\left[\begin{bmatrix} 3k & -2k \\ -2k & 3k \end{bmatrix} - \omega^2\begin{bmatrix} m & 0 \\ 0 & 2m \end{bmatrix}\right] = 0,$$

or

$$\begin{bmatrix} 3k - \omega^2 m & -2k \\ -2k & 3k - 2m\omega^2 \end{bmatrix} = 0,$$

or

$$2m^2\omega^4 - 9mk\omega^2 + 5k^2 = 0,$$

which gives the natural frequencies

$$\omega_1 = 2.775\sqrt{\frac{k}{m}}\ [s^{-1}], \qquad \omega_2 = 1.139\sqrt{\frac{k}{m}}\ [s^{-1}]$$

and the frequencies

$$f_1 = 0.442\sqrt{\frac{k}{m}}\ [\text{Hz}], \qquad f_2 = 0.181\sqrt{\frac{k}{m}}\ [\text{Hz}].$$

In general, for systems with two degrees of freedom, the equation of the natural frequency is a fourth-order equation. ▲

EXAMPLE 3.13 Determine the equation of the natural frequency for the vibrant system presented in Example 3.8.

Solution

In this particular case the matrix are

$$[M] = \begin{bmatrix} J_1 & 0 & 0 \\ 0 & J_2 & 0 \\ 0 & 0 & J_3 \end{bmatrix}, \qquad [R] = \begin{bmatrix} k_1 + k_2 & -k_2 & 0 \\ -k_2 & k_2 + k_3 & -k_3 \\ 0 & -k_3 & k_3 + k_4 \end{bmatrix}.$$

The equation of the natural frequencies is

$$\begin{bmatrix} (k_1 + k_2) - \omega^2 J_1 & -k_2 & 0 \\ -k_2 & (k_2 + k_3) - \omega^2 J_2 & -k_3 \\ 0 & -K_3 & (k_3 + k_4) - \omega^2 J_3 \end{bmatrix} = 0.$$

An equation of sixth order in ω is obtained, which through the substitution $\omega^2 = u$ leads to an equation of order 3 in u. ▲

3.5 Approximative Methods for Natural Frequencies

3.5.1 HOLZER METHOD

The elastic force in a Hooke model (Fig. 3.17) can be

$$F_{j-1} = k_{j-1}(q_{j-1} - q_j), \tag{3.38}$$

Figure 3.17
The elastic force in a Hooke model.

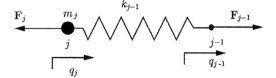

which results in the recurrence relation

$$q_j = q_{j-i} - \frac{F_{j-1}}{k_{j-1}}. \tag{3.39}$$

The equation of motion for the mass m_j is

$$m_j \ddot{q}_j = F_{j-1} - F_j. \tag{3.40}$$

With $q_j(t) = q_j \sin(\omega t + \varphi)$ one can obtain $-\omega^2 m_j q_j = F_{j-1} - F_j$, and the result is the relation of recurrence,

$$\begin{cases} q_j = q_{j-i} - \dfrac{F_{j-1}}{k_{j-1}} \\[2mm] F_j = F_{j-1} + m_j \omega^2 q_j. \end{cases} \tag{3.41}$$

Equations (3.41) are the fundamental relations of the recurrence of the Holzer method for mechanical models with lumped masses and Hooke models. In principle, the Holzer method consists in writing successive recurrence relations under the condition that a displacement or a force must be zero.

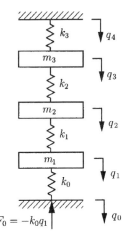

Figure 3.18
Mechanical model for Example 3.14.

$F_0 = -k_0 q_1$

EXAMPLE 3.14 Using the Holzer method, determine the natural frequencies for the mechanical model in Fig. 3.18.

Solution

Write successive Eqs. (3.41):

$$F_1 = q_1(m_1\omega^2 - k_0)$$

$$\begin{cases} q_1 = q_0 - \dfrac{F_0}{k_0} \\ F_1 = F_0 + m_1\omega^2 q_1. \end{cases}$$

But $q_0 = 0$, and $F_0 = -k_0 q_1$ results in

$$\begin{cases} q_2 = q_1 - \dfrac{F_1}{k_1} \\ F_2 = F_1 + m_2\omega^2 q_2. \end{cases}$$

Replacing F_1 as a function of q_1 and q_2 in an F_2 expression, we obtain

$$F_2 = q_1\left[m_1\omega^2 - k_0 + m_2\omega^2\left(1 - \frac{1}{k_1}(m_1\omega^2 - k_0)\right)\right]$$

$$\begin{cases} q_3 = q_2 - \dfrac{F_2}{k_2} \\ F_3 = F_2 + m_3\omega^2 q_3. \end{cases}$$

Replacing q_2 as a function of q_1 and F_2 as a function of q_1 in the expression for q_3, and then in the F_3 expression, one can obtain F_3:

$$\begin{cases} q_4 = q_3 - \dfrac{F_2}{k_2} \\ F_4 = F_3 + m_4\omega^2 q_4. \end{cases}$$

Replacing q_3 as a function of q_1 and F_3 as a function of q_1 in the expression for q_4, and $q_4 = 0$, in the end one can calculate F_4. ▲

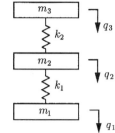

Figure 3.19
Mechanical model for Example 3.15.

EXAMPLE 3.15 Using the Holzer method, determine the natural frequencies of the mechanical model in Fig. 3.19.

Solution

Write successive Eqs. (3.41) to obtain

$$\begin{cases} q_2 = q_1 - \dfrac{F_1}{k_1} \\ F_1 = F_0 + m_1\omega^2 q_1. \end{cases}$$

But $F_0 = 0$, so $q_2 = q_1(1 - m\omega^2/k_1)$ and

$$\begin{cases} q_3 = q_2 - \dfrac{F_2}{k_2} \\ F_2 = F_1 + m_2\omega^2 q_2. \end{cases}$$

Replacing q_2 as a function of q_1 and F_2 as a function of q_1 in the expression for q_3, one obtains

$$F_3 = F_2 + m_3\omega^2 q_3.$$

Replacing F_2 as a function of q_1 and q_3 as a function of q_1 in the expression for F_3, and given that $F_3 = 0$, we obtain the expression for the natural frequencies of the mechanical system. ▲

3.5.2 APPLICATION OF THE HOLZER METHOD TO A MECHANICAL MODEL WITH ROTORS AND HOOKE MODELS

In this case, vectors of state attached to a shaft section are used. One vector of state is $[z] = [\varphi, M]^T$, where φ is the angle of rotation of the section and M is the torsion moment (torque) in some section.

Because in the zone of the rotor the diagram of moments is discontinuous, $M_i^l \neq M_i^r$, where M_i^l is the torsion moment at left and M_i^r at right (Fig. 3.20). Because the rotor is rigid, $\varphi_i^l = \varphi_i^r = \varphi_i$ (the rotation at left is equal to the rotation at right).

Vectors of state at left and right of section i are

$$[z]_i^l = \begin{bmatrix} \varphi_i^l \\ M_i^l \end{bmatrix}, \quad \text{and} \quad [z]_i^r = \begin{bmatrix} \varphi_i^r \\ M_i^r \end{bmatrix}.$$

The equation of motion of rotor i is

$$J_i\ddot{\varphi}_i = M_i^r - M_i^l, \tag{3.42}$$

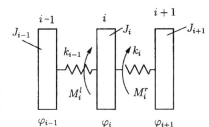

Figure 3.20
*Torsional
mechanical
model.*

and with $\varphi_i(t) = \varphi_i \sin(\omega t + \varphi)$, $M_i^l = M_i^l \sin(\omega t + \varphi)$, $M_i^r = M_i^r \sin(\omega t + \varphi)$, we obtain

$$-\omega^2 J_i \varphi_i = M_i^r - M_i^l. \tag{3.43}$$

Therefore,

$$\varphi_i^l = \varphi_i^r, \qquad M_i^r = M_i^l - \omega^2 J_i \varphi_i, \tag{3.44}$$

and the result is the recurrence relation between vectors of state,

$$[z]_i^r = [A_i][z]_i^l, \tag{3.45}$$

where

$$[A_i] = \begin{bmatrix} 1 & 0 \\ -\omega^2 J_i & 1 \end{bmatrix}.$$

For the zone between two successive rotors,

$$M_i^l = M_{i-1}^r, \qquad \varphi_i^l - \varphi_{i-1}^r = \frac{M_i^l}{k_{i-1}} = \frac{M_{i-1}^r}{k_{i-1}}. \tag{3.46}$$

There results the following recurrence relation between vectors of state at the extremity of a shaft section between two successive rotors:

$$[z]_i^l = [B_i][z]_{i-1}^r, \tag{3.47}$$

where

$$[B_i] = \begin{bmatrix} 1 & \dfrac{1}{k_{i-1}} \\ 0 & 1 \end{bmatrix}.$$

In conclusion, the recurrence relations between vectors of state are

$[z]_i^r = [A_i][z]_i^l$ in a second of rigid rotors

$[z]_i^l = [B_i][z]_{i-1}^r$ in a section between two successive rotors.

EXAMPLE 3.16 With the Holzer method, determine the natural frequencies for the mechanical model from Fig. 3.21.

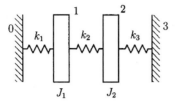

Figure 3.21
Mechanical model for Example 3.16.

Solution

We can write, successively,

$$[z]_3^l = [B_3][z]_2^r, \qquad [z]_2^r = [A_2][z]_2^l$$
$$[z]_2^l = [B_2][z]_1^r, \qquad [z]_1^r = [A_1][z]_1^l, \qquad [z]_1^l = [B_1][z]_0^r,$$

where

$$[B_3] = \begin{bmatrix} 1 & \dfrac{1}{k_3} \\ 0 & 1 \end{bmatrix}, \qquad [B_2] = \begin{bmatrix} 1 & \dfrac{1}{k_2} \\ 0 & 1 \end{bmatrix}, \qquad [B_1] = \begin{bmatrix} 1 & \dfrac{1}{k_1} \\ 0 & 1 \end{bmatrix}$$

$$[A_1] - \begin{bmatrix} 1 & 0 \\ -\omega^2 J_1 & 1 \end{bmatrix}, \qquad [A_2] = \begin{bmatrix} 1 & 0 \\ -\omega^2 J_2 & 1 \end{bmatrix}.$$

Reuniting the preceding relations, one can write

$$[z]_3^l = [B_3][A_2][B_2][A_1][B_1][z]_0^r,$$

but

$$[z]_3^l = \begin{bmatrix} 0 \\ M_3^l \end{bmatrix}, \qquad [z]_0^r = \begin{bmatrix} 0 \\ M_0^r \end{bmatrix},$$

and

$$\begin{bmatrix} 0 \\ M_3^l \end{bmatrix} = [B_3][A_2][B_2][A_1][B_1] \begin{bmatrix} 0 \\ M_0^r \end{bmatrix}.$$

With $M_0^r = 0$ one can obtain the equation for natural frequencies.   ▲

3.5.3 RAYLEIGH METHOD

The dynamic response of free undamped vibrations is an overlapping of harmonic vibrations with natural frequencies $\omega_1, \ldots, \omega_n$, resulting in

$$\begin{cases} q_1(t) = A_{11} \sin(\omega_1 t + \varphi_1) + \cdots + A_{1n} \sin(\omega_n t + \varphi_n) \\ q_2(t) = A_{21} \sin(w_1 t + \varphi_1) + \cdots + A_{2n} \sin(\omega_n t + \varphi_n) \\ \qquad\qquad\qquad \vdots \\ q_n(t) = A_{n1} \sin(\omega_1 t + \varphi_1) + \cdots + A_{nn} \sin(\omega_n t + \varphi_n). \end{cases} \tag{3.48}$$

One can denote

$$\begin{cases} A_{21} = \mu_{21} A_{11}, \ A_{31} = \mu_{31} A_{11}, \ldots, A_{n1} = \mu_{n1} A_{11} \\ A_{22} = \mu_{22} A_{12}, \ A_{32} = \mu_{32} A_{12}, \ldots, A_{n2} = \mu_{n2} A_{12} \\ \qquad\qquad\qquad\qquad \vdots \\ A_{2n} = \mu_{2n} A_{1n}, \ A_{3n} = \mu_{3n} A_{1n}, \ldots, A_{nn} = \mu_{nn} A_{1n}, \end{cases} \tag{3.49}$$

Vibration

where μ_{ik} are called the coefficients of distribution.

The vectors of distribution can be introduced:

$$[\mu]_1 = \begin{bmatrix} 1 \\ \mu_{21} \\ \vdots \\ \mu_{n1} \end{bmatrix}, \quad [\mu]_2 = \begin{bmatrix} 1 \\ \mu_{22} \\ \vdots \\ \mu_{n2} \end{bmatrix}, \dots, [\mu]_n = \begin{bmatrix} 1 \\ \mu_{2n} \\ \vdots \\ \mu_{nn} \end{bmatrix}.$$

It easy to demonstrate that the vectors of distribution $[\mu]_k$ verify the system

$$[[R] - \omega_k^2[M]][\mu]_k = [0]. \tag{3.50}$$

The natural modes of vibration associated with the natural frequency are the column vectors

$$[q]_k = \begin{bmatrix} A_{1k}\sin(\omega_k t + \varphi_k) \\ A_{2k}\sin(\omega_k t + \varphi_k) \\ \vdots \\ A_{nk}\sin(\omega_k t + \varphi_k) \end{bmatrix} = [\mu]_k A_{1k}\sin(\omega_k t + \varphi_k). \tag{3.51}$$

The kinetic energy of a system with more than one degree of freedom can be calculated using the relation

$$T = \tfrac{1}{2}[\dot{q}]^T[M][\dot{q}], \tag{3.52}$$

and the potential of elastic forces attached to Hooke models can be determined using the relation

$$V = \tfrac{1}{2}[q]^T[R][q]. \tag{3.53}$$

Corresponding to the k mode of vibration:

$$\text{Kinetic energy } T^{(k)} = \tfrac{1}{2}[\dot{q}]_k^T[M][\dot{q}]_k,$$
$$\text{Potential of elastic forces } V^{(k)} = \tfrac{1}{2}[q]_k^T[R][q]_k.$$

Replacing $[q]_k$ with Eq. (3.51), the following results are obtained:

$$T^{(k)} = \tfrac{1}{2}[\mu]_k^T[M][\mu]_k A_{1k}^2 \omega_k^2 \cos^2(\omega_k t + \varphi_k) \tag{3.54}$$

and

$$V^{(k)} = \tfrac{1}{2}[\mu]_k^T[R][\mu]_k A_{1k}^2 \sin^2(\omega_k t + \varphi_k). \tag{3.55}$$

The system is undamped; therefore, $T_{max}^{(k)} = V_{max}^{(k)}$, and there results the relation of calculus for the natural frequency,

$$\omega_k = \sqrt{\frac{[\mu]_k^T[R][\mu]_k}{[\mu]_k^T[M][\mu]_k}}. \tag{3.56}$$

The following is the methodology for working with the Rayleigh method:

■ Adopt an expression for $[\mu]_k$ and determine, using Eq. (3.56), a first value for ω_k.

■ With the natural frequency calculated in this way, introduce the torsors of inertia (forces or torques) and determine the new displacement, namely a new expression for $[\mu]_k$, which is reintroduced in Eq. (3.56). The Rayleigh method gives the minimum natural frequency superior to the real value, and maximum natural frequency inferior to the real value.

EXAMPLE 3.17 Using the Rayleigh method, determine one natural frequency for the mechanical model from Fig. 3.22, where $k_1 = k$, $k_2 = 2k$, $k_3 = k$, $m_1 = 2m$, $m_2 = 3m$, $m_3 = m$.

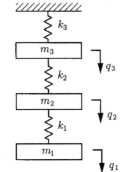

Figure 3.22
Mechanical model for Example 3.17.

Solution

The mathematical model is

$$\begin{cases} m_1\ddot{q}_1 + k_1 q_1 = k_2 q_2 = 0 \\ m_2\ddot{q}_2 - k_1 q_1(k_1 + k_2)q_2 - k_2 q_3 = 0, \\ m_3\ddot{q}_3 - k_2 q_2 + (k_2 + k_3)q_3 = 0, \end{cases}$$

where the matrix of inertia is

$$[M] = \begin{bmatrix} m_1 & 0 & 0 \\ 0 & m_2 & 0 \\ 0 & 0 & m_3 \end{bmatrix},$$

and the matrix of rigidity (stiffness) is

$$[R] = \begin{bmatrix} k_1 & -k_1 & 0 \\ -k_1 & k_1 + k_2 & -k_2 \\ 0 & -k_2 & k_2 + k_3 \end{bmatrix}.$$

The vectors of distribution is $[\mu]_k = [1, \, \mu_{2k}, \, \mu_{3k}]^T$.

Vibration

Replacing this in Eq. (3.49), one finds

$$
\omega_k = \left\{ \frac{[1,\ \mu_{2k},\ \mu_{3k}] \begin{bmatrix} k_1 & -k_1 & 0 \\ -k_1 & k_1+k_2 & -k_2 \\ 0 & -k_2 & k_2+k_3 \end{bmatrix} \begin{bmatrix} 1 \\ \mu_{2k} \\ \mu_{3k} \end{bmatrix}}{[1,\ \mu_{2k},\ \mu_{3k}] \begin{bmatrix} m_1 & 0 & 0 \\ 0 & m_2 & 0 \\ 0 & 0 & m_3 \end{bmatrix} \begin{bmatrix} 1 \\ \mu_{2k}\mu_{3k} \end{bmatrix}} \right\}^{1/2}.
$$

If the masses vibrate in phase, $[\mu]_k = [1,\ 1,\ 1]^T$, which in particular cases results in

$$
\omega = \omega_k = \sqrt{0.166\frac{k}{m}}.
$$

If two neighboring masses vibrate in phase, and the third in opposition of phase, one can have the cases

$$
[\mu]_k = [1,\ -1,\ -1]^T \quad \text{when} \quad \omega = \omega_k = \sqrt{1.5\frac{k}{m}},
$$

$$
[\mu]_k = [1,\ 1,\ -1]^T \quad \text{when} \quad \omega = \omega_k = \sqrt{0.833\frac{k}{m}}.
$$

If the neighboring masses vibrate in opposition of phase,

$$
[\mu]_k = [1,\ -1,\ 1]^T \quad \text{when} \quad \omega = \omega_k = \sqrt{2.15\frac{k}{m}}. \quad \blacktriangle
$$

3.5.4 ANALYSIS OF STABILITY OF VIBRANT SYSTEM

A vibrant system with more than one degree of freedom is a multivariable open linear system with $[F]$ or $[\phi]$ as inputs and $[q]$ as output. The matrix of transfer for the open system is

$$
[H] = [C]^T[s[I] - [A]]^{-1}[B], \tag{3.57}
$$

where matrices $[A]$, $[B]$, $[C]$ are given as functions of the inertia matrix $[M]$, damping matrix $[D]$, and stiffness matrix $[R]$.

The transfer matrix becomes

$$
[H] = [[I];\ [0]] \cdot [s[I] - [A]]^{-1} \begin{bmatrix} [0] & [0] \\ [I] & [0] \end{bmatrix}, \tag{3.58}
$$

where $[I]$ is the unit matrix and $[0]$ is the null matrix, both with the dimensions $n \times n$.

The characteristic polynomial for the open system is

$$
P(s) = \det[s[I] - [A]] \tag{3.59}
$$

or

$$
P(s) = \det \begin{bmatrix} s[I] & -[I] \\ [M]^{-1}[R] & s[I] + [M]^{-1}[D] \end{bmatrix}, \tag{3.60}
$$

which becomes

$$P(s) = \det[s^2[M] + s[D] + [R]] = 0. \tag{3.61}$$

The stability of vibrant systems with n degrees of freedom depends on the position of the roots of the polynomial Eq. (3.61). The stability criteria are algebraic (Routh, Hurwitz), or grapho-analytical criteria (Cramer, Leonhard). The use of polar diagrams (Nyquist) or Bode diagram requests a procedure to reduce the multivariable system to a monovariable system.

EXAMPLE 3.18 Determine the conditions of stability of motion for the vibrant system presented in Example 3.8.

Solution

With

$$[M] = \begin{bmatrix} J_1 & 0 & 0 \\ 0 & J_2 & 0 \\ 0 & 0 & J_3 \end{bmatrix}, \quad [D] = \begin{bmatrix} c_1 + c_2 & -c_2 & 0 \\ -c_2 & c_2 + c_3 & -c_3 \\ 0 & -c_3 & c_3 + c_4 \end{bmatrix},$$

$$[R] = \begin{bmatrix} k_1 + k_2 & -k_2 & 0 \\ -k_2 & k_2 + k_3 & -k_3 \\ 0 & -k_3 & k_3 + k_4 \end{bmatrix},$$

the characteristic polynomial becomes

$$\begin{vmatrix} \alpha_1 & -c_2 s - k_2 & 0 \\ -c_2 s - k_2 & \alpha_2 & -c_3 s - k_3 \\ 0 & -c_3 s - k_3 & \alpha_3 \end{vmatrix} = 0,$$

where $\alpha_1 = s^2 J_1 + s(c_1 + c_2) + k_1 + k_2$, $\alpha_2 = s^2 J_2 + s(c_2 + c_3) + k_2 + k_3$, $\alpha_3 = s^2 J_3 + s(c_3 + c_4) + k_3 + k_4$.

Developing this, we obtain a sixth-order polynomial, and the Hurwitz criterion can be applied to study stability.

Stability criterion Nyquist starts with the construction of the polar diagram.

The polar diagram (Fig. 3.23) has the following characteristics:

- Point C, which corresponds to $\omega = 0$, gives the specific displacement of the system in the static regime [m/N].
- The number of loops of polar diagram is the number of vibration modes (implicitly the number of degrees of freedom). For the diagram in Fig. 3.23, these are the three modes of vibration corresponding to loops 1, 2, 3.
- The value of ω that corresponds to the point on the loop placed at maximum distance from origin is the natural frequency associated with the mode of vibration that corresponds to the respective loop. For the polar diagram in Fig. 3.23, point B corresponds to one angular

frequency ω_B, giving the natural frequency ω_B that is attached to the vibration mode corresponding to loop 2.

■ The points of polar diagram in proximity to the origin give information about the behavior of the vibrant system at high frequencies of excitation.

■ The maximum deflection of the vibrant system, for each mode of vibration, is equal to the diameter of a circle that better approximates the associated loop of points of maximum distance from origin (i.e., diameter A_1A_2 of the circle associated with loop 1 is the maximum deflection for the first mode of vibration).

■ With the preceding circle diameters one can appreciate the damping level of the respective mode of vibration.

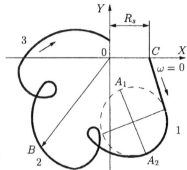

Figure 3.23
Polar diagram.

The polar diagram can be traced for a vibrant physical system using experimental data. In addition, this type of diagram indicates the linearity or nonlinearity of the vibrant system. ▲

4. Machine-Tool Vibrations

4.1 The Machine Tool as a System

The machine tool is a system (Fig. 4.1) with the following characteristics:

■ Elastic subsystem (*ES*): workpiece, tool, device, and elastic structure of the machine tool
■ Actuator subsystem (*AS*); electric motors and hydraulic motors
■ Subsystem due to the friction process and dissipation effects (*FS*)
■ Subsystem due to the cutting process (*CPS*) that includes processes from the contact between workpiece and tool

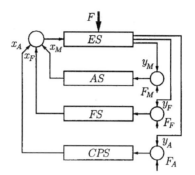

Figure 4.1
The machine-tool system.

The interconnection of the different subsystems is made with the help of the elastic subsystem (*ES*) (Fig. 4.1). The deformations of *ES* produce the variation of chip thickness, namely the variation of the cutting force (*ES*⟶*CPS*). The variation of the cutting force produces the modification of stresses and deformations of the elastic structure (*CPS*⟶*ES*).

In engineering practice, equivalent systems are used:

- Equivalent subsystem *SDE1–AS* (Fig. 4.2a). This subsystem is used for modeling the actuator's processes. As a mechanical model it has lumped masses, rotors, Hooke models, and linear dampers.

Figure 4.2
Equivalent subsystem.
(a) Subsystem SDE1–AS;
(b) subsystem SDE2–FS;
(c) subsystem SDE3–CPS.

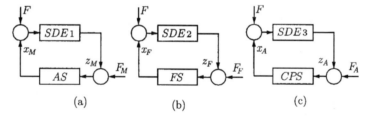

(a)   (b)   (c)

- Equivalent subsystem *SDE2–FS* (Fig. 4.2b). This subsystem is used for modeling the friction processes, with the same mechanical model as *SDE1–AS*.
- Equivalent subsystem *SDE3–CPS* (Fig. 4.2c). This subsystem is used for modeling the cutting process.

The preceding systems can be analyzed using the system presented in Fig. 4.3 where z is eliminated.

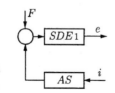

Figure 4.3
Open system.

4.2 Actuator Subsystems

Actuators can be rotating motors and linear motors. A rotating motor is associated with a torsional model (Fig. 4.4a), and a linear motor is associated with a linear displacement model (Fig. 4.4b).

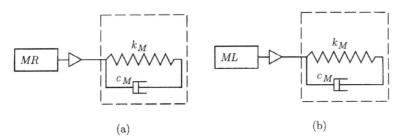

Figure 4.4
(a) Torsional model; (b) linear model.

(a)   (b)

To compute the elastic constant (spring constant) k_M and the damping coefficient c_M, the following cases will be presented.

4.2.1 ROTATING ELECTRIC MOTORS
AC Asynchronous Electric Motors
The elastic constant is

$$k_M = 2pM_{max} \text{ [Nm/rad]}, \tag{4.1}$$

where p is the number of pole pairs and M_{max} is the maximum torque in N/m.

The damping coefficient is

$$c_M = s_{cr}\omega_r J_M, \tag{4.2}$$

where ω_r is the natural frequency (in s^{-1}); $\omega_r = 2\pi f_r$, where f_r is the frequency of the alternative current (AC); and s_{cr} is the critical sliding. The critical sliding, s_{cr}, is $s_{cr} = (n_{cr} - n_0)n_0$, where n_0 is the rate of revolution for the unload function (null couple) in rev/min, and (n_{cr}) is the rate of revolution corresponding to M_{max} from the mechanical characteristic. J_M is the inertia moment for the rotor (in kg/m^2). One can compute J_M using the relation $J_M \cong 0.36G_r d_r$ (in kg/m^2), where G_r (in da/N) is the weight of rotor and d_r is the exterior diameter of the rotor (m).

DC Electric Motors
In this case the elastic constant is

$$k_M = \frac{1}{m\omega_0 T_e} \text{(Nm/rad)}, \tag{4.3}$$

where m is the slope of the static characteristic torque-sliding $(M - s)$, and ω_0 is the angular velocity for the unload function. The damping coefficient is

$$c_M = \frac{J_M}{T_e}, \tag{4.4}$$

where T_e is the electromagnetic time constant of excitation given by the relation $T_e = L_e/R_e$, L_e is the inductance of excitation, and R_e is the active resistance of excitation.

The logarithmic decrement associated with the rotating electric motor is $\delta = 1.25 \ldots 2.5$.

4.2.2 LINEAR HYDRAULIC MOTOR

For this type of motor, the elastic constant can be determined using the relation

$$k_M = \frac{Al}{2B} + \frac{DlA}{2E\delta} \ [N/m], \tag{4.5}$$

where A is the area of the active cross-section (in m^2) (Fig. 4.5), l is the displacement of the piston (the length of the active chamber, in m), B is the modulus of elasticity of the liquid (bulk modulus, in N/m^2), D is the inside diameter of the cylinder (m), E is the Young modulus of the cylinder material (N/m^2), and δ is the thickness of cylinder (m).

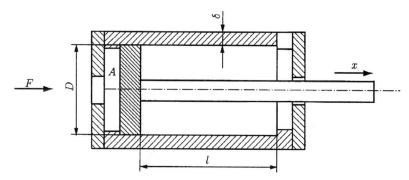

Figure 4.5
Linear hydraulic motor.

4.3 The Elastic Subsystem of a Machine Tool

4.3.1 THE ELASTIC SUBSYSTEM OF A KINEMATIC CHAIN

Consider the general case of a kinematic chain with gear, belts, and shafts. The following algorithm is given.

Step 1

Calculate the elastic constants of the shafts with negligible mass using the relation

$$k_A = \frac{GI_p}{L}, \tag{4.6}$$

where $I_p = \pi d^4/32 (m^4)$ is polar moment of inertia of the shaft with diameter d. For a tubular shaft, the polar moment of inertia is $I_p = \pi(D^4 - d^4)/32$, where D is the outside diameter and d is the inner diameter of the tube shaft,

G is the modulus of rigidity (torsional modulus of elasticity, in N/m^2), and L is the active length of the shaft.

Step 2

Calculate the moments of inertia of the rotors with respect to their geometric axes,

$$J_R = \begin{cases} \dfrac{mD^2}{8} & \text{for rotors that can be approximated with a full cylinder} \\[2ex] \dfrac{mD^2}{4} & \text{for rotors that can be approximated with a ring} \\ & \text{with mean diameter } D \end{cases} \qquad (4.7)$$

where m is the mass of the rotor in (kg).

Calculate the moments of inertia for each shaft with respect to the geometric axis of the shaft:

$$J_A = \begin{cases} \dfrac{mD^2}{8} & \text{for full section shafts} \\[2ex] \dfrac{m(D^2 - d^2)}{8} & \text{for tubular shafts} \end{cases} \qquad (4.8)$$

For rotors placed at the extremity of the shaft, the moments of inertia can be calculated using

$$J_R' = J_R + \frac{1}{6} J_A. \qquad (4.9)$$

Step 3

For each gear, calculate a supplementary elastic constant k_s that considers the bending deformation

$$k_s = \frac{rF_t}{\varphi_r}, \qquad (4.10)$$

where r is the pitch radius, F_t is the tangential force from gearing, and φ_r is the relative supplementary rotation of the gear that takes into consideration the bending.

If P is the power of the shaft and ω is the angular velocity of the shaft, then

$$k_s = \frac{P}{\omega \varphi_r}. \qquad (4.11)$$

The rotation φ_r can be calculated using

$$\varphi_r = \frac{1}{r}[(f_{1T} + f_{2T}) + (f_{1R} + f_{2R}) \tan(\alpha + \varphi)], \qquad (4.12)$$

where f_{1T}, f_{2T} are the deflections in the tangential direction of the gear for the two shafts, f_{1R}, f_{2R} are the deflections in the radial direction of the gear for the two shafts, and φ is the angle of friction of gearing (for steel, $\tan \varphi \cong 0.1$).

This results in

$$k_s = \frac{Pr}{\omega[(f_{1T} + f_{2T}) + (f_{1R} + f_{2R})\tan(\alpha + \varphi)]}. \tag{4.13}$$

The computed value of k_s is added to the value of k for the gearing case.

Step 4

Compute the equivalent length for each torsional singularity using the relation

$$l_e = \frac{GI_p}{k}\,(\text{m}), \tag{4.14}$$

where GI_p is the torsional stiffness of the shaft on which the singularity is placed, and k is the elastic constant of the singularity.

For n torsional singularities placed on the shaft, the total equivalent length is

$$L_e = \sum_{i=1}^{n} l_{ei}. \tag{4.15}$$

The length of calculus for each shaft is given by

$$Lc = L + Le, \tag{4.16}$$

where L is the active length of the shaft.

The elastic constant is computed using the relation

$$k_A' = \frac{GI_p}{Lc}\,(\text{Nm/rad}). \tag{4.17}$$

Step 5

Select a shaft of reference and reduce the inertia moments of rotors placed on other shafts with respect to the reference shaft. Reduction to reference shaft of an inertia moment J_k corresponding to a rotor placed on another shaft is given by

$$J_r = J_k i^2, \tag{4.18}$$

where $i = \omega/\omega_r$ is the transmission ratio between the shaft with the rotor and the reference shaft (ω_r is the angular velocity of the reference shaft, and ω is the angular velocity of the shaft where the rotor is placed). The reduced procedure is applied for all the rotors.

Step 6

Reduce to the reference shaft the elastic constants and the lengths of calculus for all the other shafts. One can write

$$L_{c,r} = \frac{L_c}{i^2}, \tag{4.19}$$

where L_c is the length of calculus for a shaft, $L_{c,r}$ is length of calculus reduced at the reference shaft, and i is the transmission ratio. For the elastic constant the reduction is made with the relation

$$k_r = ki^2. \tag{4.20}$$

Step 7

For each active section of the shaft, determine the damping coefficients c. The reduction of the damping coefficients to the reference shaft is given by the relation

$$c_r = ci^2. \tag{4.21}$$

Step 8

Reduce the outside torques M (that act on a rotor placed on other shafts than the reference shaft) to the reference shaft:

$$M_r = Mi. \tag{4.22}$$

Using step 1 to step 8 one may obtain an equivalent mechanical model with rotors placed on the reference shaft. The reduction of the rotors (and therefore the reduction of the degrees of freedom) is made (Fig. 4.6) using the relations

$$J'_{i-1} = J_{i+1} + J_i \frac{k_i}{k_{i+1} + k_i} \tag{4.23}$$

$$J'_{i+1} = J_{i+1} + J_i \frac{k_{i+1}}{k_{i+1} + k_i} \tag{4.24}$$

$$k'_i = \frac{k_i k_{i+1}}{k_i + k_{i+1}} \tag{4.25}$$

$$c'_i = \frac{c_i c_{i+1}}{c_i + c_{i+1}}. \tag{4.26}$$

Step 9

For the mechanical model obtained after step 7 or step 8 one can write the mathematical model. A complex kinematic chain can be transformed into a

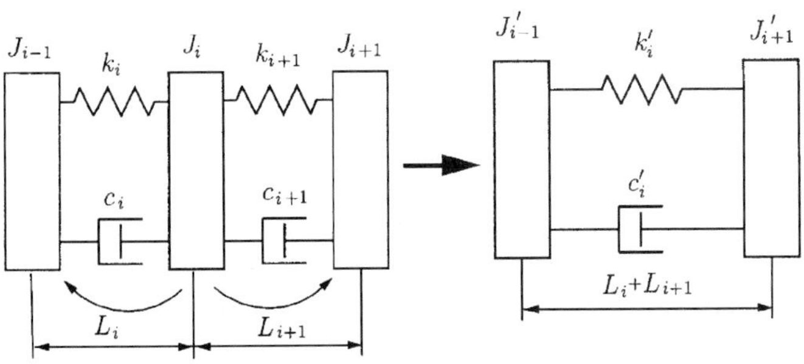

Figure 4.6
Reduction of number of rotors (and therefore of degrees of freedom).

simpler model as shown in Fig. 4.7. A point of ramification P as a rotor with inertia moment $J_P = 0$ and rotation q_P can be considered.

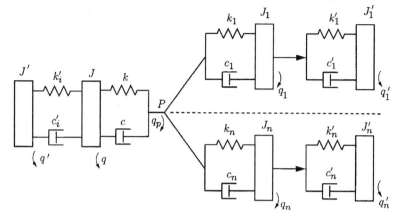

Figure 4.7
Mechanical model of the kinematic chain.

One can write the equations

$$
\begin{cases}
J_P \ddot{q}_P = 0 = -k(q_P - q) - c(\dot{q}_P - \dot{q}) - \sum_{j=1}^{n}[k_j(q_P - q_j) \\
\qquad + c_j(\dot{q}_P - \dot{q}_j)] \\
J\ddot{q} = -k(q - q_P) - c(\dot{q} - \dot{q}_P) - k'(q - q') - c'(\dot{q} - \dot{q}') \\
\qquad + M(t) \\
J_1 \ddot{q}_1 = -k_1(q_1 - q_P) - c_1(\dot{q}_1 - \dot{q}_P) - K_1'(q_1 - q_1') \\
\qquad - c_1'(\dot{q}_1 - \dot{q}_1') + M_1(t) \\
\vdots \\
J_n \ddot{q}_n = -k_n(q_n - q_P) - c_n(\dot{q}_n - \dot{q}_P) - k_n'(q_n - q_n') \\
\qquad - c_n'(\dot{q}_n - \dot{q}_n') + M_n(t).
\end{cases}
\tag{4.27}
$$

EXAMPLE 4.1 A gear train with two stages is shown in Fig. 4.8. The gear train has an asynchronous electric motor EM, an elastic clutch with bolts C, the spur gears R_1, R_2, R_3, R_4, and the rotor R with an exterior driven torque $M_R(t) = M_0 \sin(\omega t)$. The exterior driven torque M_R acts on the rotor R. The gears R_1, R_2, R_3, R_4 and the rotor R are fixed on the shafts I, II, III with the keys K_1, K_2, K_3, K_4, and K_5. Analyze the system (the torsional vibration) using a reduced system with three degrees of freedom.

Given data: Power on the shafts P_1, P_2, P_3 (W), input rotation n_1 (rev/min), AC asynchronous electric motor with p pole pairs, maximum torque M_{max} (N/m), rotation for the unload function (null couple) n_0 (rev/min), rotation corresponding to M_{max} from mechanical characteristic n_{cr}, frequency of the alternative current f_r (AC), moment of fly wheel GD^2 (N/m^2). Shaft characteristics are given in Table 4.1.

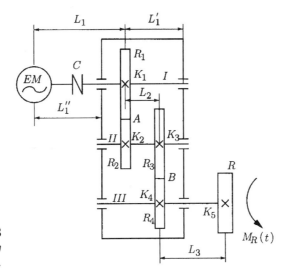

Figure 4.8
*Gear train with
two stages.*

Table 4.1 *Shaft Characteristics*

Shaft number	Diameter (m)	Section type	Transverse elasticity modulus (N/m²)
I	d_I	Full round	G
II	d_{II}	Full round	G
III	d_{III}	Full round	G

The clutch is with bolts and has the elastic dynamic constant $k_d = 3$. The rotor R has the mass M and the exterior diameter d_R. Gear characteristics are shown in Table 4.2.

Table 4.2 *Gear Characteristics*

Gear	Number of teeth	Width (m)	Pitch circle radius (m)	Deformation of a teeth pair, k_{rd} (m³/N)	Mass (kg)
R_1	N_1	l_1	r_1	$k = 6 \times 10^{-13}$	m_1
R_2	N_3	$l_2 = l_1$	r_2	$k = 6 \times 10^{-13}$	m_2
R_3	N_3	l_3	r_3	$k = 6 \times 10^{-13}$	m_3
R_4	N_4	$l_4 = l_3$	r_4	$k = 6 \times 10^{-13}$	m_4

All of the gears are cylindrical with pressure angle $\phi = 20°$. The keys are shown in Table 4.3.

Table 4.3 *Key Characteristics*

Key	Key type	Shaft diameter (m)	Length of joint (m)	Height	Number of keys
K_1	prismatic	d_I	λ_1	b_1	1
K_2	prismatic	d_{II}	λ_2	b_2	1
K_3	prismatic	d_{II}	λ_3	b_3	1
K_4	prismatic	d_{III}	λ_4	b_4	1
K_5	prismatic	d_{III}	λ_5	b_5	1

Solution

Step 1
The elastic constants of the shafts are

$$k_1 = \frac{\pi G d_I^4}{32 L_1}, \qquad k_2 = \frac{\pi G d_{II}^4}{32 L_2}, \qquad k_3 = \frac{\pi G d_{III}^4}{32 L_3} \text{(Nm/rad)}.$$

Step 2
Compute the inertia moment for electric motor

$$J'_{EM} = \frac{GD^2}{8g} \text{(kg m}^2\text{)},$$

where g is the gravitational acceleration, $g = 9.81 \text{ m/s}^2$, and D is the diameter of the rotor of the motor. The inertia moment of the clutch is neglected.

The inertia moments for the spur gears are

$$J'_{R1} = \frac{m_1 r_1 \omega_1^2}{2}, \quad J'_{R2} = \frac{m_2 r_2 \omega_2^2}{2}, \quad J'_{R3} = \frac{m_3 r_2 \omega_3^2}{2}, \quad J'_{R4} = \frac{m_4 r_4 \omega_4^2}{2} \text{ (kg m}^2\text{)}.$$

The inertia moment for the rotor is $J'_R = M d_R^2 / 8 (\text{kg m}^2)$.

Compute the inertia moments for shafts

$$J_I = \frac{M_I d_I^2}{8}, \qquad J_{II} = \frac{M_{II} d_{II}^2}{8}, \qquad J_{III} = \frac{M_{III} d_{III}^2}{8} \text{ (kg m}^2\text{)}.$$

The corrected inertia moments for extremity rotors are computed using the relations

$$J_{ME} = J'_{ME} + \tfrac{1}{6} J_I, \qquad J_{R1} = J'_{R1} + \tfrac{1}{6} J_I, \qquad J_{R2} = J'_{R2} + \tfrac{1}{6} J_{II}$$

$$J_{R3} = J'_{R3} + \tfrac{1}{6} J_{II}, \qquad J_{RA} = J'_{R4} + \tfrac{1}{6} J_{III}, \qquad J_R = J'_R + \tfrac{1}{6} J_{III}.$$

Step 3
The torsional singularities are the electric motor *EM*, the clutch *C*, the gear contacts at *A* and *B*, and the keys K_1–K_5. The elastic constants are computed for each singularity.

Vibration

For the electric motor,

$$k_{EM} = 2pM \text{ max (N m/rad)}.$$

For the spur gears,

$$k'_{R1} = \frac{l_1 r_1^2 \cos \alpha}{10 k_{rd}}, \qquad k'_{R2} = \frac{l_2 r_2^2 \cos \alpha}{10 k_{rd}}, \qquad k'_{R3} = \frac{l_3 r_3^2 \cos \alpha}{10 k_{rd}}$$

For the clutch,

$$k_C = 28.6 k_d G d_{\text{max}}^2 (\text{in N m/rad}], \text{ where } d_{max} \text{ is in meters and} G \text{ is in daN/cm}^2.$$

For the keys,

$$k_{K1} = \frac{d_I^2 \lambda_1 b_1}{10k}, \qquad k_{K2} = \frac{d_{II}^2 \lambda_2 b_2}{10k}, \qquad k_{K3} = \frac{d_{II}^2 \lambda_3 b_3}{10k}$$

$$k_{K4} = \frac{d_{III}^2 \lambda_4 b_4}{10k}, \qquad k_{K5} = \frac{d_{III}^2 \lambda_5 b_5}{10k} (\text{N m/rad}),$$

where $k = 6.4 \times 10^{-4}$ for prismatic keys.

Step 4

The angular velocities are

$$\omega_I = \frac{\pi n_I}{30}, \qquad \omega_{II} = \frac{\pi n_I N_1}{30 N_2}, \qquad \omega_{II} = \frac{\pi n_I N_1 N_3}{30 N_2 N_4} (s^{-1}).$$

The torques of the shafts are

$$M_I = \frac{P_I}{\omega_I}, \qquad M_{II} + \frac{P_{II}}{\omega_{II}}, \qquad M_{III} = \frac{P_{III}}{\omega_{III}} (\text{N m}).$$

The tangential forces in gearing are

$$\text{Gearing } R_1 - R_2 : \quad Ft_{1-2} = M_I / r_1 \text{ (N)}$$
$$\text{Gearing } R_3 - R_4 : \quad Ft_{3-4} = M_{II} / r_3 \text{ (N)}$$

The radial forces in gearing are

$$\text{Gearing } R_1 - R_2 : \quad Fr_{1-2} = Ft_{1-2} \tan \alpha \text{ (N)}$$
$$\text{Gearing } R_3 - R_4 : \quad Fr_{3-4} = Ft_{3-4} \tan \alpha \text{ (N)}$$

With these forces one can compute the deflections in tangential plane and in radial plane at the points K_1, K_2, K_3, K_4, and K_5. The following deflections are obtained:

Point K_1 (shaft I): tangential f_{IT}, radial f_{IR}
Point K_2 (shaft II): tangential f'_{IIT}, radial f'_{IIR}
Point K_3 (shaft II): tangential f''_{IIT}, radial f''_{IIR}
Point K_4 (at shaft III): tangential f_{IIIT}, radial f_{IIIR}.

With Eq. (4.12) one can calculate the coefficient of correction φ_r for each gear:

$$\varphi_{R1} = \frac{1}{r_1}[(f_{IT} + f'_{IIT}) + (f_{IR} + f'_{IIR})\tan(\alpha + \varphi)]$$

$$\varphi_{R2} = \frac{1}{r_2}[(f_{IT} + f'_{IIT}) + (f_{IR} + f'_{IIR})\tan(\alpha + \varphi)]$$

$$\varphi_{R3} = \frac{1}{r_3}[(f''_{IIT} + f_{IIIT}) + (f''_{IIR} + f_{IIIR})\tan(\alpha + \varphi)]$$

$$\varphi_{R4} = \frac{1}{r_4}[(f''_{IIT} + f_{IIIT}) + (f''_{IIR} + f_{IIIR})\tan(\alpha + \varphi)].$$

Using Eq. (4.11) or (4.10), the supplementary elastic constants are computed for each gear:

$$k''_{R1} = \frac{P_I}{\omega_I \varphi_{R1}}, \quad k''_{R2} = \frac{P_{II}}{\omega_{II} \varphi_{R2}}, \quad k''_{R3} = \frac{P_{III}}{\omega_{III} \varphi_{R3}}, \quad k''_{R4} = \frac{P_{IV}}{\omega_{III} \varphi_{R4}}, \quad \text{(Nm/rad)}.$$

The total elastic constants for each gear are

$$k_{R1} = k'_{R1} + k''_{R1}, \quad k_{R2} = k'_{R2} + k''_{R2}, \quad k_{R3} = k'_{R3} + k''_{R3}, \quad k_{R4} = k'_{R4} + k''_{R4}.$$

Step 5

The lengths for each singularity are

$$l_{EM} = \frac{GI_{PI}}{k_{ME}}, \quad l_C = \frac{GI_{PI}}{k_C}, \quad l_{K1} = \frac{GI_{PI}}{k_{K1}}$$

$$l_{K2} = \frac{GI_{PII}}{k_{K2}}, \quad l_{K3} = \frac{GI_{PII}}{kK_3}, \quad l_{K4} = \frac{GI_{PIII}}{k_{K4}}$$

$$l_{K5} = \frac{GI_{PIII}}{k_{K5}}, \quad l_{R1} = \frac{GI_{PI}}{k_{R1}}, \quad l_{R2} = \frac{GI_{PII}}{k_{R2}}$$

$$l_{R3} = \frac{GI_{PII}}{k_{R3}}, \quad l_{R4} = \frac{GI_{PIII}}{k_{R4}}$$

where

$$I_{PI} = \frac{\pi d_I^4}{32}, \quad I_{PII} = \frac{\pi d_{II}^4}{32}, \quad I_{PIII} = \frac{\pi d_{III}^4}{32}.$$

The lengths of calculus for the active section for each shaft are

$$L_I = L_1 + l_{EM} + l_C + l_{K1} + l_{R1}$$
$$L_{II} = L_2 + l_{K2} + l_{K3} + l_{R2} + l_{R3}$$
$$L_{III} = L_3 + l_{R4} + l_{K4} + l_{K5}.$$

The elastic constants can be recalculated as

$$k_I = \frac{GI_{PI}}{LI}, \quad k_{II} = \frac{GI_{PII}}{L_{II}}, \quad k_{III} = \frac{GI_{PIII}}{L_{III}} \text{ (N m/rad)}.$$

The mechanical model is shown in Fig. 4.9. The damping constants c_I, c_{II}, c_{III} can be determined with Eq. (3.7).

Vibration

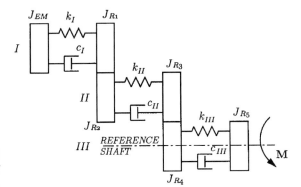

Figure 4.9
*Mechanical
model.*

Step 6

Shaft *III* is chosen as the reference shaft. The reduction of the inertia moments of rotors placed on other shafts will be calculated. The transmission ratios with respect to the reference shaft are

$$i_{1-3} = \frac{\omega_I}{\omega_{III}} = \frac{N_2 N_4}{N_1 N_3}, \qquad i_{2-3} = \frac{\omega_2}{\omega_{III}} = \frac{N_4}{N_3}.$$

The relations of reduction for the inertia moments are

$$J_{EM}^R = J_{EM}\,i_{1-3}^2, \qquad J_{R1}^R = J_{R1}\,i_{1-3}^2, \qquad J_{R2}^R = J_{R2}\,i_{2-3}^2, \qquad J_{R3}^R = J_{R3}\,i_{2-3}^2.$$

The following notations are used:

$$J_1 = J_{EM}^R, \qquad J_2 = J_{R1}^R + J_{R2}^R, \qquad J_3 = J_{R3}^R + J_{R4}^R, \qquad J_4 = J_{R5}^R.$$

Step 7

The elastic constants k_I and k_{II} are reduced to the reference shaft:

$$k_1 = k_I\,i_{1-3}^2, \qquad k_2 = k_{II}\,i_{2-3}^2, \qquad k_3 = k_{III}.$$

Step 8

The damping coefficients c_I and c_{II} are reduced to reference shaft:

$$c_1 = c_I \cdot i_{1-3}^2, \qquad c_2 = c_{II} \cdot i_{2-3}^2, \qquad c_3 = c_{III}.$$

Step 9

The external torque M acts on the reference shaft. The mechanical model reduced to the reference shaft is shown in Fig. 4.10. The mechanical model has four degrees of freedom. To simplify the calculations the number of degrees of freedom is reduced to three eliminating the rotor J_3. A new mechanical model is obtained (Fig. 4.11). The mathematical model corresponding to the new mechanical model, shown in Fig. 4.11, is

$$\begin{cases} J_1 \ddot{Q}_1 = -k_1(Q_1 - Q_2) - c_1(\dot{Q}_1 - \dot{Q}_2) \\ J_2' \ddot{Q}_2 = -k_1(Q_2 - Q_1) - c_1(\dot{Q}_2 - \dot{Q}_1) - k_{23}(Q_2 - Q_3) - c_{23}(\dot{Q}_2 - \dot{Q}_3) \\ J_4' \ddot{Q}_3 = -k_{23}(Q_3 - Q_2) - c_{23}(\dot{Q}_3 - \dot{Q}_2) + M_0 \sin(\omega t). \end{cases}$$

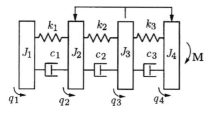

Figure 4.10
Reduced mechanical model.

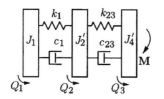

Figure 4.11
Mechanical model with three degrees of freedom.

The kinematic chain just analyzed is specific to a gear box and a feed box of a machine tool.

To obtain a linear motion, the following kinematic chains can be used: screw nuts (Fig. 4.12a), screw nuts with rolling element (Fig. 4.12b), pinion rack (Fig. 4.12c), and linear hydraulic motor (Fig. 4.12d).

For the mechanical model shown in Fig. 4.12e, m is the mass of the system and $x = q$ is the linear displacement. The elastic constant k for the usual cases is given in Table 4.4 and the coefficient c can be calculated with the relation

$$c = \frac{\delta}{\pi}\sqrt{km}\,(\text{N s/m}),$$

where δ is the logarithmic decrement. ▲

4.3.2 ELASTIC SYSTEM OF SHAFTS

The mechanical model of flexural vibration of linear shafts was presented in Example 3.5 using the method of the coefficients of influence. The mathematical model is given by Eq. (3.10), and can be completed with internal damping. Thus, to each lumped mass a linear damping is attached, $-c_i\dot{q}_i$, $i = 1, 2, \ldots, n$. The mathematical model becomes

$$\sum_{i=1}^{n}\alpha_{ji}(m_i\ddot{q}_i + c_i\dot{q}_i) + q_j = \sum_{i=1}^{n} n_{i=1}\alpha_{ji}F_i, \qquad j = 1, 2, \ldots, n. \qquad (4.28)$$

A shaft can be modeled as a beam with lumped masses with the following specifications.

A

The shaft is modeled as a beam with constant section and the constant stiffness for bending is EI. If the variation of diameter is less than 25% or the variation in length is less than 10%, the hypothesis of constant section of shaft is admissible. If the variation of the section is greater than 25%, an

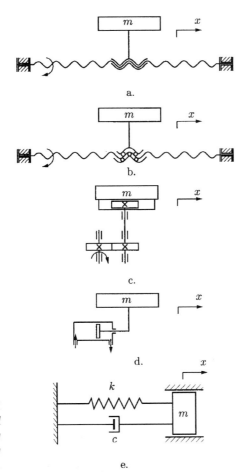

Figure 4.12
Mechanical model of a kinematic chain with screw nuts.

equivalent geometric inertia moment I is computed. For example, for the shaft shown in Fig. 4.13, the equivalent inertia moment is

$$I = \frac{I_1 I_2}{I_1 + \dfrac{(l_i + l_3)}{l^3}(I_2 - I_1)}.$$

(4.29)

B: The Lumped Masses Model Location

B_1: The lumped masses are located in the sections of the shaft where the rotors are. For example, for the main shaft of a lathe, milling machine, or boring machine (Fig. 4.14), the gear R_1 is the lumped mass m_1' and the rotor R_2 is the lumped mass m_2'.

B_2: *The lumped mass values.* The lumped masses, without the shaft mass, are numerically equal to the masses of the rotors. For the system represented in Fig. 4.14, $m_1' = m_{R1}$, $m_2' = m_{R2}$.

B_3: *The transformation of the mass shaft into lumped masses.* The calculation of the additional masses m_i'' can be achieved by two methods:

Table 4.4

Kinematic chain type	Conventional representation	Elastic constants k (N/m)	Notations
Screw nuts (sliding friction)	Fig. 4.12a	$k = \dfrac{\pi E d_m^2}{4}$	E is Young's modulus (N/m^2) d_m is mean diameter of screw (m) l is active length (m)
Screw nuts with rolling element without pretension	Fig. 4.12b	$k = 53 \times 10^{-6} z \sqrt{\dfrac{\Delta \sin^5 \alpha \cos^5 \lambda}{k_1}}$ $\Delta = 3.8 \sqrt[3]{\dfrac{F_a^2}{d_1 z^2}}$ $k_1 = m u_1 \sqrt[3]{\dfrac{2r_2 - d_1}{E^2 d_1 r_2}}$	z is the number of active rolling elements Δ is the contact deformation (μm) F_a is the axial force (daN) d_1 is the diameter of the rolling ball α is the contact angle of the rolling ball λ is the lead angle of the screw r_2 is the radius of profile for helix race m_1 is a coefficient F_0 is the pretension force
Screw nuts with rolling element with pretension	Fig. 4.12b	$k = \dfrac{1.5 \times 10^{-3} z \sin^2 \lambda \sqrt[3]{F_0}}{\delta}$ $\Delta = \dfrac{1.4 F_a}{z \sqrt[3]{F_0 d_1}}$	
Pinion–rack	Fig. 4.12c	$k = 10^{-4} K_A l \cos \alpha$ $K_A = 1.6 \times 10^3$ (daN/mm^2) for steel rack with normal teeth $K_A = 2.8 \times 10^3$ (daN/mm^2) for steel rack with helix teeth	l is the width of the pinion α is the pressure angle K_A is the specific pressure for teeth contact (daN/mm^2)

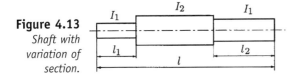

Figure 4.13
Shaft with variation of section.

1. The application of the general theory of the equivalent dynamic system with lumped masses with the elastic solid. The method is difficult to apply in current calculus.
2. The calculation of a coefficient of reduction α_i such as $m_i'' = \alpha_i m$, where m is mass of the shaft.

EXAMPLE 4.2 Determine the coefficient of reduction α_i for the mechanical models shown in Fig. 4.15. The mass of the shaft is m.

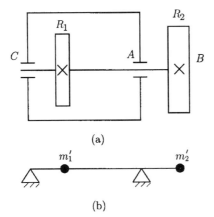

Figure 4.14
The main shaft of a lathe, milling machine, or boring machine. (a) Physical model; (b) mechanical model with lumped masses.

Vibration

Solution

The elastic constants are determined using Table 4.4. The natural frequency modes for some continuous systems are given in Table 4.5, where ρ is specific mass and A is the area of the section.

The natural frequencies are computed with the relation

$$\omega = \sqrt{\frac{k}{m}}.$$

For the model in Fig. 4.15a, one can write

$$\omega = \sqrt{\frac{k_a}{m''}} = \sqrt{\frac{3EI(a+b)}{a^2 b^2 m''}}.$$

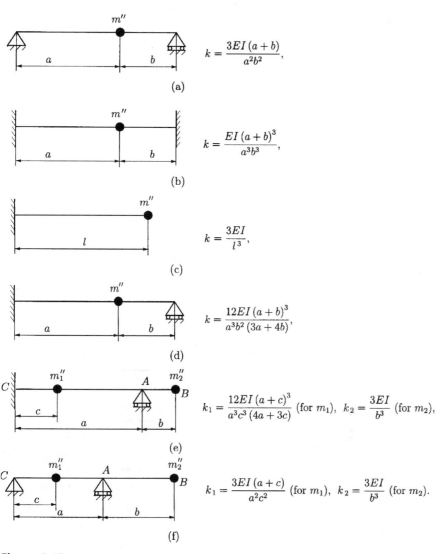

Figure 4.15 *Mechanical models and elastic constant with respect to lumped masses.*

Equalizing the preceding equation with the natural frequency ω_1 ($n = 1$), position 1 in Table 4.5, gives

$$\sqrt{\frac{3EI(a+b)}{a^2b^2m''}} = \sqrt{\frac{EI}{\rho A}} \cdot \frac{\pi^2}{(a+b)^2}.$$

But $m'' = \alpha m$ and $\rho A = m/(a+b)$. The coefficient of reduction will be

$$\alpha = \frac{3(a+b)^4}{\pi^4 a^2 b^2}.$$

Table 4.5

No.	Model	Natural frequencies	Notations
1.	EI l	$\omega_n = \sqrt{\dfrac{EI}{\rho A}} \cdot \dfrac{n^2 \pi^2}{l^2}, \; n = 1, 2, 3 \dots$	ρ, specific mass; A, cross-sectional area
2.	EI l	$\omega_1 = \sqrt{\dfrac{EI}{\rho A}} \cdot \left(\dfrac{4.73}{l}\right)^2$ $\omega_2 = \sqrt{\dfrac{EI}{\rho A}} \cdot \left(\dfrac{7.853}{l}\right)^2$	ρ, specific mass; A, cross-sectional area
3.	EI l	$\omega_1 = \sqrt{\dfrac{EI}{\rho A}} \cdot \left(\dfrac{3.927}{l}\right)^2$ $\omega_2 = \sqrt{\dfrac{EI}{\rho A}} \cdot \left(\dfrac{7.069}{l}\right)^2$	ρ, specific mass; A, cross-sectional area
4.	EI l	$\omega_1 = \sqrt{\dfrac{EI}{\rho A}} \cdot \left(\dfrac{1.875}{l}\right)^2$ $\omega_2 = \sqrt{\dfrac{EI}{\rho A}} \cdot \left(\dfrac{4.694}{l}\right)^2$	ρ, specific mass; A, cross-sectional area

For Fig. 4.15b one can write

$$\omega = \sqrt{\frac{k_b}{m''}} = \sqrt{\frac{EI(a+b)^3}{a^3 b^3 \alpha m}} = \sqrt{\frac{WI(a+b)}{m}} \left[\frac{4.73}{a+b}\right]^2,$$

and from position 3 in Table 4.5,

$$\alpha = \frac{(a+b)^6}{4.73^4 a^3 b^3}.$$

For Fig. 4.15c,

$$\alpha = \sqrt{\frac{k_c}{m''}} = \sqrt{\frac{3EI}{L^3 \alpha m}} = \sqrt{\frac{EIL}{m}} \left[\frac{1.875}{a+b}\right]^2,$$

and from position 4 in Table 4.5,

$$\alpha = \frac{3}{1.875^4} = 0.2427.$$

For Fig. 4.15d, one can use the same procedure.

For Fig. 4.15e, one can calculate two coefficients of reduction α_1 with respect to m_1'' and α_2 with respect to m_2''. If m_1 is the mass of the AC section, α_1 is determined with the relation from Fig. 4.15a, where $a \to c$, and $b \to a$:

$$\alpha_1 = \frac{3(a+c)^4}{\pi^4 a^2 c^2}, \qquad m_1'' = \alpha_1 m_1.$$

If m_2 is the mass of the AB section, α_2 is determined using the relation from Fig. 4.15e, and one can write $\alpha_2 = 0.2427$ and $m_2'' = \alpha_2 m_2$.

The situation presented in Fig. 4.15f can be calculated using the same procedure. ▲

4.4 Elastic System of Machine-Tool Structure

The machine tool is considered as an assembly of rigid bodies connected together with Hooke models (elastic models) and linear viscous dampers. The mathematical model for the mechanical model is determined using the Lagrange method. Therefore, it is essential to determine the elastic constants k and the damping coefficients c.

4.4.1 MOUNTING OF A MACHINE TOOL ON A FOUNDATION

The elastic constant k (Fig. 4.16) can be calculated for the following situations:

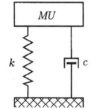

Figure 4.16
Mounting of a machine tool on a foundation.

A.1. Mounting of the machine tool with a steel spring:

$$k = \frac{zGd^4}{8D^3 n}\,(\text{N/m}). \qquad (4.30)$$

where z is the number of supports, G is the shear modulus (N/m^2), d is the wire diameter (m), D is the mean coil diameter (m), and n is the number of active coils.

A.2. Mounting of the machine tool on rubber plates or damping carpet:

$$k = \frac{E_d A}{h}. \qquad (4.31)$$

where A is the contact area (m^2), h is the thickness of the damping element (m), and E_d is the dynamic elasticity modulus of rubber (N/m^2).

A.3. Mounting of the machine tool on a concrete foundation:

$$k = A \cdot c_s \, [\text{N/m}]. \qquad (4.32)$$

Here A is the contact area (m^2), and c_s is the coefficient of elastic contraction of soil (N/m^3).

Vibration

The damping coefficient c is computed using

$$c = \frac{\delta}{\pi} \sqrt{km},\qquad(4.33)$$

where m is the mass of the machine tool and δ is the logarithmic decrement, which has different values depending on foundation type.

4.4.2 SLIDER JOINTS

The elastic constant can be calculated using

$$k = \frac{A}{k_0}\qquad (\text{N/m}),\qquad(4.34)$$

where A is the contact area (m^2) and k_0 is an elastic constant that depends on the length of the slider. In Eq. (4.34) the deflection of the slider was neglected. The mechanical model of the system can be determined as follows.

Sliding Support

Using a trapezoidal distribution of the contact pressure (Fig. 4.17a), one can obtain a suitable model. The constant k is placed on the resultant force **F** direction of the contact pressure. The coefficient k_A corresponds to the action mechanism and q_3 corresponds to a rotation of the Hooke model. The center

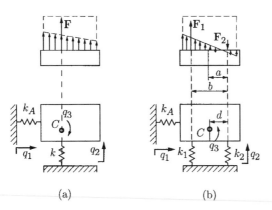

Figure 4.17
*Sliding support.
(a) Trapezoidal
repartition
(distribution) of
contact pressure;
(b) triangular
repartition of
contact pressure.*

(a)   (b)

of stiffness C is in the direction of the force **F**.

In the case of triangular distribution of the contact pressure (Fig. 4.17b), **F**$_1$ and **F**$_2$ forces are the resultants of two triangular repartitions. The center of stiffness has the position given by

$$d = \frac{k_1 a(b-a)}{k_1(b-a) + k_2 b}.\qquad(4.35)$$

To specify the values of the elastic constants k_A, k_1, k_2, the type and the geometry of the slider must be given.

Rolling Slider

The slider is modeled as a system with six degrees of freedom (Fig. 4.18). To specify the values of the elastic constants k_A, k_1, k_2, k_3, k_4, k_5, the type (balls or rolls) and the geometry of the slider must be given.

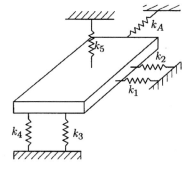

Figure 4.18
Rolling slide modeled as a system with six degrees of freedom.

4.5 Subsystem of the Friction Process

4.5.1 DRY FRICTION

In this category are dry sliding friction and mixed friction, namely, Coulombian friction. The lubricant film is present between surfaces, and the hydrodynamic phenomenon is present.

The friction force in the stationary regime is

$$F_f = \mu(k\delta),\qquad(4.36)$$

where μ is the friction coefficient, which depends on the material, relative velocity, pressure of contact, surfaces status, lubricant, temperature, etc.; k is the elastic constant to deflection in the perpendicular direction of the half joint surfaces; and δ is the perpendicular deflection

$$F_f = \frac{\mu A}{k_0}\delta.\qquad(4.37)$$

In the transitory regime, the transfer function is

$$H_F(s) = \frac{c_f}{T_F s + 1} = \frac{F_f}{\delta},\qquad(438)$$

where $c_f = \mu k$ is the stationary characteristic of the contact friction, and T_F is a time constant that is determined on the basis of the characteristic of the transitory regime. Equation (4.38) is valid in following conditions: the natural frequency of the variable mass $f = 25, \ldots, 30$ (Hz), and the variation of the relative velocity $\Delta v_r = 1, \ldots, 1.5$ (mm/s). In the case of other values a nonlinear model will be used.

4.5.2 ROLLING AND SLIDING FRICTION

The friction moment in a stationary regime is

$$M_f = r \cdot F_f = F_0 r + \mu_r N, \tag{4.39}$$

where F_0 is the constant component of the friction force, r is the radius of the rolling body (ball or roll), μ_r is the friction coefficient for rolling, and N is the normal push force. The transfer function is similar to Eq. (4.38), but some restrictive conditions are applied (c_f has lower values and T_F has greater value). At the start, F_0 can increase by 15–20% in the case of rolling slide, due to dynamic stick–slip phenomena.

EXAMPLE 4.3 Analyze the mechanical model shown in Fig. 4.19 for a slider with a mass m.

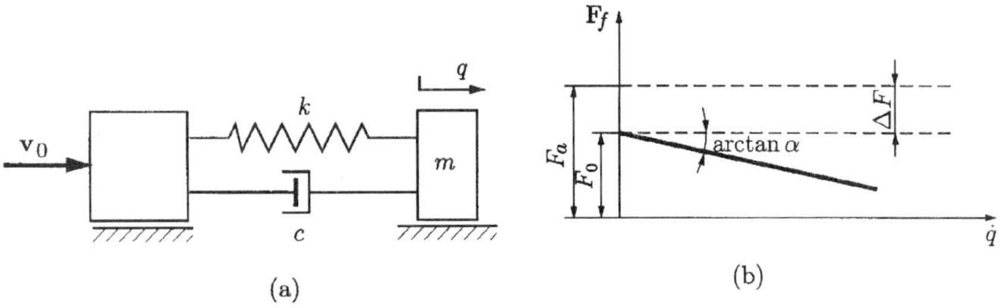

Figure 4.19
Mechanical model of a slider motion.

Solution

The friction force is

$$F_f = \begin{cases} F_a, \text{ adherence force,} & \dot{q} \equiv 0 \text{ (at rest)} \\ F_0 - \alpha\dot{q}, & \dot{q} > 0 \text{ (motion).} \end{cases} \tag{4.40}$$

The following forces act on the mass m:

The elastic force: $-k(q - v_0 t)$;
The friction force with guide-slide: $F_0 - \alpha\dot{q}$;
The damping force: $-c(\dot{q} - v_0)$.

The equation of motion is

$$m\ddot{q} = -k(q - v_0 t) - (F_0 - \alpha\dot{q}) - c(\dot{q} - v_0), \tag{4.41}$$

or

$$m\ddot{q} + (c - \alpha)\dot{q} + kq = kv_0 t + cv_0 - F_0. \tag{4.42}$$

The Cauchy problem is $q(0) = 0$, $\dot{q}(0) = 0$, and the mass starts from rest. It is considered that at $t = 0$ the damping force cv_0 is equal to the adherence force F_a: $cv_0 = F_a$. Therefore, $cv_0 - F_0 = F_a - F_0 = \Delta F$.

The solution of the mathematical model given by Eq. (4.22) is

$$q(t) = e^{-\delta t}\left\{\left(\frac{m\delta v_0}{k} - \frac{\Delta F}{k}\right)\cos(\omega t) + \left[\frac{mv_0}{k\omega}(\delta^2 - \omega^2) - \frac{\delta\Delta F}{\omega k}\right]\sin(\omega t)\right\}$$
$$+ v_0 t + \frac{\Delta F}{k} - \frac{2m\delta v_0}{k}, \qquad (4.43)$$

where

$$\delta = \frac{c - \alpha}{2m}, \qquad \omega = \sqrt{\frac{k}{m} - \delta^2}, \qquad \Delta F = F_a - F.$$

The velocity and the acceleration are

$$v = \dot{q} = e^{-\delta t}\left[-v_0\cos(\omega t) + \left(\frac{\Delta F}{m\omega} - \frac{\omega v_0}{\delta}\right)\sin(\omega t)\right] + v_0 \qquad (4.44)$$

$$a = \ddot{q} = e^{-\delta t}\left[\frac{\Delta F}{m}\cos(\omega t) + \left(\frac{k v_0}{m\omega} - \frac{\delta\Delta F}{m\omega}\right)\sin(\omega t)\right]. \qquad (4.45)$$

Functions of the value v_0, two motion regimes can appear:

Case 1: $v_0 \le v_{0cr}$

See Fig. 4.20. For $t \in [0, \quad T_1]$, the mass m is moved and at $t = T_1$ the acceleration is null or negative $a(T_1) \le 0$. From $t = T_1$ to $t = T_1 + T_2$, the mass m is at rest. At the moment $t = T_1 + T_2$, the acceleration is $a(T_1 + T_2) = \Delta F/m$, and a new cycle of motion starts. The motion is the discontinuous "stick–slip" phenomenon.

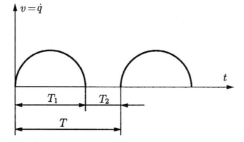

Figure 4.20
Discontinuous motion (stick–slip).

Case 2: $v_0 > v_{0cr}$

See Fig. 4.21. The motion is continuous and the velocity of the mass tends to reach the value v_0 in stationary regime.

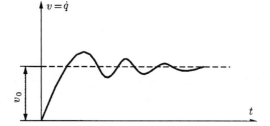

Figure 4.21
Continuous motion.

To find the value of v_{0cr}, one can follow these steps:

Set $v(T_1) = 0$ and $a(T_1) = 0$.
From Eqs. (4.45) and (4.45) with $v_0 = v_{0cr}$, the system obtained is

$$
\begin{cases}
\left[e^{\delta T_1} - \cos(\omega T_1) - \dfrac{\delta}{\omega} \sin(\omega T_1) \right] v_{0cr} + \left[\dfrac{1}{m\omega} \sin(\omega T_1) \right] \Delta F = 0, \\
\left[\dfrac{k}{m\omega} \sin(\omega T_1) \right] v_{0cr} + \left[\dfrac{1}{m} \cos(\omega T_1) - \dfrac{\delta}{m\omega} \sin(\omega T_1) \right] \Delta F = 0.
\end{cases}
\tag{4.46}
$$

The determinant of the system must be equal to zero:

$$
e^{\delta T_1} \left[\cos(\omega T_1) - \frac{\delta}{\omega} \sin(\omega T_1) \right] = 1.
\tag{4.47}
$$

The value of T_1 is replaced in one of the relations of Eqs. (4.46) to obtain the critical velocity v_{0cr}.

For v_{0cr} one can use the approximating relation

$$
v_{0cr} = \frac{\Delta F}{\sqrt{4\pi m k \theta}},
\tag{4.48}
$$

where $\theta = (c - \alpha)/2\sqrt{km}$.

The time T_2 of rest for mass m, in the condition of "stick–slip" is determined from the condition

$$
q(T_1) = v_{0cr}(T_1 + T_2) = q(T_1 + T_2),
\tag{4.49}
$$

which gives the time

$$
T_2 = -\frac{1}{\omega} e^{\delta T_1} \sin(\omega T_1).
\tag{4.50}
$$

The period of "stcik-slip" is $T = T_1 + T_2$.   ▲

4.6 Subsystem of Cutting Process

In general the cutting process is stable if the chips are continuous. For fragmentary chips and built-up edges an unstable cutting process can appear. For dynamic modeling of the cutting process, one can consider constant cutting parameters (speed v and feed f_r). In this case the cutting force depends only on the variation of depth of cut a_p (Fig. 4.22). If F_a is the cutting force, then in the stationary regime $F_a = F_a(y)$ (a nonlinear function). Considering a linear function, one can write

$$
F_a(u) = k_a u,
\tag{4.51}
$$

where $k_a = F'(y_0)$ is a stationary characteristic of the cutting process (N/mm) and $u = y - y_0$ is the relative displacement between the tool and the workpiece. The characteristic k_a is $k_a = bp$, where b is the width of the chip (mm) and p is the specific pressure (N/mm^2).

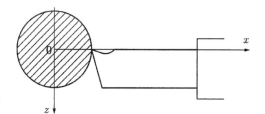

Figure 4.22
Cutting process.

For the transitory regime, with variation of the cutting force due to the variation of depth of cut and the variation of the tool geometry hypothesis, one can write

$$H_a(s) = \frac{F(u)}{u} = \frac{k_a}{T_a s + 1}, \tag{4.52}$$

where T_a is a constant that depends on the stability of the cutting process, cutting speed, depth of cut, etc.

Like the friction process, the cutting process is a self-excited vibration generator. The top edge of the tool can describe a closed trajectory $AmBnA$ that can be approximated with an ellipse (Fig. 4.23).

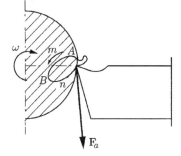

Figure 4.23
Self-excited vibration phenomenon.

One can model the tool with Hooke models, k_1 and k_2, in the direction of the main axes of rigidity with a reference frame at the center of rigidity C_0 as shown in Fig. 4.24. If $C_0 C$ is the elastic displacement of C_0 with q_1 and q_2 the components on the $C_0 Y$ and $C_0 Z$ directions, one can write the cutting force as

$$F_a = F_0 - ru, \tag{4.53}$$

where u is the relative displacement between the tool and the workpiece and r is a proportionality coefficient. If the $C_0 YY$ and $C_0 Z$ axes are the main directions of rigidity, then the coefficients of influence are

$$\alpha_{xx} = \alpha_{11} = \frac{1}{k_1}, \qquad \alpha_{yy} = \alpha_{22} = \frac{1}{k_2}, \qquad \alpha_{xy} = \alpha_{12} = \alpha_{yx} = \alpha_{21} = 0.$$

Vibration

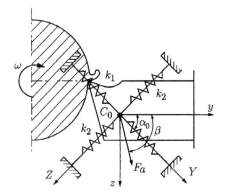

Figure 4.24
Tool model with Hooke models.

The equations of motion are (the damping is neglected)

$$\begin{cases} q_1 = -\alpha_{11} m \ddot{q}_1 - \alpha_{11} r u \cos(\beta - \alpha_0) \\ q_2 = -\alpha_{22} m \ddot{q}_2 - \alpha_{22} r u \cos(\beta - \alpha_0) \end{cases}, \tag{4.54}$$

where $u = q_1 \cos \alpha_0 - q_2 \sin \alpha_0$.

The mathematical model becomes

$$\begin{cases} m \ddot{q}_1 + [k_1 + r \cos \alpha_0 \cos(\beta - \alpha_0)] q_1 - r \sin \alpha_0 \cos(\beta - \alpha_0) q_2 = 0 \\ m \ddot{q}_2 + r \cos \alpha_0 \sin(\beta - \alpha_0) q_1 + [k_2 - r \sin \alpha_0 \sin(\beta - \alpha_0)] q_2 = 0 \end{cases}. \tag{4.55}$$

One can denote the following:

The matrix of inertia

$$M_m \begin{bmatrix} m & 0 \\ 0 & m \end{bmatrix},$$

The matrix of rigidity

$$C = \begin{bmatrix} k_1 + r \cos \alpha_0 \cos(\beta - \alpha_0) & -r \sin \alpha_0 \cos(\beta - \alpha_0) \\ r \cos \alpha_0 \sin(\beta - \alpha_0) & k_2 - r \sin \alpha_0 \sin(\beta - \alpha_0) \end{bmatrix}.$$

The characteristic polynomial is

$$P(s) = \det[s^2 [M_m] + [C]] = 0,$$

or

$$s^4 + (\alpha_1 + \alpha_2 + b - a)s^2 + (\alpha_1 \alpha_2 - a\alpha_1 + b\alpha_2) = 0, \tag{4.56}$$

where $a = \sin \alpha_0 \sin(\beta - \alpha_0)$, $b = \cos \alpha_0 \cos(\beta - \alpha_0)$, $k_1/r = \alpha_1$.

One can use the notations

$$u = s^2$$

$$M = \frac{\alpha_1 + \alpha_2 + b - a}{2} > 0$$

$$\Delta = \frac{1}{4}[(\alpha_1 + \alpha_2 + b - a)^2 - 4(\alpha_1 \alpha_2 - a\alpha_1 + b\alpha_2)].$$

The following cases are obtained:

Case 1: $\Delta > 0$, $M > \sqrt{\Delta}$, u_1, $u_2 < 0$; consequently, the characteristic equation of Eq. (4.56) has four imaginary roots. The dynamic response of the system is the overlap of two harmonic vibrations.

Case 2: $\Delta > 0$, $M < \sqrt{\Delta}$, $u_1 > 0$, $u_2 < 0$, consequently, the characteristic equation of Eq. (4.56) has two real roots (one negative and the other positive) and two imaginary roots. The dynamic response of the system is composed of one harmonic vibration, one unperiodic vibration with increasing amplitude in time, and one unperiodic vibration with decreasing amplitude in time. The motion is unstable.

Case 3: $\Delta < 0$, u_1, u_2 are complex roots; consequently, the roots of characteristic equations are complex, type $s_{1,2,3,4} = \pm e \pm jf$, quasiperiodics, one with increasing amplitude in time and the other with decreasing amplitude (Figs. 4.25, 4.26). It is the case of self-excited vibrations.

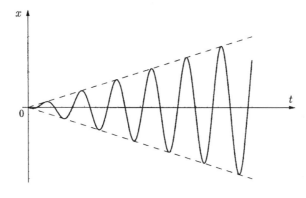

Figure 4.25
Motion with increasing amplitude in time.

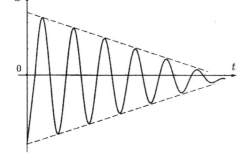

Figure 4.26
Motion with decreasing amplitude in time.

Case 4: $\Delta = 0$, u_1, $u_2 < 0$. The motion is unstable.

In conclusion, the self-vibration of the tool–workpiece system appears when

$$\left(\frac{k_1 + k_2}{r} + \cos\beta\right)^2 - 4\left[\frac{k_1 k_2}{r^2} - \frac{k_1}{r}\sin\alpha_0\sin(\beta - \alpha_0) + \frac{k_2}{r}\cos(\beta - \alpha_0)\right] < 0.$$

$$(4.57)$$

Vibration

The preceding equation can be rewritten as

$$r^2 \cos^2 \beta + 2[(k_1 + k_2) \cos \beta + 2k_1 \sin \alpha_0 \sin(\beta - \alpha_0)]r$$
$$- 4k_2 \cos \alpha_2 \cos(\beta - \alpha_0)r + (k_1 - k_2)^2 < 0 \tag{4.58}$$

References

1. D. Baganaru and P. Rinderu, *Mechanical System Vibrations*. Lotus, Craiova, 1998.

2. C. F. Beards, *Engineering Vibration Analysis with Application to Control Systems*. Halsted Press, New York, 1996.

3. G. Buzdugan, L. Fetcu, and M. Rades, *Mechanical Vibrations*, EDPG, Bucharest, 1979.

4. M. Buculei and D. Baganaru, *Mechanical Vibrations*. University of Craiova Press, Craiova, Romania, 1980.

5. A. Dimarogonas, *Vibration for Engineering*. Prentice-Hall, Upper Saddle River, NJ, 1996.

6. D. I. Inman, *Engineering Vibration*. Prentice-Hall, Englewood Cliffs, NJ, 1994.

7. C. Ispas and F. P. Simion, *Machine Vibrations*. Editura Academiei, Bucharest, 1986.

8. D. E. Newland, *An Introduction to Random Vibrations, Spectral and Wavelet Analysis*. Longman Group Limited, London, 1995.

9. W. Nowacki, *Dynamics of Elastic Systems*. John Wiley & Sons, New York, 1963.

10. M. Radoi, E. Denicu, and D. Voiculescu, *Elements of Mechanical Vibration*. University of Bucharest Press, Bucharest, 1970.

11. G. Silas and L. Brindeu, *Vibroimpacting Systems*. Editura tehnica, Bucharest, 1986.

12. W. T. Thomson, *Theory of Vibration with Applications*. Prentice-Hall, Englewood Cliffs, NJ, 1981.

13. W. T. Thomson and M. D. Dahleh, *Theory of Vibration with Applications*. Prentice-Hall, Englewood Cliffs, NJ, 1998.

14. B. H. Tongue, *Principles of Vibration*. Oxford University Press, Oxford, 1996.

Vibration

7 Principles of Heat Transfer

ALEXANDRU MOREGA

Department of Electrical Engineering, "Politehnica" University of Bucharest, Bucharest 6-77206, Romania

Inside

Heat transfer is present in almost any industrial and natural process. It suffices to mention the vital domain of energy generation and conversion, for instance energy generation through fission or fusion, the combustion processes of fossil combustibles, or the magnetohydrodynamic power generation, where numerous heat transfer problems occur. There are many heat transfer problems related to solar energy conversion systems for heating and air conditioning purposes and for electrical energy production. Heat transfer processes also influence the performance of propulsion systems, such as jet or internal combustion engines. They are present in many domains, for instance in the design of common water heating systems, incinerators, energy conversion through cryogenic systems, cooling of electronic devices and circuits, electrical machines and drives, and may imply all known heat transfer mechanisms — conduction, convection and radiation — which are then present through out the entire technological chain and design stages. In many circumstances the solutions to thermal optimization problems regarding the maximization of heat transfer rates are crucial to maintaining the thermal stability of materials and devices that may have to work under extreme thermal conditions.

At a broader scale, heat transfer processes are important factors in air and water pollution and strongly influence local and global climate.

Another class of collateral yet major problems is the thermal pollution associated with thermal residual exhaustion — such as cooling towers in thermal centrals.

Classical thermodynamics is concerned with the *initial* and *final equilibrium states* of the physical systems in between which these evolve, without a particular focus on the actual dynamics of the processes through which they take place. *Heat transfer* is a science related to thermodynamics, but distinctly constructed to study the specific modes of heat transfer. The basic concepts with which it operates are *heat transfer, temperature*, and *heat flux*. The intervening physical quantities are not only *thermodynamic* quantities, such as temperature, pressure, and heat, or *mechanical* quantities, such as velocity, mass flow, and shear stress, but also *specific* (heat transfer) quantities, such as *heat flux density* and *heat transfer coefficients*. A very suggestive, though not rigorous, definition was given by Poincaré [1, 5]: *heat transfer is driven by the temperature difference*, usually called *thermal gradient, that exists between the system and its surroundings*.

There are three basic heat transfer mechanisms: *conduction, convection*, and *radiation*. They may occur either individually or combined.

1. Heat Transfer Thermodynamics

The instantaneous state of a thermodynamic system is described through a set of physical quantities called thermodynamic *properties*. By definition, properties are those quantities whose numerical values do not depend on the particular path (evolution) that the system under investigation follows between the initial and the final thermodynamic equilibrium states. Quantities such as temperature and pressure are properties because their values depend strictly and solely on the instantaneous conditions under which they are measured. On the other hand, work, heat, mass transfer interactions and entropy are not properties, being related to the particular route followed by the system. It is important to notice that the properties must vary smoothly across the boundaries and interfaces. This restriction is also needed in order for the energy balance to make sense.

The first principle of thermodynamics, also called the *energy conservation law*, quantifies the energy interactions of a thermodynamic system. The *heat* that is transferred, Q (joules), and the *work* interactions, W (joules), are the two thermodynamic quantities that quantify the changes in the system energy inventory, E (joules). For an *infinitely small process* (i.e., between two infinitely close states of thermodynamic equilibrium) undergone by a closed system that interacts with another system — that may be also its surroundings — this principle may be written as

$$\delta Q - \delta W = dE, \tag{1.1}$$

where δQ is the heat interaction (outside–in) experienced by the system, and δW is the work interaction done by the system. Here δ and d denote elementary variations of the quantities that depend (are thermodynamic properties), or respectively do not depend (are not properties), on the particular path followed by the system, from the initial to the final equilibrium states. The per-time form of the balance equation (1.1) [3, 5] is obtained by dividing (1.1) through the corresponding time increment, dt, that is,

$$q - w = \frac{dE}{dt} = \dot{E}, \tag{1.2}$$

where q (watts) is the *heat transfer rate* and w (watts) is the *work rate* (power).

Since the work and the heat interactions are related to the environment, or neighboring systems, the terms in the left-hand side of (1.2) represent the *total* heat transfer and work rates. A generalized, global form of (1.2) is

$$\sum_i q_i - \sum_j w_j = \frac{dE}{dt} = \dot{E}, \tag{1.3}$$

where the sums extend over all possible heat and work "ports" (parts of the system boundary) of interaction.

Generally, the basic system for which balance equations and principles are formulated is a region of space of finite volume, called the *control volume*. When this region is reducible to a surface, the system is then called the *control surface*. These concepts are useful also in other circumstances, for instance in formulating solution methods to mathematical models related to physical problems, the so-called *control volume–based methods*.

Under the assumption of thermodynamic equilibrium (1.8), the energy balance (1.1) for a control volume may be formulated either on a finite time interval basis, that is, in terms of energy and heat variation (J) (Fig. 1.1),

$$E_{in} + E_g - E_{out} = \Delta E_{st}, \tag{1.4}$$

or on a time rate basis (W = J/S)

$$\dot{E}_{in} + \dot{E}_g - \dot{E}_{out} = \frac{dE_{st}}{dt} = \dot{E}_{st}. \tag{1.5}$$

The terms $\dot{E}_{in}$ and $\dot{E}_{out}$ are the energy variation rates due to the heat and work interactions experienced by the system, here the control volume. These

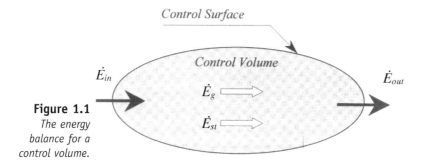

Figure 1.1
The energy balance for a control volume.

quantities represent surface or boundary interactions associated exclusively with the processes that take place at the system boundary or control surface level. The most common processes are conduction, convection, and radiation. If besides these interactions there is also mass transfer, then mechanical energies (potential and kinetic) specific to the flow may also contribute the overall energy balance.

$\dot{E}_g$ is the *internal heat generation rate* from other forms of energy, such as chemical, electrical, magnetic, or nuclear. This term is a *body source* and its magnitude is proportional to the size of the system. The quantity $\dot{E}_{st}$ is the internal energy growth rate due to the energy transfer experienced by the system through its boundary and to the internal heat generation.

When the system interacts with its surroundings—which sometimes may conveniently be seen as a second system or the complement of the first system with respect to the entire space—exclusively through heat transfer, the control volume balance equation (1.5) reduces to the control surface balance

$$\dot{E}_{in} = \dot{E}_{out}, \tag{1.6}$$

an equation that is valid under both steady and unsteady conditions. Considering the three possible heat transfer mechanisms (Fig. 1.2), it follows that for the unit boundary surface

$$q''_{cond} - q''_{conv} - q''_{rad} = 0. \tag{1.7}$$

The temperature T (kelvins) is a *primitive* quantity, in the sense that it is not introduced by utilizing other, previously defined thermodynamic quantities. Temperature does not depend on the particular path between the initial and final states; hence, it is a thermodynamic property. With respect to it, one defines the condition of *thermodynamic equilibrium* for two closed systems, A and B, as

$$T_A = T_B, \tag{1.8}$$

that is, the systems must have the same temperature. Consequently, there exists no heat transfer between systems that are in thermodynamic equilibrium with each other.

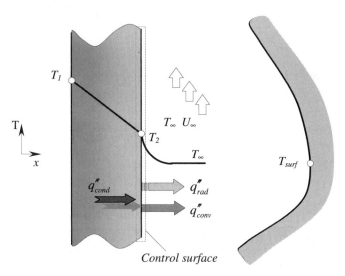

Figure 1.2
The mechanisms of heat transfer.

A method for measuring the temperature of a system consists of thermally contacting it with a *thermometer*—a *standard* system that, among other restrictions, is not to perturb the thermodynamic state of the system under investigation, while still remaining sensitive to the thermal interaction (heat transfer) experienced by the system. The conversion relation from the Kelvin to the Celsius absolute temperature scales is $T(°C) = T(K) - 273.15$.

The Second principle of thermodynamics evidences the different nature of the quantities in (1.3)—heat and work are not properties, whereas energy is a property. For a closed system, that undergoes an elementary process, this principle may be written as

$$\sum_i \frac{q_i}{T_i} \le \frac{dS}{dt},\qquad(1.9)$$

where S (joules) is the *entropy* of the system, a new thermodynamic property. The ratio q_i/T_i(W/K) is the *entropy transfer rate* at the port (boundary section) i, whose thermodynamic equilibrium temperature is T_i, and where q_i is the corresponding heat interaction rate. Here, as in (1.3), the sum extends over the heat transfer ports of the system.

A qualitative analysis of (1.3) reveals two fundamental aspects:

- The entropy transfer is not related to the work interaction (that is, work is *null entropy interaction*)
- Any thermal process, or heat interaction, is accompanied by an entropy transfer, q_i/T_i, *of the same sign as the heat that is transferred.*

Remarkably, these aspects are not evidenced by the first principle (1.1), which makes no distinction between the two types of thermodynamic

Heat Transfer

interactions, heat and work: Apparently, they contribute equally to the change in the system energy.

The classical definition of the *work interaction* is *the force that acts upon the system, dotted (vector scalar product) with the elementary displacement of its point of application.* In the same framework, the *thermal interaction* represents *the energy interaction driven by the temperature difference between the system and its surroundings.*

The specific heat at constant pressure, c_P, and **the specific heat at constant volume**, c_V, are two thermodynamic parameters that may be utilized to characterize the heat transfer processes.

The specific heat at constant pressure may be defined and evaluated as well through a heating process at constant pressure undergone by a sample of a working substance, which in this case is the thermodynamic system. The mass of the sample, m, the heat transferred to the system, δQ, the temperature variation, dT, and the change in the system volume, dV, are measurable quantities. Their knowledge yields

$$c_P \stackrel{def}{=} \left(\frac{\delta Q}{m\, dT} \right)_P. \tag{1.10}$$

The elementary heat interaction is here $\delta Q = dU + \delta W$, where $U = E$ is the *internal energy* of the system, and the work interaction is $\delta W = PdV$, where P is the *pressure* (a property) that is kept constant throughout this particular, isobaric process. Subsequently, $\delta Q = d(U + PV) = dH$, where dH is its elementary variation of a new quantity, called *enthalpy*, H (joules), and defined through $H = U + PV$. The definition (1.10) may then be put under the alternative and sometimes more convenient form

$$c_P \stackrel{def}{=} \left(\frac{\delta h}{dT} \right)_P, \tag{1.11}$$

where $h = H/m$ (J/kg) is the *specific enthalpy*.

The *specific heat at constant volume*, c_V, may be introduced through a heating process at constant volume. Measuring the heat input, δQ, and the associated temperature increase, dT, yields

$$c_V \stackrel{def}{=} \left(\frac{\delta Q}{m\, dT} \right)_V. \tag{1.12}$$

The first principle applied to this particular process — for which the work interaction, δW, is zero — leads to the alternative definition for the specific heat at constant volume

$$c_V \stackrel{def}{=} \left(\frac{\delta u}{dT} \right)_V, \tag{1.13}$$

where $u = U/m$ (J/kg) is the *specific internal energy* of the system.

In the special case of a *pure substance*, c_P and c_V are functions of both temperature and pressure. However, there are two particular situations when these quantities are functions of temperature only:

The ideal gas limit of a substance, where c_P and c_V are related by *Mayer's Law*,

$$c_P(T) + c_V(T) = R. \qquad (1.14)$$

$R = 8.315(\text{kJ/kmol K})$ is the *universal constant* of the ideal gas of the substance under consideration and denoted by a single quantity, c.

The incompressible substances (solid and liquid), for which the two specific heats are equal,

$$c_P(T) = c_V(T) = c(T). \qquad (1.15)$$

This latter case is of interest in conduction and convection heat transfer in electrical and electronic equipment and devices.

1.1 Physical Mechanisms of Heat Transfer: Conduction, Convection, and Radiation

The heat transfer mechanisms are *conduction*, *convection*, and *radiation*. However, in many applications of practical importance only one or two of these mechanisms may actually intervene.

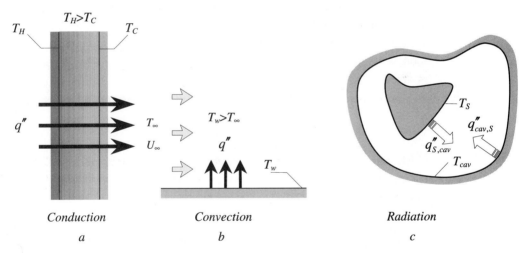

Figure 1.3 *(a) Conduction, (b) convection, and (c) radiation heat transfer.*

Conduction heat transfer occurs in nonmoving substances, solids, liquids, or gases, that experience internal temperature gradients (Fig. 1.3a). At a microscopic scale, this process may be explained by the molecular interactions. This type of heat transfer belongs to the larger class of so-called *diffusion processes*.

The *conduction heat flux*, q''_{cond}, is defined by *Fourier's law* as

$$\boldsymbol{q}''_{cond} \overset{def}{=} -k\nabla T, \qquad (1.16)$$

where k(W/mK) is the *thermal conductivity* of the substance, and ∇T is the temperature gradient (in cartesian coordinates, $\nabla = \frac{\partial}{\partial x}\mathbf{1} + \frac{\partial}{\partial y}\mathbf{J} + \frac{\partial}{\partial z}\mathbf{k}$).

When heat is transferred by a flowing fluid then, in addition to molecular diffusion, *transport* mechanisms occur, and the process is called *thermal convection* (Fig. 1.3b). Since heat is transferred between the system (for instance, a solid body) and its surroundings via a fluid conveyor (gas or liquid), the properties of the flow (velocity field, turbulence, etc.) may significantly influence the transfer rate.

Two important convection flow types are commonly utilized in a majority of technical applications, such as the cooling of electrical machines and apparatuses, electronic devices, and circuits: *external flow*—the body is bathed by the fluid with which it thermally interacts—and the *internal flow*—within the cavities, channels, ducts, etc., formed by the bodies in between which the fluid circulates.

The external flows where the fluid velocity at the point of contact (wall) is zero are characterized by hydrodynamic and temperature boundary layers. The *hydrodynamic boundary layer* is that flow region adjacent to the wall where the velocity varies from zero (at the wall) to the free stream value, U_∞ (far away from the wall). The *temperature boundary layer* is identified as that flow region adjacent to the wall where the fluid temperature varies from the wall temperature, T_w, to the incident ("fresh") fluid temperature, T_∞. The two boundary layers grow (or "develop") downstream, in the flow direction, and their particular structure (laminar, transition, or turbulent) depends on the flow parameters, wall conditions, etc. The boundary layer concepts and the related phenomena are crucial to the analysis of heat transfer problems.

The *convection heat flux*, q''_{conv}, is quantified by *Newton's law*,

$$q''_{conv} \stackrel{def}{=} h(T_w - T_\infty), \tag{1.17}$$

which states its proportionality to the temperature difference $(T_w - T_\infty)$ that may exist between the surface (of the wall, plate, etc.) temperature, T_w, and the free stream temperature, T_∞. From the thermodynamic equilibrium condition (1.28) it follows that q''_{conv} is zero when the fluid and the body are in thermal equilibrium ($T_w = T_\infty$).

The term h (W/m^{2}K) is called the *convection heat transfer coefficient, film conductance,* or *film coefficient.* It accounts for *all* heat transfer processes that take place inside the boundary layer and, consequently, depends on the particular structure of the boundary layer, geometry, flow type, and wall properties, and on the hydrodynamic and thermal properties of the fluid. By convention, the convection heat flux rate is positive when heat is transferred from the wall to the fluid ($T_w > T_\infty$), and negative conversely.

In internal flow it is the body that guides the flow. In contrast to the external flow, where the free stream temperature, T_∞, is usually a known quantity, in the internal flow the corresponding quantity is the *bulk temperature*, T_b, which is defined with respect to a flow cross section. Consequently,

$$q''_{conv} = h(T_w - T_b). \tag{1.18}$$

The contributions of the two mechanisms — diffusion and transport — to the convection heat transfer are different in the two cases, that is, the viscous internal and external flows, and they are related to the thicknesses of the hydrodynamic and temperature boundary layers. Since in viscous flows the velocity at the wall (the solid–fluid interface) is zero, the only heat transfer mechanism is the molecular diffusion. Macroscopically, this process translates into the conduction heat flux density (2.6),

$$q''_{cond} = -k \left(\frac{\partial T_w}{\partial y} \right)_{y=0^+}, \tag{1.19}$$

where k is the thermal conductivity of the fluid and $T = T_w$ is the fluid temperature at $y = 0^+$ (on the fluid side of the wall). From (1.17), (1.18), and (1.19) it follows that, at the wall,

$$h = -\frac{k}{T_w - T_\infty} \left(\frac{\partial T}{\partial y} \right)_{y=0^+} \quad \text{(external flow)}, \tag{1.20}$$

$$h = -\frac{k}{T_w - T_b} \left(\frac{\partial T}{\partial y} \right)_{y=0^+} \quad \text{(internal flow)}. \tag{1.21}$$

Apparently, the objective of the thermal analysis concerning the convection heat interaction is finding the heat transfer coefficient, h. Figure 1.4 shows the role played by the flow and the fluid in the convection heat transfer, in terms of h [2, 4]. However, it should be mentioned that there are

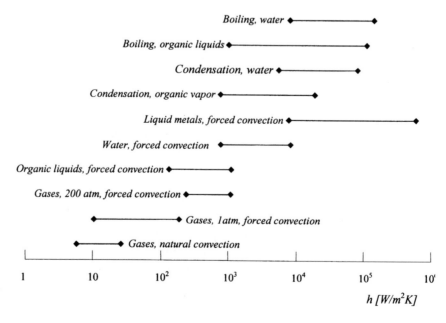

Figure 1.4
Convection heat transfer regimes.

convection heat transfer processes to which the *latent heat* associated with phase changes (liquid–gas, solid–liquid), also contributes. Two important situations — not discussed in this text — are then frequently encountered: *boiling* and *condensation* [2, 4].

The knowledge of h [Eqs. (1.20); (1.21)] necessitates the evaluation of the temperature field inside the fluid region, at the wall, and this quantity depends on the flow structure. Consequently, the flow solution — usually a nontrivial task — is a necessary stage in any consistent heat transfer problem solution, and to carry it out the heat transfer physical model has to be completed with the equations specific to the flow, namely the momentum and continuity equations.

Radiation heat transfer is conveyed by electromagnetic waves. Thermal radiation is emitted by any substance that has a finite temperature, and it is essentially a process related to the surfaces of the bodies (Fig. 1.3c). This process is of undulatory electromagnetism, photonic by nature, and it may be explained through the changes in the atomic and molecular electron configurations, which, in turn, are accompanied by a net energy transfer. Whereas conduction and convection necessitate the existence of a "working," intermediate substance, thermal radiation may exist even in the absence of the substance, that is, in vacuum.

The *Stefan–Boltzmann law* gives the maximum heat flux at which the thermal radiation may be emitted by the surface of a body, called an *ideal radiator* or *blackbody*:

$$q''_{SB} = \sigma T_S^4. \tag{1.22}$$

T_S is the absolute temperature of the body surface and σ is Stefan–Boltzmann constant, a universal constant ($\sigma = 5.67 \times 10^{-8} \, \text{W/m}^2\text{K}$). The radiative heat flux emitted by a *real* surface, called a *gray surface*, is only a fraction of that emitted by the *blackbody* (an idealized concept):

$$q''_{SB,b} = \varepsilon \sigma T_S^4, \tag{1.23}$$

where the nondimensional parameter ε is called the *emissivity* ($0 \leq \varepsilon \leq 1$). Similarly, a surface may absorb only a fraction of an incident heat flux, q''_{inc}:

$$q''_{abs} = \alpha q''_{inc}. \tag{1.24}$$

Here, α is the *absorption coefficient* ($0 \leq \alpha \leq 1$), and q''_{inc} is the incident heat flux.

A frequently encountered case of practical significance is that of a body situated in an enclosure (cavity) that interacts thermally with its confining walls. The net heat transfer rate between the surface of this body, assumed to be a gray surface of area A and temperature T_S, and the inner surface of the cavity, of temperature T_{cav}, may be evaluated by

$$q''_{rad} = \frac{q_{rad}}{A} = \sigma \varepsilon (T_S^4 - T_{cav}^4). \tag{1.25}$$

In many technical applications, a simpler, linearized form of (1.25) that is analogous to the convection heat flux definition (1.17).

$$q_{rad} = h_r A(T_S - T_{cav}) \tag{1.26}$$

$$h_r = \sigma \varepsilon (T_S + T_{cav})(T_S^2 + T_{cav}^2) \simeq \text{constant}, \tag{1.27}$$

is considered satisfactorily accurate.

If the space between the body and the cavity is filled by a fluid (i.e. not vacuum), thermal radiation may be accompanied by conduction and convection heat transfer. For an in-depth, detailed presentation of radiation heat transfer the reader is referred to, for instance, ref. 2.

1.2 Technical Problems of Heat Transfer

Heat transfer is concerned with the heat transfer interaction of the system under investigation (A), at temperature T_A, with its surrounding space (B), at temperature T_B, when $T_A \neq T_B$, that is, outside the thermal equilibrium. The intensity of this interaction is characterized by the *heat transfer rate*, $q(\text{J/s} = \text{W})$. This quantity is usually a dynamic, time dependent variable, and it depends on many factors: the initial and final states, the thermodynamic properties, and the geometry of both system (A) and its environment (B) (motion, flow, etc.). The thermodynamic equilibrium condition (1.8), on one hand, and the temperature definition, on the other hand, suggest that at a thermodynamic equilibrium state,

$$q = 0 \quad \text{for} \quad T_A = T_B. \tag{1.28}$$

Depending on the particular application, the main goal of the thermal design may be either the *inhibition* of the heat transfer (the thermal insulation problem), its *increase* (the cooling problem), or *temperature control* within prescribed, safe limits. Consequently, the following technical problems may be formulated:

> *The thermal insulation problem:* Here the temperatures of the system (A), T_A, and of its environment (B), T_B, are known, specified quantities. The key quantity of the thermal design, or optimization, is the "heat lost," and the central problem consists of leading the thermal (insulation) design so as to reduce the heat dissipation rate.
>
> *The cooling problem:* In this class of applications the heat transfer rate is prescribed. The thermal analysis is then aimed at reducing the associated temperature drop. A subsequent objective is the reduction in the entropy "production," and therefore in the "lost" work. The usual means of reaching this objective are, for instance, thermal contact improvement (reduction), and this may be achieved by selecting the type of flow for the cooling agent, the geometry, and the thermal properties of the contact surfaces—to mention only a few.

Heat Transfer

Temperature control: For a large class of problems, the thermal design objective is to maintain the system within some specific, prescribed thermal operating conditions. For instance, in order to function well, electronic devices and equipment, electrical cables, and electrical machines and apparatus must comply with specified minimum and maximum safe temperature limits. In this situation, the heat transfer rate is generally variable; hence, the system temperature is also variable.

2. Conduction Heat Transfer

To a larger or lesser extent, conduction heat transfer is present in any thermal process and in all substances: solids, liquids, and gases. At the microscopic scale, heat conduction occurs through the atomic or molecular activity of the substance, and it may be seen as a form of energy interaction, from higher energy particles to lower energy particles, via particle interactions.

An intuitive model for heat conduction is that of a gas at rest, where no macroscopic motion is perceived, that supposedly undergoes the action of an externally imposed temperature difference (Fig. 2.1). This temperature

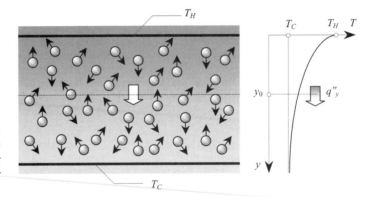

Figure 2.1
Heat conduction through atomic or molecular activity.

gradient may be produced, for instance, by two horizontal surfaces that confine the gas and that are at different, known temperatures (T_H, T_C). At any location inside the gas, the temperature is a measure of the internal energy of the gas that is due, at a microscopic scale, to the *random*, or *Brownian*, translations and vibrations of the gas molecules. In this interpretation, the higher temperatures are associated with higher molecular energies, whereas the lower temperatures correspond to lower molecular energies. The systematic "collisions" between neighboring molecules are accompanied by a net energy transfer at the molecular level, from the more "energetic" to the less energetic molecules. The heat transfer goes always *from hot to cold*, that is from the surface (zone) of high temperature to the surface (zone) of low temperature.

In the case under investigation (the upper plate is hot, T_H, and the lower plate is cold, T_C, $T_H > T_C$) the molecular collisions occur in the Oy direction, in the hot→cold direction. This phenomenon is associated with a net energy transfer in the Oy direction, and respectively to a heat flux by *molecular diffusion*.

Although in the case of liquids the molecules are much more densely packed, that is, the molecular interactions are much stronger, heat conduction is still produced by a similar mechanism. In the modern microscopic theory of heat transfer, the thermally driven *lattice waves* are responsible for this type of energy interactions [4]. If in a thermally nonconducting medium heat transfer is conveyed by these waves, in a thermally conducting medium the free-electron translational motion also contributes to the process.

2.1 The Heat Diffusion Equation

The quantitative characterization of heat transfer, in general, and that of the conduction heat transfer, in particular, relies on the evaluation of the heat transfer rate. When applied to a 1D control volume of size Δx (Fig. 2.2), the first principle yields

$$q_x - q_{x+\Delta x} - w = \frac{\partial e}{\partial t}. \tag{2.1}$$

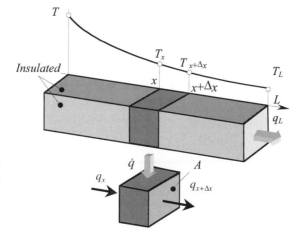

Figure 2.2
The control volume balance for the conduction heat transfer.

Here the heat transfer rate q_x is taken with respect to the positive direction of Ox, that is, from hot to cold. In this case, the single contribution to the energy inventory, e, is the internal energy, u:

$$e = (\rho A \Delta x)u. \tag{2.2}$$

$\rho(\text{kg/m}^3)$ is the *mass density* of the bar, and $A(m^2)$ is the cross-sectional area. In the case of incompressible substances, the internal energy variation is

$$du = cdT, \tag{2.3}$$

such that the right-hand side of (29) may be written as

$$\frac{\partial e}{\partial t} = (\rho cA\Delta x)\frac{\partial T}{\partial t}. \tag{2.4}$$

The work interaction, $w(\text{W/m}^3)$, which is a term in the energy balance (2.1), also accounts — in a broader sense — for interactions other than those of a mechanical nature. For instance, it also comprises the *Joule heating* related to an electrical current that may flow through the bar,

$$-w = (A\Delta x)\dot{q}. \tag{2.5}$$

Here $\dot{q}$ (W/m^3) is the *Joule heat generation rate* (2.23), and the minus sign accounts for the sense of this work interaction — *toward* the system.

From the balance equation (2.1) it results that the heat transfer rate, q_x, is proportional to the temperature drop along the control volume, Δx, that is, $q_x = C(T_x - T_{x+\Delta x})$. Empirically, it may be shown that the constant C is equal to the ratio $A/\Delta x$, which suggests the definition

$$q_x \overset{def}{=} kA\frac{\partial T}{\partial x}, \tag{2.6}$$

also called the *Fourier law*. It follows that

$$q_x'' = -k\frac{T_2 - T_1}{L} = k\frac{\Delta T}{L}, \tag{2.7}$$

where $q_x''(\text{W/m}^2)$ is the heat flux (density) rate per unit area, in the heat flow direction Ox.

Equation (2.7) shows that the heat flux is proportional to the temperature gradient in the Ox direction. The term k (W/m·K) is the *thermal conductivity* of the control volume (the bar), a property of the substance. The minus sign accounts for the heat transfer flow: *from hot to cold* $(T_1 > T_2)$.

In the particular case of a plate of finite thickness, L, with $k = \text{const}$ and under steady-state conditions, the temperature gradient is

$$\frac{dT}{dx} = \frac{T_2 - T_1}{L}. \tag{2.8}$$

This relation also suggests a convenient method of evaluating the heat flux (density) rate. If heat is transferred through a surface of size A, the *total* heat flux is then $q_x = q_x''A$.

An explicit, simpler expression for the conduction heat flux that leaves the control volume Δx may be obtained by the Taylor linearization procedure, namely

$$q_{x+\Delta x} \approx q_x + \frac{\partial q_x}{\partial x}\Delta X = -A\left[k\frac{\partial T}{\partial x} + \frac{\partial}{\partial x}\left(k\frac{\partial T}{\partial x}\right)\Delta x\right]. \tag{2.9}$$

In the limits of this first order approximation, the balance equation (2.1) for a 1D control volume becomes

$$\frac{\partial}{\partial x}\left(k\frac{\partial T}{\partial x}\right) + \dot{q} = \rho c\frac{\partial T}{\partial t}, \tag{2.10}$$

which is the (1D) *heat conduction law* or *diffusion law*.

It is instructive to recognize the terms that appear in (2.10): the *heat conduction term*, $\frac{\partial}{\partial x}\left(k\frac{\partial T}{\partial x}\right)$, the *internal heat generation term*, $\dot{q}$, and the *internal heat accumulation* or *delay* term, $\rho c\frac{\partial T}{\partial t}$. The group ρc is called the *specific heat capacity* of the substance.

The specific heat flux (or flux density) is a vector quantity. For linear, isotropic, and homogeneous media, its expression is

$$\mathbf{q}'' = q_x''\mathbf{1} + q_y''\mathbf{j} + q_z''\mathbf{k} = -k\left(\frac{\partial T}{\partial x}\mathbf{1} + \frac{\partial T}{\partial y}\mathbf{j} + \frac{\partial T}{\partial z}\mathbf{k}\right) = -k\nabla T \equiv -k\,\mathbf{grad}\,T.$$

$$(2.11)$$

The Heat Conduction Equation in the Principal Systems of Coordinates

Vector form (any coordinate system):

$$\rho c\frac{\partial T}{\partial t} = div(\overline{\overline{k}}\,\mathbf{grad}\,T) + \dot{q}$$

Cartesian (x, y, z):

$$\rho c\frac{\partial T}{\partial t} = \frac{\partial}{\partial x}(k_x\frac{\partial T}{\partial x}) + \frac{\partial}{\partial y}\left(k_y\frac{\partial T}{\partial y}\right) + \frac{\partial}{\partial z}\left(k_z\frac{\partial T}{\partial z}\right) + \dot{q}$$

Cylindrical (r, φ, z):

$$\rho c\frac{\partial T}{\partial t} = \frac{1}{r}\frac{\partial}{\partial r}(k_r r\frac{\partial T}{\partial r} + \frac{1}{r^2}\frac{\partial}{\partial \theta}\left(k_\theta\frac{\partial T}{\partial \theta}\right) + \frac{\partial}{\partial z}(k_z\frac{\partial T}{\partial z}) + \dot{q}$$

Spherical (r, φ, θ):

$$\rho c\frac{\partial T}{\partial t} = \frac{1}{r^2}\frac{\partial T}{\partial r}\left(k_r r^2\frac{\partial T}{\partial r}\right) + \frac{1}{r^2\sin\varphi}\frac{\partial}{\partial\varphi}\left(k_\varphi\frac{\partial T}{\partial\varphi}\right) + \frac{1}{r^2\sin^2\varphi}\frac{\partial}{\partial\theta}\left(k_\theta\frac{\partial T}{\partial\theta}\right) + \dot{q}$$

Unless otherwise specified, in the following sections of this chapter on conduction heat transfer the system under investigation is assumed to be a solid body.

2.2 Thermal Conductivity

Several theories try to produce explicit definitions for k in terms of the different types of thermal properties (linearity, isotropy, homogeneity). For instance, the thermal conductivity of *monatomic gases*, k, is a function of temperature only [5]:

$$k = k_0 \left(\frac{T}{T_0}\right)^n. \tag{2.12}$$

Here, $n = \frac{1}{2}$ (theoretically) and k_0 is the thermal conductivity measured at T_0 ($n \approx 0.7$ for helium). For *solid substances*, the thermal conductivity is explained by the free electron motion (the so-called *electronic gas*), k_e and by the lattice vibrations k_l; hence,

$$k = k_e + k_l. \tag{2.13}$$

However, in many practical situations a satisfactory approximation is

$$k \approx k_e. \tag{2.14}$$

The *electron–phonon* interactions, k_{ph}, and the *electron impurities*, as well as the lattice imperfections (cracks, discontinuity, etc.) k_i, contribute to the free electrons mobility reduction and may be accounted for by

$$\frac{1}{k_e} = \frac{1}{k_{ph}} + \frac{1}{k_i}, \tag{2.15}$$

and through the empirical relations

$$\frac{1}{k_{ph}} = a_f T^2, \frac{1}{k_i} = \frac{a_i}{T}, \tag{2.16}$$

where a_f and a_i are two constants, specific to a particular substance. Summing up, (41), (43), and (44) lead to the following definitions for the thermal conductivity (a function of temperature):

$$k(T) = \frac{1}{a_f T^2 + \dfrac{a_i}{T}}. \tag{2.17}$$

$k(T)$ has a maximum. Analytically,

$$k_{\max} = \frac{3}{2^{2/3}} a_f^{1/3} a_i^{2/3}, \quad \text{for} \quad T = \left(\frac{a_i}{2a_f}\right)^{1/3}. \tag{2.18}$$

A very good approximation of k for the nonpure metals is given by the *Wiedemann-Franz law* [2],

$$k \frac{\rho_e}{T} = L_0, \tag{2.19}$$

where ρ_e(S/m) is the *electrical resistivity* of the metal and $L_0 = 2.45 10^{-8}$ V$_2$/K^2 is *Lorentz's constant*.

In many applications of technical interest, within the range of working temperatures, thermal conductivity may be considered constant. Figure 2.3 [4] gives an order of magnitude image of the thermal conductivities for different substances, under normal pressure and temperature working conditions.

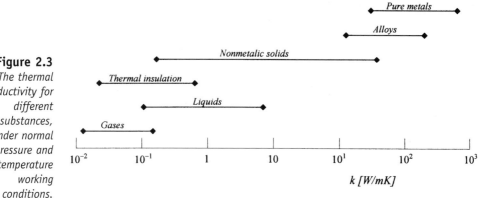

Figure 2.3
*The thermal
conductivity for
different
substances,
under normal
pressure and
temperature
working
conditions.*

Assuming that the thermal conductivity is constant, (2.10) becomes

$$\frac{\partial^2 T}{\partial x^2} + \frac{\dot{q}}{k} = \frac{1}{\alpha}\frac{\partial T}{\partial t}, \qquad (2.20)$$

where the group $\alpha = k/\rho c \, (m^2/s)$, typical of any diffusion problem, is the *thermal diffusivity*. For monatomic gases at low pressure, the mass density ρ is proportional to the ratio P/T, c_P is constant, and (2.20) becomes

$$\alpha = \alpha_0 \left(\frac{T}{T_0}\right)^{n+1} \left(\frac{P}{P_0}\right)^{-1}. \qquad (2.21)$$

Hence, α is function of temperature and pressure.

Although Fourier's law (2.6) and the heat conduction law [(2.10) and (2.11)] were introduced for solid substances, they are equally valid for liquids and gases, with the observation that here c is the specific heat at constant pressure, that is, $c_P \neq c_V$.

2.3 Initial, Boundary, and Interface Conditions

Mathematically, the macroscopic diffusion equation of heat transfer is a partial differential equation, in space and time. Consequently, the heat transfer problem is a *well-posed* problem in the Hadamard sense — that is, there *exists* a solution, which is *unique* and *depends continuously on the boundary conditions* — if consistent *initial conditions* (ICs) and *boundary conditions* (BCs) for the temperature are prescribed and if the *heat sources definitions* are specified.

The ICs — here, the temperature — need to be prescribed for the boundary *and* in the entire physical region that makes the system, and they pinpoint the initial thermodynamic state of the system.

The BCs — either the temperature (*Dirichlet* condition) or its gradient (*Neumann* condition), or a linear combination of these two quantities (*Robin* condition) — must be given for the entire boundary, at any moment (Fig. 2.4). The BCs account for the interactions of the system with its surroundings.

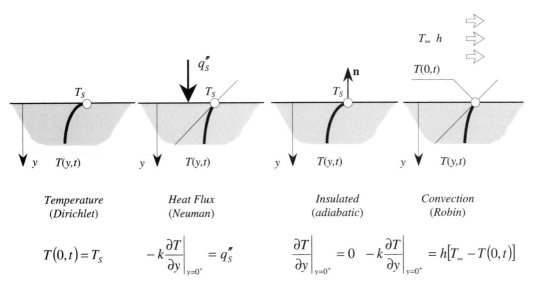

$$T(0,t)=T_S \qquad -k\frac{\partial T}{\partial y}\bigg|_{y=0^+}=q_S'' \qquad \frac{\partial T}{\partial y}\bigg|_{y=0^+}=0 \qquad -k\frac{\partial T}{\partial y}\bigg|_{y=0^+}=h[T_\infty - T(0,t)]$$

Figure 2.4 *The heat transfer boundary conditions.*

The Dirichlet condition is related to boundaries (surfaces) with prescribed temperature; the Neumann condition is related to prescribed heat flux boundaries, and the Robin condition is posed for convection and linearized radiation heat transfer boundaries. For systems with an *open* boundary, these conditions are replaced by *regularity* or *asymptotic* BCs posed on the open boundary.

If the physical region of the problem is piecewise homogeneous, that is, with piecewise constant thermal conductivity, then the interfaces between the different constituent regions are *surfaces of discontinuity* for the temperature gradient (Fig. 2.5). Mathematically, on these interfaces one specifies continuity conditions for the temperature *and* for the heat flux. The latter implies a finite "jump" in the temperature gradient. Should an interface also bear a superficial heat source (or sink), then the heat flux also has a jump, equal to the value of the source (sink). One such example is the melting/solidification interface in a phase change problem, where the latent heat of melting/solidification acts as heat source/sink.

Figure 2.5
Heat transfer through the interface that separates two solids with different thermal conductivities.

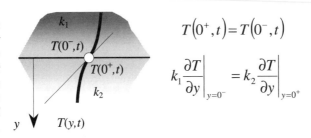

$$T(0^+,t)=T(0^-,t)$$

$$k_1\frac{\partial T}{\partial y}\bigg|_{y=0^-}=k_2\frac{\partial T}{\partial y}\bigg|_{y=0^+}$$

It should be noted that this discussion on ICs, BCs, and interface conditions is consistent with the concepts of *thermodynamic property* (whereas temperature is a property, the temperature gradient is not a property), that of *initial state* (ICs), and the *uniqueness* of the solution to the partial differential equation of the heat conduction problem (the temperature field).

2.4 Thermal Resistance

An important concept introduced through (2.8), by analogy with the electrical conduction process, is that of *thermal resistance*. Similarly to the electrical resistance, which is associated with the electrical current flow, the thermal resistance is associated with heat flow. Basically, the thermal resistance R_{th}(K/W) is defined by the ratio of temperature drop to heat flux. Depending on the particular heat transfer mechanism (conduction, convection, radiation), the thermal resistance may be

$$\text{Conduction} \quad R_{th,cond} = \frac{T_h - T_c}{q_{cond}} \frac{L}{kA}$$

$$\text{Convection} \quad R_{th,conv} = \frac{T_S - T_\infty}{q_{conv}} \frac{1}{h_{conv}A}$$

$$\text{Radiation} \quad R_{th,rad} = \frac{T_S - T_{surf}}{q_{rad}} \frac{1}{h_{rad}A}.$$

Here A is the heat transfer cross-sectional area, normal to the heat flux direction (Ox), L is the length of the heat flux tube, k is the thermal conductivity of the sample, T_h is the high temperature, T_c is the low temperature, T_S is the wall temperature of the sample, T_∞ is the ambient temperature, T_{surf} is the temperature of the radiative surface, h_{conv} is the convection heat transfer coefficient, and h_{rad} is the radiation heat transfer coefficient.

Another important derived concept is the *contact thermal resistance*. This type of thermal resistance is common to all technical applications — electrical machines and apparatuses, electronics and power electronics, etc. — wherever contact interfaces or composite media exist. The heat transfer is then accompanied by a supplementary temperature drop at these interface levels, and this phenomenon is accounted for by the *specific contact thermal resistance*, $R''_{contact}$ (Fig. 2.6).

The real, imperfect mechanical contact between parts A and B of a composite structure makes the heat flux, q''_y, flow through conduction between the contacting solid parts, q''_{cond}, and by combined convection and/or radiation through the microcavities, q''_{cavity}. The *effective* area of contact is usually small, which is particularly true for rugous surfaces. The major contribution to the heat transfer process is then left for the microcavities. The apparently abrupt temperature variation at the interface level is *not* a discontinuity (remember, temperature is a property) but a steep variation, of the order of degrees per micron.

Usually, the unwanted thermal contact resistance that exists in the case of solid bodies whose thermal conductivities exceed those of the fluid that

Heat Transfer

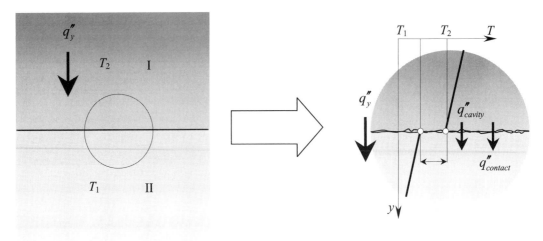

Figure 2.6   *The specific contact thermal resistance.*

fills in the contact microcavities (e.g., air) may be reduced by enhancing the effective contact area. Technically, this can be done by increasing the contact pressure, by smoothing out the rugosities, or by using a contact filler — a fluid or a soft material of a higher thermal conductivity than both A and B. However, in any situation, no solution can suppress the unwanted effect of the microcavities.

2.5 Steady Conduction Heat Transfer

2.5.1 STEADY CONDUCTION WITHOUT INTERNAL HEAT SOURCES

In this class of problems the heat flux is produced by an externally imposed temperature gradient — for example, the faces of a plate of finite thickness are connected to *heat reservoirs* that have different temperatures. Table 2.1 summarizes a few basic 1D configurations. Other common configurations are as follows:

The homogeneous pane wall (Fig. 2.7):

Equation $\quad \dfrac{d}{dx}\left(k\dfrac{dT}{dx}\right) = 0$

BCs $\qquad T(0) = T_1\, T(L) = T_2$

$$T(x) = (T_2 - T_1)\frac{x}{L} + T_1$$

$$q_x = \frac{kA}{L}(T_1 - T_2)$$

Solution $\quad q_x = \dfrac{kA}{L}(T_1 - T_2)$

$$q_x'' = kL(T_1 - T_2)$$

$$R_{tb} = \frac{1}{b_1 A} + \frac{L}{kA} + \frac{1}{b_2 A}$$

Table 2.1

Type of wall →	Plane	Cylindrical	Spherical
Energy equation	$\dfrac{d^2 T}{dx^2} = 0$	$\dfrac{1}{r}\dfrac{d}{dr}\left(r\dfrac{dT}{dt}\right) = 0$	$\dfrac{1}{r_2}\dfrac{d}{dt}\left(r^2\dfrac{dT}{dr}\right) = 0$
Temperature field	$T_1 - \Delta T\,\dfrac{x}{L}$	$T_2 - \Delta T\,\dfrac{\ln\left(\dfrac{r}{r_2}\right)}{\ln\left(\dfrac{r_1}{r_2}\right)}$	$T_1 - \Delta T\,\dfrac{1-\dfrac{r_1}{r}}{1-\dfrac{r_1}{r_2}}$
Heat flux	$k\dfrac{\Delta T}{L}$	$\dfrac{k\Delta T}{r\ln\left(\dfrac{r_2}{r_1}\right)}$	$\dfrac{k\Delta T}{r^2\left[\dfrac{1}{r_1}-\dfrac{1}{r_2}\right]}$
Heat transfer rate	$kA\dfrac{\Delta T}{L}$	$2\pi L\dfrac{k\Delta T}{r\ln\left(\dfrac{r_2}{r_1}\right)}$	$4\pi\dfrac{k\Delta T}{\dfrac{1}{r_1}-\dfrac{1}{r_2}}$
Thermal resistance	$\dfrac{L}{kA}$	$\dfrac{\ln\left(\dfrac{r_2}{r_1}\right)}{2\pi Lk}$	$\dfrac{\dfrac{1}{r_1}-\dfrac{1}{r_2}}{4\pi k}$
Critical insulation	—	$\dfrac{k}{h}$	$2\dfrac{k}{h}$

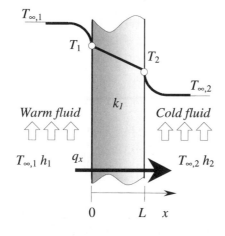

Figure 2.7
Steady heat conduction without internal heat sources within the homogeneous plane wall.

The composite plane wall (Fig. 2.8):

Equation $\dfrac{d}{dx}\left(k\dfrac{dT}{dx}\right) = 0$

BCs $T(0) = T_{\infty,1},\ T(L) = T_{\infty,2}$

Interface $k_1\left.\dfrac{dT}{dx}\right|_{x=L_a-} = k_2\left.\dfrac{dT}{dx}\right|_{x=L_a+},\ k_2\left.\dfrac{dT}{dx}\right|_{x=L_b-} = k_3\left.\dfrac{dT}{dx}\right|_{x=L_b+}$

Solution $q_x = \dfrac{T_{\infty,2} - T_{\infty,1}}{R_{tb}}$

$$R_{tb} = \dfrac{1}{b_1 A} + \dfrac{L_a}{b_1 A} + \dfrac{L_b - L_a}{k_2 A} + \dfrac{L_c - L_b}{k_3 A} + \dfrac{1}{b_2 A}$$

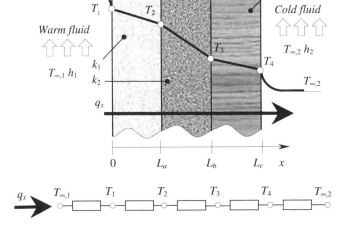

Figure 2.8
Steady heat conduction without internal heat sources within the composite plane wall.

2.5.2 STEADY CONDUCTION WITH INTERNAL HEAT SOURCE

In many circumstances, heat transfer is related to heat *generation* or *absorption*, through conversion from/to other forms of energy (e.g., chemical, nuclear, or electrical). Examples of heat generation are electrical heating, associated with the presence of electrical current; thermal effects produced by electrical and magnetic hysteresis in ferroelectric and ferromagnetic substances; exothermal chemical reactions; and nuclear fission. Examples of heat absorption processes are thermoelectric effects (Peltier, Thomson, Zeebeck [5]) and endothermal chemical reactions. From the heat transfer point of view, these internal—exo- and endothermal—processes are treated as *heat sources* and *heat sinks.*

A frequently encountered thermal effect in electrical engineering is *Joule heat generation.* The work rate (power) needed for an electrical current i (A) to pass through an electrical conductor with the electrical resistance $R_e(\Omega)$ is

$$\dot{E}_g = I^2 R_e, \tag{2.22}$$

and it dissipates as heat. Should the heat production be uniform throughout the entire conductor volume, V, then its generation rate is

$$\dot{q} = \frac{\dot{E}_g}{V} = \frac{I^2 R_e}{V} \left[\frac{W}{m^3}\right]. \tag{2.23}$$

It is important to note that the thermal resistance was introduced in the context of linear, homogeneous, and isotropic media without heat sources and, as such, its utilization in the analysis of heat transfer problems in media with heat sources may be improper.

The Plane Wall with Internal Heat Source

A simple model for the heat conduction in a region with heat sources is that of the plane wall of finite thickness $2L$, which is in contact with two thermal reservoirs that keep its faces at constant temperatures T_1 and T_2 (Fig. 2.9c). If the plate is made of a homogeneous, linear, and isotropic substance with the thermal conductivity k, and if there is a body heat source with the heat generation rate $\dot{q}$, then the mathematical model and the solution to it are

Equation $\quad \dfrac{d^2 T}{dx^2} + \dfrac{\dot{q}}{k} = 0$

BCs $\quad T(-L) = T_1, \ T(L) = T_2$

Solution $\quad T(x) = \dfrac{\dot{q}L^2}{2k}\left(1 - \dfrac{x^2}{L^2}\right)\dfrac{T_2 - T_1}{2}\dfrac{x}{L} + \dfrac{T_2 + T_1}{2}.$

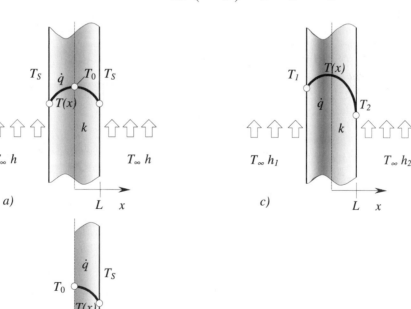

Figure 2.9
Heat conduction within the plane wall of finite thickness in a region with internal heat sources.

Should $T_1 \equiv T_2$, then (by symmetry) it results that the vertical midplane is an adiabatic surface, that is, $dT/dx|_{x=0} = 0$. Hence the problem may be reduced (Figs. 2.9a, 2.9b), and the solution is $T(x) = \dot{q}L^2/2k + T_S$.

2.6 Heat Transfer from Extended Surfaces (Fins)

By heat transfer from extended surfaces is usually understood the *global* heat transfer process—by conduction, inside a solid, finned body (inclusively the fins), and by convection and/or radiation from this one to its ambient (e.g., air). The most common applications are those where such extended surfaces are used to *enhance* the heat transfer rate from a solid body to its surrounding environment. In this context, such an extended, finned surface is called *radiator.*

For the particular cases in Fig. 2.10a there are two options for enhancing the heat transferred from the solid body to its surroundings: either by *improving* (increasing) the heat transfer coefficient h, or by *increasing* the total heat transfer area, that is, by extending it through *finning*. In many situations the first option is not affordable since it would imply a (larger) pump or even another, thermally more effective fluid (e.g., water, or dielectric fluids instead of air). Figure 2.10b shows a possible solution to the second, usually preferred option.

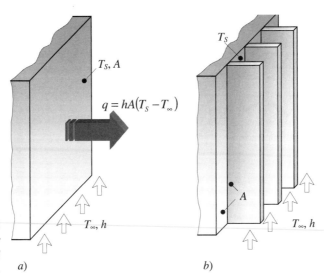

Figure 2.10
Heat transfer from extended surfaces.

A key feature that a finned structure must possess is a higher thermal conductivity than that of the substrate, or it may diminish the heat transferred rather than increasing it. Ideally, the finned structure should be made of such a material as to allow for (almost) isothermal operation, thus maximizing the heat transfer rate.

Finned surfaces are extensively used in electrical machine design, electronic and electric devices and circuits, internal combustion engines, refrigeration systems, and domestic heaters, to name only some applications. The particular design of the fins may be very different (plates, pins, tubes, etc.), depending on the particular technical application, mounting conditions, weight restrictions, fabrication technology, and cost. The radiators may be utilized either to extend the surfaces of the solid bodies through which the heat transfer takes place, or as intermediate heat transfer elements between different working fluids (heat exchangers). They may be made of fins with variable cross sections but, in any situation, they fulfill the same function: *they convey the largest part of the heat that is transferred from the finned body to its surrounding fluid environment.*

2.6.1 THE GENERAL EQUATION OF HEAT CONDUCTION IN FINS

The heat conduction equation is obtained by writing the heat transfer rate balance for a control volume. For simplicity, we shall consider that the 1D fin with variable cross section shown in Fig. 2.11 is made of a linear, isotropic, and homogeneous substance and that there is no internal heat generation. The heat balance for the dx slice is then

$$q_x = q_{x+dx} + dq_{conv}. \tag{2.24}$$

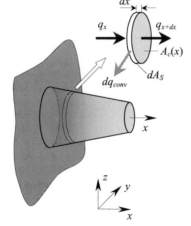

Figure 2.11
The heat transfer rate balance for a 1-D control volume within a fin of variable cross section.

Fourier's law (2.6) may be used to compute the longitudinal heat flux that enters the control volume:

$$q_x = -kA_c(x)\frac{dT}{dx}. \tag{2.25}$$

[$A_c(x)$ is the fin cross-sectional area.] Taylor's linearization scheme gives a simpler expression for the heat flux that leaves the control volume, namely,

$$q_{x+dx} = q_x + \frac{dq_x}{dx}dx, \tag{2.26}$$

which combined with (53) yields

$$q_{x+dx} = -kA_c(x)\frac{dT}{dx} - k\frac{d}{dx}\left[A_c(x)\frac{dT}{dx}\right]dx. \tag{2.27}$$

Finally, if we substitute the lateral convection heat transferred from the side wall of the control volume,

$$dq_{conv} = hdA_S(T - T_\infty), \tag{2.28}$$

and (2.25) and (2.27) in the balance equation (2.24), we obtain

$$\frac{d^2T}{dx^2} + \left(\frac{1}{A_c}\frac{dA_c}{dx}\right)\frac{dT}{dx} - \left(\frac{1}{A_c}\frac{h}{k}\frac{dA_S}{dx}\right)(T - T_\infty) = 0. \tag{2.29}$$

2.6.2 FINS WITH CONSTANT CROSS-SECTIONAL AREA

For these fins (Fig. 2.12), $A_c(x) = A_c = $ constant; the outer surface area is $A_S(x) = Px = $ const, where P is the wet perimeter of the fin cross-section; and (2.29) reduces then to

$$\frac{d^2T}{dx^2} - \frac{hP}{kA_c}(T - T_\infty) = 0, \tag{2.30}$$

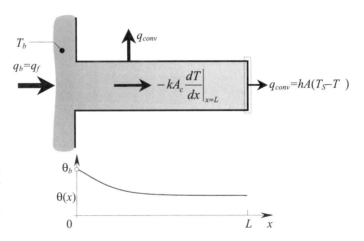

Figure 2.12
Heat transfer from fins with constant cross-section.

or, in nondimensional form,

$$\frac{d^2\theta}{dx^2} - m^2\theta = 0, \quad m^2 = \frac{hP}{kA_c}, \quad \theta(x) = T(x) - T_\infty. \tag{2.31}$$

The solution to this standard, second-order Euler ordinary differential equation is of the form

$$\theta(x) = C_1 e^{mx} + C_2 e^{-mx}, \tag{2.32}$$

where the integration constants C_1 and C_2 may be determined by imposing the boundary conditions prescribed for $x = 0$ and $x = L$. Table 2.2

Table 2.2[a]

Tip condition	Temperature, θ/θ_b	Heat transfer rate, q_f
Convection $h\theta(L) = -k\dfrac{d\theta}{dx}\Big|_{x=L}$	$\dfrac{\cosh[m(L-x)] + \dfrac{h}{mk}\sinh[m(L-x)]}{\cosh(mL) + \dfrac{h}{mk}\sinh(mL)}$	$M\,\dfrac{\sinh(mL) + \dfrac{h}{mk}\cosh(mL)}{\cosh(mL) + \dfrac{h}{mk}\sinh(mL)}$
Adiabatic $\dfrac{d\theta}{dx_{x=L}} = 0$	$\dfrac{\cosh[m(L-x)]}{\cosh(mL)}$	$M\tanh(mL)$
Temperature $\theta(L) = \theta_L$	$\dfrac{\dfrac{\theta_L}{\theta_b}\sinh(mx) + \sinh[m(L-x)]}{\sinh(mL)}$	$M\,\dfrac{\cosh[mL] - \dfrac{\theta_L}{\theta_b}}{\sinh(mL)}$
Asymptotic $\theta(L)\xrightarrow[L\to\infty]{} 0$	e^{-mx}	M

[a] $\theta = T(x) - T_\infty$; $\theta_b = \theta(0) = T(0) - T_\infty$; $m^2 = hP/kAc$; $M = \sqrt{hPkA_c}$.

summarizes some frequently encountered types of fins with isothermal bases—that is, $T(0) = T_b$, or $\theta(0) = \theta_b = T_0 - T_\infty$.

The fins' performance in enhancing the heat transferred from the finned body to its surroundings is evaluated against several quality indicators: *efficacy ε_f*, *thermal resistance, $R_{tb,f}$*; *efficiency η_f*, and *overall superficial efficiency η_{ov}*. Table 2.3 summarizes the definitions of these quantities and their actual forms for the fins listed in Table 2.2. Two common types of radiators are shown in Fig. 2.13.

Table 2.3

Effectivity	$\epsilon_f \overset{def}{=} \dfrac{q_f}{hA_{c,b}\theta_b} = \dfrac{R_{tb,b}}{R_{tb,f}}$	$\epsilon_f = \left(\dfrac{kP}{hA_c}\right)^{1/2}$ (infinite fin)
Thermal resistance	$R_{tb,f} \overset{def}{=} \dfrac{\theta_b}{q_f}$	$R_{tb,b} \overset{def}{=} \dfrac{\theta_b}{q_b}$
Efficiency	$\eta_f \overset{def}{=} \dfrac{q_f}{q_{max}} = \dfrac{q_f}{hA_f\theta_b}$	$\eta_f = \dfrac{\tanh(mL)}{mL}$ (insulated tip)
		$\eta_f = \dfrac{\tanh(mL_c)}{mL_c}$ (active tip)
Overall efficiency	$\eta_{ov} \overset{def}{=} \dfrac{q_t}{q_{max}} = \dfrac{q_t}{hA_t\theta_b}$	$\eta_{ov} = 1 - \dfrac{A_f}{A_t}(1 - \eta_f)$

$A_{c,b} = A_c(0)$, fin basis area; A_f, fin lateral area; $A_t = -A_f + A_b$, radiator total area (finned and unfinned surface).
$L_c = L + t/2$, corrected length for the active-tip fin, acceptable for $ht/k < 0.0625$.
q_f, heat flux rate transmitted by the fin; $q_t = hA_b\theta_b + hA_f\eta_f\theta_b$, total heat flux rate transmitted by the fin; q_b, heat flux rate transmitted to the fin (through the area covered by its base).
$R_{tb,b}$, convection thermal resistance (what would be without the fin).

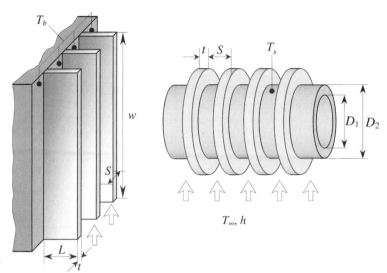

Figure 2.13
Radiators.

2.7 Unsteady Conduction Heat Transfer

In many applications heat transfer is a dynamic, time-dependent process. For instance, the onset of an electric current or the onset of a time-dependent magnetic field in an electroconductive body, or a change in the external thermal conditions of the body, are examples where the thermal steady state (if any) is reached asymptotically, through a transient regime. In these circumstances, the temperature field inside the body is obtained by solving the time-dependent energy balance equation.

2.7.1 LUMPED CAPACITANCE MODELS

When the thermal properties of the body under investigation and the thermal conditions of its surface are such that the temperature inside the body varies uniformly in time, and the body is—at any moment—almost isothermal, then the *lumped capacitance method* is a very convenient, simpler, yet satisfactory accurate tool of thermal analysis.

Let us assume that a uniformly heated, isothermal (T_i) iron chunk is immersed at $t = 0$ in a cooling fluid with $T_\infty < T_i$ (Fig. 2.14). The temperature inside the body decreases smoothly, monotonously, to eventually reaching the equilibrium value, T_∞. Heat is transferred inside the body by conduction, and by convection from the body to the surrounding fluid reservoir. If the thermal resistance of the body is small as compared to the thermal resistance of the fluid, then the heat transfer process is such that the instantaneous temperature field inside the body is uniform, which implies that the internal temperature gradients are negligibly small. The energy balance equation then takes the particular form

$$-\dot{E}_{out} = \dot{E}_{st}, \tag{2.33}$$

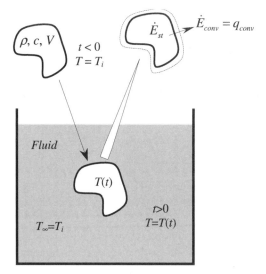

Figure 2.14
The lumped capacitance model for the cooling of a uniformly heated, isothermal iron chunk.

which means

$$-hA_S(T - T_\infty) = \rho Vc \frac{dT}{dt}, \tag{2.34}$$

or, put in nondimensional form,

$$\frac{\rho Vc}{hA_S} \frac{d\theta}{dt} = -\theta, \quad \theta = T - T_\infty. \tag{2.35}$$

Time integration from the initial state $[t = 0, \theta(0) = \theta_i]$ to the current state $[t, \theta(t)]$, that is, $(\rho Vc/hA_S) \int_{\theta_i}^{\theta} d\theta/\theta = -\int_{t_i}^{t} dt$, where $\theta_i = T_i - T_\infty$, yields

$$\frac{\rho Vc}{hA_S} \ln \frac{\theta_i}{\theta} = t, \quad \text{or} \quad \frac{\theta_i}{\theta} = \frac{T - T_\infty}{T_i - T_\infty} = e^{-(hA_S/\rho Vc)t}. \tag{2.36}$$

The group $\rho Vc/hA_S$(s) is the *thermal time constant* (seconds), and it may be re-written as

$$\tau_t = \left(\frac{1}{hA_S}\right)(\rho Vc) = R_t C_t. \tag{2.37}$$

From a practical point of view, it is particularly useful to outline the analogy that exists between the heat flux problem described by (2.36) and that of the electrical current in the $R-C$ circuit shown in Fig. 2.15.

The heat transferred to the fluid in the time span $[0, t]$,

$$Q \overset{def}{=} \int_0^t q dt = hA_S \int_0^t \theta dt, \tag{2.38}$$

is a measure of the change in the internal energy undergone by the system (body) from the initial (at $t = 0$) to the current state ($t = t$):

$$-Q = \Delta E_{st}. \tag{2.39}$$

Figure 2.15
The analogy between the heat flux problem of the lumped capacitance model and the electrical current in an R–C circuit.

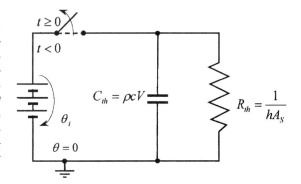

Although this result is reported here for a cooling process, such as the metallurgical process of annealing, where the internal temperature decreases (that is, $Q > 0$), the relation (2.39) is also true for heating processes, where $Q < 0$, that is, the internal energy of the body increases.

The Limits of Applicability for the Lumped Capacitance Model

It is important to recognize that, although very convenient, the lumped capacitance models have a limited validity and, subsequently, applicability criteria for them are needed.

The plate of finite thickness, L, in Fig. 2.16 is assumed to be initially isothermal, T_i. The face at $x = L$ is in contact with a fluid reservoir at T_∞ ($T_i > T_\infty$), while the face at $x = 0$ is maintained at T_i. The heat flux balance for the control surface at $x = L$ is then

$$\frac{kA}{L}(T_1 - T_2) = hA(T_2 - T_\infty), \quad \text{or} \quad \frac{T_1 - T_2}{T_2 - T_\infty} = \frac{L/kA}{1/hA} = \frac{R_{cond}}{R_{conv}} = \text{Bi}. \quad (2.40)$$

The nondimensional quantity $\text{Bi} = hL/k$ is called the *Biot number.* This group plays an important role in the evaluation of the internal conduction heat transfer processes with surface convection conditions, and it may be

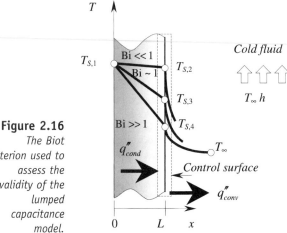

Figure 2.16
The Biot criterion used to assess the validity of the lumped capacitance model.

used to assess the validity of the lumped capacitance method for a particular case. The concept of *characteristic length*, L_c, and the Bi-criterion may be used to decide whether this assumption is valid or not. Essentially, Bi $\ll$ 1 means that the (internal) conduction thermal resistance of the body is much smaller the convection thermal resistance from this one to the fluid; hence, the lumped capacitance model may be safely used. In contrast when Bi $\gg$ 1 the (internal) conduction thermal resistance of the body is larger than the convection thermal resistance from the body to the fluid, and therefore lumped capacitance models must be used with caution.

Consequently, if Bi $= hL_c/k < 0.1$, then the lumped capacitance model is consistent. This interpretation is correct, of course, in linear, isotropic, and homogeneous substances. Figure 2.17 gives a qualitative image of the temperature field inside a plate of finite thickness for different ranges of the Bi number.

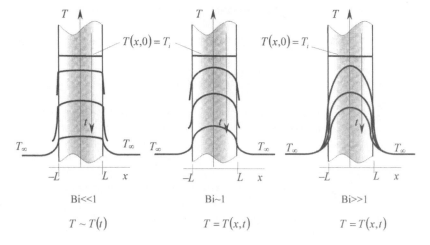

Figure 2.17
The temperature field inside a plate of finite thickness for different ranges of Biot number.

As apparent, the proper evaluation of L_c is crucial to the success of the lumped capacitance method, and for simple problems it is not too difficult to find it. For instance, in the previous problem (Fig. 2.16) $L_c = L$. For bodies of more complex geometry L_c may be taken as the size of the body in the direction of the temperature gradient (heat flux flow). Sometimes L_c is conveniently approximated by $L_c = V/A_S$, where V is the volume of the body and A_s its external surface area. This simple definition yields

$$\frac{hA_S}{\rho Vc}t = \frac{h}{\rho cL_c}t = \frac{hL_c}{k}\frac{k}{\rho c_S}\frac{1}{L_c^2}t = \left(\frac{hL_c}{k}\right)\left(\frac{\alpha}{L_c^2}t\right) = \text{Bi Fo}, \qquad (2.41)$$

where Fo $= (\alpha/L_c^2)t$ is the *Fourier number*—a nondimensional time. It should be noticed that, unlike the Bi number, Fo is *not* a constant, but rather a dynamic quantity. If we use this notation, (2.36) becomes

$$\frac{\theta}{\theta_i} = \frac{T - T_\infty}{T_i - T_\infty} = e^{-\text{Bi Fo}}. \qquad (2.42)$$

Heat Transfer

2.7.2 GENERAL CAPACITIVE THERMAL ANALYSIS

Although the Bi-criterion may be useful in deciding whether the lumped capacitance model is satisfactorily accurate, there are many situations when its validity is questionable — for instance, the presence of internal heat sources, (nonlinear) radiative heat transfer, etc.

Figure 2.18 shows a schematic of a plate whose initial temperature T_i (at $t = 0$) is such that $T_i \neq T_\infty$ and $T_i \neq T_{surf}$. The imposed heat flux, q_S'', and the

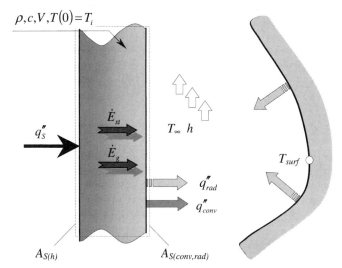

Figure 2.18
The heat flux balance for the general capacitive thermal analysis.

convection, q_{conv}'', and radiation, q_{rad}'', heat fluxes related to the body surface, $A_{S(h)}$ and $A_{S(conv,rad)}$, respectively, are assumed to be such that, globally, the total combined conduction–radiation heat flows from the body to the enclosure walls. The heat flux balance for the body (the control volume here) may be written as

$$q_S'' A_{S(h)} + \dot{E}_g - (q_{conv}'' + q_{rad}'') A_{S(conv,rad)} = \rho V c \frac{dT}{dt}, \qquad (2.43)$$

or, by using the heat flux definitions (1.17, 1.19, 1.25), as

$$q_S'' A_{S(h)} + \dot{E}_g - [h(T - T_\infty) + \sigma \varepsilon (T^4 - T_{surf}^4)] A_{S(conv,rad)} = \rho V c \frac{dT}{dt}. \qquad (2.44)$$

Although usually this nonlinear ordinary differential equation has no exact solution, in certain specific cases it may be analytically integrable. Two such circumstances are listed next.

(a) In the absence of internal heat sources $\dot{E}_g$ and imposed heat flux q_S'', if the convection heat flux is negligibly small with respect to the radiative heat flux, $q_{conv}'' \ll q_{rad}''$, then (2.44) takes the simpler form

$$\rho V c \frac{dT}{dt} + \varepsilon \sigma A_{S(rad)} (T^4 - T_{surf}^4) = 0, \qquad (2.45)$$

which is solved exactly by

$$\int_{T_i}^{T} \frac{dT}{T^4 - T_{surf}^4} = \frac{\rho Vc}{\varepsilon \sigma A_{S(rad)}} \int_0^t dt, \tag{2.46}$$

yielding an explicit definition for the time rather than for the temperature, that is,

$$t = \frac{\rho Vc}{4\varepsilon A_{S(rad)}}$$

$$\times \left\{ \ln \left| \frac{T_{surf} + T}{T_{surf} - t} \right| - \ln \left| \frac{T_{surf} + T_i}{T_{surf} - T_i} \right| + 2 \left[\tan^{-1} \left(\frac{T}{T_{surf}} \right) - \tan^{-1} \left(\frac{T_i}{T_{surf}} \right) \right] \right\}. \tag{2.47}$$

(b) If the radiation heat transfer component is negligibly small, $q''_{rad} \ll (q''_{conv}, q''_S)$, and the conduction heat transfer coefficient h is constant, then (2.44) becomes

$$\frac{d\theta'}{dt} + A\theta' = 0, \ \theta' = \theta - \frac{B}{A}, \ A = \frac{hA_{S(conv)}}{\rho Vc}, \ B = \frac{q''_S A_{S(b)} + \dot{E}_g}{\rho Vc}, \tag{2.48}$$

admitting the analytic solution

$$\frac{\theta'}{\theta'_i} = e^{-At}, \text{ or } \frac{T - T_\infty}{T_i - T_\infty} = e^{-At} + \frac{B/A}{T_i - T_\infty} \left[1 - e^{-At} \right]. \tag{2.49}$$

2.7.3 UNSTEADY HEAT CONDUCTION DRIVEN BY TEMPERATURE GRADIENTS

Outside the limits of validity for the lumped capacitance approach, the thermal problem may be solved by integrating (2.44), which may imply the solution to the domain effects due to temperature gradients. A simple, introductory model is the 1D heat transfer conduction problem of a plate of finite thickness L, made of a linear, isotropic, and homogeneous thermo-conductive substance k, and without internal heat sources. At $t = 0$ the face $x = L$, assumed to be initially isothermal, T_i, is exposed to a fluid reservoir of temperature T_∞, while the face $x = 0$ is thermally insulated, that is, no heat flux is crossing it. The heat transferred from the plate to the fluid is conveyed by conduction, inside the body, and by convection, within the fluid. The latter process is characterized by the constant convection heat transfer

Heat Transfer

coefficient h. The mathematical model and solution for the unsteady, 1D conduction heat transfer process inside the plate are then

Dimensional form: (2.50a)

$$(\text{PDE}) \rightarrow \quad \frac{\partial T}{\partial t} = \alpha \frac{\partial^2 T}{\partial x^2}, \, \alpha = \frac{k}{\rho c}$$

$$(\text{ICs}) \rightarrow \quad T(x, 0) = T_i$$

$$(\text{BCs}) \rightarrow \quad \begin{cases} \dfrac{\partial T}{\partial x}\big|_{x=0} = 0 \\[2mm] -k \dfrac{\partial T}{\partial x}\big|_{x=L} = h[T(L, t) - T_\infty]. \end{cases}$$

Nondimensional form: (2.50b)

$$(\text{PDE}) \rightarrow \quad \frac{\partial \tilde{\theta}}{\partial \text{Fo}} = \frac{\partial^2 \tilde{\theta}}{\partial \tilde{x}^2}$$

$$(\text{ICs}) \rightarrow \quad \tilde{\theta}(\tilde{x}, 0) = 1$$

$$(\text{BCs}) \rightarrow \quad \begin{cases} \dfrac{\partial \tilde{\theta}}{\partial \tilde{x}}\bigg|_{\tilde{x}=0} = 0 \\[2mm] \dfrac{\partial \tilde{\theta}}{\partial \tilde{x}}\bigg|_{\tilde{x}=L} = -\text{Bi}\, \tilde{\theta}(1, \text{Fo}) \end{cases}$$

$$\left(\text{here } \tilde{\theta} = \frac{\theta}{\theta_i} = \frac{T - T_\infty}{T_i - T_\infty}, \, \tilde{x} = \frac{x}{L}, \, \tilde{t} = \frac{\alpha t}{L^2} = \text{Fo} \right).$$

Qualitatively, the solution to the dimensional problem (2.50) is a function of several quantities — space, time, ICs, BCs, and material properties — whereas the solution to the nondimensional problem (2.50) depends only on $\tilde{x}$, Fo, and Bi:

$$T = T(x, t, T_i, T_\infty, L, k, h, \alpha), \quad \tilde{\theta} = \tilde{\theta}(\tilde{x}, \text{Fo}, \text{Bi}). \tag{2.51}$$

For a given geometry, the nondimensional temperature has the merit of a *universal function* of $\tilde{x}$, Fo, and Bi, rather than a function of all ICs, BCs, and material properties.

The 1D problem [(2.50a), (2.50b); i.e., $L \ll (H, W)$] may be solved exactly by the *variable separation* method [6], yielding

$$\tilde{\theta} = \sum_{n=1}^{\infty} C_n e^{-\zeta_n^2 \text{Fo}} \cos(\zeta_n \tilde{x}), \quad C_n = \frac{4 \sin \zeta_n}{2\zeta_n + \sin(2\zeta_n)}, \quad \zeta_n \tan \zeta_n = \text{Bi}. \tag{2.52}$$

For Fo ≥ 2 (that is, at "large" times t) this infinite series may be satisfactorily well represented by only its first term,

$$\tilde{\theta} \cong \tilde{\theta}_0 \cos(\zeta_1 x^*), \quad \tilde{\theta}_0 = C_1 e^{-\zeta_1^2 \text{Fo}}, \tag{2.53}$$

where $\tilde{\theta}_0$ stands for the midplate temperature. Table 2.4 [2, 7] gives several numerical values for ζ_1 and C_1, for different Bi numbers. An important consequence that follows is that *the temperature history at any location inside the plate repeats identically the history of temperature in the middle of the plate.*

Table 2.4

Bi	Plane wall		Infinite cylinder		Sphere	
	ζ_1(rad)	C_1	ζ_1(rad)	C_1	ζ_1(rad)	C_1
0.01	0.0998	1.0017	0.1412	1.0025	0.1730	1.0030
0.02	0.1410	1.0033	0.1995	1.0050	0.2445	1.0060
0.03	0.1732	1.0049	0.2439	1.0075	0.2989	1.0090
0.04	0.1987	1.0066	0.2814	1.0099	0.3450	1.0120
0.05	0.2217	1.0082	0.3142	1.0124	0.3852	1.0149
0.1	0.3111	1.0160	0.4417	1.0246	0.5423	1.0298
0.5	0.6533	1.0701	0.9408	1.1143	1.1656	1.1441
1.0	0.8603	1.1191	1.2558	1.2071	1.5708	1.2732
5.0	1.3138	1.2402	1.9898	1.5029	2.5704	1.7870
10.0	1.4289	1.2620	2.1795	1.5677	2.8363	1.9249
50.0	1.5400	1.2727	2.3572	1.6002	3.0788	1.9962
100.0	1.5552	1.2731	2.3809	1.6015	3.1102	1.9990

In many practical situations it is important to know the total heat transferred from the plate to the fluid, Q, from the initial moment ($t = 0$) to the current time ($t > 0$). The balance equation

$$E_{in} - E_{out} = \Delta E_{st}, \ E_{out} = Q, \ \Delta E_{st} = E(t) - E(0) \tag{2.54}$$

then yields

$$Q = -[E(t) - E(0)], \text{ or } Q = -\int_V \rho c[T(r, t) - T_i] \, dv. \tag{2.55}$$

It is convenient to put this result too in a nondimensional form by using the same scaling:

$$\tilde{Q} = \frac{Q}{Q_{ref}} = -\int_V \frac{T(r, t) - T_i}{T_i - T_\infty} \frac{dv}{V} = \int_{\tilde{V}} (1 - \tilde{\theta}) d\tilde{v}. \tag{2.56}$$

In the limits of the approximation (2.53) first, the right-hand side may be integrated to yield

$$\tilde{Q} = \frac{Q}{Q_{ref}} = 1 - \frac{\sin \zeta_1}{\zeta_1} \tilde{\theta}_0, \tag{2.57}$$

where $\tilde{\theta}_0$ may be obtained from (2.53) second, whereas the coefficients ζ_1 and C_1 can be identified in Table 2.4.

The graphical representations of these results [8, 9] the so-called "Heisler charts" — were used intensively in the unsteady heat conduction analysis. First, Fig. 2.19 [8] gives the midplate temperature, $T(0, t)$ at any moment, and then Fig. 2.20 [8] gives the corresponding temperature (at that time) at any other location within the plate. If T_0 is known for a set of values of Fo and Bi, then the corresponding temperature distribution inside the plate may be evaluated at any moment. The procedure may be reversed in order to find the time needed for the plate face to reach a prescribed temperature.

Heat Transfer

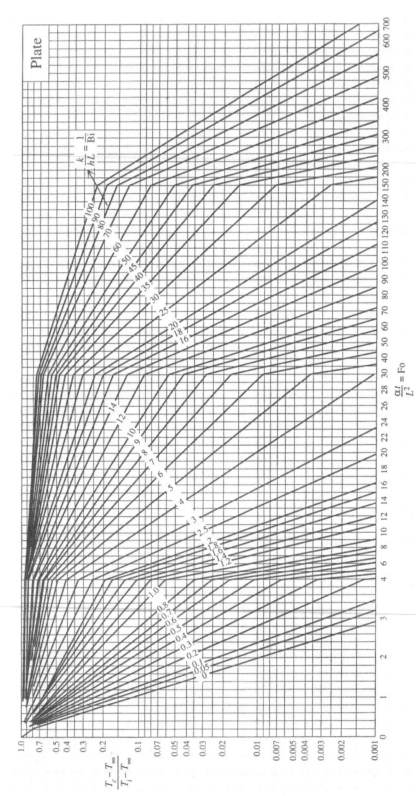

Figure 2.19 Temperature history in the midplane of a plate immersed suddenly in a fluid of a different temperature (L = plate half-thickness) — after Heisler, used with permission from A. Bejan, Heat Transfer, John Wiley, 1993, Fig. 4.7, pp. 159 and 160.

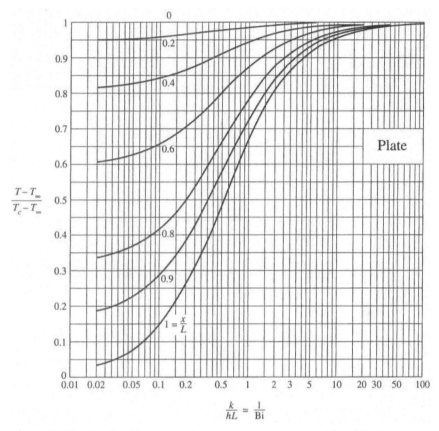

$\dfrac{T-T_\infty}{T_c-T_\infty}$

$\dfrac{k}{hL} = \dfrac{1}{\mathrm{Bi}}$

Figure 2.20 *Relationship between the temperature in any plane (x) and the temperature in the midplane (x = 0, Fig. 2.19) of a plate immersed suddenly in a fluid of a different temperature (L = plate half-thickness) — after Heisler, used with permission from A. Bejan, Heat Transfer, John Wiley, 1993, Fig. 4.8, p. 160.*

The absence of Fo from (2.57) indicates that the temperature history at any point in the plate corresponds to the midplate temperature history. This feature is a direct consequence of the assumption Fo $\geq$ 0.2, which is accurate for "large" times as compared to the initial phases of the transient process.

Figure 2.21 [9] charts the heat transferred [Eq. (2.57)]. Here, the non-dimensional temperature drop is put in terms of Fo and Bi exclusively. The results obtained for the plate of finite thickness may be also utilized for the thermal analysis of plate with one face (x = 0) thermally insulated (adiabatic) and with the other face (x = L) in contact with a convecting fluid.

Other 1D Geometries

Cylindrical bar of infinite length $\left(\dfrac{L}{r_0} > 1\right)$:

$$\tilde{\theta} = \sum_{n=1}^{\infty} C_n e^{-\zeta_n^2 \mathrm{Fo}} J_0(\zeta_n \tilde{r}), \; C_n = \frac{2}{\zeta_n} \frac{J_1(\zeta_n)}{J_0^2(\zeta_n) + J_1^2(\zeta_n)}, \; \zeta_n \frac{J_1(\zeta_n)}{J_0(\zeta_n)} = \mathrm{Bi}$$

Heat Transfer

Figure 2.21
Total heat transfer between a plate and the surrounding fluid, as a func-tion of the total time of exposure, t. (After H. Gröber, S. Erk, and U. Grigull, Fundamentals of Heat Transfer, McGraw-Hill, New York, 1961), used with permission from A. Bejan, Heat Transfer, John Wiley, 1993, Fig. 4.9, p. 162.

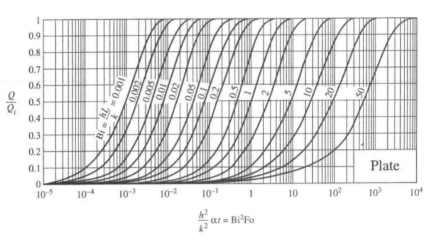

(J_0 and J_1 are Bessel functions of first type and order 0 and 1).
Sphere:

$$\tilde{\theta} = \sum_{n=1}^{\infty} C_n e^{-\zeta_n^2 \mathrm{Fo}} \frac{\sin(\zeta_n \tilde{r})}{\zeta_n \tilde{r}}, \ C_n = \frac{4[\sin(\zeta_n) - \zeta_n \cos(\zeta_n)]}{2\zeta_n - \sin(2\zeta_n)}, \ 1 - \zeta_n \cot(\zeta_n) = \mathrm{Bi}.$$

The following approximations for $\mathrm{Fo} \geq 0.2$ were indicated by Heisler:

Cylinder:

$$\tilde{\theta} = C_1 e^{-\zeta_1^2 \mathrm{Fo}} J_0(\zeta_1 \tilde{r}) = \tilde{\theta}_0 J_0(\zeta_1 \tilde{r}), \tilde{\theta}_0 = C_1 e^{-\zeta_1^2 \mathrm{Fo}}$$

$$\tilde{Q} = \frac{Q}{Q_{ref}} = 1 - \frac{2\tilde{\theta}_0}{\zeta_1} J_1(\zeta_1).$$

Sphere:

$$\tilde{\theta} = C_1 e^{-\zeta_1^2 \mathrm{Fo}} \frac{\sin(\zeta_1 \tilde{r})}{\zeta_1 \tilde{r}} = \tilde{\theta}_0 \frac{\sin(\zeta_1 \tilde{r})}{\zeta_1 \tilde{r}}, \tilde{\theta}_0 = C_1 e^{-\zeta_1^2 \mathrm{Fo}}$$

$$\tilde{Q} = \frac{Q}{Q_{ref}} = 1 - \frac{3\tilde{\theta}_0}{\zeta_1^3} [\sin(\zeta_1) - \zeta_1 \cos(\zeta_1)].$$

Table 2.4 lists the values of ζ_1 and C_1 for several Bi numbers. Figures 2.22, 2.23 [8], and 2.24 [9] represent the charts for the cylinder, and Figs. 2.25, 2.26 [8], and 2.27 [9], those for the sphere. It may be noticed that for the cylinder and sphere Bi is based on r_o rather than L.

The solutions to these unidirectional problems may be combined to construct the time-dependent solutions to more complex, 3D problems, and

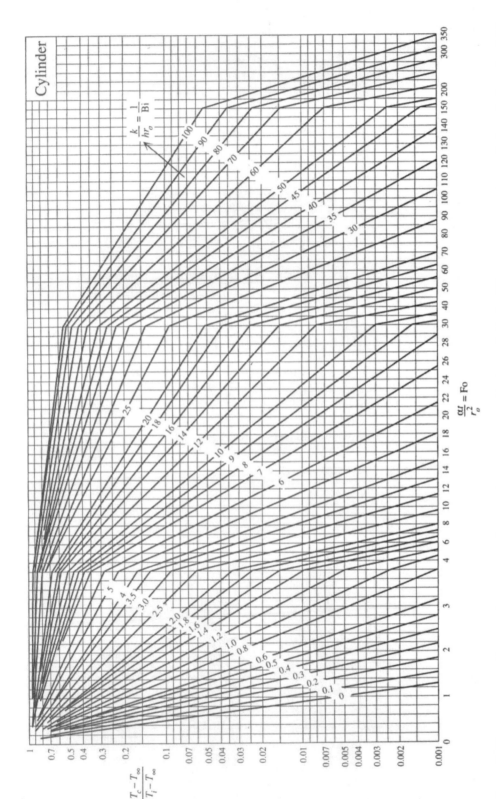

Figure 2.22 *Temperature history at the centerline of a long cylinder immersed suddenly in a fluid of a different temperature (r_0 = cylinder radius) — after Heisler, used with permission from A. Bejan, Heat Transfer, John Wiley, 1993, Fig. 4.10, pp. 164, 165.*

Figure 2.23
Relationship between the temperature at any radius (r) and the temperature at the centerline (r = 0, Figure 26) of a long cylinder immersed suddenly in a fluid of a different temperature (L = plate half-thickness) — after Heisler, used with permission from A. Bejan, Heat Transfer, John Wiley, 1993, Fig. 4.11, p. 166.

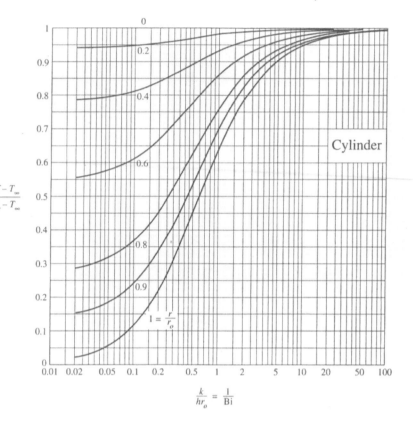

Figure 2.24
Total heat transfer between a long cylinder and the surrounding fluid, as a function of the total time of exposure, t. (After H. Gröber, S. Erk and U. Grigull, Fundamentals of Heat Transfer, McGraw-Hill, New York, 1961), used with permission from A. Bejan, Heat Transfer, John Wiley, 1993, Fig. 4.12, p. 166.

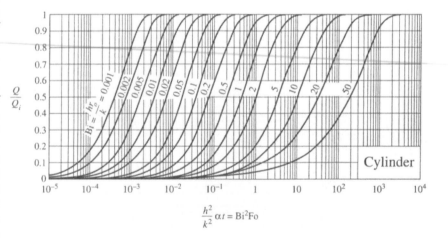

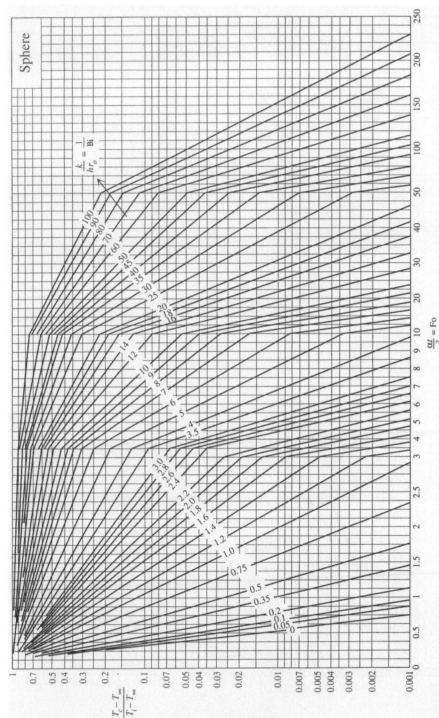

Figure 2.25 *Temperature history in the center of a sphere immersed suddenly in a fluid of a different temperature (r_0 = sphere radius) — after Heisler, used with permission from*

A. Bejan, Heat Transfer, John Wiley, 1993, Fig. 4.13, p. 167 and p. 168.

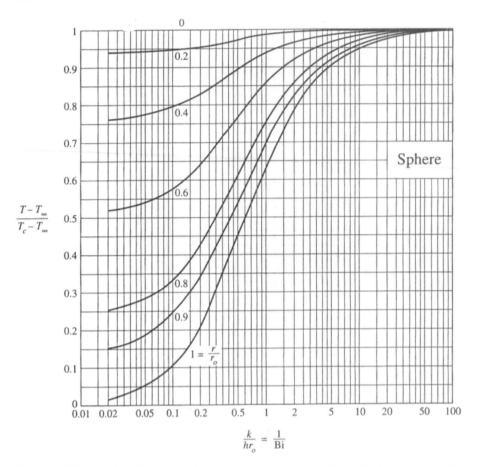

Figure 2.26 *Relationship between the temperature at any radius (r) and the temperature in the center (r = 0, Figure 29) of a sphere immersed suddenly in a fluid of a different temperature (L = plate half-thickness) — after Heisler, used with permission from A. Bejan, Heat Transfer, John Wiley, 1993, Fig. 4.14, p. 169.*

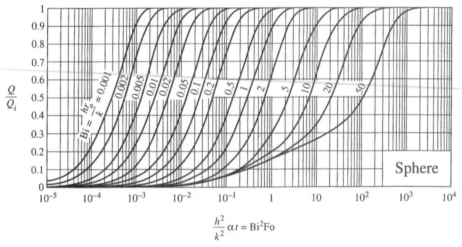

Figure 2.27 *Total heat transfer between a sphere and the surrounding fluid, as a function of the total time of exposure, t. (After H. Gröber, S. Erk and U. Grigull, Fundamentals of Heat Transfer, McGraw-Hill, New York, 1961), used with permission from A. Bejan, Heat Transfer, John Wiley, 1993, Fig. 4.15, p. 169.*

an introductory analysis may be found in ref. 2. The central idea of the method is the superposition principle that is applicable to the heat conduction problems in linear media.

2.7.4 UNSTEADY HEAT CONDUCTION IN SEMI-INFINITE SOLID BODIES

Another important 1D heat transfer problem is that of a *semi-infinite body* with a surface thermal condition — temperature, heat flux, or convection boundary conditions. Although this is a much idealized 1D problem, it may well model situations of practical interest — for instance, heat transfer from the earth's surface, or the thermal field inside a large body in the vicinity of its outer surface. The initial condition may be $T(x, 0) = T_i$, whereas the boundary condition is replaced by the prescribed, expected asymptotic behavior, $T(x, t) \xrightarrow[x \to \infty]{} T_i$. The exact solutions to the problem (2.50a) for the three types of BCs are as follows:

Specified temperature, $T(0, t) = T_S$

$$\frac{T(x, t) - T_S}{T_i - T_S} = \mathrm{erf}\left(\frac{x}{2\sqrt{\alpha t}}\right), \quad q_S''(t) = -k\frac{\partial T}{\partial x}\bigg|_{x=0} = \frac{k}{\sqrt{\pi \alpha t}}(T_S - T_i) \quad (2.58)$$

Specified heat flux, $q_S'' = q_0''$

$$T(x, t) - T_i = 2\sqrt{\alpha t}\frac{q_0''}{k}\mathrm{erf}\left[-\left(\frac{x}{2\sqrt{\alpha t}}\right)^2\right] - \frac{q_0''x}{k}\mathrm{erfc}\left(\frac{x}{2\sqrt{\alpha t}}\right) \quad (2.59)$$

Convection condition, $-k\frac{\partial T}{\partial x}\bigg|_{x=0} = h[T_\infty = T(0, t)]$

$$\frac{T(x, t) - T_i}{T_\infty - T_i} = \mathrm{erf}\left(\frac{x}{2\sqrt{\alpha t}}\right) - \mathrm{erfc}\left(\frac{x}{2\sqrt{\alpha t}} + \frac{h\sqrt{\alpha t}}{k}\right)e^{\frac{hx}{k} + \frac{h^2 \alpha t}{k^2}}. \quad (2.60)$$

The function erf η is the *Gauss error function* and erfc(η) is the *complementary error function*, erfc $\overset{def}{=} 1 - \mathrm{erf}(\eta)$. The similarity and time-dependent temperature profiles are shown in Fig. 2.28.

An interesting application where the semi-infinite model may be used is that of two contacting semi-infinite bodies that have different initial temperatures, $T_{1,i} > T_{2,i}$. If the contact thermal resistance is negligibly small then, by thermodynamic arguments, the interface must instantaneously reach the, as yet unknown, equilibrium temperature T_S, such that $T_{1,i} > T_S > T_{2,i}$. And since T_S remains unchanged throughout the entire transient regime, it follows that the temperature fields and the heat fluxes inside the bodies are of the form (2.58). The interface temperature may then be found by invoking the flux continuity condition at the interface,

$$q_{S,1}'' = q_{S,2}'', \text{ or } -\frac{k_1(T_S - T_{1,i})}{\sqrt{\pi \alpha_1 t}} = \frac{k_2(T_S - T_{2,i})}{\sqrt{\pi \alpha_2 t}}, \quad (2.61)$$

Heat Transfer

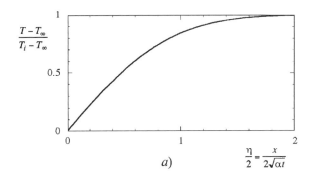

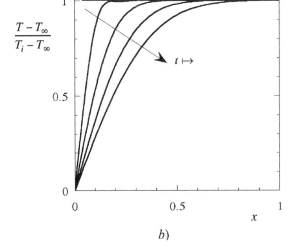

Figure 2.28
*The similarity (a)
and time-
dependent (b)
temperature
profiles for
unsteady heat
conduction in
semi-infinite
solid bodies.*

which yields

$$T_S = \frac{(\sqrt{k\rho c})_1 \, T_{1,i} + (\sqrt{k\rho c})_2 \, T_{2,i}}{(\sqrt{k\rho c})_1 + (\sqrt{k\rho c})_2}. \qquad (2.62)$$

In the preceding equations k_1 and k_2 are the thermal conductivities of the two bodies.

3. Convection Heat Transfer

3.1 External Forced Convection

In fluid media (gases and liquids), heat transfer may occur through *two* different types of mechanisms. First is the thermal diffusion, which is due to the thermal, random motion of microscopic particles. In addition, there exists an amount of heat that is convected by the *macroscopic* motion of the fluid. This motion, called *flow*, may be produced either by a source of a different, nonthermal nature (*forced convection*), by a temperature gradient (*natural* or *free convection*), or by a combination of these two types of sources (*mixed convection*). The global heat transfer is called *heat convection*, and its component produced solely by the macroscopic flow process is then called *heat convection.*

When driven by an *external* (to the heat source) *forced flow*, the convection heat transfer is called *external forced convection.*

3.1.1 THE CONVECTION HEAT TRANSFER COEFFICIENT

In the previous sections, the convection heat transfer between a solid body and the fluid within which this is immersed (air, water, etc.) was characterized by the *local heat transfer coefficient*

$$h \overset{def}{=} \frac{q''}{T_S - T_\infty},$$
(3.1)

but no particular discussion was given of the methods that may actually be used to evaluate it. Because the flow conditions may vary with respect to the body surface, it follows that h is introduced by (3.1) as a *local* quantity. The total heat transfer rate, q, is the integral of the heat transfer rate, q'', over the body surface, A_S:

$$q = \oint_{A_S} q'' dA_S.$$
(3.2)

Should the body surface be isothermal, T_S, an average heat transfer coefficient may be introduced by

$$\bar{h} \overset{def}{=} \frac{q}{A_S(T_S - T_\infty)},$$
(3.3)

which is related to h by

$$\bar{h} = \frac{1}{A_S} \oint_{A_S} h dA_S.$$
(3.4)

In the particular case of a 2D, isothermal plate of size L in the flow direction, the average heat transfer coefficient is defined as

$$\bar{h} \overset{def}{=} \frac{1}{L} \int_0^L h dx.$$
(3.5)

The central objective of convection heat transfer design is the evaluation of the local convection heat flux or of the total convection heat transfer rate. In turn, these quantities may be computed provided that the heat transfer coefficients are known. Therefore, a vast body of experimental and theoretical efforts are aimed at evaluating them. This is a difficult task since, besides the numerous parameters that may occur (fluid density, viscosity, thermal conductivity, specific heat), other factors such as the geometry of the body surface and the flow particularities also significantly contribute to the convection process. However, whatever the complexity of these dependencies, the convection heat transfer is related to the concepts of *hydrodynamic* and *thermal boundary layers.*

Heat Transfer

3.1.2 THE HYDRODYNAMIC BOUNDARY LAYER

The *boundary layer* concept was introduced by Prandtl (1904) [12, 13] to describe the shallow fluid domain that adjoins the solid wall bathed by the flow, where the velocity field varies form zero (at the wall) to the free stream velocity U_∞, and the fluid temperature varies from the wall temperature T_w to the free stream temperature T_∞.

Empirically, there is evidence that a *viscous* fluid is adherent to the rigid body surface that it "washes" (Fig. 3.1). The fluid layer between the body surface and the free flow region is called the *hydrodynamic boundary layer.* This flow domain has a particular significance in *fluid mechanics:* it is in this region that the velocity field varies from zero (at the wall) to the free stream velocity, U_∞. This flow slowing down is typical of viscous fluids, and it is due to the *shear stress,* τ, which acts parallel to the flow.

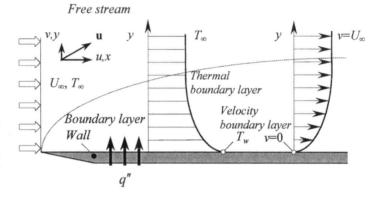

Figure 3.1
The forced convection hydrodynamic and the thermal boundary layers.

. *The boundary layer thickness* δ is a function of the local downstream coordinate x, that is, $\delta = \delta(x)$. Traditionally, the velocity field may be used to define its size in the y-direction ($u|_{u=\delta_{99}} = 0.99U_\infty$), and the boundary layer thickness is then called the *velocity boundary layer thickness* $\delta = \delta_{99}$ [12]. The wall viscous friction is evaluated by the nondimensional *wall friction coefficient,* C_f, through the wall shear stress,

$$C_f \stackrel{def}{=} \frac{\tau_w}{\rho U_\infty^2}. \tag{3.6}$$

In the case of fluids with linear, homogeneous, isotropic dynamic viscosity μ, also called *Newtonian* fluids, the wall shear stress is

$$\tau_w = \mu \frac{\partial u}{\partial y}\Big|_{y=0}. \tag{3.7}$$

3.1.3 THE THERMAL BOUNDARY LAYER

The presence of a temperature difference between the body and the fluid reservoir is responsible for the *convection* heat transfer within the fluid. However, since the viscous fluids adhere to the solid walls (the *no-slip* assumption), heat transfer at the wall–fluid interface is by pure conduction.

Consequently, on the fluid side,

$$q_w'' = -k_f \frac{\partial T}{\partial y}\bigg|_{y=0},\tag{3.8}$$

where k_f is the thermal conductivity of the *fluid*. The heat flux balance at the same interface is

$$h = -\frac{k_f \dfrac{\partial T}{\partial y}\bigg|_{y=0}}{T_w - T_\infty}.\tag{3.9}$$

The boundary layer dynamics has a direct influence upon the heat transferred between the body and the fluid. Similarly to the hydrodynamic boundary layer, and typically for convection heat transfer, in the flow domain there exists a *temperature boundary layer* (Fig. 3.1) that separates the wall (at T_w) from the thermally "fresh" fluid that is at the fluid reservoir temperature T_∞. Its size is measured by the *temperature boundary layer thickness* δ_T, defined through [10]

$$\frac{T_w - T(\delta_T)}{T_w - T_\infty} = 0.99.\tag{3.10}$$

Similarly to δ, $\delta_T = \delta_T(x)$ and it increases downstream, the fluid adjacent to the hot (cold) wall heats up (cools down) both by thermal diffusion from (to) the wall *and* by the enthalpy transported downstream (upstream) by the flow. Usually the two boundary layers have different thicknesses, $\delta \neq \delta_T$, and their ratio is equal to the fluid kinematic viscosity to thermal diffusivity ratio (the Prandtl number of the fluid).

The flow transport processes (of momentum, heat, energy, etc.) that exist inside the boundary layer are essential to convection heat transfer. Should the flow cease, then the heat convection process reduces to its diffusive, molecular component, that is, to thermal conduction.

3.1.4 LAMINAR AND TURBULENT BOUNDARY LAYER FLOWS

A first, crucial stage in the analysis of convection heat transfer consists of recognizing the particular boundary layer flow regime, or respectively the hydrodynamic boundary layer type: *laminar, transition,* or *turbulent.* Qualitatively, laminar flow is characterized by a coherent, ordered structure. Its *streamlines* appear as discernible fascicles, suggesting well-defined particle paths. In particular, the velocity field of the 2D flow represented in Fig. 3.2 has two components (u, v); the vertical component v may contribute significantly to the heat transfer process.

The specific trend of a *turbulent* boundary layer flow with high irregularity is characterized by large velocity fluctuations, which usually result in high momentum and heat transport and high mixing rates that substantially amplify the heat transfer. Unlike the envelope of a laminar boundary layer, which is a smooth surface, the envelope of a turbulent boundary layer is diffuse and irregular.

Heat Transfer

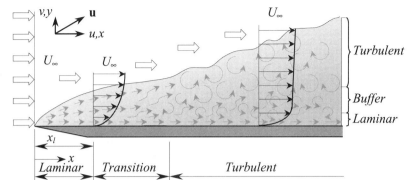

Figure 3.2
The structure of forced convection boundary layer flow: laminar, transition, and turbulent sections.

The boundary layer that develops (Fig. 3.2) is initially laminar. Small perturbations or wall roughness may trigger instabilities that eventually grow into turbulence that — somewhere downstream — destroys the regular, laminar flow structure. The region that connects the laminar section of the flow to the turbulent zone is called the *transition zone*.

The turbulent section has a complex structure that may be divided into three distinguishable layers, stratified in the y-direction. The first layer, in contact with the wall, is called the *laminar substrate* — the flow here is laminar. The second layer, called the *buffer*, is a transition zone between the laminar substrate and the outer, turbulent flow. The third layer, which is actually the *turbulent layer*, extends into and merges with the free stream, laminar flow. The velocity fields in the transition and turbulent sections have essentially 3D structures, and the associated transport phenomena are amplified through a vigorous mixing process. Figure 3.3 — after [2] — shows, qualitatively, the x-profiles of the velocity boundary layer thickness and of the heat transfer coefficient.

The main mechanisms of convection heat transfer are associated with flow transport processes; therefore, their governing equations — *mass*

Figure 3.3
A qualitative view of the velocity boundary layer thickness and of the heat transfer coefficient in the forced convection boundary layer — used with permission from A. Bejan, Heat Transfer, John Wiley, 1993, Fig. 5.16, p. 258.

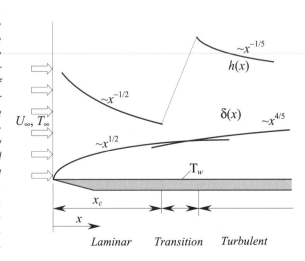

conservation and *momentum balance*—which are specific to hydrodynamics, are part of the mathematical model. To complete the analysis framework, the energy balance that describes the energy rate conservation specific to the heat transfer must be added.

3.1.5 MASS CONSERVATION (CONTINUITY)

The mass flow rates that enter and leave a fluid control volume are such that, inside the control volume, mass is neither created nor destroyed. The mass flow rate that leaves the control volume in the x-direction (Fig. 3.4) is, in the limits of the Taylor linear approximation [12],

$$\left[(\rho u) + \frac{\partial(\rho u)}{\partial x}\, dx\right] dy, \tag{3.11}$$

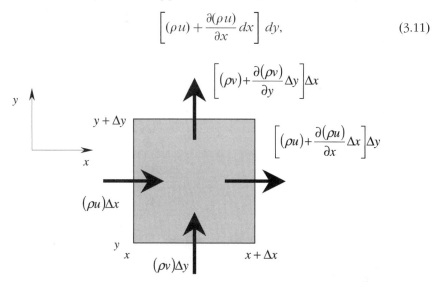

Figure 3.4
Mass conservation principle for an infinitesimal control volume in a 2D flow.

where ρ is the fluid density. This relation and a similar one for the flow in y-direction yield

$$(\rho u)dy + (\rho v)dx - \left[(\rho u) + \frac{\partial(\rho u)}{\partial x}\, dx\right] dy - \left[(\rho v) + \frac{\partial(\rho v)}{\partial y}\, dy\right] dx = 0, \tag{3.12}$$

or what is called the 2D *continuity* equation,

$$\frac{\partial(\rho u)}{\partial x} + \frac{\partial(\rho v)}{\partial y} = 0, \text{ or} \tag{3.13a}$$

$$div(\rho\mathbf{u}) = (\mathbf{u} \cdot \mathbf{grad})\rho + \rho\, div\ \mathbf{u} = 0. \tag{3.13b}$$

Here $\mathbf{u} = u\mathbf{1} + v\mathbf{j}$ is the velocity vector field, and $div\ (\mathbf{u}) \equiv \nabla \cdot \mathbf{u} = \dfrac{\partial u}{\partial x} + \dfrac{\partial v}{\partial y}$ is its *divergence*.

3.1.6 THE MOMENTUM BALANCE (NAVIER–STOKES)

Newton's law applied to the fluid control volume states that the sum of all forces that act upon the control volume is equal to the net rate of mechanical momentum relative to the control volume (Figs. 3.5, 3.6) [10, 12].

Heat Transfer

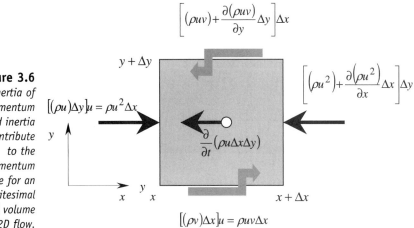

Figure 3.5
Surface and body forces that contribute to the momentum balance for an infinitesimal control volume in a 2D flow.

Figure 3.6
Inertia of momentum flows and inertia that contribute to the momentum balance for an infinitesimal control volume in a 2D flow.

There are *two* types of forces that act on the control volume: the *body forces* (X, Y — due to gravity, centrifugal motion, magnetic and electric fields, etc.), which are proportional to the control volume size, and the *superficial forces* (due to the hydrostatic pressure and to the fluid viscosity), which are proportional to the size of the surface upon which they act.

Relatively to the control volume, the mechanical effort due to the viscous flow may be decomposed uniquely into *two* components: a *normal stress (effort)* $\sigma_{i,j}$, and a *shear (tangential) stress (effort)*, $\tau_{i,j}$. The first subscript, i, indicates the orientation of the control volume surface by specifying the orientation of its normal; the second subscript, j, shows the direction of the force component. For example, for the x = surface in Fig. 3.5 the normal stress σ_{xx} corresponds to a force normal to the surface, and the tangential stress τ_{xy} corresponds to a force that acts in the y-direction, *along* the x-surface.

The arrows in Fig. 3.5 are oriented with respect to the positive direction, indicating thus the sign convention. Accordingly, the normal stress due to the hydrostatic pressure — perceived in fact as an external force — has a *compression* effect, whereas the normal stress, due to viscosity, has a *stretching* effect. The shear stress acts within the interface between adjacent fluid elements and it is a consequence of a viscous fluid flow — should the flow cease, then the shear stress vanishes. In the limits of the Taylor linear approximation, the forces that act upon the control volume are

$$F_{s,x} = \left(\frac{\partial \sigma_{xx}}{\partial x} - \frac{\partial p}{\partial x} + \frac{\partial \tau_{yx}}{\partial y} \right) dxdy \qquad (3.14a)$$

$$F_{s,y} = \left(\frac{\partial \tau_{xy}}{\partial x} + \frac{\partial \sigma_{yy}}{\partial y} - \frac{\partial p}{\partial y} \right) dxdy. \qquad (3.14b)$$

The other contributor to the momentum balance are the *net momentum fluxes* (Fig. 3.6). For instance, contributing to the x-momentum balance are the mass flow rate (ρu), that is, its associated momentum flux $(\rho u)u$, and the mass flow rate (ρv), that is, its associated momentum flux, $(\rho v)u$,

$$\frac{\partial [(\rho u)u]}{\partial x} + \frac{\partial [(\rho v)u]}{\partial y}. \qquad (3.15)$$

The x-momentum equation may now be written as

$$\frac{\partial [(\rho u)u]}{\partial x} + \frac{\partial t[(\rho v)u]}{\partial y} = \frac{\partial \sigma_{xx}}{\partial x} - \frac{\partial p}{\partial x} + \frac{\partial \tau_{yx}}{\partial y} + X. \qquad (3.16)$$

By the continuity equation (3.13a), it follows that

$$\rho \left(u \frac{\partial u}{\partial x} + v \frac{\partial u}{\partial y} \right) = \frac{\partial}{\partial x} (\sigma_{xx} - p) + \frac{\partial \tau_{yx}}{\partial y} + X. \qquad (3.17)$$

A similar expression may be obtained for the y-momentum balance, namely,

$$\rho \left(u \frac{\partial v}{\partial x} + v \frac{\partial v}{\partial y} \right) = \frac{\partial \tau_{xy}}{\partial x} + \frac{\partial}{\partial y} (\sigma_{yy} - p) + Y. \qquad (3.18)$$

The next step consists of specifying the normal and viscous shear stress in terms of the flow quantities. The normal stress produces a linear deformation, whereas the shear stress results in an angular distortion. On the other hand, the magnitude of any stress is proportional to the control volume deformation rate, which, in turn, is proportional to the fluid viscosity and to the flow gradients. For Newtonian fluids, the stress is proportional to the velocity gradients and the constant of proportionality equals the fluid viscosity. Generally, the evaluation of such dependencies is a difficult task that basically relies on empirical observations. In particular, it was found that [12]

$$\sigma_{xx} = 2\mu\frac{\partial u}{\partial x} - \frac{2}{3}\mu\left(\frac{\partial u}{\partial x} + \frac{\partial v}{\partial y}\right) \tag{3.19}$$

$$\sigma_{yy} = 2\mu\frac{\partial v}{\partial y} - \frac{2}{3}\mu\left(\frac{\partial u}{\partial x} + \frac{\partial v}{\partial y}\right) \tag{3.20}$$

$$\tau_{xy} = \tau_{yz} = \mu\left(\frac{\partial u}{\partial y} + \frac{\partial v}{\partial x}\right). \tag{3.21}$$

When utilized in (3.17) and (3.18) they yield

$$\rho\left(u\frac{\partial u}{\partial x} + v\frac{\partial u}{\partial y}\right) = -\frac{\partial p}{\partial x} + \frac{\partial}{\partial x}\left\{\mu\left[2\frac{\partial u}{\partial x} - \frac{2}{3}\left(\frac{\partial u}{\partial x} + \frac{\partial v}{\partial y}\right)\right]\right\} \tag{3.22}$$

$$\rho\left(u\frac{\partial v}{\partial x} + v\frac{\partial v}{\partial y}\right) = -\frac{\partial p}{\partial y} + \frac{\partial}{\partial y}\left\{\mu\left[2\frac{\partial v}{\partial y} - \frac{2}{3}\left(\frac{\partial u}{\partial x} + \frac{\partial v}{\partial y}\right)\right]\right\}, \tag{3.23}$$

or, in vector form,

$$\rho(\mathbf{u}\cdot\mathbf{grad})\mathbf{u} = -\mathbf{grad}p + \mu\nabla^2\mathbf{u}\left(\nabla^2 = \frac{\partial^2}{\partial x^2} + \frac{\partial^2}{\partial y^2} + \frac{\partial^2}{\partial z^2}\right). \tag{3.24}$$

It should be noted that the continuity equation (3.13b) and the momentum equation (3.14) are valid for all flows, including boundary layer flows. However, based on the specific nature of the boundary layer, simpler forms are available and, naturally, are generally preferred.

3.1.7 THE ENERGY BALANCE

The energy balance is derived also on a control volume base (Fig. 3.7) [10]. Neglecting the potential energy and its effects, the total specific energy (per unit mass) of the fluid control volume includes the *internal energy*, e, and the *kinetic energy*, $V^2/2$, where $V^2 = u^2 + v^2$ ($|\mathbf{u}| = V$ is the velocity). The internal and kinetic energies are advected through the control volume surface. The energy accumulation rate related to these transport processes in the x-direction is

$$\dot{E}_{adv,x} - \dot{E}_{adv,x+dx} = \rho u\left(e + \frac{V^2}{2}\right)dy$$

$$- \left\{\rho u\left(- + \frac{V^2}{2}\right) + \frac{\partial}{\partial x}\left[\rho u\left(e + \frac{V^2}{2}\right)dx\right]\right\}dy$$

$$= -\frac{\partial}{\partial x}\left[\rho u\left(e + \frac{V^2}{2}\right)\right]dxdy. \tag{3.25}$$

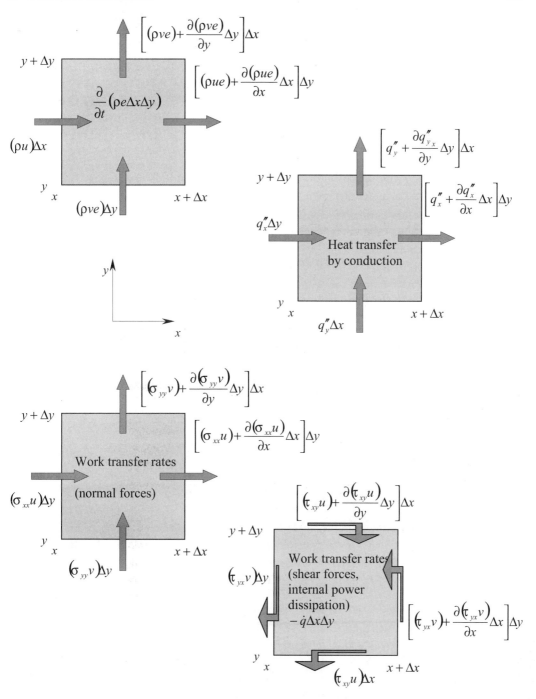

Figure 3.7 *The different contributors to the energy balance for an infinitesimal control volume in a 2D flow*

The molecular diffusivity (i.e., the thermal conduction) contributes also to the energy inventory, such that the net energy transfer rate *toward* the control volume is

$$\dot{E}_{cond,x} - \dot{E}_{cond,x+dx} = -\left(k\frac{\partial T}{\partial x}\right) dy - \left[-k\frac{\partial T}{\partial x} - \frac{\partial}{\partial x}\left(k\frac{\partial T}{\partial x}\right)\right]$$

$$= \frac{\partial}{\partial x}\left(k\frac{\partial T}{\partial x}\right) dxdy. \tag{3.26}$$

Other contributors to the energy balance include the mechanical work-type interactions, which are produced by body and surface forces. More precisely, the net rate of the mechanical work produced by the forces acting in the *x*-direction upon the control volume may be written as

$$\dot{W}_{net,x} = (Xu)\ dxdy + \frac{\partial}{\partial x}[(\sigma_{xx} - p)u]\ dxdy + \frac{\partial}{\partial y}(\tau_{yx}u)\ dxdy, \tag{3.27}$$

where the first term on the right-hand side stands for the mechanical work produced by the body force X, while the other terms are due to pressure and viscous forces.

Summing up (3.25), (3.26), (3.27) and the analogous equations for the *y*-direction yields the energy balance for the control volume:

$$\frac{\partial}{\partial x}\left[\rho u\left(e + \frac{V^2}{2}\right)\right] - \frac{\partial}{\partial y}\left[\rho v\left(e + \frac{V^2}{2}\right)\right]$$

$$= \frac{\partial}{\partial x}\left(k\frac{\partial T}{\partial x}\right) + \frac{\partial}{\partial y}\left(k\frac{\partial T}{\partial y}\right) + (Xu + Yv) - \frac{\partial}{\partial x}(pu) - \frac{\partial}{\partial y}(pv)$$

$$+ \frac{\partial}{\partial x}(\sigma_{xx}u + \tau_{xy}v) + \frac{\partial}{\partial y}(\tau_{yx}u + \sigma_{yy}v) + \dot{q} \tag{3.28}$$

Here $\dot{q}$ is the internal (body) heat generation rate inside the control volume and $e(\,J/m^3)$ is the *specific energy*. This equation accounts for the conservation of the mechanical work and thermal energy. A more convenient form is obtained by subtracting (3.27) and (3.28) multiplied by u, respectively v, from (3.28):

$$\rho u\frac{\partial e}{\partial x} + \rho v\frac{\partial e}{\partial y} = \frac{\partial}{\partial x}\left(k\frac{\partial T}{\partial x}\right) + \frac{\partial}{\partial y}\left(k\frac{\partial T}{\partial y}\right) - p\left(\frac{\partial u}{\partial x} + \frac{\partial v}{y}\right) + \mu\Phi + \dot{q}. \tag{3.29}$$

The quantity $\mu\Phi$, defined by

$$\mu\Phi = \mu\left\{\left(\frac{\partial u}{\partial y} + \frac{\partial v}{\partial x}\right)^2 + 2\left[\left(\frac{\partial u}{\partial x}\right)^2 + \left(\frac{\partial v}{\partial y}\right)^2\right] - \frac{2}{3}\left(\frac{\partial u}{\partial x} + \frac{\partial v}{\partial y}\right)^2\right\}, \tag{3.30}$$

is called *viscous dissipation*. The first term in (3.30) stands for the viscous shear stress, and the other terms are due to the normal stress. Qualitatively, viscous dissipation represents the irreversible conversion of mechanical energy in heat by viscous effects within the fluid.

In some situations, the *specific enthalpy*, $i(\,J/m^3)$,

$$i = e + \frac{p}{\rho}, \qquad (3.31)$$

is preferred rather than the energy, yielding

$$\rho u \frac{\partial i}{\partial x} + \rho v \frac{\partial i}{\partial y} = \frac{\partial}{\partial x}\left(k\frac{\partial T}{\partial x}\right) + \frac{\partial}{\partial y}\left(k\frac{\partial T}{\partial y}\right) + \mu\Phi + \dot{q}. \qquad (3.32)$$

For the *perfect gas* of a particular substance $di = cdT$; hence,

$$\rho c_p\left(u\frac{\partial T}{\partial x} + v\frac{\partial T}{\partial y}\right) = \frac{\partial}{\partial x}\left(k\frac{\partial T}{\partial x}\right) + \frac{\partial}{\partial y}\left(k\frac{\partial T}{\partial y}\right) + \mu\Phi + \dot{q}. \qquad (3.33)$$

Alternatively, for incompressible substances $c_V = c_P$ and the mass conservation equation (3.13) reduces to

$$\frac{\partial u}{\partial x} + \frac{\partial v}{\partial y} = 0. \qquad (3.34)$$

If we observe that $de = c_V dT = c_P dT$, (3.29) becomes

$$\rho c_p\left(u\frac{\partial T}{\partial x} + v\frac{\partial T}{\partial y}\right) = \frac{\partial}{\partial x}\left(k\frac{\partial T}{\partial x}\right) + \frac{\partial}{\partial y}\left(k\frac{\partial T}{\partial y}\right) + \frac{\partial}{\partial z}\left(k\frac{\partial T}{\partial z}\right) + \mu\Phi + \dot{q}. \qquad (3.35)$$

The vector form of this equation is

$$\rho c_p(\mathbf{u} \cdot \mathbf{grad})T = div(k\mathbf{grad}\,T) + \mu\Phi + \dot{q}. \qquad (3.36)$$

3.1.8 SCALE ANALYSIS RULES

Scale analysis is recommended as a first-stage, order-of-magnitude solution approach to any boundary and initial value problem. Scale analysis is often misinterpreted either as *dimensional analysis* or as the *arbitrary scaling* of a mathematical model. *Scaling* is generally aimed at stability analyses or at improving the condition number of the numerical models for specific physical models. Although it is not the direct result of scale analysis, scaling should always be considered in mathematical and numerical models.

Scale analysis is concerned with producing consistent order-of-magnitude estimates of the relevant physical quantities that occur in the equations that make up the mathematical model. When appropriately used, scale analysis may forecast the true solution within a range of approximation of utmost order one (i.e., within a few percent), as compared to more elaborate and, presumably, more accurate numerical or analytical solutions.

As example of scale analysis we shall consider the problem of a metallic plate of finite thickness $2D$, with thermal conductivity k, mass density ρ, and specific heat c, which at $t = 0$ is immersed in a warmer, high thermal conductivity fluid such that its faces instantaneously reach the equilibrium temperature $T_\infty = T_0 + \Delta T$ (Fig. 3.8) [10]. If the only quantity of interest is the *diffusion time*, that is, the time when the plate attains the final, equilibrium temperature, then a scale analysis may suffice.

Heat Transfer

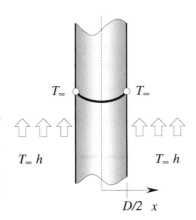

Figure 3.8
The scale analysis of the cooling process undergone by a metallic plate of finite thickness that is immersed in a warmer, high thermal conductivity fluid.

Invoking the apparent symmetry, this problem may be reduced to the half plate thickness, D, domain. The unsteady conduction heat transfer in the plate is described by the energy law, Eq. (2.20), which represents a balance between two terms of the same order of magnitude.

$$\rho c_P \frac{\partial T}{\partial t} = k \frac{\partial^2 T}{\partial x^2} = k \frac{\partial}{\partial x}\left(\frac{\partial T}{\partial x}\right)$$

$$\downarrow \qquad\qquad\qquad \downarrow \qquad\qquad\qquad\qquad (3.37)$$

$$\rho c_P \frac{\Delta T}{t} \sim \frac{k}{D}\frac{\Delta T}{D} = \frac{k\Delta T}{D^2}.$$

Hence,

$$\rho c_P \frac{\Delta T}{t} \sim \frac{k\Delta T}{D^2}, \qquad\qquad (3.38)$$

yielding

$$t \sim \frac{D^2}{\alpha}, \qquad\qquad (3.39)$$

where $\alpha = k/\rho c_P \,(\mathrm{m^2/s})$ is the *thermal diffusivity* of the metal.

Considering the analysis carried out in this example, the following scaling rules may be defined:

1. The size of the physical domain must be well defined. In the previous example the size was D.
2. Any equation expresses the balance between the scales of at least two dominant, algebraically summed terms. In the preceding example above, (3.38), the terms on the left-hand and right-hand sides are dominant, comparable in an order of magnitude sense. Generally speaking, the energy law may contain more than two terms, yet not all of them are necessarily equally important. The principles of selecting the dominant scales are given by the following rules 3–5.
3. If in the sum of two terms

$$c = a + b,$$

the order of magnitude of one of the terms is larger than that of the other one, that is,

$$O(a) > O(b),$$

then the order of magnitude of the sum is that of the leading term

$$O(a) \sim O(b).$$

The same rule applies in case of subtraction, $c = a - b$ or $c = -a + b$.

4. If the terms of the sum

$$c = a + b$$

have the same order of magnitude $O(a) \sim O(b)$, then the sum has the same order of magnitude, that is,

$$O(c) \sim O(a) \sim O(b).$$

5. The order of magnitude of the product

$$p = ab$$

is equal to the product of the orders of magnitude of the terms,

$$O(p) \sim O(a)O(b).$$

Similarly, the order of magnitude of a ratio

$$r = \frac{a}{b}$$

is equal to the ratio of the orders of magnitude of the terms,

$$O(r) \sim \frac{O(a)}{O(b)}.$$

3.1.9 STREAMLINES AND HEATLINES

Visualization has generally a special, insightful meaning in the solution to any problem. In convection processes it is important to "see" the fluid motion and, more specifically, to visualize the energy "flow." At least to this end, a particular useful vector quantity may be introduced based on the continuity equation (3.36). Under the incompressible flow assumption, the velocity field is divergence-free. Consequently, it may be conjectured that it is produced by a vector field Ψ, called the *streamfunction* [12]:

$$\mathbf{u} \stackrel{def}{=} -\mathbf{curl}\Psi \ (\mathbf{curl}\Psi \equiv \nabla \times \Psi). \tag{3.40}$$

This definition specifies only the *solenoidal* part of the streamfunction, Ψ, whereas its *potential* part is not known. However, the good news is that in 2D problems (cartesian or axially symmetric) the streamfunction is divergence-free, that is, its solenoidal part is consistently zero and, consequently, it is a well defined quantity from the point of view of the *uniqueness theorem of vector fields*. Furthermore, it follows that under such circumstances the streamfunction has only one component, orthogonal to the flow, which technically reduces it to a scalar quantity. In a 2D cartesian flow, the streamfunction $\Psi \equiv \psi\mathbf{k}$, is related to the velocity field by

$$u = \frac{\partial\psi}{\partial y}, v = -\frac{\partial\psi}{\partial x}. \qquad (3.41)$$

It may be easily verified that the ψ = const lines—called *streamlines*—are velocity field lines; hence, a very convenient method to visualize flows consists of tracing its streamlines.

Similarly, the equation that describes the steady-state convection heat transfer process without internal heat sources,

$$\rho c_p \left(u \frac{\partial T}{\partial x} + v \frac{\partial T}{\partial y} \right) = \frac{\partial}{\partial x} \left(k \frac{\partial T}{\partial x} \right) + \frac{\partial}{\partial y} \left(k \frac{\partial T}{\partial y} \right), \qquad (3.42)$$

may be rewritten as

$$\frac{\partial}{\partial x} \left(\rho c_p u T - k \frac{\partial T}{\partial x} \right) + \frac{\partial}{\partial y} \left(\rho c_p v T - k \frac{\partial T}{\partial y} \right) = 0, \qquad (3.43)$$

where we see the emergence of a new, once again divergence-free, vector field $\rho c_p \mathbf{u} T - k\mathbf{grad}\, T$. Consequently, a new vector quantity—called the *heatfunction* [10] and labeled **H**—is introduced through its **curl** part:

$$\frac{\partial H}{\partial y} = \rho c_p u T - k \frac{\partial T}{\partial x}, \frac{\partial H}{\partial x} = \rho c_p v T - k \frac{\partial T}{\partial y}. \qquad (3.44)$$

A completely analogous discussion may be conducted with respect to the solenoidal part of **H**, and with respect to the cases (cartesian, axially-symmetric) where the heatfunction is a well, and completely defined vector quantity.

When motion ceases, the heatlines coincide with the heat flux lines, which, together with the isotherms, are commonly used in the visualization of conduction heat transfer. It is important to notice that the isotherms, T = const, are meaningful in conduction heat transfer, but less so in convection heat transfer. It is only in conduction that the isotherms are locally orthogonal to the heat flow direction. Using the isotherms to visualize convection heat transfer is as inappropriate as using the P = const lines to visualize the fluid flow [10].

3.1.10 BOUNDARY LAYER HEAT TRANSFER

The hydrodynamic and energy equations that fully describe the 2D boundary layer steadystate heat transfer are as follows (Fig. 3.1) [10]:

Continuity:

$$\frac{\partial u}{\partial x} + \frac{\partial v}{\partial y} = 0. \qquad (3.45)$$

Momentum:

$$u\frac{\partial u}{\partial x} + v\frac{\partial u}{\partial y} = -\frac{1}{\rho}\frac{\partial p}{\partial x} + v\left(\frac{\partial^2 u}{\partial x^2} + \frac{\partial^2 u}{\partial y^2}\right) \tag{3.46}$$

$$u\frac{\partial v}{\partial x} + v\frac{\partial v}{\partial y} = -\frac{1}{\rho}\frac{\partial p}{\partial y} + v\left(\frac{\partial^2 v}{\partial x^2} + \frac{\partial^2 v}{\partial y^2}\right). \tag{3.46}$$

Energy:

$$u\frac{\partial T}{\partial x} + v\frac{\partial T}{\partial y} = \alpha\left(\frac{\partial^2 T}{\partial x^2} + \frac{\partial^2 T}{\partial y^2}\right), \tag{3.47}$$

where $v = \mu/\rho[\mathrm{m^2/s}]$ is the *kinematic viscosity*. The usual boundary conditions that close the physical model are

(*a*) No-slip $\quad\quad\quad\quad\quad u = 0$ ⎫
(*b*) Impermeable $\quad\quad\quad v = 0$ ⎬ at the wall
(*c*) Wall temperature $\quad T = T_0$ ⎭
(*d*) Uniform flow $\quad\quad\quad u = U_\infty$ ⎫
(*e*) Uniform flow $\quad\quad\quad v = 0$ ⎬ far from the wall
(*f*) Uniform temperature $\quad T = T_\infty$ ⎭

Boundary Layer Assumptions

The free stream structure of the flow outside the boundary layer implies [10]

$$u = U_\infty, \quad v = 0, \quad p = P_\infty, \quad T = T_\infty. \tag{3.48}$$

On the other hand, the main scales of the boundary layer are

$$x \sim L, \quad y \sim \delta, \quad \text{and} \quad u \sim U_\infty. \tag{3.49}$$

Here δ is the boundary layer thickness, and L is the downstream size (length) of the wall; therefore, the size of the boundary layer is approximately $\delta \times L$.

The *streamwise momentum equation* (3.46) indicates the balance of three terms:

$$
\begin{array}{ccc}
\text{Inertia} & \text{Pressure} & \text{Friction} \\
U_\infty\dfrac{U_\infty}{L}, v\dfrac{U_\infty}{\delta} & \dfrac{p}{\rho L} & v\dfrac{U_\infty}{L^2}, v\dfrac{U_\infty}{\delta^2}.
\end{array} \tag{3.50}
$$

By the *continuity equation* (3.45),

$$\frac{U_\infty}{L} \sim \frac{v}{\delta}, \tag{3.51}$$

and it follows that both inertia terms in (3.50) have the same order of magnitude U_∞^2/L. However, if the boundary layer is shallow, then

$$\delta \ll L, \tag{3.52}$$

and the last term in (3.50) is dominant. Therefore, $\partial^2 u/\partial x^2$ may be neglected for $\partial^2 u/\partial y^2$, which yields the simpler form

$$u\frac{\partial u}{\partial x} + v\frac{\partial u}{\partial y} = -\frac{1}{\rho}\frac{\partial p}{\partial x} + v\frac{\partial^2 u}{\partial y^2}. \tag{3.53}$$

By the same argument, that is, the boundary layer is a very shallow zone, the momentum equation in the y-direction, orthogonal to the wall, reduces to

$$u\frac{\partial v}{\partial x} + v\frac{\partial v}{\partial y} = -\frac{1}{\rho}\frac{\partial p}{\partial y} + v\frac{\partial^2 v}{\partial y^2}. \tag{3.54}$$

Although generally not used in the boundary layer analysis, (3.54) is meaningful through an important consequence: Since in the $\delta \times L$ region the pressure variation in the y-direction is negligable compared to that in x-direction, the term $\partial p/\partial x$ in (3.53) may be replaced by dP_∞/dx, which is a known quantity. The validity of this substitution may be verified by dividing the pressure differential by dx,

$$\frac{dp}{dx} = \frac{\partial p}{\partial x} + \frac{\partial p}{\partial y}\frac{dy}{dx}, \tag{3.55}$$

and by comparing the order of magnitude of the pressure gradient components, $\partial p/\partial x$ and $\partial p/\partial y$, that act as source terms in (3.53) and (3.54). The second term on the right-hand side of (3.53) and (3.54) account for *friction* (against the flow), while their left-hand sides stand for *inertia* terms. For those flow regimes where the pressure gradients are comparable to friction, (3.53), Eq. (3.54) suggests the following scale relations:

$$\frac{\partial p}{\partial x} \sim \frac{\mu U_\infty}{\delta^2}, \quad \frac{\partial p}{\partial y} \sim \frac{\mu v}{\delta^2}. \tag{3.56}$$

Further on, with the use of (3.55), they yield

$$\frac{\frac{\partial p}{\partial y}\frac{\partial y}{\partial x}}{\frac{\partial p}{\partial x}} \sim \frac{v\delta}{U_\infty L} \sim \left(\frac{\delta}{L}\right)^2 \ll 1. \tag{3.57}$$

This result was expected: In the boundary layer domain, $\delta \times L$, (3.55) reduces to

$$\frac{dP}{dx} = \frac{\partial p}{\partial x} = \frac{dP_\infty}{dx}. \tag{3.58}$$

Consequently, the *boundary layer momentum equation* is

$$u\frac{\partial u}{\partial x} + v\frac{\partial u}{\partial y} = -\frac{1}{\rho}\frac{dP_\infty}{dx} + v\frac{\partial^2 u}{\partial y^2}, \tag{3.59}$$

and, in fact, it accounts for *both* momentum equations, (3.53) and (3.54).

The boundary layer energy equation may be obtained analogously, yielding

$$u\frac{\partial T}{\partial x} + v\frac{\partial T}{\partial y} = \alpha\frac{\partial^2 T}{\partial y^2}. \tag{3.60}$$

It is important to note that this result shows the heat diffusion in x-direction exceeding the heat diffusion in y-direction.

In conclusion, the boundary layer equations are only two (3.59) and (3.60) — instead of four, the number of the complete set. Also, the absence of the second-order x-derivatives $\partial^2/\partial x^2$ makes their solution simpler. The scale analysis detailed next produces a first, simple answer that is satisfactorily accurate.

3.1.11 SCALE ANALYSIS OF THE HYDRODYNAMIC AND THERMAL BOUNDARY LAYERS

The equations (3.59) and (3.60) describe the flow and heat transfer process in the slender region of size $\delta \times L$, and may predict which of the two boundary layers is thicker, that is, which quantity, δ_T or δ, is larger, since usually $\delta \neq \delta_T$. The heat transfer boundary layer analysis is focused on two main objectives: (i) the *hydrodynamic problem*, concerned with evaluating the wall shear stress, and (ii) the *thermal problem*, targeted toward estimating the heat transfer coefficient.

(i) The *hydrodynamic problem* consists of evaluating the wall shear stress

$$\tau_w = \mu \frac{\partial u}{\partial y}\bigg|_{y=0}, \tag{3.61}$$

which, in the *velocity boundary layer*, scales as

$$\tau_w \sim \mu \frac{U_\infty}{\delta}. \tag{3.62}$$

Apparently τ_w is inversely proportional to δ: For a specified free stream flow (U_∞) of a known fluid (μ, ρ), the wall shear stress increases for a decreasing velocity boundary layer thickness, δ. As for the simpler flows, for which $dP_\infty/dx \approx 0$ (for instance the flow of a cooling fluid through a radiator made of a plate fin network), (3.59) scales as

$$\begin{array}{ccc} \text{Inertia} & \sim & \text{Friction} \\ \dfrac{U_\infty^2}{L}, \dfrac{vU_\infty}{\delta} & \sim & v\dfrac{U_\infty}{\delta^2}. \end{array} \tag{3.63}$$

It may be shown [by the continuity equation (3.45)] that the two terms on the left-hand side of (153) have the same order of magnitude; hence,

$$\delta \sim \left(\frac{vL}{U_\infty}\right)^{1/2}, \quad \text{or} \quad \frac{\delta}{L} \sim \text{Re}_L^{-1/2}. \tag{3.64}$$

The nondimensional group Re_L is the *Reynolds number* based on L, the length scale of the boundary layer.

At this point, a note on the significance of Re_L in the boundary layer theory is useful [10]. Although generally in fluid mechanics Re_L is regarded, for a given flow, as an order of magnitude estimator of the ratio of inertial to friction [12], this interpretation may not always be correct. In the boundary layer flows *always* governed by the balance inertia~friction, Re_L may reach values as high as the order of 10^4 before transition to turbulence occurs. Therefore, in boundary layer flow, the only acceptable interpretation seems to be

$$Re_L^{1/2} = \frac{\text{wall length}}{\text{boundary layer thickness}}. \tag{3.65}$$

Therefore, it is *not* Re_L but its square root, $Re_L^{1/2}$, that is meaningful: $Re_L^{1/2}$ is the *shape factor* (hence, a geometric estimator) *of the boundary layer flow region.*

An important result of the scale analysis is that δ is shown to be proportional to $L^{1/2}$. More accurate analyses confirm that along the wall $(0 < x < L)$ the velocity boundary layer thickness growth as $x^{1/2}$. As a detail, the slope of this function is infinite in the origin, that is, *at $x = 0$ the boundary layer envelope is orthogonal to the wall.*

Equation (3.64) also shows another important consequence: The velocity boundary layer assumption $\delta \ll L$ is acceptable provided $Re_L^{-1/2} \ll 1$. This result may be used as a criterion for checking the validity of a particular type of boundary layer analysis, for given circumstances. However, even if this restriction is fulfilled, there are flow domains, such as the region of size $Re_l^{-1/2}$ within the entrance region at the tip of the wall of length l, where the boundary layer assumption still *do not* apply.

By virtue of these scaling analyses, the wall shear stress is seen to scale as

$$\tau_w \sim \mu \frac{U_\infty}{L} Re_L^{1/2} \sim \rho U_\infty^2 Re_L^{-1/2}. \tag{3.66}$$

Consequently, the wall friction coefficient,

$$C_f = \frac{\tau_w}{\frac{1}{2}\rho u^2}, $$

scales as

$$C_f \sim Re_L^{-1/2} \tag{3.67}$$

which solves the first problem of the boundary layer. Numerous experiments and calculations confirm the accuracy of these results in the limit of a factor of order 1.

(ii) The *heat transfer problem* may be addressed by evaluating the convection heat transfer coefficient, h, for the temperature boundary layer of thickness δ_T [10]:

$$h \sim \frac{k(\Delta T/\delta_T)}{\Delta T} \sim \frac{k}{\delta_T}. \tag{3.68}$$

Here $\Delta T = T_w - T_\infty$ is the temperature drop within the *temperature* boundary layer of size $\delta_T \times L$.

The energy equation (3.60) is the balance of two terms: the transversal, wall-to-flow conduction heat transfer and the streamwise, convection heat transfer (enthalpy flow):

$$\text{Convection} \quad \sim \quad \text{Conduction}$$

$$u\frac{\Delta T}{L}, v\frac{\Delta T}{\delta_T} \quad \sim \quad \alpha\frac{\Delta T}{\delta_T^2}. \tag{3.69}$$

The mass conservation law (3.45) applied to the thermal boundary layer in the scaling sense leads to

$$\frac{u}{L} \sim \frac{v}{\delta_T}, \tag{3.70}$$

which reveals that the two convection scales have the same order of magnitude, namely $u(\Delta T/L)$. However, the order of magnitude of u is not necessarily U_∞! As shown by (3.69) and (3.70), the scale of u is related to the scale of δ_T; consequently, it depends on the size of δ_T *relative* to δ.

We first assume that

(1) $\delta \ll \delta T \rightarrow u \sim U_\infty$

(Fig. 3.9 after [10]). In this situation,

$$\frac{\delta_T}{L} \sim Pe_L^{-1/2} \sim Pr^{-1/2}Re_L^{-1/2}, \tag{3.71}$$

where $Pe_L = U_\infty L/\alpha$ is the nondimensional *Péclet* group. From comparing (3.64) and (3.61), it follows that the size of δ_T relative to δ is in fact the *Prandtl* number of the fluid, $Pr = v/\alpha$,

$$\frac{\delta_T}{\delta} \sim Pr^{-1/2}. \tag{3.72}$$

Consequently, the assumption $\delta \ll \delta_T$ stands only if $Pr^{1/2} \ll 1$, for fluids that fall within the range of liquid metals. In this case, the heat transfer coefficient h and *Nusselt* number, $Nu_L \overset{def}{=} h/L$, are

$$h \sim \frac{k}{L}Pr^{1/2}Re_L^{1/2}, Nu_L \sim Pr^{-1/2}Re_L^{1/2} \ (Pr \ll 1). \tag{3.73}$$

Of special practical importance are fluids with a Prandtl number of order 1 (air), or greater than 1 (water, oils, etc.). Let us assume that in this situation,

(2) $\delta \gg \delta_T \rightarrow u \sim \frac{\delta_T}{\delta} U_\infty.$

The balance *conduction*~*convection* (Fig. 3.9) is now

$$\frac{\delta_T}{L} \sim Pr^{-1/3}Re_L^{-1/2}, \text{respectively} \ \frac{\delta_T}{\delta} \sim Pr^{-1/3} \ll 1, \tag{3.74}$$

confirming that the assumption is valid for fluids with $Pr^{-1/3} \gg 1$. Subsequently, h, and Nu_L scale as

$$h \sim \frac{k}{L}Pr^{1/3}Re_L^{1/2}, Nu_L \sim Pr^{1/3}Re_L^{1/2} \ (Pr \gg 1). \tag{3.75}$$

These results are consistent within the limits of a factor of order 1, $O(1)$.

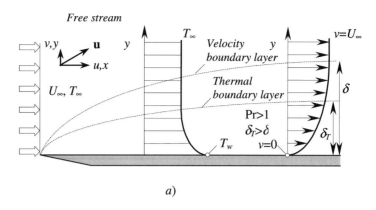

a)

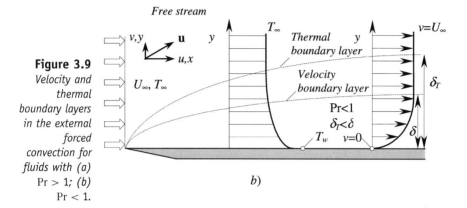

Figure 3.9
Velocity and thermal boundary layers in the external forced convection for fluids with (a) Pr > 1; (b) Pr < 1.

b)

Although straightforward, simple, and relevant, the scale analysis has several drawbacks. First, its results are meaningful only in an order of magnitude sense, more precisely in the limits of the unspecified proportionality coefficients. Another difficulty is the ambiguous significance of several quantities, such as the wall shear stress, τ_w, and the heat transfer coefficient, h: Are they local or average quantities? In any case, to be consistent with the scale analysis results, whatever other, more accurate solution method might be used, the local and average quantities it produces should be of the same order of magnitude.

3.1.12 THE INTEGRAL METHOD

This approximative method was originally introduced by Pohlhausen and von Kàrman [10], and it may be used to provide numerical values for the coefficients of proportionality left unspecified by the scale analysis. This method relies on the hypothesis that the y-profiles of u and T are not particularly relevant in the evaluation of τ_w and h.

Multiplying the continuity equation (3.13) by u and adding the result to the right hand side of (3.59), then amplifying the same continuity equation (3.13a) by T and adding this result to the left hand side of (3.60), yields

$$\frac{\partial(u^2)}{\partial x} + \frac{\partial(uv)}{\partial y} = -\frac{1}{\rho}\frac{dP_\infty}{dx} + \frac{\partial^2 u}{\partial y^2} \tag{3.76}$$

$$\frac{\partial(uT)}{\partial x} + \frac{\partial(vT)}{\partial y} = \alpha \frac{\partial^2 T}{\partial y^2}, \tag{3.77}$$

which, further integrated over the interval $y \in [0, Y]$ (Fig. 3.10), where $Y > \max\{\delta, \delta_T\}$, results in

$$\frac{d}{dx}\int_0^Y u^2\,dy + u_Y v_Y - u_0 v_0 = -\frac{Y}{\rho}\frac{dP_\infty}{dx} + v\left(\frac{\partial u}{\partial y}\right)_Y - v\left(\frac{\partial u}{\partial y}\right)_0 \tag{3.78}$$

$$\frac{d}{dx}\int_0^Y uT\,dy + u_Y T_Y - u_0 T_0 = \alpha\left[\left(\frac{\partial T}{\partial y}\right)_y - \left(\frac{\partial T}{\partial y}\right)_0\right]. \tag{3.79}$$

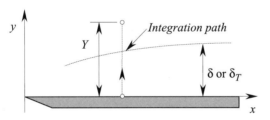

Figure 3.10
The integration path in the boundary layer integral method.

Because the free stream flow is uniform, $\partial/\partial y|_{y\geq Y} = 0$, $u_Y = U_\infty$, $T_Y = T_\infty$, and since at the wall $v_0 = 0$, it follows that v_Y may be obtained by integrating the mass conservation equation (3.13) over the boundary layer height $[0, Y]$

$$\frac{d}{dx}\int_0^Y u\,dy + v_Y - v_0 = 0. \tag{3.80}$$

Replacing this result in (3.78) and (3.79) yields the momentum and energy integral equations for the laminar boundary layer,

$$\frac{d}{dx}\int_0^Y u(U_\infty - u)\,dy = \frac{Y}{\rho}\frac{dP_\infty}{dx} + \frac{dU_\infty}{dx}\int_0^Y u\,dy + v\left(\frac{\partial u}{\partial y}\right)_0 \tag{3.81}$$

$$\frac{d}{dx}\int_0^Y u(T_\infty - T)\,dy = \alpha\left(\frac{\partial T}{\partial y}\right)_0. \tag{3.82}$$

These equations are conservation relations for the fluid control volume of size $dx \times Y$, where x is the downstream coordinate. The forces that act upon this control volume are (Fig. 3.11)

$$\text{left} \rightarrow \text{right} \begin{cases} M_x = \int_0^Y \rho u^2 \, dy \\ M_Y = U_\infty d\dot{m}, \\ P_\infty Y. \end{cases} \quad \text{where } \dot{m} = \int_0^Y \rho u \, dy,$$

$$\text{right} \rightarrow \text{left} \begin{cases} M_{x+dx} = M_x + \dfrac{dM_x}{dx} dx \\ \tau dx \\ Y[P_\infty + \dfrac{dP_\infty}{dx} dx]. \end{cases}$$

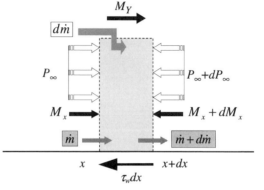

Figure 3.11
The forces that contribute to the conservation relations for the fluid control volume of size.

At this point, to integrate the energy equation (3.82) a certain $u(y)$ profile has to be assumed:

$$u = \begin{cases} U_\infty m(n), & n \in [0, 1], \\ U_\infty, & n \geq 1. \end{cases}$$

Here $m(n)$ is a *shape function*, unspecified as yet, and $n = y/\delta$. The boundary layer hypothesis, $dP_\infty/dx = 0$, may be used, yielding the following ordinary differential equation for the velocity boundary layer thickness:

$$\delta \frac{d\delta}{dx} \left[\int_0^1 m(1 - m) dn \right] = \frac{v}{U_\infty} \frac{dm}{dn} \bigg|_{n=0}. \tag{3.83}$$

Finally, by integration, the following expressions for the local velocity boundary layer thickness and for the friction coefficient,

$$\frac{\delta}{x} = a_1 \text{Re}_x^{-1/2} \tag{3.84}$$

$$C_{f,x} = \frac{\tau}{0.5\rho U_\infty^2} - a_2 \text{Re}_x^{-1/2}, \tag{3.85}$$

are obtained.

Although

$$
a_1 = \sqrt{\dfrac{2\dfrac{dm}{dn}\bigg|_{n=0}}{\int_0^1 m(1-m)\,dn}} \quad \text{and} \quad a_2 = \sqrt{2\dfrac{dm}{dn}\bigg|_{n=0}\int_0^1 m(1-m)\,dn}
$$

are functions of the particular shape function used, these results are consistent with the scale analysis conclusions. Table 3.1 [10] shows several shape functions that are frequently used, and shows their influence on the boundary layer integral solutions.

Table 3.1

Profile $m(n)$ or $m(p)$	$\dfrac{\delta}{x}\mathrm{Re}_x^{1/2}$	$C_{f.x}\mathrm{Re}_x^{1/2}$	$\mathbf{Nu\,Re}_x^{-1/2}\,\mathbf{Pr}^{-1/3}$ Uniform temperature (Pr > 1)	Uniform heat flux (Pr < 1)
$m = n$	3.46	0.577	0.289	0.364
$m = \dfrac{n(3-n^2)}{2}$	4.64	0.646	0.331	0.417
$m = \sin\dfrac{n\pi}{2}$	4.8	0.654	0.337	0.424
Similarity solution	4.92	0.664	0.332	0.458

The proportionality coefficients for the heat transfer relations that are not specified by the scale analysis may be determined in an analogous manner by assuming first a specific profile for the temperature,

$$
\frac{T_0 - T}{T_0 - T_\infty} = m(p), \quad p = \frac{y}{\delta_T} \in [0, 1]
$$

$$
T = T_\infty, \quad p > 1
$$

Recalling that the scale analysis gave evidence that the ratio $\Delta = \delta_T/\delta = \Delta(\mathrm{Pr})$ is a function only of the Prandtl number, the discussion reduces to two cases:

1. *High-Prandtl-number fluids* $(\delta_T < \delta)$: The energy equation reduces to the implicit form for $\Delta(\mathrm{Pr})$,

$$
\mathrm{Pr} = 2\frac{\dfrac{dm}{dp}\bigg|_{p=0}}{(a_1\Delta)^2}\left[\int_0^1 m(p\Delta)[1 - m(p)]\,dp\right]^{-1}, \tag{3.86}
$$

a result that is consistent with the scale analysis. If we select the simplest temperature profile, $m = p$, (3.86) leads to

$$
\Delta(\mathrm{Pr}) = \mathrm{Pr}^{-1/3}, \tag{3.87}
$$

which is identical to the scale analysis result for the fluids with $\mathrm{Pr} \gg 1$, (3.52). Table 3.1 also indicates several shape functions for the temperature and the corresponding Nu_L. A frequent choice is the cubic profile $m = p(3 - p^2)/2$, which yields

$$\Delta = \frac{\delta_T}{\delta} = 0.977 \, \mathrm{Pr}^{-1/3} \tag{3.88}$$

$$h = 0.323 \frac{k}{x} \mathrm{Pr}^{1/3} \mathrm{Re}_x^{1/2} \tag{3.89}$$

$$\mathrm{Nu}_L = \frac{hx}{k} = 0.323 \frac{k}{x} \mathrm{Pr}^{1/3} \mathrm{Re}_x^{1/2}. \tag{3.90}$$

2. *Low-Prandtl-number fluids* ($\mathrm{Pr} \ll 1$, i.e., $\Delta \gg 1$): These fluids form the liquid-metal case. Equation (3.86) is replaced by

$$\mathrm{Pr} = 2 \frac{\left.\dfrac{dm}{dp}\right|_{p=0}}{(a_1 \Delta)^2} \left[\int_0^{1/\Delta} m(p\Delta)[1 - m(p)] \, dp + \int_{1/\Delta}^1 [1 - m(p)] dp \right]^{-1}. \tag{3.91}$$

Since $\Delta \gg 1$, the second integral on the right-hand side is dominant and, if we select the profile $m = p$, yields

$$\Delta = \frac{\delta_T}{\delta} = (3\mathrm{Pr})^{-1/2}, \quad \text{or} \quad \frac{\delta_T}{x} = 2\mathrm{Pr}^{-1/2}\mathrm{Re}_x^{-1/2}, \quad \text{for } \mathrm{Pr} \ll 1. \tag{3.92}$$

The corresponding heat transfer coefficient and the Nusselt number,

$$h = \frac{k}{\delta_T} = \frac{1}{2}\frac{k}{x}\mathrm{Pr}^{1/2}, \quad \mathrm{Nu}_L = \frac{hx}{k} = \frac{1}{2}\mathrm{Pr}^{1/2}\mathrm{Re}_x^{1/2}, \tag{3.93}$$

are consistent with the scale analysis results, and are also confirmed by experiment.

3.1.13 THE SIMILARITY SOLUTIONS FOR THE HYDRODYNAMIC AND THERMAL BOUNDARY LAYERS

Exact, analytic boundary solutions were obtained by Blasius (for flow) [16] and Pohlhausen (for heat transfer) [17]. The central idea — inspired from experiment — resides in recognizing that the velocity and temperature profiles within the boundary layer region are *self-similar* with respect to the y-coordinate (orthogonal to the wall, i.e., to the flow). This suggests that there exists a set of unique, or "master" velocity and temperature profiles, such that at any location downstream (x-coordinate) the local velocity and temperature profiles are similar to them. Mathematically, it means that there exists a change of variables $(x, y) \to \eta$, where η is called the *similarity variable*, such that the velocity and temperature profiles are functions only of η. The master profiles form the object of *similarity analysis*.

For the velocity field, similarity means $u/U_\infty = function(\eta)$. Using the scaling analysis result $\delta(x) \sim x \, \mathrm{Re}_x^{-1/2}$, Blasius defined

$$\eta_{def} = \frac{y}{x}\mathrm{Re}_x^{1/2}. \tag{3.94}$$

On the other hand, it was shown (3.41) that the velocity field for a Cartesian, 2D flow is the **curl** of the streamfunction, Ψ, that is,

$$u = \frac{\partial \psi}{\partial y}, \ v = -\frac{\partial \psi}{\partial x}. \tag{3.95}$$

Consequently, it may be conjectured that

$$\frac{u}{U_\infty} = f'(\eta), \tag{3.96}$$

where the unknown function $f(\eta)$, apparently related to the streamfunction, is introduced through its derivative, $f'(\eta) = df/d\eta$, unknown as yet.

The boundary layer flow problem consists then of

- *Continuity:* $\dfrac{\partial u}{\partial x} + \dfrac{\partial v}{\partial y} = 0$

- *Momentum:* $u\dfrac{\partial u}{\partial x} + v\dfrac{\partial u}{\partial y} = v\dfrac{\partial^2 u}{\partial y^2},$ $\Rightarrow$ $\dfrac{\partial \psi}{\partial y}\dfrac{\partial^2 \psi}{\partial x \partial y} - \dfrac{\partial \psi}{\partial x}\dfrac{\partial^2 \psi}{\partial y^2} = v\dfrac{\partial^3 \psi}{\partial y^3}$

and of the following boundary conditions:

$$y = 0, \quad \Rightarrow \quad u = v = 0, \quad \Rightarrow \quad \frac{\partial \psi}{\partial y} = 0, \psi = 0$$

$$y \to \infty, \quad \Rightarrow \quad u \to U_\infty, \quad \Rightarrow \quad \frac{\partial \psi}{\partial y} \to U_\infty.$$

On the other hand, the velocity–streamfunction relations (3.95) yield

$$\psi = (U_\infty v x)^{1/2} f(\eta) \tag{3.97}$$

$$v = \frac{1}{2}\left(\frac{v U_\infty}{x}\right)^{1/2}(\eta f' - f), \tag{3.98}$$

which produce an alternative form of the momentum equation and boundary conditions, in fact the *similarity velocity boundary layer problem*:

$$2f''' + f f'' = 0 \tag{3.99}$$

$$\begin{aligned} f' = f = 0, &\quad \text{for } \eta = 0 \\ f' \to 1, &\quad \text{for } \eta \to \infty. \end{aligned} \tag{3.100}$$

Blasius solved the nonlinear ordinary differential equation (3.99) by an asymptotic method — the resulting velocity profile, obtained numerically by using a Runge–Kutta fourth-order, constant step scheme, is shown in Fig. 3.12. If we recall that $u = 0.99 U_\infty$ occurs at $\eta = 4.92$, then the velocity boundary layer thickness is

$$\frac{\delta}{x} = 4.92 \mathrm{Re}_x^{-1/2}. \tag{3.101}$$

Heat Transfer

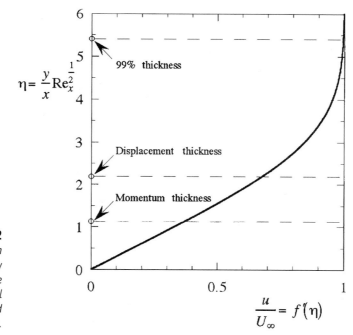

Figure 3.12
Blasius solution to the similarity velocity profile in external forced convection.

At this point it should be mentioned that other definitions for the hydrodynamic boundary layer are around, for instance,

$$\delta^* \stackrel{def}{=} \int_0^\infty \left(1 - \frac{u}{U_\infty}\right) dy, \qquad \frac{\delta^* \stackrel{f(\eta)}{=} 1.73 \mathrm{Re}_x^{1/2}}{x} \qquad \text{(displacement thickness)}$$

$$\theta \stackrel{def}{=} \int_0^\infty \frac{u}{U_\infty}\left(1 - \frac{u}{U_\infty}\right) dy, \qquad \frac{\theta \stackrel{f(\eta)}{=} 0.664 \mathrm{Re}_x^{-1/2}}{x} \qquad \text{(momentum thickness)}.$$

An important result of the similarity analysis is an estimate for the friction coefficient,

$$C_{f,x} \stackrel{f(\eta)}{=} \frac{\mu \left.\frac{\partial u}{\partial y}\right|_{y=0}}{\frac{1}{2}\rho U_\infty^2} = 2f''|_{\eta=0}\mathrm{Re}_x^{-1/2}, \tag{3.102}$$

or, using the satisfactorily good evaluation $f''|_{\eta=0} = 0.332$,

$$C_{f,x} = 0.664 \mathrm{Re}_x^{-1/2}. \tag{3.103}$$

The boundary layer *similarity heat transfer problem* for the isothermal wall ($T_w = T_0$) consists of the following partial differential equation and boundary conditions:

$$\theta'' + \frac{\mathrm{Pr}}{2}f\theta' = 0, \; \theta(\eta) = \frac{T - T_0}{T_\infty - T_0} \tag{3.104}$$

$$\theta = 0, \quad \text{for } \eta = 0 \tag{3.105}$$
$$\theta \to 1, \quad \text{for } \eta \to \infty \tag{3.106}$$

For air ($Pr = 1$), $\theta = f'$; hence, the heat transfer problem is identical to the hydrodynamic problem and, consequently, the two boundary layer envelopes coincide. Since $f(\eta)$ is a known quantity, the temperature profile may be obtained for any other Pr number. The intermediate result produced by integrating (3.104) once,

$$\theta'(\eta) = \theta'(0)e^{-\frac{Pr}{2}\int_0^\eta f(\beta)d\beta},$$ (3.107)

may be integrated again to yield

$$\theta(\eta) = \theta'(0)\int_0^\eta e^{-\frac{Pr}{2}\int_0^\gamma f(\beta)d\beta}d\gamma.$$ (3.108)

It is now apparent that $\theta(\eta)$ depends on the particular value of $\theta'(0)$—itself a function of Pr—which may be obtained by invoking the asymptotic condition (3.106),

$$\theta'(0) = \left\{\int_0^\infty e^{-\frac{Pr}{2}\int_0^\gamma f(\beta)d\beta}d\gamma\right\}^{-1}.$$ (3.109)

The evaluation of this expression is particularly important to the convection boundary layer problem since both the heat transfer coefficient

$$h = \frac{k}{x}Re_x^{1/2}\theta'(0)$$ (3.110)

and the local Nusselt number

$$Nu_x = \frac{hx}{k}Re_x^{1/2}\theta'(0)$$ (3.111)

depend on it (are functions of it).

Instead of an explicit definition for $\theta'(0)$, Pohlhausen proposed the empirical correlation

$$\theta'(0) = 0.332Pr^{1/3},$$ (3.112)

valid for $Pr > 0.5$. A different correlation has to be utilized if $Pr < 0.5$ or, alternatively, (3.104) should be solved numerically for that particular value of Pr.

Remarkably, in the limit $Pr \to 0$ (i.e., for fluids with very high thermal conductivity), the ordinary differential equation (3.104) admits an analytical solution. In this case, $f' \sim 1$, and by further differentiation it yields

$$\frac{d}{d\eta}\left(\frac{\theta''}{\theta'}\right) = -\frac{Pr}{2}f',$$ (3.113)

which is exactly integrated by

$$\theta(\eta) = erf\left(\frac{\eta}{2}\sqrt{Pr}\right).$$ (3.114)

Consequently,

$$\theta'(0) = \left(\frac{Pr}{\pi}\right)^{1/2}, \quad \underset{Pr\to 0}{Nu} = \frac{hx}{k} = 0.564Pr^{1/2}Re_x^{1/2},$$ (3.115)

in good agreement with the scale analysis results (3.73).

It is important to recognize the limits of these similarity solutions, called "exact." It suffices to analyze the behavior of the predicted v component of the velocity that, erroneously, does not vanish for $\eta \to \infty$ (in the free stream)—rather it reaches the finite value $0.86 U_\infty \mathrm{Re}_x^{-1/2}$. Since in the boundary layer theory $v/U_\infty \underset{\eta \to \infty}{\sim} \mathrm{Re}_x^{-1/2}$, the accuracy of the results increases with $\mathrm{Re}_x^{1/2}$; hence, the thinner the boundary layer the more accurate the analysis. Another difficulty occurs for $x \to 0$, which is the starting section of the boundary layer.

The analytical form of the heatfunction for the forced convection boundary layer at the warm wall is [11]

$$
\tilde{H}(\tilde{x}, \eta) = \frac{H(x, y)}{\rho c_P U_\infty (T_0 - T_\infty) L \mathrm{Re}_L^{-1/2}} = \tilde{x}^{1/2} \left[f(\eta)\theta(\eta) + \frac{2}{\mathrm{Pr}} \theta'(\eta) \right]
$$

$$
= \tilde{x}^{1/2} g_T(\eta) \quad \text{(isothermal wall)} \tag{3.116}
$$

$$
\hat{H}(\tilde{x}, \eta) = \frac{H(x, y)}{q'' L} = \tilde{x} \left[\frac{1}{2} \mathrm{Pr} f(\eta)\theta(\eta) + \theta'(\eta) \right]
$$

$$
= \tilde{x} g_F(\eta) \quad \text{(isoflux wall)}. \tag{3.117}
$$

Like the temperature field of the Pohlhausen solution, the heatfunction field depends on the Prandtl number. It is apparent that $\tilde{H}(\tilde{x}, \tilde{y})$ and $\hat{H}(\tilde{x}, \eta)$ are *not* similarity functions. However, they are the product of two terms: a similarity heatfunction, $g_T(\eta)$ or $g_F(\eta)$, and a particular function of the downstream coordinate that depends on the particular thermal boundary condition (here a warm, isothermal or constant-flux wall). An interesting feature of the similarity heatfunction $g(\eta)$ is that it clearly shows the contribution of the two mechanisms of heat transfer: transversal conduction, $\theta'(\eta)$, and convection, $f(\eta)\theta(\eta)$.

3.1.14 OTHER THERMAL CONDITIONS

The isothermal plate is the simplest and, historically, one of the oldest problems. However, in many applications other thermal conditions may occur. Although the specific quantities h and Nu differ from case to case, the boundary layer concepts and the theory introduced by Prandtl, Blasius, and Pohlhausen are, from a broader perspective, the same. Several examples are detailed next.

Wall with Unheated Starting Section

An important thermal design problem is the evaluation of the heat transferred from a wall with an unheated ($T = T_\infty$) starting section, $0 < x < x_0$ (Fig. 3.13). The solution may be found by the integral method. If we select the temperature profile $m = p(3 - p^2)/2$, and the velocity profile $m = n(3 - n^2)/2$ (Table 3.1), the energy integral equation becomes

$$
\Delta^3 + 4\Delta^2 x \frac{d\Delta}{dx} = \frac{0.932}{\mathrm{Pr}}, \tag{3.118}
$$

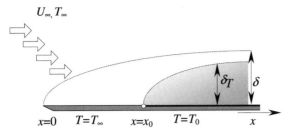

Figure 3.13
Forced convection heat transfer from a plane wall with an unheated starting section.

and it admits the solution

$$\Delta^3 = \frac{0.932}{Pr} + Cx^{-3/4}. \tag{3.119}$$

The integration constant C is obtained by applying the boundary condition: The temperature boundary layer starts at $x = x_0$, and hence,

$$\Delta = 0.977 Pr^{-1/3} \left[1 - \left(\frac{x_0}{x}\right)^{3/4} \right]^{1/3}. \tag{3.120}$$

For $x_0 = 0$ this result coincides with (3.88). The local Nusselt number is then

$$Nu_x = \frac{bx}{k} = 0.323 Pr^{1/3} Re_x^{1/2} \left[1 - \left(\frac{x_0}{x}\right)^{3/4} \right]^{1/3}. \tag{3.121}$$

Wall with Specified Nonuniform Temperature

The previous result may be used to solve a more general heat transfer problem: a nonuniform heated plane wall with a specified temperature distribution. For an isothermal source, $T_w = T_0$, located at $x_1 < x < x_2$, the temperature boundary layer may be thought of as being produced by a superposition of two temperature boundary layers, as shown in Fig. 3.14. The associated heat fluxes are as follows:

In the unheated section:

$$0 < x < x_1, \, q'' = 0.$$

In the heated section:

$$x_1 < x < x_2, \, q'' = 0.323 \frac{k}{x} Pr^{1/3} Re_x^{1/2} \frac{\Delta T}{\left[1 - \left(\frac{x_1}{x}\right)^{3/4} \right]^{1/3}}.$$

In the unheated section (by superposition):

$$x > x_2, \, q'' = 0.323 \frac{k}{x} Pr^{1/3} Re_x^{1/2} \left\{ \frac{\Delta T}{\left[1 - \left(\frac{x_1}{x}\right)^{3/4} \right]^{1/3}} + \frac{-\Delta T}{\left[1 - \left(\frac{x_2}{x}\right)^{3/4} \right]^{1/3}} \right\}.$$

Heat Transfer

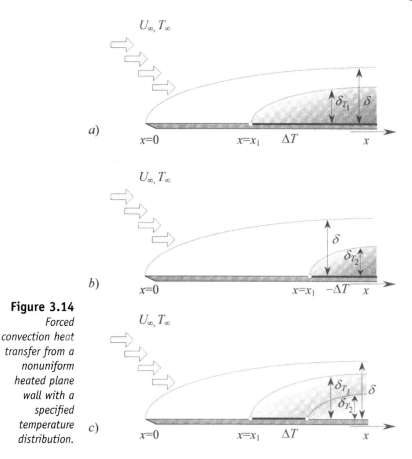

Figure 3.14
Forced convection heat transfer from a nonuniform heated plane wall with a specified temperature distribution.

This result may be generalized. For example, in the case of a heat source described by a stairlike temperature profile, ΔT_i, the heat flux is

$$q'' = 0.323 \frac{k}{x} \mathrm{Pr}^{1/3} \mathrm{Re}_x^{1/2} \sum_{i=1}^{N} \frac{\Delta T_i}{\left[1 - \left(\frac{x_i}{x} \right)^{3/4} \right]^{1/3}}, \tag{3.122}$$

where x_i is the length of the wall section i with the imposed temperature ΔT_i.

Wall with Specified Uniform Heat Flux

In many problems, such as the cooling of electronic or nuclear components, the wall heat flux rather than the temperature is the known quantity. In these applications, *overheating*, *burnout*, and *melting* are very important issues and, consequently, the evaluation of the wall temperature distribution, $T_w(x)$, is one of the main goals in thermal design.

The integral method and the profiles introduced previously may be utilized to compute the temperature drop $T_w(x) - T_\infty$ in the case when $q'' = $ const, yielding the local Nusselt number:

$$\text{Nu}_x = \frac{q''}{T_w(x) - T_\infty} \frac{x}{k} = 0.458 \text{Pr}^{1/3} \text{Re}_x^{1/2} \quad (\text{Pr} > 1). \tag{3.123}$$

A more general case is the heated plane wall with a nonuniform heat flux distribution $q''(x)$:

$$T_w(x) - T_\infty = \frac{0.623}{k} \text{Pr}^{-1/3} \text{Re}_x^{-1/2} \int_{\xi=0}^{x} \left[1 - \left(\frac{\xi}{x} \right)^{3/4} \right]^{-2/3} q''(\xi) d\xi \quad (\text{Pr} > 1). \tag{3.124}$$

3.1.15 OTHER FLOW CONDITIONS

In the boundary layer momentum balance (3.59), the horizontal component of the pressure gradient is negligibly small compared to the inertia and friction terms. This assumption is valid when the flow is parallel to the wall. If the wall is inclined with respect to the flow, let us say by an angle of $\beta/2$, then the free stream is accelerated in the x-direction, along the wall. If we neglect the laminar boundary layer, within which the viscous friction balances the inertia, the flow that engulfs an angular obstacle with the opening β may be considered inviscid. In this situation, the momentum balance is made by two terms: the friction and the pressure gradient. Consequently, the flow may be solved by methods specific to *potential* problems. The free stream velocity profile outside the boundary layer around angular obstacles is, in the limits of the potential (Laplace) theory,

$$U_\infty(x) = Cx^{m(\beta)}, C = \text{const}, m = \frac{\beta}{2\pi - \beta}. \tag{3.125}$$

By using Bernoulli's equation,

$$\frac{1}{\rho} \frac{dP_\infty}{dx} = -U_\infty \frac{dU_\infty}{dx}$$

[12], and (3.125), the momentum equation for the boundary layer around an angular obstacle (of angle β) is

$$u \frac{\partial u}{\partial x} + v \frac{\partial v}{\partial y} = \frac{m}{x} U_\infty^2 + v \frac{\partial^2 u}{\partial y^2}. \tag{3.126}$$

Falkner and Skan [18] found that this problem admits a similarity solution with an additional parameter, m (Blasius is a particular case, for $m = 0$). Analogously to Pohlhausen's method and using the Falkner–Skan solution to the flow, Eckert [19] found a similarity solution for the convection heat transfer problem. The friction coefficient in this case is $C_{f,x} \sim \text{Re}_x^{-1/2}$; Nusselt number is less sensitive to the longitudinal pressure gradient and the scale relation $\text{Nu} \sim \text{Pr}^{1/3} \text{Re}_x^{-1/2}$ is accurate for a larger range of β values. Eckert computed the Nu $Re_x^{-1/2}$ group as a function of Pr and β. Figure 3.15 [10] shows how, by employing an appropriate scaling, the $NuPr^{-1/3}Re_x^{-1/2}$ group appears as a function of only β.

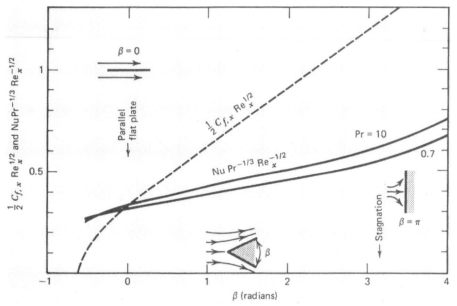

Figure 3.15 *Local friction and heat transfer in laminar boundary layer flow along an isothermal wedge-shaped body — used with permission from A. Bejan, Convection Heat Transfer, John Wiley, 1984, Fig. 2.10, p. 57*

3.2 Internal Forced Convection

In many applications convection heat transfer between a solid body and a fluid occurs through an *internal* (to the heat source) *forced flow*. The main quantities of interest are, again, the *friction coefficient*, the *longitudinal* (streamwise) *pressure drop*, and the *heat transfer coefficient* that evaluates the heat transferred to the flow. Similarly to the external forced convection problem, where the Prandtl–Pohlhausen analysis was shown to provide simpler mathematical models, the interest in simpler, yet satisfactorily accurate hydrodynamic and heat transfer equations is appealing here as well. The main concepts specific to the internal forced convection are the *fully developed flow* and *fully developed temperature field*. They are direct consequences of the hydrodynamic and temperature boundary layers that develop in external forced convection [21]. Similarly to the boundary layer that separates the external flow into two regions (a free stream flow and a boundary layer flow), the concept of "fully developed" helps distinguishing the *entrance region flow* from the *fully developed region flow*.

3.2.1 THE HYDRODYNAMIC ENTRANCE LENGTH

The flow in the 2D channel of Fig. 3.16, made of two parallel plates, has a uniform velocity entrance profile, U. As in external forced convection, boundary layers are expected to develop at top and bottom walls on a downstream length x sufficiently small compared to the size of the inter-plates channel spacing, D. When δ reaches the limiting value $D/2$, the two boundary layers start interacting. Consequently, the channel flow may be divided into two regions: an *entrance (or developing) region*—where the boundary layers do not interact (are distinct) and where there is a central free stream flow unperturbed by the presence of the boundary layers—and a *fully developed region*, where the central free stream flow ceases to exist for a central channel flow, and where the boundary layers merge into interacting.

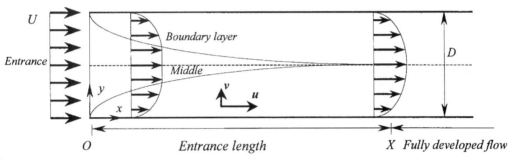

Figure 3.16 *2D channel flow with uniform velocity entrance profile: the entrance (or developing) and the fully developed regions.*

A straightforward order-of-magnitude evaluation of the entrance length size, X, may be obtained by using Blasius's solution, $\delta/x = 4.92\mathrm{Re}_x^{-1/2}$ (3.81), and the constraint $\delta(X) = D/2$:

$$\frac{X}{D}\mathrm{Re}_D^{-1} = 0.0103. \tag{3.127}$$

The more accurate integral solution for the entrance region, due to Sparrow and Crawford [20], accounts for the acceleration of the central flow caused by the growing compression exerted by the developing boundary layers. This effect is illustrated in Fig. 3.16: As the boundary layers accumulate stagnant fluid at the walls, the free stream velocity, U_c, has to increase in order to preserve the $\rho U D$ mass flow through any cross section. The momentum integral equation (3.81), with U_∞ replaced by U_c, $Y = \delta(x)$ and with the pressure gradient, dP_∞/dx, replaced by dP/dx—a function of $U_c(x)$, as suggested by Bernoulli's equation $\rho(U_c^2/2) + P = \mathrm{const}$—yields

$$U_c\frac{dU_c}{dx} + \frac{1}{\rho}\frac{dP}{dx} = 0. \tag{3.128}$$

Eliminating dP/dx between (3.128) and (3.81) results in

$$\frac{d}{dx}\int_0^\delta (U_c - u)u\,dy + \frac{dU_c}{dx}\int_0^\delta (U_c - u)\,dy = \nu\frac{\partial u}{\partial y}\bigg|_{x=0}. \tag{3.129}$$

On the other hand, the mass conservation law yields

$$\int_0^\delta \rho u\,dy + \int_0^{\frac{D}{2}} \rho U_c\,dy = \rho U\frac{D}{2}. \tag{3.130}$$

Sparrow's integral approach is based on the boundary layer profile

$$\frac{u}{U_c} = 2\frac{y}{\delta} - \left(\frac{y}{\delta}\right)^2, \tag{3.131}$$

which yields

$$\frac{x}{D}\mathrm{Re}_D^{-1} = \frac{3}{160}\left(9\frac{U_c}{U} - 2 - 7\frac{U}{U_c} - 16\ln\frac{U_c}{U}\right), \frac{\delta(x)}{D/2} = 3\left[1 - \frac{U}{U_c(x)}\right]. \tag{3.132}$$

If we recall that at the boundary layer merging station, X, the boundary layer thickness is constrained, $\delta(X) = D/2$, it follows that

$$U_c(X) = \frac{3}{2}U, \frac{X}{D}\mathrm{Re}_D^{-1} = 0.0065. \tag{3.133}$$

This solution is smaller by 37% than the scale analysis solution (3.127) [10]. The basic conclusion of the two analyses (scale and integral) is that the laminar entrance length is proportional to $D\mathrm{Re}_D$, where the proportionality coefficient is of order 10^{-2}.

Schlichting [15] solved this problem by using two asymptotic expansions for the laminar entrance region (for the beginning and for the end, respectively), and obtained

$$\frac{X}{D}\mathrm{Re}_D^{-1} = 0.04. \tag{3.134}$$

The different structures of the entrance and fully developed regions is also apparent in the wall shear stress, $\tau_w(x)$, as suggested by the friction coefficient

$$C_{f,x} \stackrel{\text{def}}{=} \frac{\tau_w(x)}{\frac{1}{2}\rho U^2},$$

which, by Sparrow's analysis is

$$C_{f,x}\mathrm{Re}_D = \frac{8}{3}\frac{U_c}{U}\left(1 - \frac{U}{U_c}\right)^{-1}. \tag{3.135}$$

Figure 3.17 [10] shows $C_{f,x}\mathrm{Re}_D$ computed by using Blasius's solution, $\delta/x = 4.92\mathrm{Re}_x^{-1/2}$, where U_∞ is replaced by U in the definitions of $C_{f,x}$ and Re_x. The friction coefficient variation indicates, as expected, the existence of two distinct boundary layers. In the fully developed region $\tau_w(x)$, and hence $C_{f,x}$, ceases to depend on x, because the velocity profile $u(x, y)$ itself is no longer a function of x.

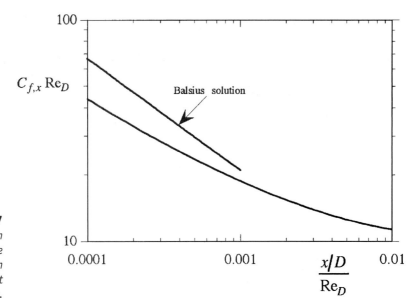

Figure 3.17
Skin friction coefficient in the entrance region of a parallel-duct flow.

The entrance region problem for a cylindrical, circular duct may be treated analogously [10].

3.2.2 FULLY DEVELOPED FLOW

The stationary forms of the mass conservation equation,

$$\frac{\partial u}{\partial x} + \frac{\partial v}{\partial y} = 0, \tag{3.136}$$

and momentum balance equation,

$$u\frac{\partial u}{\partial x} + v\frac{\partial u}{\partial y} = -\frac{1}{\rho}\frac{\partial P}{\partial x} + v\left(\frac{\partial^2 u}{\partial x^2} + \frac{\partial^2 u}{\partial y^2}\right) \tag{3.137}$$

$$u\frac{\partial v}{\partial x} + v\frac{\partial v}{\partial y} = -\frac{1}{\rho}\frac{\partial P}{\partial y} + v\left(\frac{\partial^2 v}{\partial x^2} + \frac{\partial^2 v}{\partial y^2}\right), \tag{3.137}$$

may be substantially simplified by the following argument: At the coordinate $x \sim L$, that is, in the fully developed region, $y \sim D$ and $u \sim U$. The mass conservation law (3.136) suggests the following scaling for the transversal component of the velocity:

$$v \sim \frac{DU}{L}. \tag{3.138}$$

Hence, the fully developed region is situated far enough downstream (with respect to the entrance region) that the order of magnitude (scale) of v is negligibly small. Consequently, mass conservation yields

$$v = 0, \frac{\partial u}{\partial x} = 0. \tag{3.139}$$

Heat Transfer

It should be noticed that the scale $y \sim D$ (constraint) is consistent for the fully developed region, whereas the scaling $x \sim \delta$ is appropriate for the entrance region. The first scaling corresponds to a variable velocity profile in the entire cross section; the second scaling refers to the case of a δ-thick region that grows with the x-coordinate.

The y-momentum equations (3.137) and (3.139) lead to

$$\frac{\partial P}{\partial y} = 0,$$

which indicates that, mathematically, P is a function of x only. This conclusion is similar to that obtained previously, in the external boundary layer problem: *in the fully developed region the pressure is constant in any cross section of the flow.* In virtue of this conclusion, the x-momentum equation (3.137a) becomes

$$\frac{dP}{dx} = \mu \frac{\partial^2 u}{\partial y^2} = \text{const} \tag{3.140}$$

Since, on one hand, $P = P(x)$ and, on the other hand $u = u(y)$, each term in (3.140) has to be constant. Solving (3.140) with wall no-slip boundary conditions, $u|_{y=\pm \frac{D}{2}} = 0$ yields the *Hagen-Poiseuille solution* [22, 23] to the fully developed flow between parallel plates,

$$u = \frac{3}{2} U \left[1 - \left(\frac{y}{D/2} \right)^2 \right], U = \frac{D^2}{12\mu} \left(-\frac{dP}{dx} \right). \tag{3.141}$$

Here, the y-origin is taken at the channel longitudinal axis. The velocity profile is parabolic and the velocity is proportional to the pressure drop per unit duct length.

For a duct of an arbitrary cross section, (3.140) is replaced by

$$\frac{dP}{dx} = \mu \nabla^2 u = \text{const}, \tag{3.142}$$

which, in the particular case of a cylindrical circular cross-section duct of radius r_0, becomes

$$\frac{dP}{dx} = \mu \left(\frac{\partial^2 u}{\partial r^2} + \frac{1}{r} \frac{\partial u}{\partial r} \right). \tag{3.143}$$

The solution to this problem with the no-slip wall boundary condition is

$$u = 2U \left[1 - \left(\frac{r}{r_0} \right)^2 \right], U = \frac{r_0^2}{8\mu} \left(-\frac{dP}{dx} \right). \tag{3.144}$$

The forms (3.141) and (3.144) are the simplest results for fully developed channel flow in a constant cross-section duct. Generally, solving the Poisson problem (3.142) is considerably more complicated.

For Hagen–Poiseuille flows, it is customary to define the Reynolds number as $\mathrm{Re}_D = UD/\nu$. Since in this case the governing balance is between the longitudinal (downstream) pressure gradient and the viscous friction (opposed to the flow), this group is a measure of the ratio of pressure force to friction force, which scales as

$$-\frac{\dfrac{dP}{dx}}{\mu\dfrac{\partial^2 u}{\partial r^2}} = O(1).$$ (3.145)

3.2.3 THE HYDRAULIC DIAMETER AND THE PRESSURE DROP

Although in the fully developed region the mass flow rate $\dot{m}$ and the pressure drop ΔP are proportional, their analytic relation, $\dot{m}(\Delta P)$, is not that simple, and its identification forms the object of substantial experimental work.

The longitudinal momentum equation for a control volume in the fully developed region (Fig. 3.18, after [10]) is

$$A\Delta P = \tau_w pL,$$ (3.146)

Figure 3.18
Contributors to the longitudinal momentum balance for a fluid control volume in the fully developed region.

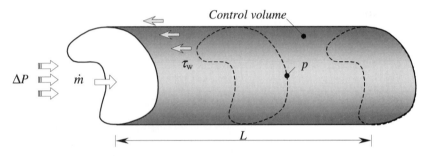

where p is the "wet" perimeter. In the fully developed region it is customary to replace the wall shear stress, τ_w, by the *friction factor*, f, a nondimensional quantity defined [similarly to $C_{f,x} = \tau_w(x)/\frac{1}{2}\rho U^2$] by

$$f \overset{\text{def}}{=} \frac{\tau_w}{\frac{1}{2}\rho U^2}.$$ (3.147)

However, unlike $C_{f,x}$, which was defined for the entrance region, f does *not* depend on x. Consequently, the pressure drop in the fully developed region is computed by

$$\Delta P = f\frac{pL}{A}\left(\frac{1}{2}\rho U^2\right).$$ (3.148)

The inverse of the ratio p/A is the characteristic size of the duct cross-section, that is,

$$r_h = \frac{A}{p}, \quad \text{(hydraulic radius), or } D_h = 4r_h = \frac{4A}{p} \quad \text{(hydraulic diameter).}$$

(3.149)

Table 3.2 after [10], gives the hydraulic diameter for several types of ducts of cross sections with reduced degrees of asymmetry, all with the same

Table 3.2

Cross section	Hydraulic diameter
Circular	$D_h = a$
Square	$D_h = a$
Triangular	$D_h = a$
Rectangular $(4:1)$	$D_h = \frac{8}{5}a$
Parallel plates	$D_h = 2a$

equivalent hydraulic diameter, D_h. For cross sections of higher degrees of asymmetry, the hydraulic diameter is of the order of magnitude of the smallest cross-sectional dimension.

By (3.148), the pressure drop is

$$\Delta P = f \frac{4L}{D_h}\left(\frac{1}{2}\rho U^2\right),\tag{3.150}$$

and it may be evaluated provided f is a known quantity. For Hagen–Poiseuille flows, the solutions to (3.141) and (3.144) are

Parallel plates (D = inter-plates spacing) $\rightarrow$ $f = \dfrac{24}{\mathrm{Re}_{D_h}}$, $D_h = 2D$,

Circular tube (D = tube diameter) $\rightarrow$ $f = \dfrac{16}{\mathrm{Re}_{D_h}}$, $D_h = D$.

Table 3.3 [10] summarizes the friction factor, f, for ducts of different cross-section types, in laminar flows ($\mathrm{Re}_{D_h} < 2000$).

Table 3.3

Cross section	$f\mathrm{Re}_{D_b}$	$\dfrac{\pi D_b^2}{4}\dfrac{1}{A_{duct}}$	$\mathrm{Nu}_D = bD_b/k$	
			q'' uniform	T_0 uniform
(triangle, 60°)	13.3	0.605	3	2.35
(square)	14.2	0.785	3.63	2.89
(circle)	16	1	4.364	3.66
(rectangle, 4a)	18.3	1.26	5.35	4.65
(parallel plates)	24	1.57	8.235	7.54
(parallel plates, Insulated)	24	1.57	5.385	4.86

The friction factor, f, may be computed provided the solution to (3.143) is known. For instance, for a duct of rectangular cross section $a \times b$ (Fig. 3.19),

$$\frac{dP}{dx} = \mu\left(\frac{\partial^2 u}{\partial y^2} + \frac{\partial^2 u}{\partial z^2}\right) = \text{const} \tag{3.151}$$

The analytic solution to this Poisson equation may be obtained, for example, by the *variable separation* method. However, a simpler yet satisfactorily accurate solution may be derived by assuming the velocity profile

$$u(y, z) = u_0\left[1 - \left(\frac{y}{a/2}\right)^2\right]\left[1 - \left(\frac{z}{b/2}\right)^2\right], \tag{3.152}$$

Heat Transfer

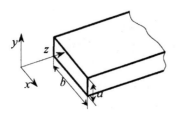

Figure 3.19
Duct of rectangular cross section.

where u_0 is the maximum velocity (in the axis). The problem then reduces to finding u_0.

It is obvious that $u(y, z)$ thus obtained does not identically match the true solution to the problem. The maximum value of the velocity results by imposing that (3.152) verifies (3.151) integrated (averaged) over the duct cross section, namely,

$$ab \frac{dP}{dx} = \mu \int_{-\frac{a}{2}}^{\frac{a}{2}} \int_{-\frac{b}{2}}^{\frac{b}{2}} \left(\frac{\partial^2 u}{\partial y^2} + \frac{\partial^2 u}{\partial z^2} \right) dz dy.$$

(3.153)

Consequently,

$$ab \frac{dP}{dx} = -\frac{16}{3} \mu u_0 \left(\frac{b}{a} + \frac{a}{b} \right),$$

(3.154)

and using the definition of the average,

$$U \overset{def}{=} \frac{1}{ab} \int_{-\frac{a}{2}}^{\frac{a}{2}} \int_{-\frac{b}{2}}^{\frac{b}{2}} u dz \, dy,$$

(3.155)

yields

$$u_0 = \frac{9}{4} U.$$

(3.156)

Finally, this result produces the following expression for the friction factor:

$$f = \frac{a^2 + b^2}{(a + b)^2} \frac{24}{\text{Re}_{D_b}}, \quad \text{where} \quad D_b = \frac{2ab}{a + b}.$$

(3.157)

3.2.4 HEAT TRANSFER IN THE FULLY DEVELOPED REGION

The key problem of heat transfer in duct flows is to determine the relationship between the wall-to-stream temperature drop and its associated heat transfer rate. For flow in a circular duct of radius r_0 and with average longitudinal velocity U, the mass flow rate is $\dot{m} = \rho \pi r_0^2 U$ (Fig. 3.20, [10]). The heat transfer rate from the wall to the stream should equal the change in the enthalpy of the stream. To verify this we write the first principle for a control volume of length dx,

$$q'' 2\pi r_0 dx = \dot{m} (h_{x+dx} - h_x).$$

(3.158)

In the perfect gas limit of the fluid ($dh = c_P dT_m$), or in the limit of an incompressible flow and under a negligible pressure variation ($dh = c_P dT_m$), this balance equation implies

$$\frac{dT_m}{dx} = \frac{2}{r_0} \frac{q''}{\rho c U},$$

(3.159)

where, for incompressible flows, c_P was replaced by c. The temperature T_m that appears in the energy balance written for the control volume is the *average temperature of the stream,* and it indicates that the actual tempera-

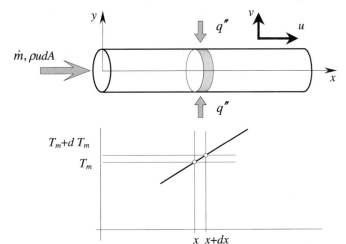

Figure 3.20

Heat transfer in the fully developed region for flow in a cylindrical duct.

ture field in the stream is *not* uniform. Its definition results from the first principle applied to a stream tube, that is, $q'' 2\pi r_0 \, dx = d \int_A \rho u c_p T dA$, as

$$T_m \overset{\text{def}}{=} \frac{1}{\rho c_p U} \frac{1}{A} \int_A \rho u c_p T dA. \tag{3.160}$$

For ducts with uniform-temperature walls, this means

$$T_m = \frac{1}{U} \frac{1}{\pi r_0^2} \int_A u T dA. \tag{3.161}$$

As the temperature varies in every cross section of the duct, there exists a wall-to-stream temperature drop $\Delta T = T_w - T_m$. The heat transfer coefficient is then

$$h \overset{\text{def}}{=} \frac{q''}{T_w - T_m} = \frac{k \left. \dfrac{\partial T}{\partial r} \right|_{r=r_0}}{\Delta T}, \tag{3.162}$$

where q'' is positive if the heat transfer is wall $\leftrightarrow$ stream.

3.2.5 THE FULLY DEVELOPED TEMPERATURE PROFILE

The heat transfer coefficient may be evaluated provided the temperature field $T(x, y)$, and hence the energy boundary value problem for the specified boundary conditions, is solved *first*. For example, for the stationary laminar flow in a straight circular duct, the energy equation is

$$\frac{1}{\alpha} \left(u \frac{\partial T}{\partial x} + v \frac{\partial T}{\partial y} \right) = \frac{\partial^2 T}{\partial r^2} + \frac{1}{r} \frac{\partial T}{\partial r} + \frac{\partial^2 T}{\partial x^2}. \tag{3.163}$$

In the hydrodynamic fully developed region, $v = 0$ and $u = u(r)$, which implies

$$\frac{u(r)}{\alpha} \frac{\partial T}{\partial x} = \frac{\partial^2 T}{\partial r^2} + \frac{1}{r} \frac{\partial T}{\partial r} + \frac{\partial^2 T}{\partial x^2}. \tag{3.164}$$

This equation indicates the balance

$$Axial\ convection \sim Radial\ conduction,\ Axial\ conduction$$

with the following scales:

$$\overbrace{\frac{U}{\alpha}\left(\frac{q''}{D\rho c_P U}\right)}^{\text{Convection}} \sim \overbrace{\underbrace{\frac{\Delta T}{D^2}}_{\text{radial}}, \underbrace{\frac{1}{x}\left(\frac{q''}{D\rho c_P U}\right)}_{\text{Longitudinal}}}^{\text{Conduction}}. \tag{3.165}$$

Note that en route to this scaling relation we used the relation

$$\frac{\partial T}{\partial x} \sim \frac{q''}{D\rho c_P U}, \tag{3.166}$$

introduced by (3.159).

Apparently, the radial conduction is a central term in (3.165) — *without its contribution there is no heat transfer associated to the internal flow.* The scales (3.165) multiplied by $D^2/\Delta T$ and the definition of the heat transfer coefficient, $h = q''/\Delta T$, yield

$$\overbrace{\frac{hD}{k}}^{\text{Convection}} \sim \overbrace{\underbrace{1}_{\text{Radial}}, \underbrace{\left(\frac{hD}{k}\right)^2\left(\frac{\alpha}{DU}\right)^2}_{\text{Longitudinal}}}^{\text{Conduction}}. \tag{3.167}$$

The bottom line to this analysis is that for large Péclet numbers, Pe_D, the axial conduction heat transfer may be negligibly small, and the energy equation becomes

$$\frac{u(r)}{\alpha}\frac{\partial T}{\partial x} = \frac{\partial^2 T}{\partial r^2} + \frac{1}{r}\frac{\partial T}{\partial r} \left(Pe_D = \frac{UD}{\alpha} \gg 1\right). \tag{3.168}$$

Furthermore, from the balance

$$Axial\ convection \sim Radial\ conduction,$$

it follows that the Nusselt number is a *constant of order 1*,

$$Nu_D = \frac{hD}{k} \sim 1 \left(Pe_D = \frac{UD}{\alpha} \gg 1\right). \tag{3.169}$$

The temperature profile produced by this analysis corresponds to fully developed flow. It represents the temperature distribution downstream from the two entrance regions (X, X_T), where both u and T are developing.

At this point it is important to note that, in the literature, the fully developed temperature profile is defined through

$$\frac{T_w - T}{T_w - T_m} = \phi\left(\frac{r}{r_0}\right), \tag{3.170}$$

where T, T_w, and T_m may be functions of x. This analytic form for $T(x, r)$ is a consequence of $Nu_D \sim 1$, so,

$$Nu_D = \frac{hD}{k} = \frac{q''}{T_w - T_m}\frac{D}{k}, \quad \text{or} \quad Nu_D = D\frac{\left.\frac{\partial T}{\partial r}\right|_{r=r_0}}{T_w - T_m} \sim 1. \tag{3.171}$$

Consequently, the variation of $\partial T/\partial r|_{r=r_0}$ with respect to x is identical to that of $T_w(x) - T_m(x)$, and since $\partial T/\partial r$ is a function of x and r, it follows that

$$\mathrm{Nu}_D = \frac{\partial T/\partial (r/r_0)}{T_w(x) - T_m(x)} = f_1\left(\frac{r}{r_0}\right) = O(1), \tag{3.172}$$

which, further integrated with respect to r/r_0, yields

$$T\left(x, \frac{r}{r_0}\right) = (T_0 - T_m)f_2\left(\frac{r}{r_0}\right) + f_3(x). \tag{3.173}$$

Here f_1, f_2, and f_3 are functions of r/r_0 and x.

3.2.6 DUCTS WITH UNIFORM HEAT FLUX WALLS

When q'' is not a function of x, the ordinary differential equation (3.168) admits an analytical solution,

$$T(x, r) = T_w(x) - \frac{q''}{b}\phi\left(\frac{r}{r_0}\right), \quad \text{hence} \quad \frac{\partial T}{\partial x} = \frac{dT_w}{dx}, \quad \text{or} \quad \frac{dT_w}{dx} = \frac{dT_m}{dx}. \tag{3.174}$$

By virtue of (3.159),

$$\frac{\partial T}{\partial x} = \frac{2}{r_0}\frac{q''}{\rho c_p U} = \text{const} \tag{3.175}$$

Consequently, the temperature at a particular location is a linear function of x, and its slope is proportional to q'' (Fig. 3.21), after [10]. On the other hand, the r-variation of T, respectively $\phi(r/r_0)$, may be found by solving the energy equation for the thermally fully developed flow. Using (3.168), the temperature profile (3.174) and the Hagen–Poiseuille velocity profile (3.144) yield

$$-2\frac{bD}{k}(1 - r^{*2}) = \frac{d^2\phi}{dr^{*2}} + \frac{1}{r^*}\frac{d\phi}{dr^*}, \quad r^* = \frac{r}{r_0}, \tag{3.176}$$

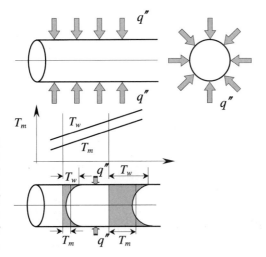

Figure 3.21
Heat transfer in the fully developed region for a cylindrical duct with uniform heat flux walls.

with $\mathrm{Nu}_D = hD/k$ now emerging explicitly. Integrating this equation twice and using the boundary condition $\phi'|_{r^*=0} = \text{finite}$ results in

$$\phi(r^*, \mathrm{Nu}_D) = C_2 - 2\mathrm{Nu}_D \left[\left(\frac{r^*}{2}\right)^2 - \left(\frac{r^{*2}}{4}\right)^2 \right]. \tag{3.177}$$

The integration constant C_2 may be found by using (3.174), (3.177), and the condition $T|_{r^*=1} = T_w$, namely,

$$T = T_w - (T_w - T_m)\mathrm{Nu}\left(\frac{3}{8} - \frac{r^{*2}}{2} + \frac{r^{*4}}{8}\right). \tag{3.178}$$

The average temperature drop $T_w - T_m$ (3.160) is then

$$T_w - T_m = \frac{1}{\pi r_0^2 U}\int_0^{2\pi}\int_0^{r_0}(T_w - T)urdrd\theta = 4\int_0^1 (T_w - T)(1 - r^{*2})r^*dr^*. \tag{3.179}$$

Consequently,

$$1 = 4\mathrm{Nu}_D\int_0^1\left(\frac{3}{8} - \frac{r^{*2}}{2} + \frac{r^{*4}}{8}\right)(1 - r^{*2})r^*dr^* = \frac{11}{48}\mathrm{Nu}_D, \tag{3.180}$$

which means that the Nusselt number for the thermally fully developed Hagen–Poiseuille region is

$$\mathrm{Nu}_D = \frac{48}{11}. \tag{3.181}$$

This result is in good agreement with the scale analysis (3.169).

Table 3.3 gives the Nusselt number, $\mathrm{Nu}_D = hD_h/k$, for different types of ducts, obtained by integrating the energy equation

$$\frac{u}{\alpha}\frac{\partial T}{\partial x} = \nabla^2 T, \tag{3.182}$$

where the longitudinal temperature gradient, $\partial T/\partial x$, may be explicitly obtained via a balance equation of type (3.159), that is,

$$\frac{dT_m}{dx} = \frac{q'}{\rho c_P A U} = \text{const} \tag{3.183}$$

q' is the heat transfer rate per duct unit length, an x-independent quantity. Generally, for the noncircular ducts, the wall temperature on a circumferential outline, T_w, is assumed constant and, subsequently, the heat flux is nonuniform in the x-direction.

3.2.7 DUCTS WITH ISOTHERMAL WALLS

Figure 3.22, after [10], shows a qualitative sketch of the temperature profile for a duct with an isothermal, T_w, wall. If the average temperature of the flow at coordinate x_1 in the fully developed region is T_1, then the heat transfer is

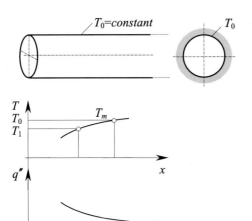

Figure 3.22
Heat transfer in the fully developed region for a cylindrical duct with isothermal walls.

driven by the temperature difference $T_w - T_1$ where the flow temperature is increasing downstream, monotonously. Consequently, the heat flux

$$q''(x) = b[T_w - T_m(x)] \tag{3.184}$$

exhibits the same trend. On the other hand, by scale analysis (3.169), b was shown to be constant. Now, eliminating q'' between (3.184) and (3.159) and then integrating it yields

$$\frac{T_0 - T_m(x)}{(T_0 - T_1)} = e^{-[\alpha(x-x_1)/r_0^2 U]\mathrm{Nu}_D}, \tag{3.185}$$

which clearly indicates an exponential decrease with respect to x. The Nu_D number that appears in (3.185) may be computed by solving the energy equation (3.168), where the temperature gradient $\partial T/\partial x$ is written as

$$\frac{\partial T}{\partial x} = \frac{\partial}{\partial x}[T_0 - \phi(T_0 - T_m)] = \phi\frac{dT_m}{dx}.$$

Merging this result, the Hagen–Poiseuille solution, and the temperature profile $T = T_0 - \phi(T_0 - T_m)$ into the energy equation (3.164) leads to the nondimensional form

$$-2\,\mathrm{Nu}_D(1 - r^{*2})\phi = \frac{d^2\phi}{dr^{*2}} + \frac{1}{r^*}\frac{d\phi}{dr^*}, \tag{3.186}$$

which, if we observe that the sign of $\phi(r^*)$ is reversed, is similar to (3.176). The corresponding boundary conditions may be

$$\text{Axial symmetry:} \quad \left.\frac{d\phi}{dr^*}\right|_{r^*=0} = 0, \tag{3.187}$$

$$\text{Isothermal wall:} \quad \phi|_{r^*=1} = 0.$$

Heat Transfer

If we apply the heat transfer coefficient definition (3.162), the Nusselt number then results as

$$\mathrm{Nu}_D = -2\frac{d\phi}{dr^*}\bigg|_{r^*=1}. \tag{3.188}$$

The solution $\phi(r^*, \mathrm{Nu})$ to the problem (3.186), (3.187) and the definition (3.188) may be found, for instance, by a fixed-point iterative procedure using a starting guess value for Nu_D that is then successively improved. The final result,

$$\mathrm{Nu}_D = 3.66, \tag{3.189}$$

is in good agreement with the scale analysis result (3.169). Table 3.3 also gives the Nusselt numbers for several common types of ducts.

3.2.8 HEAT TRANSFER IN THE ENTRANCE REGION
The previous results are valid for laminar internal forced flow, when both velocity and temperature are fully developed, that is, for $x > \max\{X, X_T\}$. The length X_T is that particular value of the x-coordinate where δ_T reaches the value of the hydraulic diameter.

The scale analysis may be used to produce order of magnitude estimates, and Fig. 3.23a shows, qualitatively, the influence of Pr on the scaling $\delta_T(X_T) \sim D_h$. As seen previously, the ratio δ/δ_T increases monotonously with Pr; hence, X/X_T has to vary conversely.

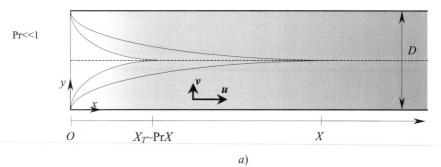

a)

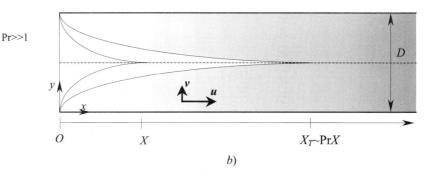

b)

Figure 3.23
The internal forced convection heat transfer in the Entrance Region for fluids with (a) Pr $\ll$ 1, *(b)* Pr $\gg$ 1.

Heat Transfer

For fluids with $\Pr \ll 1$, in virtue of (3.71), δ_T grows faster than δ,

$$\delta_T(x) \sim x\Pr^{-1/2}\mathrm{Re}_x^{-1/2}, \tag{3.190}$$

and, if we consider that at the temperature entrance region limit $x \sim X_T$ and $\delta_T \sim D_h$, it follows that

$$X_T\Pr^{-1/2}\mathrm{Re}_{X_T}^{-1/2} \sim D_h, \quad \text{or} \quad \left(\frac{X_T}{D_h}\right)^{1/2}(\mathrm{Re}_{D_h}\Pr)^{-1/2} \sim 1. \tag{3.191}$$

Alternatively, this result may also be put in the form

$$\frac{X_T}{D_h}(\mathrm{Re}_{D_h}\Pr)^{-1} \sim \text{const}, \tag{3.192}$$

where the constant value was identified, empirically, as being "approximately 0.1."

A similar scale analysis for the hydrodynamic problem leads to

$$\left(\frac{X_T}{D_h}\right)^{1/2}\mathrm{Re}_{D_h}^{-1/2} \sim 1. \tag{3.193}$$

For fluids with $\Pr \gg 1$ (water, oils, etc.), $\delta_T \nsim D_h$, although in the entrance region the temperature boundary layer grows more slowly than the velocity boundary layer. In this situation the velocity profile extends over D_h (Fig. 3.23). Hence, in the temperature boundary layer the scale of u is U, and it may be shown that $\delta_T(x) \sim x\mathrm{Re}_x^{-1/2}\Pr^{-1/2}$, that is, a result that is identical to the one obtained for fluids with $\Pr \ll 1$. The scaling relations (3.193) and (3.191) lead to

$$\frac{X_T}{X} \sim \Pr, \tag{3.194}$$

a valuable, general conclusion that is valid for any Pr.

The local Nusselt number in the thermally developing region ($x \ll X_T$) scales as

$$\mathrm{Nu}_D = \frac{bD_h}{k} \sim \frac{q''}{\Delta T}\frac{D_h}{k} \sim \frac{D_h}{\delta_T} \sim \left(\frac{x/D_h}{\mathrm{Re}_{D_h}\Pr}\right)^{-1/2}, \tag{3.195}$$

and similarly to the δ_T (3.190) scale, it is valid for any Pr. This result was validated by other, more accurate solutions.

3.3 External Natural Convection

Natural, or *free* convection occurs when the fluid flows "by itself" becauses of its density variation with the temperature, and not because of imposed, external means (e.g., a pump). For example, in a stagnant fluid reservoir that is in contact with a warmer, vertical wall (the heat source here), the fluid layer that contacts the wall is heated by the wall through thermal diffusion, and it becomes lighter as its density decreases. Consequently, this warmer fluid layer is entrained into a slow, upward motion that, provided the fluid reservoir is large enough, does not perturb the fluid away from the wall. Because the hydrostatic pressure in the stagnant fluid reservoir decreases with altitude, a control volume of fluid conveyed in this ascending motion—in fact, a wall jet—expands while traveling upward.

By virtue of the mass conservation principle (the reservoir contains a finite amount of fluid), the fluid control volume will eventually return to the bottom of the warm wall, entrained by a descending stream. In this closing sequence of its travel the fluid control volume is cooled and compressed (its density is increased) by the increasing hydrostatic pressure.

Summarizing, the fluid control volume may be seen as a system that undergoes a cyclic sequence of heating, expansion, cooling, and compression processes, which is in fact the classical thermodynamic work-producing cycle (Fig. 3.24) [2]. Here, unlike the classical thermodynamic cycles, the heating and expansion, on one hand, and the cooling and compression, on the other hand, are (respectively) simultaneous processes—neither at constant volume (the control volume expands/compresses while heating up/cooling down) nor isobaric (the hydrostatic pressure varies continuously with the altitude). The work potential produced by this cycle is "consumed" through internal friction between the fluid layers, which are in relative motion.

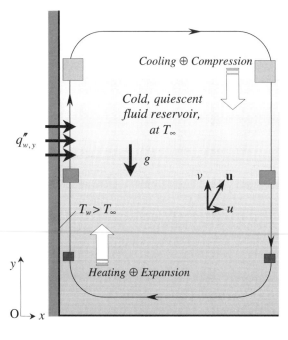

Figure 3.24
External natural convection: The fluid control volume acts as a system that undergoes a cyclic sequence of heating — the classical thermodynamic work-producing cycle.

In natural convection heat is transferred from the heat source (e.g., the warm, vertical wall) to the adjacent fluid layer by thermal diffusion, then by convection and diffusion within the fluid reservoir. When the fluid reservoir that freely convects the heat is external to the heat source, the convection heat transfer is called *external*.

3.3.1 THE THERMAL BOUNDARY LAYER

The fluid region where the temperature field varies from the wall temperature to the reservoir temperature and where, in fact, motion exists is called the *thermal (temperature) boundary layer* (Fig. 3.25).

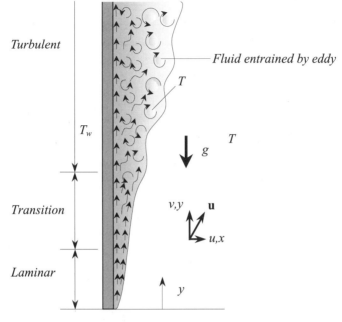

Figure 3.25
The structure of the natural convection boundary layer flow: laminar, transition, and turbulent sections.

The central object of the thermal analysis is, again, to evaluate the heat transferred from the wall to the reservoir, and the bottom line to it is finding the heat transfer coefficient

$$h_y \overset{def}{=} \frac{q''_{w,y}}{T_w - T_0} = -\frac{k\frac{\partial T}{\partial x}\Big|_{x=0}}{T_w - T_0}. \tag{3.196}$$

Here T_0 is the reservoir temperature far away from the wall, T_w is the wall temperature, k is the thermal conductivity of the fluid, and $q''_{w,y}$ is the wall heat flux rate in the y-direction (horizontal).

The *mass conservation equation* is

$$\frac{\partial u}{\partial x} + \frac{\partial v}{\partial y} = 0, \tag{3.197}$$

the *momentum equation* is

$$\rho\left(u\frac{\partial u}{\partial x} + v\frac{\partial u}{\partial y}\right) = -\frac{\partial P}{\partial x} + \mu\left(\frac{\partial^2 u}{\partial x^2} + \frac{\partial^2 u}{\partial y^2}\right) \tag{3.198}$$

$$\rho\left(u\frac{\partial v}{\partial x} + v\frac{\partial v}{\partial y}\right) = -\frac{\partial P}{\partial y} + \mu\left(\frac{\partial^2 v}{\partial x^2} + \frac{\partial^2 v}{\partial y^2}\right) - \rho g, \tag{3.198}$$

and the energy equation is

$$u\frac{\partial T}{\partial x} + v\frac{\partial T}{\partial y} = \alpha\left(\frac{\partial^2 T}{\partial x^2} + \frac{\partial^2 T}{\partial y^2}\right) \tag{3.199}$$

These form the partial differential equation part of the mathematical model of this boundary layer problem.

As for the forced convection boundary layer, the natural convection boundary layer consists of a *temperature boundary layer* of thickness δ_T, and a *hydrodynamic boundary layer*, of thickness δ.

The temperature gradient at the wall scales as

$$\left.\frac{\partial T}{\partial x}\right|_{x=0} \sim \frac{\Delta T}{\delta_T}, \quad \text{where} \quad \Delta T = T_w - T_0, \tag{3.200}$$

and, since the thermal boundary layer is a slender region, it makes sense to assume that

$$\delta_T \ll y. \tag{3.201}$$

This means that, in the governing equations, the second-order derivatives with respect to y may be neglected with respect to the second-order derivatives with respect to x.

Another important feature revealed by the scaling is that pressure does not vary significantly across the δ_T region, that is,

$$P(x, y) \simeq P(y) = P_0(y). \tag{3.202}$$

Furthermore, if we observe that the pressure distribution in the reservoir is essentially hydrostatic, the pressure gradient in the y-direction may be replaced by

$$\frac{dP_0}{dy} = -\rho_0 g. \tag{3.203}$$

All these derivations yield the following simplified forms:

$$\rho\left(u\frac{\partial v}{\partial x} + v\frac{\partial v}{\partial y}\right) = \mu\frac{\partial^2 v}{\partial x^2} + (\rho_\infty - \rho)g, \tag{3.204}$$

$$u\frac{\partial T}{\partial x} + v\frac{\partial T}{\partial y} = \alpha\frac{\partial^2 T}{\partial x^2}. \tag{3.205}$$

These form the *reduced set of boundary layer equations.*

As shown by the thermodynamic equation of state $\rho = \rho(T, P)$, the density of the fluid is a function of temperature and pressure. Consequently, by Taylor expansion, it follows that

$$\rho \simeq \rho_0 + (T - T_0)\left.\frac{\partial \rho}{\partial T}\right|_P + (P - P_0)\left.\frac{\partial \rho}{\partial P}\right|_T + \cdots. \tag{3.206}$$

Experimental evidence indicates that the pressure dependence is negligibly small compared to the temperature dependence; hence, the first-order linear, approximation of the density is

$$\rho \simeq \rho_0 + \rho_0 \beta(T - T_\infty) = \rho_0[1 - \beta(T - T_0)]. \qquad (3.207)$$

Here $\beta(\text{K}^{-1}$, a constant) is the *coefficient of volumetric expansion*.

It is important to note that this linear approximation is valid only when $\beta(T - T_0) \ll 1$, that is, when the departure from the reference density $\rho_0(T_0, P_0)$ (recorded in the reservoir) is sufficiently small. When the temperature dependence of the fluid properties within the boundary layer region is significant, for a better agreement of the theoretical model with experiment the fluid properties should be evaluated at $(T_w - T_0)/2$, called the *film temperature*.

In the limits of this linear approximation (3.207), the boundary layer momentum equation (3.204) becomes

$$\underbrace{u\frac{\partial v}{\partial x} + v\frac{\partial v}{\partial y}}_{\text{Inertia}} = \underbrace{v\frac{\partial^2 v}{\partial x^2}}_{\text{Friction}} + \underbrace{g\beta(T - T_0)}_{\text{Buoyancy}} \qquad (3.208)$$

where $v = \mu/\rho_0$ is the *kinematic viscosity*. By the scaling argument (3.201) the temperature corrections for ρ are discarded, except for the buoyancy term — its cancellation here would suppress the only source of motion. The linearization of $\rho(T, P)$ and the (3.208) form of the momentum equation are known as the *Oberbeck–Boussinesq approximation* [24, 25].

3.3.2 THE SCALE ANALYSIS

An order of magnitude analysis of 3.196 shows that the heat transfer coefficient scales as

$$h_f \sim \frac{k}{\delta_T}. \qquad (3.209)$$

By the same approach, the following balances are identified:

Mass conservation (3.197):

$$\frac{u}{\delta T} \sim \frac{v}{y}. \qquad (3.210)$$

Momentum balance (3.208):

$$u\frac{v}{\delta_T}, v\frac{v}{y} \sim v\frac{v}{\delta_T^2}, g\beta\Delta T. \qquad (3.211)$$

Energy balance (3.205):

$$u\frac{\Delta T}{\delta_T}, v\frac{\Delta T}{y} \sim \alpha\frac{\Delta T}{\delta_T^2}. \qquad (3.212)$$

Heat Transfer

The mass conservation scaling relation (3.210) may be used to produce a simpler form of the energy balance (3.217),

$$\underbrace{v\frac{\Delta T}{y}}_{\text{Convection (enthalpy)}} \sim \underbrace{\alpha\frac{\Delta T}{\delta_T^2}}_{\text{Conductiona (diffusion)}};\tag{3.213}$$

and of the momentum balance (3.211),

$$\underbrace{v\frac{v}{y}}_{\text{Inertia}} \sim \underbrace{v\frac{v}{\delta_T^2}}_{\text{Friction}}, \underbrace{g\beta\Delta T}_{\text{Buoyancy}}.\tag{3.214}$$

It is important to note that in natural convection, the leading term (i.e., the source of motion) is the buoyancy term. The other terms, inertia and friction, act as opposing body forces.

Buoyancy–Friction Balance (High-Pr Fluids)
In this case,

$$\frac{v^2}{y} < v\frac{v}{\delta_T^2},\tag{3.215}$$

and the momentum scaling (3.214) reduces to

$$v\frac{v}{\delta_T^2} \sim g\beta\Delta T,\tag{3.216}$$

which leads to the following order-of-magnitude relations:

$$u \sim \frac{\alpha}{y}\mathrm{Ra}_y^{1/4}, v \sim \frac{\alpha}{y}\mathrm{Ra}_y^{1/2}, \delta_T \sim y\,\mathrm{Ra}_y^{-1/4}.\tag{3.217}$$

The *Rayleigh group*, that emerges,

$$\mathrm{Ra}_y = \frac{g\beta(T_w - T_0)y^3}{\alpha v},\tag{3.218}$$

has the same order of magnitude as the thermal boundary layer aspect ratio y/δ_T. The last relation in (3.217) may be used to estimate the order of magnitude of the heat transfer coefficient (3.209),

$$b_y \sim \frac{k}{y}\mathrm{Ra}_y^{1/4},\tag{3.219}$$

and, further on, the local Nusselt number,

$$\mathrm{Nu}_y = \frac{b_y y}{k} \sim \mathrm{Ra}_y^{1/4}.\tag{3.220}$$

Finally, by using (3.217) second and third, the scaling relation (3.215) translates into

$$\alpha < v, \quad \text{or} \quad 1 < \mathrm{Pr}.\tag{3.221}$$

Hence, the buoyancy–friction balance occurs in fluids with high Prandtl numbers ($\mathrm{Pr} \gtrsim 1$).

Another important result of this analysis is that the velocity profile in natural convection has *two* length scales: the distance from the wall to the stagnant reservoir (δ),

$$\delta \sim y\mathrm{Ra}_y^{-1/4}\mathrm{Pr}^{1/2}, \tag{3.222}$$

and the distance from the wall to the location of the peak velocity (δ_T). The ratio of these scales is of the order of the square root of the fluid Prandtl number,

$$\frac{\delta}{\delta_T} \sim \mathrm{Pr}^{1/2} > 1. \tag{3.223}$$

Buoyancy–Inertia Balance (Low-Pr Fluids)

For low-Pr fluids the momentum balance is

$$\frac{v^2}{y} \sim g\beta\Delta T \left(\frac{v^2}{y} > v\frac{v}{\delta_T^2}\right), \tag{3.224}$$

and, consequently,

$$u \sim \frac{\alpha}{y}\mathrm{Bo}^{1/4}, \, v \sim \frac{\alpha}{y}\mathrm{Bo}^{1/2}, \delta_T \sim y\mathrm{Bo}^{-1/4}, \tag{3.225}$$

where

$$\mathrm{Bo} = \mathrm{Ra}_y\mathrm{Pr} \tag{3.226}$$

is the *Boussinesq group*.

The local Nusselt number now scales as

$$\mathrm{Nu}_y = \frac{h_y y}{k} \sim \mathrm{Bo}^{1/4}, \tag{3.227}$$

and substituting v and δ_T scales in the inequality $v_2/y > vv/\delta_T^2$ yields $< \mathrm{Pr} \gtrsim 1$.

In the buoyancy–inertia balance regime, the wall jet has the same thickness as the thermal boundary layer, δ_T. The region adjacent to the wall where the velocity profile goes from zero (at the wall) to its peak value (at δ_S) is called the *shear layer*. Its scale is

$$\delta_S \sim y\left(\frac{\mathrm{Ra}_y}{\mathrm{Pr}}\right)^{-1/4} = y\mathrm{Gr}_y^{-1/4}, \tag{3.228}$$

which confirms that δ_S is smaller than δ_T:

$$\frac{\delta_S}{\delta_T} \sim \mathrm{Pr}^{1/2} < 1. \tag{3.229}$$

The nondimensional group Gr in (3.228) is the *Grashof number.* This quantity is usually introduced through

$$\mathrm{Gr}_y \overset{def}{=} \frac{\mathrm{Ra}_y}{\mathrm{Pr}} = \frac{g\beta(T_w - T_0)y^3}{v^2}. \tag{3.230}$$

These scaling results are valid provided the shear layer *and* the thermal boundary layer are *slender,* namely,

$$1 > \frac{y}{\delta_S} \sim \left(\frac{\mathrm{Ra}_y}{\mathrm{Pr}}\right)^{1/4} \quad \text{and} \quad 1 > \frac{y}{\delta_T} \sim (\mathrm{Ra}_y\mathrm{Pr})^{1/4}$$

hold. The second restriction is stronger, and it implies that $\mathrm{Ra}_y > \mathrm{Pr}^{-1}$, which is in fact the condition that validates the scaling results. Several wall thermal conditions are discussed next.

3.3.3 VERTICAL WALL WITH UNIFORM TEMPERATURE

The dimensionless coefficients of order 1 left unspecified by the scale analysis may be found by more accurate methods, for instance, the similarity analysis.

For high-Pr fluids the *similarity variable η* (§3.3.13) may be introduced by scaling the coordinate orthogonal to the wall with the thermal boundary layer thickness, respectively,

$$\eta = \frac{x}{y\mathrm{Ra}_y^{-1/4}}. \tag{3.231}$$

The v–velocity profile (3.217) second,

$$G(\eta, \mathrm{Pr}) = \frac{v}{(\alpha/y)\mathrm{Ra}_y^{1/2}}, \tag{3.232}$$

the streamfunction $\left(u = \dfrac{\partial \psi}{\partial y}, v = -\dfrac{\partial \psi}{\partial x}, \text{ i.e., } G = -\dfrac{dF}{d\eta} \right)$

$$F(\eta, \mathrm{Pr}) = \frac{\psi}{\alpha\mathrm{Ra}_y^{1/4}}, \tag{3.233}$$

and the temperature profile for the isothermal wall, T_w,

$$\theta(\eta, \mathrm{Pr}) = \frac{T - T_0}{T_w - T_0}, \tag{3.234}$$

are the similarity forms of the heat transfer quantities.

Rewriting the energy (3.205) and momentum (3.208) equations in terms of ψ instead of u and v, and then replacing x, y, and ψ by η, F, and θ, yields the similarity equations of the natural convection thermal boundary layer (for $(\mathrm{Pr} \gtrsim 1)$ fluids):

$$\frac{1}{\mathrm{Pr}}\left(\frac{1}{2}F'^2 - \frac{3}{4}FF'' \right) = -F''' + \theta \tag{3.235}$$

$$\frac{3}{4}F\theta' = \theta''. \tag{3.236}$$

The boundary conditions that close the model are

$$\eta = 0 \Rightarrow \begin{cases} F = 0 & (u = 0, \text{ impermeable wall}) \\ F' = 0 & (v = 0, \text{ no-slip}) \\ \theta = 1 & (T = T_w, \text{ isothermal wall}) \end{cases}$$

$$\eta \to \infty \Rightarrow \begin{cases} F' \to 0 & (v = 0, \text{ stagnant fluid reservoir}) \\ \theta \to 0 & (T = T_w, \text{ isothermal fluid reservoir}). \end{cases}$$

The similarity temperature and velocity profiles along a vertical wall for laminar natural convection are shown in Fig. 3.25.[1] The local Nusselt number produced by this analysis,

$$\mathrm{Nu}_y \overset{def}{=} \frac{h_y y}{k} = \frac{-k \left.\frac{\partial T}{\partial x}\right|_{x=0} y}{(T_w - T_\infty) k} = -\left.\frac{d\theta}{d\eta}\right|_{x=0} \mathrm{Ra}_y^{1/4}, \tag{3.237}$$

indicates that the local heat flux, $q''_{w,y}$, varies as $y^{-1/4}$. This result is approximated within 0.5% by

$$\mathrm{Nu}_y = 0.503 \left(\frac{\mathrm{Pr}}{\mathrm{Pr} + 0.986\mathrm{Pr}^{1/2} + 0.492} \right)^{1/4} \mathrm{Ra}_y^{1/4}, \tag{3.238}$$

which covers the entire range of Prandtl numbers. The asymptotic values predicted by (3.238), namely,

$$\mathrm{Nu}_y = \begin{cases} 0.503\mathrm{Ra}_y^{1/4}, & \mathrm{Pr} \gg 1 \\ 0.600(\mathrm{Ra}_y \mathrm{Pr})^{1/4}, & \mathrm{Pr} \ll 1, \end{cases}$$

are in good agreement with the order-of-magnitude analysis (3.220), (3.227).

A quantity of importance to the heat transfer analysis, the *average* or *overall Nusselt number*, is defined by

$$\overline{\mathrm{Nu}}_y \overset{def}{=} \frac{\overline{h}_y y}{k} = \frac{\overline{q}''_{w,y}}{(T_w - T_0)} \frac{y}{k}, \tag{3.239}$$

where $\overline{q}''_{w,y}$ is the wall heat flux averaged over $[0, y]$ and $\overline{q}''_{w,y} y = q'_{w,y}$. The $\overline{\mathrm{Nu}}_y$ that corresponds to (3.238) is

$$\overline{\mathrm{Nu}}_y = 0.671 \left(\frac{\mathrm{Pr}}{\mathrm{Pr} + 0.986\mathrm{Pr}^{1/2} + 0.492} \right)^{1/4} \mathrm{Ra}_y^{1/4}. \tag{3.240}$$

For air ($\mathrm{Pr} = 0.72$), this correlation gives

$$\overline{\mathrm{Nu}}_y = 0.517\mathrm{Ra}_y^{1/4}. \tag{3.241}$$

3.3.4 TRANSITION TO TURBULENCE

Under specified thermal conditions and for a given fluid, the boundary layer flow is laminar for y small enough that the corresponding Ra_y is below a certain *critical* value (Fig. 3.26). Traditionally [4], the threshold limit is $\mathrm{Ra}_y \sim 10^9$, regardless of the particular Pr number of the working fluid. This criterion, which does not depend on Pr, was shown [26] to be better represented by the condition $\mathrm{Gr}_y \sim 10^9$ for all fluids within the range $10^{-3} \lesssim \mathrm{Pr} \lesssim 10^3$.

---

[1] The velocity and temperature similarity profiles shown in Fig. 3.26 were computed by the same approach that was used previously for the solution to the Blasius–Pohlhausen, external forced convection problem (§3.1.13). Equations (3.237) and (3.240) were utilized to determine $\left.\frac{d\theta}{d\eta}\right|_{x=0}$.

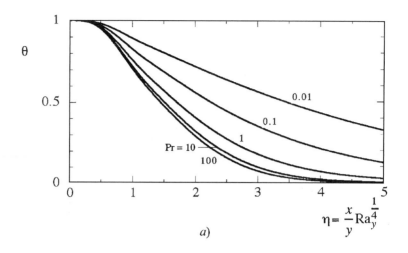

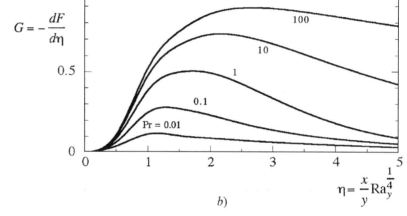

Figure 3.26
The similarity velocity and temperature profiles in laminar natural convection.

Churchill and Chu [27] recommend an empirical correlation for the overall Nusselt number,

$$
\overline{\mathrm{Nu}}_y = \left\{ 0.825 + \frac{0.387\mathrm{Ra}_y^{1/6}}{\left[1 + \left(\dfrac{0.492}{\mathrm{Pr}}\right)^{9/16}\right]^{8/27}} \right\}^2, \tag{3.242}
$$

that holds for $(10^{-1} \lesssim \mathrm{Ra}_y \lesssim 10^{12})$, and it covers the *entire* Rayleigh number range (laminar, transition, and turbulence). The physical properties that appear should be evaluated at the film temperature, $(T_w + T_0)/2$. For air $(\mathrm{Pr} = 0.72)$, (3.242) assumes the simpler form

$$
\overline{\mathrm{Nu}}_y = (0.825 + 0.325\mathrm{Ra}_y^{1/6})^2. \tag{3.243}
$$

A more accurate correlation for the laminar regimes $Gr_y < 10^9$) was found by Churchill and Chu [27]:

$$\overline{Nu}_y = 0.68 + \frac{0.67 Ra_y^{1/4}}{\left[1 + \left(\dfrac{0.492}{Pr}\right)^{9/16}\right]^{4/9}} \tag{3.244}$$

$$\overline{Nu}_y = 0.68 + 0.515 Ra_y^{1/4} \quad (Pr = 0.72). \tag{3.245}$$

3.3.5 VERTICAL WALL WITH UNIFORM HEAT FLUX

The relations (3.205) and (3.208) are general. Therefore, they may be used for the uniformly heated wall, $q_w'' = \text{const.}$ In the laminar regime of high Prandtl number fluids, the local Nusselt number scales as $Nu_y \sim Ra_y^{1/4}$, or

$$\frac{q_w''}{\Delta T(y)} \frac{y}{k} \sim \left[\frac{g\beta \Delta T(y) y^3}{\alpha \nu}\right]^{1/4} \Rightarrow \left[\frac{q_w''}{\Delta T(y)} \frac{y}{k}\right]^5 \sim \frac{g\beta q_w'' y^4}{\alpha \nu k} = Ra_y^{*1/5}, \tag{3.246}$$

where $\Delta T(y) = T_w(y) - T_0$ and Ra_y^* is the Rayleigh number defined with the heat flux (the known quantity here). Apparently, $\Delta T(y)$ varies as $y^{1/5}$. The similarity solution for the uniform heat flux [28], which is fitted satisfactorily well by

$$Nu_y \cong 0.616 Ra_y^{*1/5} \left(\frac{Pr}{Pr + 0.8}\right)^{1/4}, \tag{3.247}$$

confirms this result.

The following formulas for $\overline{Nu}_y$, defined based on the wall-averaged temperature difference $T_w(y) - T_0$, are recommended in ref. 29:

$$\left.\begin{array}{l} Nu_y = 0.6 Ra_y^{*1/5} \\[2mm] \overline{Nu}_y = 0.75 Ra_y^{*1/5} \end{array}\right\} \quad \text{laminar, } 10^5 < Ra_y^* < 10^{13} \tag{3.248a}$$

$$\left.\begin{array}{l} Nu_y = 0.568 Ra_y^{*0.22} \\[2mm] \overline{Nu}_y = 0.645 Ra_y^{*0.22} \end{array}\right\} \quad \text{turbulent, } 10^{13} < Ra_y^* < 10^{16}. \tag{3.248b}$$

For air ($Pr = 0.72$), they give

$$Nu_y = 0.55 Ra_y^{*1/5} \text{ (laminar)} \tag{3.248c}$$

$$\overline{Nu}_y = 0.75 Ra_y^{*1/5} \text{ (turbulent).} \tag{3.248d}$$

Churchill and Chu [27] proposed another correlation for the overall Nusselt number, valid for all Rayleigh and Prandtl numbers:

$$\overline{Nu}_y = \left\{0.825 + \frac{0.387 Ra_y^{1/6}}{\left[1 + \left(\dfrac{0.437}{Pr}\right)^{9/16}\right]^{8/27}}\right\}^2. \tag{3.249}$$

Heat Transfer

The Rayleigh number here is based on the y-averaged temperature difference, $\overline{T}_w - T_\infty$. For air (Pr = 0.72) this correlation predicts

$$\overline{Nu}_y = \left[0.825 + 0.328 Ra_y^{1/6}\right]^2, \tag{3.250}$$

which yields an asymptotic formula for high Rayleigh numbers $Ra_y > 10^{10}$, namely,

$$\overline{Nu}_y \cong 0.107 Ra_y^{1/6}. \tag{3.251}$$

These relations may be rewritten in terms of the flux Rayleigh number, Ra_y^*, by using the conversion $Ra_y = Ra_y^*/Nu_y$.

3.3.6 OTHER EXTERNAL NATURAL CONVECTION CONFIGURATIONS

Thermally stratified fluid reservoir (Fig. 3.27):

$$b = \frac{\Delta T_{max} - \Delta T_{min}}{\Delta T_{max}}$$

$$\overline{Nu}_H = \frac{q''_{w,H}}{\Delta T_{max}}\frac{H}{k} \qquad \overline{Nu}_H = f(b, Pr)Ra_H^{1/4} \ [2]$$

$$Ra_H = \frac{g\beta \Delta T_{max} H^3}{\alpha v}.$$

Inclined walls (Fig. 3.28, after [2]):

Isothermal wall, T_w : $Ra_y = \dfrac{(g\cos\phi)\beta\Delta T_{w-\infty}y^3}{\alpha v}$

Isoflux wall, q''_w : $Ra_y^* = \dfrac{(g\cos\phi)\beta q''_w y^4}{\alpha v k}.$

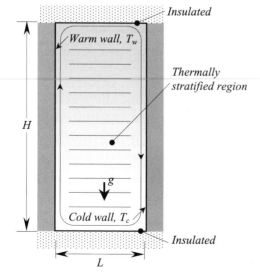

Figure 3.27
The natural convection flow in an enclosure heated differentially from the sides. The left wall is warm, at T_H, and the right wall is cold, at T_c, $T_H > T_c$.

y
g
ϕ
gcosϕ
Warm wall
Cold wall
x
Warm wall
Cold wall

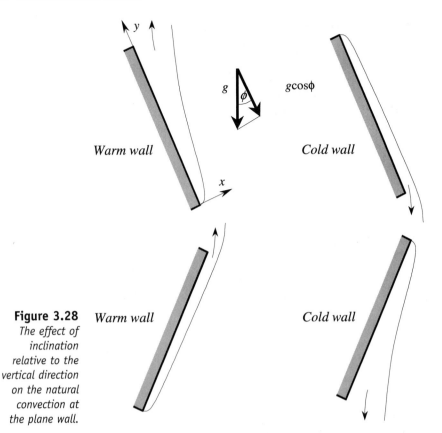

Figure 3.28 *The effect of inclination relative to the vertical direction on the natural convection at the plane wall.*

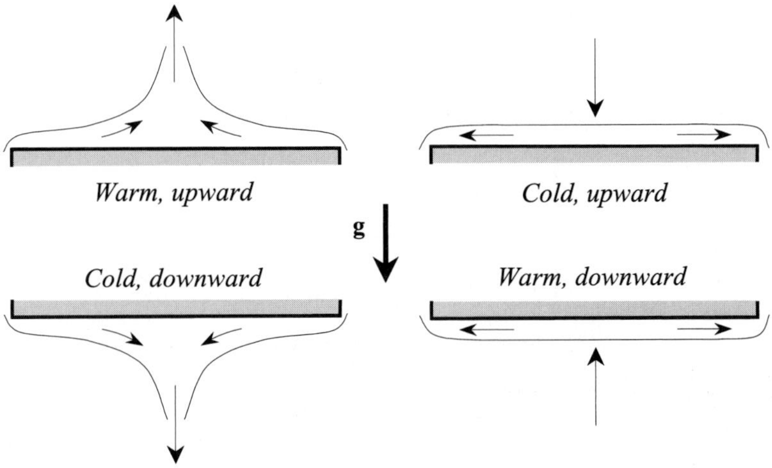

Warm, upward
Cold, upward
g
Cold, downward
Warm, downward

Figure 3.29 *Natural convection at the horizontal plane wall.*

Horizontal isothermal walls (Fig. 3.29):

Hot surface facing upward, or cold surface facing downward:

$$\overline{\mathrm{Nu}} = \begin{cases} 0.54\mathrm{Ra}_L^{1/4} & (10^4 < \mathrm{Ra}_L < 10^7) \\ 0.15\mathrm{Ra}_L^{1/4} & (10^7 < \mathrm{Ra}_L < 10^9). \end{cases}$$

Hot surface facing downward, or cold surface facing upward:

$$\overline{\mathrm{Nu}}_L = 0.27\mathrm{Ra}_L^{1/4} \quad (10^5 < \mathrm{Ra}_L < 10^{10}).$$

Immersed bodies:

Horizontal cylinder [27]:

$$\overline{\mathrm{Nu}}_D = \left\{ 0.6 + \frac{0.387\mathrm{Ra}_D^{1/6}}{\left[1 + \left(\dfrac{0.559}{\mathrm{Pr}} \right)^{9/16} \right]^{8/27}} \right\}^2 \quad (10^{-5} < \mathrm{Ra}_D < 10^{12}, \text{ any Pr})$$

$$\overline{\mathrm{Nu}}_D = \frac{q''_{w,D}}{\Delta T} \frac{D}{k}$$

$$\mathrm{Ra}_D = \frac{g\beta\Delta T D^3}{\alpha v}.$$

Sphere [31] :

$$\overline{\mathrm{Nu}}_D = \left\{ 2 + \frac{0.589\mathrm{Ra}_D^{1/4}}{\left[1 + \left(\dfrac{0.469}{\mathrm{Pr}} \right)^{9/16} \right]^{4/9}} \right\}^2 \quad (\mathrm{Ra}_D < 10^{11}, \mathrm{Pr} \gtrsim 0.7).$$

Vertical cylinder [35] :

Thick cylinder, $\delta_T \ll D, D < H$

(boundary layer theory, §3.3.3).

Thin cylinder [31]

$$\begin{cases} \overline{\mathrm{Nu}}_H = \dfrac{4}{3} \left[2 + \dfrac{7\mathrm{Ra}_H\mathrm{Pr}}{5(20 + 21\,\mathrm{Pr})} \right]^{1/4} + \dfrac{4(272 + 315\,\mathrm{Pr})}{35(64 + 63\,\mathrm{Pr})}\dfrac{H}{D} & (\text{laminar}) \\[4mm] \overline{\mathrm{Nu}}_H = \dfrac{\bar{h}H}{k}, \mathrm{Ra}_H = \dfrac{g\beta\Delta T H^3}{\alpha v} \\[3mm] \bar{h}, \text{ wall averaged heat transfer coefficient} \end{cases}$$

Sparrow and Ansari [32]

$$\overline{\mathrm{Nu}}_D = 0.775\mathrm{Ra}_D^{0.208}$$

$[H = D, \mathrm{Ra}_D > 1.4 \times 10^4, \text{ for air}]$

Other immersed bodies:

Lienhard [33]

$$\overline{Nu}_l \underset{Pr \gtrsim 0.7}{=} 0.52 Ra_l^{1/4} \begin{cases} \overline{Nu}_l = \dfrac{\overline{h}l}{k}, \ Ra_H = \dfrac{g\beta\Delta T l^3}{\alpha\nu} \\ \overline{h}, \text{ wall averaged } h \\ l, \text{ the distance traveled by the boundary} \\ \quad \text{layer fluid} \end{cases}$$

Yovanovich [34]

$$\begin{cases} \overline{Nu}_l = \overline{Nu}_l^0 + \dfrac{0.67 G_l Ra_l^{1/4}}{\left[1 + \left(\dfrac{0.492}{Pr}\right)^{9/16}\right]^{4/9}} \\ \text{or} \\ \overline{Nu}_l \cong 3.47 + 0.51 Ra_l^{1/4} \end{cases} \begin{cases} \overline{Nu}_l = \dfrac{\overline{h}_l}{k} \\ \overline{Nu}_l^0 \rightarrow \text{conduction limit} \\ Ra_l = \dfrac{g\beta\Delta T l^3}{\alpha\nu} \\ 0 < Ra_l < 10^8, \\ G_l, \text{ geometric factor} \\ l = A^{1/2}, \\ \overline{h}, \text{ wall averaged } h. \end{cases}$$

3.4 Internal Natural Convection

One of the simplest problems of internal natural convection is the vertical channel formed by two heated plates (Fig. 3.30), such as vertically mounted electronic printed boards in an electronic package, or vertical plate fins. For the beginning, we will assume that the plates are isothermal, at T_w, and that the fluid reservoir outside the channel is also isothermal, at T_∞, such that $T_w > T_\infty$. The fluid that penetrates the channel from below warms up in contact with the walls, and the thermal boundary layers that develop upward, along the plates, entrain the fluid in an ascending, free convection flow. Naturally, since the heat is removed from the channel via a fluid (gas or liquid), the flow properties (velocity field, turbulence, etc.) do significantly influence its transfer rate. The natural convection channel flow is characterized by two length scales: the channel height, H, and the interplate spacing, L.

Two important flow configurations are commonly encountered:

The wide channel limit: The interplate spacing is large enough for the thermal boundary layers to touch each other, and the temperature boundary layer is thinner than the interplate spacing, $\delta_T < L$ (Fig. 3.29). For the fluids with $Pr \gtrsim 1$, the wide channel limit means

$$\frac{L}{H} > Ra_H^{-1/4} \quad \text{or} \quad \frac{L}{H} > Ra_L^{-1}. \tag{3.52}$$

The heat transfer rate may be accurately computed with the single-wall formulas (3.242)–(3.243) or (3.248)–(3.251).

Heat Transfer

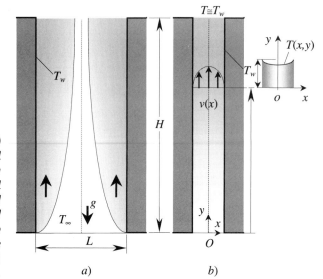

Figure 3.30
*Internal natural
convection in
the vertical
channel formed
by two heated
plates. The two
limits: (a) wide
channel; (b)
narrow channel.*

The narrow channel limit: The interplate spacing is small enough such that the hydrodynamic boundary layers interact into a Hagen–Poiseuille-like flow (Fig. 3.29b) over almost the entire channel height.

In the narrow-channel limit the flow is thermally fully developed over nearly the entire channel height, and for fluids with $\mathrm{Pr} \simeq 1$ it is also hydrodynamically fully developed—it is essentially a vertical flow with $u = 0$, $\partial v/\partial y = 0$. As for the external flows, where $dP_\infty/dy = -\rho_\infty g$, here $\partial P/\partial y = \mathrm{const} = dP/dy$. Furthermore, because both ends of the channel are open $dP/dy = dP_\infty/dy = -\rho_\infty g$.

If we notice that the fluid temperature at the channel outlet is nearly equal to the wall temperature, the momentum equation (3.208) may be simplified to

$$v\frac{\partial^2 v}{\partial x^2} \simeq -\frac{g\beta(T - T_\infty)}{v} = \mathrm{const}, \tag{3.253}$$

which is equivalent to the momentum equation (3.140) for the parallel-plate channel forced convection. However, the source of motion is here $-g\beta(T_w - T_\infty)/v$, rather than $(1/\mu)dP/dx$. The solution to (3.253) with no-slip boundary conditions is the parabolic profile

$$v(x) = \frac{g\beta \Delta T L^2}{v}\left[1 - \left(\frac{x}{L/2}\right)^2\right], \Delta T = T_w - T_\infty. \tag{3.254}$$

This profile yields the vertical mass flow rate per unit length

$$\dot{m}' = \int_{-L/2}^{L/2} \rho v\, dx = \frac{\rho g\beta \Delta T L^3}{12v}. \tag{3.255}$$

As the longitudinal pressure gradient that drives the flow, $dP/dy = -\rho g\beta(T - T_\infty)$, is independent of the channel height H, the velocity profile and the mass flow rate given by (3.254) and (3.255) are also independent of H.

The total heat carried out of the chimney by the stream is

$$q' = \dot{m}' c_P (T - T_\infty) = \frac{\rho g \beta c_P (\Delta T)^2 L^3}{12\nu}. \tag{3.256}$$

It follows that the average Nusselt number is

$$\overline{\mathrm{Nu}}_H \stackrel{\text{def}}{=} \frac{\bar{q}'' H}{\Delta T k} = \frac{1}{24} \mathrm{Ra}_L, \tag{3.257}$$

where $\bar{q}'' = q'/2H$ is the channel-averaged heat flux. This result—valid in the narrow-channel limit—is correct, provided the local wall to stream temperature drop is less than the wall to fresh fluid temperature drop, namely,

$$T_w - T(x, y) < T_w - T_\infty. \tag{3.258}$$

The scaling relation behind this inequality is $\bar{q}'' L/k < \Delta T$, and it yields a valuable criterion,

$$\mathrm{Ra}_L < \frac{H}{L}, \tag{3.259}$$

that may be used to validate the narrow-channel limit assumption, in a general chimney analysis.

Table 3.4 (after ref. 2) lists the average Nusselt number for several types of chimney flows, in the narrow-channel limit.

3.4.1 ENCLOSURES HEATED DIFFERENTIALLY FROM THE SIDES

An important category of internal free convection is concerned with flows confined to cavities. Figure 3.27 depicts the classical problem of the natural convection flow in an enclosure heated differentially from the sides (the left wall is warm, at T_H, and the right wall is cold, at T_c, $T_H > T_c$). The working fluid is entrained into a circular stream—upward at the warm wall, and downward at the cold wall. A vast body of literature is devoted to this area of investigation—several general trends are summarized in ref. 2.

For square enclosures there are, as for the channel free flows, two limiting cases: the *wide cavity* case, when the thermal boundary layer thickness is smaller than the cavity width, $\delta_T < L$; and the *narrow cavity* case, when the cavity is thin enough for the thermal boundary layers to interact.

In the first case (the wide cavity), the theory developed for external natural convection may be utilized to predict the heat transfer properties. For instance, [36] recommends the Berkovsky–Polevikov correlation

$$\overline{\mathrm{Nu}}_H = \frac{\bar{q}'' H}{k \Delta T} = 0.22 \left(\frac{\mathrm{Pr}}{0.2 + \mathrm{Pr}} \mathrm{Ra}_H \right)^{0.28} \left(\frac{L}{H} \right)^{0.09}, \tag{3.260}$$

for $H/L \in (2, 10)$, $\mathrm{Pr} < 10^5$, $\mathrm{Ra}_H < 10^{13}$.

In the opposite limit (the narrow cavity, $L/H < \mathrm{Ra}_H^{-1/4}$), if the enclosure is tall enough, then the heat transfer between the side walls approaches the

Table 3.4

Cross section	$\overline{\text{Nu}}/\text{Ra}_{D_b}$
	$\dfrac{1}{106.4}$
	$\dfrac{1}{113.6}$
	$\dfrac{1}{128}$
	$\dfrac{1}{192}$

pure conduction regime limit. The same conclusion is valid for shallow enclosures, with distinct jets along the top and bottom walls [2]. A theoretical solution to the average Nusselt number, $\overline{\text{Nu}}_H$, in the boundary layer regime for $\text{Pr} \gtrsim 1$ fluids was indicated by [37]

$$\overline{\text{Nu}}_H = \frac{\bar{q}'' H}{k \Delta T} = 0.25 \text{Ra}_H^{2/7} \left(\frac{H}{L}\right)^{1/7}. \tag{3.261}$$

When the heat flux is specified rather than ΔT, the Rayleigh number may be rewritten as $\text{Ra}_H^* = \text{Ra}_H \overline{\text{Nu}}_H = g\beta^4 q''/\alpha\nu k$, such that Eq. (3.261) becomes

$$\overline{\text{Nu}}_H = 0.34 \text{Ra}_H^{*2/9} \left(\frac{H}{L}\right)^{1/9}. \tag{3.262}$$

Apparently, as pointed out in ref. 2, the heat transfer correlations developed for systems with isothermal walls apply satisfactorily to the same configurations, but with uniform heat flux walls, provided the corresponding Rayleigh number is based on the heat flux.

3.4.2 ENCLOSURES HEATED DIFFERENTIALLY, FROM BELOW

In this case, the driving, vertical temperature gradient (Fig. 3.31, after [2]) must exceed a *critical* value for convection to occur. This condition is usually

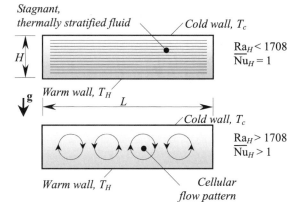

Figure 3.31
The natural convection flow in an enclosure heated differentially, from below.

put in terms of a critical Rayleigh number [38], which for shallow enclosures is

$$\mathrm{Ra}_H = \frac{g\beta(T_H - T_c)H^3}{\alpha v} \gtrsim 1708. \tag{3.263}$$

Below this limit, heat transfer is by pure conduction and the temperature varies linearly in the vertical direction. Immediately above this threshold value, counterrotating two-dimensional rolls appear, indicating the onset of the so-called *Bénard* convection. If Ra_H is further increased, then the flow pattern changes to three-dimensional cells of hexagonal shape [2]. For enclosures that are not shallow, depending on the particular length to height aspect ratio, this threshold may depart from the critical value (3.263) and the flow structure may be different.

From the heat transfer design point of view, it is important to provide adequate conditions for the system to work beyond the thermal convection onset limit, so as to augment the heat transfer into exceeding the pure conduction limit through the convection transport process. A measure of this augmentation, for the shallow enclosures, is given by the following correlation for the Nusselt number [39]:

$$\overline{\mathrm{Nu}}_H = 0.069\mathrm{Ra}_H^{1/3}\mathrm{Pr}^{0.074}.$$

This expression is validated by experiments within the range $3 \times 10^5 < \mathrm{Ra}_H < 7 \times 10^9$ [2].

3.4.3 INCLINED ENCLOSURES HEATED DIFFERENTIALLY, FROM THE SIDES

Figure 3.32 shows the impact of the orientation of the differentially heated cavity. The recommended correlations for the overall Nusselt number,

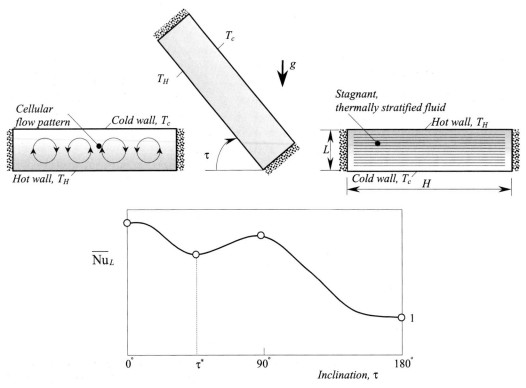

Figure 3.32 *The effect of the inclination angle on the heat transfer rate and flow pattern in an enclosure heated differentially, from opposite side walls.*

$\overline{\mathrm{Nu}}_L(\tau)$, for various ranges of the inclination angle τ are [36, 2]

$$
\overline{\mathrm{Nu}}(\tau) = \begin{cases}
1 + [\overline{\mathrm{Nu}}_L(90^\circ) - 1](\sin\tau)^{1/4} & \tau \in (90^\circ, 180^\circ) \\[6pt]
\overline{\mathrm{Nu}}_L(90^\circ)(\sin\tau)^{1/4}, & \tau \in (\tau^*, 90^\circ) \\[6pt]
\left[\dfrac{\overline{\mathrm{Nu}}_L(90^\circ)}{\mathrm{Nu}_L(0^\circ)}(\sin\tau^*)^{1/4}\right]^{\tau/\tau^*}, & \begin{cases} \tau \in (0^\circ, \tau^*) \\ \dfrac{H}{L} < 10 \end{cases} \\[14pt]
1 + 1.44\left(1 - \dfrac{1708}{\mathrm{Ra}_L\cos\tau}\right)^* & \begin{cases} \tau \in (0^\circ, \tau^*) \\ \dfrac{H}{L} > 10 \end{cases} \\[14pt]
\times\left[1 - \dfrac{1708 \times (\sin 1.8\tau)^{1.6}}{\mathrm{Ra}_L\cos\tau}\right] \\[14pt]
+ \left[\left(\dfrac{\mathrm{Ra}_L\cos\tau}{5830}\right)^{1/3} - 1\right]^*,
\end{cases}
$$

Here, the quantities marked with ()* have to be set to zero when negative.

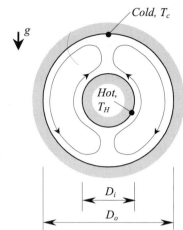

Figure 3.33
Natural convection flow in the annular space between horizontal, coaxial cylinders heated differentially.

3.4.4 OTHER CONFIGURATIONS HEATED DIFFERENTIALLY

Annular space between horizontal, coaxial cylinders, Fig. 3.33 [2, 40]:

$$q'\left[\frac{W}{m}\right] \cong \frac{2.425(T_i - T_0)k}{\left[1 + \left(\dfrac{D_i}{D_0}\right)^{3/5}\right]^{5/4}} \left(\frac{\Pr \operatorname{Ra}_{D_i}}{0.861 + \Pr}\right)^{1/4} \quad \text{(laminar regime, all Pr)}$$

for

$$\operatorname{Ra}_{D_i} = \frac{g\beta(T_i - T_0)D_i^3}{\alpha v} \lesssim 10^7 \text{ and } \operatorname{Ra}_{D_i} < \left(\frac{D_0}{D_0 - D_i}\right)^4.$$

Annular space between concentric spheres [2, 40]:

$$q[W] \cong \frac{2.325(T_i - T_0)kD_i}{\left[1 + \left(\dfrac{D_i}{D_0}\right)^{7/5}\right]^{5/4}} \left(\frac{\Pr \operatorname{Ra}_{D_i}}{0.861 + \Pr}\right)^{1/4} \quad \text{(laminar regime, all Pr)}.$$

References

1. H. Poincaré, *Thermodynamique.* pp. 66–68. Georges Carré, Paris, 1982.

2. A. Bejan, *Heat Transfer.* Wiley, New York, 1993.

3. C. Trusdell, Irreversible heat engines and second law of thermodynamics. *Lett. Heat Mass Transf.* **3**, 267–290 (1976).

4. F. P. Incropera and D. DeWitt, *Fundamentals of Heat and Mass Transfer.* Wiley, New York, 1990.

5. A. Bejan, *Advanced Engineering Thermodynamics.* Wiley, New York, 1988.

Heat Transfer

6. H. S. Carslow and J. C. Jaeger, *Conduction of Heat in Solids*, 2nd ed. Oxford University Press, London, 1959.

7. P. J. Schneider, *Conduction of Heat Transfer*. Addison-Wesley, Reading, MA, 1955.

8. M. P. Heisler, Temperature charts for induction and constant temperature heating. *Trans. ASME* **69**, 227–236 (1947).

9. H. Gröber, S. Erk and U. Grignall, *Fundamental of Heat Transfer*. McGraw-Hill, New York, 1961.

10. A. Bejan, *Convection Heat Transfer*. Wiley, New York, 1984.

11. A. M. Morega and A. Bejan, Heatline visualization of forced convection laminar boundary layers. *Int. J. Heat Transfer* **36**, 3957–3966 (1993).

12. R. W. Fox and A. T. McDonald, *Introduction to Fluid Mechanics*, 3rd ed. Wiley, New York, 1985.

13. L. Prandtl, Fluid motion with very small friction (in German). *Proc. of the Third International Congress on Mathematics, Heidelberg*, 1904.

14. L. Prandtl, *Essentials of Fluid Dynamics, with Applications to Hydraulics, Aeronautics, Meteorology and other Subjects*. Blackie & Sons Limited, London and Glasgow, 1957.

15. H. Schlichting, *Boundary Layer Theory*, 4th ed. McGraw-Hill, New York, 1960.

16. H. Blasius, Grenzschichten in Flussigkeiten mit kleiner Reibung. *Z. Math. Phys.* **56**, 1 (1908).

17. E. Pohlhausen, Der Warmeaustausch zwischen festen Korpern und Flussigkeiten mit kleiner Reibung und kleiner Warmeleitung. *Z. Angew. Math. Mech.* **1**, 115–121 (1921).

18. V. M. Faulkner and S. W. Skan, Some approximate solutions of the boundary layer equations, *Phil. Mag.* **12**, 865 (1931).

19. E. R. G. Eckert, *VDI-Forschungsh.* **416**, 1–24 (1942).

20. E. M. Sparrow and M. E. Crawford, Analysis of laminar forced convection heat transfer in the entrance region of flat rectangular ducts. *National Advisory Committee for Aeronautics, Technical Note* 3331 (1955).

21. W. M. Kays and M. E. Crawford, *Convective Heat and Mass Transfer*, 2nd ed. McGraw-Hill, New York, 1960.

22. G. Hagen, über die Bewegung des Wassers in engen zylindrischen Röhren. *Pogg. Ann.* **46**, 423 (1839).

23. J. Poiseuille, Récherches expérimentales sur le mouvement des liquides dans les tubes de très petits diamètres. *Compt. Rendu.* **11**, 961, 1041 (1840).

24. A. Oberbeck, über die Wärmeleitung der Flüssigkeit bei Berücksichtigung der Strömungen infolge von Temperaturdifferenzen. *Ann. Phys. Chem.* **7**, 271–292 (1879).

25. J. Boussinesq, *Theorie Analyltique de la Chaleur*, Vol. 2. Gauthier-Villars, Paris, 1903.

26. A. Bejan and J. L. Lage, The Prandtl number effect on the transition in natural convection along a vertical surface. *J. Heat Transfer* **122** (1990).

27. S. W. Churchill and H. H. S. Chu, Correlating equations for laminar and turbulent free convection from a vertical plate. *Int. J. Heat Mass Transfer* **18**, 1323–1329 (1975).

28. E. M. Sparrow and J. L. Gregg, Laminar free convection from a vertical plate with uniform surface heat flux, *Trans. ASME* **78**, 435–440 (1956).

29. G. C. Vliet and C. K. Liu, An experimental study of turbulent natural convection boundary layers, *J. Heat Transfer* **91**, 517–531 (1969).

30. G. C. Vliet and D. C. Ross, Turbulent natural convection on upward and downward facing inclined heat flux surfaces. *J. Heat Transfer* **97**, 549–555 (1975).

31. S. W. Churchill, Free convection around immersed bodies, Section 2.5.7 in *Heat Exchanger Design Handbook* (E. U. Schlünder, ed.) Hemisphere, New York, 1983.

32. E. M. Sparrow and M. A. Ansari, A refutation of King's rule for multi-dimensional external natural convection. *Int. J. Heat Mass Transfer* **26**, 1357–1364 (1983).

33. J. H. Lienhard, On the commonality of equations for natural convection from immersed bodies. *Int. J. Heat Mass Transfer* **26**, 2121–2123 (1983).

34. M. M. Yovanovich, On the effect of shape, aspect ratio and orientation upon natural convection from isothermal bodies of complex shape. *ASME HTD* **82**, 121–129 (1987).

35. E. J. LeFevre and J. A. Ede, Laminar free convection from the outer surface of a vertical cylinder, *Proc. Ninth. Int. Congr. Appl. Mech., Brussels* **4**, 175–183 (1956).

36. I. Catton, Natural convection in enclosures, *6th Int. Heat Transfer Conf., Toronto, 1978*, **6**, 13–43 (1979).

37. S. Kimura and A. Bejan, The boundary layer natural convection regime in a rectangular cavity with uniform heat flux from the side, *J. Heat Transfer* **106**, 98–103 (1984).

38. A. Pellew and R. V. Southwell, On maintained convection motion in a fluid heated from below, *Proc. Royal Soc.* **A176**, 312–343 (1940).

39. S. Globe and D. Dropkin, Natural convection heat transfer in liquids confined by two horizontal plates and heated from below, *J. Heat Transfer* **81**, 24–28 (1959).

40. G. D. Raithby and K. G. T. Hollands, A general method of obtaining approximate solutions to laminar and turbulent free convection problems. *Adv. Heat Transfer* **11**, 265–315 (1975).

Heat Transfer

8 Fluid Dynamics

NICOLAE CRACIUNOIU AND BOGDAN O. CIOCIRLAN

*Department of Mechanical Engineering,
Auburn University, Auburn, Alabama 36849*

Inside

1. Fluids Fundamentals

1.1 Definitions

The branches of applied mechanics that study the behavior of fluids at rest or in motion are called *fluid mechanics* and *hydraulics*.

Substances capable of flowing and taking the shape of containers are called *fluids*. The following characteristics of fluids apply:

- Fluids can be classified as liquids or gases.
- Liquids are incompressible, occupy definite volumes, and have free surfaces.
- Gases are compressible and expand until they occupy all portions of the container.
- Fluids cannot sustain shear or tangential forces when in equilibrium.
- Fluids exert some resistance to change of form.

1.2 Systems of Units

1.2.1 INTERNATIONAL SYSTEM OF UNITS

The fundamental units in the international system (SI) are meter (m), kilogram (kg), and second (s), corresponding respectively to the following fundamental mechanical dimensions: length, mass, and time. The unit of force is the newton (N) derived from Newton's second law: force (N) = mass (kg) × acceleration (m/s^2). Therefore, $1\,N = 1\,kg \cdot m/s^2$. Other units are m^3 for unit volume, kg/m^3 for unit density, joule ($1\,J = 1\,N\,m$) for work and energy, and pascal ($1\,Pa = 1\,N/m^2$) for pressure or stress. The temperature unit is the degree Celsius (C) and the unit of the absolute temperature is the kelvin (K).

1.2.2 BRITISH ENGINEERING SYSTEM OF UNITS

The fundamental units in this system (called the FPS system) are foot (ft), pound (lb), and second (sec), corresponding to the length, force, and time fundamental mechanical dimensions. Other units are ft^3 for the unit volume, ft/sec^2 for unit acceleration, ft-lb for unit work, and lb/ft^2 for unit pressure. The unit for mass is called the slug. The slug can be derived from Newton's second law applied for freely falling mass, namely weight (lb) = mass (slugs) × g (32.2 ft/sec^2. then, mass (slugs) = weight (lb)/g (32.2 ft/sec^2). Thus, 1 slug = lb-sec^2/ft. The temperature unit is the degree Fahrenheit (F) or, on the absolute scale, the degree Rankine (R).

1.3 Specific Weight

The weight of a unit volume of a fluid is called the *specific* or *unit weight* denoted by γ.

The specific weight of liquids can be considered as constant for practical applications, whereas the specific weight of gases can be calculated by using the *equation of state*

$$\frac{pv}{T} = R, \tag{1.1}$$

where p is the absolute pressure, v the volume per unit weight, T the absolute temperature, and R the gas constant. The gas constant is given by

$$R = \frac{R_0}{Mg}, \tag{1.2}$$

where R_0 is the universal gas constant and Mg the molar weight. If we substitute $v = 1/\gamma$ in Eq. (1.1), the expression for the specific weight of gases is obtained:

$$\gamma = \frac{p}{RT}. \tag{1.3}$$

The units of γ are lb/ft^3 or N/m^3.

1.4 *Viscosity*

The amount of resistance of a fluid to a shearing force can be determined by the property of the fluid called *viscosity*. The *Newtonian fluids* are the fluids for which there is a proportionality between the shear stress and the rate of shear strain. The proportionality is given by

$$\tau = \mu \frac{dV}{dy}, \tag{1.4}$$

where τ is the shear stress, μ the *absolute* or *dynamic viscosity*, V the strain, dV/dy the rate of shear strain, and y the distance between two imaginary parallel layers in the fluid, and the strain V is measured in between the layers (Fig. 1.1). The units of μ are Pa s or lb-sec/ft^2.

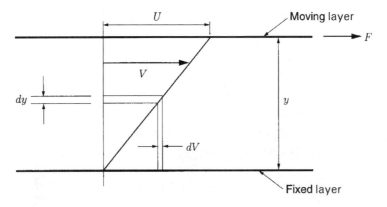

Figure 1.1
Typical system used to develop the expression for the viscosity of a fluid.

The *coefficient of kinematic viscosity* is another viscosity coefficient defined as

$$v = \frac{\mu}{\rho} = \frac{\mu g}{\gamma},$$ (1.5)

where μ is the absolute viscosity and ρ the mass density. The units of v are $\text{m}^2/2$ or $\text{ft}^2/2$.

Viscosities of liquids are not affected by the pressure changes but decrease with an increase in temperature. The absolute viscosity of gases increases with a temperature increase, but it is not affected by pressure changes. For a constant temperature, the kinematic viscosity of gases varies inversely with pressure, since the specific weight of gases changes with pressure changes.

1.5 Vapor Pressure

The *vapor pressure* is the pressure created by the vapor molecules when evaporation takes place within an enclosed space. It increases with temperature increase.

1.6 Surface Tension

The *surface tension* σ is defined as

$$\sigma = \frac{\Delta F}{\Delta L},$$ (1.6)

where ΔF is the elastic force transverse to any element length ΔL in the surface. The units of σ are N/m or lb/ft.

1.7 Capillarity

If a segment of a capillary tube is submerged in a liquid, the liquid in the capillary tube can rise or fall. This effect is called *capillarity* and is caused by the surface tension. The rise or fall depends on the relative magnitudes of the cohesion and the adhesion of the liquid to the walls of the capillary tube. Liquids rise in the tube when adhesion is greater than cohesion and fall in the tube when the cohesion is greater than adhesion. The capillarity is important when the diameter of the tube is greater than $\frac{3}{8}$ inch or 10 mm.

1.8 Bulk Modulus of Elasticity

The *bulk modulus of elasticity* E is defined as the ratio of the change in unit pressure to the corresponding volume change per unit of volume and expresses the compressibility of a fluid. E is given by

$$E = \frac{dp}{-dv/v}.$$ (1.7)

The minus sign is inserted to give a positive value for E when the pressure dp increases and the fractional volume dv/v decreases. The units of E are Pa or lb/in^2.

1.9 Statics

Pressure is perhaps the most used quantity in fluid statics. The fluid pressure acts normal to any plane and is transmitted with equal intensity in all directions. In a liquid, the pressure is the same at any point located on the same horizontal plane. For the measurement of the pressure, various gages above or below the atmospheric pressure are used. *Vacuum* is a term used to indicate a space with a pressure less than atmospheric pressure. The average atmospheric pressure, called the standard atmospheric pressure, is equal to 14.7 psi, 101.3 kPa, 760 mm of mercury, or 1 atmosphere.

The pressure can be expressed as

$$p = \frac{dF}{dA}, \tag{1.8}$$

where F is the force acting on a surface of area A. If F is uniformly distributed over the area A, then

$$p = \frac{F}{A}. \tag{1.9}$$

The units of pressure are lb/ft^2 (psf), lb/in^2 (psi), or Pa (N/m^2).

Considering two points at different levels in a liquid, h_1 and h_2, respectively, the difference in pressure between the points is given by

$$p_2 - p_1 = \gamma(h_2 - h_1), \tag{1.10}$$

where γ is the unit weight of the liquid and $h_2 - h_1$ the difference in elevation. Taking $h_1 = 0$, that is, one point located at the surface of the liquid, and assuming $h_2 = h$ is positive downward, Eq. (1.10) becomes

$$p = \gamma h. \tag{1.11}$$

The above equations are valid as long as γ is constant. The elevation h in Eq. (1.11) is also called the *pressure head* and represents the height of a column of homogeneous fluid that will produce a given intensity of pressure. Therefore,

$$h = \frac{p}{\gamma}. \tag{1.12}$$

For small changes in elevation dh, Eq. (1.11) can be written as

$$dp = -\gamma dh. \tag{1.13}$$

The negative sign means that as h increases, being positive upward, the pressure decreases.

There are two ways to express the pressure measurements, that is, absolute pressure and gage pressure. Absolute pressure uses absolute zero as its base. Gage pressure uses standard atmospheric pressure as its base. For example, if a fluid pressure is 3.8 kPa above standard atmospheric pressure, its gage pressure is equal to 3.8 kPa and its absolute pressure is equal to $3.8 + 101.3 = 105.1$ kPa. A device for measuring atmospheric pressure is the *barometer*. It consists of a tube more than 762 mm in length inserted in an open container of mercury with a closed tube end at the top and an open tube end at the bottom. The mercury extends from the container up into the tube. At sea level, the mercury rises in the tube to a height of approximately 762 mm. The level of mercury rises or falls as atmospheric pressure changes. Direct reading of the mercury level gives the pressure head of mercury, which can be converted to pressure by using Eq. (1.11).

To measure the pressure of other fluids, devices such as *piezometers* or *manometers* can be used.

1.10 Hydrostatic Forces on Surfaces

1.10.1 FORCE EXERTED ON A PLANE AREA

The force exerted by a liquid on a plane area is given by

$$F = \gamma h_{cg} A, \tag{1.14}$$

where γ is the specific weight of the liquid, h_{cg} the depth of the center of gravity of the area, and A the area. The line of action of the force passes through the center of pressure, which is given by

$$y_{cp} = \frac{I_{cg}}{y_{cg} A} + y_{cg}, \tag{1.15}$$

where I_{cg} is the moment of inertia of the area about its center of gravity axis.

1.10.2 FORCE EXERTED ON A CURVED SURFACE

In this case, the hydrostatic force has a horizontal component and a vertical component. The horizontal component on a curved surface is equal to the normal force on the vertical projection of the surface. It acts through the center of pressure for the vertical projection. The vertical projection is equal to the weight of the volume of liquid above the area. The volume can be real or imaginary. The vertical projection passes through the center of gravity of the volume.

1.10.3 HOOP OR CIRCUMFERENTIAL TENSION

Hoop (*circumferential*) *tension* is produced by the exerted internal pressure in the walls of a cylinder.

The *longitudinal stress* in thin-walled cylinders ($t < 0.1$; d, t is the wall thickness, d the cylinder diameter) closed at the ends is equal to half the hoop tension.

1.10.4 HYDROSTATIC FORCES ON DAMS

Some safety factors should be considered when one checks for dam stability. The stability of the dam can be affected by the following factors:

- Large hydrostatic forces that cause a tendency for the dam to slide horizontally and overturn it around the downstream edge
- Hydrostatic uplift along the bottom of the dam caused by water seeping under the dam

Thus, the safety factors are those against sliding and against overturning. Also, the pressure intensity on the base must be checked.

1.11 Buoyancy and Flotation

Buoyancy and *flotation* are based on Archimedes' principle, which states that the force, called the *buoyant force*, that lifts (buoys) upward a floating or submerged body in a fluid is equal to the weight of the fluid that would be in the volume displaced by the fluid. In other words, there is a balance between the weight of the floating body and the buoyant force. The point located at the center of gravity of the displaced fluid is called the *center of buoyancy*. The buoyant force acts through this point.

Some buoyancy applications include determination of irregular volumes, specific gravities of liquids, and naval architectural design.

To address the stability problem of submerged and floating bodies, the following principles apply:

- In order for a submerged body to be stable, the body's center of gravity must lie below the center of buoyancy of the displaced liquid.
- In order for a submerged body to be in neutral equilibrium for all positions, the body's centers of gravity and buoyancy must coincide.
- In order for a floating cylinder or sphere to be stable, the body's center of gravity must lie below the center of buoyancy.
- The stability of other floating objects depends on whether a righting or overturning moment is developed when the centers of gravity and buoyancy move out of vertical alignment because of the changing of position of the center of buoyancy.

1.12 Dimensional Analysis and Hydraulic Similitude

1.12.1 DIMENSIONAL ANALYSIS

The mathematics of dimensions of quantities is called *dimensional analysis*. The physical relationships among quantities can be expressed by equations. Within these equations, absolute numerical and dimensional equality must exist. By manipulating the physical relationships, they can be reduced to fundamental quantities, such as force F or mass M, length L, and time T. A typical application includes the following items:

- Converting one system of units to another
- Developing equations

- Reducing the required number of variables
- Establishing the principles of model design

1.12.2 HYDRAULIC MODELS

In general, hydraulic models can be either *true models* or *distorted models*. True models have almost all characteristics of the prototype reproduced to scale (*geometric similitude*) and fulfill the design constrains (*kinematic* and *dynamic similitude*).

1.12.3 GEOMETRIC SIMILITUDE

If the ratios of all corresponding dimensions in model and prototype are similar, we say that *geometric similitude* exists. Examples of such ratios are

$$\frac{L_{\text{model}}}{L_{\text{prototype}}} = L_{\text{ratio}} \text{ or } \frac{L_m}{L_p} = L_r \tag{1.16}$$

and

$$\frac{A_{\text{model}}}{A_{\text{prototype}}} = \frac{L_{\text{model}}^2}{L_{\text{prototype}}^2} = L_{\text{ratio}}^2 = L_r^2. \tag{1.17}$$

1.12.4 KINEMATIC SIMILITUDE

If

- The paths of homologous moving particles are geometrically similar and
- The ratios of the velocities of homologous particles are equal

then we say that *kinematic similitude* exists. Examples of useful ratios are

$$\text{Velocity} \quad \frac{V_m}{V_p} = \frac{\dfrac{L_m}{T_m}}{\dfrac{L_p}{T_p}} = \frac{L_m}{L_p} \div \frac{T_m}{T_p} = \frac{L_r}{T_r} \tag{1.18}$$

$$\text{Acceleration} \quad \frac{a_m}{a_p} = \frac{\dfrac{L_m}{T_m^2}}{\dfrac{L_p}{T_p^2}} = \frac{L_m}{L_p} \div \frac{T_m^2}{T_p^2} = \frac{L_r}{T_r^2} \tag{1.19}$$

$$\text{Discharge} \quad \frac{Q_m}{Q_p} = \frac{\dfrac{L_m^3}{T_m}}{\dfrac{L_p^3}{T_p}} = \frac{L_m^3}{L_p^3} \div \frac{T_m}{T_p} = \frac{L_r^3}{T_r}. \tag{1.20}$$

1.12.5 DYNAMIC SIMILITUDE

If the ratios of all homologous forces in model and prototype are the same, than we say that *dynamic similitude* exists. The dynamic similitude exists between systems with geometric and kinematic similitude.

Newton's second law of motion $\sum F_x = Ma_x$ forms a basis for the required conditions for complete similitude. The forces in Newton's equation can be viscous forces, pressure forces, gravity forces, surface tension forces, or elasticity forces. The following relationship between forces acting on model and prototype is obtained:

$$\frac{\sum \text{Forces}_m}{\sum \text{Forces}_p} = \frac{M_m a_m}{M_p a_p}. \tag{1.21}$$

1.12.6 USEFUL RATIOS

Inertial force ratio:

$$F_r = \frac{\text{force}_{\text{model}}}{\text{force}_{\text{prototype}}} = \frac{M_m a_m}{M_p a_p} = \frac{\rho_m L_m^3}{\rho_p L_p^3} \times \frac{L_r}{T_r^2} = \rho_r L_r^2 \left(\frac{L_r}{T_r}\right)^2$$

$$F_r = \rho_r L_r^2 V_r^2 = \rho_r A_r V_r^2. \tag{1.22}$$

Equation (1.22) expresses the general law of dynamic similarity between model and prototype, also known as *Newtonian equation.*

Inertia–pressure force ratio (Euler number):

$$\frac{Ma}{pA} = \frac{\rho L^3 \times \dfrac{L}{T^2}}{pL^2} = \frac{\rho L^4 \dfrac{V^2}{L^2}}{pL^2} = \frac{\rho L^2 V^2}{pL^2} = \frac{\rho V^2}{p}. \tag{1.23}$$

Inertia–viscous force ratio (Reynolds number):

$$\frac{Ma}{\tau A} = \frac{Ma}{\mu \left(\dfrac{dV}{dy}\right) A} = \frac{\rho L^2 V^2}{\mu \left(\dfrac{V}{L}\right) L^2} = \frac{\rho VL}{\mu}. \tag{1.24}$$

Inertia–gravity force ratio:

$$\frac{Ma}{Mg} = \frac{\rho L^2 V^2}{\rho L^3 g} = \frac{V^2}{Lg}. \tag{1.25}$$

The square root of this ratio, $V/\sqrt{Lg}$, is known as the *Froude number.*

Inertia–elasticity force ratio (Cauchy number):

$$\frac{Ma}{EA} = \frac{\rho L^2 V^2}{EL^2} = \frac{\rho V^2}{E}. \tag{1.26}$$

The square root of this ratio, $V/\sqrt{E/\rho}$, is known as the *Mach number.*

Inertia–surface tension ratio (Weber number):

$$\frac{Ma}{\sigma L} = \frac{\rho L^2 V^2}{\sigma L} = \frac{\rho L V^2}{\sigma}.$$ (1.27)

Time ratios:

$$T_r = \frac{L_r^2}{v_r}$$ (1.28)

$$T_r = \sqrt{\frac{L_r}{G_r}}$$ (1.29)

$$T_r = \sqrt{L_r^3 \times \frac{\rho_r}{\sigma_r}}$$ (1.30)

$$T_r = \frac{L_r}{\sqrt{\dfrac{E_r}{\rho_r}}}.$$ (1.31)

1.13 Fundamentals of Fluid Flow

Unlike solids, the elements of a flowing fluid can move at different velocities and can be subjected to different accelerations. The following principles apply in fluid flow:

- *The principle of conservation of mass*, from which the equation of continuity is developed
- *The principle of kinetic energy*, from which some flow equations are derived
- *The principle of momentum*, from which equations regarding the dynamic forces exerted by flowing fluids can be established

1.13.1 PROPERTIES

Fluid flow can be characterized as *steady* or *unsteady*, *uniform* or *nonuniform*, *laminar* or *turbulent*, one-dimensional, two-dimensional, or three-dimensional, and *rotational* or *irrotational*.

If the direction and magnitude of the velocity at all points in the fluid are identical, the flow is called *true one-dimensional*. It is also acceptable when the single dimension is taken along the central streamline of the flow and when the velocities and the accelerations normal to the streamline can be neglected. For this case, the average values of velocity, pressure, and elevation are used to model the flow as a whole.

If the fluid particles flow in planes or parallel planes and the streamline patterns are identical in each plane, the flow is called *two-dimensional*.

Irrotational flow is that flow in which no shear stresses occur. Therefore, no torque exist, and thus the particles do not rotate about their center of mass.

1.13.2 STEADY FLOW

If the velocity of succesive fluid particle at any point in the fluid is the same at successive moments of time, the flow is called *steady flow*. The velocity is constant with respect to time ($\partial V / \partial t = 0$), but it may vary at different points or with respect to distance. The other fluid variables do not vary with time, that is, $\partial p / \partial t = 0$, $\partial \rho / \partial t = 0$, $\partial Q / \partial t = 0$, etc. A steady flow can be *uniform* or *nonuniform*.

If the fluid variables change with time ($\partial V / \partial t \neq 0$), the flow is called *unsteady flow*.

1.13.3 UNIFORM FLOW

If the magnitude and the direction of the velocity do not vary with respect to distance ($\partial V / \partial s = 0$), the flow is called *uniform flow*. Therefore, the other fluid variables, such as y, p and Q, do not change with distance.

If the fluid variables change with distance ($\partial V / \partial s \neq 0$), the flow is called *nonuniform flow*.

1.13.4 STREAMLINES

The imaginary curves drawn through a fluid to show the direction of motion for various sections of the flow are called *streamlines*. The velocity vectors are always tangent to the streamlines and, therefore, there is no flow across a streamline at any point.

1.13.5 STREAMTUBES

A group of streamlines that bound an elementary portion of a flowing fluid is called a *streamtube*. For small cross-sectional areas of a streamtube, the velocity of the center of the cross section can be taken as the average velocity of the section as a whole.

1.13.6 EQUATION OF CONTINUITY

The *equation of continuity* is obtained from the principle of conservation of mass. For steady flow, the principle of conservation of mass becomes

$$\rho_1 A_1 V_1 = \rho_2 A_2 V_2 = \text{const},\tag{1.32}$$

or

$$\gamma_1 A_1 V_1 = \gamma_2 A_2 V_2,\tag{1.33}$$

that is, the mass of fluid passing all sections in a stream of fluid per unit time is the same. If the fluid is *incompressible* ($\gamma_1 = \gamma_2$), Eq. (1.33) yields

$$Q = A_1 V_1 = A_2 V_2 = \text{const},\tag{1.34}$$

where A_1 and A_2 are the cross-sectional areas of the stream at sections 1 and 2, respectively, and V_1 and V_2 are respectively the velocities of the stream at the same sections. Commonly used units of flow are cubic feet per second (cfs), gallons per minute (gpm), or million gallons per day (mgd).

For steady two-dimensional incompressible flow, the continuity equation is

$$A_{n_1} V_1 = A_{n_2} V_2 = A_{n_3} V_3 = \text{const}, \qquad (1.35)$$

where A_n terms are the areas normal to the respective velocity vectors.

1.13.7 FLOW NETS

A flow net consists of the following:

- A system of streamlines spaced so that the rate of flow q is the same between each succesive pair of lines
- Another system of lines normal to the streamlines spaced so that the distance between normal lines equals the distance between adjacent streamlines.

The flow nets are drawn to show the flow patterns in cases of two dimensional and three-dimensional flow. Although an infinite number of streamlines are required to completely describe a flow under a given set of boundary conditions, in practice, a small number of streamlines are used if acceptable accuracy is obtained.

1.13.8 ENERGY AND HEAD

Three forms of energy are usually considered in fluid flow problems, namely *potential, kinetic,* and *pressure energy.*

Potential energy (PE) is the energy possessed by an element of fluid due to its elevation above a reference datum (Fig. 1.2). PE is given by

$$\text{PE} = Wz, \qquad (1.36)$$

where W is the weight of the considered element and z the distance where the element is located with respect to the datum.

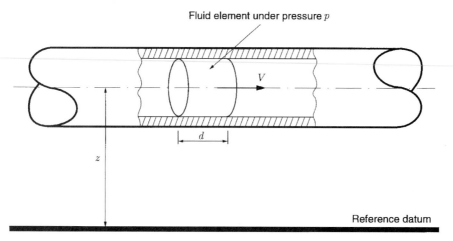

Figure 1.2

Kinetic energy (KE) is the energy possessed by an element of fluid due to its velocity. The following expression can be used to calculate KE:

$$KE = \frac{1}{2} m V^2. \tag{1.37}$$

Here m is the mass of the element and V its velocity. If the mass m is expressed as $m = W/g$, where g is the gravitational acceleration, Eq. (1.37) becomes

$$KE = \frac{1}{2} \frac{W V^2}{g}. \tag{1.38}$$

Pressure energy or *flow energy* (FE) is defined as the work needed to force the element of fluid across a certain distance against the pressure. The following expression applies:

$$FE = p\, A\, d. \tag{1.39}$$

Here p is the pressure, A is the cross-sectional area, and d the distance over which the work is done in moving the element of fluid. The term $A\, d$ is in fact the volume of the element, $A\, d = W/\gamma$, where γ is the specific weight of the fluid. Therefore,

$$FE = p\frac{W}{\gamma}. \tag{1.40}$$

The total energy E is the sum of PE, KE, and FE, and from Eqs. (1.36), (1.38), and (1.40), we obtain

$$E = Wz + \frac{1}{2}\frac{WV^2}{g} + p\frac{W}{\gamma}. \tag{1.41}$$

Each term in Eq. (1.41) can be expressed in ft-lb or N m. In fluid mechanics and hydraulic problems, it is customary to work with energy expressed as a *head*, that is, the amount of energy per unit weight of fluid. The units for head are ft-lb/lb or N m/N of fluid. Mathematically, these units are ft and m.

To express the total energy (E) as a head (H), Eq. (1.41) can be divided by the weight of the fluid W, which gives

$$H = z + \frac{V^2}{2g} + \frac{p}{\gamma}, \tag{1.42}$$

where z is known as the *elevation head*, $V^2/2g$ as the *velocity head*, and p/γ as the *pressure head*.

1.13.9 ENERGY EQUATION

The *energy equation* is derived by applying the principle of energy to fluid flow. In the direction of flow, the principle of energy is summarized by the general equation

Energy section 1 + Energy added − Energy lost − Energy extracted

$$= \text{Energy section 2}. \tag{1.43}$$

Fluid Dynamics

For the case of steady flow of incompressible fluids, Eq (1.43) becomes

$$\left(\frac{p_1}{\gamma} + \frac{V_1^2}{2g} + z_1\right) + H_A - H_L - H_E = \left(\frac{p_2}{\gamma} + \frac{V_2^2}{2g} + z_2\right). \qquad (1.44)$$

This equation is known as the *Bernoulli equation.*

1.13.10 VELOCITY HEAD

The kinetic energy per unit weight at a particular point is called *velocity head.* The true kinetic energy can be calculated by integrating the differential kinetic energies from streamline to streamline and considering the *kinetic energy correction factor* α. α is applied to the $V_{av}^2/2g$, and it is given by

$$\alpha = \frac{1}{A}\int_A \left(\frac{v}{V}\right)^3 dA, \qquad (1.45)$$

where V is the average velocity in the cross section, v is the velocity at any point in the cross section, and A is the area of the cross section. Studies indicated that $\alpha = 1.00$ for uniform distribution of velocity, $\alpha = 1.02$ to 1.15 for turbulent flows, and $\alpha = 2.00$ for laminar flow.

1.13.11 POWER

Power is given by the following relationships:

$$\text{Power } P = \gamma QH = \text{lb/ft}^3 \times \text{ft}^3/\text{sec} \times \text{ft-lb/lb} = \text{ft-lb/sec,}$$

$$\text{horsepower} = \frac{\gamma QH}{550},$$

or

$$P = \text{N/m}^3 \times \text{m}^3/\text{s} \times \text{N} \cdot \text{m/N} = \text{N} \cdot \text{m/s} = \text{watts(W).}$$

2. Hydraulics

Hydraulic systems are installed because they enabled the designer to significantly magnify and/or transfer forces. Hydraulic components in earth-moving machines allow relatively large forces to be applied at locations remote from the engine with small additional weight and complexity. For these machines, such as presses and mining machines, an engine-driven pump powers the thrust cylinders and/or the torque motors at remote locations under the control of a single operator at a central location.

2.1 Absolute and Gage Pressure

In order to select the proper pumps and reservoirs, the following quantities are of interest:

- *Absolute pressure* is that pressure measured with respect to an absolute vacuum pressure
- *Gage pressure* is the difference in pressure between the pressure being measured and the ambient pressure

The ambient pressure is assumed to be the atmospheric pressure. In general, the atmospheric pressure at a particular location depends on the altitude and the weather. It is assumed that the atmospheric pressure is 14.5 psi in the English system of units and 1 bar ($= 10^5$ Pa) in the SI system.

The other units for pressure are psia, denoting the *absolute pressure in pounds per square inch*, psig, denoting the *gage pressure in pounds per square inch*, bar, denoting the gage pressure in SI units; and the height of a water column or the height of a column of mercury that produces the specified pressure at its base. Henceforth, the height of a mercury column will refer to the absolute pressure only.

Table 2.1 lists the conversion factors among several systems of units.

Table 2.1 *Equivalents between Pressure Units*

psia	in H$_2$O	in Hg
1	27.7	2.04
0.49	13.6	1
0.036	1	0.073
14.7	407.2	29.9

2.2 Bernoulli's Theorem

Bernoulli's theorem in its simplest form can be developed by applying the conservation of energy in a nonviscous, incompressible fluid.

The kinetic energy of a volume of fluid of mass m moving with velocity v is given by

$$\mathrm{KE} = \frac{mv^2}{2},\qquad(2.1)$$

and the potential energy is given by

$$\mathrm{PE} = mgz,\qquad(2.2)$$

where z is the elevation above a reference position, and g is the gravity acceleration.

The pressure energy is

$$P_e = pV,\qquad(2.3)$$

where p is the pressure and V is the reference volume.

The total energy E of the reference volume of fluid is

$$\frac{mv^2}{2} + mgz + pV = EV. \tag{2.4}$$

If we divide by the reference volume, V, Eq. (1.4) may be written as

$$\frac{\rho v^2}{2} + \rho gz + p = E, \tag{2.5}$$

where $\rho = m/V$ is the density.

Equation (1.5) is valid for any points 1 and 2 of the circuit. Therefore,

$$\frac{\rho(v_1^2 - v_2^2)}{2} + \rho g(z_1 - z_2) + p_1 + p_2 = 0. \tag{2.6}$$

Equation (2.6) is the standard form of Bernoulli's theorem.

In the form given in Eq. (2.6), the pressure may be either gage pressure or absolute pressure. The units, however, must be consistent. If g is in ft/s^2, then v must be in ft/s, p must be in lb/ft^2, ρ must be in lb/ft^3, and z must be in ft. In the SI system if g is in m/s^2, then v must be in m/s, p must be in N/m^2, ρ must be in kg/m^3, and z must be in m.

This formula is particularly important in the design of hydraulic systems because it clearly shows the relation between pressure and flow velocity in a hydraulic line. If one is increased, the other one should be decreased.

EXAMPLE 2.1 A pressure gage mounted at station 1 in a fluid line with an internal diameter of 2.0 in reads 320.0 psig for a fluid flow of 15.0 gpm (gal/min). The fluid is a hydraulic oil of density 35.2 lb/ft^3. It passes through a reducer to a pressure hose with an internal diameter of 1.5 in, and to a gear motor of a robot arm that moves 20 in above and below the level of station 2 as shown in Fig. 2.1. Find the gage pressure at stations 2, 3, and 4. Energy losses at the reducer are negligible. ▲

Solution

Using Eq. (2.6), we get

$$\frac{\rho v_1^2}{2} + p_1 + \rho gz_1 = \frac{\rho v_2^2}{2} + p_2 + \rho gz_2,$$

where $p_1 = 320.0 \times 12^2 = 46{,}080$ psfg.

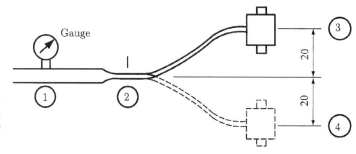

Figure 2.1
Pressure line 1, reduced section 2, and torque motor at position 3 and 4.

The rate of change of volume $\partial V/\partial t$ is denoted by $\dot{Q}$ and is measured in units of gallons per minute or liters per minute:

$$1 \text{ gallon } (1 \text{ Gal}) = 231 \text{ in}^3.$$

The velocities can be found by using

$$v_1 = \frac{\dot{Q}}{A_1}, v_2 = \frac{\dot{Q}}{A_2}, \text{ and } v_2 = v_1 \frac{A_1}{A_2},$$

along with the numerical values for velocities,

$$v_1 = \frac{15(231)}{60(12)^3} \frac{(12)^2}{\pi(1.0)^2} = 1.531 \text{ ft/s}, v_2 = 7.353 \frac{\pi(0.5)^2}{\pi(0.75)^2} = 2.723 \text{ ft/s}.$$

Since there is no change in elevation between stations (1) and (2), the z terms cancel out in Eq. (1.6). Therefore, by solving Bernoulli's equation for p_2 at station (2.2), we obtain

$$p_2 = \frac{\rho}{2}(v_1^2 - v_2^2) + p_1$$

$$= \frac{35.2}{2(32.2)}(1.531^2 - 2.723^2) + 46,080 = 46,082.798 \text{ psfg} = 320.018 \text{ psig}.$$

Because of the elevation at station (3),

$$p_3 = p_2 + \rho g(z_2 + z_3) = 46,080 + 35.2\left(0 - \frac{20}{12}\right) = 46,025.833 \text{ psfg}$$

$$= 319.623 \text{ psig}.$$

At station (4), the pressure is

$$p_4 = 46080 + 35.2\left(\frac{20}{12}\right) = 46138.667 \text{ psfg} = 320.407 \text{ psig}. \quad \blacktriangle$$

2.3 Hydraulic Cylinders

Some types of hydraulic cylinders are shown in Fig. 2.2. Types presented in Figs. 2.2a and 2.2b are the most common. Since both cylinders have two ports, one located at the head end and the other one at the cap end, the difference between single acting and double acting cylinders is whether fluid pressure is delivered under external control to both ends or to just one end. The piston and the rod in the single-acting cylinder can be extended by forcing the fluid in the port located at the cap end. To drain from that port, an external force can be applied as the rod is retracted. The port at the head end can be used to admit air or fluid as the rod is retracted. Single-acting cylinders can also be retracted hydraulically and extended mechanically. Double-acting cylinders have ports at each end of the cylinder and, thus, the piston and rod can be moved hydraulically in either direction. The double rod cylinder, shown in Fig. 2.2c, is a cylinder having a single piston and a piston rod extending from each end. These types are used when the equality of forces and speeds must be independent of travel direction. For the spring

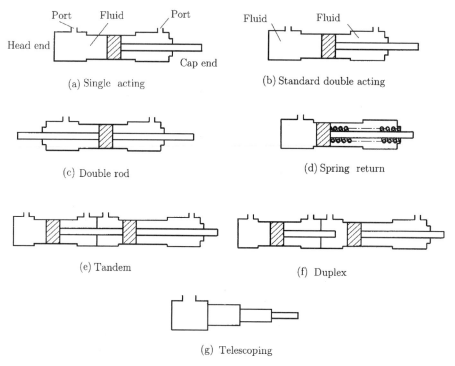

Figure 2.2 *Types of hydraulic cylinders.*

return cylinder shown in Fig. 2.2d, the returning to the initial position is due to the spring force. In Figs. 2.2e and 2.2f are shown the tandem and duplex cylinders, respectively, used for operation in two directions. A telescoping cylinder is shown in Fig. 2.2g. Terminology for the major parts of a hydraulic cylinder is shown in Fig. 2.3, where 1 is the cap end, 2 is the head end, 3 is the piston rings, 4 is the piston rod, 5 and 9 are the ports, 6 is the cylinder body, 8 is the rod gland seals, and 10 is the rod gland bushing.

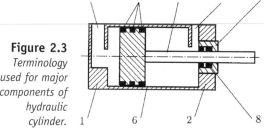

Figure 2.3
Terminology used for major components of hydraulic cylinder.

If the inertial forces are high, the cap end of the cylinder undergoes a shock when the piston and rod touch the cap. This shock can be reduced by installing a hydraulic cushion as shown in Fig. 2.4. The hydraulic cushion consists of a cushion spear, 10; a needle valve, 8; and a secondary drain line from the needle valve. As the piston, 7, approaches the end cap, the tapered portion of the cushion spear partially closes the larger port at the cap end and slows the flow from the cylinder, 3, thus decelerating the piston. When the cushion spear finally closes the larger port, the draining of the remaining fluid is slowed further and it is diverted through the small needle valve orifice. The cushion sleeve 6 performs the same function at the head end 2. In Fig. 2.4, is a check valve, and 9 is the floating cushion bushing.

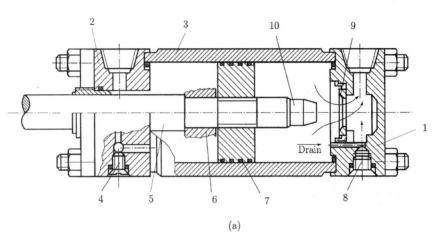

(a)

Figure 2.4
Hydraulic cylinder. (a)Cushion sleeve to cushion motion toward the head end; cushion spear for motion toward the cap end. (b) Varieties of cushion spears.

(b)

Let us consider two cylinders connected by a hydraulic line as shown in Fig. 2.5, and let force F_1 act upon cylinder 1. If the cross-sectional area of the piston is A_1, the pressure required in the fluid to hold piston 1 in equilibrium is

$$p_1 A_1 = F_1. \tag{2.7}$$

The force needed to hold piston 2 in equilibrium is given by ($p_1 = p_2 = p$)

$$p_1 A_2 = F_2, \tag{2.8}$$

since the pressure is unchanged throughout a stationary fluid.

Upon elimination of p between Eq. (2.7) and Eq. (2.8), it is found that

$$F_2 = \frac{A_2}{A_1} F_1. \tag{2.9}$$

Fluid Dynamics

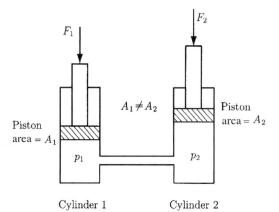

Figure 2.5

Simple hydraulic system to show force, area, and cylinder displacement relations.

If force F_2 moves piston 2, then piston 1 must also move if none of the incompressible fluid leaves or enters the system. If no energy is lost, the work done by piston 2 must be equal to that done by piston 1. Thus,

$$F_1 x_1 = F_2 x_2, \tag{2.10}$$

where x_1 and x_2 denote the displacements of pistons 1 and 2, respectively. Substituting F_2 in Eq. (2.10) from Eq. (2.9) yields

$$x_1 = \frac{A_2}{A_1} x_2. \tag{2.11}$$

The increased force on piston 1 is obtained at the expense of increased motion of piston 2. Equation (2.9) will not hold during motion because of pressure loss in the hydraulic lines and cylinders. That fact will be discussed in the following sections. However, Eq. (2.9) does hold once the pistons and fluid come to rest. Equation (2.10) is an approximation to the actual motion and force relations because of energy losses due to viscosity and turbulence. These losses are usually negligible, when compared to the energy being transmitted from cylinder to cylinder.

2.4 Pressure Intensifiers

In Fig. 2.6 is shown a pressure intensifier, also known as an intensifier. In the figure, 1 is the inlet port, 2 and 3 are the outlet ports, and 4 is the cylinder house. The pistons in the two cylinders of different diameters are connected mechanically. Since the forces on the two pistons are equal, it follows that if they are held in equilibrium by the pressurized fluid in each cylinder, the pressures must be related according to

$$p_1 A_1 = F_1 = F_2 = p_2 A_2, \tag{2.12}$$

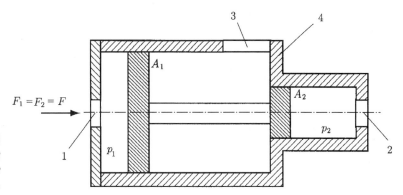

Figure 2.6
Schematic of a pressure intensifier.

so that

$$p_2 = p_1 \frac{A_1}{A_2} \qquad (2.13)$$

holds in the steady-state condition, that is, when the pistons are not accelerated or decelerated.

2.5 Pressure Gages

The pressure gages on machines are either mechanical or electrical.

The Bourdon gage, shown in Fig. 2.7, works on the principle that pressure in a curved tube will tend to straighten it. Thus, as shown in the figure, pressure acts equally on every square inch area in the tube, 1. Since the surface S_1 on the outside of the curve is greater than the surface area S_2 on the shorter radius, the force acting on S_1 is greater than the force acting on S_2. When the pressure is applied, the tube straightens out until the difference in force is balanced by the elastic resistance of the material of the tube. The tube is bent into a circular arc and it becomes oval in cross-section. Therefore, it tends to straighten more easily under pressure. The tube works by differential areas, since the area on which the pressure acts outward is greater than the area on which the pressure acts inward. The open end of the tube passes through the socket 2, which is threaded so that the gage can be screwed into an opening in the hydraulic system. The closed end of the tube, 9, is linked to a pivoted segment gear, 3, meshed with a small gear, 4, to which a pointer, 5, is attached. Beneath the pointer there is a scale, 6, reading in pounds per square inch. The gage is calibrated against known pressures to ensure accurate readings. Under pressure the tube tends to straighten and the segment moves around its pivot, 8, rotating the gear and the pointer. The pointer assembly is usually pressed on the shaft in such manner that it is removable for resetting when the gage is calibrated against a master unit. Electronic pressure indication may be obtained by means of a piezoelectric sensor that transforms a pressure-induced force into an electric charge that is electronically transformed by a coupler into a voltage that is proportional to the pressure. The active element in a piezoelectric sensor is the piezoelectric

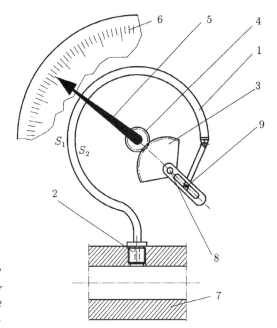

Figure 2.7
Major parts for Bourdon tube gage.

crystal, whose structure causes it to produce opposite electrical charges on its upper and lower surfaces when it is subjected to a compressive force perpendicular to those surfaces.

2.6 Pressure Controls

The first hydraulic component in the pressure line is a unit designed to control and protect the pump and drive unit from damage due to over-pressurization. This unit is called the pressure control valve. The most commonly used overload protectors or system pressure control valves are either simple direct-pressure-operated or more elaborate compound or pilot-operated relief valves. Although the basic functions of both types are similar, the methods and limits of operation vary considerably.

2.6.1 DIRECT-OPERATED RELIEF VALVES

Three types of simple direct-operated relief valves are shown in Fig. 2.8. An adjustable spring force on the disk (Fig. 2.8a), cone (Fig. 2.8b), or ball (Fig. 2.8c), seals the inlet from the outlet as long as the inlet pressure cannot overcome the spring force. The effective area for each type is denoted by A. The area A multiplied by the system pressure gives the force that pushes against the spring force that holds the valve closed. The conditions discussed at this point indicate only the pressure at which the valve begins to open or crack: To fully relieve the system pressure, we must provide for a volume of flow through the relief valve. At the *cracking pressure*, a highly restricted and very minor flow is allowed. Since the pressure is a result of the resistance to the flow, the system pressure will continue to rise after the cracking pressure

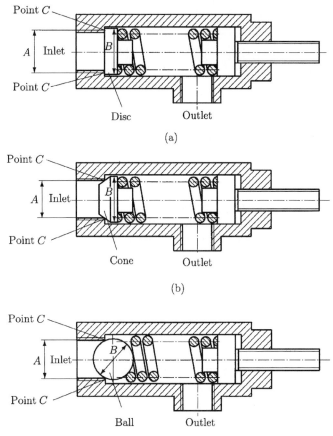

Point C

A | Inlet

Point C

Disc — Outlet

(a)

Point C

A | Inlet

Point C

Cone — Outlet

(b)

Point C

A | Inlet

Point C

Ball — Outlet

(c)

Figure 2.8

Types of simple direct-operating relief valve. (a) Direct-acting relief valve with disk poppet; (b) direct-acting relief valve with cone poppet; (c) direct-acting relief valve with ball poppet.

is reached. It will rise until the valve opening is large enough to allow sufficient decrease in flow resistance to the lower system pressure up to the desired level. At this point, the relief valve is said to operate at its *full-flow pressure*. The spring rate (the action of the spring) will further complicate the opening of the valve. As the valve seat cracks, the upward movement will compress the spring. Any upward movement of the valve will produce further spring compression and, thus, will increase the mechanical resistance of the valve until it is fully open. It will thus be faced with a full-flow pressure setting well above our cracking pressure setting and will attain a condition of *pressure override*. Protection of the system requires that the full-flow pressure setting be the one used. As a result, because of the cracking the oil leaks into the return or into the outlet line, decreasing the effective available volume to a point below the actual system relief setting. It is not uncommon for direct-acting relief valves that cracking occurs at a pressure value less than 80% of the full-flow pressure setting of the valve. Therefore, the full pumped volume is available only if the system pressure is less than 80% of the maximum operating pressure.

Direct-operated relief valves show a definite tendency to open and close rapidly, or *chatter*, owing to pressure pulsations. As a result system pressure variations may exist. In the process of chattering, the valve seal parts are rapidly damaged to the point that they will constantly leak, producing erratic control.

Figure 2.8a shows a very poor design. The flat surface is prevented from sealing by the slightest bit of contamination, thus giving very poor control. Figures 2.8b and 2.8c show examples of practical sealing methods. The tapered or rounded surface of the valve, by seating on the sharp edge of the orifice, will seal very well with minor wear on both parts. Hardening by heat treatment, the valve or the seat or both will extend the valve's life.

Direct-operated relief valves are considered to be *fail-safe*. Worn or broken parts allow excessive leakage, causing the pressure loss. In all three illustrations (Fig. 1.8), the valve area is denoted by B and, above the sealing point, it is considerably larger than the effective opening area denoted by A. Since area B is greater than area A, any pressure at the outlet port will be amplified by the ratio of area B to area A. This ratio must be added to the spring setting of the valve.

2.6.2 DIRECT-OPERATED SPOOL-TYPE PRESSURE CONTROLS

An improved design of the direct-operated relief valve is illustrated in Figs. 2.9a–2.9c. Instead of using the cone, ball, or disk shown in Fig. 2.8, a closely fitted spool is considered to open and close the outlet port of the valve. The sealing method employed in this type of pressure control valve is the same as that used for the spool-type, directional control valves on most hydraulic systems. The adjustment of the cracking pressure is accomplished by increasing or decreasing the compression of the spring, 3. The system pressure is piped into either port marked A and is transmitted to piston 2 via chamber C. As the system pressure on piston 2 exceeds the spring force 3, the spool 1 will move up and allow a flow of oil from inlet to outlet.

The pressure control characteristics of this valve will be very similar to those found with the simple direct-operating valve: a cracking pressure, a full-flow pressure, and, owing to the spring rate, a certain amount of pressure override. There may also occur the action termed "chatter" in the description of the simple direct-acting relief valve. However, this open-and-close movement with a spool-type valve will not damage the valve sealing surface. This action with a spool-type valve will be referred to as a *throttling action*. The spool may consistently vary the size of the outlet opening that is exposed to the pressure chamber and thus relieve or bypass only as much oil as is required to reduce the system pressure to the adjusted level. In the event the system pressure continues to rise, the spool will cease to throttle and assume the full-open position. At this point, the full-flow pressure condition of the valve occurs.

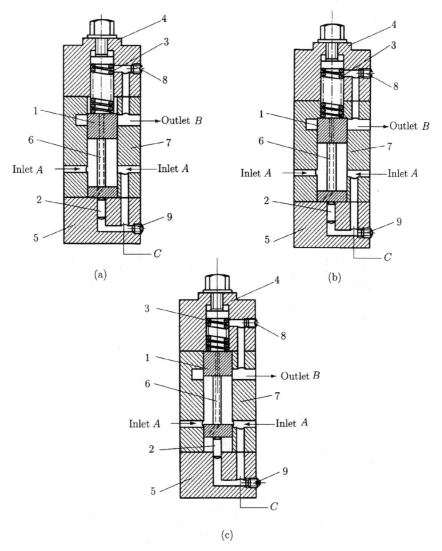

Figure 2.9
Spool-type relief valve shown (a) closed; (b) throttled; (c) fully open.

The life of the spool-type valve is superior to the life of the simple direct-operated valve. This is due to the decrease in wear on the sealing surfaces. The control action is also softer or cushioned. Figures 2.9a to 2.9c are referred to as *hydro-cushioned valves*, indicating a soft or cushioned action. This type of valve is designed with the piston 2 having an area equal to one-eighth the area of the main valve spool. This feature allows the control spring 3 to be much smaller and more sensitive than found in simple direct-operated valves. It also allows much larger volumes to be efficiently handled without the undesirable, erratic, and cumbersome springs needed for the simple direct-acting type. Opening and closing pressures are much closer. Pressure override is generally due to the spring rate encountered by continued compression of the adjustment spring as the spool moves upward to open the outlet port.

Fluid Dynamics

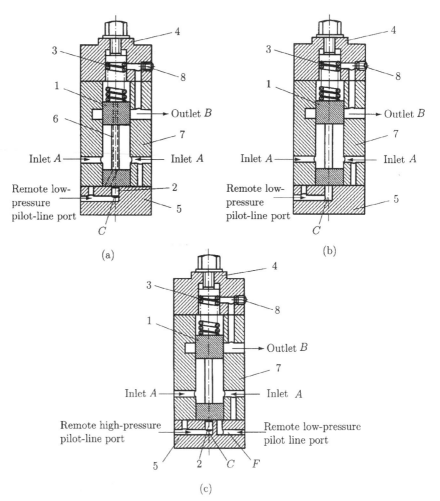

Figure 2.10

Spool-type relief valve controlled remotely. (a) high pressure; (b) low pressure; (c) dual pressure.

The spool-type valve may be controlled remotely. Figures 2.10a to 2.10c show the methods of remote control connection. In Fig. 2.10a, the bottom cap containing piston 2 is 180°. As a result, chamber C is no longer open to the inlet chamber A. Chamber C is now connected to a remote source of pressure. The remote pressure will now control the valve exactly as described in Fig. 2.9 when the system inlet pressure is the controlling medium. In Fig. 2.10b the control piston 2, shown in Fig. 2.10a, has been removed. The discussion of Fig. 2.9 indicated that piston 2 has an area equal to one-eighth of the bottom of spool 1. With the arrangement shown in Fig. 2.10b, a remote pressure that is one-eighth of that required in Fig. 2.10a causes the system in Fig. 2.10b to operate. If the system of Fig. 2.10a operates at 800 lb/in², the system of Fig. 2.10b requires only 100 lb/in² to operate. The ratio of areas inversely affects the pressure at which the spool operates. The greater the effective area, the lower the pressure needed to move the spool.

Figure 2.10c, by the addition of chamber F, allows independent operation of the valve as Fig. 2.10a or Fig. 2.10b without mutual interference.

Because of this feature, this valve is a relief valve that operates at two different pressures. It can be controlled from two remote and independent sources. Note that, in order to operate this type of valve as shown in Figs. 2.10b and 2.10c, the internal drain orifice D in Fig. 2.10a is not used. The cushion of spool 1 can still be achieved by the oil from the remote valve-operating source.

2.6.3 SEQUENCE VALVES

Figure 2.11 shows a valve unit adapted for pressure control of hydraulic actions so that one function cannot be exerted until the pressure of another function has reached a predetermined value. When performing in such a manner, the valve is called a sequence valve. Sequence valves must all be externally drained, since the system pressure is available at the outlet. The only difference between the valves shown in Figs. 2.11a and 2.11b is the method of valve actuation. The valve in Fig. 2.11a is directly operated by the system pressure at the inlet port of the valve, whereas the valve in Fig. 2.11b is adapted to be operated remotely by a pressure completely independent of the pressure at the inlet of the valve.

2.6.4 SEQUENCE VALVE FOR REVERSE FREE FLOW

Figure 2.12 shows a revised valve to allow pressure-controlled flow in one direction and uncontrolled free flow in the reverse direction. A simple check valve, 4, between the inlet port and the outlet port allows this condition. The presence of system pressure at the outlet port during operation requires that this valve be externally drained. Rotation of the bottom end cap 2 offers a choice between direct system operation, as shown in Fig. 2.12a, or remote independent operation, as shown in Fig. 2.12b.

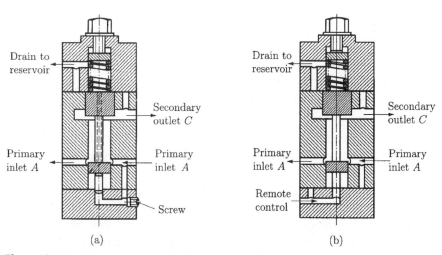

Figure 2.11 *Sequence valve. (a) Directly operated, externally drained; (b) remotely operated, externally drained. (Adapted from McNickle, 1966.)*

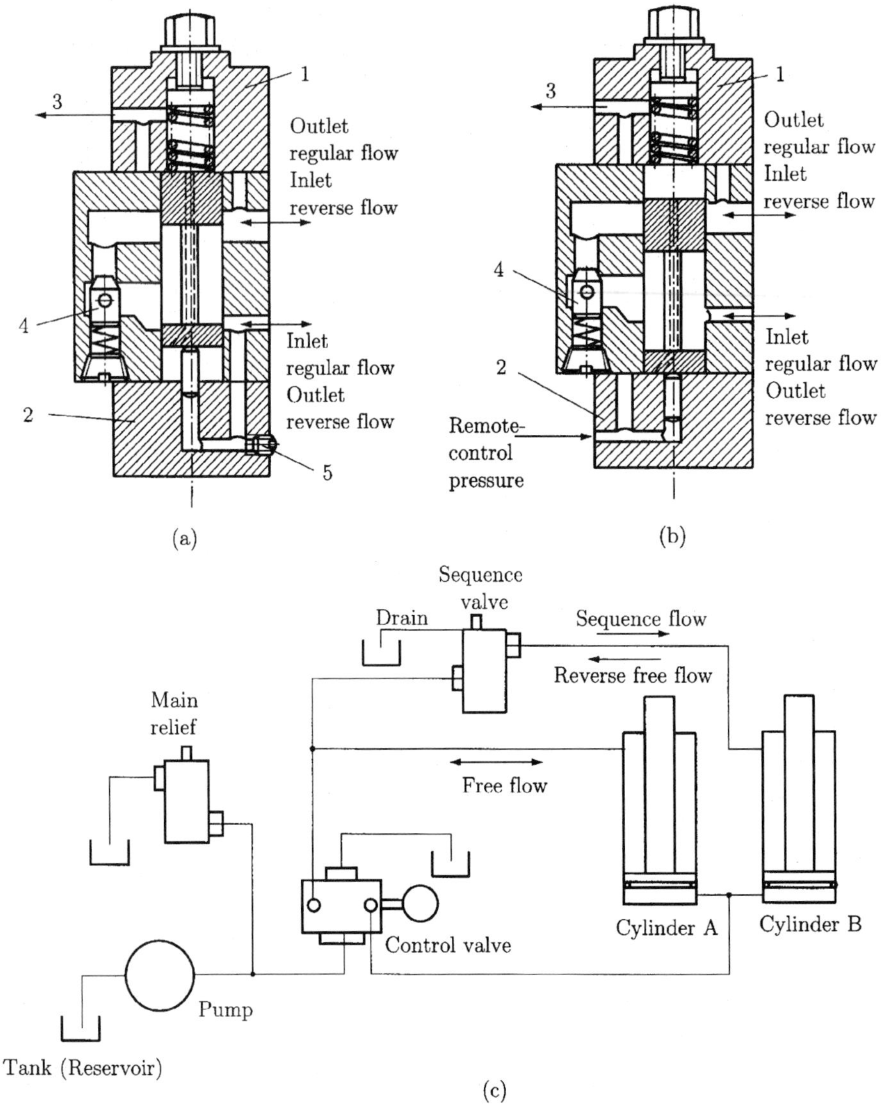

Figure 2.12 *Sequence valve for reverse free flow. (a) Directly operated, externally drained. Valve shown open in sequenced position. (b) Remotely operated, externally drained. Valve shown closed to regular flow. (c) Application drawing showing a typical circuit with sequence valve. (Adapted from McNickle, 1966.)*

The application drawing is shown incorporating the directly operated valve in Fig. 2.12c. With this circuit, both cylinder *A* and cylinder *B* extend at the same time, but cylinder *B* does not retract until cylinder *A* has fully retracted. One port of the directional control valve is connected unrestrictedly to the piston end of each cylinder. With this circuit the cylinders extend at the same time. The second port of the directional valve is directly connected to the rod end of cylinder *A*. The sequence valve is teed into this line at its inlet port, and the outlet port is connected to the rod end of cylinder *B*. As pressure is directed to this port of the directional control valve, cylinder

A retracts. Cylinder *B* does not move until cylinder *A* has completed its retraction and until the system pressure has increased to the point at which the sequence valve is opened, allowing flow from its inlet to the outlet and to cylinder *B*. The check valve in the sequence valve allows the passage of exhaust oil from the end of cylinder *B* as both cylinders are extended.

2.6.5 COUNTERBALANCE VALVES

Figure 2.13 shows an adaptation of the valve presented in the previous section used to control the operation of a hydraulic cylinder. The valve is now used in the exhaust line. Figure 2.13 is identical to Fig. 2.12, with the difference that the valve presented in Fig. 2.13 is internally drained (dashed line). This valve can be directly operated as shown in Fig. 2.13a or remotely operated as shown in Fig. 2.13b.

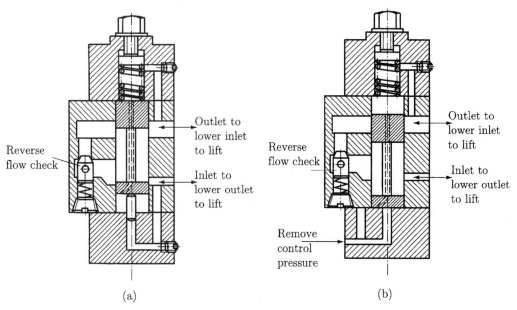

Figure 2.13 *Counterbalance valve. (a) Directly operated, internally drained, reverse free flow. (b) Remotely controlled, internally drained, reverse free flow. (Adapted from McNickle, 1966.)*

An application for this type can be similar with the application presented in Fig 2.12c if we consider just one cylinder lifting a heavy load. A counterbalance valve is installed in the line supplying the rod end of the lift cylinder so that the inlet of the counterbalance valve is piped to the cylinder and the outlet is piped to one port of the directional control valve. Under these conditions the free-flow characteristic of the counterbalance valve allows unrestricted lifting of the load. Lowering the load does not require overcoming the pressure setting of the counterbalance valve. After the counterbalance valve has opened, oil is allowed to flow through the counterbalance valve through the directional control and back to the reservoir. The setting of the counterbalance valve can be adjusted to suit the load and ensure "no-drift" holding and smooth lowering of the load.

Fluid Dynamics

2.6.6 COMPOUND RELIEF VALVES

A typical compound relief valve is shown in Fig. 2.14. This type of valve can also be referred to as a pilot-operated relief valve. Figure 2.14a indicates that hydraulic pressure at the inlet acts on the bottom of the main spool 1 and, by passing through orifice A, it also acts on the top side of spool 1. Oil passage through chamber B makes the system pressure available to the pilot valve 3. Note that pilot valve 3 is a simple direct-acting relief valve that is held on its seat by the adjustment of the force on spring 4. Spool 1 is maintained in the closed position by spring 2 and by the system pressure on its top side. Note that the area of the top side of spool 1 that is exposed to system pressure is slightly larger than the bottom area exposed to the same system pressure. This slight difference in area ensures more positive sealing and makes this valve's cracking pressure 90 to 95% of the full-flow pressure.

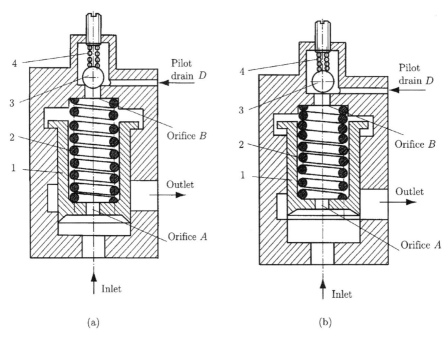

(a)   (b)

Figure 2.14 *Compound relief valve. (a) Externally drained, shown closed. (b) Valve shown open. (Adapted from McNickle, 1966.)*

A cycle of operation with this valve would start as the system pressure, conveyed to the pilot valve 3 by passages A and B, overcomes the force of spring 4 and forces the pilot 3 open. Oil is now free to flow out of the drain D at very low pressure and return to the reservoir. The system flow continues through orifice A, but note that orifice B is larger than orifice A. As a result, the pressure on the top side of spool 1 drops below the system pressure, and spool 1 moves up and opens the outlet port. As long as the pilot 3 remains open, the size difference between orifice A and orifice B maintains a condition of hydraulic unbalance and holds the valve open. A decrease of pressure allows the pilot 3 to close and quickly equalize the pressure on the top and bottom of spool 1. The force of spring 2 and the difference in area between the top side and the bottom side of spool 1 will quickly move the spool 1 down and close the outlet port.

Figure 2.15 shows another version of the compound-type relief valve. This valve is referred to as a balanced-piston-type relief valve. The operating functions in Fig. 2.15a are basically the same as those in Fig. 2.14. The main difference is in the method of flow. From the pilot valve 2, oil is returned to the reservoir. The pilot drain D is a hole or orifice passing through the main spool 1 directly into the outlet port. This feature requires fewer hydraulic lines, but it also allows outlet back pressure to adversely affect pilot operations. It is important that the outlet lines be unrestricted to ensure

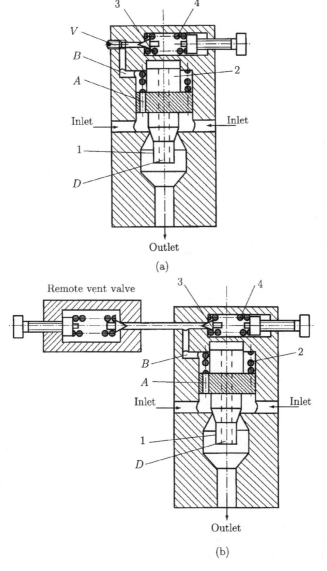

Figure 2.15
*Version of the compound-type relief valve.
(a) Valve shown closed.
(b) Compound relief valve with remote venting valve in use. Valve shown open or vented.*

minimum back pressure. Orifice B in Fig. 2.15a is placed differently, as is the pilot valve. The port V in Fig. 2.15 offers a new method of control for the compound-type unit. If port V is allowed to be open to the atmosphere or, by a simple valve, to the reservoir, the pressure on the top side of the main spool 1 will be relieved and the spool will immediately move upward and open. This practice is called *venting* and offers an auxiliary or additional method of instantly relieving the system pressure without altering or affecting the unvented operating setting of the valve. A simple direct-operating relief valve, manually operated, will handle venting functions and further increase the flexibility of this unit's operation. Figure 2.15b shows a typical arrangement that allows manually controlled venting and/or automatic system pressure operation. The vent valve is small since it is required to handle minor volumes.

2.6.7 COMPOUND-TYPE SEQUENCE VALVES

Figures 2.16a and 2.16b show the revised compound relief valve. Figure 2.16a illustrates the unit designated as the Y type. Since the outlet chamber becomes the secondary port exposed to the system pressure when the valve opens, an external bleed line E (Fig. 2.16a) must be used. The pilot chamber is no longer opened at the center orifice D of spool 1. The orifice D is now used to ensure complete hydraulic balance of spool 1 when the unit is in its sequenced position. The operation of the valve in Fig. 2.16a starts when the system pressure at the inlet port passing through orifice A reaches the level required to unseat pilot piston 3. The pressure drop through orifice A causes a reduced pressure in chamber B and allows spool 1 to move upward and open, the outlet or secondary port. Oil passing through orifice D in the center of spool 1 is now opened to system pressure, and the effective areas on both

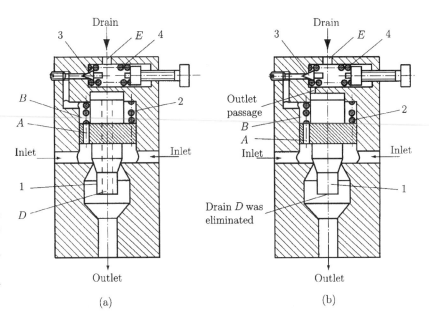

Figure 2.16
Compound-type sequence valve.
(a) Type Y sequence valve.
(b) Type X sequence valve.

sides of valve 1 are equal. The continued flow and pressure drop through orifice A maintains a lower pressure in chamber B, the valve remaining open. In the event that the inlet pressure decreases, pilot piston 3 closes, the pressure in chamber B rises, and the valve closes.

The valve presented in Fig. 2.16a is dependent on the system pressure at the point of operation. Figure 2.16b shows another modification of the basic compound unit. This model is identified as the X type. For this type an open passage between the pilot chamber and the top of spool 1 is used. The orifice D through the center of spool 1 is eliminated.

Operation of the X type is different from any type previously discussed. The purpose of this design is to fill the main circuit of a system with oil before flow to the outlet or secondary circuit is allowed. As the main circuit becomes full, the pressure rises at the inlet of Fig.2.16b. The flow through orifice A causes the opening of pilot 3, and, as a result, spool 1 opens. As the valve opens, the full area of the bottom spool 1 is exposed to system pressure. Since the small guide area of the top side spool 1 is opened to the pilot drain chamber, the spool 1 is hydraulically unbalanced because of the differential area. It is required that the system pressure be sufficient to overcome the very light force of spring 2 to remain open. This X type does not close until the system pressure has decreased nearly to zero. This valve's main purpose is therefore limited to controlling the sequence of flow as a hydraulic system is put into operation.

2.6.8 PRESSURE-REDUCING VALVES

A low-pressure, low-volume flow in addition to the main system high-pressure, high-volume flow is required by some hydraulic systems. The extra pump can be eliminated by the use of a pressure-reducing valve to supply the small flow at reduced pressure.

Figure 2.17 shows an X-model pressure-reducing valve. The X-valve combines the features of the direct-operated valve type with those of the compound-type valve. This valve incorporates a pilot 1 to control the action of the main spool 3, thus being a compound valve. The pressure-actuated spool 3 seals because of its close fit to the main body and because of a sliding action that opens and closes the outlet or reduced-pressure port. As system flow begins, the inlet is supplied with oil at the main pressure-control-valve setting. The flow to the outlet or reduced-pressure port is transmitted through orifice C, which is a narrow space between the reduced-diameter section of spool 3 and the main body. The fluid under pressure passes through chamber D and exerts a force on the bottom area of spool 3. A very small orifice E carries the pressurized oil through the center of spool 3 into chamber A. The areas of both ends of spool 3 are equal and under the same pressure so that a state of hydraulic balance exists. Spool 3 is thus held down by the force of spring 4. Since a reduced pressure at the outlet is desired, pilot 1 is adjusted to open at a pressure considerably lower than the pressure available at orifice C. The orifice E has a smaller area than chamber

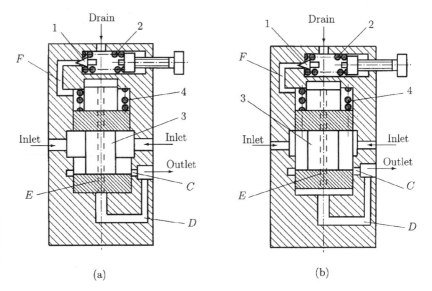

Figure 2.17
*Pressure-
reducing valve.
(a) Valve shown
static. (b) Valve
shown
operating.*

D, and the flow of the oil from E to F causes a pressure drop. The pressure in chamber F and on the top of spool 3 is lower than the pressure in chamber D or the pressure on the bottom of spool 3. Hydraulic unbalance occurs now; spring 4 is overcome and spool 3 is moved upward. As spool 3 moves upward, orifice C is reduced in size, opposing the flow, and a pressure drop is created between the inlet and outlet ports. Port C will thus be consistently changed to increase or decrease the resistance to the flow in order to maintain a constant reduced pressure at the outlet. As the flow from the outlet port increases in response to an increased low-pressure flow demand, the spool will move downward and open orifice C. As flow is diminished, orifice C will be closed. The maximum pressure available at the outlet is the sum of the forces of spring 2 and spring 4.

This valve has three critical situations. Orifice E is very small and it can be very easily plugged by minute foreign bodies. A constant flow through orifice E to the drain port of the pilot valve is needed to maintain a constant dependable reduced pressure. Orifice F must remain completely open. The pilot drain must have a free, unrestricted, unshared line to the reservoir. The final critical area of this valve is the close tolerance required between spool 3 and the bore of the main body.

2.7 Flow-Limiting Controls

2.7.1 CHECK VALVES

The simple check valve limits the flow to one direction. Figure 2.18a shows a simple check valve of the right-angle type in closed position, in which the flow from outlet to inlet is not allowed. A round poppet A is placed in the inlet port by the force of spring B. System pressure at the inlet port acts on the bottom of poppet A, compressing spring B, and opening the valve to

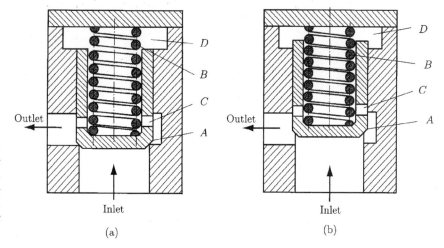

Figure 2.18
Simple check valve. Right-angle check valve shown (a) closed and (b) open.

allow the flow from inlet to outlet. The orifice C in poppet A serves as a drain for chamber D. It also exposes the top side of the poppet to the prevailing pressure of the outlet side when it is closed and of the inlet side when it is open. The system pressure need only overcome the force of spring B to hold the valve open. The pressure drop through this valve, from inlet to outlet, is thus equal to the force of spring B when the valve is properly sized with respect to flow volume. Figure 2.18b shows the valve in the opened position.

There are situations in hydraulic circuit design when it is desirable to have the automatic single-flow feature of the simple check valve for only a portion of the time and at any given time to be able to allow flow in either direction. This situation occurs in working with load-lifting devices. The normal single-flow characteristic allows the load to be lifted at any time and automatically held. It is also required that the ability to lower the load be included in the design. A pilot-operated check valve will adequately perform this function.

Figure 2.19 illustrates a pilot-operated check valve. In Fig. 2.19a, the check valve has a portion constructed in a manner similar to the simple check valve in Fig. 2.18. A pilot piston D with a stem E and a pilot pressure port for external connection have been added.

Figure 2.19b shows the valve when inlet pressure is high enough to overcome the force of spring B. Pilot pressure is still $0 \ lb/in^2$, and thus the inlet pressure acts on the top side of piston D and holds the pilot stem E downward. The valve acts as the conventional check valve in Figs. 2.19a and 2.19b.

Let us assume that we need to have the flow from the outlet port to the inlet port. A load has been lifted by allowing flow from inlet to outlet, and that it is now time to lower the load.

The application of an independent external pressure to the pilot port will move piston D upward, allowing flow from outlet to inlet, thus lowering the

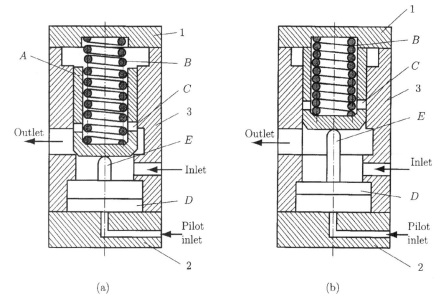

Figure 2.19
Pilot-operated check valve shown (a) closed and (b) with pilot actuated to allow constant flow or reverse flow.

(a)   (b)

load. To maintain the valve in an opened position, the following relation is required:

$$P_P \times D_B > Fs_B + P_C \times D_T$$

or

$$F_P > Fs_B + F_i$$

because

$$P \times A = F$$

where P_p is the pilot pressure, D_B is the bottom area D, F_{SB} is the spring B force, P_i is the inlet pressure, D_T is the top area D, F_P is the pilot force, and F_i is the inlet force. Also P is the pressure, A is the area, and F is the force.

2.7.2 PARTIAL-FLOW-LIMITING CONTROLS

For a hydraulic cylinder, the speed in one direction can be controlled if a simple needle valve is located in the exhaust port of the cylinder. This is referred to as a meter-out application. The exhaust pressure of a hydraulic cylinder is relatively stable and, thus, it maintains a reasonably accurate flow rate control with a simple needle valve. Figure 2.20a shows a simple needle-valve meter-out control. The unit depicted in Fig. 2.20 has the additional feature of allowing the flow to pass through a check valve B in one direction, thus being unaffected by the adjustment of the metering valve A. The metered flow direction of these valves is usually indicated by an arrow on the external surface of the unit. A fine adjustment thread on the stem of valve A provides a precise control of the flow. An adjustment of A is minor while the valve is subjected to system pressure. Excessive looseness of the locknut for valve A and excessive turning of valve A may damage the small valve seal. If large adjustments are needed, they are best accomplished at 0 lb/in^2.

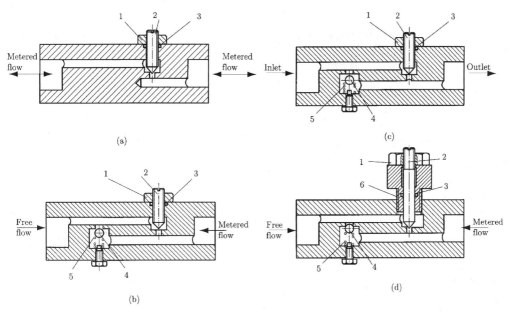

Figure 2.20 *Single-needle valve flow control. (a) Metered flow in both directions. (b) Check valve for reverse flow. (c) Reverse flow check valve shown open. (d) Valve constructed to allow adjustment while under pressure. (Adapted from McNickle, 1966.)*

The flow control illustrated in Fig. 2.20d is far superior to those shown in Figs. 2.20a to 2.20c. The unit shown in Fig. 2.20d is adjustable at any time, even while under maximum pressure.

2.8 Hydraulic Pumps

Although many hydraulic pumps and motors appear to be interchangeable in that they operate on the same principles and have similar parts, they often have design differences that make their performances better as either motors or pumps. Moreover, some motors have no pump counterparts. In this chapter only positive-displacement pumps are considered (those pumps that deliver a particular volume of fluid with each revolution of the input drive shaft). This terminology is used to distinguish them from centrifugal pumps and turbines.

2.8.1 GEAR PUMPS

The simplest type of these pumps is the gear pump, shown in Fig. 2.21, in which the fluid is captured in the spaces between the gear teeth and the housing as the gears rotate. Flow volume is controlled by controlling the speed of the drive gear. Although these pumps may be noisy unless well designed, they are simple and compact.

2.8.2 GEROTOR PUMPS

Another version of the gear pump is the gerotor, whose cross section is presented schematically in Fig. 2.22. The internal gear has one fewer tooth

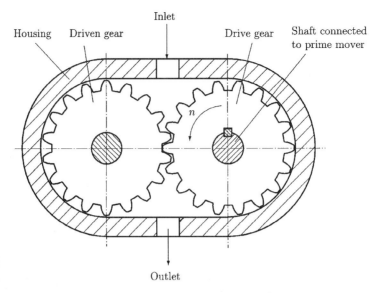

Figure 2.21
Gear pump.

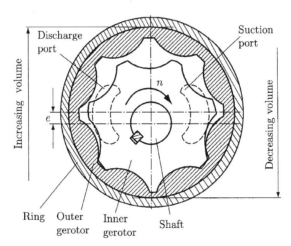

Figure 2.22
Gerotor pump.

than the external gear, which causes its axis to rotate about the axis of the external gear. The geometry is such that on one side of the internal gear the space between the inner and outer gerotor increases for one-half of each rotation and on the other side it decreases for the remaining half of the rotation. It consists of three basic parts: the ring, the outer gerotor, and the inner gerotor. The number of the teeth varies, but the outer gerotor always has one more tooth than the inner gerotor.

The figure shows the two kidney-shaped ports, namely, the suction port and discharge port. The axis around which the inner element rotates is offset by the amount *e* from the axis of the outer gerotor, which, driven by the inner gerotor, rotates within the ring.

2.8.3 VANE PUMPS

Figure 2.23a shows the sketch of a vane pump. The drive shaft center line is displaced from the housing center line, having uniformly spaced vanes mounted in radial slots so that the vanes can move radially inward and outward to always maintain contact with the housing. Fluid enters through port plates, shown in Fig. 2.23b, at each end of the housing. The advantages of vane pumps over gear pumps are that they can provide higher pressures and variable output without the need to control the speed of the prime mover (electric motor, diesel engine, etc). The design modification required for a variable volume output from a pump having a circular interior cross section is that of mounting the housing between end plates so that the axis of the cylinder in which the vane rotates may be shifted relative to the axis of the rotor, as shown in Fig. 2.24. The maximum flow is obtained when they are displaced by the maximum distance (Fig. 2.24a), and zero flow is obtained when the axis of the rotor and the housing tend to coincide (Fig. 2.24b).

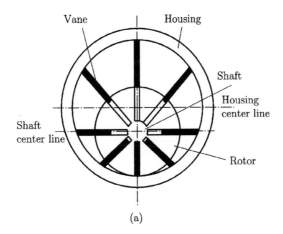

(a)

Figure 2.23
*Vane pump. (a)
Cross section
shown
schematically.
(b) Port plate
with inlet port
and outlet port.*

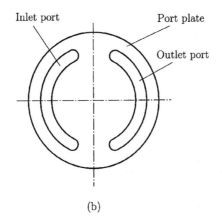

(b)

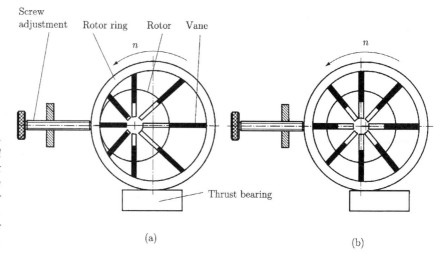

Figure 2.24
Cross-sectional schematic variable vane pump (a) for maximum flow, and (b) for minimum flow.

2.8.4 AXIAL PISTON PUMP

An axial piston pump is shown in Fig. 2.25. The major components are the swashplate, the axial pistons with shoes, the cylinder barrel, the shoeplate, the shoeplate bias spring, and the port plate. The shoeplate and the shoeplate bias spring hold the pistons against the swashplate, which is held stationary while the cylinder barrel is rotated by the prime mover. The cylinder, the shoeplate, and the bias spring rotate with the input shaft, thus forcing the pistons to move back and forth in their respective cylinders in the cylinder barrel. The input and output flows are separated by the stationary port plate with its kidney-shaped ports. Output volume may be controlled by changing the angle of the swashplate. As angle α between the normal to the swashplate and the axis of the drive shaft in Fig. 2.26b goes to zero, the flow volume decreases. If angle α increases (Fig. 2.26a), the volume also increases. Axial piston pumps with this feature are known as overcenter axial piston pumps.

2.8.5 PRESSURE-COMPENSATED AXIAL PISTON PUMPS

For these pumps the angle α of the swashplate is controlled by a spring-loaded piston that senses the pressure at a selected point in the hydraulic system. As the pressure increases, the piston can decrease α in an effort to decrease the system pressure, as illustrated in Fig. 2.26b. Pressure compensation is often used with overcenter axial piston pumps in hydrostatic transmissions to control the rotational speed and direction of hydraulic motors.

2.9 Hydraulic Motors

Hydraulic motors differ from pumps because they can be designed to rotate in either direction, can have different seals to sustain high pressure at low rpm, or can have different bearings to withstand large transverse loads so

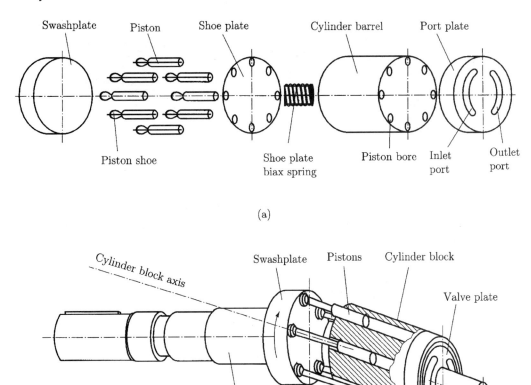

Figure 2.25 *Axial piston pump. (a) Overcenter axial pump without drive shaft shown. (b) Basic parts for axial piston pump.*

they can drive sprockets, gears, or road wheels on vehicles. A rotating valve that distributes the pressure to the pistons in sequence causes the output shaft to rotate in the desired direction. The pistons are mounted in a block that holds the pistons perpendicular to the rotor. Each piston slides laterally on a flat surface inside the housing as it applies a force between the flat portion of the housing and the eccentric rotor. Figure 2.27 shows a radial piston pump with the pistons, 2, arranged radially around the rotor hub, 1. The rotor with the cylinders and the pistons are mounted with an eccentricity in the pump house 3. The pistons, which can slide within the cylinders with a special seal system, pull and then push the fluid (the arrows on the figure) through a central valve 4.

Orientation of the block relative to the housing is maintained by means of an Oldham coupling. The schematic principle of operation of an Oldham coupling is presented in Fig. 2.28. The main parts are the end plate 1, coupling plate 2, and block 3, which contains the pistons and the eccentric

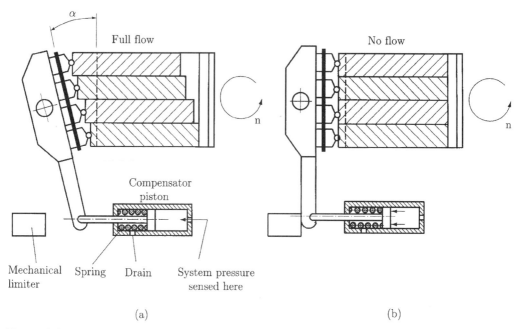

Figure 2.26 *Simplified schematic of the operation of the compensator piston in controlling the angle of the swashplate to control output flow rate. (a) Large displacement for full flow. (b) Zero displacement for no flow.*

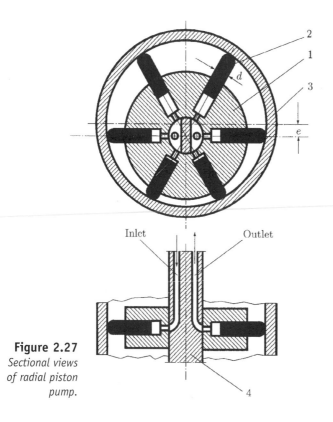

Figure 2.27
*Sectional views
of radial piston
pump.*

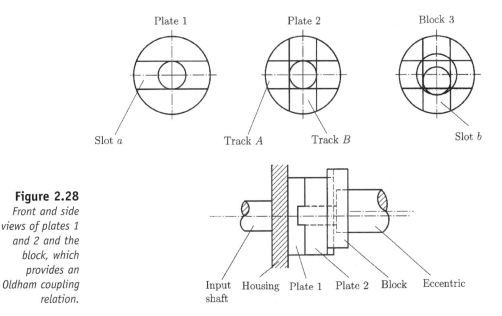

Figure 2.28
Front and side views of plates 1 and 2 and the block, which provides an Oldham coupling relation.

portion of the shaft. Slot *a* is cut into plate 1 and accepts track *A*, which is part of plate 2. Track *B* is perpendicular to track *A* and is located on the opposite side of plate 2 from track *A*. Slot *b* is cut into the block and accepts track *B*. Thus, any displacement of the block relative to plate 1, which is attached to the housing, can be decomposed into components parallel to tracks *A* and *B*. As the shaft turns the center of the eccentric and of the block, it will describe a circle about the center of the housing, but the block itself will not rotate. Pistons, block, and housing, therefore, will always maintain their proper orientation relative to one another.

2.10 Accumulators

An accumulator is a tank that accumulates and holds fluid under pressure. Accumulators are used to maintain the pressure in the presence of fluctuating flow volume, to absorb the shock when pistons are abruptly loaded or stopped, as in the case of planers, rock crushers, or pressure rollers, or to supplement pump delivery in circuits where fluid can be stored during other parts of the cycle. The bladder-type accumulator is presented in Fig. 2.29a. This design incorporates a one-piece cylindrical shell with semicircular ends to better withstand system pressure. A rubber bladder or separator bag is installed inside the outer shell. It is this bladder that, when filled with a gas precharge, supplies the energy to expel stored liquid at the desired time. A poppet valve is supplied in the lower end of the accumulator to prevent the bladder from being damaged by entering the fluid port assembly. This poppet is held open by a spring but is closed once the accumulator precharge extends the bladder and causes it to contact the top of the

Fluid Dynamics

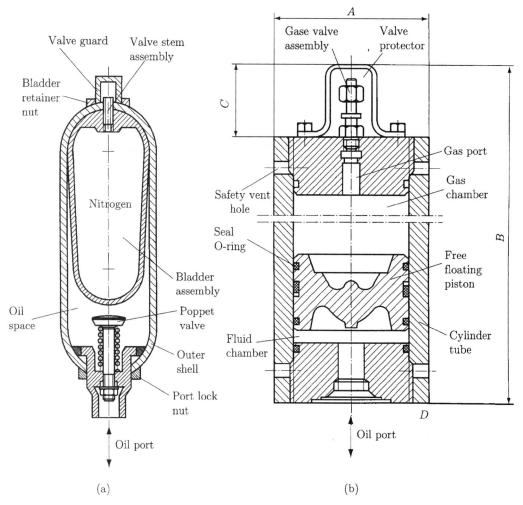

Figure 2.29 *Internal construction of accumulators. (a) Bladder-type accumulator. (b) Piston-type accumulator.*

poppet. The gas used for accumulator service is an inert gas such as nitrogen. Also, a small valve is used to fill the bladder with gas. This valve is similar to those used to fill auto tires. A locknut is provided to anchor this valve and the bladder to the shell.

The piston-type accumulator is shown in Fig. 2.29b. Note that the piston type resembles a conventional hydraulic cylinder minus the piston rod. This configuration can be identified as free or floating piston operation. The gas precharge is on one side of the piston, and the system oil is on the other. Two seals are indicated on the piston head. Therefore, the two rings keep the piston head from cocking, but actual sealing is accomplished by the seal ring on the gas side of the piston head.

2.11 Accumulator Sizing

Accumulator size depends on the amount of fluid to be stored and the means used to supply pressure to the fluid stored in the accumulator. If a weight above a piston is used, the accumulator must be large enough to hold the fluid and the volume of the weight and piston. When gas pressure is used, either in a bladder or above a piston, the sizing of the accumulator requires that we consider the behavior of the gas as it is being compressed by the incoming fluid. The gas equation is considered to be polytropic and includes isothermal and reversible adiabatic changes as special cases if the appropriate value of the exponent is selected. An isothermal process is one in which the compression is slow enough for the temperature of gas to remain constant. An adiabatic process is one that is so rapid that no heat is lost and the temperature rises accordingly. The polytropic gas equation is

$$pV^n = p_o V_o^n. \tag{2.14}$$

If the accumulator volume is V_f and pressure is p_f, when it is filled with the desired amount of fluid, and V_e and p_e are the volume and pressure when it is empty, the volume of stored fluid in V_s is

$$V_s = V_e - V_f. \tag{2.15}$$

From Eq. (2.14) it follows that

$$p_f V_f^n = p_e V_e^n, \tag{2.16}$$

so that upon solving Eq. (2.16) for V_f and substituting into Eq. (1.15), the required volume V_e of the accumulator is given by

$$V_e = \frac{V_s}{1 - \left(\dfrac{p_e}{p_f}\right)^{1/n}} \beta \tag{2.17}$$

where the value of n is given in Fig. 2.30 and β is an experimentally measured factor given by

$$\beta = 1.24, \text{ for bladder-type accumulators}$$

$$\beta = 1.11, \text{ for piston-type accumulators}$$

2.12 Fluid Power Transmitted

For calculating the power transmitted to a particular unit, it is necessary to know the functional formula for power,

$$P = Fv, \tag{2.18}$$

where F is the force, and v is the velocity.

The force F can be written as

$$F = pA, \tag{2.19}$$

where p is the pressure and A is the cross-sectional area.

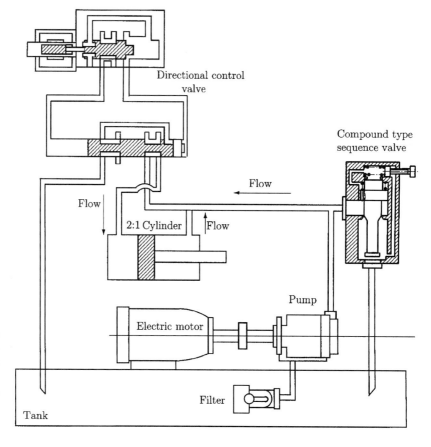

Figure 2.30
Schematic for regenerative cylinder circuit. (Adapted from Orthwein, 1990.)

If L is the distance traveled in time t by a point that moves with the fluid flowing through the hose or cylinder of cross-sectional area A, the power required to move that fluid is

$$P = pA\frac{\partial L}{\partial t} = pAv = p\frac{\partial V}{\partial t}. \tag{2.20}$$

The rate of change of time $\partial V/\partial t$ is denoted by $\dot{Q}$ in units of gallons per minute or liters per minute.

1 gallon (1 gal) $= 231$ in^3, and 1 horsepower (hp) $=$ (ft lb/min)/33,000.

2.13 Piston Acceleration and Deceleration

To analyze piston behavior, we consider piston velocity and acceleration as a function of the system parameters.

The velocity is the volumetric flow rate divided by the cross-sectional area A_c of the cylinder. Thus,

$$v_r = \frac{\dot{Q}}{A_c}, \tag{2.21}$$

where v_r denotes the velocity of the piston and rod.

During the motion, the rod and piston are accelerating, and the equilibrium equation is

$$m = \frac{d^2x}{dt^2} = pA_c - (F_r + f),$$ (2.22)

where F_r is the force opposing the motion of the piston and f is the friction and fluid losses at the exhaust ports.

If the pressure is constant, which is a reasonable approximation if the lines and fittings are large enough to produce only negligible pressure losses, one may integrate Eq. (2.22) to get

$$\frac{dx}{dt} = \frac{pA_c - (F_r + f)}{m} t$$ (2.23)

when the piston starts from rest.

When set equal to the piston maximum steady-state velocity v_r, the time needed to accelerate to velocity v_r is

$$t_a = \frac{mv_r}{pA_c - (F_r + f)}.$$ (2.24)

The distance required for the piston to reach this velocity may be calculated by integrating Eq. (2.23) with respect to time and using the condition that the motion started from $x = 0$ to obtain

$$x_a = \frac{mv_r^2}{2[pA_c - (F_r + f)]} = \frac{t_a v_r}{2},$$ (2.25)

Here m is the total accelerating mass; that is, the piston, the rod, and any mass being accelerated by the piston and rod. If the stroke of the cylinder is less than x_a, the piston will accelerate over the entire stroke.

The time needed by the piston to accelerate, move at constant velocity, and decelerate may be estimated by using the relationship

$$t_1 = t_a + t_d + \frac{s - (x_a + x_d)}{v_r},$$ (2.26)

where s is the stroke length, t_d is the deceleration time, and x_d is the deceleration distance.

Hydraulic pistons usually move relatively slowly because the ratio between hose and cylinder cross-sectional areas is generally small and because large line losses are associated with large velocities. Acceleration times and distances are negligible.

2.14 Standard Hydraulic Symbols

The time and effort to draw and modify design drawings for hydraulic systems (Fig. 2.30) can be greatly reduced by employing a set of standard design symbols to denote hydraulic components. Two different conventions have been accepted for joining and crossing hydraulic lines. They are presented in Fig. 2.31a. The main hydraulic lines are drawn as solid lines; pilot lines are drawn as long dashes; exhaust and drain line are drawn as short dashes. Check valves are drawn as in Fig. 2.31b, where flow is allowed from A to B, but not from B to A. Figures 2.32 and 2.33 show other activation symbol, and Fig. 2.34 shows the symbolic circuit for the regenerative cylinder, initially shown in Fig. 2.30.

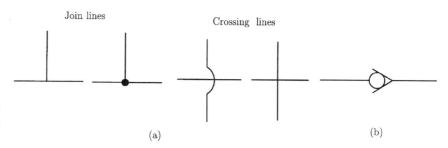

Figure 2.31
(a) Joining and (b) crossing hydraulic lines.

2.15 Filters

Filters are used in hydraulic circuits to remove foreign matter without adding appreciably to pressure loss in the circuit. Normally only one filter is added to most hydraulic systems on machines, unless a particular component is especially sensitive to dirt and must have extra protection. Filters are added in the reservoir, in the pump intake from the reservoir, or in the return line to the reservoir. Motivation for the first two choices is that the pump is usually the most expensive single component in the system and that foreign matter tends to collect in the reservoir because the flow velocity is low — it acts as a settling tank. The disadvantage of this location is that if the filter becomes clogged, it can starve the pump and cause extensive damage. This possibility may be largely eliminated by placing the filter in the return line, but with the risk of damaging the filter by forcing large particles through it. A pilot-operated check valve may be used to bypass a clogged filter, but at the expense of circulating foreign matter that should have been filtered out. Another alternative is to stop the system when the pressure across a filter exceeds a limiting value. Particulate matter is described in terms of its largest dimension in micrometers (microns), where a micrometer is 1×10^{-6} meters. Filters are classified in terms of their ability to entrap these particles by means of a β value. The symbol β is immediately followed by a number that denotes the diameter of the particles involved according to the relation

$$\beta_d = \frac{\text{Number of particles of diameter } d \text{ upstream from filter}}{\text{Numbers of particles of diameter } d \text{ downstream from the filter}}.$$

Most fluid filters are not rated for particles less than 3 μm; β may be taken as 1.0 to $d < 3$.

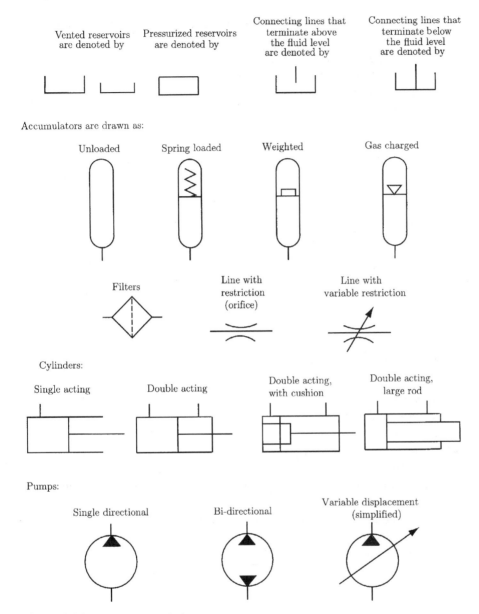

Figure 2.32 *Activation symbols.*

2.16 Representative Hydraulic System

A simple hydraulic circuit to provide bidirectional control of a hydraulic cylinder is shown in Fig. 2.35. It includes a motor with a clutch between the motor and pump, a filter in the motor intake lane from the reservoir, and a manually operated directional valve.

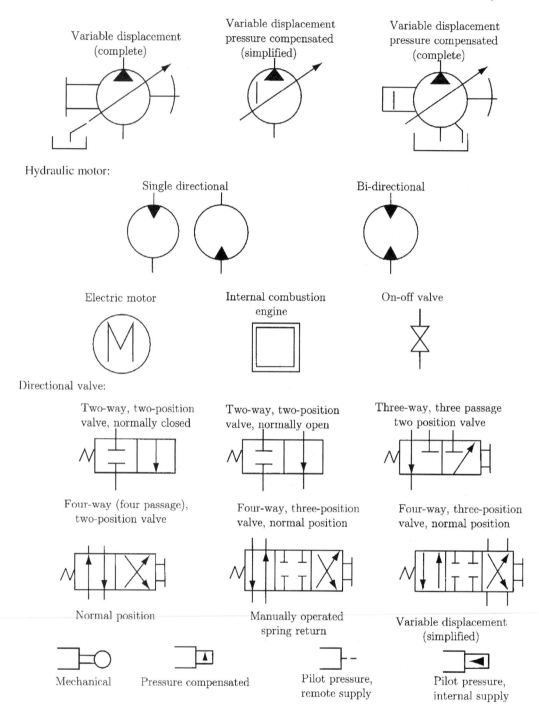

Figure 2.33 *Other activation symbols.*

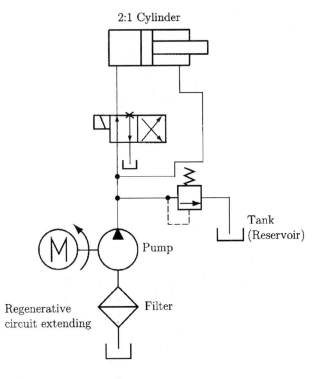

Figure 2.34
Symbolic circuit for regenerative cylinder circuit schematic shown in Fig. 2.30. (Adapted from Orthwein, 1990.)

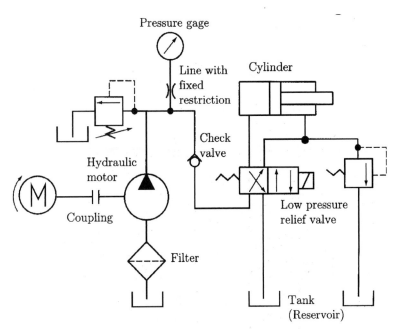

Figure 2.35
Representative hydraulic circuit.

References

1. R. W. Fox and A. T. McDonald, *Introduction to Fluid Mechanics.* John Wiley & Sons, New York, 1998.

2. R. V. Giles, J. B. Evert, and C. Liu, *Fluid Mechanics and Hydraulics.* McGraw-Hill, New York, 1994.

3. R. M. White, *Fluid Mechanics.* McGraw-Hill, New York, 1999.

4. W. C. Orthwein, *Machine Component Design.* West Publishing Company, St. Paul, MN, 1990.

5. J. J. Pippenger and T. G. Hicks, *Industrial Hydraulics*, 2nd ed. McGraw-Hill, New York, 1970.

6. E. F. Brater and H. W. King, *Handbook of Hydraulics.* McGraw-Hill, New York, 1980.

7. M. E. Walter Ernst, *Oil Hydraulic Power and Its Industrial Applications.* McGraw-Hill, New York, 1949.

8. R. P. Lambeck, *Hydraulic Pumps and Motors: Selection and Application for Hydraulic Power Control Systems.* Dekker, New York, 1983.

9. R. P. Benedict, *Fundamentals of Pipe Flow.* Wiley & Sons, New York, 1977.

10. W. G. Holzbock, *Hydraulic Power and Equipment.* Industrial Press, New York, 1968.

11. L. S. McNickle, Jr., *Simplified Hydraulics.* McGraw-Hill, New York, 1966.

9 Control

MIRCEA IVANESCU

*Department of Electrical Engineering,
University of Craiova, Craiova 1100, Romania*

Inside

This chapter, "Control," is an introduction to automation for technical students and engineers who will install, repair, or develop automatic systems in an industrial environment. It is intended for use in engineering technology programs at the postsecondary level, but it is also suitable for use in industrial technology curricula, as well as for in-service industrial training programs. Industrial managers, application engineers, and production personnel who want to become familiar with control systems or to use them in production facilities will find this chapter useful.

The text requires an understanding of the principles of mechanics and fluid power, as well as a familiarity with the basics of mathematics. Although not essential, a good knowledge of the principles of physics is also helpful.

1. Introduction

Control engineering is concerned with the automation of processes in order to provide useful economic products. These processes can be conventional systems, such as chemical, mechanical, or electrical systems, or modern complex systems such as traffic-control and robotic systems. Control engineering is based on the foundation of feedback theory and linear system analysis. The aim of the control system is to provide a desired system response.

In order to be controlled, a process can be represented by a block, as shown in Fig. 1.1. The input–output relationship represents the cause and effect relationship of the process. The simplest system of automation is the open-loop control system, which consists of a controller that provides the input size for the process (Fig. 1.2).

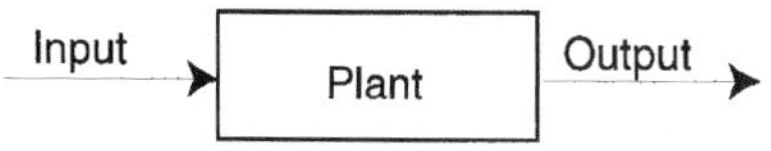

Figure 1.1
Plant.

Input → Plant → Output

Figure 1.2
*Open-loop
control system.*

A closed-loop control system (Fig. 1.3) uses a feedback signal consisting of the actual value of the output. This value is compared with the prescribed or desired input, and the result of the comparison defines the system error. This size is amplified and used to control the process by a controller. The controller acts in order to reduce the error between the desired input and the actual output. Moreover, several quality criteria are imposed in order to obtain a good evolution of the global system.

1.1 A Classic Example

One of the most popular examples of a feedback control system is the water-level float regulator (Fig. 1.4). The output of the system is defined by the

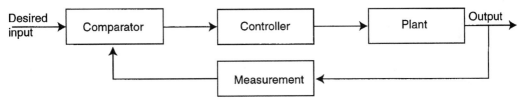

Figure 1.3   *Closed-loop control system.*

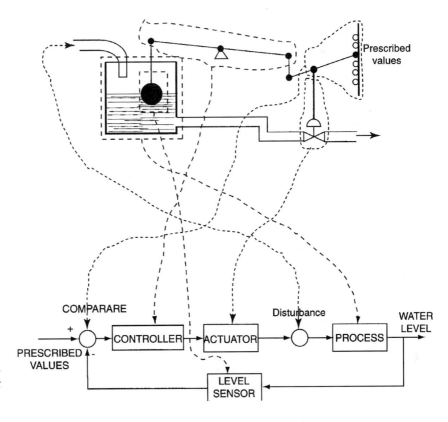

Figure 1.4
*Water-level float
regulator.*

water level measured by a float that controls, by a mechanical system, the valve that, in turn, controls the water flow out of the tank. The plant is represented by the water tank, and the water flowing into the tank represents a disturbance of the system. If the water level increases, the float moves up and, by a mechanical system representing the controller, initiates the opening of the valve. If the water level decreases, the float moves down and initiates the closing of the valve. The size of the closing or opening can be adjusted from a reference panel that provides the prescribed values. The system operates as a negative feedback control system because a difference is obtained between the prescribed value and the output; the water level and the variations of the water level are eliminated by compensation through the valve functions.

This system represents one of the simplest control systems. Effort is necessary in order to eliminate transient oscillations and to increase the accuracy of the control system.

2. Signals

The differential equations associated with the components of control systems (Appendix A) indicate that the time evolution of the output variable $x_0(t)$ is a function of the input variable $x_i(t)$. In order to obtain the main characteristics of these elements it is necessary to use standard input signals.

(a) *The impulse function* $\delta(t)$: The unit impulse is based on a rectangular function $f(t)$ such as

$$f_\varepsilon(t) = \begin{cases} 1/\varepsilon, & 0 \leq t \leq \varepsilon \\ 0, & t > \varepsilon, \end{cases} \tag{2.1}$$

where $\varepsilon > 0$. As ε approaches zero, the function $f_\varepsilon(t)$ approaches the impulse function $\delta(t)$, where

$$\int_0^\infty \delta(t)\,dt = 1 \tag{2.2}$$

$$\int_0^\infty \delta(t - \alpha)f(t)\,dt = f(\alpha). \tag{2.3}$$

The impulse input is useful when one considers the convolution integral for an output $x_0(t)$ in terms of an input $x_i(t)$,

$$x_0(t) = \int_0^t h(t - \tau)x_i(\tau)\,d\tau = \mathscr{L}^{-\infty}\{\mathscr{H}(\textstyle\int)\mathscr{X}_{\backslash}(\textstyle\int)\}, \tag{2.4}$$

where $\mathscr{H}(s)$, $\mathscr{X}_i(s)$ are the Laplace transforms of $h(t)$, $x_i(t)$, respectively (Appendix B). If the input is the impulse function $\delta(t)$,

$$x_i(t) = \delta(t) \tag{2.5}$$

$$x_0(t) = \int_0^t h(t - \tau)\delta(\tau)\,d\tau, \tag{2.6}$$

and we have

$$x_0(t) = b(t). \tag{2.7}$$

(b) *The step input:* This standard signal is defined as

$$x_i(t) = \begin{cases} a, & t > 0 \\ 0, & t < 0. \end{cases} \tag{2.8}$$

The Laplace transform is

$$X_i(s) = \frac{a}{s}. \tag{2.9}$$

This signal is shown in Fig. 2.1.

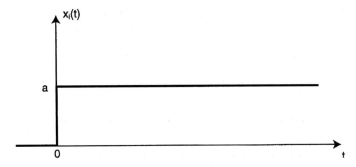

Figure 2.1
Step input.

(c) *The ramp input:* The standard test signal has the form (Fig. 2.2)

$$x_i(t) = \begin{cases} at, & t > 0 \\ 0, & t < 0 \end{cases} \tag{2.10}$$

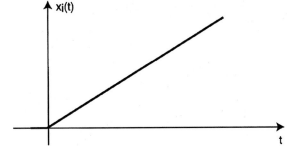

Figure 2.2
Ramp input.

and the Laplace transform

$$X_i(s) = \frac{a}{s^2}. \tag{2.11}$$

(d) *The sinusoidal input signal:* The standard form is

$$x_i(t) = a \sin \omega t, \quad \omega = \frac{2\pi}{T}, \tag{2.12}$$

where a is the amplitude, ω is the frequency of the signal (Fig. 2.3), and T is the period. This signal is used when we analyze the response of the system when the frequency of the sinusoid is varied. So, several performance measures for the frequency response of a system. The Laplace transform is

$$X_i(s) = \frac{a\omega}{s^2 + \omega^2}. \qquad (2.13)$$

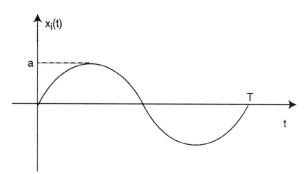

Figure 2.3
Sinusoidal input.

3. Transfer Functions

All the elements of the control system are unidirectional information trans-mission elements:

$$Y(s) = \frac{X_o(s)}{X_i(s)}. \qquad (3.1)$$

The transfer function of the linear system is defined as the ratio of the Laplace transform of the output variable $X_o(s)$ to the Laplace transform of the input variable $X_i(s)$, with all the initial conditions assumed to be zero. The transform function is defined only for linear and constant parameter systems.

3.1 Transfer Functions for Standard Elements

Transfer functions for standard elements are the following:

(a) *Proportional element.* For this element, the output is proportional to the input,

$$x_o(t) = K\, x_i(t) \qquad (3.2)$$
$$X_o(s) = K\, X_i(s), \qquad (3.3)$$

and the transfer function (Fig. 3.1) will be

$$Y(s) = \frac{X_o(s)}{X_i(s)} = K. \qquad (3.4)$$

Figure 3.1
*Transfer
function.*

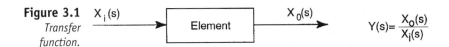

$$Y(s) = \frac{X_0(s)}{X_i(s)}$$

(b) *Integrating element.* The output is defined by the integral of the input,

$$x_o(t) = K_I \int x_i(t)\,dt, \tag{3.5}$$

or

$$\dot{x}_o(t) = K_I x_i(t), \tag{3.6}$$

$$sX_o(s) = K_I X_i(s) \tag{3.7}$$

and the transfer function

$$Y(s) = \frac{X_o(s)}{X_i(s)} = \frac{K_I}{s}. \tag{3.8}$$

(c) *Differentiating element.* This element is defined by

$$x_o(t) = K_D \dot{x}_i(t) \tag{3.9}$$

$$X_o(s) = K_D s X_i(s) \tag{3.10}$$

with the transfer function

$$Y(s) = \frac{X_o(s)}{X_i(s)} = K_D s. \tag{3.11}$$

(d) *Mixed elements.* From standard elements, we can obtain new transfer functions:

■ Proportional-integrating element (PI):

$$(s) = K + \frac{K_I}{s} \tag{3.12}$$

■ Proportional-differentiating element (PD):

$$Y(s) = K + K_D s \tag{3.13}$$

■ Proportional-integrating–differentiating element (PID):

$$Y(s) = K + \frac{K_I}{s} + K_D s \tag{3.14}$$

3.2 Transfer Functions for Classic Systems

We consider the linear spring–mass–damper system described in Appendix A.1, Eq. (A1.2), with zero initial conditions

$$Ms^2 X_o(s) + k_f s X_o(s) + k X_o(s) = A X_i(s). \tag{3.15}$$

The transfer function will be

$$Y(s) = \frac{X_o(s)}{X_i(s)} = \frac{A}{Ms^2 + k_f s + k}. \qquad (3.16)$$

In a similar way, for the linear approximated rotational spring mass damper system, we have

$$Y(s) = \frac{\Theta(s)}{T(s)} = \frac{1}{Js^2 + k_f s + (k + Mgl/2)}. \qquad (3.17)$$

4. Connection of Elements

In order to represent a system with several variables under control, an interconnecting of elements is used. Each element is represented by a block diagram. The block diagram is a unidirectional block that represents the transfer function of its variables. For example, the block diagram of the linear spring–mass–damper element is represented as in Fig. 4.1. This representation defines the relationships between the inputs, the system pressure $p(t)$, and the output, the position $z(t)$.

Figure 4.1
Block diagram of the linear spring–mass–damper element.

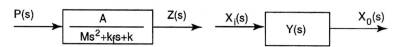

In order to represent a complex system, an interconnecting of blocks is used. This representation offers a better understanding of the contribution of each variable than is possible to obtain directly from differential equations.

(a) *Cascade connection*: In this case, the output of the first element is also the input in the second element (Fig. 4.2),

$$X_{i2}(s) = X_{o1}(s), \qquad (4.1)$$

Figure 4.2
Cascade connection.

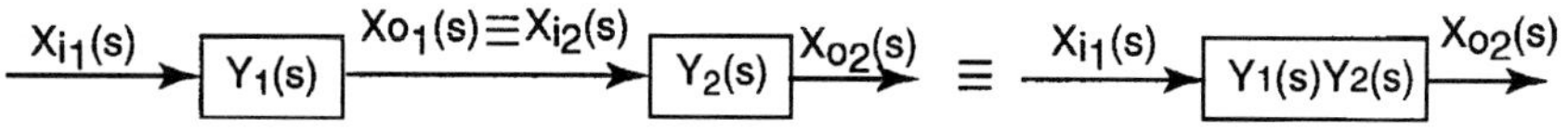

but

$$Y_1(s) = \frac{X_{o1}(s)}{X_{i1}(s)}, \qquad Y_2(s) = \frac{X_{o2}(s)}{X_{i2}(s)}, \qquad (4.2)$$

and for the overall system,

$$Y(s) = \frac{X_o(s)}{X_i(s)} = \frac{X_{o1}(s)}{X_{i1}(s)} \cdot \frac{X_{o2}(s)}{X_{i2}(s)} = Y_1(s) \cdot Y_2(s). \qquad (4.3)$$

(b) *Parallel connection*: For this structure, the input is the same for both elements, and the output is defined by summing of outputs:

$$X_{i1}(s) = X_{i2}(s) = X_i(s) \qquad (4.4)$$
$$X_o(s) = X_{o1}(s) + X_{o2}(s) \qquad (4.5)$$

and

$$Y(s) = \frac{X_o(s)}{X_i(s)} = \frac{X_{o1}(s)}{X_{i1}(s)} + \frac{X_{o2}(s)}{X_{i2}(s)} = Y_1(s) + Y_2(s). \qquad (4.6)$$

(c) *Negative feedback connection*: This system is represented in Fig. 4.3. The relations that describe the system are (Fig. 4.4)

$$E(s) = X_i(s) - X_f(s) \qquad (4.7)$$

$$\frac{X_o(s)}{Y_1(s)} = X_i(s) - X_o(s) \cdot Y_2(s). \qquad (4.8)$$

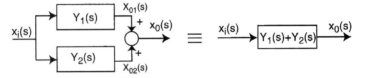

Figure 4.3
Parallel connection.

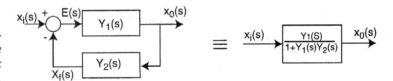

Figure 4.4
Negative feedback connection.

The transfer function will be

$$Y(s) = \frac{X_o(s)}{X_i(s)} = \frac{Y_1(s)}{1 + Y_1(s) \cdot Y_2(s)}. \qquad (4.9)$$

Relation (4.9) is particularly important because it represents the closed-loop transfer function.

(d) *Complex connection*: A complex structure can contain a number of variables under control. This system can be described by a set of equations represented as Laplace transforms:

$$X_{01}(s) = Y_{11}(s) \cdot X_{i1}(s) + Y_{12}(s) \cdot X_{i2}(s) + \cdots + Y_{1m}(s) \cdot X_{im}(s)$$
$$X_{02}(s) = Y_{21}(s) \cdot X_{i1}(s) + Y_{22}(s) \cdot X_{i2}(s) + \cdots + Y_{2m}(s) \cdot X_{im}(s),$$
$$\vdots$$
$$X_{0n}(s) = Y_{n1}(s) \cdot X_{i1}(s) + Y_{n2}(s) \cdot X_{i2}(s) + \cdots + Y_{nm}(s) \cdot X_{im}(s),$$
$$(4.10)$$

where $Y_{ij}(s)$ is the transfer function relating the ith output variable to the jth input variable (Fig. 4.5).

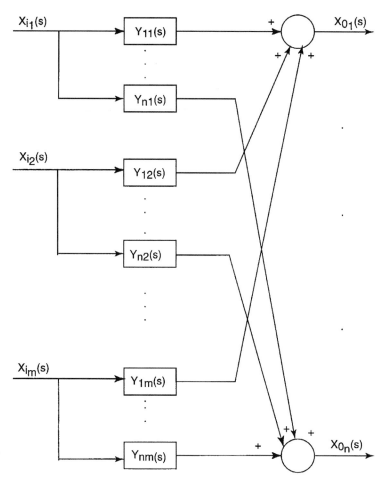

Figure 4.5
Block diagram for a complex structure.

5. Poles and Zeros

In Section 4 we introduced the negative feedback connection as the main connection in the control systems. This connection is also known as a closed-loop connection or closed-loop control system (Fig. 5.2). The presence of feedback assures the improvement of control quality.

An open-loop control system is shown in Fig. 5.1. The main difference between the open- and closed-loop control systems is the generation and utilization of the error signal. The error signal is defined as the difference between the input variable and the feedback variable,

$$e(t) = x_i(t) - x_f(t) \qquad (5.1)$$

$$E(s) = X_i(s) - X_f(s). \qquad (5.2)$$

Figure 5.1

An open-loop control system.

The closed-loop system (Fig. 5.2) operates so that the error is reduced to a minimum value. We consider that a control system with good performance ensures that

$$\lim_{t \to \infty} e(t) = 0. \tag{5.3}$$

We demonstrated that the transfer function for closed-loop systems is

$$Y(s) = \frac{Y_1(s)}{1 + Y_1(s) \cdot Y_2(s)}. \tag{5.4}$$

From (5.2) and (5.4) we easily obtain

$$E(s) = \frac{1}{1 + Y_1(s) \cdot Y_2(s)} \cdot X_i(s), \tag{5.5}$$

which defines the error signal as a function of the input variable.

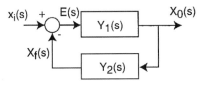

Figure 5.2

A closed-loop control system.

If we consider $Y_2(s) = 1$ (direct negative feedback system),

$$Y(s) = \frac{Y_1(s)}{1 + Y_1(s)}, \tag{5.6}$$

where $Y_1(s)$ is called "the direct-way transfer function" or "open-loop transfer function." This transfer function can be written as

$$Y_1(s) = \frac{Q(s)}{R(s)}, \tag{5.7}$$

where $Q(s)$, $R(s)$ are two polynomials.

In the relation (5.7), the denominator polynomial $R(s)$, when set equal to zero, is called the characteristic equation because the roots of this equation determine the character of the system. The roots of this characteristic equation are called the *poles* or *singularities* of the system.

The roots of the numerator polynomial $Q(s)$ in (5.4) are called the *zeros* of the system. The complex frequency s-plane plot of the poles and zeros graphically portrays the character of the system. $Y_1(s)$ can be rewritten as

$$Y_1(s) = \frac{K(s + z_1)(s + z_2) \cdots (s + z_m)}{s^l(s + p_1)(s + p_2) \cdots (s + p_n)}, \tag{5.8}$$

where K is a constant and $-z_1, -z_2, \ldots, -z_m$ are the zeros of the $Y_1(s), -p_1, -p_2, \ldots, -p_n$ are the poles, and $p = 0$ represents an l-order pole of $Y_1(s)$.

If $l = 0$, relation (5.8) defines a type-zero transfer function; for $l = 1$, we have a type-one transfer function; etc. For example, let us consider the translational mechanism in Fig. 5.3. The dynamic model is defined by the equation

$$F(t) = G + m\ddot{z}(t) + k_f\dot{z}(t), \tag{5.9}$$

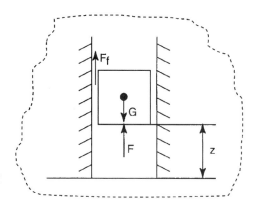

Figure 5.3
Translational mechanism.

where k_f is the friction constant, F is the input variable, and z is the output variable. If we neglect the component G, the transfer function for this element can be considered as

$$Y_m(s) = \frac{Z(s)}{F(s)} = \frac{1}{s(ms + k_f)}. \tag{5.10}$$

The relation (5.10) represents a type-one transfer function in which we find two poles,

$$p_1 = 0$$
$$p_2 = -\frac{k_f}{m}. \tag{5.11}$$

A simple closed-loop control system can be introduced, in which the direct way contains a power amplifier of the force F (Fig. 5.4) with a gain factor. The function for the direct way is

$$Y_1(s) = \frac{Z(s)}{E(s)} = \frac{k_G}{s(ms + k_f)}, \tag{5.12}$$

Figure 5.4
Closed-loop control for a translational mechanism.

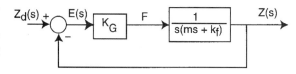

and the transfer function for the closed-loop system will be

$$Y(s) = \frac{k_G}{ms^2 + k_f s + k_G}.$$ (5.13)

6. Steady-State Error

The steady-state error is the error generated after the transient response (determined by the input variation or a disturbance signal) has decayed, leaving only the continuous response.

6.1 Input Variation Steady-State Error

Let us consider the closed-loop control system from Fig. 5.2, in which we assume $Y_2(s) = 1$ (the direct negative feedback connection, Fig. 6.1). From Eq. (5.5) we obtain

$$E(s) = \frac{1}{1 + Y_1(s)} \cdot X_i(s).$$ (6.1)

Figure 6.1
Direct negative feedback connection.

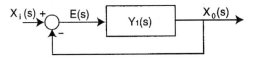

In order to calculate the steady-error E_s, we use the final-value theorem (Appendix A.2),

$$E_s = \lim_{t \to \infty} e(t) = \lim_{s \to 0} [sE(s)].$$ (6.2)

From Eq. (6.1), results

$$E_s = \lim_{s \to 0} \left[s \frac{1}{1 + Y_1(s)} X_i(s) \right].$$ (6.3)

If we consider a unit step input, $X_i(s) = 1/s$,

$$E_s = \lim_{s \to 0} \left[\frac{1}{1 + Y_1(s)} \right].$$ (6.4)

We consider $Y_1(s)$ in general form Eq. (5.8). For a type-zero system, the steady-state error is

$$E_s = \frac{1}{1 + Y_1(0)}, \tag{6.5}$$

$$E_s = \frac{1}{1 + \dfrac{K \prod\limits_{i=1}^{m} z_i}{\prod\limits_{i=1}^{n} p_i}}. \tag{6.6}$$

For a type-one system or more ($l \geq 1$),

$$Y_1(0) = \infty, \tag{6.7}$$

and

$$E_s = 0. \tag{6.8}$$

For a ramp input, $X_i(s) = 1/s^2$, we have

$$E_s = \lim_{s \to 0} \left[s \frac{1}{1 + Y_1(s)} \cdot \frac{1}{s^2} \right] \tag{6.9}$$

$$E_s = \lim_{s \to 0} \frac{1}{s Y_1(s)}. \tag{6.10}$$

If $Y_1(s)$ defines a type-zero system ($l = 0$),

$$E_s = \infty. \tag{6.11}$$

For a type-one system,

$$E_s = \frac{\prod\limits_{i=1}^{n} p_i}{K \prod\limits_{i=1}^{m} z_i}. \tag{6.12}$$

If $l > 1$,

$$E_s = 0. \tag{6.13}$$

6.2 Disturbance Signal Steady-State Error

A disturbance signal $d(t)$ is an unwanted signal that affects the system's output signal. Feedback in control systems is used to reduce the effect of disturbance input.

We will consider the closed-loop control system from Fig. 6.1, and we assume that the disturbance signal is applied on the direct path of the system (Fig. 6.2). We consider that the disturbance signal point allows the decomposition of the transfer function $Y_1(s)$ in two blocks, $Y_1'(s)$, $Y_1''(s)$. Let us

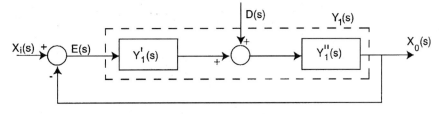

Figure 6.2
Disturbance in the control system.

assume the case in which a steady state for $x_i(t) = 0$, $d(t) = 0$ is obtained. Of course, $x_o(t) = 0$,

$$X_i(s) = 0$$
$$D(s) = 0 \tag{6.14}$$
$$X_0(s) = 0.$$

Now, we consider a disturbance signal (the input signal is $X_i(s) = 0$),

$$X_o(s) = Y_1'''(s)D(s) + Y_1'(s) \cdot Y_1''(s)E(s), \tag{6.15}$$

where

$$E_o(s) = -X_o(s) \tag{6.16}$$
$$Y_1'(s) \cdot Y_1''(s) = Y_1(s) \tag{6.17}$$

From Eqs. (6.15)–(6.17), we obtain

$$X_o(s) = \frac{Y_1'''(s)}{1 + Y_1(s)} D(s). \tag{6.18}$$

We can define the transfer function due to the disturbance as

$$Y_{OD}(s) = \frac{Y_1'''(s)}{1 + Y_1(s)}. \tag{6.19}$$

The steady-state error signal, in this case, will be $(x_i(t) = 0)$

$$|e_D(t)| = |x_o(t)| \tag{6.20}$$

$$E_D(s) = \frac{Y_1'''(s)}{1 + Y_1(s)} D(s). \tag{6.21}$$

If the disturbance signal is a unit step, $D(s) = 1/s$, we obtain

$$E_D(s) = \frac{1}{s} \frac{Y_1'''(s)}{1 + Y_1(s)}. \tag{6.22}$$

If we consider that $Y_1'''(s)$ is a type-zero transfer function and $Y_1(s)$ is a type-one transfer function, we obtain

$$E_{DS} = \lim_{t \to \infty} e_D(t) = \lim_{s \to 0} [sE_D(s)] \tag{6.23}$$

$$E_{DS} = \lim_{s \to 0} \frac{Y_1'''(s)}{1 + Y_1(s)}, \tag{6.24}$$

but

$$\lim_{s \to 0} Y_1(s) = \infty, \tag{6.25}$$

so that

$$E_{DS} = 0. \tag{6.26}$$

If we consider that $Y_1''(s)$ is a type-one transfer function, from Eq. (6.24) we obtain

$$E_{DS} = \lim_{s \to 0} \frac{Y_1''(s)}{1 + Y_1'(s) \cdot Y_1''(s)} \tag{6.27}$$

or

$$E_{DS} = \lim_{s \to 0} \frac{1}{\dfrac{1}{Y_1''(s)} + Y_1'(s)}, \tag{6.28}$$

but

$$\lim_{s \to 0} \frac{1}{Y_1''(s)} = 0, \tag{6.29}$$

and, from Eq. (6.28),

$$E_{DS} = \lim_{s \to 0} \frac{1}{Y_1'(s)}. \tag{6.30}$$

In order to obtain $E_{DS} = 0$ it is necessary that

$$\lim_{s \to 0} Y_1'(s) = \infty. \tag{6.31}$$

Relation (6.31) shows that the disturbance steady-state error approaches zero if $Y_1'(s)$ represents a type-one transfer function.

The transient response for a unit step disturbance signal is represented in Fig. 6.3. If $Y_1'(s)$ is a type-zero transfer function, then $E_{DS} \neq 0$. (Fig. 6.4).

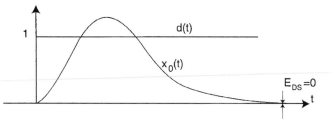

Figure 6.3
Transient response for a type-one transfer function $Y_1'(s)$.

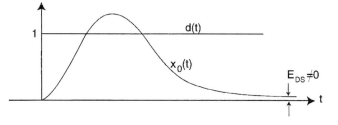

Figure 6.4
Transient response for a type-zero transfer function $Y_i'(s)$.

For example, let us reconsider the translational mechanism presented in Fig. 5.3 and 5.4. If we neglect the gravitational component, the system defines a type-one transfer function

$$Y_1(s) = \frac{k_G}{s(ms + k_f)},$$ (6.32)

and the steady-state error will be

$$E(s) = \lim_{s \to 0} \left[s \frac{s(ms + k_f)}{ms^2 + k_f s + k_G} \cdot Z_d(s) \right].$$ (6.33)

For the step input,

$$z_d(t) = z_d^*$$
$$Z_d(s) = \frac{z_d^*}{s},$$ (6.34)

we obtain

$$E(s) = 0,$$ (6.35)

and for the ramp input,

$$z_d(t) = k_d t$$
$$Z_d(s) = \frac{k_d}{s^2}.$$ (6.36)

From (6.33) results

$$E(s) = \frac{k_d k_f}{k_G}.$$ (6.37)

We now reconsider the dynamic model (5.9) in which we do not neglect the gravitational component G:

$$F(t) = G + m\ddot{z}(t) + k_f \dot{z}(t).$$ (6.38)

If we use the Laplace transform, we obtain

$$Z(s) = \frac{1}{s(ms + k_f)} F(s) - \frac{1}{s(ms + k_f)} G(s).$$ (6.39)

We can remark that the gravitational component can be interpreted as a disturbance of the system (Fig. 6.5).

Figure 6.5
Disturbance in the closed-loop control for translational mechanism.

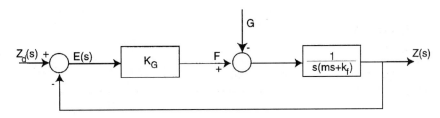

For an input $z_d = 0$, from Eqs. (6.20) and (6.21) we obtain

$$E_D(s) = -\frac{1}{ms^2 + k_f s + k_G}\, G(s), \qquad (6.40)$$

but $G(s) = G/s$ and from Eqs. (6.24) and (6.40),

$$|E_{DS}| = \frac{G}{k_G}. \qquad (6.41)$$

7. Time-Domain Performance

The time-domain performance is important because control systems are time-domain systems. This means that time performance is the most important performance for control systems. Time-domain performance is usually defined in terms of the response of a system to the test input signals: step unit or ramp input. The standard step response of a control is shown in Fig. 7.1.

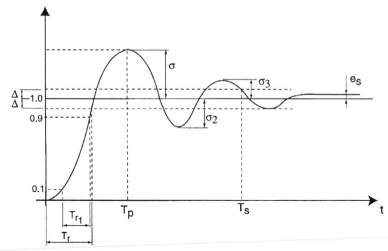

Figure 7.1
Unit step response of a system.

We can define the following parameters:

(a) *The overshoot* is defined as the difference between the peak value M of the time response and the standard input,

$$\sigma = M - 1, \qquad (7.1)$$

or, more generally,

$$\sigma = M - x_d^*, \qquad (7.2)$$

where x_d^* is the amplitude of the step input. Frequently, this parameter is defined as the percent overshoot,

$$\sigma_p = \frac{M - x_d^*}{x_d^*} \cdot 100. \tag{7.3}$$

(b) *The peak time* is defined as the time required for the system to reach the peak value.

(c) *The rise time* T_r is the time required for the system to reach the input value for the first time. These parameters are normally used for underdamped systems. For overdamped systems another parameter T_{r_1} is used that represents the rise time between $0.1x_d^*$ and $0.9x_d^*$.

(d) *The settling time* T_s is defined as the time required for the system to settle within a certain percentage Δ of the input amplitude x_d^*.

(e) *The damping order* δ is defined as

$$\delta = 1 - \frac{\sigma_3}{\sigma}, \tag{7.4}$$

where σ_3 is the amplitude of the third oscillation (Fig. 7.1). In order to obtain good performance it is necessary that

$$\delta \geq \delta_p, \tag{7.5}$$

where δ_p is the prescribed value (of course, $\delta_p \leq 1$).

All these parameters are defined in terms of the response of the system to the test input. The same procedure can be used if we analyze the response of the system to the disturbance test (Fig. 7.2).

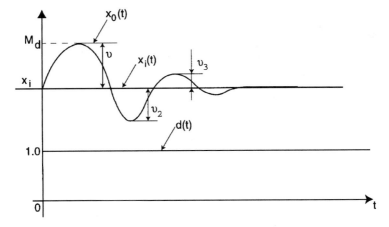

Figure 7.2
Unit step disturbance response of a system.

(f) *The disturbance overshoot* is defined as the difference between the peak value M_d of the output signal and the input (constant value) when a unit step disturbance [$d(t) = 1$] is applied:

$$\vartheta = M_d - x_i. \tag{7.6}$$

(g) *The disturbance damping order* δ is defined as

$$\delta = 1 - \frac{\vartheta_3}{\vartheta}, \tag{7.7}$$

where ϑ_3 is the third amplitude of the oscillation for a unit step disturbed system.

It is very important for the control system analysis to establish the relationship between the representation of a linear system in terms of the location of the poles and zeros of its transfer function, and its time-domain response to a unit step. This relationship will be developed in the following sections, but now we will try to present this problem.

Let us consider a closed-loop system with the transfer function

$$Y(s) = \frac{X_0(s)}{X_i(s)} = \frac{Y_1(s)}{1 + Y_1(s)},$$

where $Y_1(s)$ has the form (5.8). We assume that the roots of the characteristic equation of $Y(s)$ (the poles) are σ_i, $i = 1, \ldots, m$ (simple poles) and $-\alpha_k \pm j \cdot \omega_k$, $k = 1, \ldots, n$ (complex conjugate pairs). For a unit step input, the output of the system can be written as a partial fraction expansion [8]:

$$X_0(s) = \frac{1}{s} + \sum_{i=1}^{n} \frac{A_i}{s + \sigma_i} + \sum_{k=1}^{n} \frac{B_k}{s^2 + 2\alpha_k s + (\alpha_k^2 + \omega_k^2)}. \tag{7.8}$$

Then the inverse Laplace transform can be obtained as a sum of terms:

$$x_0(t) = 1 + \sum_{i=1}^{m} A_i e^{-\sigma_i \cdot t} + \sum_{k=1}^{n} \frac{B_k}{\omega_k} e^{-\alpha_k \cdot t} \sin \omega_k \cdot t. \tag{7.9}$$

The transient response contains a steady-state output, exponential terms, and damped sinusoidal terms. It is clear that, in order for the response to be stable, the real parts of the roots σ_i and α_k must be negative.

For example, we consider the closed-loop control system from Fig. 5.4. The closed-loop transfer function is

$$Y(s) = \frac{k_G}{s^2 m + s k_f + k_G}. \tag{7.10}$$

This relation can be rewritten as

$$Y(s) = \frac{\omega_n^2}{s^2 + 2\zeta\omega_n s + \omega_n^2}, \tag{7.11}$$

where ω_n is the natural frequency,

$$\omega_n^2 = \frac{k_G}{m}, \tag{7.12}$$

and ζ is the dimensionless damping ratio,

$$\zeta = \frac{k_f}{2\sqrt{m \cdot k_G}}. \tag{7.13}$$

The roots of the characteristic equation are

$$s_{1,2} = -\zeta\omega_n \pm \omega_n\sqrt{\zeta^2 - 1}. \tag{7.14}$$

When $\zeta > 1$, the roots are real and the system is defined as *overdamped*. For $\zeta < 1$, the roots are complex and conjugates and the system is called *underdamped*:

$$s_{1,2} = -\zeta\omega_n \pm j\omega_n\sqrt{1-\zeta^2}. \tag{7.15}$$

For $\zeta = 1$, we have double roots and the system is defined as *critical damping*:

$$s_1 = s_2 = -\zeta\omega_n. \tag{7.16}$$

If we consider a unit step input, the transient response of Eq. (7.11) is [4, 9]

$$x_0(t) = 1 - \frac{e^{-\zeta\omega_n t}}{\sqrt{1-\zeta^2}} \cdot \sin\left(\omega_n\sqrt{1-\zeta^2} \cdot t + \tan^{-1}\left(\frac{\sqrt{1-\zeta^2}}{\zeta}\right)\right). \tag{7.17}$$

The form of the transient response is illustrated in Fig. 7.3.

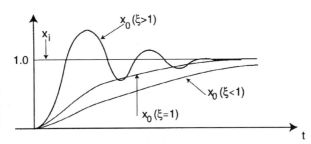

Figure 7.3
Response of the closed-loop control for the translational mechanical system.

8. Frequency-Domain Performances

The frequency response of a system is defined as the steady-state response of the system to a sinusoidal input signal. For a linear system, the output signal will be a sinusoidal signal that differs from the input only in amplitude and phase angle. The frequency transfer function is obtained by replacing s with $j\omega$. This transfer function is a function of the complex variable $j\omega$,

$$Y(j\omega) = Y(s)|_{s=j\omega} \tag{8.1}$$

$$Y(j\omega) = P(\omega) + jQ(\omega), \tag{8.2}$$

where

$$\begin{cases} P(\omega) = \text{Re}[Y(j\omega)] \\ Q(\omega) = \text{Im}[Y(j\omega)]. \end{cases} \tag{8.3}$$

The transfer function can be also represented by a magnitude $M(\omega)$ and a phase $\Psi(\omega)$,

$$Y(j\omega) = M(\omega) \cdot e^{j \cdot \Psi(\omega)}, \tag{8.4}$$

where

$$\Psi(\omega) = \tan^{-1}\left(\frac{Q(\omega)}{P(\omega)}\right) \tag{8.5}$$

$$|M(\omega)|^2 = [P(\omega)^2 + [Q(\omega)]^2]. \tag{8.6}$$

8.1 The Polar Plot Representation

The polar plot representation is the graphical representation of $Y(j\omega)$. The polar plot is obtained as the locus of the real and imaginary parts of $Y(j\omega)$ in the polar plane. The coordinates of the polar plot are the real and imaginary parts of $Y(j\omega)$. For example, we reconsider the open-loop system for a translational mechanism (5.13), $Y_1(s) = k_G/s(ms + k_f)$. The transfer function can be rewritten as

$$Y_1(s) = \frac{k}{s(\tau s + 1)}, \tag{8.7}$$

where

$$k = \frac{k_G}{k_f}$$
$$\tau = \frac{m}{k_f}. \tag{8.8}$$

The frequency transfer function $Y_1(j\omega)$ will be

$$Y_1(s)|_{s=j\omega} = Y_1(j\omega) = \frac{k}{-\omega^2 + j\omega}. \tag{8.9}$$

From Eqs. (8.3)–(8.6), we obtain

$$P_1(\omega) = \frac{-k\omega^2\tau}{\omega^4\tau^2 + \omega^2}$$
$$Q_1(\omega) = \frac{-\omega k}{\omega^4\tau^2 + \omega^2} \tag{8.10}$$

and

$$M_1(\omega) = \frac{k}{(\omega^4\tau^2 + \omega^2)^{1/2}} \tag{8.11}$$

$$\Psi_1(\omega) = -\tan^{-1}\left(\frac{-1}{\omega\tau}\right). \tag{8.12}$$

We find several values of $M(\omega)$ and $\Psi(\omega)$:

For $\omega = 0$,

$$M_1(0) = \infty$$
$$\Psi(0) = -\frac{\pi}{2};$$

For $\omega = 1/\tau$,

$$M_1\left(\frac{1}{\tau}\right) = \frac{\sqrt{2}}{2} k\tau$$

$$\Psi\left(\frac{1}{\tau}\right) = -\frac{3\pi}{4};$$

For $\omega = \infty$,

$$M_1(\infty) = 0$$

$$\Psi(\infty) = -\pi.$$

The polar plot of $Y_1(j\omega)$ is shown in Fig. 8.1.

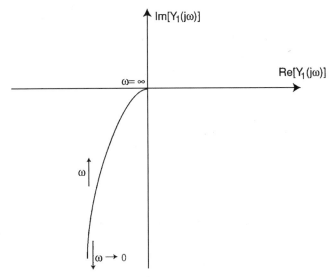

Figure 8.1
Polar plot for open-loop system of translational mechanism.

8.2 The Logarithmic Plot Representation

Logarithmic plots or *Bode plots* are based on logarithmic graphical plots for the magnitude and phase angle,

$$\log Y_1(j\omega) = \log M_1(\omega) + j\Psi_1(\omega). \tag{8.13}$$

The logarithm of $M_1(\omega)$ is normally expressed in terms of

$$\log M_1(\omega) \rightarrow 20 \log M_1(\omega), \tag{8.14}$$

where the units are decibels (dB). The logarithmic gain in decibels and the angle $\Psi_1(\omega)$ can be plotted versus the frequency ω.

In order to analyze this method we will use the model offered by the translational mechanism,

$$Y_1(j\omega) = \frac{k}{j\omega(j\omega\tau + 1)} = M_1(\omega) \cdot e^{j\cdot\Psi_1(\omega)}. \tag{8.15}$$

The transfer function can be rewritten as

$$Y_1(j\omega) = Y'(j\omega) \cdot Y''(j\omega) \cdot Y'''(j\omega), \tag{8.16}$$

where

$$Y'(j\omega) = M_1'(\omega) \cdot e^{j \cdot \Psi'(\omega)} = k \tag{8.17}$$

$$Y''(j\omega) = M_1''(\omega) \cdot e^{j \cdot \Psi''(\omega)} = \frac{1}{j\omega} \tag{8.18}$$

$$Y'''(j\omega) = M_1'''(\omega) \cdot e^{j \cdot \Psi'''(\omega)} = \frac{1}{j\omega\tau + 1}. \tag{8.19}$$

By using the logarithmic plots, the magnitude and angle can be easily plotted from the partial transfer functions $Y'(j\omega)$, $Y''(j\omega)$, $Y'''(j\omega)$:

$$20\log M_1(\omega) = 20\log M_1'(\omega) + 20\log M_1''(\omega) + 20\log M_1'''(\omega), \tag{8.20}$$

$$\Psi_1(\omega) = \Psi_1'(\omega) + \Psi_1''(\omega) + \Psi_1'''(\omega). \tag{8.21}$$

Examining the relations (8.17)–(8.19), we obtain

$$20\log M_1''(\omega) = 20\log k \tag{8.22}$$

$$20\log M_1''(\omega) = 20\log\left|\frac{1}{j\omega}\right| = -20\log\omega \tag{8.23}$$

$$20\log M_1'''(\omega) = 20\log\left|\frac{1}{j\omega\tau + 1}\right| = -20\log(1 + \omega^2\tau^2)^{1/2}. \tag{8.24}$$

The relation (8.22) defines the logarithmic gain as a constant, and the phase angle is zero (Fig. 8.2):

$$\Psi_1'(\omega) = 0. \tag{8.25}$$

In order to plot the magnitude versus frequency in a Bode diagram, we consider that the use of a logarithmic scale of frequency is the most judicious choice. In this case, the frequency axis is marked by $\log\omega$.

The logarithmic magnitude $M_1''(\omega)$ from Eq. (8.23) will be represented by a straight line. The slope of the line can be computed by choosing an interval

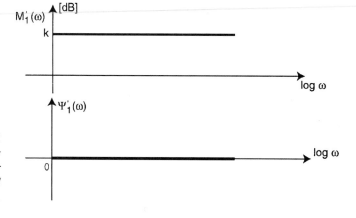

Figure 8.2
Frequency diagrams for $M_1'(\omega)$ and $\Psi_1'(\omega)$.

of two frequencies with a ratio equal to 10. This interval is called a decade. Let us consider two frequencies, $\omega_2, \omega_1, \omega_2 = 10 \cdot \omega_1$. The slope will be

$$20 \log M_1''(\omega_2) - 20 \log M_1''(\omega_1) = -20 \log \omega_2 + 20 \log \omega_1$$
$$= -20 \log \frac{\omega_2}{\omega_1} = -20 \text{ dB/decade}. \qquad (8.26)$$

The phase plot is obtained easily from Eq. (8.18):

$$\Psi_1''(\omega) = -\frac{\pi}{2}. \qquad (8.27)$$

An approximate representation of $Y_1'''(j\omega)$ can be obtained if we analyze the frequency domain around the value

$$\omega_b = \frac{1}{\tau}, \qquad (8.28)$$

called the *break frequency*. For small frequencies, $\omega \ll 1/\tau$, the relation (8.24) can be rewritten as

$$20 \log M_1'''(\omega) = 0. \qquad (8.29)$$

For large frequencies, $\omega \gg 1/\tau$, we have

$$20 \log M_1'''(\omega) = -20 \log \omega \tau. \qquad (8.30)$$

The relation (8.30) defines a straight line with the slope

$$20 \log M_1'''(\omega_2) - 20 \log M_1'''(\omega_1) = -20 \log \omega_2 \tau + 20 \log \omega_1 \tau$$
$$= -20 \log 10 = -20 \text{ dB/decade}. \qquad (8.31)$$

The approximate plot of magnitude is represented in Fig. 8.4 by the solid line, and the exact curve by the dashed line. The phase angle $\Psi_1'''(\omega)$ is obtained from Eq. (8.19) (Fig. 8.3) as

$$\Psi_1'''(\omega) = -\tan^{-1}(\omega \tau). \qquad (8.32)$$

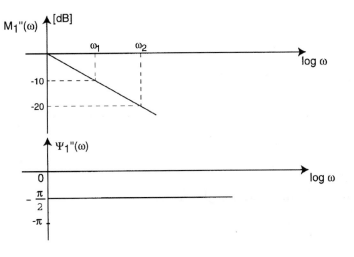

Figure 8.3
Frequency diagrams for $M_1''(\omega)$ and $\Psi_1''(\omega)$.

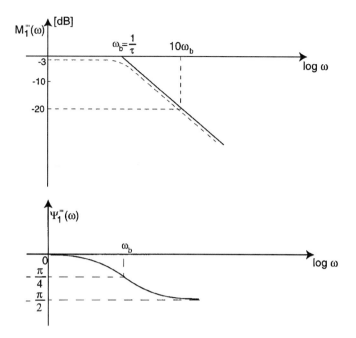

Figure 8.4
Frequency diagrams for $M_1'''(\omega)$ and $\Psi_1'''(\omega)$.

For the break value $\omega_b = 1/\tau$; we have $20 \log M_1'''(1/\tau) = -20 \log 2 = -3$ dB, $\Psi_1'''(1/\tau) = -\tan^{-1}(1) = -\pi/4$. The Bode plots for $Y_1(j\omega)$ can be obtained by adding the plots of the partial transfer functions. The result is presented in Fig. 8.5.

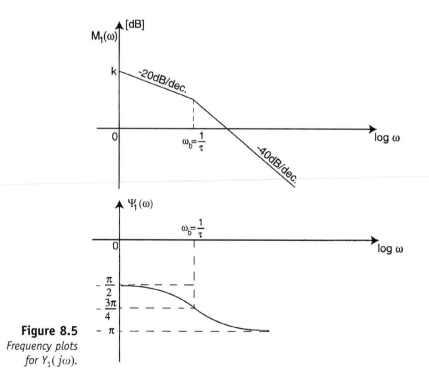

Figure 8.5
Frequency plots for $Y_1(j\omega)$.

8.3 Bandwidth

The *bandwidth* is an important parameter that defines the quality of the closed-loop control system,

$$Y(j\omega) = \frac{Y_1(j\omega)}{1 + Y_1(j\omega)},$$ (8.33)

and we consider $M(\omega)$, $\Psi(\omega)$ the magnitude and the phase of the closed-loop system, respectively:

$$Y(j\omega) = M(\omega)e^{j \cdot \Psi(\omega)}.$$ (8.34)

The general form of $M(\omega)$ is represented in Fig. 8.6.

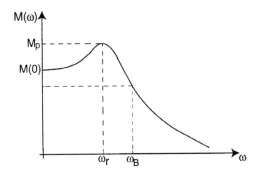

Figure 8.6
Magnitude $M(\omega)$ for a closed-loop system.

The maximum magnitude is M_p and the frequency is called the *resonant frequency* ω_r. The bandwidth is defined as the domain of frequency for which

$$M(\omega) \geq \frac{\sqrt{2}}{2} M(0).$$ (8.35)

The associated frequency is ω_b, and if we consider $M(0) = 1$, then

$$20 \log M(\omega_B) = 20 \log \frac{\sqrt{2}}{2} \cong -3 \text{ dB}.$$ (8.36)

If the open-loop transfer function has a pole in the origin, then $\lim_{\omega \to 0} Y_1(j\omega) = \infty$, and

$$\lim_{\omega \to 0} Y(j\omega) = 1,$$ (8.37)

then

$$M(0) = 1.$$ (8.38)

If $Y_1(s)$ does not have a pole in origin, then

$$M(0) < 1.$$ (8.39)

For example, we reconsider the closed-loop control system from Fig. 5.4 (translational mechanism closed-loop control).

From (7.11)–(7.13), we have the transfer function

$$Y(s) = \frac{\omega_n^2}{s^2 + 2\zeta\omega_n s + \omega_n^2}$$

$$Y(j\omega) = \frac{1}{1 - \left(\dfrac{\omega}{\omega_n}\right)^2 + 2\zeta\dfrac{\omega}{\omega_n}j}. \tag{8.40}$$

The logarithmic magnitude $M(\omega)$ and the phase angle $\Psi(\omega)$ will be

$$20\log M(\omega) = -10\log\left(\left(1 - \left(\frac{\omega}{\omega_n}\right)^2\right)^2 + 4\zeta\frac{\omega^2}{\omega_n^2}\right) \tag{8.41}$$

$$\Psi(\omega) = -\tan^{-1}\left(\frac{2\zeta\dfrac{\omega}{\omega_n}}{1 - \dfrac{\omega^2}{\omega_n^2}}\right). \tag{8.42}$$

We will approximate these functions in two frequency domains. For $\omega \ll \omega_n$, we obtain

$$20\log M(\omega) \cong 0 \tag{8.43}$$

$$\Psi(\omega) \cong 0. \tag{8.44}$$

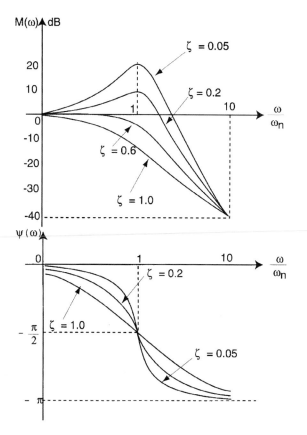

Figure 8.7
Frequency plots for a closed-loop control system.

For $\omega \gg \omega_n$, we neglect several terms in (8.41) and

$$20 \log M(\omega) \simeq -40 \log \frac{\omega}{\omega_n}, \tag{8.45}$$

which determines a curve with a slope of $-40\,\text{dB/decade}$. Also, the phase angle will be

$$\Psi(\omega) \simeq -\pi. \tag{8.46}$$

The magnitude asymptotes meet the 0-dB axis for $\omega = \omega_n$. The difference between the actual magnitude curve and the asymptotic approximation is a function of the damping ratio ζ. The maximum value of the frequency response M_p occurs at the resonant frequency ω_r. When the damping ratio ζ approaches zero, the resonant frequency ω_r approaches the natural frequency ω_n (Fig. 8.7).

9. Stability of Linear Feedback Systems

A stable system is defined as a system with a bounded system response. If the system is subjected to a bounded input or disturbance and the response is bounded in magnitude, the system is said to be stable.

A first result for the linear closed-loop system stability was obtained in Section 7. For a unit step input, the output was written as

$$X_o(s) = \frac{1}{s} + \sum_{i=1}^{m} \frac{A_i}{s + \sigma_i} + \sum_{k=1}^{n} \frac{B_k}{s^2 + 2\alpha_k s + (\alpha_k^2 + \omega_k^2)}, \tag{9.1}$$

and the time response is obtained as a sum of terms,

$$x_o(t) = 1 + \sum_{i=1}^{m} A_i e^{-\sigma_i t} + \sum_{k=1}^{n} \frac{B_k}{\omega_k} e^{-\sigma_k t} \sin \omega_k t. \tag{9.2}$$

Clearly, in order to obtain a bounded response, the real part of the characteristic roots σ_i and α_k must be negative.

We may conclude that "a necessary and sufficient condition that a feedback system be stable is that all the poles of the system transfer function have negative real parts." Of course, the main methods, investigate the stability by determining the characteristic roots. There are also other methods that do not require the determination of the roots but use only the polynomial coefficients of the characteristic equations or the frequency transfer function.

9.1 The Routh–Hurwitz Criterion

The Routh–Hurwitz criterion is a necessary and sufficient criterion for linear system stability. This criterion is based on the ordering of the coefficients of the characteristic equation [4, 8, 9, 17, 18]

$$\Delta(s) = a_n s^n + a_{n-1}s^{n-1} + \cdots + a_1 s + a_0 = 0 \tag{9.3}$$

into an array as follows:

$$
\begin{bmatrix}
s^n & a_n & a_{n-2} & a_{n-4} & \cdots \\
s^{n-1} & a_{n-1} & a_{n-3} & a_{n-5} & \cdots \\
s^{n-2} & b_{n-1} & b_{n-3} & b_{n-5} & \cdots \\
s^{n-3} & c_{n-1} & c_{n-3} & c_{n-5} & \cdots \\
s^{n-4} & d_{n-1} & d_{n-3} & d_{n-5} & \cdots \\
\vdots & & & &
\end{bmatrix}.
$$

Here,

$$b_{n-1} = -\frac{1}{a_{n-1}}\begin{vmatrix} a_n & a_{n-2} \\ a_{n-1} & a_{n-3} \end{vmatrix}, \qquad b_{n-3} = -\frac{1}{a_{n-1}}\begin{vmatrix} a_n & a_{n-4} \\ a_{n-1} & a_{n-5} \end{vmatrix}$$

$$c_{n-1} = -\frac{1}{b_{n-1}}\begin{vmatrix} a_{n-1} & a_{n-3} \\ b_{n-1} & b_{n-3} \end{vmatrix}, \qquad c_{n-3} = -\frac{1}{b_{n-1}}\begin{vmatrix} a_{n-1} & a_{n-5} \\ b_{n-1} & b_{n-5} \end{vmatrix}$$

$$d_{n-1} = -\frac{1}{c_{n-1}}\begin{vmatrix} b_{n-1} & b_{n-3} \\ c_{n-1} & c_{n-3} \end{vmatrix}, \ldots.$$

The Routh–Hurwitz criterion requires that all the elements of the first column be nonzero and have the same sign. The condition is both necessary and sufficient. For example, we consider the characteristic equation of a third-order system [8, 9, 18]

$$s^3 + (\lambda + 1)s^2 + (\lambda + \mu - 1)s + (\mu - 1) = 0. \tag{9.4}$$

The coefficient array is

$$
\begin{bmatrix}
s^3 & 1 & \lambda + \mu - 1 \\
s^2 & \lambda + 1 & \mu - 1 \\
s & \dfrac{\lambda(\lambda + \mu)}{\lambda + 1} & 0 \\
s^0 & \mu - 1 & 0
\end{bmatrix}.
$$

The necessary and sufficient conditions will be

$$
\begin{aligned}
&C_1: \lambda > -1 \\
&C_2: \lambda(\lambda + \mu) > 0 \\
&C_3: \mu > 1.
\end{aligned}
\tag{9.5}
$$

These conditions are presented in Fig. 9.1. The final domain for the condition $C = C_1 \cap C_2 \cap C_3$ is shown in Fig. 9.1b.

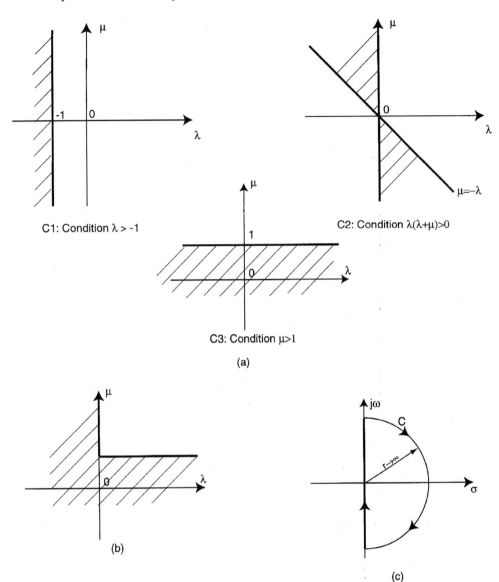

Figure 9.1 *(a) Partial conditions. (b) Condition $C = C_1 \cap C_2 \cap C_3$. (c) Nyquist contour.*

9.2 The Nyquist Criterion

We consider the closed-loop control system transfer function

$$Y(s) = \frac{Y_1(s)}{1 + Y_1(s)}. \tag{9.6}$$

The characteristic equation of this system is

$$\Delta(s) = 1 + Y_1(s). \tag{9.7}$$

For a system to be stable, all the zeros of $\Delta(s)$ must be lie in the left-hand s-plane. In order to investigate the positions of equation characteristic roots, a special contour C in the s-plane, which encloses the entire right-hand s-plane, is chosen. This contour is called the Nyquist contour (Fig. 9.1). The contour passes along the $j\omega$-axis from $-j\infty$ to $+j\infty$ and is completed by a semicircular path of radius r, where r approaches infinity. For this contour C, the corresponding contour D in the $\Delta(s)$-plane will encircle the origin in a clockwise direction N times (see Appendix C),

$$N = Z - P, \tag{9.8}$$

where Z, P represent the number of zeros and poles, respectively, of the characteristic equation $\Delta(s)$. Of course, the number of poles P of $\Delta(s)$ is equal to the number of poles of the open-loop transfer function $Y_1(s)$.

The Nyquist criterion prefers to operate by the complex function

$$Y_1(s) = \Delta(s) - 1, \tag{9.9}$$

instead of $\Delta(s)$, which also represents the transfer function of the open-loop control system. In this case the number of clockwise encirclements of the origin (Appendix C) becomes the number of clockwise encirclements of the -1 point in the plane $Y_1(s)$. We know that the stability of the closed-loop system requires that the number of zeros Z in the right-hand plane (equal to the number of poles of the closed-loop system) should be zero:

$$N = -P. \tag{9.10}$$

Now, we can formulate the Nyquist criterion as follows [6, 9]:

A necessary and sufficient condition for the stability of the closed-loop control system is that, for the contour D in the $Y_1(s)$-plane, the number of counterclockwise encirclements of the $(-1, 0)$ point be equal to the number of poles of $Y_1(s)$ from the right-hand s-plane.

In the particular case when the open-loop system $Y_1(s)$ is stable, the number of poles of $Y_1(s)$ in the right-hand s-plane is zero ($P = 0$); the Nyquist criterion requires that the contour D in the $Y_1(s)$-plane not encircle the $(-1, 0)$ point.

We will illustrate the Nyquist criterion by several examples. First, let us consider the thermal heating system presented in Fig. 9.2.

The model of a tank system containing a heated liquid is defined by the transfer function [9, 16]

$$Y_H(s) = \cfrac{1}{c_t s + \left(Q c_s + \cfrac{1}{R_t} \right)} = \frac{\theta(s)}{I(s)}, \tag{9.11}$$

where $\theta = \theta_o - \theta_e$ defines the system output and represents the temperature difference between the temperature of the fluid out and the environmental temperature, I is the electrical current that ensures heating of the system, and c_t, Q, c_s, R_t represent the thermal capacitance, the fluid flow rate (constant),

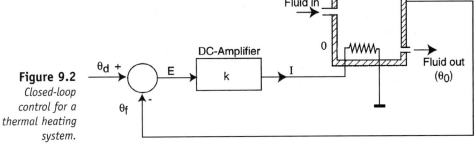

Figure 9.2
Closed-loop control for a thermal heating system.

the specific heat of the fluid, and the thermal resistance of insulation, respectively. The direct control path contains a DC amplifier with k-gain.

The transfer function of the open-loop system is

$$Y_1(s) = k\frac{1}{c_t s + \left(Qc_s + \dfrac{1}{R_t}\right)}, \tag{9.12}$$

which can be rewritten as

$$Y_1(s) = \frac{k_1}{\tau_1 s + 1}, \tag{9.13}$$

where

$$k_1 = \frac{k}{Qc_s + \dfrac{1}{R_t}}$$

$$\tau_1 = \frac{c_t}{Qc_s + \dfrac{1}{R_t}}. \tag{9.14}$$

The transfer function of $Y_1(s)$ has a pole $p_1 = -1/\tau_1$ so that the open-loop system is stable. The mapping contour of $Y_1(s)$ is a circle (Fig. 9.3) that does not encircle the point $(-1, 0)$ (see Appendix A.3). Of course, the closed-loop system is stable.

Figure 9.3
Mapping contour for thermal heating system.

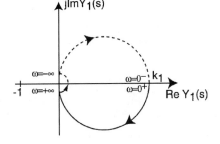

A second example is offered by the closed-loop control of a translational mechanism (Sections, Fig 5.4). The transfer function of the open-loop system is

$$Y_1(s) = \frac{k}{s(ms + k_f)},\qquad(9.15)$$

or

$$Y_1(s) = \frac{k_1}{s(\tau s + 1)},\qquad(9.16)$$

where

$$\begin{aligned}k_1 &= \frac{k}{k_f}\\ \tau &= \frac{m}{k_f}.\end{aligned}\qquad(9.17)$$

We can remark the presence of two poles,

$$\begin{aligned}p_1 &= -\frac{1}{\tau}\\ p_2 &= 0.\end{aligned}\qquad(9.18)$$

In this case, the Nyquist contour C contains an infinitesimal semicircle contour of radius $r_1 \to 0$ in order to satisfy the condition that the contour cannot pass through the pole of origin. The mapping contour by $Y_1(s)$ is obtained from the rules discussed in Appendix A.3. This is presented in Fig 9.4b.

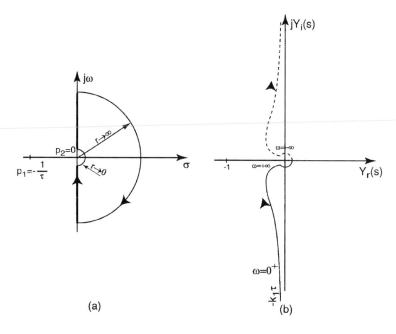

Figure 9.4
(a) Nyquist contour. (b) Mapping contour for translational mechanism.

(a)

(b)

When s traverses the large semicircle of the Nyquist contour, ω varies from $+\infty$ to $-\infty$, and we have

$$Y_1(s)|_{s=r_2 e^{j\phi}} \cong \frac{k_1}{\tau r_2^2} e^{-2j\phi}, \tag{9.19}$$

$$\lim_{r_2 \to \infty} Y_1(s) \cong \lim_{r_2 \to \infty} \left| \frac{k_1}{\tau r_2^2} \right| \cdot e^{-2j\phi}. \tag{9.20}$$

It is clear that ϕ changes from $\phi = \pi/2$ at $\omega = +\infty$ to $\phi = -\pi/2$ at $\omega = -\infty$ so that the angle change of $Y_1(s)$ is from $-2\pi/2 = -\pi$ at $\omega = +\infty$ to $2\pi/2 = \pi$ at $\omega = -\infty$. The magnitude of the contour is defined by an infinitesimal circle of radius $k_1/\tau r_2^2$, where r_2 is infinite.

For the small semicircular detour around the pole $p_2 = 0$, the mapping can be approximated by

$$Y_1(s)|_{s=r_1 e^{j\phi}} \cong \frac{k_1}{r_1 e^{j\phi}} \tag{9.21}$$

$$\lim_{r_1 \to 0} Y_1(s)|_{s=r_1 e^{j\varphi}} \cong \lim_{r_1 \to 0} \frac{k_1}{r_1} e^{-j\phi}. \tag{9.22}$$

The angle of $Y_1(s)$ has a value from $\pi/2$ at $\omega = 0^-$ to $-\pi/2$ at $\omega = 0^+$. The magnitude will be infinite.

If we now consider the portion of the contour C from $\omega = 0^+$ to $\omega = +\infty$, we will have

$$Y_1(s) = -k_1 \frac{\omega\tau + j}{\omega[(\omega\tau)^2 + 1]} = Y_r(s) + jY_i(s). \tag{9.23}$$

For $\omega \to 0$ we obtain

$$\lim Y_r(\omega) = -k_1 \tau$$
$$\lim Y_i(\omega) = -\infty. \tag{9.24}$$

The same procedure can be used for the portion $\omega = -\infty$ to $\omega = 0^-$, and we obtain a symmetrical polar plot.

If we analyze the stability of the closed-loop system, we see that the open- loop system is at the limit of stability ($Z = 0$ and a pole at the origin), but the closed-loop system has a mapping contour that does not encircle the -1 point, so that the system is always stable.

Let us consider the same system, the translational mechanism, for which the driving system is represented by a solenoid (Fig. 9.5). A solenoid is an electrical system that converts direct current (DC) electrical energy into translational mechanism energy. The transfer function is

$$Y_s = \frac{F(s)}{V(s)} = \frac{k_s}{R_s + sL_s}, \tag{9.25}$$

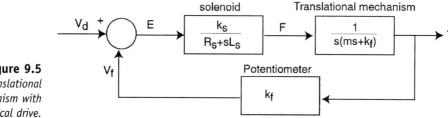

Figure 9.5
A translational mechanism with electrical drive.

where R_s and L_s are electrical system parameters, and k_s represents a coefficient of the mechanical system. A feedback path is obtained by a potentiometer,

$$Y_f(s) = \frac{V(s)}{Z(s)} = k_f^1. \tag{9.26}$$

The open-loop transfer function is

$$Y_1(s) = \frac{k_s}{s(ms + k_f)(R_s + sL_s)}, \tag{9.27}$$

and the closed-loop transfer function will be

$$Y(s) = \frac{Y_1(s)}{1 + Y_1(s)Y_f(s)}. \tag{9.28}$$

We write

$$Y_1'(s) = Y_1(s)Y_f(s) = \frac{k'}{s(\tau_1 s + 1)(\tau_2 s + 1)}, \tag{9.29}$$

where

$$\tau_1 = \frac{m}{k_f}$$

$$\tau_2 = \frac{L_s}{R_s} \tag{9.30}$$

$$k' = \frac{k_s}{k_f^1 R_s}.$$

If we use the same procedure as that discussed in the previous example, we can use the Nyquist contour from Fig. 9.4a.

We see that

$$\lim_{r_1 \to 0} Y_1'(s)|_{s=r_1 e^{j\varphi}} \cong \lim_{r_1 \to 0} \frac{k'}{r_1} e^{-j\phi}. \tag{9.31}$$

The magnitude approaches infinity and the angle of the $Y_1'(s)$-plane contour changes from $-\pi/2$ at $\omega = 0^+$ to $\pi/2$ at $\omega = 0^-$.

Also, for $-\infty < \omega < +\infty$, we have

$$\lim_{r_2 \to \infty} Y_1'(s)|_{s=r_2 e^{j\varphi}} \cong \frac{k'}{r_2^3} e^{-3j\phi}, \tag{9.32}$$

so that the $Y_1'(s)$ angle varies from $-3\pi/2$ at $\omega = +\infty$ to $3\pi/2$ at $\omega = -\infty$. The magnitude of $Y_1'(s)$ approaches zero. The mapping in the $Y_1'(s)$-plane contour is shown in Fig. 9.6.

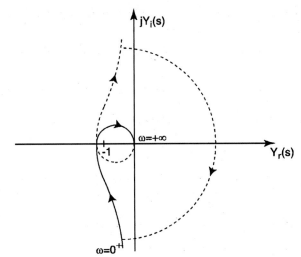

Figure 9.6

Mapping contour for the translational mechanism with electrical drive.

We see that it is possible that the contour may encircle the -1 point. In this case the number of encirclements is equal to two and the system becomes unstable. In order to improve the quality of the motion, we will compute the point where $Y_1'(s)$ intersects the real axis. First, we calculate the frequency ω_c of the intersection point from the condition

$$Y_{1i}'(\omega) = 0, \tag{9.33}$$

where we considered

$$Y_1'(j\omega) = Y_{1r}'(\omega) + jY_{1i}'(\omega). \tag{9.34}$$

From Eqs. (9.33) and (9.29) we get [4, 9] the critical frequency

$$\omega_c = \frac{1}{\sqrt{\tau_1 \tau_2}}. \tag{9.35}$$

The intersection point coordinate is obtained as

$$Y_{1r}'(\omega_c) = \frac{-k'\tau_1\tau_2}{\tau_1 + \tau_2}. \tag{9.36}$$

The stability condition requires

$$\frac{-k'\tau_1\tau_2}{\tau_1 + \tau_2} > -1. \tag{9.37}$$

The relation (9.37) enables us to impose certain constraints on the gain factor k',

$$k' < \frac{\tau_1 + \tau_2}{\tau_1\tau_2}. \tag{9.38}$$

If the gain factor k' has the limit value

$$k'_c = \frac{\tau_1 + \tau_2}{\tau_1 \tau_2}, \tag{9.39}$$

the closed-loop system will have a pole on the $j\omega$-axis, and the system is at the limit of stability. As k' decreases below this limit value, stability increases and the margin between the new gain k' and k'_c is a measure of the relative stability.

This measure of relative stability is called the gain margin and is defined as the reciprocal of the gain $|Y'_1(j\omega)|$ at the frequency at which the phase angle is π:

$$\frac{1}{d} = \frac{1}{|Y'_1(\omega_c)|}. \tag{9.40}$$

It can also be defined in decibels:

$$20 \log\left(\frac{1}{d}\right) = -20 \log |Y'_1(\omega_c)|. \tag{9.41}$$

Another measure of the relative stability can be defined [6, 9, 20] by the phase margin (w) as the phase angle through which the $Y'_1(j\omega)$ plot must be rotated in order that the unity magnitude $|Y'_1(j\omega)| = 1$ point should pass through the $(-1, 0)$ point in the $Y'_1(s)$-plane. This index is called the phase margin and is equal to the additional phase log required before the system becomes unstable.

9.3 Stability by Bode Diagrams

In the previous section we established that the gain and phase margins are a measure of relative stability. But the gain and phase margins are easily evaluated from the Bode diagram.

The critical point for stability is defined for

$$|Y_1(j\omega_c)| = 1, \tag{9.42}$$

which corresponds to [see Eq. (8.4)]

$$|M(\omega_c)|_{dB} = 0 \tag{9.43}$$

$$\Psi(\omega_c) = k\pi \tag{9.44}$$

where k is an odd number.

From the Bode diagram, the gain margin will be

$$d = M(\omega_t), \tag{9.45}$$

where ω_t is the frequency for which

$$\Psi(\omega_t) = -\pi, \tag{9.46}$$

and the phase margin w is

$$w = |\Psi(\omega_c)| - \pi. \tag{9.47}$$

If the polar plot for $Y_1(j\omega)$ approaches the critical point $(-1, 0)$, the system is at the limit of stability, the logarithmic magnitude is 0 dB, and the phase angle is π on the Bode diagram. Let us now consider the translational mechanism with electrical drive discussed in the preceding section. From Eq. (9.29) we have

$$Y_1(j\omega) = \frac{k'}{j\omega(j\omega\tau_1 + 1)(j\omega\tau_2 + 1)}. \qquad (9.48)$$

We assume $k' = 1$, $\tau_1 = 0.2$, $\tau_2 = 0.1$. The Bode characteristics are presented in Fig 9.7.

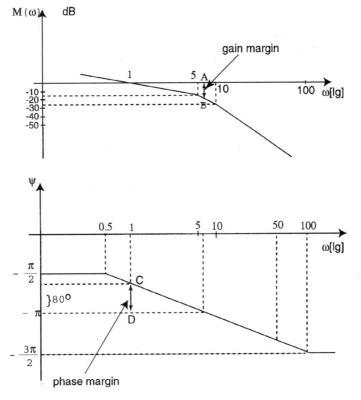

Figure 9.7
Bode characteristics for the translational mechanism with electrical drive.

The critical frequency is $\omega_c = 1$. The gain margin is estimated by the segment AB, $d \approx 20$ dB, and the phase margin is evaluated by the segment CD, $w \approx -80°$.

10. Design of Closed-Loop Control Systems by Pole-Zero Methods

In the preceding sections we analyzed the design and adjustment of the system parameters in order to provide a desirable quality of the control system. But often it is not simple to adjust a parameter of a technological process that has a complex configuration. It is preferable to reconsider the structure of the control system and to introduce new structure components that allow a better selection and adjustment of the parameters for the overall system.

These new structure components are called controllers.

10.1 Standard Controllers

In Fig. 10.1 we present a feedback control system in which a controller ensures the quality of the control system. The adjustment of the controller parameters in order to provide suitable performance is called compensation.

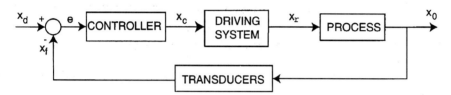

Figure 10.1 *A feedback control system with controller.*

The transfer function of the controller is designated as

$$Y_c(s) = \frac{X_c(s)}{E(s)}, \tag{10.1}$$

where $E(s)$ is the system error and $X_c(s)$ defines the output of the controller. This variable acts as the input for the second component, the driving system, which represents an interface between the controller and the mechanical process:

$$Y_D(s) = \frac{X_p(s)}{X_c(s)}. \tag{10.2}$$

If we suppose that the transfer function of the mechanical system (process) is $Y_p(s)$, the open-loop system has the transfer function $Y_1(s) = Y_c(s) \cdot Y_D(s) \cdot Y_p(s)$ and the closed-loop control system

$$Y(s) = \frac{Y_c(s) \cdot Y_D(s) \cdot Y_p(s)}{1 + Y_c(s) \cdot Y_D(s) \cdot Y_p(s) \cdot Y_T(s)}, \tag{10.3}$$

where $Y_T(s)$ represents the transfer function of the transducer on the feedback path.

In order to facilitate the selection of the best control structure, several types of standard controllers are used.

(a) The *P controller* (proportional controller) is defined by the equation

$$x_c(t) = K_p \cdot e(t). \tag{10.4}$$

This controller provides a proportional output as a function of the error

$$Y_c(s) = K_p. \tag{10.5}$$

(b) *I controller* (integration controller):

$$x_c(t) = \frac{1}{T_i} \int e(t)\,dt \tag{10.6}$$

$$Y_c(s) = \frac{1}{T_i s}. \tag{10.7}$$

(c) *PI controller* (proportional-integration controller):

$$x_c(t) = K_p\left(e(t) + \frac{1}{T_i}\int e(t)\,dt\right) \tag{10.8}$$

$$Y_c(s) = K_p\left(1 + \frac{1}{T_i s}\right). \tag{10.9}$$

(d) *PD controller* (proportional-derivative controller):

$$x_c(t) = K_p\left(e(t) + T_d\frac{de(t)}{dt}\right) \tag{10.10}$$

$$Y_c(s) = K_p(1 + T_d s). \tag{10.11}$$

(e) *PDD$_2$ controller* (proportional-derivative-derivative controller):

$$x_c(t) = K_p\left(T_{d_1} T_{d_2}\frac{d^2 e(t)}{dt^2} + (T_{d_1} + T_{d_2})\frac{de(t)}{dt} + e(t)\right) \tag{10.12}$$

$$Y_c(s) = K_p(1 + T_{d_1} s)(1 + T_{d_2} s). \tag{10.13}$$

(f) *PID controller* (proportional-integration-derivative controller):

$$x_c(t) = K_p\left(e(t) + \frac{1}{T_i}\int(t)\,dt + T_d\frac{de(t)}{dt}\right) \tag{10.14}$$

$$Y_c(s) = K_p\left(1 + \frac{1}{T_i s} + T_d s\right). \tag{10.15}$$

The design of a control system requires the selection of a type of controller, the arrangement of the system structure, and then the selection and adjustment of the controller parameters in order to obtain a set of desired performances.

If the theoretical design of the controller requires a transfer function more complex than those of PID or PDD$_2$ controllers, it is preferable to connect several types of standard controllers that can achieve the desired performance.

10.2 P-Controller Performance

We consider the transfer function of an open-loop system (5.8) rewritten in the form

$$Y_1(s) = \frac{K}{s^l}\cdot\frac{Q(s)}{R(s)}, \tag{10.16}$$

where $l = 0, 1, 2$ ($l > 2$ determines the instability of the system) and $Q(s)$, $R(s)$ are polynomials with coefficients of s^0 equal to 1:

$$\frac{Q(0)}{R(0)} = 1.$$

From Eq. (10.16) we obtain

$$K = \lim_{s \to 0}[s^l \cdot Y_1(s)]. \tag{10.17}$$

If we consider the transfer function for $l = 0$ (a type-zero system) in Eq. (5.8),

$$Y_1(s) = \frac{A \cdot \prod_{i=1}^{m}(s + z_i)}{\prod_{j=1}^{n}(s + p_j)}, \tag{10.18}$$

where $-z_i$, $-p_j$ represent the zeros and poles of the open-loop system, respectively, we obtain

$$K = \frac{A \cdot \prod_{i=1}^{m} z_i}{\prod_{j=1}^{n} p_j}. \tag{10.19}$$

For a unit step input, from Eq. (6.6) we get the steady error

$$E(s) = \frac{1}{1 + \dfrac{A \cdot \prod_{i=1}^{m} z_i}{\prod_{j=1}^{n} p_j}}. \tag{10.20}$$

We will now analyze closed-loop systems. First, we consider the transfer function of a thermal heating system (9.12)–(9.14),

$$Y_1(s) = \frac{k_1}{\tau_1 s + 1},$$

with the closed-loop transfer function

$$Y(s) = \frac{X_0(s)}{X_i(s)} = \frac{Y_1(s)}{1 + Y_1(s)} = \frac{k_1^*}{s + p_1}, \tag{10.21}$$

where

$$\begin{cases} p_1 = -\dfrac{1 + k_1}{\tau_1}, \\[2mm] k_1^* = \dfrac{k_1}{\tau_1}, \end{cases} \tag{10.22}$$

and k_1, τ_1 are defined in Eq. (9.14).

The transient response for a unit step input $x_i(t)$ will be

$$X_0(s) = Y(s) \cdot X_i(s)$$

$$X_0(s) = \frac{k_1^*}{s + p_1} \cdot \frac{1}{s}. \tag{10.23}$$

Expanding Eq. (10.23) in a partial fraction expansion, we obtain

$$X_0(s) = \frac{c_0}{s} + \frac{c_1}{s + p_1}, \tag{10.24}$$

where

$$c_0 = [s \cdot X_0(s)]|_{s=0} = \frac{k_1^*}{p_1} \tag{10.25}$$

$$c_1 = [(s + p_1) \cdot X_0(s)]|_{s=-p_1} = \frac{-k_1^*}{p_1}. \tag{10.26}$$

Then, the relation (10.24) becomes

$$x_0(t) = \frac{k_1^*}{p_1} - \frac{k_1^*}{p_1} e^{-p_1 \cdot t}. \tag{10.27}$$

The transient response is presented in Fig. 10.2. It is composed of the steady-state output k_1^*/p_1 and an exponential term $(k_1^*/p_1)e^{-p_1 t}$. The steady-state error will be

$$E_s = 1 - x_0(\infty) = 1 - \frac{k_1^*}{p_1}. \tag{10.28}$$

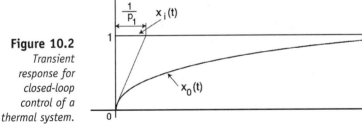

Figure 10.2
Transient response for closed-loop control of a thermal system.

We remark that when p_1 approaches the origin ($|p_1|$ decreases), the time constant $1/p_1$ and also the duration of the transient response increase. It is clear that a fast transient response requires a large p_1 that will determine the increase of the steady-state error. As a second case, we consider the closed-loop transfer function for a translational mechanism (Fig. 5.4). The open-loop transfer function is given by Eq. (9.16). The closed-loop transfer function will be

$$Y(s) = \frac{\omega_n^2}{s^2 + 2\zeta\omega_n s + \omega_n^2}, \tag{10.29}$$

where the natural frequency ω_n and damping ratio ζ are

$$\begin{cases} \omega_n^2 = \dfrac{k_1}{\tau_1} \\[2mm] \zeta = \dfrac{1}{\tau_1 \omega_n}, \end{cases} \tag{10.30}$$

and k_1, τ_1 are given in Eq. (9.17). The system poles are (Fig. 10.3)

$$p_{1,2} = -\zeta\omega_n \pm \sqrt{1-\zeta^2}. \tag{10.31}$$

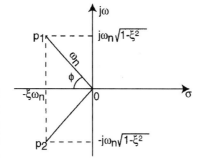

Figure 10.3
Poles of a closed-loop for a translational mechanism.

In Section 6, we obtained that the steady-state error is

$$E_s = 0 \tag{10.32}$$

for a unit step input (6.8), and

$$E_s = \frac{2\zeta}{\omega_n}, \tag{10.33}$$

for a ramp input (6.12). The *overshoot* of the transient response can be obtained by using the identity

$$s^2 + 2\zeta\omega_n s + \omega_n^2 = (s + \zeta\omega_n)^2 + (\omega_n\sqrt{1-\zeta^2})^2. \tag{10.34}$$

From Eq. (10.29),

$$X_0(s) = Y(s) \cdot X_i(s) = \frac{1}{s} - \frac{s + 2\zeta\omega_n}{s^2 + 2\zeta\omega_n s + \omega_n^2}, \tag{10.35}$$

or

$$X_0(s) = \frac{1}{s} - \left(\frac{s + \zeta\omega_n}{(s + \zeta\omega_n)^2 + (\omega_n\sqrt{1-\zeta^2})^2} \right)$$
$$+ \left(\frac{\zeta}{\sqrt{1-\zeta^2}} \cdot \frac{\omega_n\sqrt{1-\zeta^2}}{(s + \zeta\omega_n)^2 + (\omega_n\sqrt{1-\zeta^2})^2} \right). \tag{10.36}$$

The inverse Laplace transform of Eq. (10.36) will give

$$x_0(t) = 1 - \frac{e^{-\zeta \omega_n t}}{\sqrt{1 - \zeta^2}} \cdot \sin\left(\omega_n \sqrt{1 - \zeta^2}\, t + \tan^{-1}\left(\frac{\sqrt{1 - \zeta^2}}{\zeta}\right)\right). \tag{10.37}$$

The transient response is shown in Fig. 10.4. The maximum value of the time response is obtained for

$$\frac{dx_o(t)}{dt} = 0. \tag{10.38}$$

We obtain the values of time for which $x_0(t)$ achieves the extremes [4]

$$t_{ex} = \frac{k\pi}{\omega_n \sqrt{1 - \zeta^2}}; \qquad k = 0, 1, 2, \ldots. \tag{10.39}$$

For $k = 0$ we obtain the absolute minimum value at $k = 0$; for $k = 1$ we obtain the peak value time

$$T_p = \frac{\pi}{\omega_n \sqrt{1 - \zeta^2}}. \tag{10.40}$$

If we substitute T_p in Eq. (10.37) we obtain the overshoot

$$\sigma = e^{-\pi \zeta / \sqrt{1 - \zeta^2}}. \tag{10.41}$$

We see that for $\zeta = 0$, the overshoot is 100 (the system is at the limit of stability) and for $\zeta > 0.85$ the overshoot approaches zero.

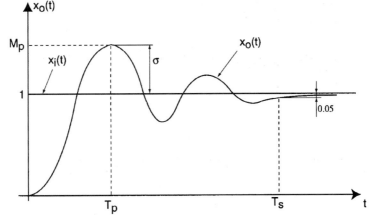

Figure 10.4
Transient response of a closed-loop control for a translational mechanism.

The *settling time* (T_s) is defined as the time required for the system to settle within a certain percentage δ of the input amplitude. From Eq. (10.36) we obtain the condition [4]

$$\frac{e^{-\zeta \omega_n T_s}}{\sqrt{1 - \zeta^2}} = \delta, \tag{10.42}$$

and

$$T_s = \frac{\ln(\delta\sqrt{1-\zeta^2})}{-\zeta\omega_n}. \tag{10.43}$$

The *bandwidth* (ω_B) was discussed in Section 8. From Eqs. (8.35), (8.40), and (8.29) we obtain

$$\frac{\omega_n^2}{\sqrt{(\omega_n^2 - \omega_B^2)^2 + (2\zeta\omega_n\omega_B)^2}} = \frac{\sqrt{2}}{2}. \tag{10.44}$$

The bandwidth ω_B will be

$$\omega_B = \omega_n\sqrt{1 - 2\zeta^2 + \sqrt{2 - 4\zeta^2 + 4\zeta^4}}. \tag{10.45}$$

For example, for $\zeta = 0.5$,

$$\omega_B \cong 1.27\omega_n, \tag{10.46}$$

and for $\zeta = 0.7$,

$$\omega_B \cong \omega_n. \tag{10.47}$$

10.3 Effects of the Supplementary Zero

We consider a closed-loop control system as in Fig. 10.1 where the controller is defined by a PD transfer function (10.11). We assumed that the controlled process is represented by the translational mechanism (9.16). The closed-loop transfer function will be

$$Y_{PD}(s) = \frac{\frac{\omega_n}{z} \cdot (s + z)}{s^2 + 2\zeta\omega_n s + \omega_n^2}, \tag{10.48}$$

where z is the zero introduced by the PD controller,

$$z = -\frac{1}{T_D}, \tag{10.49}$$

and

$$\omega_n^2 = \frac{k_p}{\tau}$$

$$\zeta = \frac{1 + k_p T_D}{2\tau\omega_n}. \tag{10.50}$$

The transfer function of the open-loop control system from Fig. 10.5 represents a type-one system, so that the steady-state error will be

$$E_s = 0, \tag{10.51}$$

for a unit step input. For a ramp input signal, we obtain from Eq. (6.10)

$$E_s = \lim_{s \to 0}\left[\frac{1}{s \cdot Y_{1_{PD}}(s)}\right], \tag{10.52}$$

Figure 10.5
A closed-loop control system with PD controller.

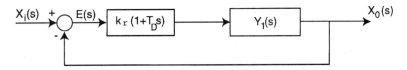

where

$$Y_{1_{PD}}(s) = \frac{Y_{PD}(s)}{1 - Y_{PD}(s)},$$

$$Y_{1_{PD}}(s) = \frac{\frac{\omega_n^2}{z} \cdot (s + z)}{s\left(s + \left(2\zeta\omega_n - \frac{\omega_n^2}{z}\right)\right)}. \qquad (10.53)$$

Substituting $Y_{1_{PD}}(s)$ in Eq. (10.52), we obtain

$$E_{S_{PD}} = \frac{2\zeta - \frac{\omega_n}{z}}{\omega_n}. \qquad (10.54)$$

It is clear that the steady-state error decreases by the value $\lambda = \omega_n/z$. If we cancel the effect of the zero, $z \to \infty$, the PD steady-state error approaches the P steady-state error,

$$E_{S_{PD}} \to E_{S_P} = \frac{2\zeta}{\omega_n}.$$

From Eq. (10.54) we also have the condition

$$2\zeta > \frac{\omega_n}{z}. \qquad (10.55)$$

In order to analyze the transient response, we will rewrite (10.48) as

$$Y_{PD}(s) = \left(1 + \frac{s}{z}\right) Y_P(s), \qquad (10.56)$$

where $Y_P(s)$ represents the closed-loop transfer function with a P controller discussed in the preceding section,

$$Y_P(s) = \frac{\omega_n^2}{s^2 + 2\zeta\omega_n s + \omega_n^2}. \qquad (10.57)$$

For a unit step input, the output $x_0(t)$ will be

$$X_0(s) = Y_P(s) \cdot X_i(s) + \frac{1}{z} s Y_P(s) X_i(s). \qquad (10.58)$$

The inverse Laplace transformation of (10.58) will give

$$x_{0_{PD}}(t) = x_{0_P}(t) + \frac{1}{z} \frac{dx_{0_P}(t)}{dt}, \qquad (10.59)$$

where $x_{0_{PD}}$, x_{0_P} denote the output signal for a PD controller or a P controller in the control system, respectively. It is clear that the overshoot of this system will be increased by the term $(1/z) \cdot (dx_{0_P}(t)/dt$ (Fig. 10.6).

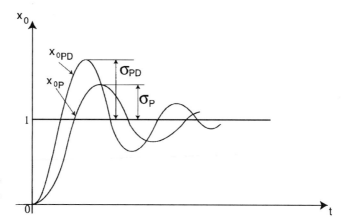

Figure 10.6
Transient response with PD controller.

From Eqs. (10.37) and (10.59) we obtain

$$x_{0_{PD}}(t) = 1 - e^{-\zeta\omega_n t} \frac{\sqrt{\lambda^2 - 2\zeta\lambda + 1}}{\sqrt{1-\zeta^2}} \sin(\omega_n\sqrt{1-\zeta^2}\,t + \gamma), \qquad (10.60)$$

where

$$\gamma = \tan^{-1} \frac{\sqrt{1-\zeta^2}}{\zeta - \lambda} \qquad (10.61)$$

$$\lambda = \frac{\omega_n}{z}. \qquad (10.62)$$

The maximum value is obtained by

$$\frac{dx_{0_{PD}}(t)}{dt} = 0, \qquad (10.63)$$

which enables us to calculate the time [4]:

$$T_{P_{PD}} = \frac{\pi - (\gamma - \varphi)}{\omega_n\sqrt{1-\zeta^2}}. \qquad (10.64)$$

We remark that if $z \to \infty$, $\lambda \to 0$, $\gamma = \varphi$ and the value of (10.64) is the same as that determined for the P-controller (10.40).

The overshoot will be

$$\sigma_{PD} = \sqrt{\lambda^2 - 2\zeta\lambda + 1} \cdot e^{-\zeta\cdot\pi - (\gamma-\varphi)/\sqrt{1-\zeta^2}}. \qquad (10.65)$$

The settling time $T_{s_{PD}}$ can be determined by using the condition

$$e^{-\zeta\omega_n T_{s_{PD}}} \cdot \frac{\sqrt{\lambda^2 - 2\zeta\lambda + 1}}{\sqrt{1-\zeta^2}} = 0.05. \qquad (10.66)$$

If we develop Eq. (10.66) and consider the settling time T_{s_P} defined by (10.42), (10.43), we obtain

$$T_{s_{PD}} = \frac{1}{\zeta\omega_n} \cdot \ln\sqrt{\lambda^2 - 2\zeta\lambda + 1} + T_{s_P}. \qquad (10.67)$$

In general, the values of ζ and λ verify the conditions (10.55)

$$0 < \zeta < 1$$
$$0 < \lambda < 2\zeta. \tag{10.68}$$

In these cases

$$\ln \sqrt{\lambda^2 - 2\zeta\lambda + 1} < 0, \tag{10.69}$$

so that

$$T_{S_{PD}} < T_{S_p}. \tag{10.70}$$

We can conclude that a PD controller does not increase the duration of the transient response. Let us now analyze the bandwidth ω_B. From Eq. (10.48) we obtain

$$20 \log M(\omega) = 20 \log \frac{\dfrac{\omega_n^2}{z}\sqrt{\omega^2 + z^2}}{\sqrt{(\omega_n^2 - \omega^2) + (2\zeta\omega_n\omega)^2}}. \tag{10.71}$$

The Bode diagram of magnitude $M(\omega)$ is presented in Fig. 10.7. The zero z introduces a new break frequency

$$\omega_z = z, \tag{10.72}$$

and a straight line with slope $+20\,\mathrm{dB/decade}$. The last line, determined by the denominator expression, will have a slope $-20\,\mathrm{dB/decade}$. We can compare the $M(\omega)$ Bode diagram defined by Eq. (10.71) with a typical $M(\omega)$ Bode diagram described by the relation (8.40).

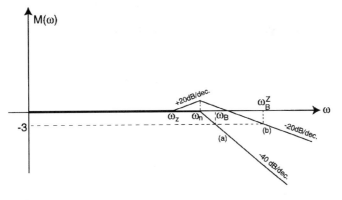

Figure 10.7
Bode diagram of $M(\omega)$: (a) for $Y(j\omega)$ defined by (8.40); (b) for $Y(j\omega)$ defined by (10.71).

Examining both diagrams, we see that

$$\omega_B^z > \omega_B. \tag{10.73}$$

In ref. 8, is proven that the increasing of the bandwidth is limited at

$$\omega_B < \omega_B^z < 2\omega_B. \tag{10.74}$$

10.4 Effects of the Supplementary Pole

In the preceding section we discussed the effects of the zeros introduced by the PD controller on the performance of closed-loop control systems.

We will analyze the effects of the poles that are added to the transfer function of the direct path by integrating controllers. We assume that the new closed-loop transfer function has the form

$$Y_I(s) = \frac{k_I}{(s^2 + 2\zeta\omega_n s + \omega_n^2)(s + p^*)}, \tag{10.75}$$

where p^* is the new pole and k_I is chosen as

$$k_I = \omega_n^2 p, \tag{10.76}$$

in order to obtain the condition

$$|Y_I(0)| = M(0) = 1. \tag{10.77}$$

The open-loop transfer function (10.53) is

$$Y_{1_I}(s) = \frac{\omega_n^2 p^*}{s(s^2 + (2\zeta\omega_n + p^*)s + (2\zeta\omega_n p^* + \omega_n^2))}. \tag{10.78}$$

The steady-state error E_{s_I} will be, for a unit step input,

$$E_{s_I} = 0, \tag{10.79}$$

and for a ramp input (2.10),

$$E_{s_I} = \frac{2\zeta + \dfrac{\omega_n}{p^*}}{\omega_n}, \tag{10.80}$$

which determines an increase of the steady-state error by the value ω_n/p^*. If we cancel the pole effect, $p \to \infty$, the steady-state error achieves the value of the P-controller steady-state error, $2\zeta/\omega_n$. The effects of the pole p^* are insignificant if the pole approaches the origin.

The transient response for a unit step input will be

$$X_{0_I}(s) = Y_I(s) \cdot \frac{1}{s} = \frac{\omega_n^2 p^*}{s(s + p^*)(s^2 + 2\zeta\omega_n s + \omega_n^2)}. \tag{10.81}$$

The partial fraction expansion of (10.81) is

$$X_{0_I}(s) = \frac{1}{s} + \frac{C_1}{s + p_1} + \frac{C_2}{s + p_2} + \frac{C_3}{s + p^*}, \tag{10.82}$$

where p_1, p_2 are the conjugate complex poles of $s^2 + 2\zeta\omega_n s + \omega_n^2$. The inverse Laplace transform of (10.82) has the form

$$x_0(t) = 1 + C_1 e^{-p_1 \cdot t} + C_2 e^{-p_2 \cdot t} + C_3 e^{-p^* \cdot t}, \tag{10.83}$$

where the last two terms represent the damped oscillation (for $0 < \zeta < 1$) of the system determined by the two conjugate complex poles. The second term $C_3 e^{-p^* \cdot t}$ represents a new exponential oscillation. The amplitude of this oscillation can be calculated by multiplying by the denominator factor of (10.82) corresponding to C_3 and setting s equal to the root

$$C_3 = [(s + p^*)X_{0_j}(s)]_{s=-p^*}.$$ (10.84)

Alternatively, the equation may be written as

$$C_3 = -\frac{\omega_n^2}{p^{*^2} - 2\zeta\omega_n p^* + \omega_n^2}.$$ (10.85)

For C_1, C_2 we can use the same procedure

$$C = [(s + p_1)X_{0_j}(s)]_{s=-p_1} = -\frac{p_2}{p_2 - p_1}$$ (10.86)

$$C_2 = [(s + p_2)X_{0_j}(s)]_{s=-p_2} = \frac{p_1}{p_2 - p_1}.$$ (10.87)

If we evaluate the relation (10.85) we see that $p^{*^2} - 2\zeta\omega_n p^* + \omega_n^2$ is always negative for $0 < \zeta < 1$ so that

$$C_3 < 0.$$ (10.88)

This inequality indicates that the pole p^* has a favorable influence on the transient response because it contributes to the diminution of the oscillation component.

In order to analyze the bandwidth ω_B^P, we will represent the $M(\omega)$ Bode diagram (Fig. 10.8). It is clear that the bandwidth ω_B^P is decreased by introducing the pole $-p^*$:

$$\omega_B^P < \omega_B.$$ (10.89)

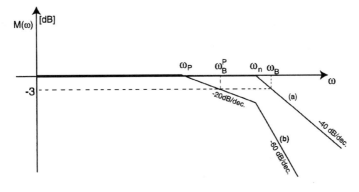

Figure 10.8
Bode diagram of $M(\omega)$: (a) for $Y(j\omega)$ defined by (8.40); (b) for $Y(j\omega)$ defined by (10.75).

10.5 Effects of Supplementary Poles and Zeros

In order to illustrate the characteristics and advantages of introducing poles and zeros, we will consider the closed-loop transfer function

$$Y_{PZ}(s) = \frac{\dfrac{p^*}{z^*}\omega_n^2(s + z^*)}{(s^2 + 2\zeta\omega_n s + \omega_n^2)(s + p^*)},$$ (10.90)

Control

where $-p^*$, $-z^*$ represent the new pole and zero and the p^*/z^* coefficient ensures the condition

$$|Y_{PZ}(0)| = 1. \tag{10.91}$$

The steady-state error will be

$$E_{S_{PZ}} = 0 \tag{10.92}$$

for a unit step input, and for a ramp input,

$$E_{S_{PZ}} = \frac{2\zeta}{\omega_n} + \left(\frac{1}{p^*} - \frac{1}{z^*}\right). \tag{10.93}$$

In this case, it is possible to improve the steady-state error if

$$\frac{1}{p^*} < \frac{1}{z^*} \tag{10.94}$$

or

$$p^* > z^* \tag{10.95}$$

and p^*, z^* approaches the origin (Fig. 10.9). The transient response for a unit step input is obtained as in Eq. (10.81),

$$X_{0_{PZ}}(s) = \frac{\dfrac{p^*}{z^*}\omega_n^2(s + z^*)}{s(s^2 + 2\zeta\omega_n s + \omega_n^2)(s + p^*)}, \tag{10.96}$$

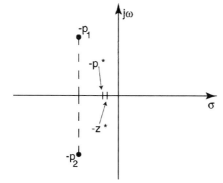

Figure 10.9
An s-plane plot of the poles and zeros.

which can be rewritten as

$$X_{0_{PZ}}(s) = \frac{1}{s} + \frac{C_1^*}{s + p_1} + \frac{C_2^*}{s + p_2} + \frac{C_3^*}{s + p^*}, \tag{10.97}$$

where $-p_1$, $-p_2$ are conjugate complex poles. The inverse Laplace transform of Eq. (10.97) will be

$$x_{0_{PZ}}(t) = 1 + C_1^* e^{-p_1 \cdot t} + C_2^* e^{-p_2 \cdot t} + C_3^* e^{-p^* \cdot t}, \tag{10.98}$$

where

$$C_1^* = [(s + p_1)X_{0_{PZ}}(s)]_{s=-p_1}, \qquad (10.99)$$

$$C_2^* = [(s + p_2)X_{0_{PZ}}(s)]_{s=-p_2} \qquad (10.100)$$

$$C_3^* = [(s + p^*)X_{0_{PZ}}(s)]_{s=-p^*}, \qquad (10.101)$$

or

$$C_1^* = -\frac{p_2}{p_2 - p_1} \cdot \frac{p^*}{z^*} \cdot \frac{z^* - p_1}{p^* - p_1} \qquad (10.102)$$

$$C_2^* = \frac{p_1}{p_2 - p_1} \cdot \frac{p^*}{z^*} \cdot \frac{z^* - p_2}{p^* - p_2} \qquad (10.103)$$

$$C_3^* = \frac{\frac{p^*}{z^*}\omega_n^2(z^* - p^*)}{-} p^*(p^{*^2} - 2\zeta\omega_n p^* + \omega_n^2). \qquad (10.104)$$

If we compare C_1^*, C_2^* with C_1, C_2 from Eqs. (10.86) and (10.87), respectively, we obtain

$$C_1^* = C_1 \frac{p^*}{z^*} \cdot \frac{z^* - p_1}{p^* - p_1} \qquad (10.105)$$

$$C_2^* = C_2 \frac{p^*}{z^*} \cdot \frac{z^* - p_2}{p^* - p_2}. \qquad (10.106)$$

If the new zero $-z^*$ and pole $-p^*$ verify the condition (see Fig. 10.9)

$$p^* \cong z^* \qquad (10.107)$$

or

$$\frac{p^*}{z^*} \cong 1, \qquad (10.108)$$

and

$$p^* \cong z^* \ll |-p_1| = |-p_2| \cong \omega_n, \qquad (10.109)$$

then

$$\left|\frac{z^* - p_1}{p^* - p_1}\right| = \left|\frac{z^* - p_2}{p^* - p_2}\right| \cong 1. \qquad (10.110)$$

From Eqs. (10.105), (10.106), and (10.110), we obtain

$$C_1^* \cong C_1 \qquad (10.111)$$

$$C_2^* \cong C_2. \qquad (10.112)$$

We conclude that the introduction of the new pole and zero $-p^*$, $-z^*$ does not influence the first two transient components of $x_{0_{PZ}}(t)$. The last component $C_3^* e^{-p^* \cdot t}$ can be analyzed from the relation (10.104) and the condition (10.108). Therefore C_3^* can be rewritten as

$$C_3^* \cong \frac{\frac{p^*}{z^*}\omega_n^2(z^* - p^*)}{-p^*\omega_n^2},$$

or

$$C_3^* \cong -1 + \frac{p^*}{z^*}. \tag{10.113}$$

Because the pole $-p^*$ and the zero $-z^*$ verify the condition (10.95), from (10.113) an alteration of the transient response results, which determines the increase of the overshoot. In order to improve the transient response, we can introduce the constraint

$$C_3^* \leq \Delta\sigma, \tag{10.114}$$

where, for a typical application, $\Delta\sigma$ is chosen as

$$\Delta\sigma = 0.01 \div 0.05, \tag{10.115}$$

which determines an overshoot variation between 1% and 5%. The bandwidth ω_B^{PZ} is not modified because the effects of the pole $-p^*$ that determines the break frequency

$$\omega_B^P = p^* \tag{10.116}$$

are compensated by the effects of the zero $-z^*$ with a break frequency

$$\omega_B^Z = z^*, \tag{10.117}$$

in the conditions for which the relation (10.107) is verified.

10.6 Design Example: Closed-Loop Control of a Robotic Arm

Consider the control system for a rotational robotic arm (Fig. 10.10), where $Y_C(s)$ represents the controller transfer function and the robotic arm is described by the transfer function

$$Y_{ARM}(s) = \frac{k_A}{s(s + \tau_A)}. \tag{10.118}$$

Equation (10.118) is easily obtained from the dynamic model described in Appendix A.1 (A.1.4) in which the gravitational term is neglected. We assume that the parameters identify the following values for k_A, τ_A [1, 4]:

$$k_A = 15$$
$$\tau_A = 95.$$

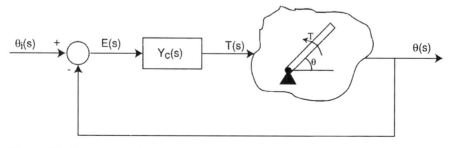

Figure 10.10 *Rotational robotic arm control system.*

First, let us try a P-controller with the transfer function

$$Y_C(s) = k_P. \tag{10.119}$$

In this case, the open-loop transfer function will be

$$Y_1(s) = \frac{15k_P}{s(s+95)}. \tag{10.120}$$

The control system requires the following performance:

Overshoot:

$$\sigma[\%] \leq 7.5\%. \tag{10.121}$$

Steady-state error:

$$E_s = 0, \tag{(10.122)}$$

for unit step input, and

$$E_s[\%] \leq 2\% \tag{(10.123)}$$

for ramp input.
Bandwidth:

$$\omega_B \leq 100. \tag{10.124}$$

The last condition of the bandwidth allows the estimation of the damping ratio. From Eq. (10.41) we obtain

$$\zeta = 0.636. \tag{10.125}$$

If we use the pole representation from Fig. 10.3, we obtain the pole phase angle

$$\varphi = \cos^{-1}\zeta = 50°30'. \tag{10.126}$$

The condition (10.123) of the ramp steady-state error determines the natural frequency from Eq. (10.33),

$$\frac{2\zeta}{\omega_n} \leq 0.02.$$

Therefore, we obtain

$$\omega_n \geq 63.5, \tag{10.127}$$

but the relation (10.45) requires

$$\omega_B = 1.1\omega_N, \tag{10.128}$$

and from the condition (10.124),

$$\omega_N \leq 91. \tag{10.129}$$

The inequalities (10.127) and (10.129) define the natural frequency domain ω_n,

$$6.5 \leq \omega_N \leq 91. \tag{10.130}$$

For a closed-loop transfer function of type (10.29), the open-loop transfer function $Y_1(s)$ has the form (10.53)

$$Y_1(s) = \frac{\omega_n^2}{s(s + 2\zeta\omega_n)}.$$
(10.131)

From the denominator expressions of relations (10.130) and (10.131), we obtain

$$2\zeta\omega_n = 95,$$

or

$$\omega_n \cong 75.$$
(10.132)

This value of ω_n verifies the condition (10.130) and can be adopted as the optimum value of the natural frequency (Fig. 10.11).

Figure 10.11
Optimal pole distribution for a closed control system of a robotic arm with P controller.

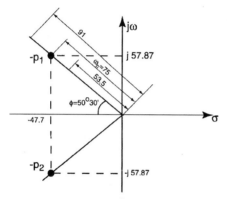

The numerator expressions of the same relations enable us to obtain

$$15k_P = \omega_n^2.$$
(10.133)

Then, the gain coefficient of the P controller will be

$$k_P = 375.$$
(10.134)

The pole distribution of the closed-loop system is (Fig. 10.11)

$$p_1 = -\zeta\omega_n + j\omega_n\sqrt{1 - \zeta^2} = -47.7 + j57.87$$
$$p_2 = -47.7 - j57.87.$$
(10.135)

We can conclude that a closed-loop control system for a robotic arm with P controller satisfies all the conditions (10.120)–(10.123), and the system parameters are

$$\omega_n = 75$$
$$\zeta = 0.636$$
$$\varphi = 50°36'.$$

Let us consider that the mechanical parameters of the arm define a transfer function by the form

$$Y_{ARM}(s) = \frac{10}{s(s+0.1)}, \qquad (10.136)$$

and the closed-loop control performances are as follows:

Overshoot:

$$\sigma[\%] \leq 7.5\%. \qquad (10.137)$$

Steady-state error:

$$E_s = 0 \quad \text{for unit step input} \qquad (10.138)$$
$$E_s[\%] = 2\% \quad \text{for ramp input.} \qquad (10.139)$$

Bandwidth:

$$\omega_B \leq 50. \qquad (10.140)$$

In this case, the conditions (10.127), (10.129) require

$$\zeta = 0.636$$
$$\omega_n \geq 63.5, \qquad (10.141)$$

but, from Eq. (10.45),

$$\omega_B = 1.1\omega_n. \qquad (10.142)$$

It is clear that the condition (10.140) can not be verified. In this case we propose introducing an *additional pole and zero* p^*, z^*, in the transfer function of the system. First, we divide the overshoot in two parts,

$$\sigma = \sigma^* + \sigma_{PZ}, \qquad (10.143)$$

where σ_{PZ} is the overshoot determined by the additional pole and zero. We estimate σ_{PZ} as

$$\sigma_{PZ} = 0.03. \qquad (10.144)$$

Then σ^* determined by the main poles will be

$$\sigma^* = 0.45. \qquad (10.145)$$

From Eq. (10.41) we obtain

$$\zeta \cong 0.7. \qquad (10.146)$$

and

$$\varphi = \cos^{-1}\zeta = 45°. \qquad (10.147)$$

In the preceding section we established that the influence of an additional pole and zero on the bandwidth is negligible, and for $\zeta = 0.07$ we have

$$\omega_B \cong \omega_n.$$

From the condition (10.140), we impose

$$\omega_B = \omega_n = 50. \qquad (10.148)$$

It follows that the main pole positions will be defined by (Fig. 10.12)

$$p_1 = -35 + j35$$
$$p_2 = -35 - j35.$$

(10.149)

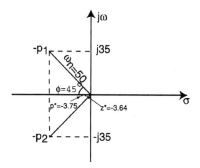

Figure 10.12
Pole-zero distribution for a closed control system of a robotic arm with complex structure controller.

We can remark that these pole positions do not verify the ramp steady-state error (10.139). Indeed, we have

$$E_s = \frac{2\zeta}{\omega_n} = 0.028$$
$$E_s[\%] = 2.8\%.$$

(10.150)

From Eqs. (10.93) and (10.139), the effects of the new pole and zero on the ramp steady-state error determine the condition

$$\frac{1}{z^*} - \frac{1}{p^*} = \frac{0.4}{50},$$

(10.151)

and the overshoot σ_{PZ} from Eq. (10.144) determines a new equation [(10.113)–(10.115)],

$$\frac{p^*}{z^*} = 1.03.$$

(10.152)

Solving p^*, z^* from Eqs. (10.151) and (10.152), we have

$$p^* = 3.75$$
$$z^* = 3.64.$$

(10.153)

From Eq. (10.90), the closed-loop transfer function will be

$$Y_{PZ}(s) = \frac{2523(s + 3.64)}{(s^2 + 70s + 2450)(s + 3.75)}.$$

(10.154)

The open-loop transfer function of (10.154) is obtained by the form

$$Y_{1_{PZ}}(s) = \frac{2523(s + 3.64)}{s(s + 71.07)(s + 2.67)},$$

(10.155)

but

$$Y_{1_{PZ}}(s) = Y_C(s)Y_{ARM}(s) = Y_C(s)\frac{10}{s(s+0.1)}. \tag{10.156}$$

From the relations (10.155) and (10.156) we obtain the transfer function of the controller,

$$Y_C(s) = \frac{253.3(s+3.64)(s+0.1)}{(s+71.07)(s+2.67)}. \tag{10.157}$$

This transfer function defines a complex structure controller that can be obtained by a cascade connection of standard controllers (PD) and compensator networks.

11. Design of Closed-Loop Control Systems by Frequential Methods

In Section 8, frequency-domain performance was discussed and the main advantages for designing in this field were specified. These results will be used to deduce the transfer function of the controller in a closed-loop control system and to adjust its parameters in order to satisfy the system performance. We will discuss this procedure by examining a typical model, a second-order system, described by a transfer function

$$Y_1(s) = \frac{k_1}{s(\tau_1 s + 1)}, \tag{11.1}$$

which defines the dynamic behavior of the translational mechanism. We assume that the following performances are imposed:

Settling time:

$$T_s \leq T_{simp}; \tag{11.2}$$

Overshoot:

$$\sigma \leq \sigma_{imp}; \tag{11.3}$$

Steady-state error:

$$E_s = 0 \quad \text{for a unit step input} \tag{11.4}$$

$$E_s \leq E_{simp} \quad \text{for a ramp input.} \tag{11.5}$$

The first step is the same as the one we discussed in the previous section. We will try to identify the position of the main poles. The condition (11.3) and the relation (10.41) enable us to calculate the damping ratio ξ, and the condition (11.2) introduced in the relation (10.43) determines the natural frequency ω_n.

Then, we can estimate the transfer function of the closed-loop system that satisfies the first two conditions (11.2) and (11.3),

$$Y(s) = \frac{\omega_n^2}{s^2 + 2\xi\omega_n s + \omega_n^2},\tag{11.6}$$

or, in the frequency domain,

$$Y(j\omega) = \frac{\omega_n^2}{(\omega_n^2 - \omega^2) + j2\xi\omega_n\omega}.\tag{11.7}$$

The magnitude plot $20\log|Y(j\omega)|$ is presented in Fig. 11.1 (curve a). This plot has a break point at frequency ω_n, and high frequency is represented by a straight line with a slope of $-40\,\mathrm{dB/decade}$.

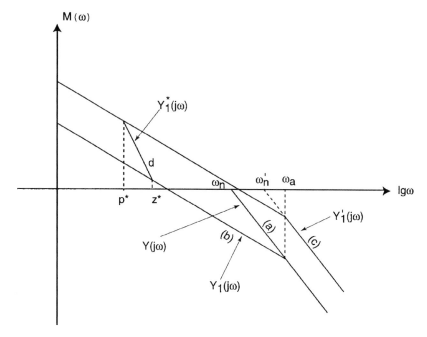

Figure 11.1
Bode plots for the design by frequency methods.

From the transfer function (11.7) we obtain the open-loop transfer function

$$Y_1(j\omega) = \frac{\omega_n^2}{j\omega(j\omega + 2\xi\omega_n)},\tag{11.8}$$

which has the same representation at high frequency as the closed-loop transfer function (11.7) (a straight line with slope $-40\,\mathrm{dB/decade}$), but the break frequency is (curve b, Fig. 11.1)

$$\omega_a = 2\xi\omega_n.\tag{11.9}$$

The transfer function (11.8) defines a type-one system, which determines a steady-state error $E_s = 0$ for a unit step input so that the condition (11.4) is verified.

In order to solve the condition of steady-state error for a ramp input (11.5), we rewrite the relation (11.8) as

$$Y_1(j\omega) = \frac{\omega_n/2\xi}{j\omega\left(\dfrac{j\omega}{2\xi\omega_n}+1\right)}. \tag{11.10}$$

The numerator expression defines a gain factor k,

$$k = \frac{\omega_n}{2\xi} = \frac{1}{E_s}. \tag{11.11}$$

If condition (11.5) is not satisfied, we can increase the gain k_p of the controller, so that the overall gain of the open loop system is

$$k' k_c = \frac{1}{E_{simp}}, \tag{11.12}$$

where k_c defines the critical value of the gain that satisfies the condition (11.5). Of course, this new gain modifies the damping ratio ξ' and the natural frequency ω'_n;

$$k' = \frac{\omega'_n}{2\xi'}, \tag{11.13}$$

but

$$2\xi\omega_n = 2\xi'\omega'_n = \omega_a. \tag{11.14}$$

The new transfer function is represented by curve c in Fig. 11.1:

$$Y_1'(j\omega) = \frac{k'}{j\omega\left(\dfrac{j\omega}{2\xi\omega_n}+1\right)}. \tag{11.15}$$

The new natural frequency ω'_n can be evaluated by the intersection of the $|Y_1'(j\omega)|$ high-frequency plot (the slope $-40\,\text{dB/decade}$) and the ω-axis. We can remark that

$$\omega'_n > \omega_n, \tag{11.16}$$

which determines the damping ratio

$$\xi' < \xi, \tag{11.17}$$

which can increase the prescribed value of the overshoot. In order to eliminate these difficulties, we introduce a cascade network that must have a frequency response of the same magnitude as the type $Y_1'(j\omega)$ (curve c) for small frequencies while, for medium frequencies, having a frequency response of the magnitude of the type $Y_1(j\omega)$ (curve b). In this case, we ensure the steady-state performance ($t \to \infty$ or $\omega \to 0$) and transient performance for the frequencies $\omega \sim \omega_n$. This network will introduce a zero z^* and a pole p^*. The magnitude plot is presented in Fig. 11.2.

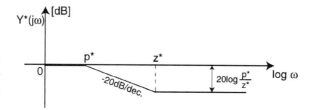

Figure 11.2
|Y*(jω)| plot for a compensation network.

The transfer function is defined by

$$Y_1^*(j\omega) = \frac{\dfrac{1}{z^*}j\omega + 1}{\dfrac{1}{p^*}j\omega + 1} = \frac{p^*}{z^*}\frac{j\omega + z^*}{j\omega + p^*}, \tag{11.18}$$

where

$$p^* < z^*. \tag{11.19}$$

The overall open-loop transfer function will be

$$Y_1^*(j\omega) = \frac{k'\left(\dfrac{1}{z^*}j\omega + 1\right)}{j\omega\left(\dfrac{j\omega}{2\xi\omega_n} + 1\right)\left(\dfrac{1}{p^*}j\omega + 1\right)}. \tag{11.20}$$

The magnitude plot of $Y_1^*(j\omega)$ is presented in Fig. 11.1 curve d. We see that if we make a good selection of the coefficients p^* and z^*, we can satisfy all the performances for steady and transient states.

12. State Variable Models

The state variable method represents an attractive method for the analysis and design of control systems based on reconsidering the dynamic models of the systems described by differential equations. Thus, these methods represent time-domain techniques, in which the response and description of a system are given in terms of time t. The time-domain methods can be readily used for nonlinear systems, for time-varying control systems for which one or more of the parameters of the system may vary as a function of time, for multivariable systems (the systems with several inputs and outputs) etc. In this sense, these methods represent stronger techniques than the classical methods of the Laplace transform or frequency response.

State variables are those variables that determine the future behavior of a system when the present state and the input signals are known.

The state variables are represented by a state vector

$$x = [x_1, x_2, \ldots, x_n]^T, \tag{12.1}$$

where the components $x_1, x_2, \ldots, x_n$ define the system state variables. The state of the system is described by a set of first-order differential equations [5, 6, 8, 9, 18] written in terms of the state variables

$$\dot{x}_1 = a_{11}x_1 + a_{12}x_2 + \cdots + a_{1n}x_n + b_{11}u_1 + \cdots + b_{1m}u_m$$
$$\dot{x}_2 = a_{21}x_1 + a_{22}x_2 + \cdots + a_{2n}x_n + b_{21}u_1 + \cdots + b_{2m}u_m$$
$$\vdots$$
$$\dot{x}_n = a_{n1}x_1 + a_{n2}x_2 + \cdots + a_{nn}x_n + b_{n1}u_1 + \cdots + b_{nm}u_m, \tag{12.2}$$

where the new variables $u_1, u_2, \ldots, u_m$ represent the input signals.

Equation (12.2) can be rewritten in matrix form,

$$\dot{x} = Ax + Bu, \tag{12.3}$$

where

$$A = \begin{bmatrix} a_{11} & a_{12} & \cdots & a_{1n} \\ a_{21} & a_{22} & \cdots & a_{2n} \\ \vdots & & & \\ a_{n1} & a_{n2} & \cdots & a_{nn} \end{bmatrix} \tag{12.4}$$

$$B = \begin{bmatrix} b_{11} & \cdots & b_{1m} \\ \vdots & & \\ b_{n1} & \cdots & b_{nn} \end{bmatrix},$$

and

$$u = [u_1, u_2, \ldots, u_m]^T \tag{12.5}$$

defines the input vector of the system.

The initial state of the system is defined by the vector

$$x_0 = [x_1(t_0), x_2(t_0), \ldots, x_n(t_0)]^T. \tag{12.6}$$

The state variables are not all readily measurable or observable. The variables that can be measured represent the output variables. They are defined by the matrix equation

$$y = Cx + Du, \tag{12.7}$$

where C and D are $(p \times n), (p \times m)$ constant matrices and y is the output vector

$$y = [y_1, y_2, \ldots, y_p]^T. \tag{12.8}$$

In order to illustrate the concept of the state variables, we can use several examples.

The first example is represented by the linear spring–mass–damper mechanical system (Fig. 12.1). From Appendix A we obtain the differential equation that describes the behavior of this system,

$$M\ddot{z} + k_f \dot{z} + kz = AP. \tag{12.9}$$

Figure 12.1
*The linear
spring–mass–
damper
mechanical
system.*

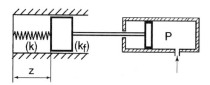

The state variables that can define this system rigorously are the position and the velocity. We can write

$$x_1 = z$$
$$x_2 = \dot{z},$$
(12.10)

the system state variables. The input is pressure $p = u$. Equation (12.10) can be rewritten as

$$\dot{x}_1 = x_2$$
$$\dot{x}_2 = -\frac{k_f}{M}x_2 - \frac{k}{M}x_1 + \frac{A}{M}u.$$
(12.11)

In the matrix form (12.3), we will have

$$A = \begin{bmatrix} 0 & 1 \\ -\dfrac{k}{M} & -\dfrac{k_f}{M} \end{bmatrix}$$
(12.12)

$$B = \frac{A}{M}.$$

We assume that only the position is measurable, so that we have for the output

$$y = Cx,$$
(12.13)

where

$$y = x_1,$$
$$C = [1 \quad 0].$$
(12.14)

The second example is represented by a coupled spring–mass system shown in Fig. 12.2.

The dynamic model is described by the differential equations

$$\begin{cases} m_1\ddot{z}_1 + k_1(z_1 - z_2) = F \\ m_2\ddot{z}_2 + k_f\dot{z}_2 + k_2z_2 - k_1(z_1 - z_2) = 0. \end{cases}$$
(12.15)

Figure 12.2
*The coupled
spring–mass
system.*

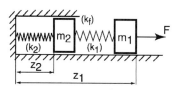

We define the state vector as

$$x = [x_1, x_2, x_3, x_4]^T,$$

where

$$x_1 = z_1$$
$$x_2 = \dot{z}_1$$
$$x_3 = z_2$$
$$x_4 = \dot{z}_2,$$

the input is

$$u = F,$$

and the output variables are represented by positions z_1, z_2

$$y = [y_1, y_2]^T.$$

Equations (12.3) and (12.7) will have the

$$A = \begin{bmatrix} 0 & 1 & 0 & 0 \\ -\dfrac{k_1}{m_1} & 0 & \dfrac{k_1}{m_1} & 0 \\ 0 & 0 & 0 & 1 \\ \dfrac{k_1}{m_1} & 0 & -\left(\dfrac{k_f}{m_2} + \dfrac{k_1}{m_1}\right) & -\dfrac{k_f}{m_2} \end{bmatrix} \qquad (12.16)$$

$$B = \begin{bmatrix} 0 \\ \dfrac{1}{m_1} \\ 0 \\ 0 \end{bmatrix}$$

$$C = \begin{bmatrix} 1 & 0 & 0 & 0 \\ 0 & 0 & 1 & 0 \end{bmatrix}. \qquad (12.17)$$

The mathematical model offered by the matrix equations (12.3) and (12.7) is called in the literature [9, 18] "the input–state–output" model.

In matrix form, the solution of Eq. (12.3) can be written as an exponential function [8, 9, 18]:

$$x(t) = \exp(At)x(0) + \int_0^t \exp[A(t - \tau)]Bu(\tau)d\tau. \qquad (12.18)$$

The Laplace transform of this relation has the form

$$X(s) = [sI - A]^{-1}x(0) + [sI - A]^{-1}BU(s), \qquad (12.19)$$

where $X(s)$, $U(s)$ are the Laplace transforms of the state and input vectors, and

$$[sI - A]^{-1} = \phi(s) \qquad (12.20)$$

is the Laplace transform of

$$\phi(t) = \exp A(t). \qquad (12.21)$$

$\phi(t)$ is called the transition matrix.

The solution of the state equation can be rewritten as

$$x(t) = \phi(t)x(0) + \int_0^t \phi(t-\tau)Bu(\tau)d\tau. \tag{12.22}$$

If the initial conditions $x(0)$, the input $u(t)$, and the transition matrix $\phi(t)$ are known, we can calculate the time response $x(t)$. Thus, finding the transition matrix is a very important issue.

If we consider the input $u = 0$, we obtain

$$x(t) = \phi(t)x(0), \tag{12.23}$$

or

$$X(s) = \phi(s)x(0). \tag{12.24}$$

We can rewrite Eq. (12.23) by components:

$$\begin{bmatrix} x_1(t) \\ x_2(t) \\ \vdots \\ x_n(t) \end{bmatrix} = \begin{bmatrix} \phi_{11}(t) & \phi_{12}(t) & \cdots & \phi_{1n}(t) \\ \phi_{21}(t) & & \cdots & \phi_{2n}(t) \\ \vdots & & & \\ \phi_{n1}(t) & \phi_{n2}(t) & \cdots & \phi_{nn}(t) \end{bmatrix} \begin{bmatrix} x_1(0) \\ x_2(0) \\ \vdots \\ x_n(0) \end{bmatrix} \tag{12.25}$$

From this equation we see that the matrix coefficient $\phi_{ij}(t)$ is the response of the ith state variable due to an initial condition on the jth state variable when there are zero initial conditions for all the other states,

$$\phi_{ij}(t) = x_i(t) \left|\begin{array}{l} \\ x_j(0) = 1 \\ x_k(0) = 0, \qquad \forall k \neq j \\ u(t) = 0 \end{array}\right. \tag{12.26}$$

or by the Laplace transform

$$\Phi_{ij}(s) = X_i(s) \left|\begin{array}{l} \\ x_j(0) = 1 \\ x_k(0) = 0, \qquad \forall k \neq j \\ U(s) = 0 \end{array}\right. \tag{12.27}$$

There are several techniques that allow the evaluation of the matrix coefficients $\phi_{ij}(t)$ or $\phi_{ij}(s)$ [18]. In order to illustrate these methods, we will use the signal flow diagram of the system presented in Appendix A.4. We will develop this procedure for the linear spring–mass–damper mechanical system (Fig. 12.1) described by Eq. (12.11). The signal diagram flow in Laplace variable is shown in Fig. 12.3.

We note, therefore, that in order to determine the matrix coefficients, it is necessary to evaluate the $X_i(s)$, changing the initial conditions $x_i(0)$. Thus, the coefficient $\phi_{11}(s)$ is obtained from the initial conditions $x_1(0) = 1$; $x_2(0) = 0$.

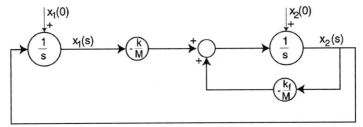

Figure 12.3
The signal-flow diagram for the linear spring–mass–damper mechanical system.

From Fig. 12.3 we can easily obtain

$$X_1 = \frac{1}{s}(1 + X_2)$$

$$X_2 = \frac{1}{s}\left(-\frac{k}{M}X_1 - \frac{k_f}{M}X_2\right),$$

then

$$\phi_{11}(s) = X_1(s) = \frac{s + \dfrac{k_f}{M}}{s^2 + \dfrac{k_f}{M}s + \dfrac{k}{M}}. \tag{12.28}$$

If we repeat this procedure for all matrix coefficients, we obtain

$$\phi_{12}(s) = \frac{1}{s^2 + \dfrac{k_f}{M}s + \dfrac{k}{M}}$$

$$\phi_{21}(s) = \frac{s + \dfrac{k_f}{M}}{s^2 + \dfrac{k_f}{M}s + \dfrac{k}{M}} \tag{12.29}$$

$$\phi_{22}(s) = \frac{s}{s^2 + \dfrac{k_f}{M}s + \dfrac{k}{M}}.$$

The transition matrix $\phi(t)$ is obtained by the inverse Laplace transforms of $\phi_{ij}(s)$. The stability of the state variable models can be easily studied by analyzing matrix A. Indeed, the unforced system has the form

$$\dot{x} = Ax, \tag{12.30}$$

which gives an exponential solution of $x(t)$ (12.18). It has been proven [8, 9, 17, 18] that the stability of the system (12.30) is obtained by solving the characteristic equation

$$\det(\lambda I - A) = 0. \tag{12.31}$$

The placement of the characteristic equation roots, the A eigenvalues in the left-hand part, will determine the system stability. For example, if we consider the linear spring–mass–damper system from Fig. 12.1 with $k/M = 2$, $k_f/M = 3$, from Eq. (12.12) we obtain

$$A = \begin{bmatrix} 0 & 1 \\ -2 & -3 \end{bmatrix}.$$

The characteristic equation will be

$$\det \begin{bmatrix} \lambda & -1 \\ 2 & \lambda + 3 \end{bmatrix} = 0,$$

which has the roots $\lambda_1 = -2$, $\lambda_2 = -1$. The system is stable.

13. Nonlinear Systems

In the preceding sections we have discussed the analysis methods and design techniques of systems for which linear models are valid. However, in the control system there are many nonlinearities whose discontinuous nature does not allow linear approximation. These nonlinearities include Coulomb friction, saturation, dead zones, and hysteresis and are found in a great number of models in control engineering. Their effects cannot be derived by linear methods, and nonlinear analysis techniques must be developed to predict a system's performance in the presence of these inherent nonlinearities.

13.1 Nonlinear Models: Examples

Nonlinearities in the mechanical systems can be classified as inherent (natural) and intentional (artificial).

Inherent nonlinearities are those that are produced in natural ways. Examples of inherent nonlinearities include centripetal forces in rotational motion, and Coulomb friction between contacting surfaces. Artificial nonlinearities are introduced by the designer in order to improve system performance. We offer, as typical examples, the nonlinear control laws, the adaptive control law, and the sliding control.

Nonlinearities can also be classified [2, 3, 12] in terms of their mathematical properties as continuous and discontinuous. The discontinuous nonlinearities cannot be locally approximated by linear functions, for example, hysteresis or saturation.

In this section we will present several typical nonlinearities and nonlinear models.

1. The gravitational pendulum with rotational spring–mass–damper mechanical system is presented in Fig. 13.1. The dynamic model is described by the differential equation

$$J\ddot{\theta} + k_f\dot{\theta} + k\theta + Mg\frac{l}{2}\sin\theta = T, \qquad (13.1)$$

where the nonlinearity is defined by the gravitational component $Mg\sin\theta$.

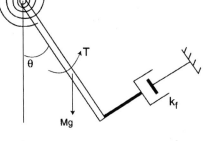

Figure 13.1
Gravitational pendulum with rotational spring–mass–damper system.

2. The nonlinear mass–damper–spring system is presented in Fig. 13.2. The dynamic equation of the free system is [12]

$$M\ddot{x} + b\dot{x}|\dot{x}| + kx + k_1x^3 = 0, \qquad (13.2)$$

where $b\dot{x}|\dot{x}|$ represents the nonlinear damping and $(kx + k_1x^3)$ represents the nonlinear spring.

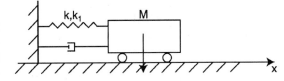

Figure 13.2
Nonlinear mass–damper–spring system.

3. The hydraulic actuator used for the linear positioning of a mass is shown in Fig. 13.3 [8]. An input displacement x moves the control value, and thus fluid passes into the upper part of the cylinder and the piston is moved. When the input is small, its increase leads to a corresponding (often proportional) increase of the output, the piston displacement. But when the input reaches a certain level, its further increase produces little or no increase of the output. The output simply stays around its maximum value. The device is said to be in saturation (Fig. 13.4).

4. Transmission systems frequently offer a nonlinearity termed backlash [12]. It is caused by the small gaps that exist in transmission mechan-

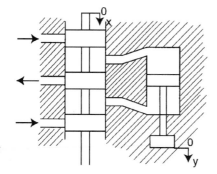

Figure 13.3
Hydraulic actuator.

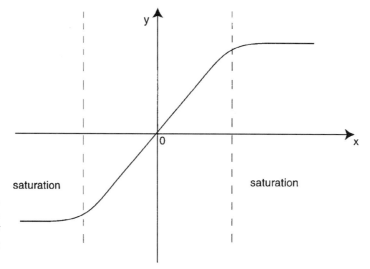

saturation

saturation

Figure 13.4
Saturation nonlinearity of the hydraulic actuator.

isms. These gaps are determined by the unavoidable errors in manufacturing and assembly. As a result of the gaps, when the driving gear (Fig. 13.5) rotates a smaller angle than the gap, the driven gear does not move at all, which corresponds to the dead-zero (OA segment in Fig. 13.5); after contact has been established between the two gears, the driver gear follows the rotation of the driving gear in a linear fashion (AB segment). When the driving gear rotates in the reverse direction by a distance of two gaps, the driven gear does not move (BC segment). After the contact between the two gears is established, the driven gear follows the rotation of the driving gear (CD segment). The overall nonlinearity is presented in Fig. 13.5 [12].

5. The two-axis planar articulated robot is an example of the complexity of the dynamic model of this class of mechanical structures. Let us consider the planar robotic structure in Fig. 13.5b. Applying the Denavit–Hartenberg algorithm, we obtain the differential equations that describe the system [1, 13],

$$\tau_1 = a_1 \ddot{q}_1 + a_2 \ddot{q}_2 + b_1 \dot{q}_1 \dot{q}_2 + b_1 \dot{q}_2^2 + c_1$$
$$\tau_2 = a_3 \ddot{q}_1 + a_4 \ddot{q}_2 + b_3 \dot{q}_1^2 + c_2,$$

(13.3)

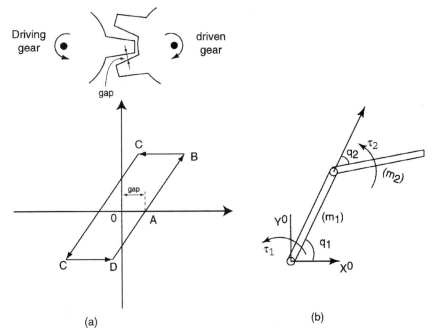

Figure 13.5
(a) A backlash nonlinearity in a transmission mechanism. (b) A two-axis planar articulated robot.

where q_1, q_2 represent the generalized coordinates of the motions; τ_1, τ_2, define the input torques; a_1, a_2, a_3, a_4, b_1, b_2, b_3, c_1, c_2 are nonlinear coefficients of the motion parameters [1],

$$a_1 = m_1 l_1^2 + J_1 + m_2(l_1^2 + l_{c_2}^2 + 2l_1 l_{c_2} \cos q_2) + J_2$$
$$a_2 = a_3 = m_2 l_1 l_{c_2} \cos q_2 + m_2 l_{c_2}^2 + J_2$$
$$a_4 = m_2 l_{c_2}^2 + J_2$$
$$b_2 = b_3 = \frac{b_1}{2} = m_2 l_1 l_{c_2} \sin q_2$$
$$c_1 = m_1 l_{c_1} g \cos q_1 + m_2 g[l_{c_2} \cos(q_1 + q_2) + l_1 \cos q_1]$$
$$c_2 = m_2 g l_{c_2} \cos(q_1 + q_2);$$

and m_1, m_2, l_1, l_2, J_1, J_2, l_{c_1}, l_{c_2} represent the parameters of the mechanical structure.

13.2 Phase Plane Analysis

The phase plane method is concerned with the graphical study of second-order autonomous systems described by [2, 3, 12]

$$\dot{x}_1 = f_1(x_1, x_2)$$
$$\dot{x}_2 = f_2(x_1, x_2), \tag{13.4}$$

where x_1, x_2 represent the system state variables and f_1, f_2, are nonlinear functions of the states.

The state space of the x_1, x_2 defines a plane called the *phase plane*. A solution $x(t)$ of Eq. (12.1) defines a phase plane trajectory, and a family of these trajectories represents a phase portrait of the system [2].

An important concept in phase plane analysis is that of a *singular point*. A singular point is an equilibrium point in the phase plane, which implies the conditions

$$\dot{x}_1 = 0$$
$$\dot{x}_2 = 0. \tag{13.5}$$

From conditions (13.5) and Eqs. (13.4) we obtain the equilibrium relations

$$f_1(x_1, x_2) = 0$$
$$f_2(x_1, x_2) = 0. \tag{13.6}$$

From Eq. (13.6) we obtain the values x_1, x_2 that define the equilibrium point.

There are several techniques for generating phase plane portraits, by using analytical graphical [2] and numerical methods [3] based on computers. The analytical methods are based on the behavior of nonlinear systems similar to a linear system around each equilibrium point. Consider x^0 the equilibrium point and we can define a vicinity around x^0,

$$x = x^0 + \mu y. \tag{13.7}$$

In this vicinity, Eqs. (13.4) can be rewritten as

$$y = Ay + \dot{g}(\mu, x_1^0, x_2^0), \tag{13.8}$$

where

$$\lim_{\mu \to 0} g(\mu, x_1^0, x_2^0) = 0. \tag{13.9}$$

The matrix A is the Jacobian of f_1, f_2, and it has the form [3]

$$A = \begin{bmatrix} \dfrac{\partial f_1}{\partial x_1} & \dfrac{\partial f_1}{\partial x_2} \\ \dfrac{\partial f_2}{\partial x_1} & \dfrac{\partial f_2}{\partial x_2} \end{bmatrix} \mu = 0 = \begin{bmatrix} a_{11} & a_{12} \\ a_{21} & a_{22} \end{bmatrix}$$
$$x_1 = x_1^0 \tag{13.10}$$
$$x_2 = x_2^0.$$

The characteristic equation of the matrix A has the form

$$\Delta: \lambda^2 + a_1 \lambda + a_0 = 0, \tag{13.11}$$

where

$$a_1 = -(a_{11} + a_{12}) = -trA$$
$$a_0 = \det A. \tag{13.12}$$

The trajectories in the vicinity of this singularity point can display different characteristics, depending on the values of the characteristic equation roots, λ_1, λ_2. We have the following cases [12]:

■ A *stable node* is obtained when both eigenvalues are negative. In this case $x_1(t) = x(t)$ and $x_2(t) = \dot{x}(t)$ converge to zero exponentially (Fig. 13.6a).

■ An *unstable node* is obtained when both eigenvalues are positive, $\lambda_1 > 0$, $\lambda_2 > 0$, and $x(t)$, $\dot{x}(t)$ diverge from zero exponentially (Fig. 13.6b).

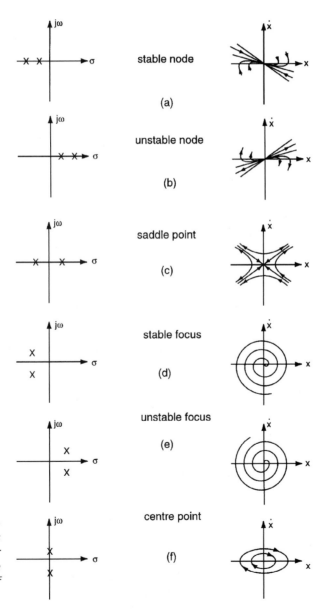

Figure 13.6
Phase portraits of nonlinear systems in the vicinity of singular points.

- *A saddle point* corresponds to the case when $\lambda_1 < 0$, $\lambda_2 < 0$. Now, the system trajectory determined by λ_2 will diverge to infinity (Fig. 13.6c).
- A *stable focus* is obtained when both eigenvalues are conjugate complex and the real parts are negative. The trajectories converge to the origin but encircle the origin one or more times (Fig. 13.6d).
- *An unstable focus* is determined by the case when both eigenvalues have positive real parts. The trajectories encircle the origin and diverge to infinity (Fig. 13.6e).
- *A center point* is produced when both eigenvalues have real parts equal to zero (Fig. 13.6f). The trajectories are ellipses.

We consider as an example the mechanical system presented in Fig. 13.7. This system consists of a rigid beam with a rotational spring around a center pivot and a solid ball rolling along a groove in the top of the beam. The control problem is to position the ball in the desired position $\theta_d = 0$, by using a torque applied to the beam as a control input at the pivot.

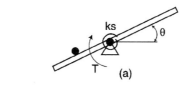

Figure 13.7

Control system for the mechanical system of ball and beam.

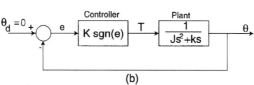

The dynamic model of the mechanical system can be approximated by

$$J\ddot{\theta} + k_s\theta = T, \qquad (13.13)$$

where k_s is the spring constant. We neglect the mass of the ball.

The nonlinearity of the control system is determined by the controller, which is a bang-bang controller:

$$T = -k\,\mathrm{sgn}(e). \qquad (13.14)$$

The unforced system is described by

$$\ddot{\theta} + \frac{k_s}{J}\theta = 0, \qquad (13.15)$$

but

$$\ddot{\theta} = \frac{d\dot{\theta}}{dt} = \frac{d\dot{\theta}}{d\theta}\frac{d\theta}{dt} = \dot{\theta}\frac{d\dot{\theta}}{d\theta}, \qquad (13.16)$$

so that Eq. (13.15) can be rewritten as

$$\dot{\theta}\,d\dot{\theta} + \frac{k_s}{J}\,\theta\,d\theta = 0. \tag{13.17}$$

Integrating this equation yields

$$\dot{\theta}^2 + \frac{k_s}{J}\,\theta^2 = c. \tag{13.18}$$

The characteristic equation eigenvalues are

$$\lambda_{1,2} = \pm j\sqrt{\frac{k_s}{J}}. \tag{13.19}$$

Therefore, the phase trajectories are a family of ellipses and the singular point is a center point (Fig. 13.8). These cases represent the behavior of nonlinear systems in the vicinity of singular points similarly to approximated linear systems, but nonlinear systems can have more complicated behavior in terms of limit cycles.

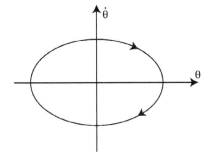

Figure 13.8
Phase portrait for the unforced system of ball and beam.

In the phase plane, a limit cycle is defined as an isolated closed curve. Depending on the trajectories in the vicinity of the limit cycle, there are the following types of limits cycles [12]:

- *Stable limit cycles* at which all trajectories in the vicinity of the limit cycle converge to it.
- *Unstable limit cycles* where all trajectories in the vicinity of the limit cycle diverge from it.
- *Semistable limit cycles* where some of the trajectories in the vicinity of the limit cycle converge to it and others diverge from it (Fig. 13.9).

13.3 Stability of Nonlinear Systems

We reconsider the nonlinear system described by Eq. (13.4) in the general form

$$\dot{x} = f(x, u, t), \tag{13.20}$$

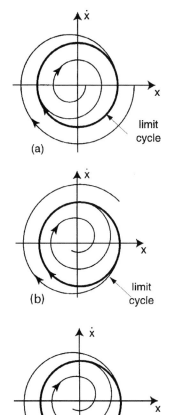

Figure 13.9
Limit cycles:
(a) stable,
(b) unstable,
(c) semistable.

where u is the system input and $x^0 = x(0)$ is the initial condition. The nonlinear system (13.20) is said to be autonomous if f does not depend explicitly on time:

$$\dot{x} = f(x, u). \tag{13.21}$$

Otherwise, the system is called nonautonomous [12]. We define the equilibrium points by relations (13.5) and (13.6). If we now consider a constant input $u(t) = u^*$, we can define an equilibrium point x^* of the system (13.21) associated with the input $u(t) = u^*$, a point in the state space that verifies the condition [3, 12]

$$f(x^*, u^*) = 0. \tag{13.22}$$

It is evident that $\dot{x}(t) = 0$ at each equilibrium point. Thus, if $x(0) = x^0$ is an equilibrium point, then $x(t) = x^0$ for $t \geq 0$.

An autonomous nonlinear system can have no equilibrium points, one equilibrium point, or multiple equilibrium points. We consider as an example the dynamic model of the one-axis robot (Fig. 13.10) [13, 21],

$$\left(\frac{m}{3} + m_l\right)a^2\ddot{q} + g\left(\frac{m}{2} + m_l\right)a\cos q + b(\dot{q}) = T, \tag{13.23}$$

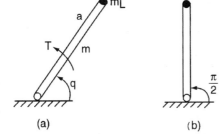

Figure 13.10
(a) One-axis robot.
(b) Equilibrium point.

(a)   (b)

where m, m_l define the arm and load mass, respectively, q is the generalized coordinate of the system, and T is the input torque. Equation (13.23) can be written in terms of state variables

$$\begin{aligned} x_1 &= q \\ x_2 &= \dot{q}, \end{aligned} \tag{13.24}$$

as

$$\dot{x}_1 = x_2,$$

$$\dot{x}_2 = \frac{1}{a^2\left(\dfrac{m}{3} + m_2\right)}\left(-g\left(\frac{m}{2} + m_l\right)a\cos x_1 - b(x_2) + T\right). \tag{13.25}$$

If the input torque is $T = u^* = $ constant, the equilibrium point $x_1^* = q^*$ is obtained from the condition

$$\cos x_1^* = \frac{u^*}{ag\left(\dfrac{m}{2} + m_l\right)}, \tag{13.26}$$

because the term $b(x_2)$ satisfies the relation

$$b(0) = 0. \tag{13.27}$$

If the system is unforced, $u^* = 0$, we obtain (Fig. 13.10b)

$$x_1^* = \frac{\pi}{2}. \tag{13.28}$$

Definition and knowledge of the equilibrium points are the essential elements in the interpretation of the asymptotic nonlinear system stability. An equilibrium point x^* of the system (13.21) is asymptotically stable [12] if and only if for each $\varepsilon > 0$ there is a $\delta > 0$ such that if

$$\|x(0) - x^*\| < \delta, \tag{13.29}$$

then

$$\|x(t) - x^*\| < \varepsilon, \qquad \text{for } t \to 0, \tag{13.30}$$

and

$$x(t) \to x^*, \qquad \text{for } t \to \infty.$$

Thus, if an equilibrium point x^* is asymptotically stable, then any solution that starts out sufficiently close to x^* stays close in the sense that $\|x(t) - x^*\|$ remains small and the solution asymptotically approaches x^* in the limits as $t \to \infty$.

13.4 Liapunov's First Method

Let x^* be an equilibrium point of the system (13.21), for the input $u^* = \text{constant}$, and let $J(x^*)$ be the Jacobian matrix of $f(x, u^*)$ evaluated at $x = x^*$. Let λ_k be the eigenvalue of $J(x^*)$,

$$\det[\lambda I - J(x^*)] = 0. \tag{13.31}$$

Then x^* is asymptotically stable if the real part of each eigenvalue is negative:

$$\text{Re } \lambda_k < 0; \qquad 0 \leq k \leq n. \tag{13.32}$$

Liapunov's first method represents a sufficient condition for asymptotic stability, but it is not a necessary condition.

For example [13], we can reconsider the one-axis robot system from Fig. 13.10, where, for simplicity, we assume that the friction is purely viscous,

$$b(x_2) = b_1 x_2, \tag{13.33}$$

and the input is constant,

$$T = u^*. \tag{13.34}$$

From Eqs. (13.24) and (13.25) we obtain the Jacobian

$$J(x) = \begin{bmatrix} 0 & 1 \\ \dfrac{ga}{c}\left(\dfrac{m}{2} + m_l\right)\sin x_1^* & -\dfrac{b_1}{c} \end{bmatrix}, \tag{13.35}$$

where

$$c = a^2\left(\dfrac{m}{3} + m_l\right).$$

The characteristic equation of $J(x^*)$ is obtained from (13.31),

$$\lambda^2 + \dfrac{b_1}{c}\lambda - \dfrac{g\left(\dfrac{m}{2} + m_l\right)a\sin x_1^*}{c} = 0, \tag{13.36}$$

and the eigenvalues will be

$$\lambda_{1,2} = -\dfrac{b_1}{2c} \pm \dfrac{1}{2}\left(\dfrac{b_1^2}{c^2} + \dfrac{4g\left(\dfrac{m}{2} + m_l\right)a\sin x_1^*}{c}\right)^{1/2}. \tag{13.37}$$

If we analyze the discriminant of Eq. (13.37), assigning typical values to the arm parameters and noting that the coefficient b_1 has small values, it is clear that we have the eigenvalues in the left half of the complex plane if

$$\sin x_1^* = \sin q^* < 0, \tag{13.38}$$

which represents the domain

$$-\pi < q < 0. \tag{13.39}$$

We conclude that the asymptotic stability domain corresponds to the positions under the horizontal axis.

13.5 Liapunov's Second Method

This method is based on a fundamental physical observation: If the total energy of a mechanical system is continuously dissipated, then the system must reach an equilibrium point. Thus, the stability of a system can be studied by examining a scalar function, that is, an energy or Liapunov function [2, 3, 12].

A Liapunov function is a function $V(x)$ that satisfies the following properties:

1. $V(x)$ has a continuous derivative
2. $V(0) = 0$; $\tag{13.40}$
3. $V(x) > 0$ for $x \neq 0$.

Properties 2 and 3 define this function as a positive-definite function. Liapunov's second method is a direct method based on the finding of a Liapunov function.

Let V be a Liapunov function. Then x^* (the equilibrium point of the system) is asymptotically stable if the system has the following solutions:

$$\dot{V}(x(t)) \leq 0$$
$$\dot{V}(x(t)) \equiv 0 \quad \text{if } x(t) \equiv 0. \tag{13.41}$$

Condition 1 indicates that the values of $V(x(t))$ do not increase along the solutions of the system. Condition 2 indicates that there is a single solution for which $V(x(t))$ remains constant, $x(t) = 0$.

Then, because $V(x(t))$ does not increase and does not stay constant, it must decrease. Therefore, $V(x(t)) \to 0$ for $t \to \infty$ and $x(t) \to 0$ for $t \to \infty$.

In order to evaluate $\dot{V}(x)$, we calculate

$$\dot{V}(x) = \frac{\partial V}{\partial x} \dot{x}, \tag{13.42}$$

and from Eq. (13.21) we obtain

$$\dot{V}(x) = \frac{\partial V}{\partial x} f(x, u^*). \tag{13.43}$$

In order to illustrate this method we will reconsider the control of the mechanical system of ball and beam. We rewrite Eq. (13.13) in the form

$$\ddot{\theta} + a\theta = cu, \tag{13.44}$$

where

$$a = \frac{k_s}{J}$$

$$c = \frac{1}{J}.$$

We define the state variables x_1, x_2 as

$$x_1 = \theta$$
$$x_2 = \dot{\theta}. \tag{13.45}$$

We introduce a nonlinear complex controller that controls both variables (Fig. 13.11), the position $x_1 = \theta$ and the velocity $x_2 = \dot{\theta}$. The control law is assumed to be

$$T = -k'x_2 - b\,\text{sgn}(x_1 + kx_2), \tag{13.46}$$

where k', b, k are constants.

Figure 13.11
The control system for the mechanical system of ball and beam, with nonlinear complex controller.

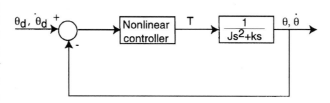

The state equations of this system are obtained from relations (13.44)–(13.46):

$$\dot{x}_1 = x_2$$
$$\dot{x}_2 = -ax_1 - k'cx_2 - bc\,\text{sgn}(x_1 + kx_2). \tag{13.47}$$

We choose a Liapunov function as

$$V = \frac{1}{2}\sigma^2, \tag{13.48}$$

where

$$\sigma = x_1 + kx_2. \tag{13.49}$$

From Eq. (13.42) we obtain

$$\dot{V} = \sigma\dot{\sigma} = \sigma(\dot{x}_1 + k\dot{x}_2). \tag{13.50}$$

Substituting $\dot{x}_1, \dot{x}_2$ from Eq. (13.47) in $\dot{V}$, we have

$$\dot{V} = \sigma[x_2 + k(-ax_1 - k'cx_2 - bc\,\text{sgn}(\sigma))]. \tag{13.51}$$

If we choose the gain coefficient k' as

$$k' = \frac{1 - k^2 a}{kc}, \tag{13.52}$$

from (13.51) we obtain

$$\dot{V} = \sigma[-ka(x_1 + kx_2) - bc\,\text{sgn}(\sigma)], \tag{13.53}$$

or

$$\dot{V} = -ka\sigma^2 - bc|\sigma|. \tag{13.54}$$

Condition 1 of Liapunov's second method is satisfied. For the second condition, we note that

$$\dot{V}(x_1(t), x_2(t)) \equiv 0 \tag{13.55}$$

requires

$$\sigma(t) \equiv 0. \tag{13.56}$$

Then

$$\begin{aligned} x_1(t) &\equiv 0 \\ x_2(t) &\equiv 0. \end{aligned} \tag{13.57}$$

We conclude that the equilibrium point $x_1^* = 0$, $x_2^* = 0$ is asymptotically stable.

14. Nonlinear Controllers by Feedback Linearization

As in the analysis of nonlinear control systems, there is no general method for designing nonlinear controllers. Several methods and techniques applicable to particular classes of nonlinear control problems are presented in the literature [1–3, 12, 13].

One of the most attractive methods is feedback linearization. Feedback linearization techniques determine a transformation of the original system models into equivalent models of a simpler form.

Consider the system defined by the equation

$$\dot{x} = f(x, u). \tag{14.1}$$

Feedback linearization is solved in two steps [2, 12].

First, one finds the input-state transformation

$$u = g(x, w), \tag{14.2}$$

so that the nonlinear system dynamics are transformed into equivalent linear time-invariant dynamics

$$\dot{x} = Ax + bw. \tag{14.3}$$

The second step is to determine a linear technique in order to obtain a good placement of the poles

$$w = -kx. \tag{14.4}$$

This method, also called input-state linearization, is simply applied for a special class of nonlinear systems described by the so-called *companion form* [12],

$$\begin{aligned}
\dot{x}_1 &= x_2 \\
\dot{x}_2 &= x_3 \\
&\vdots \\
\dot{x}_{n-1} &= x_n \\
\dot{x}_n &= f(x) + b(x)u,
\end{aligned} \tag{14.5}$$

where u is a scalar control input and f and b are nonlinear functions of the state. In order to cancel the nonlinearities and impose a desired linear dynamics we can use particular transformations of (14.2) by the form

$$u = \frac{1}{b(x)}(w - f(x)), \tag{14.6}$$

where we assume that

$$b(x) \neq 0, \tag{14.7}$$

for $x \in X$-state space.

In this case, we obtain a linear model

$$\begin{aligned}
\dot{x}_1 &= x_2 \\
\dot{x}_2 &= x_3 \\
&\vdots \\
\dot{x}_{n-1} &= x_n \\
\dot{x}_n &= w.
\end{aligned} \tag{14.8}$$

If we introduce a control law

$$w = -k_0 x - k_1 \dot{x} - \cdots - k_{n-1} x^{(n-1)}, \tag{14.9}$$

the closed-loop control system (14.8), (14.9) will have the characteristic equation

$$s^2 + k_{n-1} s^{n-1} + \cdots + k_0 = 0, \tag{14.10}$$

and we can choose the coefficients $k_0, k_1, \ldots, k_{n-1}$ such that all the roots of Eq. (14.10) are strictly in the left half complex plane.

As a first example, we reconsider the dynamic model of a nonlinear mass–damper–spring system (13.1), which can be rewritten as

$$M\ddot{x} + b\dot{x}|\dot{x}| + kx + k_1 x^3 = F. \tag{14.11}$$

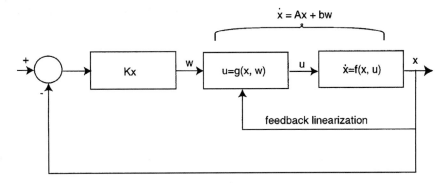

Figure 14.1
Input-state linearization.

By introducing the state variables x_1, x_2, and the control u (Fig. 14.1) we will have the companion form

$$\dot{x}_1 = x_2$$
$$\dot{x}_2 = \frac{1}{M}(-bx_2|x_2| - kx_1 - k_1x_1^3) + \frac{1}{M}u. \qquad (14.12)$$

A feedback linearization is obtained by

$$u = M(bx_2|x_2| + kx_1 + k_1x_1^3 + w), \qquad (14.13)$$

and the linear control can be chosen as

$$w = -\alpha_1 x_1 - \alpha_2 x_2.$$

In this case, we obtain the closed-loop system in the form (Fig. 14.2)

$$\dot{x}_1 = x_2$$
$$\dot{x}_2 = -\alpha_1 x_1 - \alpha_2 x_2, \qquad (14.14)$$

and a good selection of coefficients α_1, α_2 enables us to obtain the desired performance.

Figure 14.2
Nonlinear controller for the control problem of the nonlinear mass–damper–spring system.

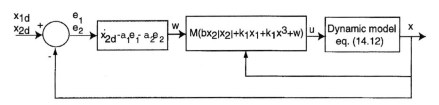

For a desired behavior defined by x_{1_d}, x_{2_d}, $\dot{x}_{2_d}$, we can choose the control law as

$$w = \dot{x}_{2_d} - \alpha_1 e_1 - \alpha_2 e_2, \qquad (14.15)$$

and the characteristic equation will have the form

$$\ddot{e} + \alpha_2 \dot{e} + \alpha_1 e = 0. \qquad (14.16)$$

The second example is offered by the control problem of the *two-axis planar articulated robot* (Fig. 14.3). The dynamic model of the robot was presented by Eq. (13.3), and they can be rewritten as

$$A(q)\ddot{q} + B(q)\dot{q} + c(q) = \tau, \tag{14.17}$$

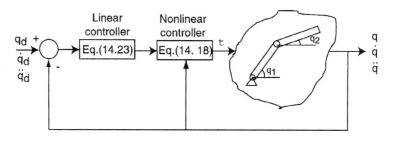

Figure 14.3
Nonlinear controller for a two-axis planar articulated robot.

where

$$q = [q_1, q_2]^T$$

$$A(q) = \begin{bmatrix} a_1(q) & a_2(q) \\ a_3(q) & a_4(q) \end{bmatrix}$$

$$B(q) = \begin{bmatrix} b_1\dot{q}_2 & b_2\dot{q}_2 \\ b_3\dot{q}_1 & 0 \end{bmatrix}$$

$$c(q) = \begin{bmatrix} c_1(q) \\ c_2(q) \end{bmatrix}$$

$$\tau = [\tau_1 \quad \tau_2]^T.$$

It is known that the inertial matrix $A(q)$ is invertible [13, 21] so that we can propose a nonlinear control

$$\tau = A(q)w + B(q)\dot{q} + c(q), \tag{14.18}$$

where w is the new input vector,

$$w = [w_1 w_2]^T. \tag{14.19}$$

If we define by

$$q^* = [q, \dot{q}, q^*]^T \tag{14.20}$$

and

$$q_d^* = [q_d, \dot{q}_d, \ddot{q}_d]^T \tag{14.21}$$

the desired values of the position, velocity, and acceleration for each arm, then the error system will be

$$q_d^* - q^* = [q_d - q, \dot{q}_d - \dot{q}, \ddot{q}_d - \ddot{q}]^T. \tag{14.22}$$

The linear control is assigned the form

$$w = \ddot{q}_d - \alpha_1(\dot{q}_d - \dot{q}) - \alpha_0(q_d - q). \tag{14.23}$$

From Eqs. (14.17), (14.18), and (14.23) we obtain

$$(\ddot{q}_d - \ddot{q}) + \alpha_1(\dot{q}_d - \dot{q}) + \alpha_0(q_d - q) = 0, \tag{14.24}$$

which defines the linear dynamics of the error $(q_d - q)$. A good selection of α_1, α_0 enables us to obtain the desired performance.

15. Sliding Control

15.1 Fundamentals of Sliding Control

The methods discussed in the preceding sections require a good knowledge of system parameters and therefore suffer from sensitivity to errors in the estimates of these parameters. Thus, modeling inaccuracies can have strong adverse effects on nonlinear control. Therefore, any practical design must ensure system robustness in conditions in which a model's imprecision is a reality.

One of the simplest approaches to robust control is the so-called *sliding control*. The basic idea is that the control signal changes abruptly on the basis of the state of the system. A control system of this type is also referred to as a *variable-structure system* [1, 12, 14].

Let us consider a dynamic system in a companion form,

$$\begin{aligned}
\dot{x}_1 &= x_2 \\
\dot{x}_2 &= x_3 \\
&\vdots \\
\dot{x}_n &= f(x) + b(x)u.
\end{aligned} \tag{15.1}$$

In Eqs. (15.1), the functions $f(x)$ and $b(x)$ are not known exactly, but we do know the sign and are bounded by known, continuous functions of x.

The control issue is to get the state x to track a specific time-varying state

$$x_d = [x_d, \dot{x}_d, \dots, x_d^{(n-1)}]^T, \tag{15.2}$$

where

$$x_d(0) = x(0) = x^0. \tag{15.3}$$

Let e be the tracking error,

$$e^* = [x - x_d, \dot{x} - \dot{x}_d, \dots, x^{(n-1)} - x_d^{(n-1)}]^T, \tag{15.4}$$

or

$$e^* = [e, \dot{e}, \dots, e^{(n-1)}]^T. \tag{15.5}$$

Let us define a surface

$$\sigma(x) = P \cdot e^* + \dot{e}^*, \tag{15.6}$$

where P can be any positive-definite matrix. For example, P can be a diagonal matrix with positive diagonal elements

$$P = \mathrm{diag}(p_1, p_2, \ldots, p_n). \tag{15.7}$$

The set of all x such that

$$\sigma(x) = 0 \tag{15.8}$$

is a $(2n - 1)$-dimensional subspace or hyperplane that is called the *switching surface.*

The switching surface divides the state space into two regions. If $\sigma(x) > 0$, then we are on one side of the switching surface and the control law has one form; if $\sigma(x) < 0$, then we are on the other side of the switching surface and the control law will have a different form. Thus, the control changes structure when the state of the system crosses the switching surface.

For the case of $n = 1$, the second system $\sigma(x) = 0$ corresponds to a line through the origin with a slope of $-p_1$ (Fig. 15.1):

$$\sigma(x) = p_1 e + \dot{e} = 0. \tag{15.9}$$

Our objective is to develop a control law that will drive the system to the switching surface in a finite time and then constrain the system to stay on the switching surface.

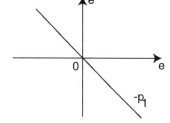

Figure 15.1
Switching line for a second-order system.

When the system is operating on the switching surface, we say that it is in the *sliding mode* (Fig. 15.2).

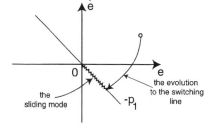

Figure 15.2
Trajectories for a sliding control.

The closed-loop control system for sliding mode control is presented in Fig. 15.3.

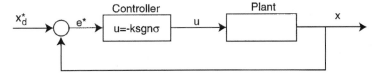

Figure 15.3
A sliding control system.

The controller ensures the evolution to the switching line. When the trajectory penetrates the switching line, the control law switches and directs the solution back toward the switching line, then switches again and keeps the motion on the switching line.

One of the simplest control laws can be

$$u = -k \, \text{sgn}(\sigma),\tag{15.10}$$

but the control law can be more complex because of the complexity and nonlinearity of the system.

On the switching line, in theory, the oscillations about the switching surface have zero amplitude and infinite frequency. However, in practice they have a small amplitude and a high frequency depending on the controller performance.

When the system is in the sliding mode, $\sigma(x) = 0$. Then, from Eq. (15.9),

$$p_1 e + \dot{e} = 0.\tag{15.11}$$

We can conclude that, in the sliding mode, the error is independent of the system parameters. This property defines the robustness of sliding control systems.

In order to illustrate the sliding control, we again consider the control issue of the "*ball and beam.*" We cancel the rotational spring and assume that the dynamic model of the system is determined by

$$J\ddot{\Theta} = T,\tag{15.12}$$

or

$$\begin{aligned}\dot{x}_1 &= x_2 \\ \dot{x}_2 &= cu,\end{aligned}\tag{15.13}$$

with

$$c = \frac{1}{J}.$$

We consider that the desired values of the state variables are x_{1d}, x_{2d}. Then the errors will be defined by

$$\begin{aligned}e_1 &= x_{1d} - x_1 \\ e_2 &= x_{2d} - x_2.\end{aligned}\tag{15.14}$$

If we adopt, for simplicity, that x_{1d}, x_{2d}, are equal to zero, Eq. (15.3) can be rewritten in terms of e_1, e_2 as

$$\dot{e}_1 = e_2$$
$$\dot{e}_2 = -cu. \tag{15.15}$$

Let the switching line be defined by

$$\sigma(e) = pe_1 + e_2 = 0. \tag{15.16}$$

First, we propose a bang-bang controller of the form (Fig. 15.3)

$$u = U \, \text{sgn}(\sigma), \tag{15.17}$$

or

$$u = \begin{cases} +U, & \text{if } \sigma > 0 \\ -U, & \text{if } \sigma < 0. \end{cases} \tag{15.18}$$

The phase trajectory can be obtained from Eq. (15.15) rewritten in the form

$$\ddot{e} = -cu. \tag{15.19}$$

Integrating this equation we obtain (see Section 13)

$$\dot{e}^2 = -2cu + d, \tag{15.20}$$

where d is a constant. This relation defines a family of parabolas. The phase portrait of this control is presented in Fig. 15.4.

The trajectory starts from the initial point for $u = +U$, and the evolution is represented by the parabola arc that emerges from the initial point. When the trajectory penetrates the switching line, the control u changes at $u = -U$ and the new trajectory is represented by another parabola arc. The motion continues until the trajectory penetrates the switching line again and a new change of the control is produced, etc.

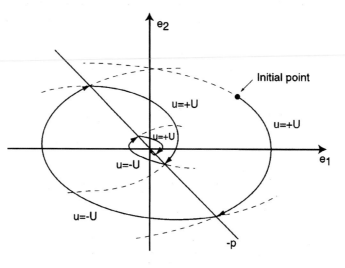

Figure 15.4
Phase portrait of the sliding control for $u = \text{sgn}(\sigma)$ for the ball and beam problem.

We can note that when the trajectory reaches the origin, the motion is stable, although the initial system (15.13) is at the limit of stability (an oscillation) determined by the two poles $s_1 = s_2 = 0$.

We reconsider Eq. (15.15), but we suppose that the control law is described by

$$u = -k'e_1 + k'' \operatorname{sgn}(\sigma)|\sigma|. \qquad (15.21)$$

The closed-loop system is (Fig. 15.5)

$$\begin{aligned} \dot{e}_1 &= e_2 \\ \dot{e}_2 &= ck'e_1 - ck'' \operatorname{sgn}(\sigma)|\sigma|. \end{aligned} \qquad (15.22)$$

Figure 15.5
The closed-loop control system with a complex control law: $u = k'e_1 + k'' \operatorname{sgn}(\sigma)|\sigma|$ for the ball and beam problem.

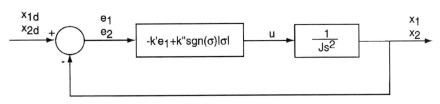

In order to prove the system's stability, we consider the Liapunov function

$$V = \frac{1}{2}\sigma^2, \qquad (15.23)$$

where

$$\sigma = pe_1 + e_2. \qquad (15.24)$$

Now, we can evaluate

$$\dot{V} = \sigma\dot{\sigma} = \sigma(p\dot{e}_1 + \dot{e}_2). \qquad (15.25)$$

Substituting $\dot{e}_1$, $\dot{e}_2$ in the last relation, we obtain

$$\dot{V} = \sigma(pe_2 + ck'e_1 - ck'' \operatorname{sgn}(\sigma)|\sigma|). \qquad (15.26)$$

We choose k', so that

$$ck' = p^2.$$

In this case, we have

$$\dot{V} = \sigma(p\sigma - ck'' \cdot \operatorname{sgn} \sigma|\sigma|),$$

or

$$\dot{V} = -(-p + ck'')\sigma^2. \qquad (15.27)$$

If we choose

$$ck'' > p, \qquad (15.28)$$

we obtain the asymptotic stability of the system,

$$\dot{V} \leq 0. \tag{15.29}$$

The solution $\dot{V} \equiv 0$ is obtained only for $\sigma \equiv 0$ or $e_1(t) \equiv 0$, $e_2(t) \equiv 0$.

These examples define the sliding control, a control method in which several parameters of the controller are modified abruptly. For this reason, this method is also known in the literature as the *variable structure control*.

15.2 Variable Structure Systems

We wish to extend the previous results to another control problem class in which several parameters of the process, of the plant, itself are modified abruptly in order to satisfy the control performances. We must make clear the differences between the control system discussed in the previous section and this method. In the conventional sliding control the controller parameters were modified; now we wish to change the components of the process of the system.

In order to analyze this method, we will discuss a particular example. We consider a nonforced second-order system defined by the differential equation

$$\ddot{x} + 2\zeta\omega_n\dot{x} + \omega_n^2 x = 0, \tag{15.30}$$

If the desired value is $x_d = 0$, the dynamic of the error

$$e = x_d - x \tag{15.31}$$

can be obtained as

$$\ddot{e} + 2\zeta\omega_n\dot{e} + \omega_n^2 e = 0. \tag{15.32}$$

In this case, we consider that the variable parameter, which can offer a control solution by the principle of variable structure, is the damping ratio ζ. The solution of this differential equation is well-known [(7.17)]:

$$e(t) = \frac{e^{-\zeta\omega_n t}}{\sqrt{1 - \zeta^2}} \sin\left(\omega_n\sqrt{1 - \zeta^2}\, t + \tan^{-1}\left(\frac{\sqrt{1 - \zeta^2}}{\zeta}\right)\right). \tag{15.33}$$

The phase portrait for $\zeta < 1$ in the plane $(\dot{e}, e)$ is presented in Fig. 15.6. We can note the damped oscillation form of the trajectory.

In order to establish the system switching law, we introduce a switching line

$$\sigma(e) = pe_1 + e_2 = 0. \tag{15.34}$$

We assume that the motion of the system is started at point A and the evolution is determined by a damping ratio $\zeta < 1$. When the trajectory penetrates the switching line, we try to find the control law of the damping ratio ζ in order to obtain an evolution along the switching line, toward the origin.

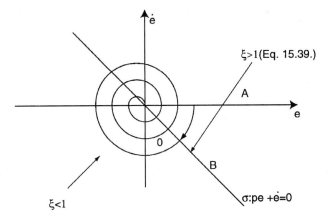

Figure 15.6
*The phase
portrait of a
second-order
system with ζ
control.*

From Eq. (15.32) we obtain

$$\ddot{e} = -2\zeta\omega_n\dot{e} - \omega_n^2 e, \tag{15.35}$$

but

$$\ddot{e} = \frac{d\dot{e}}{dt},$$

and dividing by $\dot{e}$ gives

$$\frac{d\dot{e}/dt}{de/dt} = \frac{-2\zeta\omega_n\dot{e} - \omega_n^2 e}{\dot{e}}, \tag{15.36}$$

or

$$\frac{d\dot{e}}{de} = -2\zeta\omega_n - \omega_n^2 \frac{e}{\dot{e}}. \tag{15.37}$$

However, from Eq. (15.34) we have

$$\frac{d\dot{e}}{de} = \frac{e}{\dot{e}} = -\rho. \tag{15.38}$$

Substituting (15.38) in (15.37) we obtain

$$\rho = \frac{-2\zeta\omega_n\rho - \omega_n^2}{-\rho}.$$

This relation enables us to determine the critical value of the damping
ratio for which the system evolution is on the switching line:

$$\zeta^* = \frac{\rho^2 + \omega_n^2}{2\omega_n\rho}. \tag{15.39}$$

In this case, the control system requires the following sequential
procedure:

Step 1: A conventional control is utilized and the system motion is
produced by the trajectory segment AB.

Step 2: When the trajectory penetrates the switching line, the damping ratio is increased such that

$$\zeta \geq \zeta^*. \tag{15.40}$$

The motion occurs on the switching line, toward the origin (Fig. 15.6).

We will use this procedure. It is clear that changing the damping ratio when the system is moving requires a special technology. The following example will try to illustrate the problem of the variable structure systems. The *ER rotational damper system* represents a rotational mechanism in which the damping ratio is controlled by electrorheological fluids (ER fluids).

ER fluids are emulsions formed by mixing an electrically polarizable solid material in powder form in a nonconducting liquid medium [15, 16]. When the suspension is subjected to a DC electric field, the fluid behaves as a solid (Fig. 15.7), the viscosity increasing significantly. The mechanism is presented in Fig. 15.8.

Figure 15.7
Viscosity control of ER fluids by an electric field.

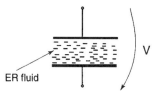

Figure 15.8
ER rotational damper mechanism.

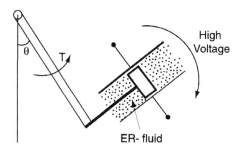

The dynamic model can be approximated by

$$J\ddot{\Theta} + k_f\dot{\Theta} = T, \tag{15.41}$$

where the ER fluid viscosity modifies the parameter k_f. If we assume a simple amplifier with k-gain in the open-loop system, we obtain

$$Y_1(s) = \frac{\omega_n^2}{s(s + 2\zeta\omega_n s)}, \tag{15.42}$$

where ζ is the damping ratio,

$$\zeta = \frac{k}{2\sqrt{J}}, \tag{15.43}$$

and the natural frequency is

$$\omega_n = \frac{k}{J}.$$ (15.44)

The closed-loop control system will have the transfer function

$$Y(s) = \frac{\omega_n^2}{s^2 + 2\zeta\omega_n s + \omega_n^2}.$$ (15.45)

The variable structure system is presented in Fig. 15.9. We note two control loops. The first is the conventional feedback control, and the second is the variable structure control of the damping ratio.

Figure 15.9
*Variable
structure system
for the ER
rotational
damper
mechanism.*

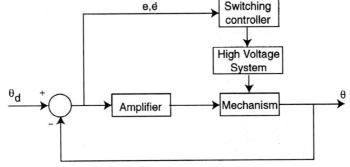

A switching controller determines the intersection between the trajectory and the switching line. A high-voltage system ensures the changing of the damping ratio as in the relation (15.39).

The phase portrait is shown in Fig. 15.10.

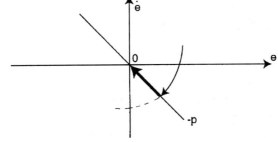

Figure 15.10
*Phase portrait
for a variable
structure system.*

A. Appendix

A.1 Differential Equations of Mechanical Systems

All the elements of the control system are single-way information transmission elements. For these elements we can establish an "input" and an "output."

In order to describe the behavior of an element in the control system, it is necessary to define an equation that expresses the output as a function of the input. Because the systems under consideration are dynamic, the descriptive equations are usually differential equations.

The differential equations describing the dynamic performance of a physical system are obtained by utilizing the physical laws of the process.

The general form of a linear differential equation is

$$\cdots + a_3 \dddot{x}_0(t) + a_2 \ddot{x}_0(t) + a_1 \dot{x}_0(t) + a_0 x_0(t) = b_1 x_i(t) + b_1 \dot{x}_i(t) + \cdots,$$

$$(A.1.1)$$

where $x_i(t)$, $x_0(t)$ denote the input and output variables, respectively, as functions of time. In this equation, x_i, x_0 can be physical variables, such as translational velocity, angular velocity, pressure, or temperature, but they do not represent the same physical dimension. For example, if x_i is a force, x_0 can be the translational velocity.

Equation (A.1.1) represents a simplified form of the mathematical model of a physical system. Practically, the complexity of the system requires a complex mathematical model defined by nonlinear differential equations. The analysis of these systems is very complicated. But, a great majority of physical systems are linear within some range of the variables, so we can accept the model (A.1.1) as a linear, lumped approximation for the majority of elements that we will analyze.

For a complex mechanical structure, the dynamic equations can be obtained by using Newton's second law of motion.

A summary of differential equation for lumped, linear elements is given in Table A.1.1 [8, 9, 16]. The symbols used are defined in Table A.1.2.

If we consider the mechanical system shown in Fig. A.1.1, we obtain

$$AP(t) - k_f \dot{z}(t) - kz(t) = M\ddot{z}(t),$$

where the active force is composed from

- $AP(t)$, the pressure force
- $k_f \dot{z}(t)$, the translational damper force,
- $-kz(t)$, the translational spring force,
- $M\ddot{z}(t)$, the inertial force

or

$$M\ddot{z}(t) + k_f \dot{z}(t) + kz(t) = AP(t), \qquad (A.1.2)$$

where A is the piston area, k_f is the friction constant, and k is the spring constant (we neglect the gravitational term M_g).

If we use conventional notation for the input, $x_i = P$, and for the output, $x_0 = z$, we obtain

$$M\ddot{x}_0(t) + k_f \dot{x}_0(t) + kx_0(t) = Ax_i(t), \qquad (A.1.3)$$

Table A.1.1 *Differential Equations for Ideal Elements*

Physical element	Describing equation	Symbol
Translational spring	$v = \dfrac{1}{K}\dfrac{dF}{dt}$	
Rotational spring	$\omega = \dfrac{1}{K}\dfrac{dT}{dt}$	
Fluid inertia	$P = I\dfrac{dQ}{dt}$	
Translational mass	$F = M\dfrac{dv}{dt}$	
Rotational mass	$T = J\dfrac{d\omega}{dt}$	
Fluid capacitance	$Q = C_f\dfrac{dP}{dt}$	
Translational damper	$F = fv$	
Rotational damper	$T = f\omega$	

Table A.1.2 *Symbols for Physical Quantities*

F, force
T, torque
Q, fluid volumetric flow rate
v, translational velocity
ω, angular velocity
P, pressure
M, mass
J, moment of inertia
C_f, fluid capacitance
f, viscous friction

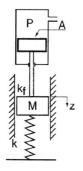

Figure A.1.1
Linear spring–mass–damper mechanical system.

which represents a second-order linear constant-coefficient differential equation.

For the rotational spring–mass–damper mechanical system from Fig. A.1.2, we obtain a similar equation, but if we do not neglect the gravitational component, we will have

$$J\ddot{\theta}(t) + k_f\dot{\theta}(t) + k\theta(t) + Mg\frac{l}{2}\sin\theta(t) = T(t). \tag{A.1.4}$$

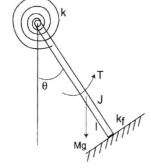

Figure A.1.2
The rotational spring–mass–dampler mechanical system.

This equation defines a second-order nonlinear differential equation. A linear approximation of the nonlinear term can be obtained if we use the Taylor series expansion about the operating point x^* [8, 9, 17],

$$h(x) = h(x^*) + \frac{dh}{dx}\Big|_{x=x^*}\frac{(x - x_0^*)}{1!} + \frac{d^2h}{dx^2}\Big|_{x=x^*}\frac{(x - x_0^*)^2}{2!} + \cdots, \tag{A.1.5}$$

and we use only the first two terms (neglecting the higher-order terms),

$$h(x) \approx h(x^*) + \frac{dh}{dx}\Big|_{x=x^*}\frac{(x - x_0^*)}{1!}. \tag{A.1.6}$$

We consider the operating point, the equilibrium point $\theta^* = 0$,

$$Mg\frac{l}{2}\sin\theta \approx Mg\frac{l}{2}\sin 0 + Mg\frac{l}{2}\cos(\theta - 0),$$

$$Mg\frac{l}{2}\sin\theta \approx Mg\frac{l}{2}\theta. \tag{A.1.7}$$

The linear approximation of (A.1.3) will be

$$J\ddot{\theta}(t) + k_f\dot{\theta}(t) + \left(k + Mg\frac{l}{2}\right)\theta(t) = T(t). \qquad \text{(A.1.8)}$$

Of course, this approximation is reasonably accurate if the variations $\Delta\theta$ are sufficiently small about the operating point $\theta^* = 0$.

A.2 The Laplace Transform

The Laplace transform for a function $f(t)$ is defined by [6, 8, 9, 17]

$$F(s) = \int_0^\infty f(t)e^{-st}\,dt. \qquad \text{(A.2.1)}$$

Normally, we write

$$F(s) = L\{f(t)\}. \qquad \text{(A.2.2)}$$

The Laplace transform (A.2.1) for a function $f(t)$ exists if the transformation integral converges. Therefore,

$$\int_0^\infty |f(t)|e^{-\sigma_1 t}\,dt < \infty, \qquad \text{(A.2.3)}$$

for some real, positive σ_1. If the magnitude of $f(t)$ is $|f(t)| < Me^{\alpha t}$ for all positive t, the integral will converge for $\sigma_1 > \alpha$. σ_1 is defined as the abscissa of absolute convergence.

The inverse Laplace transform is

$$f(t) = \frac{1}{2\pi j}\int_{\sigma-j\infty}^{\sigma+j\infty} F(s)e^{st}\,ds. \qquad \text{(A.2.4)}$$

The Laplace variable s can be considered to be the differential operator

$$s = \frac{d}{dt}, \qquad \text{(A.2.5)}$$

and

$$\frac{1}{s} = \int_{0+}^t dt. \qquad \text{(A.2.6)}$$

The Laplace transform has the advantage of substituting the differential equations by the algebraic equations. A list of some important Laplace transforms is given in Table A.2.1.

A.3 Mapping Contours in the s-Plane

A transfer function is a function of complex variable $s = \sigma + j\omega$. The function $Y(s)$ is itself complex and can be defined as

$$Y(s) = Y_r + jY_i. \qquad \text{(A.3.1)}$$

It can be represented on a complex $Y(s)$-plane with coordinates Y_r, Y_i.

Table A.2.1 *Laplace Transforms*

$f(t)$	$F(s)$
Impulse function $\delta(t)$	1
1	$\dfrac{1}{s}$
t^n	$n!/s^{n+1}$
e^{-at}	$\dfrac{1}{s+a}$
$e^{-at}f(t)$	$F(s+a)$
$\sin \omega t$	$\dfrac{\omega}{s^2 + \omega^2}$
$\cos \omega t$	$\dfrac{s}{s^2 + \omega^2}$
$1 - e^{-at}$	$\dfrac{a}{s(s+a)}$
$\dfrac{1}{(b-a)}(e^{-at} - e^{-bt})$	$\dfrac{\omega}{(s+a)(s+b)}$
$e^{-at} \sin \omega t$	$\dfrac{1}{(s+a)^2 + \omega^2}$
$e^{-at} \cos \omega t$	$\dfrac{s+a}{(s+a)^2 + \omega^2}$
$\dfrac{\omega_n}{\sqrt{1-\xi^2}} e^{-\xi\omega_n t} \sin \omega_n \sqrt{1-\xi^2}\, t, \quad \xi < 1$	$\dfrac{\omega_n^2}{s^2 + 2\xi\omega_n s + \omega_n^2}$
$\dfrac{d^k f(t)}{dt^k}$	$s^k F(s) - s^{k-1}\dot{f}(0^+) - \cdots - f^{(k-1)}(0^+)$
$\displaystyle\int_{-\infty}^{t} f(t)\,dt$	$\dfrac{F(s)}{s} + \dfrac{\int_{-\infty}^{0} f(t)\,dt}{s}$

If the s variable varies in the s-plane on a contour C, $Y(s)$ will define another contour D in the $Y(s)$-plane. Thus, the contour C has been mapped by $Y(s)$ onto a contour D. On this contour, for each value of the $Y(s)$ function will correspond a value of the complex variable s.

The direction of traversal of the s-plane contour is shown by arrows on the contour (Fig. A.3.1). Then, a similar traversal occurs on the $Y(s)$-plane contour D, as we pass beyond C in order, as shown by the arrows. By convention [8, 9, 18, 19] the area within a contour to the right of the transverse of the contour is considered to be the area enclosed by the

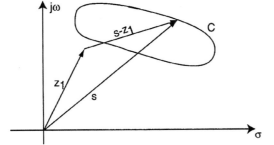

Figure A.3.1
*Contour C in
s-plane.*

contour. Therefore, we will assume the clockwise traversal of a contour to be positive and the area enclosed within the contour to be on the right.

We consider the transfer function

$$Y(s) = A(s - z_1), \tag{A.3.2}$$

where A is a constant and z_1 defines a zero of $Y(s)$,

$$s = z_1 + Re^{j\varphi},$$

where R, φ are variables,

$$s - z_1 = Re^{j\varphi}.$$

If the contour C in the s-plane encircles the zero z_1, this is equivalent to a rotation of the $(s-z_1)$ vector by 2π in the case when the corresponding contour D in the $Y(s)$-plane encircles the origin in a clockwise direction. If the contour C does not encircle the zero z_1, the angle of $s-z_1$ is zero when the traversal is in a clockwise direction along the contour and the contour D does not encircle the origin (Fig. A.3.2).

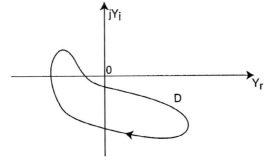

Figure A.3.2
*Contour D in
Y(s)-plane.*

Now, we consider the transfer function

$$Y(s) = \frac{A}{s - p_1}, \tag{A.3.3}$$

where p_1 defines a pole of $Y(s)$

$$s = p_1 + Re^{j\varphi},$$

Then

$$Y(s) = \frac{A}{R}e^{-j\varphi}. \tag{A.3.4}$$

We assume that the contour C encircles the pole p_1. Considering the vectors s as shown for a specific contour C (Fig. A.3.3) we can determine the angles as s traverses the contour. Clearly, as the traversal is in a clockwise direction along the contour, the traversal of D in the $Y(s)$-plane is in the opposite direction. When s traverses along C a full rotation of 2π rad, for the $Y(s)$-plane we will have an angle -2π rad (Fig. A.3.4).

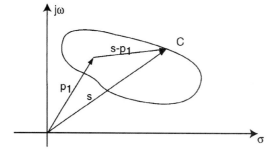

Figure A.3.3
Contour C in s-plane with a pole p_1.

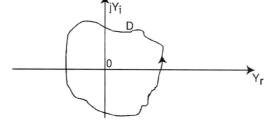

Figure A.3.4
Contour D in Y(s)-plane.

We can generalize these results. If a contour C in the s-plane encircles Z zeros and P poles of $Y(s)$ as the traversal is in a clockwise direction along the contour, the corresponding contour D in the $Y(s)$-plane encircles the origin of the $Y(s)$-plane

$$N = Z - P$$

times in a clockwise direction. The resultant angle of $Y(s)$ will be

$$\Delta\phi = 2\pi Z - 2\pi P. \tag{A.3.5}$$

The case in which the contour C encircles $Z = 3$ zeros and $P = 1$ pole in the s-plane is presented in Figs. A.3.5 and A.3.6. The corresponding contour D in the $Y(s)$-plane encircles the origin two times in a clockwise direction.

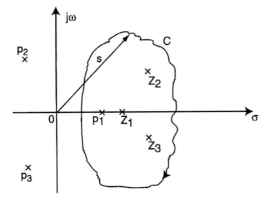

Figure A.3.5
Contour C for
Z = 3 and
P = 1.

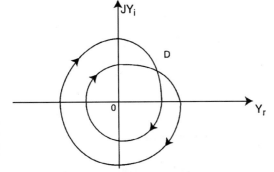

Figure A.3.6
Contour D for
N = Z and
P = 2.

One of the most interesting mapping contours in the s-plane is from the Nyquist contour. The contour passes along the $j\omega$-axis from $-j\infty$ to $+j\infty$ and is completed by a semicircular path of radius r (Fig. A.3.7). We choose the transfer function $Y_1(s)$ as

$$Y_1(s) = \frac{k}{\tau s + 1}, \qquad (A.3.6)$$

which has the pole $p_1 = -1/\tau$ real and negative. In this case, the Nyquist contour does not encircle a pole.

Figure A.3.7
(a) Nyquist
contour and a
pole $p_1 = -1/\tau$.
(b) Mapping
contour for
$Y_1(s) =$
$k/(\tau s + 1)$.

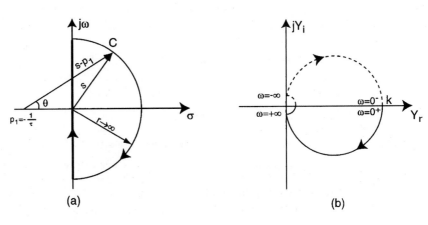

From (A3.4) we have

$$Y_1(s) = \frac{k}{\tau R} e^{-j\theta}. \tag{A.3.7}$$

When s traverses the semicircle C with $r \to \infty$ from $\omega = +\infty$ to $\omega = -\infty$, the vector $Y_1(s)$ with the magnitude $k/\tau R$ has an angle change from $-\pi/2$ to $+\pi/2$.

When s traverses the positive imaginary axis, $s = j\omega$ ($0 < \omega < \infty$), the mapping is represented by

$$Y(s) = \frac{k}{1 + j\omega\tau} = k\frac{1 - j\omega\tau}{1 + \omega^2\tau^2}, \tag{A.3.8}$$

which represents a semicircle with diameter k (Fig. A.3.7b, the solid line).

The portion from $\omega = -\infty$ to $\omega = 0^-$ is mapped by the function

$$Y_1(s)|_{s=-j\omega} = Y_1(-j\omega) = k\frac{1 + j\omega\tau}{1 + \omega^2\tau^2}. \tag{A.3.9}$$

Thus, we obtain the complex conjugate of $Y_1(j\omega)$, and the plot for the portion of the polar plot from $\omega = -\infty$ to $\omega = 0^-$ is symmetrical to the polar plot from $\omega = +\infty$ to $\omega = 0^+$ (Fig. A.3.7b).

A.4 The Signal Flow Diagram

A signal flow diagram is a representation of the relationship between the system variables. The signal flow diagram consists of unidirectional operational elements that are connected by the unidirectional path segments. The operational elements are integration, multiplication by a constant, multiplication of two variables, summation of several variables, etc. (Fig. A.4.1).

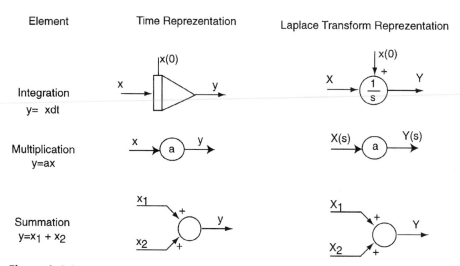

Figure A.4.1 *Signal flow diagram elements.*

These functions are often sufficient to develop a simulation model of a system.

For example, we consider a dynamic model described by the differential equation

$$\ddot{x} + a_1\dot{x} + a_2 x = u. \tag{A.4.1}$$

Using the notations

$$x = x_1$$
$$\dot{x} = x_2,$$

Eq. (A.4.1) can be rewritten as

$$\dot{x}_1 = x_2$$
$$\dot{x}_2 = -a_1 x_2 - a_2 x_1 + u. \tag{A.4.2}$$

The signal flow diagram of (A.4.1) is presented in Fig. A.4.2. The diagram has two representations, one for time-domain variables and one for the Laplace transform representation.

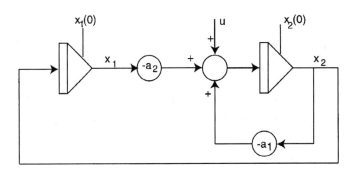

(a)

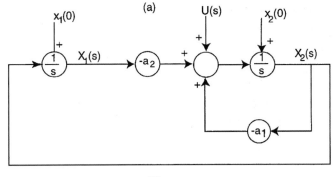

(b)

Figure A.4.2
Signal flow diagrams (a) in the time domain, (b) in the Laplace variable domain.

References

1. R. J. Schilling, *Fundamentals of Robotics—Analysis and Control*. Prentice Hall, Englewood Cliffs, NJ, 1990.

2. P. G. Darzin, *Nonlinear Systems*. Cambridge University Press, 1992.

3. I. E. Gibson, *Nonlinear Automatic Control*. McGraw-Hill, New York, 1963.

4. S. Călin, *Automatic Regulators*. Ed. Did-Ped, Bucharest, 1967.

5. E. Kamen, *Introduction to Signals and Systems*. Macmillan, New York, 1990.

6. B. C. Kuo, *Automatic Control Systems*. Prentice Hall, Englewood Cliffs, NJ, 1990.

7. T. Yoshikawa, *Foundation of Robotics*. M.I.T. Press, Cambridge, MA, 1990.

8. G. I. Thaler, *Automatic Control Systems*. West Publishing, St. Paul, MN, 1990.

9. R. G. Dorf, *Modern Control Systems*. 6th ed. Addison-Wesley, Reading, MA, 1992.

10. B. W. Niebel, *Modern Manufacturing Process Engineering*. McGraw-Hill, New York, 1989.

11. S. C. Jacobsen, Control strategies for tendon driven manipulators. *IEEE Control Systems*, Vol. 10, Feb., 23–28 (1990).

12. I. I. E. Slotine and Li Weiping, *Applied Nonlinear Control*. Prentice-Hall International, New York, 1991.

13. M. Asada and I. I. E. Slotine, *Robot Analysis and Control*. Wiley-Interscience, New York, 1986.

14. H. Bühler, *Réglage par mode de glissement*. Presses Polytechniques Romandes, Lausanne, 1986.

15. M. Ivanescu and V. Stoian, A distributed sequential controller for a tentacle manipulator, in *Computational Intelligence* (Bernd Reusch, ed.), pp. 232–238. Springer Verlag, Berlin, 1996.

16. R. C. Rosenberg and D. C. Karnopp, *Introduction to Physical System Design*. McGraw-Hill, New York, 1986.

17. R. I. Smith and R. C. Dorf, *Circuits, Devices and Systems*, 5th ed. John Wiley & Sons, New York, 1991.

18. W. L. Brogan, *Modern Control Theory*. Prentice Hall, Englewood Cliffs, NJ, 1991.

19. R. E. Ziemer, *Signals and Systems*, 2nd ed. Macmillan, New York, 1989.

20. C. L. Phillps and R. D. Harbor, *Feedback Control Systems*. Prentice Hall, Springer Verlag, New York, 1988.

21. R. L. Wells, Control of a flexible robot arm. *IEEE Control Systems*, Vol. 10, Jan., 9–15 (1990).

Appendix Differential Equations and Systems of Differential Equations

HORATIU BARBULESCU

Department of Mechanical Engineering,
Auburn University, Auburn, Alabama 36849

Inside

1. Differential Equations

1.1 Ordinary Differential Equations: Introduction

1.1.1 BASIC CONCEPTS AND DEFINITIONS

Operatorial Equation

Let X, Y be arbitrary sets and $f: X \rightarrow Y$ a function defined on X with values in Y. If $y_0 \in Y$ is given, and $x \in X$ must be found so that

$$f(x) = y_0, \tag{1.1}$$

then it is said that an *operatorial equation* must be solved. A *solution* of Eq. (1.1) is any element $x \in X$ that satisfies Eq. (1.1). The sets X, Y can have different algebraical and topological structures: linear spaces, metrical spaces, etc. If f is a linear function, that is, $f(\alpha x_1 + \beta x_2) = \alpha f(x_1) + \beta f(x_2)$, and if X and Y are linear spaces, then Eq. (1.1) is called a *linear equation*. If Eq. (1.1) is a linear equation and $y_0 = \theta_Y$ (the null element of space Y), then Eq. (1.1) is called a *linear homogeneous equation*.

Differential Equation

An equation of the form (1.1) for which X and Y are sets of functions is called a *functional equation*. A functional equation in which is implied an unknown function and its derivatives of some order is called a *differential equation*. The maximum derivation order of the unknown function is called the *order of the equation*. When the unknown function depends on a single independent variable, the equation is termed an *ordinary differential equation* (or, more briefly; a *differential equation*). If the unknown function depends on more independent variables, the corresponding equation is called a *partial differential equation*. *The general form* of a differential equation of order n is

$$F(t, x, x', x'', \ldots, x^{(n)}) = 0, \tag{1.2}$$

where t is the independent variable, $x = x(t)$ is the unknown function, and F is a function defined on a domain $D \subseteq \mathbf{R}^{n+2}$ ($\mathbf{R}$ is the set of real numbers). It is called a *solution of* Eq. (1.2) *on the interval* $I = (a, b) \subset \mathbf{R}$, a function $\varphi = \varphi(t)$ of $C^n(I)$ class [i.e., $\varphi(t)$ has continuous derivatives until n-order], which has the following properties:

1. $(t, \varphi(t), \varphi'(t), \ldots, \varphi^{(n)}(t)) \in D, \quad \forall t \in I$
2. $F(t, \varphi(t), \varphi'(t), \ldots, \varphi^{(n)}(t)) = 0, \quad \forall t \in I$

If the function F can be explicated with the last argument, it then yields

$$x^{(n)} = f(t, x, x', \ldots, x^{(n-1)}), \tag{1.3}$$

which is called the *normal form* of the n-order equation.

Cauchy's Problem

Let $G \subset \mathbf{R}^{n+1}$ be the definition domain of the function f from Eq. (1.3) and $(t_0, x_0^0, x_1^0, \ldots, x_{n-1}^0) \in G$. If the solution of Eq. (1.3) satisfies the initial conditions

$$x(t_0) = x_0^0, \qquad x'(t_0) = x_1^0, \ldots, x^{(n-1)}(t_0) = x_{n-1}^0, \qquad (1.4)$$

then the problem of finding that solution of Eq. (1.3) is called the *Cauchy's problem* for Eq. (1.3). The Cauchy's problem for Eq. (1.2) is formulated analogously. *The general solution* of Eq. (1.3) is a function family depending on the independent variable and on n arbitrary independent constants $\varphi = \varphi(t; c_1, c_2, \ldots, c_n)$, and which satisfies the conditions

1. $\varphi(t; c_1, c_2, \ldots, c_n)$ is a solution for Eq. (1.2) on an interval I_c
2. For any initial conditions (1.4), there could be determined the values $c_1^0, c_2^0, \ldots, c_n^0$ of the constants $c_1, c_2, \ldots, c_n$ so that $\varphi(t; c_1^0, c_2^0, \ldots, c_n^0)$ is the solution for the Cauchy's problem with the conditions (1.4)

A solution obtained from the general solution for particular constants $c_1, c_2, \ldots, c_n$ is called a *particular solution*. A *singular solution* is a solution that cannot be obtained from the general solution.

EXAMPLE 1.1 Consider the following equation:

$$x'^2 + x^2 - 1 = 0. \qquad (1.5)$$

This is a differential equation of the first order. Expressing it with x', we obtain

$$x' = \sqrt{1 - x^2} \qquad \text{and} \qquad x' = -\sqrt{1 - x^2}; \qquad (1.6)$$

hence two equations of normal form. Let us consider the first equation. This is of the form of Eq. (1.3) with $n = 1$, the right-hand side function being $f(x) = \sqrt{1 - x^2}$ and defined on $G = \mathbf{R} \times (-1, 1)$. In its expression, the independent variable t does not appear explicitly. The function $\varphi(t) = \sin t$ is derivable, and substituting it in the equation, we find the equality $\cos t = |\cos t|$. This is an identity on any interval I_k of the form $I_k = (-(\pi/2) + 2k\pi, (\pi/2) + 2k\pi)$, $k \in \mathbf{Z}$ ($\mathbf{Z}$ is the set of integer numbers). Then, on each of these intervals the function $\varphi(t) = \sin t$ is the solution for equation $x' = \sqrt{1 - x^2}$. Now, consider the functions family $\varphi(t; c) = \sin(t + c)$, $c \in \mathbf{R}$. Let us set the interval $I_c = (-(\pi/2 - c), (\pi/2) + c)$. For $t \in I_c$, $t + c \in (-(\pi/2), (\pi/2))$ and $\varphi(t; c) = \sin(t + c)$ is the solution on I_c for the equation $x' = \sqrt{1 - x^2}$. If $(t_0, x_0) \in G = \mathbf{R} \times (-1, 1)$ settled from the condition $\varphi(t_0, c) = x_0$, then $\sin(t_0 + c) = x_0$ and the value of c obtained from this condition and denoted by c_0 is $c_0 = \arcsin x_0 - t_0$. The function $\varphi(t, c_0) = \sin(t + c_0)$ is a solution for the equation and satisfies the initial condition, so it is the general solution. ▲

Remark 1.1

The constant functions $\varphi_1(t) \equiv 1$ and $\varphi_2(t) \equiv -1$ are solutions on $\mathbf{R}$ for the equation $x' = \sqrt{1 - x^2}$, but they cannot be obtained from the general solution for any particular constant c, and hence there are singular solutions. ▲

1.1.2 SYSTEMS OF DIFFERENTIAL EQUATIONS

A system of differential equations is constituted by two or more differential equations. A *system* with n *differential equations of the first order in normal form* is a system of the form

$$
\begin{cases}
x_1' = f_1(t, x_1, x_2, \ldots, x_n) \\
x_2' = f_2(t, x_1, x_2, \ldots, x_n) \\
\qquad \cdots \\
x_n' = f_n(t, x_1, x_2, \ldots, x_n),
\end{cases}
\tag{1.7}
$$

$f_i \colon I \times D \subset \mathbf{R} \times \mathbf{R}^n$. A *solution of the system* of equations (1.7) on the interval $J \subseteq I$ is an assembly of n functions $(\varphi_1(t), \varphi_2(t), \ldots, \varphi_n(t))$ derivable on J and that, when substituted with the unknowns $x_1, x_2, \ldots, x_n$, satisfies Eqs. (1.7) in any $t \in J$.

Initial Conditions

Consider that $t_0 \in J$ and $(x_1^0, x_2^0, \ldots, x_n^0) \in D$ are settled. The conditions

$$
x_1(t_0) = x_1^0, \qquad x_2(t_0) = x_2^0, \ldots, x_n(t_0) = x_n^0
\tag{1.8}
$$

are called *initial conditions* for the system of equations (1.7).

Vectorial Writing of the System of Equations (1.7)

If the column vector $X(t)$ is written as

$$
X(t) = \begin{Bmatrix} x_1(t) \\ x_2(t) \\ \cdots \\ x_n(t) \end{Bmatrix},
$$

with

$$
f(t, X) = \begin{Bmatrix} f_1(t, x_1, x_2, \ldots, x_n) \\ f_2(t, x_1, x_2, \ldots, x_n) \\ \cdots \\ f_n(t, x_1, x_2, \ldots, x_n) \end{Bmatrix},
$$

then the system of equations (1.7) can be written in the form

$$
X'(t) = f(t, X)
\tag{1.9}
$$

and the initial conditions (1.8) are

$$
X(t_0) = X_0,
\tag{1.10}
$$

where

$$X_0 = \left\{ \begin{array}{c} x_1^0 \\ x_2^0 \\ \ldots \\ x_n^0 \end{array} \right\}. \tag{1.11}$$

Relation between a System of First-Order Differential Equations with n Equations and an Equation of Order n

Consider the equation of order n

$$u^{(n)} = f(t, u, u', \ldots, u^{(n-1)}), \tag{1.12}$$

with the initial condition

$$u(t_0) = u_0^0, \qquad u'(t_0) = u_1^0, \ldots, u^{(n-1)}(t_0) = u_{n-1}^0. \tag{1.13}$$

Using the notation

$$u = x_1, \qquad u' = x_2, \ldots, u^{(n-2)} = x_{n-1}, u^{(n-1)} = x_n, \tag{1.14}$$

by derivation we find the system

$$\left\{ \begin{array}{l} x_1' = x_2 \\ x_2' = x_3 \\ \ldots \\ x_{n-1}' = x_n \\ x_n' = f(t, x_1, x_2, \ldots, x_n), \end{array} \right. \tag{1.15}$$

which is of the form of Eqs. (1.7). If the vector

$$X(t) = \left\{ \begin{array}{c} x_1(t) \\ x_2(t) \\ \ldots \\ x_n(t) \end{array} \right\}$$

is a solution for the system of equations (1.15), then the function $u(t) = x_1(t)$ is a solution for Eq. (1.12). Conversely, if $u(t)$ is a solution of Eq. (1.12), then

$$X(t) = \left\{ \begin{array}{c} x_1(t) \\ x_2(t) \\ \ldots \\ x_n(t) \end{array} \right\}$$

with $x_1, x_2, \ldots, x_n$ given by Eq. (1.13) is a solution of the system of equations (1.15). In summary, finding the solution of an n-order differential equation is equivalent to finding the solution of an equivalent system with n equations of the first order. The situation is the same for Cauchy's problem. We conclude that finding the solution of a system with superior-order equations is equivalent to finding the solution of a system with more equations of first order.

1.1.3 GEOMETRICAL MEANING

Let us consider a first-order equation in the normal form

$$y' = f(x, y). \tag{1.16}$$

In the plane xOy, the quantity $y' = dy/dx$ represents for the curve $y = y(x)$ the slope of the tangent at that curve. The differential equation (1.16) thus establishes a relation between the coordinates of a point and the slope of the tangent to the solution graph at that point. It defines a direction field, and the problem of finding solutions (the integration of the differential equation) is reduced to the problem of finding curves, called *integral curves*, which have the property that the tangent directions in any of their point coincide with the field direction. The Cauchy's problem for Eq. (1.16) with the condition $y(x_0) = y_0$ means to find the integral curve that passes through the point $M_0(x_0, y_0)$. The *isocline* is the locus of points $M(x, y)$ in the definition domain of function f for which the tangents to the integral curves have the same direction. The equation

$$f(x, y) = k, \tag{1.17}$$

where $k \in \mathbf{R}$, gives the points in which the field has the direction k. For $k = 0$ will be obtained the extreme points for integral curves. The general solution of Eq. (1.16) represents a curve family depending on a parameter, and it is called a *complete integral.*

The Differential Equation of a Curve Family

Let us consider the curve family

$$y = \varphi(x, a), \tag{1.18}$$

where a is a real parameter. It can be considered that Eq. (1.18) represents the complete integral of the differential equation that must be found. Derivation of Eq. (1.18) with respect to x yields

$$y'(x) = \frac{\partial \varphi(x, a)}{\partial x}. \tag{1.19}$$

Eliminating the parameter a from Eqs. (1.18) and (1.19), we find a relation of the form

$$F(x, y, y') = 0, \tag{1.20}$$

which is called *the differential equation of the family* Eq. (1.18). If the curve family is given in the form

$$\Phi(x, y, a) = 0, \tag{1.21}$$

then the differential equation of this family is obtained by eliminating the parameter a from the equations

$$\begin{cases} \Phi(x, y, a) = 0 \\ \dfrac{\partial \Phi}{\partial x} + \dfrac{\partial \Phi}{\partial y} y' = 0. \end{cases} \tag{1.22}$$

In the case when the curves family depends on n parameters

$$\Phi(x, y, a_1, a_2, \ldots, a_n) = 0,$$

the differential equation of this family is obtained by eliminating the parameters $a_1, a_2, \ldots, a_n$ from the above equation and the other $n - 1$ equations obtained by derivation with respect to x, successively until the nth-order. The differential equation of a curve family has as a general solution the given family itself.

Differential Equations

Isogonal Trajectories

Let us consider the plane curve family

$$\Phi(x, y, a) = 0, \tag{1.23}$$

where a is a real parameter. A curve in the plane xOy is called *isogonal* to the family Eq. (1.23) if it intersects the curves of the family Eq. (1.23) at the same angle α. When $\alpha = \pi/2$, the curve is called *orthogonal*. The isogonal curve family Eq. (1.23) with $\tan \alpha = k$ is obtained as follows:

- Find the differential equation of the family Eq. (1.23).
- In that differential equation, substitute y' by $(y' - k)/(1 + ky')$, if $\alpha \neq \pi/2$ or, $-1/y'$ when $\alpha = \pi/2$.
- Find the general solution of the differential equation obtained; this is the isogonal family of Eq. (1.23).

EXAMPLE 1.2 Consider the differential equation $y' = -x/2y$.

(a) Make an approximation construction of the integral curves, using the isoclines.
(b) Find the general solution (complete integral).
(c) Find the orthogonal family of the complete integral. ▲

Solution

(a) The differential equation is of the form $y' = f(x, y)$, with $f(x, y) = -x/2y$. The definition domain of function f is $D = \mathbf{R} \times (-\infty, 0) \cup \mathbf{R} \times (0, \infty)$. The equation of isoclines (1.17) gives $-x/2y = k$; hence $y = -x/2k, k \in \mathbf{R} \backslash \{0\}$. The corresponding isoclines to the values of k

$$-1, \quad -\tfrac{1}{2}, \quad -\tfrac{1}{4}, \quad \tfrac{1}{4}, \quad \tfrac{1}{2}, \quad 1$$

are the straight lines

$$y = \tfrac{1}{2}x, \quad y = x, \quad y = 2x, \quad y = -2x, \quad y = -x, \quad y = -\tfrac{1}{2}x,$$

presented in Fig. 1.1.

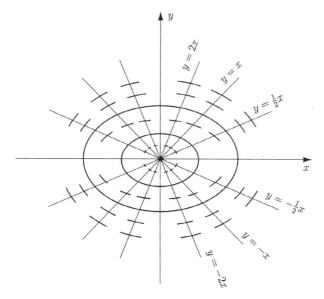

Figure 1.1
The correspond-ing isoclines
$(y = \frac{1}{2}x, y = x,$
$y = 2x,$
$y = -2x,$
$y = -x,$
$y = -\frac{1}{2}x).$

(b) The equation can be written as

$$x + 2yy' = 0$$

or

$$x\,dx + 2y\,dy = 0.$$

Then,

$$d(\tfrac{1}{2}x^2 + y^2) = 0.$$

Consider that $x^2/2 + y^2 = a^2$, with a^2 an arbitrary positive constant. The general solution (the complete integral) is a parabolas family. The differential equation of this is $y' = -x/2y$. Substituting y' by $-1/y'$ in this equation yields the differential equation of the isogonal family $-1/y' = -x/2y$. This equation is written as

$$xy' = 2y, \quad x\,dy = 2y\,dx, \quad \frac{dy}{y} = 2\frac{dx}{x}, \quad d(\ln|y|) = 2\,d(\ln|x|).$$

This yields $\ln|y| = 2\ln|x| + c$. The arbitrary constant c could be chosen in the form $c = \ln b$, $b > 0$. The isogonal curve family is $|y| = bx^2$, which represents a family of parabolas. In Fig. 1.2 are represented two isogonal families.

1.1.4 PHENOMENA INTERPRETED BY MEANS OF DIFFERENTIAL EQUATIONS

Differential equations make it possible to study some finite-determinist and differentiable phenomena. Determinist phenomena are those processes whose future evolution state is uniquely determined by the state of present

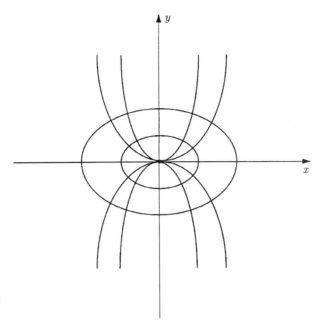

Figure 1.2
Two isogonal families.

conditions. Finite phenomena are those processes that need a finite number of parameters for their correct description. Differentiable phenomena are those processes in which the functions used for their description are derivable (up to some order). Next, some examples of modeling by means of differential equations are presented.

Phenomenon of Growth (or Decay)

In the study of some phenomena of growth from economy, biology, etc. [For instance, the growth (decay) of production, population of a race, material quantity], there appear differential equations of the form

$$\frac{df(t)}{dt} = k(t)f(t), \tag{1.24}$$

where $k(t) > 0$ in the case of a growth phenomenon and $k(t) < 0$ when it is the curve of a decaying phenomenon. For example, if in a study of the evolution of a certain species, we denote by $f(t)$ the number of individuals at the moment t, by n and m the coefficient of birth rate and death rate, respectively, then given the assumption that the population is isolated (i.e., there is no immigration or emigration), the variation rate of the population $f'(t)$ will be given by

$$\frac{df(t)}{dt} = (n - m)f(t). \tag{1.25}$$

If we take into account the inhibiting effect of crowding, we obtain the equation

$$f'(t) = (n - m)f(t) - \alpha f^2(t), \tag{1.26}$$

where α is a positive constant that is very small compared to $(n - m)$.

A Mathematical Model of Epidemics

Let us consider a population of n individuals and a disease spreading by direct touch. At a moment t, the population is composed of three categories:

- $x(t) =$ the number of individuals not infected
- $y(t) =$ the number of infected individuals that are not isolated (are free)
- $z(t) =$ the number of infected individuals that are isolated (under observation)

It is natural to presume that the infection rate $-x'(t)$ is proportional to $x \cdot y$ and the infected individuals become isolated at a rate that is proportional to their number, y. This yields the system

$$\begin{cases} x' = -\beta xy, \\ y' = \beta xy - \gamma y, \\ x + y + z = n. \end{cases} \qquad (1.27)$$

The Dog Trajectory

A man walks on a line Oy with the uniform velocity v. At the moment $t = 0$ he is at point O, and he calls his dog, which at that moment is at point A at the distance $OA = a$. The dog runs to the master with the uniform velocity $v_1 = kv$ ($k > 0$), the velocity being always oriented to the master (Fig. 1.3). Find the equation of the dog trajectory and the time at which it will reach its master. Discuss.

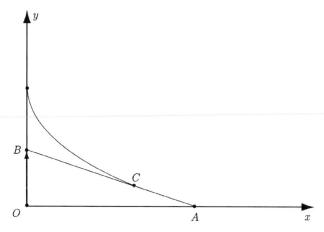

Figure 1.3
The dog trajectory.

The Basic Dynamical Equations

Consider that M is a material point of mass m that moves in $\mathbf{R}^3$ under the action of a force F [which usually depends on time, on the position $r(t)$ of the point M, and on the motion velocity dr/dt. Applying Newton's second law of dynamics, we obtain

$$m\frac{d^2r}{dt^2} = F\left(t, r, \frac{dr}{dt}\right),\tag{1.28}$$

where $r(t)$ is the vector with components $x(t)$, $y(t)$, $z(t)$. The scalar transcription of Eq. (1.28) leads to the system

$$\begin{cases} m\dfrac{d^2x}{dt^2} = f_1(t, x, y, z, x'(t), y'(t), z'(t)) \\[2mm] m\dfrac{d^2y}{dt^2} = f_2(t, x, y, z, x'(t), y'(t), z'(t)) \\[2mm] m\dfrac{d^2z}{dt^2} = f_3(t, x, y, z, x'(t), y'(t), z'(t)), \end{cases}\tag{1.29}$$

where f_1, f_2, f_3 are the components of F.

The Problem of the Second Cosmic Velocity

The goal of this problem is to find the velocity v_0 of the vertical launching of a body as it escapes the influence of the earth's gravitational attraction. If we use the law of universal attraction and Newton's second law, the equation of motion is

$$m\ddot{r}(t) = -k\frac{mM}{r^2(t)},\tag{1.30}$$

where $r(t)$ is the distance from the center of the earth to the center of the body, m is the mass of the body, M is the mass of the earth, and k is the constant of universal attraction.

Equation of an Electrical Oscillatory Circuit

Let us consider an electrical circuit composed of an inductance L, a resistor R, and a capacitor C, having a voltage U. The laws of electricity yield the differential equation

$$LI''(t) + RI'(t) + \frac{1}{C}I(t) = f(t),\tag{1.31}$$

where $I(t)$ is the intensity and $f(t) = U'(t)$.

Equation of a Mechanical Oscillator

The equation of motion of a material point with mass m, which moves on the Ox axis under the action of an elastic force $F = -\omega^2 x$, is

$$m\ddot{x} + \omega^2 x = 0.\tag{1.32}$$

Taking into account the existence of a friction force proportional to the exterior velocity $f(t)$, we find the differential equation

$$m\ddot{x} + b\dot{x} + \omega^2 x = f(t),\tag{1.33}$$

which is called the *equation of vibration*.

Directions of the Normal Stresses in a Plane Problem of Elasticity Theory
These directions are defined by the differential equation

$$\left(\frac{dy}{dt}\right)^2 + \frac{\sigma_x - \sigma_y}{\tau_{xy}}\left(\frac{dy}{dt}\right) - 1 = 0, \tag{1.34}$$

where σ_x, σ_y, τ_{xy} are the stresses (assumed known).

Deflection of a Beam
To find the critical load for the deflection of a beam with one end joined and the other one cantilevered, we use the differential equation

$$\frac{d^2 w}{dx^2} + \frac{P}{EI} w = \frac{H}{EI} x. \tag{1.35}$$

Rotating Shaft
If the shaft rotates with the angular speed ω, then

$$\frac{d^4 y}{dx^4} - \frac{\gamma A \omega^2}{gEI} y = 0, \tag{1.36}$$

where ω is the angular velocity, A is the area of the transverse section of the shaft, γ is the specific mass of the material, g is the gravitational acceleration, and y is the deflection at point x.

1.2 Integrable Types of Equations

A differential equation is *integrable by quadratures* if the general solution of the equation can be expressed in an explicit or implicit form that may contain quadratures (i.e., indefinite integrals).

1.2.1 FIRST-ORDER DIFFERENTIAL EQUATIONS OF THE NORMAL FORM
Equations with Separable Variables
An *equation with separable variables* is a first-order differentiable equation of the form

$$x'(t) = p(x)q(t), \tag{1.37}$$

where q is a continuous function defined on the interval $(a_1, a_2) \subset \mathbf{R}$, and p is a continuous nonzero function, on the interval $(b_1, b_2) \subset \mathbf{R}$. If we separate the variables (dividing by $p(x)$) and integrate, the solution is of the form

$$\int \frac{dx}{p(x)} = \int q(t) + C. \tag{1.38}$$

The solution of Eq. (1.37) with the initial condition

$$x(t_0) = x_0 \tag{1.39}$$

is given by the equality

$$\int_{x_0}^{x} \frac{du}{p(u)} = \int_{t_0}^{t} q(s)ds.$$

Remark 1.2

If $p(x) = 0$ has solutions $x = c_k$, then they are singular solutions. An equation of the form

$$M(x)N(y)dx + P(x)Q(y)dy = 0 \tag{1.40}$$

is also an equation with separable variables. Let us assume that the functions that appear are continuous and $N(y) \neq 0$, $P(x) \neq 0$. Dividing by $N(y)P(x)$ yields

$$\frac{M(x)}{P(x)} dx + \frac{Q(y)}{N(y)} dy = 0, \tag{1.41}$$

with the general solution

$$\int \frac{M(x)}{P(x)} dx + \int \frac{Q(y)}{N(y)} dy = C. \tag{1.42}$$

EXAMPLE 1.3 Consider the equation $dx/dt = at^{\alpha}(x^{\beta_1} + bx^{\beta_2})$; $a, b, \alpha \in \mathbf{R}$, $\beta_1, \beta_2 \in \mathbf{Q}$. The equation is written in the form $dx/(x^{\beta_1} + bx^{\beta_2}) = at^{\alpha}$; $(x^{\beta_1} + bx^{\beta_2} \neq 0)$. If β_1, $\beta_2 \in \mathbf{Z}$, in the first term is necessary to integrate by decomposing a rational function into simple fractions. If $\beta_1, \beta_2 \in \mathbf{Q}$, $\beta_1 = n_1'/n_1$, $\beta_2 = n_2'/n_2$, we can make the replacement $x = y^r$, where r is a common multiple of numbers n_1 and n_2. Finally, we will obtain a rational function. ▲

Remark 1.3

If the equation $x^{\beta_1} + bx^{\beta_2} = 0$ has as solutions $x = k$, then these are singular solutions of the initial equation.

Particular Cases

1. $b = 0$, $\beta_1 = 1$, $\displaystyle\int \frac{dx}{x} = \ln|x|$

2. $b = 0$, $\beta_1 \neq 1$, $\displaystyle\int \frac{dx}{x^{\beta_1}} = \frac{x^{\beta_1+1}}{\beta_1 + 1}$

3. $\beta_1 = 1$, $\beta_2 = 0$, $\displaystyle\int \frac{dx}{x+b} = \ln|x+b|$

4. $\beta_1 = 2$, $\beta_2 = 0$, $b > 0$, $\displaystyle\int \frac{dx}{x^2 + b} = \frac{1}{\sqrt{b}} \arctan\frac{x}{\sqrt{b}}$

5. $\beta_1 = 2$, $\beta_2 = 0$, $b < 0$, $\displaystyle\int \frac{dx}{x^2 + b} = \frac{1}{2\sqrt{-b}} \ln\left|\frac{x - \sqrt{-b}}{x + \sqrt{-b}}\right|$

6. $\beta_1 = 1$, $\beta_2 = \dfrac{1}{2}$.

Making the replacement $x = y^2$ yields

$$\int \frac{dx}{x + b\sqrt{x}} = \int \frac{2y\,dy}{y^2 + by} = 2\int \frac{dy}{y + b} = 2\ln|y + b| = 2\ln|\sqrt{x} + b|.$$

APPLICATION 1.1 The *relaxation phenomenon* (the decrement in time of the stresses of a piece under a constant deformation and a constant temperature) is described by the differential equation $\dot{\sigma}/E + \sigma/\eta = 0$, in which σ represents the stress in the transverse sections of the piece; $\dot{\sigma} = d\sigma/dt$ is the derivative of stress with respect to time; E is the modulus of elasticity of the material (constant); and η is the viscidity coefficient (constant). Determine the solution $\sigma = \sigma(t)$. ▲

Solution

The equation is separable and can be written as

$$\frac{\dot{\sigma}}{E} = -\frac{\sigma}{\eta}; \quad \frac{\dot{\sigma}}{\sigma} = -\frac{E}{\eta}; \quad \frac{d\sigma}{\sigma} = -\frac{E}{\eta}\,dt; \quad \int \frac{d\sigma}{\sigma} = -\int \frac{E}{\eta}\,dt + C;$$

$$\ln \sigma = -\frac{E}{\eta}t + \ln c_1,$$

where $c_1 > 0$. (The arbitrary constant C is of the form $C = \ln c_1$). From the last equality, the general solution is

$$\sigma(t) = c_1 e^{-(E/\eta)t}.$$

Considering the condition $\sigma(0) = \sigma_0$ yields $c_1 = \sigma_0$, and the solution is

$$\dot{\sigma}(t) = \sigma_0 e^{-(E/\eta)t}.$$

APPLICATION 1.2 The rotation φ of a console is given by the differential equation

$$\frac{d\varphi}{dx} = \frac{pl^2}{2EI_0} \cdot \frac{a\dfrac{x^2}{l^2} + b\dfrac{x^3}{l^3}}{a + 3b\dfrac{x}{l}},$$

where a, b, p, EI_0 are prescribed nonzero constants. In the fixed end ($x = l$), the rotation is zero.

Solution

The preceding equation has separated variables. The function φ is given by the integral

$$\varphi = \frac{pl^2}{2EI_0} \int \frac{a\dfrac{x^2}{l^2} + b\dfrac{x^3}{l^3}}{a + 3b\dfrac{x}{l}}\,dx + C.$$

After the replacement $x/l = y$, $dx = l\,dy$,

$$\varphi = \frac{pl^2}{2EI_0}\int\frac{ay^2 + by^3}{a + 3by}\,dy + C = \frac{pl^3}{6EI_0}\int\frac{3by^3 + 3ay^2}{3by + a}\,dy + C$$

$$= \frac{pl^3}{6EI_0}\int\left(y^2 + \frac{2a}{3b}y - \frac{2a^2}{9b^2} + \frac{2a^3}{9b^2}\frac{1}{3by + a}\right)dy + C$$

$$\varphi = \frac{pl^3}{6EI_0}\left[\frac{y^3}{3} + \frac{a}{3b}y^2 - \frac{2a^2}{3b^2}y + \frac{2a^3}{27b^3}\ln(3by + a)\right] + C$$

$$\varphi(x) = \frac{pl^3}{18EI_0}\left[\left(\frac{x}{l}\right)^3 + \frac{a}{b}\left(\frac{x}{l}\right)^2 - \frac{2a^2}{b^2}\left(\frac{x}{l}\right) + \frac{2a^3}{9b^3}\ln\left(3b\frac{x}{l} + a\right)\right] + C.$$

The condition $\varphi(l) = 0$ yields

$$C = -\frac{pl^3}{18EI_0}\left[1 + \frac{a}{b} - \frac{2a^2}{b^2} + \frac{2a^3}{9b^3}\ln(3b + a)\right],$$

and then the solution is

$$\varphi(x) = \frac{pl^3}{18EI_0}\left[\left(\frac{x}{l}\right)^3 - 1 + \frac{a}{b}\left(\frac{x^2}{l^2} - 1\right) - \frac{2a^2}{b^2}\left(\frac{x}{l} - 1\right) + \frac{2a^3}{9b^3}\ln\frac{3b\frac{x}{l} + a}{3b + a}\right]. \quad \blacktriangle$$

APPLICATION 1.3 *Equation of Radioactive Disintegration*

The disintegration rate of a radioactive substance is proportional to the mass of that substance at the time t, namely, $x(t)$. The differential equation of disintegration is $x'(t) = -\alpha x(t)$, where α is a positive constant that depends on the radioactive substance. Determine the disintegration law and the halving time.

Solution

The differential equation $dx/dt = -\alpha x(t)$ is a separable equation. Separating the variables, dividing by $x(t)$, and integrating gives $dx/x = -\alpha\,dt$; $\int dx/x = -\alpha\int dt + C$. Then, $\ln x(t) = -\alpha t + \ln c_1$ (C was chosen as $\ln c_1$) and the general solution $x(t) = c_1 e^{-\alpha t}$. From the initial condition, $x(t_0) = x_0$, $x_0 = c_1 e^{-\alpha t_0}$, which yields $c_1 = x_0 e^{\alpha t_0}$. The solution is

$$x(t) = x_0 e^{-\alpha(t-t_0)},$$

where x_0 is the substance quantity at the time t_0. The halving time is the time period T after which the substance quantity is reduced by half, $x(t_0 + T) = \frac{1}{2}x_0$. From this condition results

$$x_0 e^{-\alpha T} = \tfrac{1}{2}x_0, \quad\text{or}\quad e^{-\alpha T} = \tfrac{1}{2}.$$

Thus,

$$-\alpha T = -\ln 2,$$

or

$$T = \frac{1}{\alpha}\ln 2.$$

Remark 1.4

Radioactive disintegration is a decaying phenomenon. ▲

APPLICATION 1.4 *Newton's Law of Cooling*

The rate at which a body is cooling is proportional to the difference of the temperatures of the body and the surrounding medium. It is known that the air temperature is $U_1 = 10°C$ and that during $T = 15$ minutes the body is cooled from $U_2 = 90°C$ to $U_3 = 50°C$. Find the law for the changing body temperature with respect to time.

Solution

If we denote the time by t and the body temperature by $U(t)$, then $dU/dt = k(U - U_1)$, where k is the proportionality factor. Separating the variables gives $dU/(U - U_1) = k\,dt$. Taking integrals of the left- and right-hand sides gives

$$\int \frac{dU}{U - U_1} = k \int dt + C,$$

or $\ln(U - U_1) = kt + \ln c_1$. Hence, $U - U_1 = c_1 e^{kt}$. Then $U = U_1 + c_1 e^{kt}$. To find the constants c_1 and k, we use the conditions of the problem,

$$U(0) = U_2 \qquad \text{and} \qquad U(T) = U_3.$$

Hence, $U_2 = U_1 + c_1$. Then $c_1 = U_2 - U_1$ and $U_3 = U_1 + (U_2 - U_1)e^{kT}$. Thus, $e^{kT} = (U_3 - U_1)/(U_2 - U_1)$, or $e^k = [(U_3 - U_1)/(U_2 - U_1)]^{1/T}$ and $U(t) = U_1 + (U_2 - U_1)[(U_3 - U_1)/(U_2 - U_1)]^{t/T}$. Substituting the values $U_1 = 10°C$, $U_2 = 90°C$, $U_3 = 50°C$, $T = 15$ min gives

$$U = 10 + 80(\tfrac{1}{2})^{t/15}. \quad ▲$$

Remark 1.5

In Application 1.4, it was assumed that the proportionality factor is constant. Sometimes it is supposed that it depends linearly on time, $k = k_0(1 + \alpha t)$. In this case,

$$\frac{dU}{dt} = k_0(1 + \alpha t)(U - U_1)$$

or

$$\frac{dU}{U - U_1} = k_0 \int (1 + \alpha t)dt + C$$

$$\ln(U - U_1) = k_0\left(t + \frac{\alpha t^2}{2}\right) + \ln c_1; \quad U - U_1 = c_1 e^{k_0[t + (\alpha t^2/2)]}.$$

Thus,

$$U(t) = U_1 + c_1 e^{k_0[t+(\alpha t^2/2)]}.$$

Using

$$U(0) = U_2$$

yields

$$U_2 = U_1 + c_1,$$

or

$$c_1 = U_2 - U_1.$$

Next is used

$$U(T) = U_3,$$

or

$$U_3 = U_1 + (U_2 - U_1)e^{k_0[T+(\alpha T^2/2)]}.$$

Thus,

$$e^{k_0[T+(\alpha T^2/2)]} = \frac{U_3 - U_1}{U_2 - U_1},$$

or

$$e^{k_0} = \left(\frac{U_3 - U_1}{U_2 - U_1}\right)^{1/[T+(\alpha T^2/2)]}.$$

Finally,

$$U(T) = U_1 + (U_2 - U_1)\left(\frac{U_3 - U_1}{U_2 - U_1}\right)^{(2t+\alpha t^2)/(2T+\alpha T^2)}. \quad \blacktriangle$$

APPLICATION 1.5 *The Emptying of a Vessel*

Study the law of leakage of water from a vessel that has the shape of a rotation surface about a vertical axis, with a hole A in the bottom part. Study the following particular cases:

(a) The vessel has a hemisphere shape of radius R
(b) The vessel has a truncated cone shape with the small base as bottom, the radii R_1, R_2, and height H
(c) The vessel has a truncated cone shape with the large base as bottom, the radii R_1, R_2, and height H
(d) The vessel has a cone shape with the vertex at bottom
(e) The vessel has a cylinder shape $\quad \blacktriangle$

Solution

In hydrodynamics is deduced an expression of the form $v = k\sqrt{h}$ that determines the leakage velocity through a hole at depth h from the free surface of the liquid. The equation of median curve of the form $r = r(h)$ is assumed to be known. The volume of water that leaks in elementary time dt is evaluated in two different ways. The liquid leaks through the hole and fills a cylinder with base A and height $v\,dt$; hence, $dV = Av\,dt = Ak\sqrt{h}\,dt$. On the other side, the height of liquid in the vessel will descend by dh; the differential volume that leaks is $dV = -\pi r^2\,dh$. Introducing into equations the two expressions of dV gives the differential equation with separable variables $-\pi r^2\,dh = Ak\sqrt{h}\,dt$. Separating the variables yields

$$dt = -\frac{\pi}{Ak}\,\frac{r^2(h)}{\sqrt{h}}\,dh,$$

and after solving the integral from the expression, we find

$$t = -\frac{\pi}{Ak}\int \frac{r^2(h)}{\sqrt{h}}\,dh + C.$$

From the condition $h(0) = H$, the constant C is determined.

(a) In the case of a spherical shape (Fig. 1.4), it can be written that $r^2 = h(2R - h)$. Then,

$$t = -\frac{\pi}{Ak}\int \frac{h(2R - h)}{\sqrt{h}}\,dh + C$$

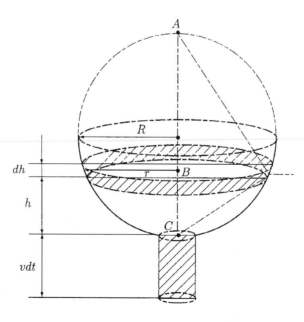

Figure 1.4
The emptying of a spherical vessel.

or

$$t = -\frac{\pi}{Ak}\left[2R\int \sqrt{h}\,dh - \int h^{3/2}\,dh\right] + C$$
$$= -\frac{\pi}{Ak}\left[\frac{4}{3}Rh^{3/2} - \frac{2}{5}h^{5/2}\right] + C.$$

Using the condition

$$h(0) = H$$

yields

$$C = \frac{\pi}{Ak}\left[\frac{4}{3}RH^{3/2} - \frac{2}{5}H^{5/2}\right],$$

and hence $t = \frac{\pi}{Ak}\left[\frac{4}{3}R(H^{3/2} - h^{3/2}) - \frac{2}{5}(H^{5/2} - h^{5/2})\right]$. The time T for

which $h(T) = 0$ is $T = \frac{\pi}{Ak}H^{3/2}\left(\frac{4}{3}R - \frac{2}{5}H\right)$. For $H = R$ (the hemisphere

is full), $T = \left(\frac{14}{15}\right)\frac{\pi R^{5/2}}{Ak}$.

(b) From the geometry of a cone (Fig. 1.5),

$$\frac{r - R_1}{h} = \frac{R_2 - R_1}{H} \qquad \text{and} \qquad r = R_1 + \frac{R_2 - R_1}{H}h.$$

Then,

$$\frac{r^2}{\sqrt{h}} = \frac{R_1^2}{\sqrt{h}} + \frac{2R_1(R_2 - R_1)}{H}\sqrt{h} + \left(\frac{R_2 - R_1}{H}\right)^2 h^{3/2},$$

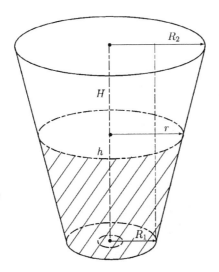

Figure 1.5
The emptying of a truncated-cone vessel with the small base at the bottom.

and substituting this in the expression of t, after integration, yields

$$t = -\frac{\pi}{Ak}\left[2R_1^2\sqrt{b} + \left(\frac{4}{3}\right)\frac{R_1(R_2 - R_1)}{H}b^{3/2} + \left(\frac{2}{5}\right)\frac{R_2 - R_1^2}{H}b^{5/2}\right] + C.$$

Using the condition

$$b(0) = H,$$

we find that

$$C = \frac{\pi}{Ak}\left[2R_1^2\sqrt{H} + \left(\frac{4}{3}\right)\frac{R_1(R_2 - R_1)}{H}H^{3/2} + \left(\frac{2}{5}\right)\frac{R_2 - R_1^2}{H}H^{5/2}\right],$$

and hence,

$$t = \frac{\pi}{Ak}\left[2R_1^2(\sqrt{H} - \sqrt{b}) + \left(\frac{4}{3}\right)\frac{R_1(R_2 - R_1)}{H}(H^{3/2} - b^{3/2})\right.$$
$$\left. + \left(\frac{2}{5}\right)\frac{R_2 - R_1^2}{H}(H^{5/2} - b^{5/2})\right].$$

The condition $b(T) = 0$ implies

$$T = \frac{\pi\sqrt{H}}{Ak}\left[2R_1^2 + \frac{4}{3}R_1(R_2 - R_1) + \frac{2}{5}(R_2 - R_1)^2\right].$$

(c) From Fig. 1.6, $(r - R_1)/(H - b) = (R_2 - R_1)/H$ and yields,

$$r = R_2 + \frac{(R_1 - R_2)}{H}b. \quad \blacktriangle$$

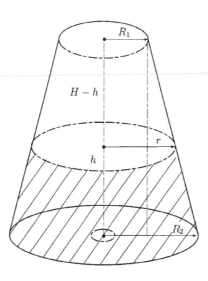

Figure 1.6
The emptying of a truncated-cone vessel with the large base at the bottom.

Remark 1.6

If in the expression of r from case (b) we replace R_1 by R_2, we find the expression of r from case (c). Consequently, the expressions of t and T for case (c) will be obtained from the corresponding expressions obtained in (b), in which R_1 is replaced by R_2 and R_2 by R_1:

$$t = \frac{\pi}{Ak}\left[2R_2^2(\sqrt{H} - \sqrt{b}) + \left(\frac{4}{3}\right)\frac{R_2(R_1 - R_2)}{H}(H^{3/2} - b^{3/2}) \right.$$
$$\left. + \frac{5}{2}\left(\frac{R_1 - R_2}{H}\right)^2 (H^{5/2} - b^{5/2}) \right]$$

$$T = \frac{\pi\sqrt{H}}{Ak}\left[2R_2^2 + \frac{4}{3}R_2(R_1 - R_2) + \frac{2}{5}(R_1 - R_2)^2 \right]. \ \blacktriangle$$

Remark 1.7

If we compare the expressions of T in cases (b) and (c), denoting by T' the expression in case (c), then

$$T' - T = \frac{\pi\sqrt{H}}{Ak}\left[2(R_2^2 - R_1^2) + \frac{4}{3}R_2 R_1 - \frac{4}{3}R_2^2 - \frac{4}{3}R_1 R_2 + \frac{4}{3}R_1^2 \right.$$
$$\left. + \frac{2}{5}(R_1 - R_2)^2 - \frac{2}{5}(R_2 - R_1)^2 \right] = \frac{\pi\sqrt{H}}{Ak}\frac{2}{3}(R_2^2 - R_1^2),$$

or

$$T' = T + \frac{2}{3}\frac{\pi\sqrt{H}}{Ak}(R_2^2 - R_1^2).$$

(d) It is obtained from case (b), taking $R_1 = 0$, $R_2 = R$. Hence,

$$t = \frac{2\pi R^2}{5AkH^2}(H^{5/2} - b^{5/2}) \quad \text{and} \quad T = \frac{2\pi R^2}{5Ak}\sqrt{H}.$$

(e) It is obtained from case (b), taking $R_1 = R_2 = R$. Then,

$$t = \frac{2\pi R^2}{Ak}(\sqrt{H} - \sqrt{b}) \quad \text{and} \quad T = \frac{2\pi R^2}{Ak}\sqrt{H}. \ \blacktriangle$$

Equations That Can Be Reduced to Separable Equations (Equations with Separable Variables)

Equations of the form

$$x' = f(at + bx) \tag{1.43}$$

are reducible to equations with separable variables by the replacement of functions

$$u = at + bx. \tag{1.44}$$

Indeed, $\dfrac{du}{dt} = a + b\dfrac{dx}{dt} = a + bf(u)$; hence,

$$\frac{du}{a + f(u)} = dt \quad \text{and} \quad t = \int \frac{du}{a + bf(u)} + C.$$

EXAMPLE 1.4 Solve the equation $x'(2t + 2x + 1) = (t + x - 1)^2$. ▲

Solution

Let us make the notation $t + x = u$ and then $1 + \dfrac{dx}{dt} = \dfrac{du}{dt}$, or

$$1 + \frac{(u - 1)^2}{2u + 1} = \frac{du}{dt} \Rightarrow \frac{u^2 + 2}{2u + 1} = \frac{du}{dt}$$

$$\Rightarrow \frac{2u + 1}{u^2 + 2} \, du = dt$$

$$\Rightarrow \int \frac{2u + 1}{u^2 + 2} \, du = \int dt + C.$$

If we separate the first member in two integrals,

$$\int \frac{2u}{u^2 + 2} \, du + \int \frac{du}{u^2 + 2} = t + c \Rightarrow \ln(u^2 + 2) + \frac{1}{\sqrt{2}} \arctan \frac{u}{\sqrt{2}} = t + c.$$

The general solution is

$$\ln(t^2 + x^2 + 2tx + 2) + \frac{1}{\sqrt{2}} \arctan \frac{t + x}{\sqrt{2}} = t + c. ▲$$

APPLICATION 1.6 A body with mass m is acted on by a force proportional to time (the proportionality factor is equal to k_1). In addition the body experiences a counteraction by the medium that is proportional to the velocity of the body (the proportionality factor being equal to k_2). Find the law of the body's motion. ▲

Solution

The differential equation of motion is $m \dfrac{dv}{dt} = k_1 t - k_2 v$. Denoting $k_1 t - k_2 v = u$, we find (after derivation with respect to t) $k_1 - k_2 \dfrac{dv}{dt} = \dfrac{du}{dt}$. Multiplying by m and taking into account the replacement $k_1 m - k_2 u = m \dfrac{du}{dt}$, which is an equation with separable variables, $du/(k_1 m - k_2 u) = \dfrac{1}{m} dt$. After integration,

$$\int \frac{du}{k_1 m - k_2 u} = \frac{1}{m} \int dt + C \Rightarrow -\frac{1}{k_2} \ln |k_1 m - k_2 u| = \frac{t}{m} + C.$$

The initial condition $v(0) = 0$ results in $u(0) = 0$; hence, $-\dfrac{1}{k_2} \ln |k_1 m| = C$.

Replacing the value of C yields $-\dfrac{1}{k_2} \ln |k_1 m - k_2 u| = \dfrac{t}{m} - \dfrac{1}{k_2} \ln |k_1 m|$.

Multiplying by $(-k_2)$, $\ln |k_1 m - k_2 u| = \ln |k_1 m| - \dfrac{k_2}{m} t$; hence, $k_1 m - k_2 u = k_1 m e^{-k_2 t/m} \Rightarrow k_2 u = k_1 m - k_1 m e^{-k_2 t/m}$. Replacing u by its expression depending on v gives

$$k_2 k_1 t - k_2^2 v = k_1 m - k_1 m e^{-k_2 t/m} \Rightarrow v(t) = \frac{k_1 m}{k_2^2} e^{-k_2 t/m} + \frac{k_1}{k_2} t - \frac{k_1 m}{k_2^2}.$$

To find the dependence of displacement on time, we use the equality

$$v(t) = \frac{ds(t)}{dt} \quad \text{or} \quad s(t) = \int v(t)dt + C; \quad s(0) = 0.$$

This yields

$$s(t) = \int \left(\frac{k_1 m}{k_2^2} e^{-k_2 t/m} + \frac{k_1}{k_2} t - \frac{k_1 m}{k_2^2} \right) dt + C$$

$$= -\frac{k_1 m^2}{k_2^3} e^{-k_2 t/m} + \frac{k_1}{2k_2} t^2 - \frac{k_1 m}{k_2^2} t + C$$

$$s(0) = s_0 \Rightarrow s_0 = C - \frac{k_1 m^2}{k_2^3};$$

hence,

$$C = s_0 + \frac{k_1 m^2}{k_2^3} \quad \text{and} \quad s(t) = s_0 + \frac{k_1 m^2}{k_2^3} - \frac{k_1 m}{k_2^2} t + \frac{k_1}{2k_2} t^2 - \frac{k_1 m^2}{k_2^3} e^{-k_2 t/m}. \quad \blacktriangle$$

Homogeneous Equations

A differential equation of the first order,

$$x' = f(t, x), \tag{1.45}$$

is called a *homogeneous equation* if the function $f(t, x)$ is homogeneous with degree zero, that is,

$$f(\lambda t, \lambda x) = f(t, x), \quad \forall \lambda \neq 0. \tag{1.46}$$

If $\lambda = 1/t$ ($t \neq 0$), then $f(t, x) = f(1, x/t)$, which means that the function f depends only on the ratio x/t. After replacement,

$$\frac{x}{t} = u, \tag{1.47}$$

and making the notation $f(1, u) = \varphi(u)$, $x = tu$, $\dfrac{dx}{dt} = u + t\dfrac{du}{dt}$ and Eq. (1.45) becomes $u + t\dfrac{du}{dt} = \varphi(u)$, which is an equation with separable variables, $\dfrac{du}{dt} = (\varphi(u) - u)/t$. This equation is defined on domains of the form $(-\infty, 0) \times (u_1, u_2)$ or $(0, \infty) \times (u_1, u_2)$, where u_1, u_2 are two consecutive zeros of function $\varphi(u) - u$. Separating the variables and integrating,

$$\frac{du}{\varphi(u) - u} = \frac{dt}{t}.$$

The general solution is

$$\int \frac{du}{\varphi(u) - u} = \ln|t| + C. \tag{1.48}$$

Remark 1.8

If u_k are solutions of the equation

$$\varphi(u) - u = 0, \tag{1.49}$$

then the straight lines $x = u_k t$ are singular solutions of Eq. (1.45).

Remark 1.9

If $f(x, y) = P(x, y)/Q(x, y)$, with P and Q homogeneous polynomials that have the same degree, then Eq. (1.45) is a homogeneous equation.

EXAMPLE 1.5 Find the general solution of equation $y/y' = x + \sqrt{x^2 + y^2}$. ▲

Solution

The equation is written in the form $y\dfrac{dx}{dy} = x + \sqrt{x^2 + y^2}$, or

$$\frac{dx}{dy} = \frac{x}{y} + \sqrt{\frac{x^2}{y^2} + 1}.$$

The replacement $\dfrac{x}{y} = u$ yields $x = yu$; the general solution is

$$\frac{dx}{dy} = u + y\frac{du}{dy} \Rightarrow u + y\frac{du}{dy} = u + \sqrt{u^2 + 1} \Rightarrow \frac{du}{\sqrt{u^2 + 1}} = \frac{dy}{y} \Rightarrow \int \frac{du}{\sqrt{u^2 + 1}}$$

$$= \int \frac{dy}{y} + \ln c \Rightarrow \ln(u + \sqrt{u^2 + 1}) = \ln y + \ln c \Rightarrow u + \sqrt{u^2 + 1}$$

$$= cy \Rightarrow \frac{x}{y} + \sqrt{\frac{x^2}{y^2} + 1} = cy \Rightarrow \sqrt{x^2 + y^2} = cy^2 - x \Rightarrow x^2 + y^2$$

$$= c^2 y^4 - 2cxy^2 + x^2 \Rightarrow c^2 y^2 = 2cx + 1. ▲$$

APPLICATION 1.7 *Parabolic Mirror*

Find a mirror such that light from a point source at the origin O is reflected in a beam parallel to a given direction.

Solution

Consider the plane section of the mirror (Fig. 1.7). Consider that the ray of light OP strikes the mirror at M and is reflected along MR, parallel to the x-axis. If MT is the tangent in M and α, i and r are the angles indicated, $i = r$ by the optical law of reflection, and $r = \alpha$ by geometry. Hence, $\alpha = i$ and $|OT| = |OM|$; $|OT| = |PT| - x$, $MP/PT = \tan \alpha = y' \Rightarrow |MP| = y'|PT| \Rightarrow y = y'|PT| \Rightarrow |OT| = |y/y'| - x$; $|OM| = \sqrt{x^2 + y^2}$. The differential equation is

$$\left|\frac{y}{y'}\right| - x = \sqrt{x^2 + y^2} \Rightarrow \frac{y}{y'} = x + \sqrt{x^2 + y^2}.$$

This is the equation from the previous example. Its general solution is the family of parabolas $c^2 y^2 = 2cx + 1$. If we state the condition, for example, $y(-1) = 0$, we obtain $c = \frac{1}{2}$, and the solution is $x = \frac{1}{4}y^2 - 1$. ▲

APPLICATION 1.8 *The Problem of the Swimmer*

To cross a river, a swimmer starts from a point P on the bank. He wants to arrive at a point Q on the other side. The velocity v_1 of the running water is

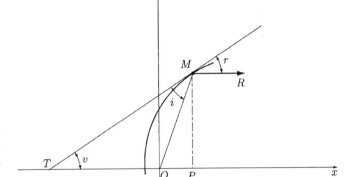

Figure 1.7
Reflection of a beam in a parabolic mirror.

k_1, and the velocity v_2 of the swimmer's motion is constant k_2. Find the trajectory described by the swimmer, knowing that the relative velocity is always directed to Q. ▲

Solution

Consider that M is the swimmer's position at time t (Fig. 1.8). The components of the absolute velocity on the two axes Ox and Oy are

$$\begin{cases} \dfrac{dx}{dt} = k_1 - k_2 \dfrac{x}{\sqrt{x^2 + y^2}} \\ \dfrac{dy}{dt} = -k_2 \dfrac{y}{\sqrt{x^2 + y^2}}. \end{cases}$$

Dividing these equalities gives the differential equation of the demanded trajectory,

$$\frac{dx}{dy} = \frac{x}{y} - k\sqrt{\frac{x^2}{y^2} + 1}; \quad k = \frac{k_1}{k_2}.$$

This is a homogeneous equation, and after the replacement $x = yu$ and $dx/dy = u + y(du/dy)$, the equation becomes $y(du/dy) = -k\sqrt{u^2 + 1}$, or $du/\sqrt{u^2 + 1} = -k(dy/y)$. Integration gives

$$\ln(u + \sqrt{u^2 + 1}) = -k \ln y + \ln c (c > 0) \quad \text{or} \quad u + \sqrt{u^2 + 1} = cy^{-k}.$$

This then yields

$$u = \frac{1}{2}\left(\frac{c}{y^k} - \frac{y^k}{c}\right).$$

Returning to x and y, we find

$$x = \frac{1}{2}y\left(\frac{c}{y^k} - \frac{y^k}{c}\right).$$

The condition for the trajectory to pass through $P(x_0, y_0)$ yields $c = y_0^{k-1}(x_0 + \sqrt{x_0^2 + y_0^2})$. The condition for trajectory to pass through Q is written as

$$\lim_{y \to 0} \frac{1}{2} y \left(\frac{c}{y^k} - \frac{y^k}{c} \right) = 0,$$

and it is possible if $k < 1$. ▲

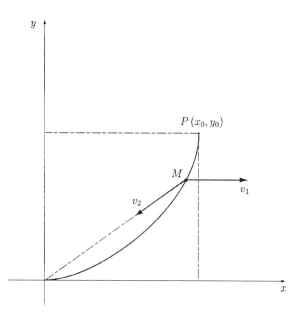

Figure 1.8
Trajectory of the swimmer.

Remark 1.10

For $k_1 = 0$, $k = 0$, and the trajectory has the equation $x = \dfrac{x_0}{y_0} y$, that is, the linear segment between P and Q. ▲

Equations That Can Be Reduced to Homogeneous Equations
Consider the equation

$$x' = f\left(\frac{a_1 t + b_1 x + c_1}{a_2 t + b_2 x + c_2} \right) \quad a_i, \ b_i, \ c_i \in \mathbf{R}; \quad i = 1, 2, \tag{1.50}$$

to which will be attached the algebraic system

$$\begin{cases} a_1 t + b_1 x + c_1 = 0 \\ a_2 t + b_2 x + c_2 = 0. \end{cases} \tag{1.51}$$

If

$$a_1 b_2 - a_2 b_1 \neq 0, \tag{1.52}$$

then the system of equations (1.51) has a unique solution (t_0, x_0). Let us change the variables:

$$\begin{cases} x = y + x_0 \\ t = s + t_0. \end{cases} \tag{1.53}$$

Because $dx/dt = dy/ds$, Eq. (1.50) becomes

$$\frac{dy}{ds} = f\left(\frac{a_1 s + b_1 y}{a_2 s + b_2 y}\right) = f\left(\frac{a_1 + b_1 \dfrac{y}{s}}{a_2 + b_2 \dfrac{y}{s}}\right) \overset{not}{=} \varphi\left(\frac{y}{s}\right),$$

which is a homogeneous equation. If $a_1 b_2 - a_2 b_1 = 0$, then $a_1/a_2 = b_1/b_2 = k$; hence,

$$\frac{a_1 t + b_1 x + c_1}{a_2 t + b_2 x + c_2} = \frac{a_1 t + b_1 x + c_1}{k(a_1 t + b_1 x) + c_2},$$

and if we make the notation $y = a_1 t + b_1 x$, Eq. (1.50) may be written as

$$\frac{dy}{dt} = b_1 f\left(\frac{y + c_1}{ky + c_2}\right) + a_2,$$

which is an equation with separable variables.

EXAMPLE 1.6 Integrate the equation $\dfrac{dx}{dt} = \dfrac{t - x + 3}{t + x - 1}$. ▲

Solution

The system

$$\begin{cases} t - x + 3 = 0 \\ t + x - 1 = 0 \end{cases} \tag{1.54}$$

has the unique solution $t_0 = -1$, $x_0 = 2$. Let us change the variables:

$$\begin{cases} t = s - 1 \\ x = y + 2. \end{cases}$$

Because $dx/dt = dy/ds$, the initial equation becomes

$$\frac{dy}{ds} = \frac{s - y}{s + y} = \frac{1 - \dfrac{y}{s}}{1 + \dfrac{y}{s}},$$

which is a homogeneous equation. The replacement $y = su$, $dy/ds = u + s(du/ds)$ leads to

$$u + s\frac{du}{ds} = \frac{1 - u}{1 + u} \quad \text{or} \quad s\frac{du}{ds} = \frac{1 - 2u - u^2}{1 + u}.$$

Separating the variables and integrating gives

$$\frac{(u + 1)du}{u^2 + 2u - 1} + \frac{ds}{s} = 0 \Rightarrow \frac{1}{2}\ln|u^2 + 2u - 1| + \ln|s| = c$$

$$\Rightarrow s^2(u^2 + 2u - 1) = c_1.$$

Returning to the initial variables yields $y^2 + 2ys - s^2 = c_1$. Hence, the general solution is $(x - 2)^2 + 2(t + 1)(x - 2) - (t + 1)^2 = c_1$. ▲

Remark 1.11

From $u^2 + 2u - 1 = 0$, we obtain $u_{1,2} = -1 \pm \sqrt{2}$ and $(x - 2)/(t + 1) = -1 \pm \sqrt{2}$ as singular solutions.

EXAMPLE 1.7 Integrate the equation

$$\frac{dx}{dt} = \frac{2t + 2x + 4}{t + x - 1}. \quad \blacktriangle$$

Solution

The right-hand member

$$f(t, x) = \frac{2(t + x) + 4}{t + x - 1}$$

is a function only of $t + x$. Changing the function, $t + x = y$ yields $1 + \dfrac{dx}{dt} = \dfrac{dy}{dt}$; hence,

$$\frac{dy}{dt} = 1 + \frac{2y + 4}{y - 1} \Rightarrow \frac{dy}{dt} = 3\frac{y + 1}{y - 1} \Rightarrow \frac{y - 1}{y + 1} dy = 3dt.$$

Integrating, $y + 2\ln|y + 1| = 3t + c$, hence, the general solution is

$$x - 2t + 2\ln|t + x + 1| = c. \quad \blacktriangle$$

Total Differential Equations

A first-order equation of the form

$$b_1(t, x)dt + b_2(t, x)dx = 0 \tag{1.55}$$

is said to be a *total (or exact) differential equation* if its lefthand side is a total differential of some function $H(t, x)$,

$$b_1 dt + b_2 dx \equiv dH \equiv \frac{\partial H}{\partial t} dt + \frac{\partial H}{\partial x} dx. \tag{1.56}$$

This equality is possible if and only if

$$\frac{\partial b_1}{\partial x} = \frac{\partial b_2}{\partial t} \tag{1.57}$$

should hold in some range D of variables t and x. The general integral of Eq. (1.55) is of the form

$$H(t, x) = c \quad \text{or} \quad \int_{t_0}^{t} b_1(\tau, x_0)d\tau + \int_{x_0}^{x} b_2(t, s)ds = c. \tag{1.58}$$

EXAMPLE 1.8 Solve the equation $(2t - 3xt^2)dt + (x^2 + x - t^3)dx = 0$. $\quad \blacktriangle$

Solution

Let us verify that this equation is a total differential equation: $\partial b_1/\partial x = -3t^2$; $\partial b_2/\partial t = -3t^2$, so that $\partial b_1/\partial x = \partial b_2/\partial t$; that is, the condition (1.57) is fulfilled. The general integral is $\int_{t_0}^t (2\tau - 3\tau^2 x_0)d\tau + \int_{x_0}^x (s^2 + s - t^3)ds = c \Rightarrow$

$$\tau^2|_{t_0}^t - \tau^3|_{t_0}^t x_0 + \left(\frac{s^3}{3} + \frac{s^2}{2}\right)|_{x_0}^x - t^3 s|_{x_0}^x = c \Rightarrow t^2 - t_0^2 - t^3 x_0 + t_0^3 x_0 + \frac{x^3}{3} + \frac{x^2}{2}$$

$$-\frac{x_0^3}{3} - \frac{1}{2}x_0^2 - t^3 x + t^3 x_0 = c \Rightarrow t^2 + \frac{x^3}{3} + \frac{x^2}{2} - t^3 x = c_1, \text{ where}$$

$$c_1 = c + t_0^2 - t_0^3 x_0 + \frac{x_0^3}{3} + \frac{1}{2}x_0^2. \quad \blacktriangle$$

The Integrating Factor

In some cases, where Eq. (1.55) is not a total differential equation, it may be possible to select a function $\mu(t, x)$ that, after multiplying the left-hand side, turns into a total differential $dH = \mu b_1\, dt + \mu b_2\, dx$. This function $\mu(t, x)$ is called an *integrating factor*. From the definition of the integrating factor, $\frac{\partial}{\partial x}(\mu b_1) = \frac{\partial}{\partial t}(\mu b_2)$, or $b_1 \frac{\partial \mu}{\partial x} - b_2 \frac{\partial \mu}{\partial t} = \left(\frac{\partial b_2}{\partial t} - \frac{\partial b_1}{\partial x}\right)\mu$. Hence,

$$b_1 \frac{\partial \ln \mu}{\partial x} - b_2 \frac{\partial \ln \mu}{\partial t} = \frac{\partial b_2}{\partial t} - \frac{\partial b_1}{\partial x}, \tag{1.59}$$

and a partial differential equation has been obtained. The integrating factor is relatively easy to find in the following cases.

1. If $\mu = \mu(t)$, then $\frac{\partial \mu}{\partial x} = 0$, and Eq. (1.59) will take the form

$$\frac{d \ln \mu}{dt} = \frac{\frac{\partial b_1}{\partial x} - \frac{\partial b_2}{\partial t}}{b_2} = \varphi(t). \tag{1.60}$$

Then,

$$\ln \mu = \int \varphi(t)dt. \tag{1.61}$$

2. If $\mu = \mu(x)$, then $\frac{\partial \mu}{\partial t} = 0$, and Eq. (1.59) will take the form

$$\frac{d \ln \mu}{dx} = \frac{\frac{\partial b_2}{\partial t} - \frac{\partial b_1}{\partial x}}{b_1} = \psi(x). \tag{1.62}$$

Then,

$$\ln \mu = \int \psi(x)dx. \tag{1.63}$$

EXAMPLE 1.9 Solve the equation $(x^2 + 3tx)dt + (tx + t^2)dx = 0$. $\blacktriangle$

Solution

The derivatives are

$$
\begin{cases}
\dfrac{\partial b_1}{\partial x} = 2x + 3t, \\[2mm]
\dfrac{\partial b_2}{\partial t} = x + 2t,
\end{cases}
$$

with $\dfrac{\partial b_1}{\partial x} \neq \dfrac{\partial b_2}{\partial t}$, but

$$
\dfrac{\dfrac{\partial b_1}{\partial x} - \dfrac{\partial b_2}{\partial t}}{b_2} = \dfrac{x+t}{t(x+t)} = \dfrac{1}{t}.
$$

This satisfies condition (1.61), and $\ln \mu = \int (1/t)dt \Rightarrow \ln \mu = \ln t \Rightarrow \mu = t$. Multiplying the given equation by t, we find the total differential equation $(x^2 t + 3t^2 x)dt + (t^2 x + t^3)dx = 0$, whose general solution is given by the equality $\int_{t_0}^{t}(x_0^2 \tau + 3\tau^2 x_0)d\tau + \int_{x_0}^{x}(t^2 s + t^3)ds = c$. After the evaluation of integrals and reducing the similar terms, the general solution is

$$
t^2 x^2 + 2t^3 x = c_1, \quad c_1 \text{ being an arbitrary constant.} \quad \blacktriangle
$$

EXAMPLE 1.10 Solve the equation $(tx + 2t)dt + (t^2 + x^2)dx = 0$. $\blacktriangle$

Solution

We can write that $\dfrac{\partial b_1}{\partial x} = t; \dfrac{\partial b_2}{\partial t} = 2t \Rightarrow \dfrac{\partial b_1}{\partial x} \neq \dfrac{\partial b_2}{\partial t}$, but

$$
\left(\dfrac{\partial b_2}{\partial t} - \dfrac{\partial b_1}{\partial x}\right)/b_1 = \dfrac{t}{t(x+2)} = \dfrac{1}{x+2} = \psi(x).
$$

The equation has an integrable factor $\mu = \mu(x)$:

$$
\ln \mu = \int \dfrac{dx}{x+2} \Rightarrow \ln \mu = \ln |x+2| \Rightarrow \mu = x+2.
$$

Multiplying the given equation by $x + 2$, we find the equation with total differentials

$$
(tx^2 + 4tx + 4t)dt + (xt^2 + x^3 + 2t^2 + 2x^2)dx = 0.
$$

With Eq. (1.58), after the evaluation of integrals and reducing similar terms, the general solution is

$$
2t^2 + \dfrac{t^2 x^2}{2} + \dfrac{x^4}{4} + 2t^2 x + \dfrac{2}{3}x^3 = c_1. \quad \blacktriangle
$$

EXAMPLE 1.11 Solve the equation $(3x^2 - t)dt + (2x^3 - 6tx)dx = 0$ with an integrating factor of the form $\mu = \varphi(t + x^2)$. $\blacktriangle$

Differential Equations

Solution

Set $t + x^2 = y$. Then $\mu = \mu(y)$, and consequently,

$$\frac{\partial \ln \mu}{\partial t} = \frac{d \ln \mu}{dy}\left(\frac{\partial y}{\partial t}\right) = \frac{d \ln \mu}{dy}, \qquad \frac{\partial \ln \mu}{\partial x} = \frac{d \ln \mu}{dy}\left(\frac{\partial y}{\partial x}\right) = 2x\frac{d \ln \mu}{dy}.$$

Equation (1.59) used to find the integrating factor will be of the form

$$(2xb_1 - b_2)\frac{d \ln \mu}{dy} = \frac{\partial b_2}{\partial t} - \frac{\partial b_1}{\partial x} \qquad \text{or} \qquad \frac{d \ln \mu}{dy} = \frac{\dfrac{\partial b_2}{\partial t} - \dfrac{\partial b_1}{\partial x}}{2xb_1 - b_2}.$$

Since $b_1 = 3x^2 - t$, $b_2 = 2x^3 - 6tx$, then

$$\frac{\dfrac{\partial b_2}{\partial t} - \dfrac{\partial b_1}{\partial x}}{2xb_1 - b_2} = \frac{-3}{t + x^2} = \frac{-3}{y},$$

and hence $d \ln \mu / dy = -3/y$, or $\mu = y^{-3}$, that is, $\mu = 1/(t + x^2)^3$, and multi-plying the given equation by $\mu = 1/(t + x^2)^3$ yields

$$\frac{3x^2 - t}{(t + x^2)^3} dt + \frac{2x^3 - 6tx}{(t + x^2)^3} dx = 0.$$

This is a total differential equation, and according to Eq. (1.58), its general integral is

$$\int_{t_0}^{t} \frac{3x_0^2 - \tau}{(\tau + x_0^2)^3} d\tau + \int_{x_0}^{x} \frac{2s^3 - 6ts}{(t + s^2)^3} ds = c \qquad \text{or,} \qquad (t + x^2)^2 c = t - x^2. \quad \blacktriangle$$

Linear Differential Equations of First Order

A first-order linear differential equation is an equation that is linear in the unknown function and its derivative. A linear equation has the form

$$\frac{dx}{dt} + a(t)x = b(t), \tag{1.64}$$

where $a(t)$ and $b(t)$ will henceforth be considered continuous functions of t in the domain in which it is required to integrate Eq. (1.64). If $b(t) \equiv 0$, then Eq. (1.64) is called homogeneous linear. In this case,

$$\frac{dx}{dt} + a(t)x = 0, \tag{1.65}$$

which is an equation with separable variables, $dx/x = -a(t)dt$, and inte-grating,

$$\ln |x| = -\int a(t)dt + \ln c_1, \qquad c_1 > 0 \Rightarrow x = ce^{-\int a(t)dt}, \qquad c \neq 0. \tag{1.66}$$

This is the general solution of Eq. (2.29). Also, it satisfies the initial condition $x(t_0) = x_0$, given by

$$x(t) = x_0 e^{-\int_{t_0}^{t} a(\tau)d\tau}.$$

The general solution of a nonhomogeneous Eq. (1.65) can be found by the *method of variation of an arbitrary constant*, which consists in finding the solution of Eq. (1.65) in the form

$$x = c(t)e^{-\int a(t)dt}, \tag{1.67}$$

where $c(t)$ is a new unknown function of t. Computing the derivative

$$\frac{dx}{dt} = c'(t)e^{-\int a(t)dt} - c(t)a(t)e^{-\int a(t)dt}$$

and substituting it into the original Eq. (1.64),

$$c'(t)e^{-\int a(t)dt} - a(t)c(t)e^{-\int a(t)dt} + a(t)c(t)e^{-\int a(t)dt} = b(t),$$

or

$$c'(t) = b(t)e^{\int a(t)dt}.$$

Integrating, we find

$$c(t) = \int b(t)e^{\int a(t)dt}\, dt + c_1,$$

and consequently,

$$x = e^{-\int a(t)dt}\left[c_1 + \int b(t)e^{\int a(t)dt}\, dt \right] \tag{1.68}$$

is the general solution of Eq. (1.64).

Remark 1.12

The general solution of a nonhomogeneous linear equation is the sum of the general solution of the corresponding homogeneous equation $c_1 e^{-\int a(t)dt}$ and a particular solution of the nonhomogeneous equation $e^{-\int a(t)dt}\int b(t)e^{\int a(t)dt}\, dt$ obtained from Eq. (2.32) for $c_1 = 0$.

Properties

1. The general solution of a nonhomogeneous linear Eq. (1.64) is of the form

$$x = A(t)c + B(t), \quad c \in \mathbf{R}, \tag{1.69}$$

and the general solution of the homogeneous linear Eq. (1.65) is of the form

$$x = A(t)c, \quad c \in \mathbf{R}. \tag{1.70}$$

2. If x_1 and x_2 are two particular solutions of Eq. (1.64), then the general solution is

$$x = x_1 + c(x_2 - x_1). \tag{1.71}$$

3. A linear equation remains linear whatever replacements of the independent variable $t = \varphi(\tau)$ are made [where $\varphi(\tau)$ is a differentiable function].

4. A linear equation remains linear whatever linear transformations of the sought-for function $x = \alpha(t)y + \beta(t)$ take place [where $\alpha(t)$ and $\beta(t)$ are arbitrary differentiable functions, with $\alpha(t) \neq 0$ in the interval under consideration].

EXAMPLE 1.12 Find the solution through $(0, -1)$ for the equation

$$(t^2 + 1)x' + 2tx = t^2. \quad \blacktriangle$$

Solution

Dividing both sides of equation by $t^2 + 1$ gives

$$x' + \frac{2t}{t^2 + 1}x = \frac{t^2}{t^2 + 1}.$$

The corresponding homogeneous equation of the given equation is

$$x' + \frac{2t}{t^2 + 1}x = 0,$$

and its general solution is $x = c/(t^2 + 1)$. The general solution of the nonhomogeneous equation is sought in the form $x = c(t)/(t^2 + 1)$. Substituting it into the nonhomogeneous equation yields

$$\frac{c'(t)}{t^2 + 1} - \frac{2tc(t)}{(t^2 + 1)^2} + \frac{2tc(t)}{(t^2 + 1)^2} = \frac{t^2}{t^2 + 1}; \quad c'(t) = t^2, \quad c(t) = \frac{t^3}{3} + c_1.$$

Hence, the general solution is

$$x = \frac{c_1}{t^2 + 1} + \frac{t^3}{3(t^2 + 1)}.$$

Using the condition $x(0) = -1$, then $c_1 = -1$. The solution of the initial value problem is

$$x = \frac{1}{t^2 + 1}\left(\frac{t^3}{3} - 1\right). \quad \blacktriangle$$

EXAMPLE 1.13 Solve the equation $y' = y/(2y \ln y + y - x). \quad \blacktriangle$

Solution

This equation is linear if one considers x as a function of y,

$$\frac{dx}{dy} = \frac{2y \ln y + y - x}{y}, \quad \text{or} \quad \frac{dx}{dy} + \frac{x}{y} = 2 \ln y + 1. \text{ Using Eq. (1.68), we find that}$$

$$x = e^{-\int (1/y) dy}\left[c_1 + \int (2 \ln y + 1)e^{\int (1/y) dy}\, dy\right],$$

$$x = \frac{1}{y}\left[c_1 + \int (2 \ln y + 1)y\, dy\right],$$

and finally,

$$x = \frac{c}{y} + y \ln y.$$

EXAMPLE 1.14 Find the solution of an equation satisfying the indicated condition $2xy' - y = 1 - (2/\sqrt{x})$, $y \to -1$ as $x \to +\infty$. ▲

Solution

The general solution of equation $y' - \dfrac{1}{2x}y = \dfrac{1}{2x} - \dfrac{1}{x\sqrt{x}}$ is

$$y = e^{\int (1/2x)dx}\left[c + \int \left(\frac{1}{2x} - \frac{1}{x\sqrt{x}} \right) e^{-\int (1/2x)dx} \right]$$

$$\Rightarrow y = \sqrt{x}\left[c + \int \left(\frac{1}{2x} - \frac{1}{x\sqrt{x}} \right) \frac{1}{\sqrt{x}}\, dx \right] = \sqrt{x}\left[c + \frac{1}{x} - \frac{1}{\sqrt{x}} \right]$$

$$= c\sqrt{x} + \frac{1}{\sqrt{x}} - 1 \; \lim_{x\to\infty} \left(c\sqrt{x} + \frac{1}{\sqrt{x}} - 1 \right) = -1 \Rightarrow c = 0.$$

The solution with the indicated condition is

$$y = \frac{1}{\sqrt{x}} - 1.$$

APPLICATION 1.9 *The Motion of Parachutes*

A body with mass m is dropped with an initial velocity v_0 from some height. It is required to establish the law of velocity v variation as the body falls, if in addition to the force of gravity the body is acted upon by the decelerating force of the air, proportional to the velocity (with constant k). ▲

Solution

By Newton's second law, $m\dfrac{dv}{dt} = F$, where $\dfrac{dv}{dt}$ is the acceleration of the moving body and F is the force acting on the body in the direction of motion. This force is the resultant of two forces: the force of gravity mg, and the force of air resistance $-kv$, which has the minus sign because it is in the opposite direction to the velocity. Then, $m\dfrac{dv}{dt} = mg - kv$, or $\dfrac{dv}{dt} + \dfrac{k}{m}v = g$. The solution of this linear equation is $v = ce^{-kt/m}[c + \int ge^{kt/m}dt] \Rightarrow v(t) = ce^{-kt/m} + gm/k$. Taking into account that $v(0) = v_0$, then $v_0 = c + mg/k$; $c = v_0 - mg/k$, and $v(t) = (v_0 - mg/k)e^{-kt/m} + mg/k$. From this formula it follows that for sufficiently large t, the velocity v depends slightly on v_0. It will be noted that if $k = 0$ (the air resistance is absent or so small that it can be neglected), we obtain a result familiar for physics,

$$v = v_0 + gt.$$

Remark 1.13

The equation $m\dfrac{dv}{dt} = mg - kv$ could be also solved as an equation with separable variables. The equation from Application 1.6 could be regarded as a first-order linear equation. ▲

APPLICATION 1.10

Consider an electrical circuit with a resistor R and a self-inductance L that satisfies the differential equation $L\dfrac{dI}{dt} + RI = E$, where E is the electromotive force. Find the electrical intensity I, t seconds after the moment of switching on, if E changes according to the sinusoidal law $E = E_0 \cos \omega t$, and $I = 0$ for $t = 0$. ▲

Solution

Using the notation $\alpha = \dfrac{R}{L}$, we obtain the linear equation $\dfrac{dI}{dt} + \alpha I = (E_0/L) \cos \omega t$. The solution is $I = e^{-\alpha t}[c + (E_0/L) \int e^{\alpha t} \cos \omega t \, dt]$. Integrating twice by parts, we obtain

$$I(t) = ce^{-\alpha t} + \frac{E_0(\omega \sin \omega t + \alpha \cos \omega t)}{L(\omega^2 + \alpha^2)}.$$

If $I(0) = 0$, then

$$c = -\frac{E_0 \alpha}{L(\omega^2 + \alpha^2)};$$

and consequently,

$$I(t) = \frac{E_0}{L(\omega^2 + \alpha^2)}(\omega \sin \omega t + \alpha \cos \omega t - \alpha e^{-\alpha t}).$$

Since t is sufficiently large, and $e^{-\alpha t}$ is a small quantity ($\alpha > 0$) and can be ignored,

$$I(t) \approx \frac{E_0}{L(\omega^2 + \alpha^2)}(\omega \sin \omega t + \alpha \cos \omega t). \quad ▲$$

Bernoulli's Equation

This equation is of the form

$$\frac{dx}{dt} + a(t)x = b(t)x^\alpha, \quad \alpha \in \mathbf{R} \backslash \{0, 1\}. \tag{1.72}$$

Using the substitution $x^{1-\alpha} = y$, Bernoulli's equation may be reduced to a linear equation. Indeed, differentiating,

$$x^{1-\alpha} = y, \tag{1.73}$$

and $(1 - \alpha)x^{-\alpha}\dfrac{dx}{dt} = \dfrac{dy}{dt}$. Substituting into Eq. (1.72), we find the linear equation

$$\frac{dy}{dt} + (1 - \alpha)a(t)y = (1 - \alpha)b(t). \tag{1.74}$$

EXAMPLE 1.15 Solve the equation $\dfrac{dx}{dt} - \dfrac{2t}{t^2 + 1}x = -\dfrac{t^2}{t^2 + 1}x^2$ and find the solution through $(0, -1)$. ▲

Solution

Dividing both sides of equation by x^2 gives

$$\frac{1}{x^2}\left(\frac{dx}{dt}\right) - \left(\frac{1}{x}\right)\frac{2t}{t^2+1} = -\frac{t^2}{t^2+1}.$$

We use the substitution

$$\frac{1}{x} = y, \quad -\frac{1}{x^2}\left(\frac{dx}{dt}\right) = \frac{dy}{dt}.$$

After substitution, the last equation turns into the linear equation

$$\frac{dy}{dt} + \frac{2t}{t^2+1}y = \frac{t^2}{t^2+1}$$

with the general solution

$$y = \frac{3c_1 + t^3}{3(t^2+1)}$$

(see Example 1.12). Hence, the general solution of the given equation is $x = 3(t^2+1)/(3c_1 + t^3)$. If $x(0) = -1$, then $c = -1$. It follows that

$$x = \frac{3(t^2+1)}{t^3 - 3}. \quad \blacktriangle$$

EXAMPLE 1.16   Find the solution of equation satisfying the indicated condition

$$xy' - y = \left(1 - \frac{2}{\sqrt{x}}\right)\sqrt{y}; \quad \lim_{x\to+\infty} = 1. \quad \blacktriangle$$

Solution

If we divide both sides of the equation by $\sqrt{y}$, $\dfrac{x}{\sqrt{y}}\left(\dfrac{dy}{dx}\right) - \sqrt{y} = 1 - \dfrac{2}{\sqrt{x}}$,

and set $\sqrt{y} = z$, $\dfrac{1}{2\sqrt{y}}\left(\dfrac{dy}{dx}\right) = \dfrac{dz}{dx}$, then $2x\dfrac{dz}{dx} - z = 1 - \dfrac{2}{\sqrt{x}}$, which is a

linear equation whose general solution is (see Example 1.14)

$$z = c\sqrt{x} + \frac{1}{\sqrt{x}} - 1.$$

The general solution of the given equation is

$$y = z^2 = \left(c\sqrt{x} + \frac{1}{\sqrt{x}} - 1\right)^2; \quad \lim_{x\to\infty}\left(c\sqrt{x} + \frac{1}{\sqrt{x}} - 1\right)^2 = 1 \Rightarrow c = 0;$$

hence, the required solution is

$$y = \frac{1 - 2\sqrt{x} + x}{x}. \quad \blacktriangle$$

APPLICATION
1.11

Find the law of motion for a body provided by the resistance of the medium depending on the velocity, $F = \lambda_1 v + \lambda_2 v^\alpha$, $\alpha \neq 1$.   ▲

Solution

The equation of motion is assumed having the form $m\dfrac{dv}{dt} = -\lambda_1 v - \lambda_2 v^\alpha$, or $\dfrac{dv}{dt} + \dfrac{\lambda_1}{m}v = -\dfrac{\lambda_2}{m}v^\alpha$, which is a Bernoulli equation. Dividing by v^α and making the substitution $u = v^{1-\alpha}$, we find the linear equation

$$\frac{du}{dt} + (1-\alpha)\frac{\lambda_1}{m}u = -\frac{1-\alpha}{m}\lambda_2,$$

whose complete integral is

$$u(t) = ce^{(\alpha-1)(\lambda_1/m)t} - \frac{\lambda_2}{\lambda_1} \quad \text{and} \quad v = \left(ce^{(\alpha-1)(\lambda_1/m)t} - \frac{\lambda_2}{\lambda_1}\right)^{1/(1-\alpha)}.$$

To find the constant c, we use the initial condition $v(t_0) = v_0$.

Riccati's Equation

The equation of the form

$$\frac{dx}{dt} = p(t)x^2 + q(t)x + r(t) \tag{1.75}$$

is called Riccati's equation and in the general form is not integrable by quadratures, but may be transformed into a linear equation by changing the function, if a single particular solution $x_1(t)$ of this equation is known. Indeed, using the substitution

$$x(t) = x_1(t) + \frac{1}{y(t)}, \tag{1.76}$$

we obtain

$$x'(t) = x_1'(t) - \frac{y'(t)}{y^2(t)}; \quad x^2(t) = x_1^2(t) + \frac{2x_1(t)}{y(t)} + \frac{1}{y^2(t)}.$$

Substituting into Eq. (1.75), since $x_1'(t) = p(t)x_1^2 + q(t)x_1 + r(t)$, yields the linear equation

$$y' + [2p(t)x_1(t) + q(t)]y = -p(t). \tag{1.77}$$

Finding its complete integral and substituting it into Eq. (1.76), we obtain the complete integral of Riccati's equation.

Properties

1. The general solution of Riccati's equation is of the form

$$x = \frac{A_1(t)c + B_1(t)}{A(t)c + B(t)}, \quad c \in \mathbf{R}. \tag{1.78}$$

This results from Eqs. (1.76) and (1.69).

2. If there are known four particular solutions x_i, $1 \leq i \leq 4$, of the form of Eq. (1.78), corresponding to the constants c_i, $1 \leq i \leq 4$, then

$$\frac{x_4 - x_1}{x_4 - x_2} : \frac{x_3 - x_1}{x_3 - x_2} = \frac{c_4 - c_1}{c_4 - c_2} : \frac{c_3 - c_1}{c_3 - c_2}.$$

3. If there are known three particular solutions x_1, x_2, x_3 of Riccati's equation, then the general solution is given by

$$\frac{x - x_1}{x - x_2} : \frac{x_3 - x_1}{x_3 - x_2} = c, \quad c \in \mathbf{R}. \tag{1.79}$$

4. If there are known two particular solutions x_1, x_2, then to find the solution of Eq. (1.75) means to calculate a single quadrature.

EXAMPLE 1.17 The equation

$$x' = Ax^2 + \frac{B}{t}x + \frac{C}{t^2}, \quad A, B, C = \text{constants}; \quad (B+1)^2 \geq 4AC \tag{1.80}$$

has a particular solution of the form $x_1 = a/t$, a being constant.

$$\text{Particular case:} \quad t^2 x' + t^2 x^2 = tx - 1. \quad \blacktriangle$$

Solution

We can write

$$x_1 = \frac{a}{t}; \quad x_1' = -\frac{a}{t^2},$$

and

$$-\frac{a}{t^2} = A\frac{a}{t^2} + B\frac{a}{t^2} + \frac{c}{t^2} \Leftrightarrow Aa^2 + (B+1)a + C = 0.$$

Because $(B+1)^2 \geq 4AC$, the last equation has real roots. In the particular case,

$$x' = -x^2 + \frac{1}{t}x - \frac{1}{t^2}; \quad A = -1, B = 1, C = -1 \Rightarrow (B+1)^2 - 4AC = 0.$$

For a, the equation could be written as $-Aa^2 + 2a - 1 = 0 \Rightarrow a = 1$, and $x_1 = 1/t$ is a particular solution. If substitute

$$x = \frac{1}{t} + \frac{1}{y}, \quad x' = -\frac{1}{t^2} - \frac{y'}{y^2},$$

then

$$t^2\left(-\frac{1}{t^2} - \frac{y'}{y^2}\right) + t^2\left(\frac{1}{t} + \frac{1}{y}\right)^2 = t\left(\frac{1}{t} + \frac{1}{y}\right) - 1 \quad \text{or} \quad y' - \frac{1}{t}y = 1,$$

which is a linear equation, with the general solution

$$y = e^{\int (1/t)dt}\left[c + \int e^{-\int (1/t)dt} \, dt\right];$$

hence, $y = ct + \ln t$. The general solution of the given equation is

$$x = \frac{1}{t} + \frac{1}{ct + \ln t}.$$

EXAMPLE 1.18 Solve the equation $x' + Ax^2 = Bt^m$, $A, B \in \mathbf{R}$, in the particular cases

(a) $m = 0$
(b) $m = -2$. ▲

Solution

(a) The equation $x' + Ax^2 = B$ has separable variables, $dx/(Ax^2 - B) = -dt$. The solution is

$$\int \frac{dx}{Ax^2 - B} = c - t.$$

(b) The equation is $x' + Ax^2 = B/t^2$. If we substitute

$$x = \frac{y}{t}; \quad x' = \frac{ty' - y}{t^2},$$

then $ty' = y - Ay^2 + B$, or

$$\frac{dy}{B + y - Ay^2} = \frac{dt}{t}.$$

EXAMPLE 1.19 Show that the Riccati's equation

$$x' = a\frac{x^2}{t} + \left(\frac{1}{2}\right)\frac{x}{t} + c, \quad a, c \in \mathbf{R} \tag{1.81}$$

could be transformed into an equation with separable variables, changing the function $x = y\sqrt{t}$. ▲

Solution

We can write $x = y\sqrt{t}$, $x' = y'\sqrt{t} + y\frac{1}{2\sqrt{t}}$, and the equation becomes

$$y'\sqrt{t} + \frac{1}{2\sqrt{t}}y = ay^2 + \frac{1}{2}\frac{y}{\sqrt{t}} + c, \text{ or } \sqrt{t}\frac{dy}{dt} = ay^2 + c,$$ which is an equation with separable variables. Separating the variables gives

$$\frac{dy}{ay^2 + c} = \frac{dt}{\sqrt{t}}; \quad \int \frac{dy}{ay^2 + c} = 2\sqrt{t} + k, \quad (k = \text{constant}). ▲$$

PARTICULAR EXAMPLE Integrate the equation $tx' - \frac{1}{2}x^2 - \frac{1}{2}x = \frac{1}{2}t$. It is similar to

$x' = \frac{1}{2}\left(\frac{x^2}{t}\right) + \frac{1}{2}\left(\frac{x}{t}\right) + \frac{1}{2}$. This equation is of the form of Eq. (1.81), with

$a = \frac{1}{2}$, $c = \frac{1}{2}$, and the solution is

$$\int \frac{dy}{y^2 + 1} = 4\sqrt{t} + k_1 \Rightarrow \arctan y = 4\sqrt{t} + k_1,$$

$$\Rightarrow \arctan \frac{x}{\sqrt{t}} = 4\sqrt{t} + k_1 \Rightarrow x = \sqrt{t}\tan(4\sqrt{t} + k_1). ▲$$

1.2.2 ELEMENTARY TYPES OF EQUATIONS NOT SOLVED FOR DERIVATIVE

A first-order differential equation not solved for derivative is of the form

$$F(t, x, x') = 0. \tag{1.82}$$

Equations of the First Order and Degree n in x'

Consider the differential equation

$$a_0(t, x)(x')^n + a_1(t, x)(x')^{n-1} + \cdots + a_n(t, x) = 0, \tag{1.83}$$

with $a_j \in C(D)$; $D \subseteq \mathbf{R}^2$, $a_0 \neq 0$. Let us solve this equation for x'. Let

$$x' = f_1(t, x), \quad x' = f_2(t, x), \ldots, x' = f_m(t, x) \quad (m \leq n) \tag{1.84}$$

be a real solution for Eq. (1.81). The general integral equation will be expressed by a sum of the integrals

$$F_1(t, x, c_1) = 0, \quad F_2(t, x, c_2) = 0, \ldots, F_m(t, x, c_m) = 0, \tag{1.85}$$

where $F_i(t, x, c_i)$ is the integral of the equation

$$x' = f_i(t, x), \quad (i = 1, 2, \ldots, m). \tag{1.86}$$

Thus, k integral curves pass through each point of the domain where x' takes on real values.

EXAMPLE 1.20 Solve the equation $tx'^2 - 2xx' - t = 0$. ▲

Solution

Let us solve this equation for x':

$$x' = \frac{x + \sqrt{x^2 + t^2}}{t}, \quad x' = \frac{x - \sqrt{x^2 + t^2}}{t}.$$

Then we obtain two homogeneous equations that can be solved by the substitution $x/t = u$. Integrating each one gives

$$x = \frac{1}{2}\left(ct^2 - \frac{1}{c}\right) \quad \text{and} \quad x = \frac{1}{2}\left(c - \frac{t^2}{c}\right).$$

Both families of solutions satisfy the original equation.

Equations of the Form F(x') = 0

For equations of the form

$$F(x') = 0, \tag{1.87}$$

there exists at least one constant root $x' = k_i$, since Eq. (1.87) does not contain t and x (k_i is a constant). Consequently, integrating equation $x' = k_i$, then $x = k_i t + c$, or $k_i = (x - c)/t$; hence,

$$F\left(\frac{x - c}{t}\right) = 0 \tag{1.88}$$

is an integral of Eq. (1.87).

EXAMPLE 1.21 Solve the equation $(x')^5 - (x')^3 + x' + 2 = 0.$ ▲

Solution

The integral of this equation is

$$\left(\frac{x-c}{t}\right)^5 - \left(\frac{x-c}{t}\right)^3 + \frac{x-c}{t} + 2 = 0. \quad ▲$$

Equations of the Form $F(t, x') = 0$

For equations of the form

$$F(t, x') = 0, \tag{1.89}$$

let us consider the case when these equations cannot be solved for x'.

(a) If Eq. (1.89) is readily solvable for t, $t = \varphi(x')$, then it is nearly always convenient to introduce $x' = p$ as parameter. Then $t = \varphi(p)$. Differentiating this equation and replacing dx by $p\, dt$ yields $dx = p\varphi'(p)dp$, $x = \int p\varphi'(p)dp + c$. The general solution is

$$t = \varphi(p); \quad x = \int p\varphi'(p)dp + c. \tag{1.90}$$

(b) If Eq. (1.89) is not solvable (or is difficult to solve) for both t and x', but allows the expression of t and x' in terms of some parameter p,

$$t = \varphi(p), \quad x' = \psi(p).$$

Then, $dt = \varphi'(p)dp$, $dx = x'\, dt = \psi(p)\varphi'(p)dp$, and

$$x = \int \psi(p)\varphi'(p)dp + c; \quad t = \varphi(p). \tag{1.91}$$

So, the general solution of Eq. (1.89) is obtained in the parametric form Eq. (1.91).

EXAMPLE 1.22 Solve the equation $tx' = (x')^3 + (x')^2 - 1.$ ▲

Solution

Substitute $x' = p$. Then

$$t = p^2 + p - \frac{1}{p}; \quad dt = \left(2p + 1 + \frac{1}{p^2}\right)dp,$$

$$dx = x'\, dt = p\, dt = p\left(2p + 1 + \frac{1}{p^2}\right)dp; \quad x = \int\left(2p^2 + p + \frac{1}{p}\right)dp + c.$$

The general solution of the equation is

$$t = p^2 + p - \frac{1}{p}, \quad x = \frac{2}{3}p^3 + \frac{p^2}{2} + \ln|p| + c. \quad ▲$$

EXAMPLE 1.23 Solve the equation $t^{2/3} + (x')^{2/3} = a^{2/3}.$ ▲

Solution

If we set $t = a \sin^3 p$, $x' = a \cos^3 p$, then

$$dx = x' \, dt = a \cos^3 p (3a \sin^2 p) \cos p \, dp; \quad dx = (3a^2 \sin^2 p) \cos^4 p \, dp,$$

$$x = \frac{3a^2}{8} \int \sin^2 2p (1 + \cos 2p) dp + c.$$

Hence,

$$\begin{cases} x = \dfrac{3a^2}{16} \left[p - \dfrac{1}{4} \sin 4p + \dfrac{1}{3} \sin^3 2p \right] + c, \\ t = a \sin^3 p. \end{cases} \tag{1.92}$$

Equations of the Form $F(x, x') = 0$

Consider equations of the form

$$F(x, x') = 0. \tag{1.93}$$

(a) If $x' = \varphi(x)$, then $\dfrac{dx}{\varphi(x)} = dt$; $\displaystyle\int \dfrac{dx}{\varphi(x)} = t + c.$

(b) If $x = \psi(x')$; $x' = p$, and $dt = \dfrac{dx}{x'} = \dfrac{\psi'(p)}{p} dp$, or

$$\begin{cases} t = \displaystyle\int \dfrac{\psi'(p)}{p} dp + c, \\ x = \psi(t) \end{cases} \tag{1.94}$$

is the general solution.

(c) If it is difficult to solve Eq. (1.93) for x' and x, then it is advisable to introduce the parameter p and replace Eq. (1.93) by two equations, $x = \varphi(p)$ and $x' = \psi(p)$. Since $dx = x' dt$, then $dt = \dfrac{dx}{x'} = \dfrac{\varphi'(p)}{\psi(p)} dp$. Thus, in parametrical form, the desired integral curves are defined by the equations

$$\begin{cases} t = \displaystyle\int \dfrac{\varphi'(p)}{\psi(p)} dp, \\ x = \varphi(p). \end{cases} \tag{1.95}$$

EXAMPLE 1.24 Solve the equation $x = (x')^3 + (x')^2 + 20$. ▲

Solution

Substitute $x' = p$. Then

$$x = p^3 + p^2 + 20; \quad dx = (3p^2 + 2p) dp;$$

$$dt = \frac{dx}{x'} = \frac{3p^2 + 2p}{p} dp \Rightarrow dt = (3p + 2) dp.$$

Equations $t = \frac{3}{2}p^2 + 2p + c$ and $x = p^3 + p^2 + 20$ are parametrical equations for an integral curve family. ▲

EXAMPLE 1.25 Solve the equation $x^{2/5} + (x')^{2/5} = 1$. ▲

Solution

Substitute $x = \sin^5 p$, $x' = \cos^5 p$. Then

$$dt = \frac{dx}{x'} = \frac{5\sin^4 p}{\cos^4 p}\,dp; \quad t = 5\int \frac{\sin^4 p}{\cos^4 p}\,dp + c.$$

The general solution of the given differential equation in the parametrical form is

$$\begin{cases} t = 5\left[\frac{1}{3}\tan^3 p - \tan p + p\right], \\ x = \sin^5 p. \end{cases} \quad ▲$$

The Lagrange Equation

The equation

$$x = t\varphi(x') + \psi(x') \qquad (1.96)$$

is called *Lagrange's equation*. Setting $x' = p$, differentiating with respect to t, and replacing dx by $p\,dt$, this equation is reduced to a linear equation in t as a function of p,

$$x = t\varphi(p) + \psi(p); \quad p = \varphi(p) + t\varphi'(p)\frac{dp}{dt} + \psi'(p)\frac{dp}{dt},$$

or

$$[p - \varphi(p)]\frac{dt}{dp} = t\varphi'(p) + \psi'(p). \qquad (1.97)$$

Finding the solution of this last equation $t = r(p, c)$, we obtain the general solution of the original equation in parametrical form,

$$\begin{cases} t = r(p, c) \\ x = r(p, c)\varphi(p) + \psi(p). \end{cases}$$

In addition, the Lagrange equation may have some singular solution of the form $x = \varphi(c)t + \psi(c)$, where c is the root of equation $c = \varphi(c)$.

EXAMPLE 1.26 Integrate the equation $x = t(x')^2 - x'$. ▲

Solution

Let us set $x' = p$, then $x = tp^2 - p$. Differentiating, $p = p^2 + (2pt - 1)\frac{dp}{dt}$, or $p - p^2 = (2pt - 1)\frac{dp}{dt}$, whence $(p - p^2)\frac{dt}{dp} = 2pt - 1$, or

$$\frac{dt}{dp} = \frac{2p}{p(1-p)}t - \frac{1}{p(1-p)}.$$

We obtain a first-order linear equation

$$\frac{dt}{dp} = \frac{2}{1-p}t + \frac{1}{p(p-1)}.$$

Integrating this linear equation gives

$$t = e^{\int [2/(1-p)]dp}\left[c + \int \frac{1}{p(p-1)}e^{\int [2/(p-1)]dp}\,dp\right];$$

$$t = \frac{1}{(p-1)^2}[c + p - \ln|p|];$$

$$x = \frac{p^2}{(p-1)^2}[c + p - \ln|p|] - p.$$

The roots of the equation $p - \varphi(p) = 0$, that is, $p - p^2 = 0$, are $p = 0$ and $p = 1$, corresponding to $x = 0$ and $x = t - 1$, which are the singular solutions for the original equation. ▲

The Clairaut Equation

This equation has the form

$$x = tx' + \psi(x'). \tag{1.98}$$

The method of solution is the same as for the Lagrange equation. Using the substitution $x' = p$, we obtain $x = tp + \psi(p)$. Differentiating with respect to t, $p = p + [t + \psi'(p)]\dfrac{dp}{dt}$ or, $(t + \psi'(p))\dfrac{dp}{dt} = 0$, whence either $\dfrac{dp}{dt} = 0$ and hence $p = c$ or, $t + \psi'(p) = 0$. In the first case, eliminating p gives

$$x = tc + \psi(c), \tag{1.99}$$

which is a one-parameter family of integral curves. In the second case, the solution is defined by the equation

$$\begin{cases} x = tp + \psi(p), \\ t + \psi'(p) = 0. \end{cases} \tag{1.100}$$

This is the "envelope" of the family of integral curves from Eq. (1.99). The form of Eq. (1.99) is obtained directly from Eq. (1.98), replacing x' with c.

EXAMPLE 1.27 Integrate the equation $x = tx' + 2(x')^2$. ▲

Solution

Setting $x' = p$, then $x = tp + 2p^2$. Differentiating this equation with respect to t, $p = p + (t + 4p)\dfrac{dp}{dt}$, whence $(t + 4p)\dfrac{dp}{dt} = 0$. A one-parameter family of integral straight lines has the form $x = tc + 2c^2$. The envelope of this family, defined by equation $x = tc + 2c^2$ and $t + 4c = 0$, is an integral curve. Eliminating c yields

$$x = t\left(-\frac{t}{4}\right) + 2\left(\frac{t}{4}\right)^2, \quad \text{or} \quad x = -\frac{t^2}{8}.$$

(See Fig. 1.9.) ▲

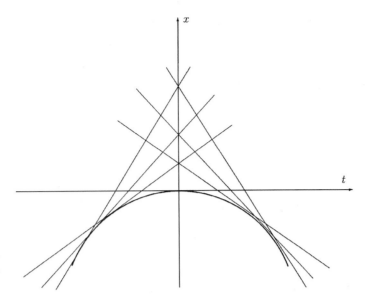

Figure 1.9
*The envelope
of family
$x = tc + 2c^2$.*

1.2.3 DIFFERENTIAL EQUATIONS OF HIGHER ORDERS
Equations of the Form $x^{(n)} = f(t)$

Consider equations of the form

$$x^{(n)} = f(t). \tag{1.101}$$

After n-fold integration, the general solution is

$$x = \int \cdots \int f(t)dt \cdots dt + c_1 \frac{t^{n-1}}{(n-1)!} + c_2 \frac{(t-t_0)^{n-2}}{(n-2)!} + \cdots + c_{n-1}t + c_n. \tag{1.102}$$

It is possible to prove by induction that the general solution of Eq. (1.101) could be written in the form

$$x = \frac{1}{(n-1)!} \int_{t_0}^{t} (t-s)^{n-1} f(s)ds + c_1 \frac{(t-t_0)^{n-1}}{(n-1)!} + c_2 \frac{(t-t_0)^{n-2}}{(n-2)!} + \cdots + c_n. \tag{1.103}$$

EXAMPLE 1.28 Find the complete integral of the equation

$$y'' = k(l - x), \quad (k, l = \text{constants})$$

and a particular solution satisfying the initial conditions $y(0) = 0$, $y'(0) = 0$.

▲

Solution

Integrating the equation gives $y' = k \int_0^x (l-x)dx + c_1 = k\left(lx - \frac{x^2}{2}\right) + c_1$;

$y = k \int_0^x \left(lx - \frac{x^2}{2}\right)dx + c_1 x + c_2$; $y = k\left(l\frac{x^2}{2} - \frac{x^3}{6}\right) + c_1 x + c_2$, the

complete integral. From the initial conditions, we find that $c_1 = 0$, $c_2 = 0$, and $y = \dfrac{k}{2}\left(l\dfrac{x^2}{2} - \dfrac{x^3}{6}\right)$ is the particular solution that satisfies the initial conditions.

APPLICATION 1.12

A cantilevered beam fixed at the extremity O is subjected to the action of a concentrated vertical force P applied to the end L of the beam, at a distance l from O (Fig. 1.10). The weight of the beam is ignored. Let us consider a cross section at the point $N(x)$. The bending moment relative to section N is equal to $M(x) = (l - x)P$. The differential equation of the bent axis of a beam is

$$\frac{y''}{(1 + y'^2)^{3/2}} = \frac{M(x)}{EJ},$$

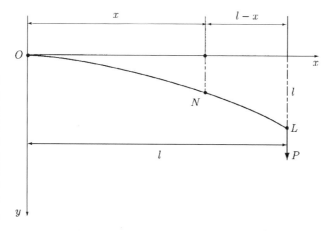

Figure 1.10
A cantilevered beam fixed at point 0 and subjected to the action of a concentrated vertical force applied to point L.

where E is the modulus of elasticity and J is the moment of inertia of the cross-sectional area of the beam, relative to the horizontal line passing through the center of gravity of the cross-sectional area, and $R = (1 + y'^2)^{3/2}/y''$ is the curvature radius of the bent beam axis. Assuming that the deformations are small and that the tangents to the bent axis form a small angle with the x-axis, the square of the small quantity y'^2 can be ignored and the differential equation of the bent beam will have the form

$$y'' = \frac{P}{EJ}(l - x).$$

The initial conditions are $y(0) = 0$, $y'(0) = 0$. The solution for this problem is (see Example 1.28)

$$y = \frac{P}{2EJ}\left(lx^2 - \frac{x^3}{3}\right).$$

Particularly, the deflection h at the extremity of the beam is

$$h = y(l) = \frac{Pl^3}{3EJ}. \quad \blacktriangle$$

Equations of the Form $F(t, x^{(n)}) = 0$

Consider equations of the form

$$F(t, x^{(n)}) = 0. \tag{1.104}$$

(a) Equation (1.104) can be solved for the second argument $x^{(n)} = f(t)$ and it is found Eq. (1.101).

(b) The equation is not solvable for $x^{(n)}$, but allows expressions for t and $x^{(n)}$ in terms of some parameter p:

$$t = \varphi(p), \quad x^{(n)} = \psi(p).$$

Then,

$$dx^{(n-1)} = x^{(n)} dt = \psi(p)\varphi'(p)dp,$$

$$x^{(n-1)} = \int \psi(p)\varphi'(p)dp + c_1 = \psi_1(p, c_1),$$

and similarly,

$$dx^{(n-2)} = x^{(n-1)} dt = \psi_1(p)\varphi'(p)dp$$

$$x^{(n-2)} = \int \psi_1(p)\varphi'(p)dp + c_2 = \psi_2(p, c_2)$$

Finally, the general solution in parametrical form is

$$\begin{cases} x = \psi_n(p, c_1, c_2, \ldots, c_n) \\ t = \varphi(p). \end{cases} \tag{1.105}$$

EXAMPLE 1.29 Solve the equation $t - e^{x''} - (x'')^2 = 0. \quad \blacktriangle$

Solution

Choosing $x'' = p$, then $t = e^p + p^2$, $dt = (e^p + 2p)dp$; $dx' = x'' dt = p(e^p + 2p)dp \Rightarrow x' = \int (pe^p + 2p^2)dp + c_1$ or $x' = pe^p - e^p + \frac{2}{3}p^3 + c_1$; $dx = x' dt = [pt^p - e^p + \frac{2}{3}p^2 + c_1](e^p + 2p)dp$ and $x = \int (pe^p - e^p + \frac{2}{3}p^3 + c_1)(e^p + 2p)dp + c_2$.

Integrating,

$$\begin{cases} x = \frac{1}{2}e^{2p}\left(p - \frac{3}{2}\right) + e^p\left(\frac{2}{3}p^3 - 2p + \frac{8}{3} + c_1\right) + \frac{4}{15}p^5 + c_1 p^2 + c_2 \\ t = e^p + p^2. \end{cases}$$

Equations of the Form $F(t, x^{(k)}, x^{(k+1)}, \ldots, x^{(n)}) = 0$

Consider an equation of the form

$$F(t, x^{(k)}, x^{(k+1)}, \ldots, x^{(n)}) = 0. \tag{1.106}$$

In this case the order of the equation may be reduced to $n - k$ by changing the variables $x^{(k)} = y$. Equation (1.105) becomes $F(t, y, y', \ldots, y^{(n-k)}) = 0$. Using this equation we find $y = f(t, c_1, c_2, \ldots, c_{n-k})$, and x is found from $x^{(k)} = f(t, c_1, c_2, \ldots, c_{n-k})$ by k-fold integration.

EXAMPLE 1.30 Find the solution of the equation $tx''' - x'' = \left(1 - \dfrac{2}{\sqrt{t}}\right)\sqrt{x''}$, satisfying the initial condition $x(1) = x'(1) = x''(1) = 0$. ▲

Solution

If we denote $x'' = y$, the Bernoulli's equation is $ty' - y = \left(1 - \dfrac{2}{\sqrt{t}}\right)\sqrt{y}$ with

the general solution (see Example 1.16) $y = \left(c\sqrt{t} + \dfrac{1}{\sqrt{t}} - 1\right)^2$. Replacing y

by $x'': x'' = \left(c\sqrt{t} + \dfrac{1}{\sqrt{t}} - 1\right)^2$. From the initial condition $x''(1) = 0$, we

obtain $c = 0$; hence, $x'' = \dfrac{1}{t} - \dfrac{2}{\sqrt{t}} + 1$; $x' = \ln t - 4\sqrt{t} + t + c_1$; $x'(1) = 0$

$\Rightarrow c_1 = 3$; hence, $x' = \ln t - 4\sqrt{t} + t + 3$, and then $x(t) = t\ln t -$

$1 - \dfrac{8}{3}t^{3/2} + \dfrac{t^2}{2} + 3t + c_2$; $x(1) = 0 \Rightarrow c_2 = \dfrac{1}{6}$. The solution for the given problem is

$$x(t) = t\ln t - \frac{8}{3}t^{3/2} + \frac{t^2}{2} + 3t - \frac{5}{6}.$$

Equations of the Form $F(x, x', x'', \ldots, x^{(n)}) = 0$

Consider equations of the form

$$F(x, x', x'', \ldots, x^{(n)}) = 0. \tag{1.107}$$

The substitution $x' = y$ makes it possible to reduce the order of the equation. In this case y is regarded as a new unknown function of x. All the derivatives $x', x'', \ldots, x^{(n)}$ are expressed in terms of derivatives of the new unknown function y with respect to

$$x, x' = \frac{dx}{dt} = y$$

$$x'' = \frac{dy}{dt} = \frac{dy}{dx}\left(\frac{dx}{dt}\right) = y\frac{dy}{dx}$$

$$x''' = \frac{d}{dt}\left(y\frac{dy}{dx}\right) = \frac{d}{dx}\left(y\frac{dy}{dx}\right)\frac{dx}{dt} = y\left[y\frac{d^2y}{dx^2} + \left(\frac{dy}{dx}\right)^2\right], \text{ etc.}$$

Substituting these expressions for $x', x'', \ldots, x^{(n)}$ in the equation yields a differential equation of order $(n - 1)$.

EXAMPLE 1.31 *Escape-Velocity Problem*

Determine the smallest velocity with which a body must be thrown vertically upward so that it will not return to the earth. Air resistance is neglected (see Eq. 1.301). ▲

Solution

Denote the mass of the earth by M and the mass of the body by m. By Newton's law of gravitation, the force of attraction f acting on the body m is $f = -k\dfrac{Mm}{r^2}$, where r is the distance between the center of the earth and center of gravity of the body, and k is the gravitational constant. The differential equation of the motion for the body is

$$m\frac{d^2r}{dt^2} = -k\frac{Mm}{r^2} \quad \text{or,} \quad \frac{d^2r}{dt^2} = -k\frac{M}{r^2}. \tag{1.108}$$

The minus sign indicates a negative acceleration. The differential Eq. (1.108) is an equation of type Eq. (1.107). This will be solved for the initial conditions

$$r(0) = R, \quad \frac{dr}{dt}(0) = v_0. \tag{1.109}$$

Here, R is the radius of the earth, and v_0 is the launching velocity. Let us make the following notations: $\dfrac{dr}{dt} = v, \dfrac{d^2r}{dt^2} = \dfrac{dv}{dt} = \dfrac{dv}{dr}\left(\dfrac{dr}{dt}\right) = v\dfrac{dv}{dr}$, where v is the velocity of motion. Substituting in Eq. (1.i08) gives $v\dfrac{dv}{dr} = -k\dfrac{M}{r^2}$. Separating variables, we find $v\,dv = -kM\dfrac{dr}{r^2}$. Integrating this equation yields $\dfrac{v^2}{2} = kM\dfrac{1}{r} + c_1$. From conditions (1.109), c_1 is found:

$$\frac{v_0^2}{2} = kM\left(\frac{1}{R}\right) + c_1$$

or

$$c_1 = -\frac{kM}{R} + \frac{v_0^2}{2},$$

and

$$\frac{v^2}{2} = kM\frac{1}{r} + \left(\frac{v_0^2}{2} - \frac{kM}{R}\right). \tag{1.110}$$

The body should move so that the velocity is always positive; hence, $v^2/2 > 0$. Since for a boundless increase of r the quantity kM/R becomes arbitrarily small, the condition $v^2/2 > 0$ will be fulfilled for any r only for the case

$$\frac{v_0^2}{2} - \frac{kM}{R} \geq 0 \quad \text{or} \quad v_0 \geq \sqrt{\frac{2kM}{k}}.$$

Hence, the minimal velocity is determined by the equation

$$v_0 = \sqrt{\frac{2kM}{R}}, \tag{1.111}$$

where $k = 6.66(10^{-8})\,\text{cm}^3/(\text{g s}^2)$, $R = 63(10^7)\,\text{cm}$. At the Earth's surface, for $r = R$, the acceleration of gravity is $g = 981\,\text{cm/s}^2$. For this reason, Eq. (1.108) yields $g = k\dfrac{M}{R^2}$, or $M = \dfrac{gR^2}{k}$. Substituting this value of M into Eq. (1.111), we find

$$v_0 = \sqrt{2gR} = \sqrt{2(981)(63)(10^7)} \approx 11.2(10^5)\,\text{cm/s} = 11.2\,\text{km/s}.$$

Equations of the Form $F(x, x'') = 0$

Consider equations of the form

$$F(x, x'') = 0. \tag{1.112}$$

This equation is a particular case of Eq. (1.107), frequently encountered in applications. The order can be reduced with

$$\frac{dx}{dt} = y, \quad \frac{d^2x}{dt^2} = \frac{dy}{dt} = \frac{dy}{dx}\left(\frac{dx}{dt}\right) = y\frac{dy}{dx}.$$

If Eq. (1.112) is readily solvable for the second argument $x'' = f(x)$, then multiplying this equation by $2x'\,dt = 2dx$ yields $d(x')^2 = 2f(x)dx$, and

$$\frac{dx}{dt} = \pm\sqrt{2\int f(x)dx + c_1}, \quad \pm\frac{dx}{\sqrt{2\int f(x)dx + c_1}} = dt$$

$$t + c_2 = \pm\int \frac{dx}{\sqrt{2\int f(x)dx + c_1}}.$$

Equation (1.112) may be replaced by its parametric representation $x = \varphi(p)$, $x'' = \psi(p)$; then $dx' = x''\,dt$ and $dx = x'\,dt$ yields $x'\,dx' = x''\,dx$ or $\frac{1}{2}d(x')^2 = \psi(p)\varphi'(p)dp \Rightarrow (x')^2 = 2\int \psi(p)\varphi'(p)dp + c_1 \Rightarrow x' = \pm\sqrt{2\int \psi(p)\varphi'(p)dp + c_1}$, and from $dx = x'\,dt$ we find dt and then t,

$$dt = \frac{dx}{x'} = \frac{\varphi'(p)dp}{\pm\sqrt{2\int \psi(p)\varphi'(p)dp + c_1}}$$

$$t = \pm\int \frac{\varphi'(p)dp}{\sqrt{2\int \psi(p)\varphi'(p)dp + c_1}} + c_2, \tag{1.113}$$

and $x = \varphi(p)$. Equations (1.113) define in parametrical form a family of integral curves.

APPLICATION 1.13

A chain of length l meters slides off from a table. At the initial instant of motion, d meters of chain hang from the table (Fig. 1.11). Study how long it will take for the whole chain to slide off. (Suppose there is no friction.) ▲

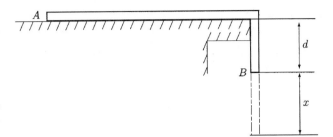

Figure 1.11
A chain that slides off a table.

Solution

If the extremity B decrease with distance x, the differential equation of motion is $gl\dfrac{d^2x}{dt^2} = g\rho(x + d)$, where ρ is the linear density. This equation is of the form $\dfrac{d^2x}{dt^2} = \dfrac{g}{l}(x + d)$. Let us make the following notation:

$$\frac{dx}{dt} = v, \qquad \frac{d^2x}{dt^2} = \frac{dv}{dt} = \frac{dv}{dx}\left(\frac{dx}{dt}\right) = v\frac{dv}{dx}.$$

Then, $v\dfrac{dv}{dx} = \dfrac{g}{l}(x + d)$. Separating the variables and integrating yields $v\,dv = \dfrac{g}{l}(x + d)dx$, $v^2 = \dfrac{g}{l}(x^2 + 2xd) + c_1$. From the initial conditions $t = 0$, $x(0) = 0$, $v(0) = 0$, we find $c_1 = 0$, and then

$$v^2 = \frac{g}{l}(x^2 + 2xd) \qquad \text{or} \qquad \frac{dx}{dt} = v = \sqrt{\frac{g}{l}}(\sqrt{x^2 + 2xd}).$$

Separating the variables again and integrating gives

$$\frac{dx}{\sqrt{x^2 + 2xd}} = \sqrt{\frac{g}{l}}\,dt \Rightarrow \int\frac{dx}{\sqrt{x^2 + 2xd}} = \sqrt{\frac{g}{l}}t + c_2$$

or $\ln(x + d + \sqrt{x^2 + 2xd}) = \sqrt{\dfrac{g}{l}}\,t + c_2$. From $x(0) = 0$, we obtain $c_2 = \ln d$; hence,

$$t = \left(\sqrt{\frac{l}{g}}\right)\ln\frac{x + d + \sqrt{x^2 + 2xd}}{d}.$$

For $x = l - d$,

$$T = \left(\sqrt{\frac{l}{g}}\right)\ln\frac{l + \sqrt{l^2 - d^2}}{d}.$$

The Equation $F(t, x, x', x'', \ldots, x^{(n)}) = 0$; Homogeneous in the Variables $x, x, \ldots, x^{(n)}$

For such an equation,

$$F(t, x, x', x'', \ldots, x^{(n)}) = 0, \tag{1.114}$$

the following relation can be established:

$$F(t, kx, kx', \ldots, kx^{(n)}) = k^m F(t, x, x', \ldots, x^{(n)}). \qquad (1.115)$$

The order of such an equation can be reduced using the substitution

$$x = e^{\int y \, dt}, \qquad (1.116)$$

where y is a new unknown function of t. Differentiating gives

$$\begin{cases} x' = y e^{\int y \, dt} \\ x'' = (y^2 + y') e^{\int y \, dt} \\ \qquad \cdots \\ x^{(k)} = \Phi(y, y', \ldots, y^{(k-1)}) e^{\int y \, dx}. \end{cases}$$

Substituting into the initial equation and observing that the factor $e^{m \int y \, dt}$ may be taken outside the sign of function F gives $f(t, y, y', \ldots, y^{(n-1)}) = 0$.

EXAMPLE 1.32 Solve the equation $txx'' - tx'^2 - xx' = 0$. ▲

Solution

$$F(t, x, x', x'') = txx'' - tx'^2 - xx', \quad F(t, kx, kx', kx'') = k^2 F(t, x, x', x'').$$

The equation is homogeneous in x, x', x''. If we use the substitution

$$x = e^{\int y \, dt}, \quad x' = y e^{\int y \, dt}, \quad y'' = (y^2 + y') e^{\int y \, dt},$$

then

$$e^{2 \int y \, dt}[ty^2 + ty' - ty^2 - y] = 0, \quad \text{or} \quad ty' - y = 0.$$

The solution of this last equation is $y = c_1 t$, and the general solution of the given equation is

$$x = e^{\int c_1 t \, dt} = e^{c_1(t^2/2) + \ln c_2}, \quad \text{or} \quad x = c_2 e^{c_1(t^2/2)}.$$

1.3 On the Existence, Uniqueness, Continuous Dependence on a Parameter, and Differentiability of Solutions of Differential Equations

1.3.1 THEOREMS OF EXISTENCE AND UNIQUENESS OF THE SOLUTION FOR THE EQUATION $\dfrac{dy}{dx} = f(x, y)$

THEOREM 1.1 On the Existence and Uniqueness of the Solution
If in the equation

$$\frac{dy}{dx} = f(x, y), \qquad (1.117)$$

function $f(x, y)$ is continuous in the rectangle $D: x_0 - a \le x \le x_0 + a$, $y_0 - b \le y \le y_0 + b$ and satisfies, in D, the Lipschitz condition

$$|f(x, y_1) - f(x, y_2)| \le L|y_1 - y_2|, \qquad (1.118)$$

where L is a constant, then exists a unique solution $y = y(x)$, $x \in [x_0 - h, x_0 + h]$ of Eq. (1.117) that satisfies the condition $y(x_0) = x_0$, where

$$h < \min\left(a, \frac{b}{M}\right), \quad M = \max |f(x, y)| \quad \text{in } D. \quad \blacktriangle \qquad (1.119)$$

Remark 1.14

The existence of a solution for Eq. (1.117) can be proved only if we assume the continuity of function $f(x, y)$.

Remark 1.15

The existence and uniqueness of solution $y = y(x)$ on the interval $(x_0 - h, x_0 + h)$ can be extended over an interval of length h_1 if the conditions of the existence and uniqueness theorem are fulfilled in the neighborhood of the new initial point $(x_0 + h, y(x_0 + h))$, by repeating the reasoning.

Remark 1.16

The existence of bounded derivative $\partial f / \partial y$ in domain D is sufficient for a function $f(x, y)$ to satisfy a Lipschitz condition in D and

$$L = \max_{D} \left| \frac{\partial f}{\partial y} \right|.$$

Remark 1.17

The solution of Eq. (1.117) with $y(x_0) = y_0$ can be found using the method of successive approximations as follows. First, construct a sequence $\{y_n(x)\}$ of functions defined by the recurrence relations

$$y_n(x) = y_0 + \int_{x_0}^{x} f(t, y_{n-1}(t))dt, \quad n = 1, 2, \ldots. \qquad (1.120)$$

The sequence $\{y_n(x)\}$ converges to an exact solution of Eq. (1.117), satisfying the condition $y(x_0) = x_0$ in some interval $(x_0 - h, x_0 + h)$. The estimative error resulting from the replacement of the exact solution $y(x)$ by the nth approximation $y_n(x)$ is given by the inequality

$$|y(x) - y_n(x)| \leq \frac{ML^n}{(n+1)!} h^{n+1}. \qquad (1.121)$$

Remark 1.18

In a quite analogous way it is possible to prove the theorem of the existence and uniqueness of the solution $y_1(x), y_2(x), \ldots, y_n(x)$ for the system of equations

$$\frac{dy_i}{dx} = f_i(x, y_1, y_2, \ldots, y_n), \quad y_i(x_0) = y_{i_0}, \quad (i = 1, 2, \ldots, n), \qquad (1.122)$$

or

$$y_i = y_{i_0} + \int_{x_0}^{x} f_i(t, y_1(t), y_2(t), \ldots, y_n(t))dt, \quad (i = 1, 2, \ldots, n). \qquad (1.123)$$

THEOREM 1.2   Consider that in the region D defined by the inequalities $x_0 - a \le x_0 + a$, $y_{i_0} - b_i \le y_i \le y_{i_0} + b_i$ $(i = 1, 2, \ldots, n)$, the right-hand side of Eq. (1.122) satisfies the following conditions:

1. All functions $f_i(x, y_1, y_2, \ldots, y_n)$, $(i = 1, 2, \ldots, n)$ are continuous and hence also bounded, $|f_i| \le M$,
2. All functions $f_i(x, y_1, y_2, \ldots, y_n)$, $(i = 1, 2, \ldots, n)$ satisfy the Lipschitz condition

$$|f_i(x, y_1, y_2, \ldots, y_n) - f_i(x, z_1, z_2, \ldots, z_n)| \le N \sum_{i=1}^{n} |y_i - z_i|.$$

Then, the system of equations (1.122) has the unique solution $y_1, y_2, \ldots, y_n$. The functions $y_1, y_2, \ldots, y_n$ are defined on the interval $x_0 - h_0 < x < x_0 + h_0$, where

$$h_0 \le \min\left(a, \frac{b_1}{M}, \ldots, \frac{b_n}{M}\right). \quad \blacktriangle \qquad (1.124)$$

THEOREM 1.3   There exists a unique solution of an n-order differential equation $y^{(n)} = f(x, y', y'', \ldots, y^{(n-1)})$ that satisfies the conditions $y(x_0) = y_0$, $y'(x_0) = y_0', \ldots, y^{(n-1)}(x_0) = y_n^{(n-1)}$, if in the neighborhood of the initial values $(x_0, y_0, y_0', \ldots, y_0^{(n-1)})$ function f is a continuous function in all its arguments and satisfies the Lipschitz condition with respect to all arguments from the second onwards. $\blacktriangle$

THEOREM 1.4   On the Continuous Dependence of a Solution on a Parameter and on the Initial Values

(a) If the right side of the differential equation

$$\frac{dx}{dy} = f(x, y, \mu) \qquad (1.125)$$

is continuous with respect to μ for $\mu_0 \le \mu \le \mu_1$ and satisfies the conditions of the theorem of existence and uniqueness, and the Lipschitz constant L does not depend on μ, then the solution $y(x, \mu)$ of this equation that satisfies condition $y(x_0) = y_0$ depends continuously on μ.

(b) If $y = y(x; x_0, y_0)$ is the solution of Eq. (1.117) satisfying the condition $y(x_0) = y_0$, then this function depends continuously on the initial values x_0 and y_0. $\blacktriangle$

THEOREM 1.5   On the Differentiability of Solutions

If in the neighborhood of a point (x_0, y_0), a function $f(x, y)$ has continuous derivatives until k order inclusive, then solution $y(x)$ of Eq. (1.117) that satisfies the initial condition $y(x_0) = y_0$ has continuous derivatives until $(k + 1)$ order inclusive, in the neighborhood of x_0. $\blacktriangle$

EXAMPLE 1.33 Find the first three successive approximations y_1, y_2, y_3 for the solution of the equation $dy/dx = 1 + xy^2$, $y(0) = 0$, in a rectangle D: $-\frac{1}{2} \le x \le \frac{1}{2}$, $-\frac{1}{2} \le y \le \frac{1}{2}$. ▲

Solution

We can write that $f(x, y) = 1 + xy^2$, $\max_D |f(x, y)| = 1 + \frac{1}{8} = \frac{9}{8}$, that is, $M = \frac{9}{8}$; $b = \min(a, b/M) = \min(\frac{1}{2}, \frac{4}{9}) = \frac{4}{9}$, $\partial f / \partial y = 2xy \Rightarrow \max_D |\partial f / \partial y| = \frac{1}{2}$, that is, $L = \frac{1}{2}$. If we choose $y_0(x) \equiv 0$, then

$$y_1 = \int_0^x f(x, y_0)dx = \int_0^x dx = x,$$

$$y_2 = \int_0^x f(x, y_1)dx = \int_0^x (1 + x^3)dx = x + \frac{x^4}{4},$$

$$y_3 = \int_0^x f(x, y_2)dx = \int_0^1 \left[1 + x\left(x + \frac{x^4}{4}\right)^2\right]dx = x + \frac{x^3}{3} + \frac{x^7}{14} + \frac{x^{10}}{160}.$$

The absolute error in the third approximation does not exceed the value

$$|y - y_3| \le \frac{ML^3}{4!}b^4 = \frac{9}{8}\left(\frac{1}{2}\right)^3\left(\frac{4}{9}\right)^4\frac{1}{4!} \approx \frac{2.286}{10^4}.$$

EXAMPLE 1.34 Find the first three successive approximations for solution of the system

$$\begin{cases} y' = z, & y(0) = 0 \\ z' = -y^2, & z(0) = \frac{1}{2}, \end{cases}$$

$$D: \begin{cases} -1 \le x \le 1 \\ -1 \le y \le 1 \\ -1 \le z \le 1. \end{cases} ▲$$

Solution

The functions $f(x, y, z) = z$ and $g(x, y, z) = -y^2$ are continuous in D. $\max_D |f(x, y, z)| = \max_D |z| = 1$; $\max_D |g(x, y, z)| = \max_D |-y^2| = 1$, that is, $M = 1$,

$$\frac{\partial f}{\partial y} = 0 \Rightarrow \max\left|\frac{\partial f}{\partial y}\right| = 0,$$

$$\frac{\partial f}{\partial z} = 1 \Rightarrow \max\left|\frac{\partial f}{\partial z}\right| = 1 \Rightarrow \text{function } f \text{ satisfies the Lipschitz condition with}$$
$$L_f = 1,$$

$$\frac{\partial g}{\partial y} = -2y \Rightarrow \left|\frac{\partial g}{\partial y}\right| = |-2y| \le 2,$$

$$\frac{\partial g}{\partial z} = 0 \Rightarrow \max\left|\frac{\partial g}{\partial y}\right| = 0 \Rightarrow \text{function } g \text{ satisfies the Lipschitz condition with}$$
$$L_g = 2.$$

Then,

$$b = \min\left(a, \frac{b_1}{M}, \frac{b_2}{M}\right) = 1.$$

There is a unique solution $y = y(x)$, $z = z(x)$, $x \in [-1, 1]$ of the initial system. Writing $y_0(x) = 0$, $z_0(x) = \frac{1}{2}$, then yields

$$y_n(x) = y_0 + \int_0^x f(t, y_{n-1}(t), z_{n-1}(t))dt = \int_0^x z_{n-1}(t)dt$$

$$z_n(x) = z_0 + \int_0^x g(t, y_{n-1}(t), z_{n-1}(t))dt = \frac{1}{2} + \int_0^x (-y_{n-1}^2(t))dt$$

$$y_1(x) = \int_0^x z_0(t)dt = \int_0^x \frac{1}{2}dt = \frac{x}{2}; \quad z_1(x) = \frac{1}{2} + \int_0^x 0 dt = \frac{1}{2}$$

$$y_2(x) = \int_0^x z_1(t)dt = \int_0^x \frac{1}{2}dt = \frac{x}{2}; \quad z_2(x) = \frac{1}{2} + \int_0^x \left(-\frac{t^2}{4}\right)dt = \frac{1}{2} - \frac{x^3}{12}$$

$$y_3(x) = \int_0^x z_2(t)dt = \int_0^x \left(\frac{1}{2} - \frac{t^3}{12}\right)dt = \frac{x}{2} - \frac{x^4}{48}; \quad z_3(x) = \frac{1}{2} + \int_0^x \left(-\frac{t^2}{4}\right)dt$$

$$= \frac{1}{2} - \frac{x^3}{12}.$$

1.3.2 EXISTENCE AND UNIQUENESS THEOREM FOR DIFFERENTIAL EQUATIONS NOT SOLVED FOR DERIVATIVE

In this section, let us consider equations of the form

$$F(x, y, y') = 0. \tag{1.126}$$

It is obvious that generally, for such equations, not one but several integral curves pass through some point (x_0, y_0), since as a rule, when solving the equation $F(x, y, y') = 0$ for y', we find several (not one) real values $y' = f_i(x, y)$, $(i = 1, 2, \ldots)$, and if each of the equations $y' = f_i(x, y)$ in the neighborhood of point (x_0, y_0) satisfies the conditions of the theorem of existence and uniqueness, then for each one of these equations there will be a unique solution satisfying the condition $y(x_0) = y_0$.

THEOREM 1.6 There is a unique solution $y = y(x)$, $x_0 - h \le x \le x_0 + h$ (with h_0 sufficiently small) of Eq. (1.126) that satisfies the condition $y(x_0) = y_0$, for which $y'(x_0) = y_0'$, where y_0' is one of the real roots of the equation $F(x_0, y_0, y') = 0$ if in a closed neighborhood of the point (x_0, y_0, y'), the function $F(x, y, y')$ satisfies the following conditions:

1. $F(x, y, y')$ is continuous with respect to all arguments.
2. The derivative $\partial F / \partial y'$ exists and is nonzero in (x_0, y_0, y').
3. There exists the derivative $\partial F / \partial y$ bounded in absolute value $|\partial F / \partial y|$. ▲

Remark 1.19

The uniqueness property for the solution of equation $F(x, y, y') = 0$, which satisfies condition $y(x_0) = y_0$, is usually understood in the sense that not more than one integral curve of equation $F(x, y, y') = 0$ passes through a given point (x_0, y_0) in a given direction.

EXAMPLE 1.35 Let us consider the problem $xy'^2 - 2yy' - x = 0$; $y(1) = 0$.

    (a) Study the application of Theorem 1.6.
    (b) Solve the problem. ▲

Solution

(a) The function $F(x, y, y') = xy'^2 - 2yy' - x$ is continuous with respect to all arguments and $x_0 = 1$, $y_0 = 0$. $F(x_0, y_0, y') = 0 \Rightarrow F(1, 0, y') = 0$, that is, $y'^2 - 1 = 0 \Rightarrow y'_{01} = 1$ and $y'_{02} = -1$. $\partial F/\partial y = 2xy' - 2y$; $\partial F/\partial y'(1, 0, 1) = 2 \neq 0$; $\partial F/\partial y'(1, 0, -1) = -2 \neq 0$. Hence, the derivative exists and is non-zero. $\partial F/\partial y = -2y'$; $|\partial F/\partial y| = |-2y'| \leq M_1$. The considered problem has a unique solution

$$y = y(x), \quad 1 - h \leq x \leq 1 + h.$$

(b) Solving the equation for y' (see Example 1.20), we obtain two equations

$$y' = \frac{y + \sqrt{y^2 + x^2}}{x}, \quad y' = \frac{y - \sqrt{y^2 + x^2}}{x}$$

with the solutions $y = \frac{1}{2}[cx^2 - (1/c)]$ and $y = \frac{1}{2}[c - (x^2/c)]$. For $y'_{01} = 1$, $x_0 = 1$, $y_0 = 0$, we obtain $0 = \frac{1}{2}[c - (1/c)] \Rightarrow c = 1$ and $y = \frac{1}{2}(x^2 - 1)$. Setting $y'_{02} = -1$, $x_0 = 1$, $y_0 = 0$ yields

$$y = \frac{1}{2}(1 - x^2).$$

EXAMPLE 1.36 For the equation $2y'^2 + xy' - y = 0$, $y(2) = -\frac{1}{2}$:

    (a) Study the application of Theorem 1.6.
    (b) Solve the problem. ▲

Solution

(a) The function $F(x, y, y') = 2y'^2 + xy' - y$ is continuous and $\partial F/\partial y = -1 \Rightarrow |\partial F/\partial y| = 1$ is bounded. $\partial F/\partial y' = 4y' + x$, $x_0 = 2$, $y_0 = -\frac{1}{2}$, $F(x_0, y_0, y') = 0 \Rightarrow 2y'^2 + 2y' + \frac{1}{2} = 0 \Rightarrow y'_0 = -\frac{1}{2}$. This yields $\partial F/\partial y'(x_0, y_0, y') = \partial F/\partial y'(2, -\frac{1}{2}, -\frac{1}{2}) = 0$. The derivative $\partial F/\partial y'$ exists, but is zero. The uniqueness condition is not fulfilled.

    (b) The equation $2y'^2 + xy' - y = 0$ has the general solution $y = cx + 2c^2$ and the singular solution $y = -(x^2/8)$. Condition $y(2) = -\frac{1}{2}$

gives $-\frac{1}{2} = 2c + 2c^2 \Rightarrow 4c^2 + 4c + 1 = 0 \Rightarrow c = -\frac{1}{2}$; $y = -\frac{1}{2}x + \frac{1}{2}$ is a solution, and $y = -(x^2/8)$ is another solution.

1.3.3 SINGULAR SOLUTIONS OF DIFFERENTIAL EQUATIONS

The set of points (x, y) at which the uniqueness of solutions for equation

$$F(x, y, y') = 0 \tag{1.127}$$

is violated is called a *singular set*. If the conditions (1) and (3) of Theorem 1.6 are fulfilled, then in the points of a singular set, the equations

$$F(x, y, y') = 0 \qquad \text{and} \qquad \frac{\partial F}{\partial y} = 0 \tag{1.128}$$

must be satisfied simultaneously. Eliminating y' from these equations, we find the equations

$$\Phi(x, y) = 0, \tag{1.129}$$

which must be satisfied by the points of the singular set. However, the uniqueness of solution of Eq. (1.127) is not necessarily violated at every point that satisfies Eq. (1.129), because the conditions of Theorem 1.6 are only sufficient for the uniqueness of solutions, but are not necessary, and hence the violation of a condition of theorem does not imply the violation of uniqueness. The curve determined by Eq. (1.129) is called a *p-discriminant curve* (PDC), since Eqs. (1.128) are most frequently written in the form $F(x, y, p) = 0$ and $\partial F/\partial p = 0$. If a branch $y = \varphi(x)$ of the curve $\Phi(x, y) = 0$ belongs to the singular set and at the same time is an integral curve, it is called a *singular integral curve*, and the function $y = \varphi(x)$ is called a *singular solution*. Thus, in order to find the singular solution of Eq. (1.127) it is necessary to find the PDC defined by the equations $F(x, y, p) = 0$, $\partial F/\partial p = 0$, to find out [by direct substitutions into Eq. (1.127)] whether there are integral curves among the branches of the PDC and, if there are such curves, to verify whether uniqueness is violated in the points of these curves or not. If the uniqueness is violated, then such a branch of the PDC is a singular integral curve. *The envelope of the family of curves*

$$\Phi(x, y, c) = 0 \tag{1.130}$$

is the curve that in each of its points is tangent to some curve of the family (1.130), and each segment is tangent to an infinite set of curves of this family. If Eq. (1.130) is the general integral of Eq. (1.127), then the envelope of the family (1.130) (if it exists) will be a singular integral curve of Eq. (1.127). The envelope forms a part of the c-discriminant curves (CDCs) determined by the system

$$\Phi(x, y, c) = 0, \quad \frac{\partial \Phi}{\partial c} = 0. \tag{1.131}$$

For some branch of a CDC defined by an envelope, it is sufficient to satisfy on it the following:

1. There exist the bounded partial derivatives

$$\left| \frac{\partial \Phi}{\partial x} \right| \leq N_1, \quad \left| \frac{\partial \Phi}{\partial y} \right| \leq N_2, \tag{1.132}$$

where N_1, N_2 are constants.

2. One of the following conditions is satisfied:

$$\left| \frac{\partial \Phi}{\partial x} \right| \neq 0 \quad \text{or} \quad \left| \frac{\partial \Phi}{\partial y} \right| \neq 0. \tag{1.133}$$

Remark 1.20

Note that these conditions are only sufficient; thus, curves involving a violation of one of conditions (1) or (2) can also be envelopes.

EXAMPLE 1.37 Find the singular solutions of the differential equation $2y'^2 + xy' - y = 0$. ▲

Solution

The following p-discriminant curves are found:

$$\begin{cases} F(x, y, p) \equiv 2p^2 + xp - y = 0 \\ \dfrac{\partial F}{\partial p} \equiv 4p + x = 0 \Rightarrow p = -\dfrac{x}{4}. \end{cases}$$

Substituting $p = -(x/4)$ into the first equation yields $y = -(x^2/8)$. Substituting $y = -(x^2/8)$ and $y' = -(x/4)$ in the given equation $2y'^2 + xy' - y = 0$, we observe that $y = -(x^2/8)$ is a solution of this equation. Let us test if the solution $y = -(x^2/8)$ is singular.

Version I

The general solution is

$$y = cx + 2c^2.$$

Let us write the conditions for the tangency of the curves $y = y_1(x)$ and $y = y_2(x)$ in the point with abscissa $x = x_0$: $y_1(x_0) = y_2(x_0)$, $y_1'(x_0) = y_2'(x_0)$. The first equation shows the ordinate coincidence of curves and, second, the slope coincidence of tangents to those curves at the point with abscissa $x = x_0$. Setting $y_1(x) = -(x^2/8)$, $y_2(x) = cx + 2c^2$, then $-(x_0^2/8) = cx_0 + 2c^2$, $-(x_0/4) = c$. Substituting $c = -(x_0/4)$ into the first equation, we find that $-(x_0/8) = -(x_0/8)$, that is, for $c = -(x_0/4)$ the first equation is identically satisfied, since x_0 is the abscissa of an arbitrary point. It can be concluded that at each of its points, the curve $y = -(x^2/8)$ is touched by some other curves of the family $y = cx + 2c^2$ (see Example 1.27). Hence, $y = -(x^2/8)$ is a singular solution for the given equation.

Version II

The c-discriminant curves are

$$\begin{cases} \Phi(x, y, c) \equiv cx + 2c^2 - y = 0 \\ \dfrac{\partial \Phi}{\partial c}(x, y, c) \equiv x + 4c = 0, \ \text{hence } c = -\dfrac{x}{4}. \end{cases}$$

Substituting $c = -(x/4)$ in the first equation yields $y = -(x^2/8)$. This is the c-discriminant curve. Making a direct substitution, we observe that a solution for the given equation is found. Then, $\partial \Phi/\partial y = -1 \neq 0$, so that one of the conditions (1.133) is fulfilled. The curve $y = -(x^2/8)$ is the envelope of the family $y = cx + 2c^2$; hence $y = -(x^2/8)$ is a singular solution for the equation $2y'^2 + xy' - y = 0$.

1.4 Linear Differential Equations

1.4.1 HOMOGENEOUS LINEAR EQUATIONS: DEFINITIONS AND GENERAL PROPERTIES

DEFINITION 1.1 An n-order differential equation is called linear if it is of the first degree in the unknown function y and its derivatives y', y', ..., $y^{(n)}$, or is of the form

$$a_0 y^{(n)} + a_1 y^{(n-1)} + \cdots + a_n y = f(x), \tag{1.134}$$

where $a_0, a_1, \ldots, a_n$ and f are given continuous functions of x and $a_0 \neq 0$ (assume $a_0 = 1$) for all the values of x in the domain in which Eq. (1.134) is considered. The function $f(x)$ is called the right-hand member of the equation. If $f(x) \neq 0$, then the equation is called a *nonhomogeneous linear equation*. But if $f(x) \equiv 0$, then the equation has the form

$$y^{(n)} + a_1 y^{(n-1)} + \cdots + a_n y = 0 \tag{1.135}$$

and is called a *homogeneous linear equation*. ▲

Remark 1.21

If the coefficients $a_i(x)$ are continuous on the interval $a \leq x \leq b$, then, in the neighborhood of any initial values

$$y(x_0) = y_0, \quad y'(x_0) = y_0', \ldots, y^{(n-1)}(x_0) = y_0^{(n-1)}, \tag{1.136}$$

where $x_0 \in (a, b)$, the conditions of the theorem of existence and uniqueness are satisfied.

DEFINITION 1.2 The functions $y_1(x), y_2(x), \ldots, y_n(x)$ are said to be *linearly dependent* in the interval (a, b) if there exist constants $\alpha_1, \alpha_2, \ldots, \alpha_n$, not all equal to zero, such that

$$\alpha_1 y_1(x) + \alpha_2 y_2(x) + \cdots + \alpha_n y_n(x) \equiv 0$$

is valid. If, however, the identity holds only for $\alpha_1 = \alpha_2 = \cdots = \alpha_n = 0$, then the functions $y_1(x), y_2(x), \ldots, y_n(x)$ are said to be linearly independent in the interval (a, b). ▲

EXAMPLE 1.38 Show that the system of functions

$$1, x, x^2, \ldots, x^n \tag{1.137}$$

is linearly independent in the interval $(-\infty, \infty)$. ▲

Solution

The identity $\alpha_0 1 + \alpha_1 x + \cdots + \alpha_n x^n \equiv 0$ may hold only for $\alpha_0 = \alpha_1 = \cdots = \alpha_n = 0$.

EXAMPLE 1.39 Show that the system of functions

$$e^{k_1 x}, e^{k_2 x}, \ldots, e^{k_n x}, \tag{1.138}$$

where $k_1, k_2, \ldots, k_n$ are different in pairs, is linearly independent in the interval $-\infty \le x \le \infty$. ▲

Solution

Suppose the contrary, that is, that the given system is linearly dependent in this interval. Then, $\alpha_1 e^{k_1 x} + \alpha_2 e^{k_2 x} + \cdots + \alpha_n e^{k_n x} \equiv 0$ on the interval $(-\infty, \infty)$, and at least one of the members $\alpha_1, \alpha_2, \ldots, \alpha_n$ is nonzero, for example $\alpha_n \ne 0$. Dividing this identity by $e^{k_1 x}$ yields $\alpha_1 + \alpha_2 e^{(k_2 - k_1)x} + \alpha_3 e^{(k_3 - k_1)x} + \cdots + a_n e^{(k_n - k_1)x} \equiv 0$. Differentiating this identity gives

$$\alpha_2(k_2 - k_1)e^{(k_2 - k_1)x} + \alpha_3(k_3 - k_1)e^{(k_3 - k_1)x} + \cdots + \alpha_n(k_n - k_1)e^{(k_n - k_1)x} \equiv 0.$$

After dividing this identity by $e^{(k_2 - k_1)x}$, we find

$$\alpha_2(k_2 - k_1) + \alpha_3(k_3 - k_1)e^{(k_3 - k_2)x} + \cdots + \alpha_n(k_n - k_1)e^{(k_n - k_2)x} \equiv 0.$$

Differentiating yields

$$\alpha_3(k_3 - k_1)(k_3 - k_2)e^{(k_3 - k_2)x} + \cdots + \alpha_n(k_n - k_1)(k_n - k_2)e^{(k_n - k_2)x} \equiv 0,$$

and then

$$\alpha_n(k_n - k_1)(k_n - k_2) \cdots (k_n - k_{n-1})e^{(k_n - k_{n-1})x} \equiv 0,$$

which is false, since $\alpha_n \ne 0$, $k_n \ne k_1$, $k_n \ne k_2, \ldots, k_n \ne k_{n-1}$ according to the condition, and $e^{(k_n - k_{n-1})x} \ne 0$.

EXAMPLE 1.40 Prove that the system of functions

$$e^{\alpha x} \sin \beta x, \quad e^{\alpha x} \cos \beta x, \quad \beta \neq 0 \tag{1.139}$$

is linearly independent in the interval $(-\infty, \infty)$. ▲

Solution

Consider the identity $\alpha_1 e^{\alpha x} \sin \beta x + \alpha_2 e^{\alpha x} \cos \beta x \equiv 0$. If we divide by $e^{\alpha x} \neq 0$, then $\alpha_1 \sin \beta x + \alpha_2 \cos \beta x \equiv 0$. Then we substitute in this identity the value of $x = 0$ to get $\alpha_1 = 0$ and hence $\alpha_1 \sin \beta x \equiv 0$; but the function $\sin \beta x$ is not identically equal to zero, so $\alpha_1 = 0$. The initial identity holds only when $\alpha_1 = \alpha_2 = 0$, that is, the given functions are linearly independent in the interval $-\infty < x < \infty$.

DEFINITION 1.3 Consider the functions $y_1(x), y_2(x), \ldots, y_n(x)$ that have derivatives of order $(n - 1)$. The determinant

$$W[y_1, y_2, \ldots, y_n] = \begin{vmatrix} y_1(x) & y_2(x) & \cdots & \cdots & y_n(x) \\ y_1'(x) & y_2'(x) & \cdots & \cdots & y_n'(x) \\ \cdots & \cdots & \cdots & \cdots & \cdots \\ y_1^{(n-1)}(x) & y_2^{(n-1)}(x) & \cdots & \cdots & y_n^{(n-1)}(x) \end{vmatrix} \tag{1.140}$$

is called the *Wronskian determinant* for these functions. Let $y_1(x)$, $y_2(x), \ldots, y_n(x)$ be a system of functions given in the interval $[a, b]$. ▲

DEFINITION 1.4 Let us set

$$\langle y_i, y_j \rangle = \int_a^b y_i(x) y_j(x) dx, \quad i, j = 1, 2, \ldots, n. \tag{1.141}$$

The determinant

$$\Gamma(y_1, y_2, \ldots, y_n) = \begin{vmatrix} \langle y_1, y_1 \rangle & \langle y_1, y_2 \rangle & \cdots & \cdots & \langle y_1, y_n \rangle \\ \langle y_2, y_1 \rangle & \langle y_2, y_2 \rangle & \cdots & \cdots & \langle y_2, y_n \rangle \\ \cdots & \cdots & \cdots & \cdots & \cdots \\ \langle y_n, y_1 \rangle & \langle y_n, y_2 \rangle & \cdots & \cdots & \langle y_n, y_n \rangle \end{vmatrix} \tag{1.142}$$

is called the *Grammian* of the system of functions $\{y_k(x)\}$. ▲

THEOREM 1.7 If a system of functions $y_1(x), y_2(x), \ldots, y_n(x)$ is linearly dependent in the interval $[a, b]$, then its Wronskian is identically equal to zero in this interval. ▲

THEOREM 1.8 *Liouville's Formula*

Consider that $y_1(x), y_2(x), \ldots, y_n(x)$ are solutions of the homogeneous linear Equations (1.135), and $W(x) = W[y_1(x), y_2(x), \ldots, y_n(x)]$ is the Wronskian determinant. Then

$$W(x) = W(x_0)e^{-\int_{x_0}^{x} a_1(x)dx}. \tag{1.143}$$

Equation (1.143) is called *Liouville's formula.* ▲

THEOREM 1.9 If the Wronskian $W(x) = W[y_1(x), y_2(x), \ldots, y_n(x)]$, formed for the solutions $y_1, y_2, \ldots, y_n$ of the homogeneous linear Eqs. (1.135), is not zero for some value $x = x_0$ on the interval $[a, b]$, where the coefficients of the equation are continuous, then it does not vanish for any value of x on that interval. ▲

THEOREM 1.10 If the solutions $y_1, y_2, \ldots, y_n$ of Eq. (1.135) are linearly independent on an interval $[a, b]$, then the Wronskian W formed for these solutions does not vanish at any point of the given interval. ▲

THEOREM 1.11 For a system of functions $y_1(x), y_2(x), \ldots, y_n(x)$ to be linearly dependent, it is necessary and sufficient that the Grammian be zero.

Let us write the homogeneous linear Eq. (1.135) as

$$L[y] = 0, \tag{1.144}$$

where

$$L[y] = y^{(n)} + a_1(x)y^{(n-1)} + \cdots + a_n(x)y. \tag{1.145}$$

Then $L[y]$ will be termed a *linear differential operator.* ▲

THEOREM 1.12 The linear differential operator L has the following two basic properties:

1. $L[cy] = cL[y]$, $c = $ constant.
2. $L[y_1 + y_2] \equiv L[y_1] + L[y_2]$. ▲

Remark 1.22

The linearity and homogeneity of Eq. (1.135) are retained for any transformation of the independent variable $x = \varphi(t)$, where $\varphi(t)$ is an arbitrary n-times differentiable function, with $\varphi'(t) \neq 0$, $t \in (a, b)$.

Remark 1.23

The linearity and homogeneity are also retained in a homogeneous linear transformation of the unknown function $y(x) = \alpha(x)z(x)$.

THEOREM 1.13 The totality of all solutions of Eq. (1.135) is an n-dimensional linear space. A basis in such a space is a fundamental system of solutions, that is, any family of n linearly independent solutions of Eq. (1.135). ▲

COROLLARY 1.1 (for Theorem 1.13)

The general solution for $a \leq x \leq b$ of the homogeneous linear equation (1.135) with the coefficients $a_i(x)$, $(i = 1, 2, \ldots, n)$ continuous on the interval $a \leq x \leq b$ is the linear combination

$$y = \sum_{i=1}^{n} c_i y_i \tag{1.146}$$

of n linearly independent (on the same interval) partial solutions y_i $(i = 1, 2, \ldots, n)$ with arbitrary constant coefficients. ▲

THEOREM 1.14 If a homogeneous linear equation $L[y] = 0$ with real coefficients $a_i(x)$ has a complex solution $y(x) = u(x) + iv(x)$, then the real part of this solution $u(x)$ and its imaginary part $v(x)$ are separately solutions of that homogeneous equation. ▲

Remark 1.24

Knowing one nontrivial particular solution y_1 of the homogeneous linear Eq. (1.135), it is possible with the substitution

$$y = y_1 \int u \, dx, \quad u = \left(\frac{y}{y_1} \right)' \tag{1.147}$$

to reduce the order of the equation and retain its linearity and homogeneity. Knowing k linearly independent (on the interval $a \leq x \leq b$) solutions $y_1, y_2, \ldots, y_k$ of a homogeneous linear equation, it is possible to reduce the order of the equation to $(n - k)$ on the same interval $a \leq x \leq b$.

Remark 1.25

If two equations have the form

$$y^{(n)} + a_1(x)y^{(n-1)} + \cdots + a_n(x)y = 0 \tag{1.148}$$
$$y^{(n)} + b_1(x)y^{(n-1)} + \cdots + b_n(x)y = 0, \tag{1.149}$$

where the functions $a_i(x)$ and $b_i(x)$, $(i = 1, 2, \ldots, n)$ are continuous on the interval $a \leq x \leq b$ and have a common fundamental system of solutions, $y_1, y_2, \ldots, y_n$, then the equations coincide. This means that $a_i(x) \equiv b_i(x)$, $(i = 1, 2, \ldots, n)$ on the interval $a \leq x \leq b$.

Remark 1.26

Consider that the system of functions $y_1(x), y_2(x), \ldots, y_n(x)$, with all derivatives to the n-order inclusively, is linearly independent in the interval $[a, b]$. Then, the equation

$$\begin{vmatrix} y_1(x) & y_2(x) & \cdots & y_n(x) & y(x) \\ y_1'(x) & y_2'(x) & \cdots & y_n'(x) & y'(x) \\ \cdots & \cdots & \cdots & \cdots & \cdots \\ y_1^{(n)}(x) & y_2^{(n)}(x) & \cdots & y_n^{(n)}(x) & y^{(n)}(x) \end{vmatrix} = 0, \qquad (1.150)$$

with $y(x)$ an unknown function, will be a linear differential equation for which the functions $y_1(x), y_2(x), \ldots, y_n(x)$ are a fundamental system of solutions. The coefficient of $y^{(n)}(x)$ in Eq. (1.150) is the Wronskian $W[y_1, y_2, \ldots, y_n]$ of this system.

Remark 1.27

(a) If

$$a_0(x) + a_1(x) + \cdots + a_n(x) = 0, \qquad (1.151)$$

then $y_1 = e^x$ is a particular solution of the equation

$$a_0(x)y^{(n)} + a_1(x)y^{(n-1)} + \cdots + a_n(x)y = 0. \qquad (1.152)$$

(b) If

$$\sum_{k=0}^{n} (-1)^k a_{n-k}(x) = 0, \qquad (1.153)$$

then $y_1 = e^{-x}$ is a particular solution of Eq. (1.152).
(c) If all coefficients $a_0(x), a_1(x), \ldots, a_n(x)$ are polynomials, then the equation can have a polynomial as a particular solution.
(d) If

$$a_{n-1}(x) + x a_n(x) = 0, \qquad (1.154)$$

then $y_1 = x$ is a particular solution of Eq. (1.152).

EXAMPLE 1.41 Solve the problem

$$xy''' - y'' - xy' + y = 0; \quad y(2) = 0, \quad y'(2) = 1, \quad y''(2) = 2. \quad \blacktriangle$$

Solution

Using Remark 1.27, we can find the following:

(a) $x - 1 - x + 1 = 0 \Rightarrow y_1 = e^x$ is a particular solution.
(b) $-x - 1 + x + 1 = 0 \Rightarrow y_2 = e^{-x}$ is a particular solution.

(c) $-x + x(1) = 0 \Rightarrow y_3 = x$ is a particular solution. The Wronskian is

$$W[y_1, y_2, \ldots, y_n] = \begin{vmatrix} e^x & e^{-x} & x \\ e^x & -e^{-x} & 1 \\ e^x & e^{-x} & 0 \end{vmatrix} = 2x \neq 0 \qquad \text{for } x \neq 0.$$

The system $\{e^x, e^{-x}, x\}$ is a fundamental system of solutions on $(0, \infty)$. The general solution of the equation $xy''' - y'' - xy' + y = 0$ is

$$y(x) = c_1 e^x + c_2 e^{-x} + c_3 x.$$

Using the initial conditions yields

$$\begin{cases} e^2 c_1 + e^{-2} c_2 + 2c_3 = 0 \\ e^2 c_1 - e^{-2} c_2 + c_3 = 1 \\ e^2 c_1 + e^{-2} c_2 = 2. \end{cases}$$

The solution of this system is $c_1 = 2e^{-2}$, $c_2 = 0$, $c_3 = -1$. Then, the solution of the equation with the initial condition is

$$y(x) = 2e^{-2} e^x - x.$$

EXAMPLE 1.42 Find the general solution of the equation

$$(2x - 3)y''' - (6x - 7)y'' + 4xy' - 4y = 0, \qquad x \in [2, \infty). \quad \blacktriangle$$

Solution

Using Remark 1.27, we find the particular solutions $y_1 = e^x$, $y_2 = x$ with the Wronskian

$$W[y_1, y_2] = \begin{vmatrix} e^x & x \\ e^x & 1 \end{vmatrix} = e^x(1 - x) \neq 0, \qquad x \in [2, \infty).$$

The solution y_1, y_2 is linearly independent on $[2, \infty)$. Using Eq. (1.147) gives

$$y' = y_1' \int u\,dx + y_1 u = e^x \int u\,dx + e^x u$$

$$y'' = y_1'' + 2y_1' u + y_1 u' = e^x \int u\,dx + 2e^x u + e^x u'$$

$$y''' = y_1''' + 3y_1'' u + 3y_1' u' + y_1 u'' = e^x \int u\,dx + 3e^x u + 3e^x u' + e^x u'',$$

and the equation takes the form $(2x - 3)u'' - 2u' - (2x - 5)u = 0$. With the substitution $u = u_1 \int v\,dx$, $u_1 = (y_2/y_1)'$, it is possible to reduce the order of the equation:

$$(2x - 3)u'' - 2u' - (2x - 5)u = 0.$$

Indeed,

$$u_1 = \left(\frac{x}{e^x}\right)' = (xe^{-x})' = (1-x)e^{-x}$$

$$u = u_1 \int v\,dx = (1-x)e^{-x} \int v\,dx$$

$$u' = u_1' \int v\,dx + u_1 v = (x-2)e^{-x} \int v\,dx + (1-x)e^{-x}v$$

$$u'' = u_1'' \int v\,dx + 2u_1'v + u_1v' = (3-x)e^{-x} \int v\,dx + 2(x-2)e^{-x}v$$
$$+ (1-x)e^{-x}v',$$

and the equation takes the form $-(2x^2 - 5x + 3)v' + (4x^2 - 12x + 10)v = 0$, or

$$\frac{v'}{v} = \frac{4x^2 - 12x + 10}{2x^2 - 5x + 3}.$$

After integration, we obtain

$$v(x) = \frac{e^{2x}(2x-3)}{(x-1)^2}.$$

Substituting this expression in $u = u_1 \int v\,dx$ gives

$$u(x) = (1-x)e^{-x} \int \frac{e^{2x}(2x-3)}{(x-1)^2}\,dx$$

$$= (1-x)e^{-x}\left[\int e^{2x}\left(\frac{2}{x-1} - \frac{1}{(x-1)^2}\right)dx\right]$$

$$= (1-x)e^{-x}\left[\int \frac{2e^{2x}}{x-1}\,dx + \int\left(\frac{1}{x-1}\right)' e^{2x}\,dx\right]$$

$$= (1-x)e^{-x}\left[\int \frac{2e^{2x}}{x-1}\,dx + \frac{1}{x-1}e^{2x}\,dx - \int \frac{2e^{2x}}{x-1}\,dx\right] = -e^x,$$

and $y = y_1 \int u(x)\,dx = e^x \int (-e^x)\,dx = -e^{2x}$. If $y = -e^{2x}$ is a solution of a homogeneous equation, then $y_3(x) = e^{2x}$ is also a solution of that equation. The system of functions $\{e^x, x, e^{2x}\}$ is a fundamental system ($W[y_1 y_2, y_3] = -(2x-3) \neq 0$, $\forall x \in [2, \infty)$), and the general solution is

$$y(x) = c_1 e^x + c_2 x + c_3 e^{2x}.$$

EXAMPLE 1.43 Find the homogeneous linear differential equation, knowing that the fundamental system of solutions is x, e^x, e^{2x}. ▲

Solution

Using Remark 1.26, this equation is

$$
\begin{vmatrix}
x & e^x & e^{2x} & y \\
1 & e^x & 2e^{2x} & y' \\
0 & e^x & 4e^{2x} & y'' \\
0 & e^x & 8e^{2x} & y'''
\end{vmatrix} = 0,
$$

that is,

$$
y''' \begin{vmatrix}
x & e^x & e^{2x} \\
1 & e^x & 2e^{2x} \\
0 & e^x & 4e^{2x}
\end{vmatrix} - y'' \begin{vmatrix}
x & e^x & e^{2x} \\
1 & e^x & 2e^{2x} \\
0 & e^x & 8e^{2x}
\end{vmatrix} + y' \begin{vmatrix}
x & e^x & e^{2x} \\
0 & e^x & 4e^{2x} \\
0 & e^x & 8e^{2x}
\end{vmatrix} = 0,
$$

and finally,

$$
(2x - 3)y''' - (6x - 7)y'' + 4xy' - 4y = 0.
$$

1.4.2 NONHOMOGENEOUS LINEAR EQUATIONS

A nonhomogeneous linear differential equation is of the form

$$
a_0(x)y^{(n)} + a_1(x)y^{(n-1)} + \cdots + a_n(x)y = g(x).
$$

If $a_0(x) \neq 0$ for the interval of variation of x, then, after division by $a_0(x)$, we find

$$
y^{(n)} + p_1(x)y^{(n-1)} + \cdots + p_n(x)y = f(x). \tag{1.155}
$$

This equation is written briefly as (see Eq. (1.145))

$$
L[y] = f(x). \tag{1.156}
$$

If, for $a \leq x \leq b$ all the coefficients $p_i(x)$ in Eq. (1.155) and $f(x)$ are continuous, then it has a unique solution that satisfies the conditions

$$
y^{(k)}(x_0) = y_0^{(k)} \quad (k = 0, 1, \ldots, n - 1), \tag{1.157}
$$

where $y_0^{(k)}$ are any real numbers and x_0 is any number in the interval $a \leq x \leq b$. The properties of the linear operator L yield the following.

THEOREM 1.15 The sum $\tilde{y} + y_1$ of the solution $\tilde{y}$ of the nonhomogeneous equation

$$
L[y] = f(x) \tag{1.158}
$$

and of the solution y_1 of the corresponding homogeneous equation $L[y] = 0$ is a solution of the nonhomogeneous Eq. (1.156). ▲

THEOREM 1.16 *The Principle of Superposition*
If y_i is a solution of equation $L[y] = f_i(x)$ $(i = 1, 2, \ldots, m)$, then $y = \sum_{i=1}^{m} \alpha_i y_i$ is a solution of the equation

$$
L[y] = \sum_{i=1}^{m} \alpha_i f_i(x), \tag{1.159}
$$

where α_i are constants. ▲

Remark 1.28

This property also holds true for $m \to \infty$ if the series $\sum_{i=1}^{m} \alpha_i y_i$ converges and admits an n-fold differentiation.

THEOREM 1.17 If the equation $L[y] = U(x) + iV(x)$ [where all the coefficients $p_i(x)$ and functions $U(x)$ and $V(x)$ are real] has a solution $y = u(x) + iv(x)$, then the real part of the solution $u(x)$ and the imaginary part $v(x)$ are, respectively, solutions for

$$L[y] = U(x), \quad L[y] = V(x). \quad \blacktriangle \qquad (1.160)$$

THEOREM 1.18 The general solution on the interval $a \le x \le b$ of the equation $L[y] = f(x)$ with continuous (on the same interval) coefficients $p_i(x)$ and $f(x)$ is equal to the sum of the general solution $\sum_{i=1}^{n} c_i y_i$ of the corresponding homogeneous equation and of some particular solution $\tilde{y}$ of the nonhomogeneous equation. Hence, the integration of a nonhomogeneous linear equation reduces to give one particular solution of the equation and to integrate the corresponding homogeneous linear equation. $\quad \blacktriangle$

The Method of Variation of Parameters (Lagrange's Method)

If to choose a particular solution of the nonhomogeneous equation is difficult, but the general solution of the corresponding homogeneous equation $y = \sum_{i=1}^{n} c_i y_i$ is found, then it is possible to integrate the nonhomogeneous linear equation by the method of parameter variation. To apply this method, it is assumed that the solution of the nonhomogeneous Eq. (1.155) has the form

$$y = c_1(x)y_1 + c_2(x)y_2 + \cdots + c_n(x)y_n. \qquad (1.161)$$

Since the choice of functions $c_i(x)$, $(i = 1, 2, \ldots, n)$ has to satisfy only one equation, (1.155), it is required for these n functions $c_i(x)$ to satisfy some other $(n-1)$ equations

$$\sum_{i=1}^{n} c_i'(x)y_i^{(k)} = 0, \quad k = 0, 1, \ldots, n-2.$$

With the conditions (1.164), Eq. (1.155) takes the form

$$\sum_{i=1}^{n} c_i' y_i^{(n-1)} = f(x).$$

To summarize, the functions $c_i(x)$, $(i = 1, 2, \ldots, n)$ are determined from the system of n linear equations

$$\begin{cases} c_1'(x)y_1 + c_2'(x)y_2 + \cdots + c_n'(x)y_n = 0 \\ c_1'(x)y_1' + c_2'(x)y_2' + \cdots + c_n'(x)y_n' = 0 \\ \quad \cdots \\ c_1'(x)y_1^{(n-2)} + c_2'(x)y_2^{(n-2)} + \cdots + c_n'(x)y_n^{(n-2)} = 0 \\ c_1'(x)y_1^{(n-1)} + c_2'(x)y_2^{(n-1)} + \cdots + c_n'(x)y_n^{(n-1)} = f(x), \end{cases} \qquad (1.162)$$

with a nonzero determinant of the system

$$\begin{vmatrix} y_1 & y_2 & \cdots & \cdots & y_n \\ y_1' & y_2' & \cdots & \cdots & y_n' \\ \cdots & \cdots & \cdots & \cdots & \cdots \\ y_1^{(n-1)} & y_2^{(n-1)} & \cdots & \cdots & y_n^{(n-1)} \end{vmatrix} = W[y_1, y_2, \ldots, y_n] \neq 0.$$

Knowing all $c_i'(x) = \varphi_i(x)$ from Eq. (1.162) yields (using quadratures) that

$$c_i(x) = \int \varphi_i(x) dx + c_i,$$

where c_i are arbitrary constants. Substituting the obtained values for $c_i(x)$ in Eq. (1.161), we can find the general solution of Eq. (1.155).

EXAMPLE 1.44 Solve the equation $y'' - y = f(x)$. ▲

Solution

The corresponding homogeneous equation is $y'' - y = 0$, and its fundamental system of solutions is $y_1 = e^x$, $y_2 = e^{-x}$. Let us find the general solution of the given equation by the method of parameter variation $y = c_1(x)e^x + c_2(c)e^{-x}$. From the following system,

$$\begin{cases} c_1'(x)e^x + c_2'(x)e^{-x} = 0 \\ c_1'(x)e^x - c_2'(x)e^{-x} = f(x) \end{cases}$$

yields

$$\begin{cases} c_1'(x) = \frac{1}{2}e^{-x}f(x) \\ c_2'(x) = -\frac{1}{2}e^x f(x). \end{cases}$$

After integration we find

$$\begin{cases} c_1(x) = \frac{1}{2}\int_0^x e^{-t}f(t)dt + c_1 \\ c_2(x) = -\frac{1}{2}\int_0^x e^t f(t)dt + c_2. \end{cases}$$

If we substitute these values of $c_1(x)$ and $c_2(x)$ in the expression for y, the general solution of the given equation is

$$y(x) = c_1 e^x + c_2 e^{-x} + \frac{1}{2}\int_0^x [e^{x-t} - e^{-(x-t)}]f(t)dt.$$

Cauchy's Method for Finding a Particular Solution of a Nonhomogeneous Linear Equation $L[y(x)] = f(x)$

Consider the nonhomogeneous linear equation

$$L[y(x)] = f(x). \tag{1.163}$$

In this method it is assumed that we know the solution $K(x, t)$ (dependent on a single parameter) of the corresponding homogeneous equation $L[y(x)] = 0$, and the solution satisfies the conditions

$$K(t, t) = K'(t, t) = \cdots = K^{(n-2)}(t, t) = 0 \quad \text{and} \quad K^{(n-1)}(t, t) = 1.$$

$$(1.164)$$

In this case,

$$y_p(x) = \int_{x_0}^{x} K(x, t) f(t) dt \qquad (1.165)$$

will be a particular solution of Eq. (1.156), and the solution satisfies the zero initial conditions $y(x_0) = y'(x_0) = \cdots = y^{(n-1)}(x_0) = 0$. The solution $K(x, t)$, called *Cauchy's function*, may be isolated from the general solution $y = \sum_{i=1}^{n} c_i y_i(x)$ of the homogeneous equation if the arbitrary constants c_i are chosen to satisfy the conditions (1.164).

EXAMPLE 1.45 Solve the equation $y'' + y = f(x)$. ▲

Solution

The corresponding homogeneous equation is $y'' + y = 0$. It is easy to verify that its general solution is $\tilde{y}(x) = c_1 \sin x + c_2 \cos x$. The conditions (1.162) leads to the following equations:

$$\begin{cases} c_1 \sin t + c_2 \cos t = 0 \\ c_1 \cos t - c_2 \sin t = 1. \end{cases}$$

Hence, $c_1 = \cos t$, $c_2 = -\sin t$, and the solution $K(x, t)$ must be of the form

$$K(x, t) = \sin x \cos t - \cos x \sin t = \sin(x - t).$$

According Eqs. (1.165), the solution of the given equation that satisfies zero initial conditions is

$$y_p(x) = \int_{x_0}^{x} \sin(x - t) f(t) dt.$$

Remark 1.29

To find the Cauchy's function, the general solution of Eq. (1.156) and the solution satisfying the initial conditions (1.157), the procedure is as follows. First, find the fundamental system $y_1(t), y_2(t), \ldots, y_n(t)$ of the corresponding homogeneous equation $L[y] = 0$ and the Wronskian $W(t) = W[y_1(t), y_2(t), \ldots, y_n(t)]$. The Cauchy's function is

$$K(x, t) = \frac{1}{W(t)} \begin{vmatrix} y_1(t) & y_2(t) & \cdots & \cdots & y_n(t) \\ y_1'(t) & y_2'(t) & \cdots & \cdots & y_n'(t) \\ \cdots & \cdots & \cdots & \cdots & \cdots \\ y_1^{(n-2)}(t) & y_2^{(n-2)}(t) & \cdots & \cdots & y_n^{(n-2)}(t) \\ y_1(x) & y_2(x) & \cdots & \cdots & y_n(x) \end{vmatrix}. \qquad (1.166)$$

The general solution of Eq. (1.156) is

$$y(x) = c_1 y_1(x) + c_2 y_2(x) + \cdots + c_n y_n(x) + \int_{x_0}^{x} K(x, t) f(t) dt. \qquad (1.167)$$

If we set $y^{(k)}(x_0) = y_0^{(k)}$, $k = 0, 1, 2, \ldots, n-1$, then

$$c_i = \frac{W_i(x_0)}{W(x_0)}, \qquad (1.168)$$

where

$$W_i(x_0) =$$

$$\begin{vmatrix} y_1(x_0) & y_2(x_0) & \cdots & y_{i-1}(x_0) & y_0 & y_{i+1}(x_0) & \cdots & y_n(x_0) \\ y_1'(x_0) & y_2'(x_0) & \cdots & y_{i-1}'(x_0) & y_0' & y_{i+1}'(x_0) & \cdots & y_n'(x_0) \\ \cdots & \cdots & \cdots & \cdots & \cdots & \cdots & \cdots & \cdots \\ y_1^{(n-1)}(x_0) & y_2^{(n-1)}(x_0) & \cdots & y_{i-1}^{(n-1)}(x_0) & y_0^{(n-1)} & y_{i+1}^{(n-1)}(x_0) & \cdots & y_n^{(n-1)}(x_0) \end{vmatrix}.$$

$$(1.169)$$

The solution of Eq. (1.156) satisfying Eq. (1.157) is

$$y(x) = \sum_{i=1}^{n} \frac{W_i(x_0)}{W(x_0)} y_i(x) + \int_{x_0}^{x} K(x, t) f(t) dt.$$

A Physical Interpretation

In many problems the solution $y(t)$ of the equation

$$y^{(n)} + p_1(t) y^{(n-1)} + \cdots + p_n(t) y = f(t) \qquad (1.170)$$

describes the displacement of some system, the function $f(t)$ describes a force acting on the system, and t is the time. First, suppose that when $t < s$ the system is at rest, its displacement is caused by a force $f_\varepsilon(t)$ that differs from zero only in the interval $s < t < s + \varepsilon$, and the momentum of this force is $\int_s^{s+\varepsilon} f_\varepsilon(\tau) d\tau = 1$. Denote by $y_\varepsilon(t)$ the solution of equation

$$y^{(n)} + p_1(t) y^{(n-1)} + \cdots + p_n(t) y = f_\varepsilon(t).$$

Then, $y_\varepsilon(t) = \int_{t_0}^{t} K(t, s) f_\varepsilon(s) ds$; $\lim_{\varepsilon \to 0} y_\varepsilon(t) = K(t, s)$. The function $K(t, s)$ is called the *influence function* of the instantaneous momentum at time $t = s$. The solution of Eq. (1.163) with zero initial conditions in the form $y = \int_{t_0}^{t} K(t, s) f(s) ds$ indicates that the effect of a constant acting force may be retarded as the superposition of influences of the instantaneous momentum.

EXAMPLE 1.46 Solve the problem $y'' + w^2 y = f(x)$; $y(x_0) = y_0$, $y'(x_0) = y_0'$. ▲

Solution

The fundamental system of equation $y'' + w^2 y = 0$ is $y_1 = \cos wx$, $y_2 = \sin wx$. Then, the Wronskian is

$$W[y_1, y_2] = \begin{vmatrix} \cos wx & \sin wx \\ -w\sin wx & w\cos wx \end{vmatrix} = w; \quad W(t) = w.$$

The Cauchy's function is

$$K(x, t) = \frac{1}{W(t)} \begin{vmatrix} y_1(t) & y_2(t) \\ y_1(x) & y_2(x) \end{vmatrix}$$

$$= \frac{1}{w} \begin{vmatrix} \cos wt & \sin wt \\ \cos wx & \sin wx \end{vmatrix} = \frac{1}{w} \sin w(x - t)$$

$$W_1(x_0) = \begin{vmatrix} y_0 & \sin wx_0 \\ y_0' & w\cos wx_0 \end{vmatrix} = y_0 w \cos wx_0 - y_0' \sin wx_0$$

$$W_2(x_0) = \begin{vmatrix} \cos wx_0 & y_0 \\ -w\sin wx_0 & y_0' \end{vmatrix} = y_0' \cos wx_0 + wy_0 \sin wx_0$$

$$y(x) = \frac{W_1(x_0)}{W(x_0)} y_1(x) + \frac{W_2(x_0)}{W(x_0)} y_2(x) + \int_{x_0}^{x} K(x, t) f(t) dt$$

$$y(x) = \frac{1}{w} [(y_0 w \cos wx_0 - y_0' \sin wx_0) \cos wx + (y_0' \cos wx_0$$

$$+ wy_0 \sin wx_0) \sin wx] + \frac{1}{w} \int_{x_0}^{x} \sin w(x - t) f(t) dt$$

$$y(x) = y_0 \cos w(x - x_0) + \frac{1}{w} y_0' \sin w(x - x_0) + \frac{1}{w} \int_{x_0}^{x} \sin w(x - t) f(t) dt.$$

EXAMPLE 1.47 Solve the problem $x''(t) + w^2 x(t) = A\sin at$, $x(0) = x_0$, $x'(0) = v_0$. ▲

Solution

Using Example 1.46 yields

$$x(t) = x_0 \cos wt + \frac{1}{w} v_0 \sin wt + \frac{A}{w} \int_0^t \sin w(t - s) \sin as\, ds;$$

$$= \int_0^t \sin w(t - s) \sin as\, ds$$

$$= \frac{1}{2} \int_0^t \{\cos[wt - (w + a)s] - \cos[wt - (w - a)]\}\, ds$$

$$= \frac{1}{2} \left[\frac{\sin[wt - (w + a)s]}{-(w + a)} \Big|_0^t + \frac{\sin[wt - (w - a)s]}{w - a} \Big|_0^t \right]$$

$$= \frac{1}{2} \left[\frac{\sin at}{w + a} + \frac{\sin wt}{w + a} + \frac{\sin at}{w - a} - \frac{\sin wt}{w - a} \right]$$

$$= \frac{1}{w^2 - a^2} [w\sin at - a\sin wt].$$

Hence

$$x(t) = x_0 \cos wt + \frac{1}{w} v_0 \sin wt + \frac{A}{w^2 - a^2} \left(\sin at - \frac{a}{w} \sin wt \right).$$

Remark 1.30

If $a = w$, then

$$\int_0^t \sin w(t-s)\sin wsds = \frac{1}{2}\int_0^t [\cos w(t-2s) - \cos wt]ds$$

$$= \frac{1}{2}\left[\frac{\sin wt}{w} - t\cos wt\right],$$

and

$$x(t) = x_0 \cos wt + \frac{1}{w}v_0 \sin wt + \frac{A}{2w^2}[\sin wt - tw\cos wt].$$

EXAMPLE 1.48 Solve the problem

$$(x-1)y'' - xy' + y = x^2 - 2x + 1, \quad y(2) = -1, \quad y'(2) = 1. \quad \blacktriangle$$

Solution

Let us write the equation in the form

$$y'' - \frac{x}{x-1}y' + \frac{1}{x-1}y = x - 1.$$

The coefficients are continuous for $x \in (1, \infty)$. The corresponding homogeneous equation is

$$y'' - \frac{x}{x-1}y' + \frac{1}{x-1}y = 0,$$

and its fundamental system of solutions is $y_1 = x$, $y_2 = e^x$. The Wronskian is

$$W[y_1, y_2] = \begin{vmatrix} x & e^x \\ 1 & e^x \end{vmatrix} = e^x(x-1) \neq 0, \quad x \in (1, \infty), \quad W(t) = e^t(t-1).$$

The Cauchy's function is

$$K(x, t) = \frac{1}{W(t)}\begin{vmatrix} y_1(t) & y_2(t) \\ y_1(x) & y_2(x) \end{vmatrix} = \frac{1}{(t-1)e^t}\begin{vmatrix} t & e^t \\ x & e^x \end{vmatrix} = (e^{x-t}t - x)\frac{1}{t-1}.$$

The particular solution of the equation

$$y'' - \frac{x}{x-1}y' + \frac{1}{x-1}y = x - 1$$

that satisfies the initial conditions

$$y(x_0) = 0, \quad y'(x_0) = 0$$

is

$$y_p(x) = \int_{x_0}^x K(x, t)f(t)dt = \int_{x_0}^x (e^{x-t}t - x)dx,$$

$$y_p(x) = -(x+1) + (x_0 + 1)e^{x-x_0} - x^2 + xx_0 = -x^2 + x - 1 + 3e^{x-2}.$$

The general solution of the equation

$$y'' - \frac{x}{x-1}y' + \frac{1}{x-1}y = x - 1$$

is

$$y(x) = c_1 x + c_2 e^x - (x+1) + (x_0 + 1)e^{x-x_0} - x^2 + xx_0$$
$$= c_1 x + c_2 e^x - x^2 + x - 1 + 3e^{x-2},$$

and

$$W_1(x_0) = \begin{vmatrix} y_0 & y_2(x_0) \\ y_0' & y_2'(x_0) \end{vmatrix} = \begin{vmatrix} -1 & e^2 \\ 1 & e^2 \end{vmatrix} = -2e^2 \quad (x_0 = 2, y_0 = -1, y_0' = 1)$$

$$W_2(x_0) = \begin{vmatrix} y_1(x_0) & y_0 \\ y_1'(x_0) & y_0' \end{vmatrix} = \begin{vmatrix} x_0 & -1 \\ 1 & 1 \end{vmatrix} = x_0 + 1 = 3.$$

The solution for the given problem is

$$y(x) = \frac{W_1(x_0)}{W(x_0)}y_1(x) + \frac{W_2(x_0)}{W(x_0)}y_2(x) + y_p(x),$$

that is,

$$y(x) = 6e^{x-2} - x^2 - x - 1.$$

EXAMPLE 1.49 Find the general solution of the equation $x'' + 4x = \sin x + 4t$. ▲

Solution

Consider the equations $x'' + 4x = \sin t$ and $x'' + 4x = 4t$. Using Example 1.47 with $w = 2$, $a = 2$, $A = 1$, the general solution of the equation $x'' + 4x = \sin t$ is $x_1(t) = c_1 \cos 2t + c_2 \sin 2t + \frac{1}{3}\sin t - \frac{1}{6}\sin 2t$. A solution of the equation $x'' + 4x = 4t$ is $x_2(t) = t$. Using Theorem 1.16 (the principle of superposition), we find that the sum

$$x(t) = x_1(t) + x_2(t) = c_1 \cos 2t + c_2 \sin 2t + \frac{1}{3}\sin t - \frac{1}{6}\sin 2t + t$$

is the general solution for the given equation.

1.4.3 HOMOGENEOUS LINEAR EQUATIONS WITH CONSTANT COEFFICIENTS

Consider the differential equation

$$a_0 y^{(n)} + a_1 y^{(n-1)} + \cdots + a_n y = 0, \tag{1.171}$$

where $a_0, a_1, \ldots, a_n$ are real constants, $a_0 \neq 0$. To find the general solution of Eq. (1.171), the following steps are necessary:

1. Write the characteristic equation for Eq. (1.171),

$$a_0 \lambda^n + a_1 \lambda^{n-1} + \cdots + a_{n-1}\lambda + a_n = 0. \tag{1.172}$$

2. Find the roots $\lambda_1, \lambda_2, \ldots, \lambda_n$ of the characteristic equation (1.172).

3. According to the nature of the roots for the characteristic equation (1.172), write out the linearly independent particular solutions of Eq. (1.171), taking into account the following:

(a) Corresponding to each real single root λ of the characteristic equation (1.172), there is a particular solution $e^{\lambda x}$ of the differential equation (1.171).

(b) Corresponding to each single pair of complex conjugate roots $\lambda_1 = \alpha + i\beta$, $\lambda_2 = \alpha - i\beta$ of the characteristic equation (1.172), there are two linearly independent solutions $e^{\alpha x} \cos \beta x$ and $e^{\alpha x} \sin \beta x$ of the differential equation (1.171).

(c) Corresponding to each real root λ of multiplicity s of the characteristic equation (1.172), there are s linearly independent particular solutions $e^{\lambda x}, xe^{\lambda x}, x^2 e^{\lambda x}, \ldots, x^{s-1} e^{\lambda x}$ of the differential equation (1.171).

(d) Corresponding to each pair of complex conjugate roots $\lambda_1 = \alpha + i\beta$, $\lambda_2 = \alpha - i\beta$ of multiplicity s, there are $2s$ linearly independent particular solutions of the differential equation (1.171): $e^{\alpha x} \cos \beta x, xe^{\alpha x} \cos \beta x, \ldots, x^{s-1} e^{\alpha x} \cos \beta x$, and $e^{\alpha x} \sin \beta x, xe^{\alpha x} \sin \beta x, \ldots, x^{s-1} e^{\alpha x} \sin \beta x$. The number of particular solutions of the differential equation (1.171) is equal to the order of the equation. All the solutions constructed are linearly independent and make up the fundamental system of solutions of the differential Eq. (1.172).

EXAMPLE 1.50 Find the general solution of the equation $y''' - y'' - 2y' = 0$. ▲

Solution

The characteristic equation $\lambda^3 - \lambda^2 - 2\lambda = 0$, has the roots $\lambda_1 = 0$, $\lambda_2 = -1$, $\lambda_3 = 2$. Since they are real and distinct, the general solution is

$$y_{g,h}(x) = c_1 + c_2 e^{-x} + c_3 e^{2x}.$$

EXAMPLE 1.51 Find the general solution of the equation $y''' - 4y'' + 5y' - 2y = 0$. ▲

Solution

The characteristic equation $\lambda^3 - 4\lambda^2 + 5\lambda - 2 = 0$ has the roots $\lambda_1 = \lambda_2 = 1$; $\lambda_3 = 2$. The general solution is

$$y_{g,h}(x) = c_1 e^x + c_2 xe^x + c_3 e^{2x}.$$

EXAMPLE 1.52 Find the general solution of the equation $y''' + 2y'' + 2y' = 0$. ▲

Solution

The characteristic equation $\lambda^3 + 2\lambda^2 + 2\lambda = 0$ has the roots $\lambda_1 = 0$, $\lambda_2 = -1 - i$, $\lambda_3 = -1 + i$. The general solution is

$$y_{g,h} = c_1 + c_2 e^{-x} \cos x + c_3 e^{-x} \sin x.$$

EXAMPLE 1.53 Find the general solution of the equation $y^{(6)} + 8y^{(4)} + 16y'' = 0$. ▲

Solution

The characteristic equation $\lambda^6 + 8\lambda^4 + 16\lambda^2 = 0$ has the roots $\lambda_1 = \lambda_2 = 0$; $\lambda_3 = \lambda_4 = 2i$; $\lambda_5 = \lambda_6 = -2i$. The general solution is

$$y_{g,h}(x) = c_1 + c_2 x + c_3 \cos 2x + c_4 \sin 2x + c_5 x \sin 2x + c_6 x \cos 2x.$$

EXAMPLE 1.54 Differential Equation of Mechanical Vibrations
Find the general solution of the equation

$$y''(x) + p y'(x) + q y(x) = 0, \quad p > 0, \quad q > 0. \quad ▲$$

Solution

The characteristic equation $\lambda^2 + p\lambda + q = 0$ has the roots

$$\lambda_1 = -\frac{p}{2} + \sqrt{\frac{p^2}{4} - q}, \quad \lambda_1 = -\frac{p}{2} - \sqrt{\frac{p^2}{4} - q}.$$

(a) If $p^2/4 > q$, the general solution is $y(t) = c_1 e^{-\lambda_1 t} + c_2 e^{-\lambda_2 t}$ ($\lambda_1 < 0$, $\lambda_2 < 0$). From this formula it follows that $\lim_{t \to \infty} t \to \infty y(t) = 0$.
(b) Let $p^2/4 = q$. The general solution will be

$$y(t) = c_1 e^{-pt/2} + c_2 t e^{-pt/2} = (c_1 + c_2 t) e^{-pt/2}.$$

Here $y(t)$ also approaches zero as $t \to \infty$, but not so fast as in the preceding case (because of the factor $c_1 + c_2 t$).
(c) Let $p^2/4 < q$. The general solution is

$$y(t) = e^{-pt/2} \left[c_1 \cos \left(\sqrt{-\frac{p^2}{4} + q} \right) t + c_2 \sin \left(\sqrt{-\frac{p^2}{4} + q} \right) t \right],$$

or

$$y(t) = e^{-pt/2} (c_1 \cos \beta t + c_2 \sin \beta t); \quad \beta = \sqrt{-\frac{p^2}{4} + q}.$$

In the following formula, the arbitrary constants c_1 and c_2 will be replaced. Let us introduce the constants A and φ_0, connected with c_1 and c_2 by the relations $c_1 = A \sin \varphi_0$, $c_2 = A \cos \varphi_0$, where A and φ_0 are defined as follows in terms of c_1 and c_2: $A = \sqrt{c_1^2 + c_2^2}$,

$\varphi_0 = \arctan(c_1/c_2)$. Substituting the values of c_1 and c_2 in the formula of $y(t)$ yields $y(t) = Ae^{-pt/2} \sin(\beta t + \varphi_0)$.

1.4.4 NONHOMOGENEOUS LINEAR EQUATIONS WITH CONSTANT COEFFICIENTS

Consider the equation

$$a_0 y^{(n)} + a_1 y^{(n-1)} + \cdots + a_n y = f(x) \tag{1.173}$$

with $a_0, a_1, \ldots, a_n$ being real constant coefficients. From Theorem 1.18, the integration of Eq. (1.173) reduces to find one particular solution of this equation and to integrate the corresponding homogeneous linear equation. For the right-hand side $f(x)$ of Eq. (1.173), we suppose that has the following form:

$$f(x) = e^{\alpha x}[P_l(x) \cos \beta x + Q_m(x) \sin \beta x]. \tag{1.174}$$

Here $P_l(x)$ and $Q_m(x)$ are polynomials of degree l and m, respectively. In this case, a particular solution $y_{p,n}$ of Eq. (1.173) must be of the form

$$y_{p,n} = x^s e^{\alpha x}[\tilde{P}_k(x) \cos \beta x + \tilde{Q}_k(x) \sin \beta x], \tag{1.175}$$

where $k = \max(m, l)$, $\tilde{P}_k(x)$ and $\tilde{Q}_k(x)$ are polynomials of degree k of the general form with undetermined coefficients, and s is the multiplicity of the root $\lambda = \alpha + i\beta$ of the characteristic equation (if $\alpha \pm i\beta$ is not a root of the characteristic equation, then $s = 0$).

EXAMPLE 1.55 Find the general solution of the equation $y''' - y'' - 2y' = e^x$. ▲

Solution

The general solution of corresponding homogeneous equation is (see Example 1.50)

$$y_{g,h} = c_1 + c_2 e^{-x} + c_3 e^{2x}.$$

The right-hand side $f(x) = e^x$ is of the form of Eq. (1.174) with $\alpha = 1$, $\beta = 0$, $P_l(x) = 1$. But $\alpha \pm i\beta = 1$ is not a root of the characteristic equation. Hence, a particular solution $y_{p,n}$ of the given equation must be of the form $y_{p,n} = ae^x$, $a = $ constant. Then,

$$y'_{p,n} = ae^x, \quad y''_{p,n} = ae^x, \quad y'''_{p,n} = ae^x.$$

Substituting in the given equation yields $-2ae^x = e^x$, and hence $a = -\frac{1}{2}$. Consequently, the particular solution is $y_{p,n} = -\frac{1}{2}e^x$, and the general solution $y_{g,n}$ of the given equation is of the form

$$y_{g,n} = c_1 + c_2 e^{-x} + c_3 e^{2x} - \frac{1}{2}e^x.$$

EXAMPLE 1.56 Find the general solution of the equation

$$y''' - y'' - 2y' = 6x^2 + 2x. \quad ▲$$

Solution

The characteristic equation of the corresponding homogeneous equation has the roots $\lambda_1 = 0$, $\lambda_2 = -1$, $\lambda_3 = 2$, and the general solution

$$y_{g,b} = c_1 + c_2 e^{-x} + c_3 e^{2x}.$$

The right-hand side $f(x) = 6x^2 + 2x$ is of the form Eq. (1.174) with $\alpha = 0$, $\beta = 0$, $P_2(x) = 6x^2 + 2x$. Since $\alpha + i\beta = 0$ is a root of the characteristic equation, a particular solution must be of the form $y_{p,n} = x(ax^2 + bx + c) = ax^3 + bx^2 + cx$. If we substitute the expression for $y_{p,n}$ in the given equation, then $-6ax^2 + (-6a - 4b)x + (6a - 2b - 2c) = 6x^2 + 2x$, whence

$$\begin{cases} -6a = 6 \\ -6a - 4b = 2 \\ 6a - 2b - 2c = 0. \end{cases}$$

This system has the solution $a = -1$, $b = 1$, $c = -4$ and hence $y_{p,n} = -x^3 + x^2 - 4x$. The general solution of the given equation is

$$y_{g,n} = c_1 + c_2 e^{-x} + c_3 e^{2x} - x^3 + x^2 - 4x.$$

EXAMPLE 1.57 Find the general solution of the equation

$$y''' - y'' - 2y' = e^x + 6x^2 + 2x. \quad \blacktriangle$$

Solution

The general solution $y_{g,b}$ of the corresponding homogeneous equation is

$$y_{g,b} = c_1 + c_2 e^{-x} + c_3 e^{2x}.$$

To find the particular solution $y_{p,n}$ of the given equation, let us find the particular solutions of the two equations

$$y''' - y' - 2y'' = e^x \quad \text{and} \quad y''' - y'' - 2y' = 6x^2 + 2x.$$

The first has a particular solution $y_1 = -\frac{1}{2} e^x$ (see Example 1.55), and the second has a particular solution $y_2 = -x^3 + x^2 - 4x$ (see Example 1.56). Using the principle of superposition of solutions, the particular solution $y_{p,n}$ of the given equation is the sum of the particular solutions y_1 and y_2,

$$y_{p,n} = y_1 + y_2 = -\frac{1}{2} e^x - x^3 + x^2 - 4x,$$

and the general solution is

$$y_{g,n} = c_1 + c_2 e^{-x} + c_3 e^{2x} - \frac{1}{2} e^x - x^3 + x^2 - 4x.$$

EXAMPLE 1.58 Find the general solution of the equation

$$y^{(4)} + 8y'' + 16y = x \cos 2x. \quad \blacktriangle$$

Solution

The characteristic equation of the corresponding homogeneous equation is $\lambda^4 + 8\lambda^2 + 16 = 0$ and has the roots $\lambda_1 = \lambda_2 = 2i$, $\lambda_3 = \lambda_4 = -2i$. The general solution of the homogeneous equation is

$$y_{g,h} = c_1 \cos 2x + c_2 x \cos 2x + c_3 \sin 2x + c_4 x \sin 2x.$$

To find a particular solution for the nonhomogeneous equation, the procedure is as follows. Consider the equation $z^{(4)} + 8z'' + 16z = xe^{2ix}$. It can easily be seen that the right-hand side of the last equation is $x \cos 2x = Re(xe^{2ix})$. Then, $z_{p,n}$ of the last equation is found:

$$z_{p,n} = x^2(ax + b)e^{2ix} = (ax^3 + bx^2)e^{2ix}$$

$$z''_{p,n} = [6ax + 2b + 4i(3ax^2 + 2bx) - 4(ax^3 + bx^2)]e^{2ix}$$

$$z^{(4)}_{p,n} = [48ai - 24(6ax + 2b) - 32i(3ax^2 + 3bx) + 16(ax^3 + bx^2)]e^{2ix}.$$

Substituting into the equation and reducing e^{2ix} from both sides yields $-14(6ax + 2b) + 48ai = x$, whence $a = -\frac{1}{96}$, $b = -\frac{1}{64}i$, so that

$$z_{p,n} = \left(-\frac{1}{96}x^3 - \frac{i}{64}x^2\right)e^{2ix} = -\frac{1}{96}\left(x^3 + \frac{3}{2}ix^2\right)(\cos 2x + i \sin 2x)$$

$$z_{p,n} = -\frac{1}{96}\left(x^3 \cos 2x - \frac{3}{2}x^2 \sin 2x\right) - \frac{1}{96}i\left(x^3 \sin 2x + \frac{3}{2}x^2 \cos 2x\right).$$

Hence, $y_{p,n} = Re z_{p,n} = -\frac{1}{96}(x^3 \cos 2x - \frac{3}{2}x^2 \sin 2x)$. The general solution of the given equation is

$$y_{g,n} = y_{g,h} + y_{p,n}$$

$$y_{g,n} = (c_1 + c_2 x)\cos 2x + (c_3 + c_4 x)\sin 2x - \frac{1}{96}\left(x^3 \cos 2x - \frac{3}{2}x^2 \sin 2x\right).$$

1.4.5 LINEAR EQUATIONS THAT CAN BE REDUCED TO LINEAR EQUATIONS WITH CONSTANT COEFFICIENTS

(a) Euler's equations: Equations of the form

$$a_0(ax + b)^n y^{(n)} + a_1(ax + b)^{n-1} y^{(n-1)} + \cdots + a_{n-1}(ax + b)y' + a_n y = f(x),$$

$$(1.176)$$

where a_i and a, b are constants, are called *Euler's equations*. An Euler's equation can be transformed by changing the independent variable

$$ax + b = e^t (\text{or} \quad ax + b = -e^t, \quad \text{if} \quad ax + b < 0) \qquad (1.177)$$

into a linear equation with constant coefficients. Indeed, we can write that

$$\frac{dy}{dx} = \frac{dy}{dt}\left(\frac{dt}{dx}\right) = e^{-t}\frac{dy}{dt}, \frac{d^2y}{dx^2} = \frac{d}{dt}\left(e^{-t}\frac{dy}{dt}\right)\left(\frac{dt}{dx}\right) = e^{-2t}\left(\frac{d^2y}{dt^2} - \frac{dy}{dt}\right), \cdots,$$

$$\frac{d^ky}{dx^k} = e^{-kt}\left(\beta_1\frac{dy}{dt} + \beta_2\frac{d^2y}{dt^2} + \cdots + \beta_k\frac{d^ky}{dt^k}\right),$$

where β_i are constants; upon substitution into Eq. (1.177), it follows that the transformed equation will be a linear equation with constant coefficients.

(b) Consider the differential equation

$$y^{(n)} + p_1(x)y^{(n-1)} + p_2(x)y^{(n-2)} + \cdots + p_n(x)y = f(x). \tag{1.178}$$

If, by the substitution

$$t = \varphi(x), \tag{1.179}$$

Eq. (1.178) may be reduced to a linear equation with constant coefficients, that will be possible only if

$$t = c \int \sqrt[n]{p_n(x)}\, dx. \tag{1.180}$$

EXAMPLE 1.59 Find the general solution of the equation

$$x^2 y'' - 2xy' + 2y = x. \quad \blacktriangle$$

Solution

Using the substitution $x = e^t$, then $y' = e^{-t}\dfrac{dy}{dt}$, $y'' = e^{-2t}\left(\dfrac{d^2 y}{dt^2} - \dfrac{dy}{dt}\right)$ and the equation takes the form $\dfrac{d^2 y}{dt^2} - 3\dfrac{dy}{dt} + 2y = e^t$. The roots of the characteristic equation are $\lambda_1 = 1$, $\lambda_2 = 2$, and the general solution of the homogeneous equation is $y_{g,h}(t) = c_1 e^t + c_2 e^{2t}$. The right-hand side is $f(t) = e^t$, and since the number 1 is a root for the characteristic equation, the particular solution $y_{p,n}(t)$ of the nonhomogeneous equation must be of the form $y_{p,n}(t) = ate^t$. Substituting the expression for $y_{p,n}(t)$ yields $a = -1$, whence $y_{p,n}(t) = -te^t$. The general solution of the nonhomogeneous equation with constant coefficients is $y_{g,n}(t) = c_1 e^t + c_2 e^{2t} - te^t$. But since $x = e^t$,

$$y = c_1 x + c_2 x^2 - x \ln x$$

is the general solution for the initial equation.

EXAMPLE 1.60 *The Chebyshev Equation*
Find the general solution of the equation

$$(1 - x^2)y'' - xy' + n^2 y = 0, \quad x \in (-1, 1). \quad \blacktriangle \tag{1.181}$$

Solution

The equation is of the form

$$y'' - \frac{x}{1 - x^2}y' + \frac{n^2}{1 - x^2}y = 0.$$

Using Eq. (1.180) with $p_2(x) = n^2/(1 - x^2)$, yields

$$t = c \int \sqrt{\frac{n^2}{1 - x^2}}\, dx.$$

For $c = -1/n$, we obtain

$$t = \arccos x \quad \text{or,} \quad x = \cos t \quad \text{and} \quad y' = \frac{dy}{dx} = \frac{dy}{dt}\left(\frac{dt}{dx}\right) = -\frac{1}{\sin t}\left(\frac{dy}{dt}\right)$$

$$y'' = \frac{d}{dt}\left(-\frac{1}{\sin t}\frac{dy}{dt}\right)\left(\frac{dt}{dx}\right) = \frac{1}{\sin^2 t}\left(\frac{d^2 y}{dt^2}\right) - \frac{\cos t}{\sin^3 t}\left(\frac{dy}{dt}\right).$$

If we substitute into Eq. (1.181), then $(d^2 y/dt^2) + n^2 y = 0$ with the general solution $y(t) = c_1 \cos nt + c_2 \sin nt$, or $y = c_1 \cos(n \arccos x) + c_2 \sin(n \arccos x)$. The functions

$$T_n(x) = \cos(n \arccos x) \tag{1.182}$$

are called *Chebyshev polynomials.*

1.4.6 SECOND-ORDER LINEAR DIFFERENTIAL EQUATIONS
General Properties
The general form of a second-order nonhomogeneous linear equation is

$$a_0(x)y'' + a_1(x)y' + a_2(x)y = g(x), \quad a_0(x) \neq 0, \tag{1.183}$$

or, after division by $a_0(x)$,

$$y'' + p(x)y' + q(x)y = f(x). \tag{1.184}$$

The corresponding homogeneous linear equations are

$$a_0(x)y'' + a_1(x)y' + a_2(x)y = 0, \tag{1.185}$$

or

$$y'' + p(x)y' + .q(x)y = 0. \tag{1.186}$$

The Wronskian of functions y_1 and y_2 is

$$W(x) = W[y_1(x), y_2(x)] = \begin{vmatrix} y_1 & y_2 \\ y_1' & y_2' \end{vmatrix} = y_1 y_2' - y_1' y_2, \tag{1.187}$$

and Liouville's formula has the form

$$W(x) = W(x_0)e^{-\int_{x_0}^{x} p(x)dx}. \tag{1.188}$$

If y_1 is a particular solution of Eq. (1.186), to find the general solution reduces to integrating the functions. Indeed, a second particular solution of Eq. (1.186) is

$$y_2 = y_1 \int \frac{e^{-\int p(x)dx}}{y_1^2} dx, \tag{1.189}$$

and the general solution of the homogeneous Eq. (1.186) is

$$y = c_1 y_1 + c_2 y_1 \int \frac{e^{-\int p(x)dx}}{y_1^2} dx. \tag{1.190}$$

If $p(x)$ is differentiable, Eq. (1.186) can be transformed into another equation in which the first-derivative term does not occur. Indeed, if we set

$$y = \alpha(x)z; \quad \alpha(x) = e^{-\int \frac{p(x)}{2} dx};$$

(1.191)

the result of substituting in Eq. (1.184) is

$$z'' + \left(-\frac{p'(x)}{2} - \frac{p^2(x)}{4} + q(x) \right) z = 0,$$

(1.192)

where the function

$$Q(x) = -\frac{p'(x)}{2} - \frac{p^2(x)}{4} + q(x)$$

(1.193)

is called the *invariant of* Eq. (1.186). Equation (1.186) can be transformed into

$$\frac{d^2 y}{dt^2} + q(x)e^{2 \int p(x) dx} y = 0 \quad (x = x(t))$$

(1.194)

by the substitution of the independent variable

$$t = \int e^{-\int p(x) dx} dx.$$

(1.195)

Equations of the form

$$(a(x)y')' + b(x)y = 0, \quad a(x) > 0$$

(1.196)

are called *self-adjoint*. Equations (1.185) and (1.196) can be transformed into one another. Indeed, if we multiply Eq. (1.185) by the function

$$\mu(x) = \frac{1}{a_0(x)} e^{\int \frac{a_1(x)}{a_0(x)} dx},$$

(1.197)

we find

$$e^{\int \frac{a_1(x)}{a_0(x)} dx} y'' + \frac{a_1(x)}{a_0(x)} e^{\int \frac{a_1(x)}{a_0(x)} dx} y' + \frac{a_2(x)}{a_0(x)} e^{\int \frac{a_1(x)}{a_0(x)} dx} y = 0,$$

(1.198)

and if we set

$$a(x) = e^{\int \frac{a_1(x)}{a_0(x)} dx}, \quad b(x) = \frac{a_2(x)}{a_0(x)} e^{\int \frac{a_1(x)}{a_0(x)} dx},$$

(1.199)

then

$$a(x)y''(x) + a'(x)y'(x) + b(x)y = 0,$$

(1.200)

or

$$(a(x)y')' + b(x)y = 0.$$

(1.201)

Equation (1.185) is self-adjoint if

$$a_0'(x) = a_1(x),$$

(1.202)

that is, if it has the form

$$a_0(x)y'' + a_0'(x)y + b(x)y = 0.$$

(1.203)

Equation (1.192) is also self-adjoint. For the self-adjoint equation,

$$p(x)y'' + p'(x)y' + q(x)y = 0 \tag{1.204}$$

yields

$$p(x)W(x) = p(x_0)W(x_0) = \text{constant} \tag{1.205}$$

Indeed, after division by $p(x)$, this yields

$$y'' + \frac{p'(x)}{p(x)}y' + \frac{q(x)}{p(x)}y = 0$$

and the Liouville's formula (1.188) has the form

$$W(x) = W(x_0)e^{-\int_{x_0}^{x}\frac{p'(s)}{p(s)}ds} = W(x_0)e^{\ln\frac{p(x_0)}{p(x)}} = \frac{W(x_0)p(x_0)}{p(x)}. \tag{1.206}$$

The order of Eq. (1.184) can be reduced (see Eq. (1.114)), using the substitution

$$\frac{y'}{y} = z. \tag{1.207}$$

The result is the Riccati's equation

$$z' = -z^2 - p(x)z - q(x). \tag{1.208}$$

EXAMPLE 1.61   Find the general solution of the equation

$$(1 - x^2)y'' - 2xy' + 2y = 0, \quad x \in (-1, 1). \quad \blacktriangle$$

Solution

The function $y_1 = x$ is a particular solution. Then, a second particular solution is

$$y_2 = y_1 \int \frac{e^{\int -p(x)dx}}{y_1^2}dx = x \int \frac{e^{\int \frac{2x}{1-x^2}dx}}{x^2}dx = x \int \frac{dx}{(1-x^2)x^2}$$
$$= x\left(-\frac{1}{x} + \frac{1}{2}\ln\frac{1+x}{1-x}\right).$$

The general solution is [see Eq. (1.190)]

$$y = x\left[c_1 + c_2\left(-\frac{1}{x} + \frac{1}{2}\ln\frac{1+x}{1-x}\right)\right].$$

Remark 1.31

Only $y_1 = x$ is bounded for $x \to \pm 1$.

EXAMPLE 1.62

(a) Transform the Bessel equation

$$x^2 y'' + xy' + (x^2 - v^2)x = 0, \quad x > 0, \quad v \in \mathbf{R}, \tag{1.209}$$

into another equation in which the first-derivative term does not occur.

(b) Find the general solution of the equation $x^2 y'' + x y' + (x^2 - \frac{1}{4})y = 0$, $x > 0$. ▲

Solution

(a) Dividing by x^2 yields

$$y'' + \frac{1}{x} y' + \left(1 - \frac{v^2}{x^2}\right) y = 0.$$

Using Eq. (1.191) for $p = 1/x$, $q = 1 - (v^2/x^2)$, we find

$$y = e^{-\int \frac{1}{2x} dx} z = \frac{z}{\sqrt{x}}$$

$$Q(x) = -\frac{p'(x)}{2} - \frac{p^2(x)}{4} + q(x) = 1 + \frac{\frac{1}{4} - v^2}{x^2},$$

and the equation

$$z'' + \left(1 + \frac{\frac{1}{4} - v^2}{x^2}\right) z = 0.$$

(b) If $v = \frac{1}{2}$, the transformed equation is $z'' + z = 0$ with the general solution $z = c_1 \sin x + c_2 \cos x$, or $y = c_1((\sin x)/\sqrt{x}) + c_2((\cos x)/\sqrt{x})$.

Remark 1.32

Multiplying the solutions $y_1 = (\sin x)/\sqrt{x}$ and $y_2 = (\cos x)/\sqrt{x}$ by $\sqrt{2/\pi}$ yields

$$y(x) = c_1 \sqrt{\frac{2}{\pi x}} \sin x + c_2 \sqrt{\frac{2}{\pi x}} \cos x.$$

The functions

$$J_{1/2}(x) = \sqrt{\frac{2}{\pi x}} \sin x; \quad J_{-1/2}(x) = \sqrt{\frac{2}{\pi x}} \cos x$$

are called *Bessel functions* with $v = \frac{1}{2}$ and $v = -\frac{1}{2}$.

EXAMPLE 1.63

(a) By the substitution of the independent variable, transform the equation

$$(1 - x^2)y'' + 2xy' + (1 - x^2)^3 y = 0, \quad x \in (-1, 1)$$

into another equation in which the first-derivative term does not occur.

(b) Find the general solution for the given equation. ▲

Solution

(a) Using the substitution Eq. (1.195) yields

$$t = \int e^{-\int \frac{2x}{1-x^2}dx}\,dx = \int e^{\ln|x^2-1|}\,dx = \int (1-x^2)\,dx = x - \frac{x^3}{3},$$

$$y' = \frac{dy}{dx} = \frac{dy}{dt}\left(\frac{dt}{dx}\right) = (1-x^2)\frac{dy}{dt},$$

$$y'' = \frac{d}{dx}\left[(1-x^2)\frac{dy}{dt}\right] = -2x\frac{dy}{dt} + (1-x^2)^2\frac{d^2y}{dt^2}.$$

Substituting into the given equation, we find

$$(1-x^2)^3\frac{d^2y}{dt^2} + (1-x^2)^3 y = 0, \quad \text{or} \quad \frac{d^2y}{dt^2} + y = 0.$$

(b) The general solution of the last equation is

$$y = c_1\cos t + c_2\sin t,$$

or

$$y = c_1\cos\left(x - \frac{x^3}{3}\right) + c_2\sin\left(x - \frac{x^3}{3}\right).$$

EXAMPLE 1.64 The *Legendre equation*

$$(1-x^2)y'' - 2xy' + n(n+1)y = 0 \qquad (1.210)$$

is a self-adjoint equation. ▲

EXAMPLE 1.65 Transform the Bessel equation $x^2y'' + xy' + (x^2 - v^2)y = 0$ into a self-adjoint equation. ▲

Solution

Dividing by x gives

$$xy'' + y' + \left(x - \frac{v^2}{x}\right)y = 0 \quad \text{or,} \quad (xy')' + \left(x - \frac{v^2}{x}\right)y = 0,$$

which is a self-adjoint equation.

EXAMPLE 1.66 Transform the Chebyshev equation

$$(1-x^2)y'' - xy' + n^2y = 0, \quad x \in (-1, 1),$$

into a self-adjoint equation. ▲

Solution

Multiplying the given equation by the function

$$\mu(x) = \frac{1}{1-x^2}e^{\int \frac{-x}{1-x^2}dx} = \frac{1}{\sqrt{1-x^2}}$$

yields

$$\sqrt{1-x^2}\,y'' - \frac{x}{\sqrt{1-x^2}}\,y' + \frac{n^2}{\sqrt{1-x^2}}\,y = 0$$

or

$$(\sqrt{1-x^2}(y'))' + \frac{n^2}{\sqrt{1-x^2}}(y) = 0,$$

which is a self-adjoint equation.

Zeros of Solutions for Second-Order Linear Differential Equations

Let us consider the equation

$$y'' + p(x)y' + q(x)y = 0. \tag{1.211}$$

(a) If $y(x)$ is a particular solution of Eq. (1.211), then the roots (zeros) of $y(x) = 0$ are simple and isolated.

(b) If two solutions of Eq. (1.211) have in common a zero, then those solutions are linearly dependent and have all zeros in common.

(c) *Sturm's theorem*: If y_1 and y_2 are two particular linear independent solutions of Eq. (1.211), then their zeros are mutually separated.

(d) *The maximum principle*: If $q(x) < 0$, $\forall x \in (a, b)$, and y is a particular solution of Eq. (1.211), then it does not touch the positive maximal points and the negative points on (a, b).

(e) Is the following corollary:

COROLLARY 1.2 *Bilocal Problem*

If $q(x) < 0$, $\forall x \in (a, b)$, then the problem

$$\begin{cases} y'' + p(x)y' + q(x)y = f(x) \\[2mm] x(a) = x(b) = 0, \end{cases} \qquad p, q, f \in C[a, b],$$

has at most one solution. ▲

(f) *Sturm's comparison theorem*: If $Q_1(x) \le Q_2(x)$, $\forall x \in [a, b]$, Q_1, $Q_2 \in C[a, b]$ then, between two consecutive zeros of the equation $y'' + Q_1(x)y = 0$, there is at least one zero of the equation

$$y'' + Q_2(x)y = 0.$$

COROLLARY 1.3

1. The distance between two consecutive zeros t_1 and t_2 of any solution of the equation

$$y'' + w^2 y = 0 \tag{1.212}$$

is

$$t_2 - t_1 = \frac{\pi}{w}.\tag{1.213}$$

2. *Comparison equations:* Consider the equation

$$x'' + Q(t)x = 0, \quad 0 < m \le Q(t) \le M, \quad t \in [a, b], \tag{1.214}$$

and the comparison equations

$$\begin{cases} y'' + my = 0 \\ z'' + Mz = 0. \end{cases}$$

If x is a solution of Eq. (1.214) and t_1, t_2 are two consecutive zeros of this equation, then

$$\frac{\pi}{\sqrt{M}} \le t_2 - t_1 \le \frac{\pi}{\sqrt{m}}. \quad \blacktriangle \tag{1.215}$$

EXAMPLE 1.67 Consider the Bessel equation $x^2 y'' + xy' + (x^2 - v^2)x = 0$ or, transformed into an equation in which the first-derivative term does not occur,

$$z'' + \left(1 + \frac{\frac{1}{4} - v^2}{x^2}\right) z = 0.$$

If we compare with $z'' + z = 0$, the distance between two consecutive zeros is

$$t_2 - t_1 > \pi \quad \text{if } v > \tfrac{1}{2} \qquad \text{and} \qquad t_2 - t_1 < \pi \quad \text{if } v < \tfrac{1}{2}.$$

Boundary-Value Problems

Consider the following problem: Find the solution of the equation

$$\frac{d}{dx}(p(x)y') + q(x)y = f(x), \quad x \in [a, b], \tag{1.216}$$

with the conditions

$$a_0 y(a) + a_1 y'(a) = 0, \quad b_0 y(b) + b_1 y'(b) = 0. \tag{1.217}$$

Suppose that the real-valued function $p(x)$ has a continuous derivative in the interval $[a, b]$, $p(x) > 0$ for $a \le x \le b$, the real-valued function $q(x)$ is continuous in the interval $[a, b]$, a_0, a_1, b_0, b_1 are real constants, a_0, a_1 are not both zero, and b_0, b_1 are not both zero. This problem is called a *boundary-value problem*. If $f(x) \equiv 0$, then it is considered a *homogeneous boundary-value problem*. Generally, boundary-value problems are not always solvable, that is, sometimes there are no solutions, and if solutions exist, there may be several or even an infinity of solutions. The problem consisting of Eqs. (1.216) and (1.217) has at most one solution if the corresponding homogeneous boundary-value problem has only one zero solution $y(x) \equiv 0$. Indeed, if one assumes the existence of two different solutions $y_1(x)$ and $y_2(x)$ for the boundary-value problem Eqs. (1.216) and (1.217), then the function $y(x) = y_1(x) - y_2(x)$ is a solution for the *homogeneous boundary-value problem*. In solving homogeneous boundary-value problems, proceed as follows. First, find the general solution of the given differential equation $y = c_1 y_1(x) + c_1 y_2(x)$, $y_1(x)$ and $y_2(x)$ being linearly independent solutions. Then, require that this solution $y(x)$ should satisfy the given conditions (1.217). This leads to some linear system of equations that is used to find c_1, c_2. By solving this system, if possible, the solution will be found for the given boundary-value problem.

EXAMPLE 1.68 Solve the problem

$$y'' - y' = 0, \quad y(0) = 0, \quad y(1) = 0. \quad \blacktriangle$$

Solution

The general solution is $y(x) = c_1 e^x + c_2 e^{-x}$. If we set $y(0) = 0$, $y(1) = 0$, then

$$\begin{cases} c_1 + c_2 = 0 \\ c_1 e + c_2 e^{-1} = 0, \end{cases} \tag{1.218}$$

and $c_1 = c_2 = 0$. Only $y(x) = 0$ is solution for the given problem.

EXAMPLE 1.69 Solve the following boundary-value problem:

$$y'' + y' = 0; \quad y(0) = 0, \quad y(\pi) = 0. \quad \blacktriangle$$

Solution

The general solution of the differential equation is

$$y(x) = c_1 \cos x + c_2 \sin x.$$

Setting $y(0) = 0$, $y(\pi) = 0$ yields

$$\begin{cases} c_1 = 0 \\ c_2 \cdot \sin \pi = 0, \end{cases}$$

which is satisfied for an arbitrary c_2. All the functions $y = c_2 \sin x$ are solutions for given boundary-value problem.

EXAMPLE 1.70 Solve the boundary-value problem

$$y'' + y = 0; \quad y(0) - y(2\pi) = 0, \quad y'(0) - y'(2\pi) = 0. \quad \blacktriangle$$

Solution

The general solution of the differential equation is $y(x) = c_1 \cos x + c_2 \sin x$. The boundary conditions involved are $c_1 - c_1 = 0$, $c_2 - c_2 = 0$ and are satisfied anyway. All the functions $y = c_1 \cos x + c_2 \sin x$ are solutions for the given problem.

EXAMPLE 1.71 Solve the problem $y'' + y = 1; \ y(0) = 0, \ y'(\pi) = 0. \quad \blacktriangle$

Solution

The general solution of the differential equation is

$$y(x) = c_1 \cos x + c_2 \sin x + 1.$$

From the boundary conditions, $c_1 = -1$, $c_2 = 0$. Only the function $y = 1 - \cos x$ is a solution for the given problem.

Green's Function for the Boundary-Value Problem

It was presumed that there exists no nontrivial solution $y(x)$ for the homogeneous equation

$$\frac{d}{dx}(p(x)y') + q(x)y = 0 \tag{1.219}$$

satisfying the boundary conditions (1.217). This condition guarantees the existence and uniqueness of a solution for the boundary-value problem consisting of Eqs. (1.216) and (1.217). Consider that $y_1(x)$ is the solution of Eq. (1.219) with the first condition (1.217) and $y_2(x)$ is the solution of Eq. (1.219) with the second boundary condition (1.217). The functions y_1 and y_2 are linearly independent. The solution for the nonhomogeneous boundary problem consisting of Eqs. (1.216) and (1.217) must be of the form

$$y = c_1(x)y_1(x) + c_2(x)y_2(x), \tag{1.220}$$

with functions $c_1(x)$, $c_2(x)$ solutions of the system

$$\begin{cases} c_1'(x)y_1(x) + c_2'(x)y_2(x) = 0 \\ c_1'(x)y_1'(x) + c_2'(x)y_2'(x) = \dfrac{f(x)}{p(x)}. \end{cases} \tag{1.221}$$

Then

$$\begin{cases} c_1'(x) = -\dfrac{f(x)y_2(x)}{W(x)p(x)} \\ c_2'(x) = \dfrac{f(x)y_1(x)}{W(x)p(x)}, \end{cases} \tag{1.222}$$

and using Eq. (1.206) yields

$$\begin{cases} c_1(x) = -\dfrac{1}{W(a)p(a)}\displaystyle\int_a^x y_2(s)f(s)ds + c_1 \\[2mm] \quad\quad = \dfrac{1}{W(a)p(a)}\displaystyle\int_x^b y_2(t)f(t)dt + c_1 \\[2mm] c_2(x) = \dfrac{1}{W(a)p(a)}\displaystyle\int_a^x y_1(s)f(s)ds + c_2, \end{cases} \tag{1.223}$$

with c_1, c_2 constants. Involving the boundary conditions gives $c_1(b) = 0$, $c_2(a) = 0$ or $c_1 = 0$, $c_2 = 0$, and hence

$$\begin{cases} c_1(x) = \dfrac{1}{W(a)p(a)} \displaystyle\int_x^b y_2(t)f(t)dt \\[3mm] c_2(x) = \dfrac{1}{W(a)p(a)} \displaystyle\int_a^x y_1(t)f(t)dt. \end{cases} \tag{1.224}$$

If we substitute Eq. (1.224) into Eq. (1.220), then

$$y(x) = \frac{1}{W(a)p(a)}\left[\int_a^x y_2(x)y_1(t)f(t)dt + \int_x^b y_1(x)y_2(t)f(t)dt\right]. \tag{1.225}$$

If we denote by

$$G(x,t) = \frac{1}{W(a)p(a)}\begin{cases} y_1(t)y_2(x), & a \le t < x \\ y_1(x)y_2(t), & x \le t \le b, \end{cases} \tag{1.226}$$

and substitute into Eq. (1.225), the solution of the nonhomogeneous boundary-value problem of Eqs. (1.216) and (1.217) must be of the form

$$y(x) = \int_a^b G(x,t)f(t)dt. \tag{1.227}$$

The function $G(x,t)$ is called *Green's function* and has the following properties:

1. $G(x,t)$ is continuous with respect to x for fixed t and $a \le x \le b$, $a < t < b$.
2. $G(t,x) = G(x,t)$ ($G(x,t)$ is symmetrical).
3. $[G'(t+0,t) - G'(t-0,t)] = \dfrac{1}{p(t)}$.
4. For $x \ne t$, $G(x,t)$ satisfies the differential equation $\dfrac{d}{dx}(p(x)y')+q(x)y=0$ and the boundary conditions

$$\begin{cases} a_0 G(a,t) + a_1 \dfrac{\partial G}{\partial x}(a,t) = 0 \\[3mm] b_0 G(b,t) + b_1 \dfrac{\partial G}{\partial x}(b,t) = 0. \end{cases} \tag{1.228}$$

THEOREM 1.19 If the homogeneous boundary-value problem has only the trivial solution, then the Green's function exists and is unique, and the solution for the nonhomogeneous boundary value problem is Eq. (1.227). ▲

EXAMPLE 1.72 Find the Green's function of the boundary-value problem

$$y'' = f(x), \quad y(0) - y'(0) = 0, \quad y(1) - y'(1) = 0. \quad ▲ \tag{1.229}$$

Solution

The general solution of the homogeneous differential equation $y'' = 0$ is $y(x) = c_1 x + c_2$. The solution of the homogeneous equation that satisfies the first condition is $y_1(x) = c_1(x + 1)$. Indeed,

$$y(x) = c_1 x + c_2, \quad y'(x) = c_1; \quad y(0) - y'(0) = c_2 - c_1 \Rightarrow c_2 = c_1.$$

The solution of the homogeneous equation that satisfies the second condition is $y_2 = c_1 x$. Setting $c_1 = 1$, then $y_1 = x + 1$, $y_2 = x$;

$$W(x) = \begin{vmatrix} x+1 & x \\ 1 & 1 \end{vmatrix} = 1;$$

$W(0)p(0) = 1$. Using Eq. (1.226) yields

$$G(x, t) = \begin{cases} x(t+1), & 0 \le t < x \\ (x+1)t, & x \le t \le 1, \end{cases} \tag{1.230}$$

and the solution is

$$y(x) = \int_0^x x(t+1)f(t)dt + \int_x^1 (x+1)tf(t)dt. \tag{1.231}$$

EXAMPLE 1.73 Find the Green's function of the boundary-value problem

$$y'' + w^2 y = f(x); \quad y(0) = 0, \quad y(1) = 0. \quad \blacktriangle \tag{1.232}$$

Solution

The general solution of the homogeneous differential equation is

$$y(x) = c_1 \cos wx + c_2 \sin wx.$$

The solution of the homogeneous equation with the condition $y(0) = 0$ is $y_1(x) = c_2 \sin wx$. The solution of equation $y'' + w^2 y = 0$ with the condition $y(1) = 0$ is $y_2(x) = c_2(\sin wx - \tan w \cos wx)$. Indeed, $y(1) = 0 \Leftrightarrow c_1 \cos w + c_2 \sin w = 0 \Rightarrow c_1/c_2 = -\tan w \Rightarrow c_1 = -c_2 \tan w$ and $y_2(x) = c_2(\sin wx - \tan w \cos wx)$. Setting $c_2 = 1$, yields $y_1(x) = \sin wx$; $y_2(x) = \sin wx - \tan w \cos wx$; $W(x) = w \tan w$; $W(0)p(0) = w \tan w$:

$$G(x, t) = \frac{\cos w}{w \sin w} \begin{cases} \sin wt[\sin wx - \tan w \cos wx] & 0 \le t < x \\ \sin wx[\sin wt - \tan w \cos wt] & x \le t \le 1. \end{cases}$$

Hence,

$$G(x, t) = \frac{1}{w \sin w} \begin{cases} \sin wt \sin(x-1)w & 0 \le t < x \\ \sin wx \sin(t-1)w & x \le t \le 1, \end{cases} \tag{1.233}$$

and the solution of problem Eq. (1.232) is

$$y(x) = \frac{1}{w \sin w} \left[\int_0^x \sin wt \sin(x-1)wdt + \int_x^1 \sin wx \sin(t-1)wdt \right]. \tag{1.234}$$

EXAMPLE 1.74 Find the Green's function of the boundary-value problem

$$y'' - w^2 y = f(x); \quad y(-1) = y(1); \quad y'(-1) = y'(1). \quad \blacktriangle$$

Solution

The general solution of time homogeneous differential equation is

$$y(x) = c_1 e^{wx} + c_2 e^{-wx}.$$

The condition $y(-1) = y(1)$ gives $c_2 = c_1$, so that the solution of the homogeneous equation with this condition is $y_1(x) = c_1(e^{wx} + e^{-wx}) = 2c_1 \cosh wx$. For $c_1 = \frac{1}{2}$, $y_1(x) = \cosh wx$. The condition $y'(-1) = y'(1)$ gives $c_2 = -c_1$ and $y_2(x) = 2c_1 \sinh wx$, and setting $c_1 = \frac{1}{2}$, $y_2(x) = \sinh wx$,

$$W(x) = \begin{vmatrix} \cosh wx & \sinh wx \\ w \sinh wx & w \cosh wx \end{vmatrix} = w$$

$$G(x, t) = \frac{1}{w} \begin{cases} \cosh wt \cdot \sinh wx, & -1 \le t < x \\ \cosh wx \cdot \sinh wt, & x \le t \le 1. \end{cases} \tag{1.235}$$

The solution of the given problem is

$$y(x) = \frac{1}{w} \left[\int_{-1}^{1} \sinh wx \cosh wt f(t) dt + \int_{x}^{1} \cosh wx \sinh wt f(t) dt \right]. \tag{1.236}$$

Integration of Differential Equations by Means of Series

The problem of integrating the homogeneous linear equation of order n

$$p_0(x) y^{(n)} + p_1(x) y^{(n-1)} + \cdots + p_n(x) y = 0 \tag{1.237}$$

reduces to choosing n or at least $(n-1)$ linearly independent particular solutions. In more involved cases, the particular solutions are in the form of a sum of a certain series, especially often in the form of the sum of a power series or a generalized power series. The conditions under which there are solutions in the form of the sum of power series or a generalized power series are ordinarily established by the methods of functions theory. The function $\varphi(x)$ is said to be *holomorphic* (or *analytic*) in some neighborhood $|x - x_0| < \rho$ of the point $x = x_0$ if it can be represented in that neighborhood by the power series

$$\varphi(x) = \sum_{k=0}^{n} c_k (x - x_0)^k, \tag{1.238}$$

converging in the domain $|x - x_0| < \rho$. Similarly, the function $\varphi(x_1, x_2, \ldots, x_n)$ is said to be *holomorphic over all its independent variables* in some neighborhood $|x_k - x_k^{(0)}| < \rho_k$ $(k = 1, 2, \ldots, n)$ of the point $(x_1^{(0)}, x_2^{(0)}, \ldots, x_n^{(0)})$ if it can be represented by the power series

$$\varphi(x_1, x_2, \ldots, x_n) = \sum_{k=0}^{n} c_{k_1, k_2, \ldots, k_n} (x_1 - x_1^{(0)})^{k_1} (x_2 - x_2^{(0)})^{k_2} \cdots (x_n - x_n^{(0)})^{k_n}, \tag{1.239}$$

converging in the domain $|x_k - x_k^{(0)}| < \rho_k$ $(k = 1, 2, \ldots, n)$.

THEOREM 1.20

(a) If $p_0(x)$, $p_1(x)$, $p_2(x)$ are analytical functions of x in the neighborhood of the point $x = x_0$ and $p_0(x) \neq 0$, then the solutions of the equation

$$p_0(x)y'' + p_1(x)y' + p_2(x)y = 0 \qquad (1.240)$$

are also analytical functions in a certain neighborhood of the same point, and hence, the solution of Eq. (1.240) may be in the form

$$y = a_0 + a_1(x - x_0) + a_2(x - x_0)^2 + \cdots + a_n(x - x_0)^n + \cdots . \qquad (1.241)$$

(b) If Eq. (1.240) satisfies the conditions of the previous theorem, but $x = x_0$ is a zero of finite order s for the function $p_0(x)$, a zero of $(s - 1)$ order or higher for the function $p_1(x)$ (if $s > 1$), and a zero of order not lower than $(s - 2)$ for time coefficient of $p_2(x)$ (if $s > 2$), then there is at least one nontrivial solution of Eq. (1.240) in the sum form of the generalized power series

$$y = a_0(x - x_0)^k + a_1(x - x_0)^{k+1} + a_2(x - x_0)^{k+3} + \cdots$$
$$+ a_n(x - x_0)^{k+n} + \cdots , \qquad (1.242)$$

where k is a real number that may be either positive or negative. ▲

Particular Case

Let us consider the second-order differential equation

$$y'' + p(x)y' - q(x)y = 0. \qquad (1.243)$$

If the coefficients $p(x)$ and $q(x)$ of the equation can be represented as

$$p(x) = \frac{\sum_{k=0}^{\infty} a_k x^k}{x}, \quad q(x) = \frac{\sum_{k=0}^{\infty} b_k x^k}{x^2}, \qquad (1.244)$$

where the series in the numerators converge in a domain $|x| < R$ and the coefficients a_0 and b_0 are not simultaneously, zero then then Eq. (1.243) has at least one solution in the form of a generalized power series

$$y = x^v \sum_{k=0}^{\infty} c_k x^k \quad (c_0 \neq 0), \qquad (1.245)$$

which converges at least in the same domain $|x| < R$. In order to find the exponent v and the coefficients c_k, it is necessary to substitute series from Eq. (1.245) in Eq. (1.243), cancel x^v, and equate the coefficients of all powers of x to zero (the method of undetermined coefficients). Here the number v is found from the governing equation

$$v(v - 1) + a_0 v + b_0 = 0, \qquad (1.246)$$

where

$$a_0 = \lim_{x \to 0} x p(x), \quad b_0 = \lim_{x \to 0} x^2 q(x). \qquad (1.247)$$

Consider that v_1 and v_2 are the roots of the governing equation (1.246). Then there are three cases.

1. If the difference $v_1 - v_2$ is not equal to an integer or zero, then it is possible to develop two solutions of the form of Eq. (1.245),

$$y_1 = x^{v_1} \sum_{k=0}^{\infty} c_k x^k \quad (c_0 \neq 0), \quad y_2 = x^{v_2} \sum_{k=0}^{\infty} A_k x^k \quad (A_0 \neq 0).$$

2. If the difference $v_1 - v_2$ is a positive integer number, then it is possible in general to develop only one series [the solution of Eq. (1.243)],

$$y_1 = x^{v_1} \sum_{k=0}^{\infty} c_k x^k.$$

3. If Eq. (1.246) has a multiple root $v_1 = v_2$, then it is possible to construct only one series, the solution (1.245). In the first case, the constructed solutions $y_1(x)$ and $y_2(x)$ will surely be linearly independent (i.e., their ratio will not be constant). In the second and the third case, only one solution was constructed for each case. Note that if the difference $v_1 - v_2$ is a positive integer number or zero, then besides solution (1.245), Eq. (1.245) will have a solution of the form

$$y_2 = A y_1(x) \ln x + x^{v_2} \sum_{k=0}^{\infty} A_k x^k. \tag{1.248}$$

In this case, $y_2(x)$ contains an extra term of the form $A y_1(x) \ln x$, where $y_1(x)$ is given in the form of Eq. (1.248). The constant A in Eq. (1.248) may turn out to be zero, and then we will find an expression in the form of a generalized power series for y_2. We will mention another method of integrating differential equations in series, found to be easier when applied to nonlinear differential equations. Suppose that we are given the differential equation

$$y^{(n)} = f(x, y, y', \ldots, y^{(n-1)}), \tag{1.249}$$

and the initial conditions

$$y(x_0) = y_0, \quad y'(x_0) = y_0', \ldots, y^{(n-1)}(x_0) = y_0^{(n-1)}. \tag{1.250}$$

THEOREM 1.21 If the right-hand side of Eq. (1.249) is holomorphic over all its independent variables $x, y, y', y'', \ldots, y^{(n-1)}$ in a neighborhood Ω, $|x - x_0| < R$, $|y - y_0| < R_1$, $|y' - y_0'| < R_1, \ldots, |y^{(n-1)} - y_0^{(n-1)}| < R_1$, of the point

$$(x_0, y_0, y_0', \ldots, y_0^{(n-1)}),$$

then Eq. (1.249) has the unique solution

$$y(x) = y_0 + y_0'(x - x_0) + \frac{y_0'}{2!}(x - x_0)^2 + \cdots + \frac{y_0^{(n-1)}}{(n-1)!}(x - x_0)^{n-1}$$
$$+ \sum_{k=n}^{\infty} a_k(x - x_0)^k, \quad \left(a_k = \frac{y_0^{(k)}}{k!}\right),$$

(1.251)

satisfying the initial conditions (1.250) and being holomorphic in some neighborhood of the point $x = x_0$. Series (1.252) converges in the domain $|x - x_0| < \rho$, where $\rho = a[1 - e^{-b/(n+1)aM}]$. a and b are constants, satisfying the conditions $0 < a < R$, $0 < b < R$, and $M = \max_\Omega |f(x, y, y', \ldots, y^{(n-1)})|$. ▲

EXAMPLE 1.75 Find the solution of the equation $(x - 1)y'' - xy' + y = 0$ in the form of the power series. ▲

Solution

Consider that $y(x)$ is in the form of series $y(x) = \sum_{k=0}^{\infty} c_k x^k$. Then $y'(x) = \sum_{k=1}^{\infty} kc_k x^{k-1}$, $y'' = \sum_{k=2}^{\infty} k(k-1)c_k x^{k-2}$. Substituting $y(x)$, $y'(x)$, and $y''(x)$ in the given equation yields

$$\sum_{k=2}^{\infty} k(k-1)c_k x^{k-1} - \sum_{k=2}^{\infty} k(k-1)c_k x^{k-2} - \sum_{k=1}^{\infty} kc_k x^k + \sum_{k=0}^{\infty} c_k x^k = 0.$$

Gathering together the similar terms and equating the coefficients of all powers of x to zero, we obtain relations from which are found the coefficients $c_0, c_1, \ldots, c_n, \ldots$. Thus, $x^0 | c_0 - 2(1)(c_2) = 0$, $x^1 | c_1 - 1(c_1) + 2(1)(c_2) - 3(2)(c_3) = 0$, $x^2 | c_2 - 2(c_2) + 3(2)(c_3) - 4(3)(c_4) = 0, \ldots$, $x^k | c_k - kc_k + (k+1)(k)(c_{k+1}) - (k+2)(k+1)c_{k+2} = 0$. Choosing $c_0 = 1$ and $c_1 = 1$ yields

$$c_2 = \frac{1}{1(2)}, c_3 = \frac{1}{1(2)(3)}, \ldots, c_k = \frac{1}{k!}, \ldots,$$

and consequently,

$$y_1(x) = 1 + \frac{x}{1!} + \frac{x^2}{2!} + \cdots + \frac{x^n}{n!} + \cdots = e^x.$$

If we choose $c_0 = 0$ and $c_1 = 1$, then $c_2 = 0$, $c_3 = 0, \ldots, c_n = 0$, $n \geq 2$, and the solution is $y_2(x) = x$. The choice $c_0 = 1$, $c_1 = 1$ is equivalent to the initial conditions $y_1(0) = 1$, $y_1'(0) = 1$. The choice $c_0 = 0$, $c_1 = 1$ is equivalent to the initial conditions $y_2(0) = 0$, $y_2'(0) = 1$. Any solution of the given equation will be a linear combination of solutions $y_1(x)$ and $y_2(x)$.

EXAMPLE 1.76   *Legendre's Equation*
Find the solution of equation

$$(1 - x^2)y'' - 2xy' + n(n + 1)y = 0, \quad n = \text{constant},$$

(1.252)

in the form of power series. ▲

Solution

If the equation is divided by $1 - x^2$, so that the coefficient of y'' is 1, we obtain

$$p(x) = \frac{-2x}{1 - x^2}, \quad q(x) = \frac{n(n + 1)}{1 - x^2}.$$

Since these coefficients have power-series expansions valid for $|x| < 1$, the solution y also has a power-series expansion for $|x| < 1$. Let us assume that

$$y = \sum_{k=0}^{\infty} c_k x^k \quad \text{and yields} \quad y' = \sum_{k=1}^{\infty} k c_k x^{k-1}, \quad y'' = \sum_{k=2}^{\infty} k(k-1) c_k x^{k-2}.$$

Substituting in the given equation and considering the coefficient of x^k yields

$$c_{k+2}(k+2)(k+1) = c_k[k(k+1) - n(n+1)].$$

For $k \geq 0$ results

$$c_{k+2} = c_k \frac{(k-n)(k+n+1)}{(k+1)(k+2)}.$$

The coefficients for even k are determined from c_0, and those for odd n are determined from c_1. Computing the coefficients, gives as the final result

$$y = c_0 \left[1 - \frac{n(n+1)}{2!} x^2 + \frac{(n-2)n(n+1)(n+3)}{4!} x^4 \right.$$
$$\left. - \frac{(n-4)(n-2)n(n+1)(n+3)(n+5)}{6!} x^6 + \cdots \right]$$
$$+ c_1 \left[x - \frac{(n-1)(n+2)}{3!} x^3 + \frac{(n-3)(n-1)(n+2)(n+4)}{5!} x^5 - \cdots \right].$$

If n is a positive integer, either the coefficient of c_0 or the coefficient of c_1 reduces to a polynomial, depending on whether n is even or odd. Choosing c_0 and c_1 so that the polynomials have the value 1 when $x = 1$ yields the Legendre polynomials $P_n(x)$,

$$1, \quad x, \quad \frac{3}{2}x^2 - \frac{1}{2}, \quad \frac{5}{2}x^3 - \frac{3}{2}x, \quad \frac{35}{8}x^4 - \frac{15}{4}x^2 + \frac{3}{8}, \ldots.$$

EXAMPLE 1.77 *Bessel's Equation*

Solve the Bessel equation

$$x^2 y'' + xy'(x^2 - v^2)y = 0, \quad x > 0, \tag{1.253}$$

with v a given constant. ▲

Solution

The Bessel equation can be rewritten as

$$y'' + \frac{1}{x} y' + \frac{x^2 - v^2}{x^2} y = 0.$$

Then $p(x) = 1/x$, $q(x) = (x^2 - v^2)/x^2$, so that

$$a_0 = \lim_{x \to 0} xp(x) = 1, \quad b_0 = \lim_{x \to 0} x^2 q(x) = -v^2.$$

The governing equation (1.246) for ρ is

$$\rho(\rho - 1) + 1(\rho) - v^2 = 0 \quad \text{or} \quad \rho^2 - v^2 = 0, \quad \text{whence} \quad \rho_1 = v, \quad \rho_2 = -v.$$

The first particular solution of the Bessel equation must be of the form of the generalized power series $y = x^v \sum_{k=0}^{\infty} c_k x^k$. Substituting y, y', and y'' in this equation gives

$$x^2 \sum_{k=0}^{\infty} c_k(k + v)(k + v - 1)x^{k+v-2} + x \sum_{k=0}^{\infty} c_k(k + v)x^{k+v-1}$$

$$+ (x^2 - v^2) \sum_{k=0}^{\infty} c_k x^{k+v} = 0,$$

or, after some simple transformations and canceling x^v,

$$\sum_{k=0}^{\infty} [(k + v)^2 - v^2]c_k x^k + \sum_{k=0}^{\infty} c_k x^{k+2} = 0.$$

From this, equating to zero the coefficients of all powers of x yields

$$(v^2 - v^2)c_0 = 0$$
$$[(1 + v)^2 - v^2]c_1 = 0$$
$$[(2 + v)^2 - v^2]c_2 + c_0 = 0$$
$$[(3 + v)^2 - v^2]c_3 + c_1 = 0$$
$$\cdots$$
$$[(k + v)^2 - v^2]c_k + c_{k-2} = 0$$
$$\cdots$$

The first relation is satisfied for any value of the coefficient c_0. The second relation gives $c_1 = 0$, the third gives

$$c_2 = -\frac{c_0}{(2 + v)^2 - v^2} = -\frac{c_0}{2^2(1 + v)},$$

the fourth $c_3 = 0$, and the fifth

$$c_4 = -\frac{c_2}{(4 + v)^2 - v^2} = \frac{c_0}{2^4(1 + v)(2 + v)(1)(2)}.$$

It is obvious that all coefficients with odd indices are zero, $c_{2k+1} = 0$, $k = 0, 1, 2, \ldots$. The coefficients with even indices are of the form

$$c_{2k} = \frac{(-1)^k c_0}{2^{2k}(v + 1)(v + 2) \cdots (v + k)k!}, \quad k = 1, 2, \ldots.$$

To simplify further computations, assume that $c_0 = c_0/2^v \Gamma(1 + v)$, Γ being the *Euler gamma function* defined for all positive values (as well as for all complex values with a positive real part) as follows:

$$\Gamma(p) = \int_0^\infty e^{-t} \cdot t^{p-1} dp, \quad \Gamma(p + 1) = p\Gamma(p). \tag{1.254}$$

Then, the particular solution of the Bessel equation, denoted by $J_v(x)$, takes the form

$$J_v(x) = \sum_{k=0}^{\infty} \frac{(-1)^k}{k!\Gamma(k+v+1)} \left(\frac{x}{2}\right)^{2k+v}. \tag{1.255}$$

This function is called Bessel's function of the first type of order v. The second particular solution of the Bessel equation (1.253) is

$$J_{-v}(x) = \sum_{k=0}^{\infty} \frac{(-1)^k}{k!\Gamma(k-v+1)} \left(\frac{x}{2}\right)^{2k-v}, \tag{1.256}$$

derived from solution (1.255) by replacing v by $-v$, since Eq. (1.253) contains v to the even power and remains unchanged when v is replaced by $-v$. If v is not an integer, then the solutions $J_v(x)$ and $J_{-v}(x)$ are linearly independent and the general solution of the Bessel equation may be taken in the form

$$y = AJ_v(x) + BJ_{-v}(x),$$

where A and B are arbitrary constants. If v is an integer, then

$$J_{-n}(x) = (-1)^n J_n(x) \quad (n \text{ is integer})$$

and it is necessary to seek another solution instead of $J_{-n}(x)$, a solution that would be linearly independent of $J_n(x)$. Let us introduce a new function

$$Y_v(x) = \frac{J_v(x)\cos v\pi - J_{-v}(x)}{\sin v\pi}, \tag{1.257}$$

called the Bessel function of the second type and order v. The function $Y_v(x)$ is a solution of Eq. (1.253), linearly independent of $J_v(x)$. It follows that the general solution of Eq. (1.253) can be represented as

$$y = AJ_p(x) + BY_p(x),$$

A and B being arbitrary constants.

Remark 1.33

Now, let us consider the frequently occurring equation

$$x^2 y'' + xy' + (b^2 x^2 - v^2)y = 0 \quad (b \neq 0, \text{ constant}). \tag{1.258}$$

The general solution of Eq. (1.258) is

$$y = AJ_v(bx) + BY_v(bx).$$

Periodic Solutions of Linear Differential Equations

Consider the second-order nonhomogeneous linear differential equation with constant coefficients

$$y'' + p_1 y' + p_2 y = f(x), \tag{1.259}$$

$f(x)$ being a periodic function of period 2π that can be expanded into Fourier series,

$$f(x) = \frac{a_0}{2} + \sum_{n=1}^{\infty} (a_n \cos nx + b_n \sin nx). \tag{1.260}$$

The periodic solution of Eq. (1.259) must be of the form

$$y(x) = \frac{A_0}{2} + \sum_{n=1}^{\infty} (A_n \cos nx + B_n \sin nx). \tag{1.261}$$

Let us substitute series (1.261) in Eq. (1.259) and select its coefficients so that Eq. (1.259) is satisfied formally. Equating the left-hand and right-hand free terms, the coefficients of $\cos nx$ and $\sin nx$ of the obtained equation, yields

$$A_0 = \frac{a_0}{p_2}, \quad A_n = \frac{(p_2 - n^2)a_n - p_1 n b_n}{(p_2 - n^2)^2 + p_1^2 n^2}$$

$$B_n = \frac{(p_2 - n^2)b_n + p_1 n a_n}{(p_2 - n^2)^2 + p_1^2 n^2}, \quad n = 1, 2, \ldots. \tag{1.262}$$

The first equation of (1.262) gives the necessary condition for the existence of a solution of the form Eq. (1.261): If $a_0 \neq 0$, then it is necessary that $p_2 \neq 0$. Substituting Eq. (1.262) in Eq. (1.261) gives

$$y(x) = \frac{a_0}{2p} + \sum_{n=1}^{\infty} \frac{[(p_2 - n^2)a_n - p_1 n b_n] \cos nx + [(p_2 - n^2)b_n + p_1 n a_n] \sin nx}{(p_2 - n^2)^2 + p_1^2 n^2} \tag{1.263}$$

When $p_1 = 0$ and $p_2 = k^2$, with $k = 1, 2, \ldots$, a periodic solution will exist, and

$$a_k = \frac{1}{\pi} \int_0^{2\pi} f(x) \cos kx\, dx, \quad b_k = \frac{1}{\pi} \int_0^{2\pi} f(x) \sin kx\, dx. \tag{1.264}$$

For $n \neq k$ the coefficients A_n and B_n are found from Eq. (1.262) and coefficients A_k and B_k remain arbitrary, since the expression $A_k \cos kx + B_k \sin kx$ is the general solution of the corresponding homogeneous equation. If conditions (1.264) fail to hold, Eq. (1.259) has no periodic solutions. When $p_2 = 0$ and $a_0 = 0$, the coefficient A_0 remains undetermined and Eq. (1.259) has an infinite number of periodic solutions differing from one another by a constant term. If the right-hand side $f(x)$ of Eq. (1.259) is of period $2l \neq 2\pi$, then it is necessary to expand $f(x)$ using the period $2l$ and seek a solution of Eq. (1.259) in the form

$$y(x) = \frac{A_0}{2} + \sum_{n=1}^{\infty} \left(A_n \cos \frac{n\pi x}{l} + B_n \sin \frac{n\pi x}{l} \right),$$

Eq. (1.262) changing accordingly.

EXAMPLE 1.78 Find the periodic solutions of the equation

$$y'' + 2y = \sum_{n=1}^{\infty} \frac{2(-1)^{n+1}}{n} \sin nx. \quad \blacktriangle$$

Solution

We can write

$$p_1 = 0, \quad p_2 = 2 \neq k^2 \quad (k \in \mathbf{N}^*)$$

$$a_0 = 0, \quad a_n = 0, \quad b_n = \frac{2(-1)^{n+1}}{n} \quad (n = 1, 2, \ldots).$$

The solution must be in the form of the series

$$y(x) = \frac{A_0}{2} + \sum_{n=1}^{\infty} (A_n \cos nx + B_n \sin nx),$$

and finding the coefficients A_n and B_n from Eq. (1.262), then

$$A_0 = \frac{a_0}{p_2} = 0, \quad A_n = 0, \quad B_n = \frac{2(-1)^{n+1}}{n(2 - n^2)}, \quad (n = 1, 2, \ldots).$$

The periodic solution of the given equation is

$$y(x) = \sum_{n=1}^{\infty} \frac{2(-1)^{n+1}}{n(2 - n^2)} \sin nx.$$

EXAMPLE 1.79 Find the periodic solutions of the equation

$$y'' + y' = \sum_{n=1}^{\infty} \frac{2(-1)^{n+1}}{n} \sin nx. \quad \blacktriangle$$

Solution

It can be written that

$$p_1 = 0, \quad p_2 = 1 = k^2, \quad k = 1,$$
$$a_1 = 0, \quad b_1 = 2 \neq 0$$

and conditions (1.264) fail to hold. Consequently, the given equation has no periodic solutions. Indeed, the given equation may be written in the form

$$y'' + y' = \sum_{n=2}^{\infty} \frac{2(-1)^{n+1}}{n} n \sin nx + 2 \sin x,$$

and in accordance with the principle of superposition, its general solution is

$$y(x) = \tilde{y}(x) + y_p(x),$$

where $\tilde{y}(x)$ is the general solution of equation

$$y'' + y' = \sum_{n=2}^{\infty} \frac{2(-1)^{n+1}}{n} \sin nx$$

and y_p is a particular solution of equation

$$y'' + y = 2 \sin x.$$

One solution is $y_p = -2x \cos x$, which is clearly nonperiodic.

EXAMPLE 1.80 Find the periodic solutions of the equation

$$y'' + y = \sum_{n=2}^{\infty} \frac{2(-1)^{n+1}}{n} \sin nx. \quad \blacktriangle$$

Solution

It can be written that

$$p_1 = 0, \quad p_2 = 1 = 1^2.$$

The resonance terms $a_1 \cos X + b_1 \sin x$ are absent in the right-hand member. Therefore, a periodic solution exists and is determined by Eqs. (1.262) and (1.263),

$$a_0 = 0, \quad a_1 = 0, \quad b_1 = 0, \quad a_n = 0, \quad b_n = \frac{2(-1)^{n+1}}{n}, \quad n = 2, 3, \ldots,$$

$$A_0 = 0, \quad A_1 = 0, \quad A_n = 0, \quad (n = 2, 3, \ldots), \quad B_1 = 0, \quad B_n = \frac{2(-1)^{n+1}}{n(2 - n^2)},$$

$$n = 2, 3, \ldots, y_p(x) = \sum_{n=2}^{\infty} \frac{2(-1)^{n+1}}{n(2 - n^2)} \sin nx.$$

Remark 1.34

The general solution is

$$y(x) = \sum_{n=2}^{\infty} \frac{2(-1)^{n+1}}{n(2 - n^2)} \sin nx + C_1 \cos x + C_2 \sin x,$$

where C_1 and C_2 are arbitrary constants and $y(x)$ is periodic for any C_1 and C_2.

2. Systems of Differential Equations

2.1 Fundamentals

Consider the system of first-order equations

$$\begin{cases} \dfrac{dx_1}{dt} = f_1(t, x_1, x_2, \ldots, x_n) \\[2mm] \dfrac{dx_2}{dt} = f_2(t, x_1, x_2, \ldots, x_n) \\[1mm] \cdots \\[1mm] \dfrac{dx_n}{dt} = f_n(t, x_1, x_2, \ldots, x_n) \end{cases} \tag{2.1}$$

and the initial conditions

$$x_i(t_0) = x_{i_0}, \quad (i = 1, 2, \ldots, n). \tag{2.2}$$

The sufficient conditions for the existence and uniqueness of a solution of the system of Eqs. (2.1), given the initial conditions (2.2), are as follows:

1. Continuity of all functions f_i in the neighborhood of the initial values
2. fulfillment of the Lipschitz condition for all functions f_i with respect to all arguments, beginning with the second one, in the same neighborhood

Condition 2 may by replaced by a condition requiring the existence of partial derivatives bounded in the absolute value

$$\frac{\partial f_i}{\partial x_j}, \quad (i, j = 1, 2, \ldots, n).$$

The solution of the system of equations (2.1) is an n-dimensional vector function $x_1(t), x_2(t), \ldots, x_n(t)$, which will be denoted by $\mathbf{X}(t)$. Using this notation, the system of equations (2.1) may be written as

$$\frac{d\mathbf{X}}{dt} = F(t, \mathbf{X}), \tag{2.3}$$

where F is a vector function with the coordinates $f_1, f_2, \ldots, f_n$, the initial conditions are

$$\mathbf{X}(t_0) = \mathbf{X}_0, \tag{2.4}$$

and $\mathbf{X}_0$ is a n-dimensional vector with coordinates $(x_{10}, x_{20}, \ldots, x_{n0})$.

The solutions of the system of equations

$$x_1 = x_1(t), x_2 = x_2(t), \ldots, x_n = x_n(t)$$

or, briefly, $\mathbf{X} = \mathbf{X}(t)$, defines in the Euclidean space with coordinates $t, x_1, x_2, \ldots, x_n$ a certain curve called the *integral curve*.

Geometrically, the Cauchy problem with Eqs. (2.1) and (2.2) can be stated as follows: Find in the space of variables $(t, x_1, x_2, \ldots, x_n)$ an integral curve passing through a given point $(t_0, x_{10}, x_{20}, \ldots, x_{n0})$. A different interpretation of solutions is possible.

In the Euclidean space with rectangular coordinates $x_1, x_2, \ldots, x_n$, the solution $x_1 = x_1(t), x_2 = x_2(t), \ldots, x_n = x_n(t)$ defines a law of motion of some trajectory depending on the variation of parameter t, which in this interpretation will be called "time." In such interpretation, the derivative $d\mathbf{X}/dt$ will be the velocity of motion of a point, and dx_1/dt, $dx_2/dt, \ldots, dx_n/dt$ will be the coordinates of velocity of that point. Given this interpretation, which is convenient and natural in many mechanical problems, the system of Eqs. (2.1) or (2.3) is ordinarily called *dynamical*, the space with coordinates $x_1, x_2, \ldots, x_n$ is called the *phase space*, and the curve $\mathbf{X} = \mathbf{X}(t)$ is called the *phase trajectory*. At a specified instant of time t, the dynamical system of Eqs. (2.1) defines a field of velocities in the space $x_1, x_2, \ldots, x_n$. If the vector function F is dependent explicitly on t, then the field of velocities varies with time and the phase trajectories can intersect. But if the vector F or the equivalent of all functions f_i is not dependent explicitly on t, then the field of velocities is stationary and the motion will be steady. In the last case, if the conditions of the theorem of existence and uniqueness are fulfilled, then only one trajectory will pass through each point of the phase space $(x_1, x_2, \ldots, x_n)$. Indeed, in this case an infinite number of different motions $\mathbf{X} = \mathbf{X}(t + c)$, where c is an arbitrary constant, occur along each trajectory $\mathbf{X} = \mathbf{X}(t)$; this is easy to see if we make the variables changing $t_1 = t + c$, after which the dynamical system does not change the form

$$\frac{d\mathbf{X}}{dt_1} = F(\mathbf{X}),$$

and consequently $\mathbf{X} = \mathbf{X}(t_1)$ will be a solution, or, in the previous variables, $\mathbf{X} = \mathbf{X}(t + c)$. Assume that two trajectories pass through a certain point $\mathbf{X}_0$ of the phase space, $\mathbf{X}_1(t)$ passing through $\mathbf{X}_0$ at the moment of time $\overline{t_0}$ and $\mathbf{X}_2(t)$ passing through $\mathbf{X}_0$ at the moment of time $\overline{\overline{t_0}}$ ($\overline{t_0}$ and $\overline{\overline{t_0}}$ are are two values of variable t-time),

$$\mathbf{X} = \mathbf{X}_1(t) \qquad \text{and} \qquad \mathbf{X} = \mathbf{X}_2(t), \mathbf{X}_1(\overline{t_0}) = \mathbf{X}_2(\overline{\overline{t_0}}) = \mathbf{X}_0.$$

On each of the trajectories we consider the solution

$$\mathbf{X} = \mathbf{X}_1(t - t_0 + \overline{t_0}) \qquad \text{and} \qquad \mathbf{X} = \mathbf{X}_2(t - t_0 + \overline{\overline{t_0}}).$$

We find a contradiction with the theorem of existence and uniqueness, since two different solutions $\mathbf{X}_1(t - t_0 + \overline{t_0})$ and $\mathbf{X}_2(t - t_0 + \overline{\overline{t_0}})$ satisfy one and the same initial condition $\mathbf{X}(t_0) = \mathbf{X}_0$.

EXAMPLE 2.1 Show that the system of functions $x(t) = A\cos(2t + \varphi)$, $y(t) = -2A\sin(2t + \varphi)$ is the general solution for the system of equations

$$\frac{dx}{dt} = y; \qquad \frac{dy}{dt} = -4x. \quad \blacktriangle$$

Solution

In this example, the domain D is $-\infty < t < +\infty$, $-\infty < x, y < +\infty$. Substituting the functions $x(t)$ and $y(t)$ in the system gives identities in t, valid for any values of constants A and φ. Note that for the given system the conditions of the theorem of existence and uniqueness for the Cauchy problem hold in the whole domain D. Therefore, any triplet of numbers t_0, x_0, y_0 may be taken as the initial conditions $x(t_0) = x_0$, $y(t_0) = y_0$. The system is

$$\begin{cases} x_0 = A\cos(2t_0 + \varphi) \\ y_0 = -2A\sin(2t_0 + \varphi), \end{cases}$$

and A, φ are found:

$$A = \frac{1}{2}\sqrt{4x_0^2 + y_0^2}$$

$$\cos(2t_0 + \varphi) = \frac{2x_0}{\sqrt{4x_0^2 + y_0^2}}$$

$$\sin(2t_0 + \varphi) = -\frac{y_0}{\sqrt{4x_0^2 + y_0^2}}.$$

Regarding t as a parameter, we find a family of parabolas on the phase plane x, y with center at the origin of coordinates

$$4x^2 + y^2 = A^2; \quad A = \frac{1}{2}\sqrt{4x_0^2 + y_0^2}.$$

The right-hand member of the given system is not dependent on t, and then the trajectories do not intersect. Fixing A yields a definite trajectory, and for different φ there will correspond different motions along this trajectory. The equation of the trajectory $4x^2 + y^2 = A^2$ does not depend on φ so that motions for fixed A are along the same trajectory. When $A = 0$, the phase trajectory consists of a single point, called in this case the *rest point* of the system.

2.2 Integrating a System of Differential Equations by the Method of Elimination

One of the main methods for integrating a system of differential equations consists of the following: All unknown functions, except one, are eliminated from the equations of the system of equations (2.1) and from the equations obtained by the differentiation of the equations that make up the system; to find this function, a single differential equation of the higher order is obtained. Integrating the equation of higher order provides one of the unknown functions; the other unknown functions are determined from the original equations and from the equations obtained as a result of their differentiation. Next will be described more exactly the process of eliminating all unknown functions except one, say $x_1(t)$.

Let us make the assumption that all functions f_i have continuous partial derivatives up to $(n-1)$-order inclusive, with respect to all arguments. Differentiating the first equation of the system of equations (2.1) with respect to t gives

$$\frac{d^2 x_1}{dt^2} = \frac{\partial f_1}{\partial t} + \sum_{i=1}^{n} \frac{\partial f_1}{\partial x_i}\left(\frac{dx_i}{dt}\right).$$

Replacing the derivatives dx_i/dt with their expressions f_i from Eqs. (2.1) and designating for the right-hand side $F_2(t, x_1, x_2, \ldots, x_n)$ yields

$$\frac{d^2 x_1}{dt^2} = F_2(t, x_1, x_2, \ldots, x_n).$$

Differentiating this identity and then doing as before, we obtain

$$\frac{d^3 x_1}{dt^3} = F_3(t, x_1, x_2, \ldots, x_n).$$

Continuing in the same way, finally we find the system

$$\begin{cases} \dfrac{dx_1}{dt} = f_1(t, x_1, x_2, \ldots, x_n) \\[2mm] \dfrac{d^2 x_1}{dt^2} = F_2(t, x_1, x_2, \ldots, x_n) \\[2mm] \cdots \\[2mm] \dfrac{d^n x_1}{dt^n} = F_n(t, x_1, x_2, \ldots, x_n). \end{cases} \tag{2.5}$$

Suppose that the determinant

$$\frac{D(f_1, F_2, F_3, \ldots, F_{n-1})}{D(x_2, x_3, \ldots, x_n)} \neq 0. \tag{2.6}$$

Then, from the first $(n-1)$ equations of the system (2.5), we find $x_2, x_3, \ldots, x_n$, expressed in terms of the variables t, x_1, $dx_1/dt, \ldots$, $d^{n-1} x_1/dt^{n-1}$:

$$\begin{cases} x_2 = \varphi_2(t, x_1, x_1', \ldots, x_1^{(n-1)}) \\[2mm] x_3 = \varphi_3(t, x_1, x_1', \ldots, x_1^{(n-1)}) \\[2mm] \cdots \\[2mm] x_n = \varphi_n(t, x_1, x_1', \ldots, x_1^{(n-1)}). \end{cases} \tag{2.7}$$

Substituting these expressions into the last of Eqs. (2.5), we find an n-order equation to determine x_1,

$$\frac{d^n x_1}{dt^n} = \phi(t, x_1, x_1', \ldots, x_1^{(n-1)}). \tag{2.8}$$

Solving this equation yields x_1:

$$x_1 = \psi_1(t, c_1, c_2, \ldots, c_n). \tag{2.9}$$

Differentiating the expression $(n-1)$ times, we find the derivatives, $dx_1/dt, d^2 x_1/dt^2, \ldots, d^{n-1} x_1/dt^{n-1}$ as functions of $t, c_1, c_2, \ldots, c_n$. If we substitute these functions into Eqs. (2.7), then $x_1, x_2, \ldots, x_n$ are

$$\begin{cases} x_2 = \psi_2(t, c_1, c_2, \ldots, c_n) \\[2mm] x_3 = \psi_3(t, c_1, c_2, \ldots, c_n) \\[2mm] \cdots \\[2mm] x_n = \psi_n(t, c_1, c_2, \ldots, c_n). \end{cases} \tag{2.10}$$

For this solution, to satisfy the given initial conditions (2.2), it is necessary to find [from Eqs. (2.9) and (2.10)] the appropriate values of the constants $c_1, c_2, \ldots, c_n$.

Remark 2.1

If the system (2.1) is linear in the unknown functions, then Eq. (2.8) is also linear.

Remark 2.2

If the condition (2.6) is not fulfilled, then the same process may be employed, but in the place of function x_1 we take another function from $x_1, x_2, \ldots, x_n$. If condition (2.6) is not fulfilled for any choice of some functions $x_1, x_2, \ldots, x_n$ in the place of x_1, then various exceptional cases are possible.

Remark 2.3

The differential equations of a system may contain higher-order derivatives and then yield a system of differential equations of higher order.

EXAMPLE 2.2   Solve the system of equations

$$\frac{dx}{dt} = y, \quad \frac{dy}{dt} = -4x + t$$

for the initial conditions

$$x(0) = 1, \quad y(0) = \tfrac{9}{4}. \quad \blacktriangle$$

Solution

If we differentiate the first equation of the system with respect to t and substitute $dy/dt = -4x + t$ in the obtained equations, the given system is reduced to a second-order linear equation,

$$\frac{d^2x}{dt^2} + 4x = t.$$

The general solution of this equation is

$$x(t) = c_1 \cos 2t + c_2 \sin 2t + \tfrac{1}{4}t.$$

Since $y = dx/dt$, then $y = -2c_1 \sin 2t + 2c_2 \cos 2t + \tfrac{1}{4}$, and thus the general solution of the given system is

$$x = c_1 \cos 2t + c_2 \sin 2t + \tfrac{1}{4}t, \quad y = -2c_1 \sin 2t + 2c_2 \cos 2t + \tfrac{1}{4}.$$

From the initial conditions $x(0) = 1$, $y(0) = \tfrac{9}{4}$, we find $c_1 = 1$, $c_2 = 1$, and the particular solution of the given system that satisfies the initial conditions is

$$x = \cos 2t + \sin 2t + \tfrac{1}{4}t, \quad y = -2 \sin 2t + 2 \cos 2t + \tfrac{1}{4}.$$

EXAMPLE 2.3   Solve the system of equations

$$\frac{d^2 x_1}{dt^2} + a x_1 - b x_2 = 0, \quad \frac{d^2 x_2}{dt^2} - c x_1 + d x_2 = 0, \quad a, b, c, d \in \mathbf{R}_+, \quad ad > bc.$$

▲

Solution

Differentiating the first equation twice with respect to t gives

$$\frac{d^4 x_1}{dt^4} + a \frac{d^2 x_1}{dt^2} - b \frac{d^2 x_2}{dt^2} = 0,$$

but

$$\frac{d^2 x_2}{dt^2} = c x_1 - d x_2 = c x_1 - \frac{d}{b} \left(\frac{d^2 x_1}{dt^2} + a x_1 \right),$$

and we find a linear equation of the fourth order,

$$\frac{d^4 x_1}{dt^4} + (a + d) \frac{d^2 x_1}{dt^2} + (ad - bc) x_1 = 0.$$

The characteristic equation is

$$r^4 + (a + d) r^2 + (ad - bc) = 0.$$

Since $a, b, c, d \in (0, \infty)$ and $ad > bc$, in the last equation, $r^2 < 0$. If we substitute $r^2 = -\lambda$, then

$$\lambda^2 - (a + d)\lambda - (bc - ad) = 0$$

$$\lambda_{1,2} = \frac{a + d \pm \sqrt{(a - d)^2 + 4bc}}{2}.$$

The general solution of the linear equation is

$$x_1(t) = c_1 \cos \sqrt{\lambda_1} t + c_2 \sin \sqrt{\lambda_2} t + c_3 \cos \sqrt{\lambda_3} t + c_4 \sin \sqrt{\lambda_4} t.$$

In the last formula, let us replace the arbitrary constants c_1, c_2, c_3, c_4 with others. We introduce the constants A_1 and φ_1, which are related to c_1 and c_2:

$$c_1 = A_1 \sin \varphi_1, \quad c_2 = A_1 \cos \varphi_1.$$

The constants A_1 and φ_1 are defined as follows in terms of c_1 and c_2:

$$A_1 = \sqrt{c_1^2 + c_2^2}, \quad \varphi_1 = \arctan \frac{c_1}{c_2}.$$

Analogously, $A_2 = \sqrt{c_3^2 + c_4^2}$, $\varphi_1 = \arctan(c_3/c_4)$. Substituting the values of c_1, c_2, c_3 and c_4 gives

$$\begin{cases} x_1(t) = A_1 \sin(\sqrt{\lambda_1} t + \varphi_1) + A_2 \sin(\sqrt{\lambda_2} t + \varphi_2) \\ x_2(t) = \dfrac{A_1}{b}(a - \lambda_1) \sin(\sqrt{\lambda_1} t + \varphi_1) + \dfrac{A_2}{b}(a - \lambda_2) \sin(\sqrt{\lambda_2} t + \varphi_2), \end{cases}$$

which is the general solution of the given system.

2.3 Finding Integrable Combinations

The integration of the system of differential equations

$$\frac{dx_i}{dt} = f_i(t, x_1, x_2, \ldots, x_n), \quad (i = 1, 2, \ldots, n) \tag{2.11}$$

is accomplished by choosing integrable combinations. *Integrable combinations* represents a differential equation of the form

$$F\left(t, u, \frac{du}{dt}\right) = 0,$$

which is a consequence of Eqs. (2.11), using suitable arithmetic operations (addition, subtraction, multiplication, and division), but which is readily integrable: for example, an equation of the type

$$d\phi(t, x_1, x_2, \ldots, x_n) = 0.$$

One integrable combination permits one finite equation

$$\phi_1(t, x_1, x_2, \ldots, x_n) = c_1,$$

which relates the unknown functions and the independent variable. Such a finite equation is called a *first integral of the system* of equations (2.11). Thus, the first integral

$$\phi(t, x_1, x_2, \ldots, x_n) = c$$

of the system (2.11) is a finite equation that is converted into an identity for some value c if the solution of the system of equations (2.11) is substituted in the place of $x_i(t)$ $(i = 1, 2, \ldots, n)$. The left-hand member $\phi(t, x_1, x_2, \ldots, x_n)$ is also called a first integral, and then the first integral is defined as a function not identically equal to a constant, but retaining a constant value along the integral curves of the system of Eqs. (2.11). Geometrically, the first integral $\phi(t, x_1, x_2, \ldots, x_n) = c$ for fixed c may be interpreted as an n-dimensional surface in $(n + 1)$-dimensional space with the coordinates $t, x_1, x_2, \ldots, x_n$, each integral curve having a common point with the surface that lies entirely within the surface. From n integrable combinations found, the first n integrals are

$$\begin{cases} \phi_1(t, x_1, x_2, \ldots, x_n) = c_1 \\ \phi_2(t, x_1, x_2, \ldots, x_n) = c_2 \\ \quad \cdots \\ \phi_n(t, x_1, x_2, \ldots, x_n) = c_n, \end{cases} \tag{2.12}$$

and if all these integrals are independent, that is,

$$\frac{D(\phi_1, \phi_2, \ldots, \phi_n)}{D(x_1, x_2, \ldots, x_n)} \neq 0,$$

then all the unknown functions are determined from the system of equations (2.12). To find integrable combinations in solving the system of differential equations (2.11), it is sometimes convenient to use the symmetrical *form* when writing the system (2.11),

$$\frac{dx_1}{f_1(t, x_1, x_2, \ldots, x_n)} = \frac{dx_2}{f_2(t, x_1, x_2, \ldots, x_n)} = \cdots = \frac{dx_n}{f_n(t, x_1, x_2, \ldots, x_n)} = \frac{dt}{1}.$$

(2.13)

To solve the system of Eq. (2.13), one may either take pairs of relations allowing separation of the variables, or use the derived proportions

$$\frac{a_1}{b_1} = \frac{a_2}{b_2} = \cdots = \frac{a_n}{b_n} = \frac{\lambda_1 a_1 + \lambda_2 a_2 + \cdots + \lambda_n a_n}{\lambda_1 b_1 + \lambda_2 b_2 + \cdots + \lambda_n b_n},$$

(2.14)

where the coefficients $\lambda_1, \lambda_2, \ldots, \lambda_n$ are arbitrary and chosen so that the numerator should be the differential of the denominator or so that the numerator should be the total differential and the denominator equal to zero.

EXAMPLE 2.4   Solve the system

$$\frac{dx}{dt} = \frac{y}{(y-x)^2}, \quad \frac{dy}{dt} = \frac{x}{(y-x)^2}. \quad \blacktriangle$$

Solution

Dividing the first equation by the second, yields $dx/dy = y/x$ or $x\,dx = y\,dy$, whence $x^2 - y^2 = c_1$. Subtracting the second equation from the first, we find $d(x-y)/dt = 1/(y-x)$, whence $(x-y)^2 + 2t = c_2$. Thus, we find two first integrals of the given system,

$$\psi_1(t, x, y) = x^2 - y^2 = c_1, \quad \psi_2(t, x, y) = (x-y)^2 + 2t = c_2,$$

which are independent, since the Jacobian

$$\frac{D(\psi_1, \psi_2)}{D(x, y)} = \begin{vmatrix} \dfrac{\partial \psi_1}{\partial x} & \dfrac{\partial \psi_1}{\partial y} \\ \dfrac{\partial \psi_2}{\partial x} & \dfrac{\partial \psi_2}{\partial y} \end{vmatrix} = \begin{vmatrix} 2x & -2y \\ 2(x-y) & -2(x-y) \end{vmatrix} = -4(x-y)^2 \neq 0.$$

Solving the system $x^2 - y^2 = c_1$, $(x-y)^2 + 2t = c_2$ for the unknown functions, we obtain the general solution for the given systems,

$$x = \frac{c_1 + c_2 - 2t}{2\sqrt{c_2 - 2t}}, \quad y = \frac{c_1 - c_2 + 2t}{2\sqrt{c_2 - 2t}}.$$

EXAMPLE 2.5   Solve the system of equations

$$\frac{dt}{x_1 - x_2} = \frac{dx_1}{x_2 - t} = \frac{dx_2}{t - x_1}. \quad \blacktriangle$$

Solution

Adding the numerators and denominators gives

$$\frac{dt}{x_1 - x_2} = \frac{dx_1}{x_2 - t} = \frac{dx_2}{t - x_1} = \frac{dt + dx_1 + dx_2}{0}$$

(here $\lambda_1 = 1$, $\lambda_2 = 1$, $\lambda_3 = 1$). Hence $dt + dx_1 + dx_2 = 0$ or $d(t + x_1 + x_2) = 0$, and so $t + x_1 + x_2 = 0$ is a first integral of the system. Multiplying the numerators and denominators of fractions in this system gives

$$\frac{t\,dt}{tx_1 - tx_2} = \frac{x_1\,dx_1}{x_1 x_2 - x_1 t} = \frac{x_2\,dx_2}{tx_2 - x_1 x_2} = \frac{t\,dt + x_1\,dx_1 + x_2\,dx_2}{0},$$

hence,

$$t\,dt + x_1\,dx_1 + x_2\,dx_2 = 0 \qquad \text{or} \qquad d(t^2 + x_1^2 + x_2^2) = 0,$$

and so the second first integral is

$$t^2 + x_1^2 + x_2^2 = c_2.$$

Solving the system $t + x_1 + x_2 = c_1$, $t^2 + x_1^2 + x_2^2 = c_2$ for the unknown functions, we find the general solution.

2.4 Systems of Linear Differential Equations

A system of differential equations is called *linear* if it is linear in all unknown functions and their derivatives. A system of n linear equations of the first order, written in the normal form, looks like

$$\frac{dx_i}{dt} = \sum_{j=1}^{n} a_{ij}(t)x_j + f_i(t), \quad (i = 1, 2, \ldots, n). \tag{2.15}$$

The system of equations (2.15) may be compactly written in the form of one matrix equation

$$\frac{dX}{dt} = AX + F, \tag{2.16}$$

where

$$A = \begin{bmatrix} a_{11} & a_{12} & \cdots & a_{1n} \\ a_{21} & a_{22} & \cdots & a_{2n} \\ \cdots & \cdots & \cdots & \cdots \\ a_{n1} & a_{n2} & \cdots & a_{nn} \end{bmatrix}, \quad X = \begin{Bmatrix} x_1 \\ x_2 \\ \vdots \\ x_n \end{Bmatrix} \qquad (2.17)$$

$$\frac{dX}{dt} = \begin{Bmatrix} \dfrac{dx_1}{dt} \\ \dfrac{dx_2}{dt} \\ \vdots \\ \dfrac{dx_n}{dt} \end{Bmatrix}, \quad F = \begin{Bmatrix} f_1 \\ f_2 \\ \vdots \\ f_n \end{Bmatrix}. \qquad (2.18)$$

The column matrix

$$Y(t) = \begin{Bmatrix} y_1(t) \\ y_2(t) \\ \cdots \\ y_n(t) \end{Bmatrix}$$

is said to be a *particular solution* of Eq. (2.15) in the interval (a, b) if the identity

$$\frac{dY}{dt} = AY(t) + F(t)$$

holds for $a < t < b$. If all functions $a_{ij}(t)$ and $f_i(t)$ in Eqs. (2.15) are continuous on the interval $[a, b]$, then in a sufficiently small neighborhood of every point $(t_0, x_{10}, x_{20}, \ldots, x_{n0})$ where $a \leq t_0 \leq b$, the conditions of the theorem of existence and uniqueness are fulfilled and, hence, a unique integral curve of the system of equations (2.15) passes through every such point. If we define the *linear operator* L by the equality

$$L[X] = \frac{dX}{dt} - AX, \qquad (2.19)$$

then Eq. (2.16) may be written more concisely as

$$L[X] = F. \qquad (2.20)$$

If all $f_i(t) \equiv 0$, $(i = 1, 2, \ldots, n)$, or the matrix $F = 0$, then the system of equations (2.15) is called *homogeneous linear*; it is of the form

$$L[X] = 0. \qquad (2.21)$$

The solutions of the linear system have the following basic properties.

THEOREM 2.1   If X is a solution of the homogeneous linear system $L[X] = 0$, then cX, where c is an arbitrary constant, is a solution of the same system.   ▲

THEOREM 2.2   The sum $X_1 + X_2$ of two solutions X_1 and X_2 of a homogeneous linear system of equations is a solution of that system.   ▲

COROLLARY 2.1   A linear combination, $\sum_{i=1}^{n} cX_i$ with arbitrary constant coefficients of solutions $X_1, X_2, \ldots, X_n$ of a homogeneous linear system $L[X] = 0$ is a solution of that system.   ▲

THEOREM 2.3   If the homogeneous linear system of equations (2.21) with real coefficients $a_{ij}(t)$ has a complex solution $X = U + iV$, then the real and imaginary parts

$$U = \left\{ \begin{array}{c} u_1 \\ u_2 \\ \cdots \\ u_n \end{array} \right\}$$

and

$$V = \left\{ \begin{array}{c} v_1 \\ v_2 \\ \cdots \\ v_n \end{array} \right\}$$

are separately solutions of that system.   ▲

THEOREM 2.4   If the Wronskian W of solutions $X_1, X_2, \ldots, X_n$,

$$W(t) \equiv W(X_1, X_2, \ldots, X_n) = \begin{vmatrix} x_{11} & x_{12} & \cdots & x_{1n} \\ x_{21} & x_{22} & \cdots & x_{2n} \\ \cdots & \cdots & \cdots & \cdots \\ x_{n1} & x_{n2} & \cdots & x_{nn} \end{vmatrix}, \qquad (2.22)$$

of the homogeneous system of equations (2.21) with coefficients $a_{ij}(t)$ continuous on the interval $a \le t \le b$ is zero at least in one point $t = t_0$ of the interval $a \le t \le b$, then the solutions $X_1, X_2, \ldots, X_n$ are linearly dependent on that interval, and hence $W(t) \equiv 0$ on that interval. ▲

THEOREM 2.5   The linear combination $\sum_{i=1}^{n} c_i X_i$ of the homogeneous linear system of equations (2.21) with the coefficients $a_{ij}(t)$ continuous on the interval $a \le t \le b$ is the general solution of the system of equations (2.21) on that interval.   ▲

THEOREM 2.6   If $\widetilde{X}$ is a solution of the nonhomogeneous linear system of equations (2.20) and X_1 is a solution of the corresponding homogeneous system of equations (2.21), then the sum $X_1 + \widetilde{X}$ is a solution of the nonhomogeneous system of equations (2.20).   ▲

THEOREM 2.7 The general solution, on the interval $a \leq t \leq b$, of the nonhomogeneous system of equations (2.20) with the coefficients $a_{ij}(t)$ continuous on that interval is equal to the sum of solution $\sum_{i=1}^{n} c_i X_i$ of the corresponding homogeneous system and the particular solution $\tilde{X}$ of the nonhomogeneous system. ▲

THEOREM 2.8 *The Principle of Superposition*
The solution of the system of linear equations

$$L[X] = \sum_{i=1}^{n} F_i$$

$$F_i = \begin{Bmatrix} f_1 i(t) \\ f_2 i(t) \\ \cdots \\ f_n i(t) \end{Bmatrix}$$

is the sum $\sum_{i=1}^{n} X_i$ of solutions X_i of the equations

$$L[X_i] = F_i \quad (i = 1, 2, \ldots, n). ▲$$

THEOREM 2.9 If the system of linear equations

$$L[X] = U + iV$$

$$U = \begin{Bmatrix} u_1 \\ u_2 \\ \cdots \\ u_n \end{Bmatrix}$$

$$V = \begin{Bmatrix} v_1 \\ v_2 \\ \cdots \\ v_n \end{Bmatrix},$$

with real functions $a_{ij}(t), u_i(t), v_i(t), (i, j = 1, 2, \ldots, n)$, has the solution $X = \tilde{U} + i\tilde{V}$,

$$\tilde{U} = \begin{Bmatrix} \widetilde{u_1} \\ \widetilde{u_2} \\ \cdots \\ \widetilde{u_n} \end{Bmatrix}$$

and

$$\tilde{V} = \begin{Bmatrix} \widetilde{v_1} \\ \widetilde{v_2} \\ \cdots \\ \widetilde{v_n} \end{Bmatrix}$$

then the real part of the solution $\tilde{U}$ and the imaginary part $\tilde{V}$ are suspected to be solutions of the equations

$$L[X] = U \quad \text{and} \quad L[X] = V. \quad \blacktriangle$$

2.4.1 THE METHOD OF VARIATION OF ARBITRARY PARAMETERS (THE LAGRANGE METHOD)

If the general solution of the corresponding homogeneous system of equations (2.21) is known, and one cannot choose a particular solution of the system of equations (2.20), then the method of variation of parameters may be applied.

Let $X = \sum_{i=1}^{n} c_i X_i$ be the general solution of the system (2.21).

The solution of the nonhomogeneous system (2.20) must be of the form

$$X(t) = \sum_{i=1}^{n} c_i(t) X_i, \tag{2.23}$$

where $c_i(t)$ are the new unknown functions. If we substitute into the nonhomogeneous equation, we obtain

$$\sum_{i=1}^{n} c_i'(t) X_i = F.$$

This vector equation is equivalent to a system of n equations

$$\begin{cases} \sum_{i=1}^{n} c_i'(t) x_{1i} = f_1(t) \\ \sum_{i=1}^{n} c_i'(t) x_{2i} = f_2(t) \\ \qquad \cdots \\ \sum_{i=1}^{n} c_i'(t) x_{ni} = f_n(t). \end{cases} \tag{2.24}$$

All $c_i'(t)$ are determined from this system, $c_i'(t) = \varphi_i(t) \ (i = 1, 2, \ldots, n)$, whence

$$c_i(t) = \int \varphi_i(t) dt + \bar{c}_i \quad (i = 1, 2, \ldots, n).$$

The system

$$X_1 = \begin{Bmatrix} x_{11} \\ x_{21} \\ \vdots \\ x_{n1} \end{Bmatrix}, \quad X_2 = \begin{Bmatrix} x_{12} \\ x_{22} \\ \vdots \\ x_{n2} \end{Bmatrix}, \ldots, X_n = \begin{Bmatrix} x_{1n} \\ x_{2n} \\ \vdots \\ x_{nn} \end{Bmatrix}$$

of particular solutions of the homogeneous system of differential equations is said to be *fundamental* in the interval (a, b) if its Wronskian

$$W(t) \equiv W(X_1, X_2, \ldots, X_n) = \begin{vmatrix} x_{11}(t) & x_{12}(t) & \cdots & x_{1n}(t) \\ x_{21}(t) & x_{22}(t) & \cdots & x_{2n}(t) \\ \cdots & \cdots & \cdots & \cdots \\ x_{n1}(t) & x_{n2}(t) & \cdots & x_{nn}(t) \end{vmatrix} \neq 0$$

for all $t \in (a, b)$. In this case, the matrix

$$M(t) = \begin{bmatrix} x_{11}(t) & x_{12}(t) & \cdots & x_{1n}(t) \\ x_{21}(t) & x_{22}(t) & \cdots & x_{2n}(t) \\ \cdots & \cdots & \cdots & \cdots \\ x_{n1}(t) & x_{n2}(t) & \cdots & x_{nn}(t) \end{bmatrix} \tag{2.25}$$

is said to be a *fundamental matrix*. The general solution of the homogeneous linear system of equations (2.21) is

$$X(t) = M(t)c$$

$$c = \begin{Bmatrix} c_1 \\ c_2 \\ \cdots \\ c_n \end{Bmatrix}, \tag{2.26}$$

The solution of the homogeneous system

$$\frac{dX}{dt} = AX$$

that satisfies the initial condition $X(t_0) = X_0$ is

$$X(t) = M(t)M^{-1}(t_0)X_0. \tag{2.27}$$

The system of equations (2.24) may be written in the form

$$M(t)c'(t) = F(t),$$

and hence

$$c(t) = \int_{t_0}^{t} M^{-1}(s)F(s)ds + \tilde{c}.$$

The general solution of the system of equations (2.16) is

$$X(t) = M(t)\tilde{c} + M(t)\int_{t_0}^{t} M^{-1}(s)F(s)ds, \tag{2.28}$$

and the solution that satisfies $X(t_0) = X_0$ is

$$X(t) = M(t)M^{-1}(t_0)X_0 + \int_{t_0}^{t} M(t)M^{-1}(s)F(s)ds. \tag{2.29}$$

THEOREM 2.10 *Liouville's Formula*

Let $W(t)$ be the Wronskian of solutions $X_1, X_2, \ldots, X_n$ of the homogeneous system of equations (2.21). Then

$$W(t) = W(t_0)e^{\int_{t_0}^{t} \sum_{j=1}^{n} a_{jj}(s)ds}, \tag{2.30}$$

where $t_0 \in (a, b)$ is arbitrary. The homogeneous linear system of differential equations

$$\frac{dx_i}{dt} = \sum_{j=1}^{n} a_{ij}x_j \tag{2.31}$$

for which the functions $X_1, X_2, \ldots, X_n$,

$$X_k = \begin{Bmatrix} x_{1k} \\ x_{2k} \\ \cdots \\ x_{nk} \end{Bmatrix},$$

are linearly independent solutions, may be written as

$$\begin{vmatrix} \dfrac{dx_i}{dt} & \dfrac{dx_{i1}}{dt} & \dfrac{dx_{i2}}{dt} & \cdots & \dfrac{dx_{in}}{dt} \\ x_1 & x_{11} & x_{12} & \cdots & x_{1n} \\ x_2 & x_{21} & x_{22} & \cdots & x_{2n} \\ \cdots & \cdots & \cdots & \cdots & \cdots \\ x_n & x_{n1} & x_{n2} & \cdots & x_{nn} \end{vmatrix} = 0 \quad (i = 1, 2, \ldots, n). \tag{2.32}$$

EXAMPLE 2.6 Show that the system of vectors

$$X_1 = \begin{Bmatrix} 1 \\ t \end{Bmatrix}, \quad X_2 = \begin{Bmatrix} -t \\ e^t \end{Bmatrix}$$

is a fundamental system of solutions for the following system:

$$\begin{cases} \dfrac{dx_1}{dt} = \dfrac{t}{e^t + t^2} x_1 - \dfrac{1}{e^t + t^2} x_2 \\ \dfrac{dx_2}{dt} = \dfrac{e^t(1 - t)}{e^t + t^2} x_1 + \dfrac{e^t + t}{e^t + t^2} x_2. \end{cases} \quad \blacktriangle$$

Solution

The Wronskian determinant is

$$W(t) = \begin{vmatrix} 1 & -t \\ t & e^t \end{vmatrix} = e^t + t^2 \neq 0, \quad \text{for all} \quad t \in R.$$

The vector

$$X_1 = \begin{Bmatrix} 1 \\ t \end{Bmatrix}$$

has the components $x_{11}(t) = 1$, $x_{21}(t) = t$ and

$$\dfrac{dx_{11}}{dt} = 0; \quad \dfrac{t}{e^t + t^2} x_{11} - \dfrac{1}{e^t + t^2} x_{21} = \dfrac{t}{e^t + t^2} - \dfrac{t}{e^t + t^2} = 0 = \dfrac{dx_{11}}{dt}; \dfrac{dx_{21}}{dt} = 1$$

$$\dfrac{e^t(1 - t)}{e^t + t^2} x_{11} + \dfrac{e^t + t}{e^t + t^2} x_{21} = \dfrac{e^t(1 - t)}{e^t + t^2} + \dfrac{(e^t + t)t}{e^t + t^2} = \dfrac{e^t + t^2}{e^t + t^2} = 1 = \dfrac{dx_{21}}{dt}.$$

Hence, X_1 is a solution for the given system. Analogously, X_2 if a solution. $\blacktriangle$

Remark 2.4

The given system can be written as

$$\frac{dX}{dt} = A(t)X; \quad A(t) = \begin{bmatrix} \dfrac{t}{e^t + t^2} & \dfrac{-1}{e^t + t^2} \\[2mm] \dfrac{e^t(1-t)}{e^t + t^2} & \dfrac{e^t + t}{e^t + t^2} \end{bmatrix}.$$

Replacing X_1 (respectively X_2) in the equation yields

$$\frac{dX_1}{dt} = \begin{Bmatrix} 0 \\ 1 \end{Bmatrix}; \quad AX_1 = \begin{bmatrix} \dfrac{t}{e^t + t^2} & \dfrac{-1}{e^t + t^2} \\[2mm] \dfrac{e^t(1-t)}{e^t + t^2} & \dfrac{e^t + t}{e^t + t^2} \end{bmatrix} \begin{Bmatrix} 1 \\ t \end{Bmatrix} = \begin{Bmatrix} 0 \\ 1 \end{Bmatrix};$$

hence, X_1 is a solution for the given system.　▲

EXAMPLE 2.7　Find the homogeneous linear system of differential equations for which the following vectors are linearly independent solutions:

$$X_1 = \begin{Bmatrix} 1 \\ t \\ t^2 \end{Bmatrix}, \quad X_2 = \begin{Bmatrix} -t \\ 1 \\ 2 \end{Bmatrix}, \quad X_3 = \begin{Bmatrix} 0 \\ 0 \\ e^t \end{Bmatrix}. \quad ▲$$

Solution

The Wronskian determinant is

$$W(t) = \begin{vmatrix} 1 & -t & 0 \\ t & 1 & 0 \\ t^2 & 2 & e^t \end{vmatrix} = e^t(1 + t^2) \neq 0 \qquad \text{for all} \qquad t \in R.$$

Equations (2.32), in this case, are

$$\begin{vmatrix} \dfrac{dx_1}{dt} & 0 & -1 & 0 \\ x_1 & 1 & -t & 0 \\ x_2 & t & 1 & 0 \\ x_3 & t^2 & 2 & e^t \end{vmatrix} = 0, \quad \text{or} \quad \frac{dx_1}{dt} = \frac{t}{1+t^2}x_1 - \frac{1}{1+t^2}x_2$$

$$\begin{vmatrix} \dfrac{dx_2}{dt} & 1 & 0 & 0 \\ x_1 & 1 & -t & 0 \\ x_2 & t & 1 & 0 \\ x_3 & t^2 & 2 & e^t \end{vmatrix} = 0, \quad \text{or} \quad \frac{dx_2}{dt} = \frac{t}{1+t^2}x_1 + \frac{t}{1+t^2}x_2$$

and

$$
\begin{vmatrix}
\dfrac{dx_3}{dt} & 2t & 0 & e^t \\
x_1 & 1 & -t & 0 \\
x_2 & t & 1 & 0 \\
x_3 & t^2 & 2 & e^t
\end{vmatrix} = 0, \quad \text{or} \quad \frac{dx_3}{dt} = \frac{4t - t^2}{1 + t^2} x_1 + \frac{2t^2 - t^3 - 2}{1 + t^2} x_2 + x_3.
$$

We find the system

$$
\frac{dX}{dt} = A(t)X, \quad \text{where} \quad A(t) = \begin{bmatrix}
\dfrac{t}{1+t^2} & \dfrac{-1}{1+t^2} & 0 \\[2mm]
\dfrac{1}{1+t^2} & \dfrac{t}{1+t^2} & 0 \\[2mm]
\dfrac{4t-t^2}{1+t^2} & \dfrac{2t^2-t^3-2}{1+t^2} & 1
\end{bmatrix}; \quad X = \begin{Bmatrix} x_1 \\ x_2 \\ x_3 \end{Bmatrix}.
$$

EXAMPLE 2.8

The following system is considered:

$$
\begin{cases}
\dfrac{dx_1}{dt} = \dfrac{t}{1+t^2} x_1 - \dfrac{1}{1+t^2} x_2 \\[3mm]
\dfrac{dx_2}{dt} = \dfrac{1}{1+t^2} x_1 + \dfrac{t}{1+t^2} x_2 \\[3mm]
\dfrac{dx_3}{dt} = \dfrac{4t-t^2}{1+t^2} x_1 + \dfrac{2t^2-t^3-2}{1+t^2} x_2 + x_3.
\end{cases}
$$

(a) Find the general solution.
(b) Find the particular solution with the initial condition

$$
X(0) = \begin{Bmatrix} 1 \\ 1 \\ 3 \end{Bmatrix}. \quad \blacktriangle
$$

Solution

The system of vectors

$$
X_1 = \begin{Bmatrix} 1 \\ t \\ t^2 \end{Bmatrix}, \quad X_2 = \begin{Bmatrix} -t \\ 1 \\ 2 \end{Bmatrix}, \quad X_3 = \begin{Bmatrix} 0 \\ 0 \\ e^t \end{Bmatrix}
$$

is a fundamental system of solutions. The general solution is

$$
X(t) = c_1 X_1 + c_2 X_2 + c_3 X_3 = \begin{Bmatrix} c_1 - c_2 t \\ c_1 t + c_2 \\ c_1 t^2 + 2c_2 + c_3 e^t \end{Bmatrix}.
$$

The initial condition

$$X(0) = \begin{Bmatrix} 1 \\ 1 \\ 3 \end{Bmatrix}$$

gives $c_1 = 1$, $c_2 = 1$, $c_3 = 1$, and the solution that satisfies the initial condition is

$$X(t) = \begin{Bmatrix} 1 - t \\ t + 1 \\ t^2 + 2 + e^t \end{Bmatrix}, \quad \text{or} \quad \begin{cases} x_1(t) = 1 - t \\ x_2(t) = t + 1 \\ x_3(t) = 2 + t^2 + e^t. \end{cases} \quad \blacktriangle$$

EXAMPLE 2.9 Consider the system

$$\begin{cases} \dfrac{dx_1}{dt} = \dfrac{t}{1 + t^2} x_1 - \dfrac{1}{1 + t^2} x_2 + t \\[2mm] \dfrac{dx_2}{dt} = \dfrac{1}{1 + t^2} x_1 + \dfrac{t}{1 + t^2} x_2 + t^2 \\[2mm] \dfrac{dx_3}{dt} = \dfrac{4t - t^3}{1 + t^2} x_1 + \dfrac{2t - t^3 - 2}{1 + t^2} x_2 + x_3 + e^{2t}. \end{cases}$$

(a) Find the general solution.
(b) Find the particular solution with the initial condition

$$X(0) = \begin{Bmatrix} 1 \\ 1 \\ 3 \end{Bmatrix}. \quad \blacktriangle$$

Solution

The corresponding homogeneous system is that from the previous example, and its general solution is

$$X(t) = \begin{Bmatrix} c_1 - c_2 t \\ c_1 t + c_2 \\ c_1 t^2 + 2c_2 + c_3 e^t \end{Bmatrix}.$$

The general solution of the given system will be found by the method of parameter variation,

$$X(t) = \begin{Bmatrix} c_1(t) - t c_2(t) \\ c_1(t) t + c_2(t) \\ c_1(t) t^2 + 2 c_2(t) + c_3(t) e^t \end{Bmatrix}.$$

From the system

$$\begin{cases} c_1'(t) - t c_2'(t) = t \\ c_1'(t) t + c_2'(t) = t^2 \\ c_1'(t) t^2 + 2 c_2'(t) + c_3'(t) e^t = e^{2t}, \end{cases}$$

we obtain $c_1'(t) = t$, $c_2'(t) = 0$, $c_3'(t) = e^t - t^3 e^{-t}$. Integrating yields

$$c_1(t) = \frac{t^2}{2} + \tilde{c}_1, \quad c_2(t) = \tilde{c}_2, \quad c_3(t) = e^t + e^{-t}(t^3 + 3t^2 + 6t + 6) + \tilde{c}_3.$$

The general solution of the given system is

$$X(t) = \left\{ \begin{matrix} \tilde{c}_1 - \tilde{c}_2 t \\ \tilde{c}_1 t + \tilde{c}_2 \\ \tilde{c}_1 t^2 + 2\tilde{c}_2 + \tilde{c}_3 e^t \end{matrix} \right\} + \begin{bmatrix} 1 & -t & 0 \\ t & 1 & 0 \\ t^2 & 2 & e^t \end{bmatrix} \left\{ \begin{matrix} t^2/2 \\ 0 \\ e^t + e^{-t}(t^3 + 3t^2 + 6t + 6) \end{matrix} \right\}$$

$$X(t) = \left\{ \begin{matrix} \tilde{c}_1 - \tilde{c}_2 t \\ \tilde{c}_1 t + \tilde{c}_2 \\ \tilde{c}_1 t^2 + 2\tilde{c}_2 + \tilde{c}_3 e^t \end{matrix} \right\} + \left\{ \begin{matrix} \frac{1}{2} t^2 \\ \frac{1}{2} t^3 \\ \frac{1}{2} t^4 + e^{2t} + t^3 + 3t^2 + 6t + 6 \end{matrix} \right\}, \quad \blacktriangle$$

or

$$\begin{cases} x_1(t) = \tilde{c}_1 - \tilde{c}_2 t + \frac{1}{2} t^2 \\ x_2(t) = \tilde{c}_1 t + \tilde{c}_2 + \frac{1}{2} t^3 \\ x_3(t) = \tilde{c}_1 t^2 + 2\tilde{c}_2 + \tilde{c}_3 e^t + \frac{1}{2} t^4 + e^{2t} + t^3 + 3t^2 + 6t + 6. \end{cases}$$

(b) The initial condition

$$X(0) = \left\{ \begin{matrix} 1 \\ 1 \\ 3 \end{matrix} \right\}$$

yields $\tilde{c}_1 = 1$, $\tilde{c}_2 = 1$, $\tilde{c}_3 = -6$. The solution that satisfies the given initial condition is

$$\begin{cases} x_1(t) = 1 - t + \frac{1}{2} t^2 \\ x_2(t) = t + 1 + \frac{1}{2} t^3 \\ x_3(t) = e^{2t} - 6e^t + \frac{1}{2} t^4 + t^3 + 4t^2 + 6t + 8. \end{cases} \quad \blacktriangle$$

2.5 Systems of Linear Differential Equations with Constant Coefficients

A linear system with constant coefficients is a system of differential equations of the form

$$\frac{dx_i}{dt} = \sum_{j=1}^{n} a_{ij} x_j + f_i(t) \quad (i = 1, 2, \ldots, n) \tag{2.33}$$

where the coefficients a_{ij} are constants. The system (2.33) may be compactly written in the form of one matrix equation

$$\frac{dX}{dt} = AX + F, \tag{2.34}$$

where matrix A is constant.

The linear systems can be integrated by the method of elimination, by finding integrable combinations, but it is possible to find directly the fundamental system of solutions of a homogeneous linear system with constant coefficients.

For the system

$$
\begin{cases}
\dfrac{dx_1}{dt} = a_{11}x_1 + a_{12}x_2 + \cdots + a_{1n}x_n \\[2mm]
\dfrac{dx_2}{dt} = a_{21}x_1 + a_{22}x_2 + \cdots + a_{2n}x_n \\[1mm]
\qquad\qquad \cdots \\[1mm]
\dfrac{dx_n}{dt} = a_{n1}x_1 + a_{n2}x_n + \cdots + a_{nn}x_n,
\end{cases}
\tag{2.35}
$$

the solution must be of the form

$$
x_1 = s_1 e^{\lambda t}, \, x_2 = s_2 e^{\lambda t}, \ldots, x_n = s_n e^{\lambda t},
\tag{2.36}
$$

with s_i ($i = 1, 2, \ldots, n$) and λ constants. Substituting Eqs. (2.36) in Eqs. (2.35) and canceling $e^{\lambda t}$ yields

$$
\begin{cases}
(a_{11} - \lambda)s_1 + a_{12}s_2 + \cdots + a_{1n}s_n = 0 \\
a_{21}s_1 + (a_{22} - \lambda)s_2 + \cdots + a_{2n}s_n = 0 \\
\qquad\qquad \cdots \\
a_{n1}s_1 + a_{n2}s_2 + \cdots + (a_{nn} - \lambda)s_n = 0.
\end{cases}
\tag{2.37}
$$

The system of equations (2.37) has a nonzero solution when its determinant is zero,

$$
\Delta =
\begin{vmatrix}
a_{11} - \lambda & a_{12} & \cdots & a_{1n} \\
a_{21} & a_{22} - \lambda & \cdots & a_{2n} \\
\cdots & \cdots & \cdots & \cdots \\
a_{n1} & a_{n2} & \cdots & a_{nn} - \lambda
\end{vmatrix}
= 0.
\tag{2.38}
$$

Equation (2.38) is called the *characteristic equation.*

Let us consider a few cases.

2.5.1 CASE I: THE ROOTS OF THE CHARACTERISTIC EQUATION ARE REAL AND DISTINCT

Denote by $\lambda_1, \lambda_2, \ldots, \lambda_n$ the roots of the characteristic equation. For each root λ_j, write the system of equations (2.37) and find the coefficients

$$
s_{1j}, s_{2j}, \ldots, s_{nj}.
$$

The coefficients s_{ij} ($i = 1, 2, \ldots, n$) are ambiguously determined from the system of equations (2.37) for $\lambda = \lambda_j$, since the determinant of the system is zero; some of them may be considered equal to unity. Thus,

- For the root λ_1, the solution of the system of equations (2.35) is

$$
x_{11} = s_{11} e^{\lambda_1 t}, \, x_{21} = s_{21} e^{\lambda_1 t}, \ldots, x_{n1} = s_{n1} e^{\lambda_1 t}.
$$

- For the root λ_2, the solution of the system (2.35) is

$$
x_{12} = s_{12} e^{\lambda_2 t}, \, x_{22} = s_{22} e^{\lambda_2 t}, \ldots, x_{n2} = s_{n2} e^{\lambda_2 t}.
$$

. . .

■ For the root λ_n, the solution of the system (2.35) is

$$x_{1n} = s_{1n}e^{\lambda_n t}, \, x_{2n} = s_{2n}e^{\lambda_n t}, \ldots, x_{nn} = s_{nn}e^{\lambda_n t}.$$

By direct substitution into equations, the system of functions

$$\begin{cases} x_1 = c_1 s_{11} e^{\lambda_1 t} + c_2 s_{12} e^{\lambda_2 t} + \cdots + c_n s_{1n} e^{\lambda_n t} \\ x_2 = c_1 s_{21} e^{\lambda_1 t} + c_2 s_{22} e^{\lambda_2 t} + \cdots + c_n s_{2n} e^{\lambda_n t} \\ \qquad\qquad \cdots \\ x_n = c_1 s_{n1} e^{\lambda_1 t} + c_2 s_{n2} e^{\lambda_2 t} + \cdots + c_n s_{nn} e^{\lambda_n t}, \end{cases} \tag{2.39}$$

where $c_1, c_2, \ldots, c_n$ are arbitrary constants, is the *general solution for the system* of equations (2.35). Using vector notation, we obtain the same result, but more compactly:

$$\frac{dX}{dt} = AX. \tag{2.40}$$

The solution must have the form

$$X = \tilde{S}e^{\lambda t}$$

$$\tilde{S} = \begin{Bmatrix} s_1 \\ s_2 \\ \cdots \\ s_n \end{Bmatrix}.$$

The system of equations (2.37) has the form

$$(A - \lambda I)\tilde{S} = 0, \tag{2.41}$$

where I is the unit matrix. For each root λ_j of the characteristic equation $|A - \lambda I| = 0$ is determined, from Eq. (2.41), the nonzero matrix S_j and, if all roots λ_j of the characteristic equation are distinct, we obtain n solutions

$$X_1 = S_1 e^{\lambda_1 t}, \, X_1 = S_2 e^{\lambda_2 t}, \ldots, X_n = S_n e^{\lambda_n t},$$

where

$$S_j = \begin{Bmatrix} s_{1j} \\ s_{2j} \\ \cdots \\ s_{nj} \end{Bmatrix}.$$

The general solution of the system (2.35) or (2.40) is of the form

$$X = \sum_{j=1}^{n} S_j c_j e^{\lambda_j t}, \tag{2.42}$$

where c_j are arbitrary constants.

Differential Equations

2.5.2 CASE II: THE ROOTS OF THE CHARACTERISTIC EQUATION ARE DISTINCT, BUT INCLUDE COMPLEX ROOTS

Among the roots of the characteristic equation, let the complex conjugate roots be

$$\lambda_1 = \alpha + i\beta, \quad \lambda_2 = \alpha - i\beta.$$

To these roots correspond the solutions

$$\begin{cases} x_{i1} = s_{i1} e^{(\alpha+i\beta)t} & (i = 1, 2, \ldots, n) \\ x_{i2} = s_{i2} e^{(\alpha-i\beta)t} l; & (i = 1, 2, \ldots, n). \end{cases} \tag{2.43}$$

The coefficients s_{i1} and s_{i2} are determined from the system of equations (2.37). It may be shown that the real and imaginary parts of the complex solution are also solutions. Thus, we obtain two particular solutions,

$$\begin{cases} \tilde{x}_{i1} = e^{\alpha t}(\tilde{s}'_{i1} \cos \beta t + \tilde{s}'_{i2} \sin \beta t) \\ \tilde{x}_{i2} = e^{\alpha t}(\tilde{s}''_{i1} \cos \beta t + \tilde{s}''_{i2} \sin \beta t), \end{cases} \tag{2.44}$$

where $\tilde{s}'_{i1}, \tilde{s}'_{i2}, \tilde{s}''_{i1}, \tilde{s}''_{i2}$ are real numbers determined in terms of s_{i1} and s_{i2}.

2.5.3 CASE III: THE CHARACTERISTIC EQUATION HAS A MULTIPLE ROOT λ_k OF MULTIPLICITY ρ

The solution of the system of equations (2.35) is of the form

$$X(t) = (S_0 + S_1 t + \cdots + S_{\rho-1} t^{\rho-1}) e^{\lambda_s t}, \tag{2.45}$$

where

$$S_j = \begin{Bmatrix} s_{1j} \\ s_{2j} \\ \cdots \\ s_{nj} \end{Bmatrix};$$

s_{ij} are constants. Substituting Eq. (2.45) into Eq. (2.40) and requiring an identity to be found, we define the matrices S_j; some of them, including $S_{\rho-1}$ as well, may turn out to be equal to zero.

EXAMPLE 2.10 Solve the system

$$\begin{cases} \dfrac{dx_1}{dt} = -x_1 - 2x_2 \\ \dfrac{dx_2}{dt} = 3x_1 + 4x_2. \end{cases} \quad \blacktriangle$$

Solution

The characteristic equation

$$\begin{vmatrix} -1 - \lambda & -2 \\ 3 & 4 - \lambda \end{vmatrix} = 0, \quad \text{or} \quad \lambda^2 - 3\lambda + 2 = 0,$$

has the roots $\lambda_1 = 1$, $\lambda_2 = 2$. For $\lambda_1 = 1$, the system of equations (2.41) has the form

$$\begin{cases} -2s_{11} - 2s_{21} = 0 \\ 3s_{11} + 3s_{21} = 0. \end{cases}$$

Hence $s_{11} = -s_{21} =$ arbitrary $(=1)$, and

$$S_1 = \begin{Bmatrix} 1 \\ -1 \end{Bmatrix}.$$

For $\lambda_2 = 2$, the system of equations (2.41) has the form

$$\begin{cases} -3s_{12} - 2s_{22} = 0 \\ 3s_{12} + 2s_{22} = 0, \end{cases} \quad \text{or} \quad s_{22} = -\frac{3}{2} s_{12}.$$

Substituting $s_{12} = 2$, then $s_{22} = -3$ and

$$S_2 = \begin{Bmatrix} 2 \\ -3 \end{Bmatrix}.$$

The general solution (2.42) is

$$\begin{Bmatrix} x_1 \\ x_2 \end{Bmatrix} = \begin{Bmatrix} c_1 e^t + 2c_2 e^{2t} \\ -c_1 e^t - 3c_2 e^{2t} \end{Bmatrix}, \quad \text{hence} \quad \begin{cases} x_1 = c_1 e^t + 2c_2 e^{2t}, \\ x_2 = -c_1 e^t - 3c_2 e^{2t}. \end{cases} \quad \blacktriangle$$

EXAMPLE 2.11 Solve the system

$$\begin{cases} \dfrac{dx_1}{dt} = 2x_1 - x_2 \\[2mm] \dfrac{dx_2}{dt} = x_1 + 2x_2. \end{cases} \quad \blacktriangle$$

Solution

The characteristic equation

$$\begin{vmatrix} 2 - \lambda & -1 \\ 1 & 2 - \lambda \end{vmatrix} = 0, \quad \text{or} \quad \lambda^2 - 4\lambda + 5 = 0,$$

has the roots $\lambda_1 = 2 + i$, $\lambda_2 = 2 - i$.

For $\lambda_1 = 2 + i$, the system of equations (2.41) has the form

$$\begin{cases} -is_{11} - s_{12} = 0 \\ s_{11} + is_{12} = 0, \end{cases} \quad \text{or} \quad s_{12} = -is_{11};$$

hence, $s_{11} = 1$, $s_{12} = -i$, and

$$S_1 = \begin{Bmatrix} 1 \\ -i \end{Bmatrix}.$$

For $\lambda_2 = 2 - i$, the system of equations (2.41) has the form

$$\begin{cases} is_{12} - s_{22} = 0 \\ s_{12} + is_{22} = 0, \end{cases} \quad \text{or} \quad s_{22} = is_{12};$$

hence $s_{12} = 1$, $s_{22} = i$, and

$$S_2 = \left\{\begin{matrix} 1 \\ i \end{matrix}\right\}.$$

The general solution is

$$\left\{\begin{matrix} x_1 \\ x_2 \end{matrix}\right\} = c_1 S_1 e^{(2+i)t} + c_2 S_2 e^{(2-i)t} = \left\{\begin{matrix} c_1 e^{(2+i)t} + c_2 e^{(2-i)t} \\ -ic_1 e^{(2+i)t} + ic_2 e^{(2-i)t} \end{matrix}\right\}$$

$$= e^{2t} \left\{\begin{matrix} (c_1 + c_2)\cos t + i(c_1 - c_2)\sin t \\ (c_1 + c_2)\sin t - i(c_1 - c_2)\cos t \end{matrix}\right\}.$$

Taking $\tilde{c}_1 = c_1 + c_2$, $\tilde{c}_2 = i(c_1 - c_2)$, the general solution is

$$\begin{cases} x_1 = e^{2t}(\tilde{c}_1 \cos t + \tilde{c}_2 \sin t) \\ x_2 = e^{2t}(\tilde{c}_1 \sin t - \tilde{c}_2 \cos t). \end{cases} \blacktriangle$$

EXAMPLE 2.12 Solve the system

$$\begin{cases} \dfrac{dx_1}{dt} = 3x_1 - x_2 \\[2mm] \dfrac{dx_2}{dt} = x_1 + x_2. \end{cases} \blacktriangle$$

Solution

The characteristic equation

$$\begin{vmatrix} 3 - \lambda & -1 \\ 1 & 1 - \lambda \end{vmatrix} = 0, \quad \text{or} \quad \lambda^2 - 4\lambda + 4 = 0,$$

has the roots $\lambda_1 = \lambda_2 = 2$. Hence, the solution must have the form

$$\begin{cases} x_1 = (s_{10} + s_{11}t)e^{2t} \\ x_2 = (s_{20} + s_{21}t)e^{2t}. \end{cases}$$

Substituting in the given system, we obtain

$$\begin{cases} 2(s_{10} + s_{11}t) + s_{11} \equiv 3(s_{10} + s_{11}t) - s_{20} - s_{21}t \\ 2(s_{20} + s_{21}t) + s_{21} \equiv s_{10} + s_{11}t + s_{20} + s_{21}t, \end{cases}$$

whence

$$s_{21} = s_{11}, \quad s_{10} - s_{20} = s_{11}.$$

s_{10} and s_{20} remain arbitrary. If we denote these arbitrary constants by c_1 and c_2, respectively, the general solution is of the form

$$\begin{cases} x_1 = [c_1 + (c_2 - c_1)t]e^{2t} \\ x_2 = [c_2 + (c_2 - c_1)t]e^{2t}. \end{cases} \blacktriangle$$

EXAMPLE 2.13 Solve the system

$$\begin{cases} \dfrac{dx_1}{dt} = -x_1 + x_2 + x_3 \\[2mm] \dfrac{dx_2}{dt} = x_1 - x_2 + x_3 \quad \blacktriangle \\[2mm] \dfrac{dx_3}{dt} = x_1 + x_2 - x_3. \end{cases}$$

Solution

The characteristic equation is

$$\begin{vmatrix} -1 - \lambda & 1 & 1 \\ 1 & -1 - \lambda & 1 \\ 1 & 1 & -1 - \lambda \end{vmatrix} = 0, \quad \text{and} \quad \lambda_1 = 1, \quad \lambda_2 = \lambda_3 = -2.$$

Corresponding to the root $\lambda_1 = 1$ is the solution

$$X_1 = S_1 e^t = \begin{Bmatrix} \alpha_1 \\ \alpha_2 \\ \alpha_3 \end{Bmatrix} e^t.$$

The system (2.41) has the form

$$\begin{cases} -2\alpha_1 + \alpha_2 + \alpha_3 = 0 \\ \alpha_1 - 2\alpha_2 + \alpha_3 = 0 \\ \alpha_1 + \alpha_2 - 2\alpha_3 = 0. \end{cases}$$

Hence, $\alpha_1 = \alpha_2 = \alpha_3 = c_1$ are arbitrary, and

$$X_1 = \begin{Bmatrix} c_1 e^t \\ c_1 e^t \\ c_1 e^t \end{Bmatrix}.$$

Corresponding to the multiple root $\lambda_2 = \lambda_3 = -2$ is the solution

$$X_2(t) = (S_0 + S_1 t) e^{-2t} = \begin{Bmatrix} s_{10} + s_{11} t \\ s_{20} + s_{21} t \\ s_{30} + s_{31} t \end{Bmatrix} e^{-2t}$$

$$X_2'(t) = -2(S_0 + S_1 t) e^{-2t} + S_1 e^{-2t}.$$

Substituting in the system $X'(t) = AX(t)$ yields

$$(A + 2I)(S_0 + S_1 t) = S_1 \quad \text{or} \quad \begin{bmatrix} 1 & 1 & 1 \\ 1 & 1 & 1 \\ 1 & 1 & 1 \end{bmatrix} \begin{Bmatrix} s_{10} + s_{11} t \\ s_{20} + s_{21} t \\ s_{30} + s_{31} t \end{Bmatrix} = \begin{Bmatrix} s_{11} \\ s_{12} \\ s_{13} \end{Bmatrix}.$$

Hence,

$$s_{10} + s_{20} + s_{30} + (s_{11} + s_{21} + s_{31})t \equiv s_{11} = s_{12} = s_{13},$$

and

$$s_{11} = s_{12} = s_{13} = 0$$
$$s_{10} + s_{20} + s_{30} = 0.$$

The quantities s_{10} and s_{20} remain arbitrary. Denoting them by c_2 and c_3, respectively, yields

$$X_2(t) = \left\{ \begin{array}{c} c_2 e^{-2t} \\ c_3 e^{-2t} \\ -(c_2 + c_3)e^{-2t} \end{array} \right\},$$

and the general solution of the given system is

$$X(t) = \left\{ \begin{array}{c} x_1(t) \\ x_2(t) \\ x_3(t) \end{array} \right\} = X_1(t) + X_2(t) = \left\{ \begin{array}{c} c_1 e^t + c_2 e^{-2t} \\ c_1 e^t + c_3 e^{-2t} \\ c_1 e^t - (c_2 + c_3)e^{-2t} \end{array} \right\}. \quad \blacktriangle$$

EXAMPLE 2.14 Solve the system

$$\begin{cases} \dfrac{dx_1}{dt} = -x_1 + x_2 + x_3 + e^{2t} \\[2mm] \dfrac{dx_2}{dt} = x_1 - x_2 + x_3 + 1 \\[2mm] \dfrac{dx_3}{dt} = x_1 + x_2 - x_3 + t \end{cases}$$

with the initial conditions $x_1(0) = -\frac{1}{4}$, $x_2(0) = 1$, $x_3(0) = \frac{5}{4}$. $\quad \blacktriangle$

Solution

The corresponding homogeneous system

$$\begin{cases} \dfrac{dx_1}{dt} = -x_1 + x_2 + x_3 \\[2mm] \dfrac{dx_2}{dt} = x_1 - x_2 + x_3 \\[2mm] \dfrac{dx_3}{dt} = x_1 + x_2 - x_3 \end{cases}$$

has the general solutions

$$X(t) = \left\{ \begin{array}{c} c_1 e^t + c_2 e^{-2t} \\ c_1 e^t + c_3 e^{-2t} \\ c_1 e^t - c_2 e^{-2t} - c_3 e^{-2t} \end{array} \right\} = \begin{bmatrix} e^t & e^{-2t} & 0 \\ e^t & 0 & e^{-2t} \\ e^t & -e^{-2t} & -e^{-2t} \end{bmatrix} \left\{ \begin{array}{c} c_1 \\ c_2 \\ c_3 \end{array} \right\} = M(t)c,$$

where

$$M(t) = \begin{bmatrix} e^t & e^{-2t} & 0 \\ e^t & 0 & e^{-2t} \\ e^t & -e^{-2t} & -e^{-2t} \end{bmatrix}; \quad c = \left\{ \begin{array}{c} c_1 \\ c_2 \\ c_3 \end{array} \right\}.$$

We seek the solution of the nonhomogeneous system in the form

$$X(t) = M(t)c(t).$$

Substituting in the given system yields

$$M'(t)c(t) + M(t)c'(t) = A(t)M(t)c(t) + F(t),$$

or

$$M(t)c'(t) = F(t).$$

Hence,

$$c'(t) = M^{-1}(t)F(t),$$

where $M^{-1}(t)$ is the inverse of $M(t)$. Then,

$$M^{-1}(t) = \frac{1}{3}\begin{bmatrix} e^{-t} & e^{-t} & e^{-t} \\ 2e^{2t} & -e^{2t} & -e^{2t} \\ -2e^{2t} & 2e^{2t} & -e^{2t} \end{bmatrix},$$

and

$$c'(t) = \frac{1}{3}\begin{bmatrix} e^{-t} & e^{-t} & e^{-t} \\ 2e^{2t} & -e^{2t} & -e^{2t} \\ -2e^{2t} & 2e^{2t} & -e^{2t} \end{bmatrix}\begin{Bmatrix} e^{2t} \\ 1 \\ t \end{Bmatrix} = \frac{1}{3}\begin{Bmatrix} e^t + e^{-t} + te^{-t} \\ 2e^{4t} - e^{2t} - te^{2t} \\ -e^{4t} + 2e^{2t} - te^{2t} \end{Bmatrix}.$$

Integrating yields

$$c(t) = \frac{1}{3}\begin{Bmatrix} e^t - 2e^{-t} - te^{-t} + \tilde{c}_1 \\ \dfrac{1}{2}e^{4t} - \dfrac{1}{4}e^{2t} - \dfrac{t}{2}e^{2t} + \tilde{c}_2 \\ -\dfrac{1}{4}e^{4t} + \dfrac{5}{4}e^{2t} - \dfrac{t}{2}e^{2t} + \tilde{c}_3 \end{Bmatrix}.$$

The general solution of the nonhomogeneous system is

$$X(t) = \begin{Bmatrix} x_1(t) \\ x_2(t) \\ x_3(t) \end{Bmatrix} = \begin{Bmatrix} \tilde{c}_1 e^t + \tilde{c}_2 e^{-2t} + \frac{1}{2}e^{2t} - \frac{1}{2}t - \frac{3}{4} \\ \tilde{c}_1 e^t + \tilde{c}_3 e^{-2t} + \frac{1}{4}e^{2t} - \frac{1}{2}t - \frac{1}{4} \\ \tilde{c}_1 e^t - \tilde{c}_2 e^{-2t} - \tilde{c}_3 e^{-2t} + \frac{1}{4}e^{2t} - 1 \end{Bmatrix}.$$

From the initial conditions,

$$\begin{cases} \tilde{c}_1 + \tilde{c}_2 - \frac{1}{4} = -\frac{1}{4} \\ \tilde{c}_1 + \tilde{c}_3 = 1 \\ \tilde{c}_1 - \tilde{c}_2 - \tilde{c}_3 - \frac{3}{4} = \frac{5}{4}, \end{cases}$$

whence $\tilde{c}_1 = 1$, $\tilde{c}_2 = -1$, $\tilde{c}_3 = 0$. The solution of the system with the initial values is

$$\begin{cases} x_1(t) = e^t - e^{-2t} + \frac{1}{2}e^{2t} - \frac{1}{2}t - \frac{3}{4} \\ x_2(t) = e^t + \frac{1}{4}e^{2t} - \frac{1}{2}t - \frac{1}{4} \\ x_3(t) = e^t + e^{-t} + \frac{1}{4}e^{2t} - 1. \end{cases}$$ ▲

EXAMPLE 2.15 Solve the system of the second-order differential equations

$$
\begin{cases}
\dfrac{d^2 x_1}{dt^2} = a_{11}x_1 + a_{12}x_2 \\[2mm]
\dfrac{d^2 x_2}{dt^2} = a_{21}x_1 + a_{22}x_2.
\end{cases}
$$

The numerical case is

$$
\begin{cases}
\dfrac{d^2 x_1}{dt^2} = -2x_1 + 3x_2 \\[2mm]
\dfrac{d^2 x_2}{dt^2} = -2x_1 + 5x_2.
\end{cases}
$$

Solution

Again, we seek the solution in the form

$$
x_1 = s_1 e^{\lambda t}, \quad x_2 = s_2 e^{\lambda t}.
$$

Substituting these expressions into the system and canceling out $e^{\lambda t}$, we find a system of equations for determining s_1, s_2, and λ,

$$
\begin{cases}
(a_{11} - \lambda^2)s_1 + a_{12}s_2 = 0 \\
a_{21}s_1 + (a_{22} - \lambda^2)s_2 = 0,
\end{cases}
$$

or $(A - \lambda^2 I)S = 0$, where

$$
S = \begin{Bmatrix} s_1 \\ s_2 \end{Bmatrix}.
$$

Nonzero s_1 and s_2 are determined only when the determinant of the system is equal to zero,

$$
|A - \lambda^2 I| = 0.
$$

This is the characteristic equation of the given differential system. For each root λ_j of the characteristic equation, we find

$$
S_j = \begin{Bmatrix} s_{1j} \\ s_{2j} \end{Bmatrix} \quad (j = 1, 2, 3, 4).
$$

The general solution will have the form

$$
\begin{Bmatrix} x_1 \\ x_2 \end{Bmatrix} = \sum_{j=1}^{4} S_j e^{\lambda_j t} c_j,
$$

where c_j are arbitrary constants. The differential system

$$
\begin{cases}
\dfrac{d^2 x_1}{dt^2} = -2x_1 + 3x_2 \\[2mm]
\dfrac{d^2 x_2}{dt^2} = -2x_1 + 5x_2
\end{cases}
$$

has the characteristic equation

$$\begin{vmatrix} -2 - \lambda^2 & 3 \\ -2 & 5 - \lambda^2 \end{vmatrix} = 0,$$

and the roots

$$\lambda_1 = i, \quad \lambda_2 = -i, \quad \lambda_3 = 2, \quad \lambda_4 = -2.$$

For $\lambda_1 = i$ and $\lambda_2 = -i$, the system $(A - \lambda^2 I)S = 0$ yields

$$-s_1 + 3s_2 = 0 \quad \text{or} \quad s_1 = 3s_2; \quad S_1 = \begin{Bmatrix} 3 \\ 1 \end{Bmatrix}, \quad S_2 = \begin{Bmatrix} 3 \\ 1 \end{Bmatrix}.$$

For $\lambda_3 = 2$ and $\lambda_4 = -2$, it yields

$$-6s_1 + 3s_2 = 0, \quad \text{or} \quad s_2 = 2s_1 \quad \text{and} \quad S_3 = \begin{Bmatrix} 1 \\ 2 \end{Bmatrix}, \quad S_4 = \begin{Bmatrix} 1 \\ 2 \end{Bmatrix}.$$

The general solution is

$$\begin{cases} x_1 = 3c_1 e^{it} + 3c_2 e^{-it} + c_3 e^{2t} + c_4 e^{-2t} \\ x_2 = c_1 e^{it} + c_2 e^{-it} + 2c_3 e^{2t} + 2c_4 e^{-2t}. \end{cases}$$

Let us write out the complex solutions

$$x_{11} = e^{it} = \cos t + i \sin t; \quad x_{21} = e^{-it} = \cos t - i \sin t.$$

The real and imaginary parts are separated from the solutions:

$$\tilde{x}_{11} = \cos t, \quad \tilde{x}_{21} = \sin t.$$

Now, the general solution can be expressed as

$$\begin{cases} x_1(t) = 3c_1 \cos t + 3c_2 \sin t + c_3 e^{2t} + c_4 e^{-2t} \\ x_2(t) = c_1 \cos t + c_2 \sin t + 2c_3 e^{2t} + 2c_4 e^{-2t}. \end{cases}$$

References

1. G. N. Berman, *A Problem Book in Mathematical Analysis*, English translation. Mir Publishers, Moscow, 1977.

2. R. L. Burden and J. D. Faires, *Numerical Analysis*. PWS-Kent Publishing Company, Boston, 1989.

3. L. Elsgolts, *Differential Equations and the Calculus of Variations*. Mir Publishers, Moscow, 1970.

4. P. Hartman, *Ordinary Differential Equations*. Wiley, New York, 1964.

5. M. L. Krasnov, A. I. Kiselyov, and G. I. Makarenko, *A Book of Problems in Ordinary Differential Equations*, English translation. Mir Publishers, Moscow, 1981.

6. E. Kreyszig, *Advanced Engineering Mathematics*. Wiley and Sons, New York, 1988.

7. V. A. Kudryavtsev and B. P. Demidovich, *A Brief Course of Higher Mathematics*, English translation. Mir Publishers, Moscow, 1981.

8. N. Piskunov, *Differential and Integral Calculus*, Vol. II, English translation. Mir Publishers, Moscow, 1974.

9. I. S. Sokolnikoff and R. M. Redheffer, *Mathematics of Physics and Modern Engineering*. McGraw-Hill, New York, 1966.

10. Y. B. Zeldovich and A. D. Myskis, *Elements of Applied Mathematics*, English translation. Mir Publishers, Moscow, 1976.

Index